Wave Optics in Infrared Spectroscopy

Wave Optics in Infrared Spectroscopy
Theory, Simulation, and Modeling

Thomas G. Mayerhöfer
Leibniz Institute of Photonic Technology, Jena, Germany

Elsevier
Radarweg 29, PO Box 211, 1000 AE Amsterdam, Netherlands
125 London Wall, London EC2Y 5AS, United Kingdom
50 Hampshire Street, 5th Floor, Cambridge, MA 02139, United States

Copyright © 2024 Elsevier Inc. All rights are reserved, including those for text and data mining, AI training, and similar technologies.

Publisher's note: Elsevier takes a neutral position with respect to territorial disputes or jurisdictional claims in its published content, including in maps and institutional affiliations.

No part of this publication may be reproduced or transmitted in any form or by any means, electronic or mechanical, including photocopying, recording, or any information storage and retrieval system, without permission in writing from the publisher. Details on how to seek permission, further information about the Publisher's permissions policies and our arrangements with organizations such as the Copyright Clearance Center and the Copyright Licensing Agency, can be found at our website: www.elsevier.com/permissions.

This book and the individual contributions contained in it are protected under copyright by the Publisher (other than as may be noted herein).

Notices

Knowledge and best practice in this field are constantly changing. As new research and experience broaden our understanding, changes in research methods, professional practices, or medical treatment may become necessary.

Practitioners and researchers must always rely on their own experience and knowledge in evaluating and using any information, methods, compounds, or experiments described herein. In using such information or methods they should be mindful of their own safety and the safety of others, including parties for whom they have a professional responsibility.

To the fullest extent of the law, neither the Publisher nor the authors, contributors, or editors, assume any liability for any injury and/or damage to persons or property as a matter of products liability, negligence or otherwise, or from any use or operation of any methods, products, instructions, or ideas contained in the material herein.

ISBN: 978-0-443-22031-9

For information on all Elsevier publications
visit our website at https://www.elsevier.com/books-and-journals

Publisher: Candice Janco
Acquisitions Editor: Charles Bath
Editorial Project Manager: Andrea Dulberger
Production Project Manager: Nadhiya Sekar
Cover Designer: Christian Bilbow

Typeset by STRAIVE, India

Dedication

"…Und was wir an gültigen Sätzen gefunden,
Dran bleibt aller irdische Wandel gebunden…"

Chor der Toten, Conrad Ferdinand Meyer (1825–98).

Contents

Foreword xi
Preface xiii

Part I
Scalar theory

1. **What is wrong with absorbance?**

2. **Transition from the Bouguer-Beer-Lambert approximation to wave optics and dispersion theory**
   2.1 The electric field and the electric displacement — 11
   2.2 The magnetic field and the magnetic induction — 14
   2.3 Maxwell's equations in simplified form — 15
      2.3.1 1st equation—Gauss's law — 15
      2.3.2 2nd equation—Faraday's law of induction — 16
      2.3.3 3rd equation—Gauss's law for magnetism — 16
      2.3.4 4th equation—Ampere's circuital law — 16
   2.4 Deriving the wave equation — 17
   2.5 One-dimensional and harmonic waves — 17
   2.6 Harmonic molecular vibrations and the dielectric function — 21
   2.7 The Kramers-Kronig relations — 28
   2.8 The influence of absorption on the electromagnetic waves — 30
   2.9 Reflection and transmission at an interface separating two scalar media under normal incidence — 32
   2.10 Transmission through a thick slab suspended in vacuum — 34
   2.11 Transmission through a thin slab suspended in vacuum — 36
   2.12 Transmission through a layer on a substrate suspended in vacuum — 38
   2.13 Scalar and vector fields — 40
   2.14 Further reading — 44
   References — 45

3. **The electromagnetic field**
   3.1 Maxwell's relations — 47
   3.2 Boundary conditions — 48
   3.3 Energy density and flux — 49
   3.4 The wave equation — 50
   3.5 Polarized waves — 52
   3.6 Further reading — 55
   References — 55

4. **Reflection and transmission of plane waves**
   4.1 Reflection and transmission at an interface separating two scalar media under normal incidence — 57
   4.2 Reflection and transmission at an interface separating two scalar semiinfinite media under nonnormal incidence — 60
      4.2.1 *s*-Polarized light — 61
      4.2.2 *p*-Polarized light — 63
      4.2.3 Calculation of reflectance and transmittance — 65
      4.2.4 Example: Dependence of the reflectance from the angle of incidence — 66
   4.3 Reflection and transmission at an interface separating two scalar media under nonnormal incidence-absorbing media — 67
   4.4 Reflection and transmission at an interface separating two scalar media under nonnormal incidence—Total/internal reflection — 68

4.5 Reflection and transmission at an interface separating two scalar media under nonnormal incidence—Matrix formalism . . . 70
   4.5.1 Matrix formulation for *s*-polarized waves at a single interface . . . 71
   4.5.2 Matrix formulation for *p*-polarized waves at a single interface . . . 72
   4.5.3 Combined matrix formulation for waves at a single interface . . . 73
   4.5.4 A layer sandwiched by two semiinfinite media . . . 74
   4.5.5 Arbitrary number of layers . . . 76
   4.5.6 Calculating the electric field strengths of a layered medium—Coherent layers . . . 77
   4.5.7 Incoherent layers . . . 78
   4.5.8 Mixed coherent and incoherent layers . . . 80
   4.5.9 Calculating the electric field strengths of a layered medium—Mixed coherent-incoherent multilayers . . . 81
4.6 Further reading . . . 84
References . . . 84

5. Dispersion relations
5.1 Dispersion relation—Uncoupled oscillator model . . . 86
5.2 Excursus: Lorentz profile vs. Lorentz oscillator . . . 98
5.3 Excursus: Dispersion relations and Beer's approximation . . . 101
5.4 Dispersion relation—Coupled oscillator model . . . 110
5.5 Dispersion relation—Semi-empirical four-parameter models . . . 116
   5.5.1 Berreman-Unterwal model . . . 116
   5.5.2 Kim oscillator . . . 117
   5.5.3 Classical model with frequency-dependent damping constant . . . 119
   5.5.4 Classical model with complex oscillator strength . . . 120
   5.5.5 Convolution model . . . 123
5.6 Dispersion relation—Inverse dielectric function model . . . 124
5.7 Dispersion relation—Drude model . . . 130
5.8 Kramers-Kronig relations and sum rules . . . 131
   5.8.1 The basics . . . 131
   5.8.2 Determination of the optical constants directly from transmittance or reflectance . . . 140
   5.8.3 The sum rules . . . 148
   5.8.4 The dielectric and the refractive index background . . . 152

5.9 Further reading . . . 155
References . . . 156

6. Deviations from the (Bouguer-)Beer-Lambert approximation
6.1 Transmittance of a slab embedded in vacuum/air . . . 161
6.2 Transmittance of a free-standing film embedded in vacuum/air . . . 164
6.3 Reflection of a layer on a highly reflecting substrate—Transflection . . . 168
6.4 Transmission of a layer on a transparent substrate . . . 171
6.5 Attenuated total reflection . . . 178
6.6 Mixing rules . . . 182
6.7 How to correct the deviations and to obtain a wave-optics conform solution . . . 184
   6.7.1 Correction of the apparent absorbance . . . 184
   6.7.2 Dispersion analysis . . . 189
6.8 Further reading . . . 194
References . . . 195

7. Additional insights gained by wave optics and dispersion theory
7.1 Infrared refraction spectroscopy . . . 197
7.2 Surface-enhanced infrared absorption (SEIRA) . . . 201
7.3 Investigation of coupling effects . . . 207
   7.3.1 Indirect coupling . . . 207
   7.3.2 Direct coupling of oscillators . . . 214
   7.3.3 Strong coupling between vibrations and the electric field—Polaritons . . . 217
7.4 Further reading . . . 221
References . . . 221

8. 2D correlation analysis
8.1 Basics . . . 225
8.2 Smart error sum . . . 230
8.3 2T2D smart error sum . . . 236
8.4 Further reading . . . 240
References . . . 241

9. Chemometrics
9.1 Introduction . . . 243
9.2 Classical least squares (CLS) regression . . . 245
9.3 Inverse least squares (ILS) regression . . . 249
9.4 Principal component analysis (PCA)/principal component regression (PCR) . . . 249
9.5 Multivariate curve resolution (MCR)-alternating least squares (ALS) . . . 252
9.6 Further reading . . . 255
References . . . 255

10. Spectral mixing rules
- 10.1 Introduction — 257
- 10.2 Lorentz-Lorenz theory — 258
- 10.3 Maxwell-Garnett approximation — 259
- 10.4 Bruggeman approximation — 261
- 10.5 The Bergman representation — 263
    - 10.5.1 Dipole interactions and resulting polarization in many-particle systems — 263
    - 10.5.2 Basic properties of the spectral density — 267
    - 10.5.3 Percolation — 268
    - 10.5.4 Dependence of the effective dielectric function on concrete spectral densities — 269
- 10.6 Microheterogeneity and size dependence of spectral features — 275
- 10.7 Further reading — 278
- References — 279

Part II
Tensorial theory

11. What is wrong with linear dichroism theory

12. Reflection and transmission of plane waves from and through anisotropic media—Generalized 4×4 matrix formalism
- 12.1 Berreman's formalism: Maxwell equations and constitutive relations — 294
- 12.2 Berreman's formalism: Calculation of the refractive indices and the polarization directions — 295
- 12.3 Yeh's formalism: Maxwell equations and constitutive relations — 297
- 12.4 Yeh's formalism: Calculation of the refractive indices and the polarization directions — 298
- 12.5 The transfer matrix — 299
- 12.6 The treatment of singularities — 300
    - 12.6.1 Degenerate eigenvalues — 300
    - 12.6.2 Singular form of the Dynamical Matrix — 303
- 12.7 The calculation of reflectance and transmittance coefficients — 303
- 12.8 Simplifications for special cases — 304
    - 12.8.1 Nonmagnetic ($\mu=1$), dielectric anisotropic ($\varepsilon_{ij}=\varepsilon_{ji}$) material and normal incidence — 305
    - 12.8.2 Nonmagnetic ($\mu=1$), dielectric ($\varepsilon_{ij}=\varepsilon_{ji}$) monoclinic material—a-c-plane — 307
    - 12.8.3 Nonmagnetic ($\mu=1$), dielectric uniaxial ($\varepsilon_{ij}=\varepsilon_{ji}$, $\varepsilon_a=\varepsilon_b$) material — 309
    - 12.8.4 Nonmagnetic ($\mu=1$), dielectric ($\varepsilon_{ij}=\varepsilon_{ji}$) uniaxial or orthorhombic material with principal orientations — 312
    - 12.8.5 Nonmagnetic ($\mu=1$), biisotropic medium — 313
- 12.9 Further reading — 314
- Reference — 314

13. Dispersion relations—Anisotropic oscillator models
- 13.1 Cubic crystal system — 318
- 13.2 Optically uniaxial: Tetragonal, hexagonal, and trigonal crystal systems — 318
- 13.3 Orthorhombic crystals — 320
- 13.4 Monoclinic crystals — 321
- 13.5 Triclinic crystals — 321
- 13.6 Generalized oscillator models — 322
- 13.7 Further reading — 323
- References — 323

14. Dispersion analysis of anisotropic crystals—Examples
- 14.1 Optically uniaxial crystals — 325
- 14.2 Orthorhombic crystals — 327
- 14.3 Monoclinic crystals — 328
- 14.4 Excursus: Perpendicular modes — 329
- 14.5 Triclinic crystals — 337
- 14.6 Generalized dispersion analysis — 340
- 14.7 Further reading — 343
- References — 343

15. Polycrystalline materials
- 15.1 How to calculate reflectance and transmittance for random orientation — 345
- 15.2 Optical properties of randomly oriented polycrystalline materials with large crystallites compared to those consisting of small crystallites — 356
- 15.3 Large crystallites and nonrandom orientation — 362
- 15.4 Further reading — 366
- References — 366

16. **Vibrational circular dichroism**
    16.1 Introduction 367
    16.2 Calculating the spectra of chiral materials 368
    16.3 Chiral dispersion analysis 369
    16.4 Further reading 378
    References 378

Index 381

Foreword

Infrared spectroscopy is a powerful analytical technique that has revolutionized our understanding of the molecular world. It is used in a wide range of fields, including chemistry, physics, materials science, and biology, to investigate the structure, composition, and dynamics of molecules. However, to fully understand and interpret infrared spectra, it is essential to have a strong foundation in wave optics.

In his new book, *Wave Optics in Infrared Spectroscopy: Theory, Simulation, and Modelling*, Thomas G. Mayerhöfer provides a comprehensive and accessible introduction to this important topic. The book covers a wide range of materials, from the basic principles of wave optics to the more advanced concepts of dispersion and absorption in infrared spectroscopy.

One of the key strengths of the book is its focus on the physical basis of infrared spectroscopy. Mayerhöfer explains various phenomena that affect the absorption, reflection, and transmission of infrared radiation in a clear and concise manner. This allows the reader to develop a deep understanding of how infrared spectra are generated and how they can be interpreted.

Another important feature of the book is its emphasis on practical applications. Mayerhöfer discusses a variety of real-world examples of how infrared spectroscopy is used to study different types of materials, including polymers, semiconductors, and biological systems. This helps the reader see the relevance of wave optics to infrared spectroscopy and appreciate the power of this technique.

Overall, *Wave Optics in Infrared Spectroscopy: Theory, Simulation, and Modelling* is an essential resource for anyone who wants to develop a deep understanding of infrared spectroscopy. It is a well-written and informative book suitable for both students and researchers. I highly recommend it to anyone interested in this important field of science.

In addition to its strengths in theory and practice, the book also makes a number of unique and important contributions to the field of infrared spectroscopy. First, it provides a unified and consistent treatment of wave optics and infrared spectroscopy that has not been previously available. Second, the book introduces a number of new concepts and insights, such as the distinction between micro-homogeneous and micro-heterogeneous samples and the importance of considering the dispersion of the refractive index in the analysis of infrared spectra. Third, the book discusses a number of challenging and complex aspects of infrared spectroscopy in a clear and accessible manner, such as the inverse problem of extracting quantitative information from infrared spectra.

I am confident that *Wave Optics in Infrared Spectroscopy: Theory, Simulation, and Modelling* will become a standard reference work for the field of infrared spectroscopy. It is an essential resource for anyone who wants to develop a deep understanding of this important technique.

Isao Noda
Department of Materials Science and Engineering, University of Delaware,
Newark, DE, United States

Preface

When I started my scientific career nearly three decades ago, I was sure that scientific knowledge was steadily increasing over time. This seems to be a logical assumption, given that Scientists must surely be the most objective type of humans, diligently studying the findings of their predecessors and building on these insights, and that the number of scientists keeps growing inevitably over the years. Even though it is thinkable that this increase might not be perfectly steady and that there can be some hurdles, it is certainly impossible that the amount of knowledge decreases with time, is it not?

Having worked for more than 25 years in the field of infrared spectroscopy, I have to admit that I have developed serious doubts regarding my prior view. Before explaining the reasons behind, let me first introduce the term naïve realism, which I learned from the book *The Science of Storytelling* by Will Storr. Naïve realism refers to the tendency to assume that our understanding of the world accurately mirrors its objective reality without any bias or distortion. It involves overlooking the impact of our emotions, previous experiences, and cultural background on our perception, leading us to presume that others share the same perspective as ours. While this is the general definition, if you exchange "world" for "field" and "cultural" for "scientific," you probably already see where this leads. You may interpose that as scientists we are not alone. We frequently work together, exchange our view on a personal level, on conferences, and in written form. This should ensure that we all share a view of our field, which was obtained via critical discussion and therefore is objective and unbiased. In particular, we have textbooks that comprise the knowledge gained over the decades and centuries, to safeguard old knowledge and secure the increase of new knowledge. Therefore, as a young scientist, you can fully rely on these books and stand on the shoulders of your predecessors.

However, in practice it is not as simple as that. Despite our best efforts to be as objective as possible in our work as scientists, we must not forget that our human nature will always influence us. We tend to look specifically for clues that verify our current understanding of theory and quickly disregard contradictory results or opinions. The technical term for this tendency is confirmation bias. From my experience, both effects, naïve realism and confirmation bias, can lead to collective beliefs in scientific communities that do not reflect the entire knowledge, which is already present. For example, it can be forgotten that certain theories are solely an approximation, which only holds within certain limits. When these approximations are applied over long periods of time, it can happen that the awareness is lost that a certain theory is in fact just an approximation, not a universally true law. I do not want to condemn anyone—from my own experience, I know it is too easy to conform to the opinion of those around us. We are social creatures. We want to belong and therefore like to live in harmony, i.e., in agreement, with the others. This also applies to scientific communities. Sometimes little to no harm is done, but in some cases, this can lead to misinformation and wrong interpretations of results, and it can, in the long run, seriously impair the progress in this field.

So, what can you do?

Keep an open mind. Talk to and exchange with other scientists, who are not directly in your research field. Consider their perspective, and try to learn from them. In particular, in infrared spectroscopy, this is relatively easy, as I will discuss later on.

Factor in your human nature. To avoid confirmation bias, question your results and the theories you apply (this certainly also holds for the contents of this book!). Look for contradictory evidence of purpose. Bonus: when you are very critical toward your results and have questioned them rigorously, it will be very hard for anyone to surprise you with a critical question, be it your boss or someone from the audience, when giving a speech.

Be curious and explore, for example, where the equations you are using for spectra evaluation are actually coming from. Familiarize yourself with the history of your field and look for original papers. Do not disregard older literature—a more recent paper is not automatically better than an older one. In addition, try every equation in simulations to better understand it and the role of the parameters. Do not only compare experiments with simulations but also try to model them.

What makes IR spectroscopy special is that there is not a single community, but at least three different ones, and each of these communities has its own perceptions that are partly shared, but also partly contradicting those of the other communities. In addition to the molecular spectroscopists, who are mostly interested in organic or biological matter, a second community mainly wants to learn more about the physics of new inorganic or metalorganic compounds and a third one deals with determining the kind and nature of matter in space and on stellar objects. Having worked on subjects that belong to each of these communities, I found that there is very little exchange between these communities that could eliminate the theoretical discrepancies. Strangely, contemporary textbooks about infrared spectroscopy mainly originated from the first community and those are, in my opinion, not really suitable to fit the needs of the other communities.

From a historical point of view, there existed in principle two different tendencies. There is the school of William Coblentz who once stated that "Experimental observations always have some value. This is not always true of theories which are built, more or less, upon hypotheses and must stand or fall with them." In this regard, an interesting eye-opener for me was an article from Yakov M. Rabkin dealing with the translation of IR spectrometer from the labs of a few specialists into the labs of nearly every university (Technological Innovation in Science: The Adoption of Infrared Spectroscopy by Chemists, Isis 78, 31–54). In this article from 1987, Rabkin, wondering why the translation was mainly launched by companies in the USA and not by those in Germany, stated that "Accumulating spectra that no theoretical specialist in the infrared field could even begin to interpret had perhaps no place in the social structure of German science and technology." Having read the old literature about IR spectroscopy, I cannot help calling this conclusion plainly wrong. Since much of the pre-World War II literature is in German, Rabkin's statement may be excusable. In fact, I would argue that Heinrich Rubens, which I personally see as the first infrared spectroscopist, was, e.g., well aware of the fact that the law named after August Beer and Heinrich Lambert (Pierre Bouguer should be mentioned here as well) was a mere approximation. His most famous PhD student Marianus Czerny proved this, e.g., in a paper where he performed the first analysis of the NaCl spectrum based on wave optics and dispersion theory (he performed the first dispersion analysis well before the advent of computers!). Also mentioned should be Clemens Schaefer and his PhD Student Frank Matossi, who published in 1930 one of the first books about Infrared Spectroscopy (Das ultrarote Spektrum), which I would still consider as one of the most advanced one, even considering newer books.

So how is it possible that in the beginning of the third millennium several books and reviews have been published where it was not at all mentioned that the Bouguer-Beer-Lambert (BBL) law is just a sometimes very rough approximation? In addition to the factors mentioned above, one reason may be that, for the translation and economic exploitation of IR spectrometers, it may not be useful to explain that the quantitative evaluation of an IR spectrum can be a very complex and sometimes unsuccessful task. In addition, as Fourier-transform IR spectrometer took over in the beginning of the 1970s, they required a computer which made it also possible to convert a spectrum easily from transmittance or reflectance to absorbance—before the advent of this type of spectrometer, absorbance was a quantity much more rarely used. Nearly at the same time, chemometrics started its triumph, which is in its conventional form based exclusively on the BBL approximation. If you add these two reasons to what John William Strutt, the later Lord Rayleigh, said, namely, "In many departments of science a tendency may be observed to extend the field of familiar laws beyond their proper limits...," it is understandable that nowadays many PhD students working on IR spectroscopy never have heard that IR spectra contain also physical information that needs to be properly removed before a spectrum is obtained with which the chemical information can be evaluated by conventional chemometrics.

It is the main goal of this book to reunite the field and make the young generation of IR spectroscopists aware that there is more than the Bouguer-Beer-Lambert approximation and conventional chemometrics. I think that concepts like the quantum mechanical foundation of infrared spectroscopy or group theory and instrumental aspects are well introduced in other textbooks; therefore, this book will concentrate on introducing wave optics and dispersion theory to the interested reader. Therefore, it should be used as a kind of add-on. This is also how I understand the lecture series that I provide about this topic for master of photonics students at the Friedrich-Schiller-University, from which this book is derived. I hope the book somehow reflects the spirit of Paul Drude from whom it is said that he was originally skeptic about the, at his times newly introduced, Maxwell equations, but then obviously learned to value those highly. To be more accurate, my hope is it not only reflects this spirit but is also able to induce the same enthusiasm in its readers.

I will not end this preface without thanking those who helped me to realize this work. In particular, I thank Charles Bath who convinced me to publish this book with Elsevier and Andrea Dulberger who helped me along the journey to the final book. I am also deeply indebted to many of my colleagues at the Leibniz Institute of Photonic Technology with its unique scientifically stimulating environment and its scientific director Jürgen Popp who made it possible for me to return, little by little, from a six-year hiatus where I exclusively had administrative duties. My former colleague Georg Peiter, as well as Hartmut Hobert and Helga Dunken, my doctoral supervisor, have to be mentioned, who sparked my interest in dealing with infrared spectroscopy and evaluating spectra quantitatively. Some of the ideas in this book would not have come to life without my former Postdocs Sonja Höfer and Vladimir Ivanovski. Finally, I cannot express my gratitude in an appropriate quantity to my colleague Susanne Pahlow, not only for proofreading many chapters in this book but also for multiple suggestions on how to improve this manuscript and, last, but not at all least, for the beautiful cover design.

Part I

Scalar theory

# Chapter 1

What is wrong with absorbance?

What is wrong with absorbance? Since every textbook in infrared spectroscopy sees it as the fundamental quantity, it must have been checked a thousand times—nothing can be wrong with it. On the other hand, absorbance is not even mentioned once in the most seminal textbook of optics ("Principles of Optics," Born et al. [1]). Who is right? I gave (and still give) this much consideration, since I also was once firmly convinced that nothing can be wrong with absorbance. At present, I consider it as a result of a low-level theory, and it seems to me, at least in the molecular IR spectroscopy community, nowadays it has been forgotten that there exists also a high-level theory for more accurately describing light-matter interaction. One reason that the current status is what it is, with too much importance placed on absorbance, might be mostly (but not only) due to a fundamental misunderstanding that arose from a misinterpretation in connection with Fermi's Golden Rule—a strong argument for the correctness of this hypothesis can be found in a review article by Matossi [2], one of the, also forgotten, pioneers of IR spectroscopy. Accordingly, the intensity I of a light beam is decreased proportionally to the distance l when it travels through an absorbing medium that is characterized by a Napierian absorption coefficient $\alpha(\tilde{\nu})$ ($\tilde{\nu}$ is the wavenumber, the inverse of the wavelength):

$$dI = -\alpha(\tilde{\nu})I dl. \quad (1.1)$$

Actually, it was originally not the light beam intensity I that was used in this equation. Initially, it was the electric field intensity E^2:

$$dI = -\alpha(\tilde{\nu})E^2 dl. \quad (1.2)$$

If we now focus on the part of the intensity that is absorbed, I_A, relative to the initial intensity of the light beam I_0, the equation reads:

$$dI_A/I_0 \equiv dA = \alpha(\tilde{\nu})E^2 dl. \quad (1.3)$$

Here it is the absorbance! Actually, it is not. A is not the absorbance; it is the absorptance that is defined by $1-R-T$ (R and T are the specular reflectance and the transmittance, and for this formula to hold, we assume that no scattering takes place) or, put into words, the part of the intensity that is absorbed. Accordingly, absorption is proportional to the local electric field intensity. It is depending on the location, because it can be altered not only by absorption. Every interface changes it, as does wave interference! At this point, the only thinkable situation where the electric field intensity remains unchanged by such complications is when we have a strongly diluted gas (which is the case Bouguer and, later, Lambert were investigating). Under these circumstances, the local electric field intensity can be replaced by the intensity of the light since the only process that changes this intensity is absorption. In this case (and only in this case!), Eq. (1.1) can be integrated to yield absorbance A (to distinguish it from absorptance, the symbol is written in nonitalic style in the following):

$$\begin{aligned}\frac{dI}{I} &= -\alpha(\tilde{\nu})dl \rightarrow \\ -\ln\left(\frac{I}{I_0}\right) &= \alpha(\tilde{\nu})d \rightarrow \\ \mathrm{A} = -\log_{10}\left(\frac{I}{I_0}\right) &= (\log_{10}e)\cdot a(\tilde{\nu})d = a(\tilde{\nu})d\end{aligned} \quad (1.4)$$

In this equation, $a(\tilde{\nu})$ is the decadic absorption coefficient. Note that in this special case we need not take into account the wave properties of light. For samples different from a diluted gas, this is not a valid assumption. In fact, it would be quite paradox, because in quantum mechanics we allow matter to have wave properties to understand absorption. It is important to keep in mind that the application of Eq. (1.4) or use of the quantity absorbance A implies that effects resulting from the wave properties of light are neglected.

A further, perhaps even more subtle problem is introduced by the polarization of matter by light. While many learn of this effect already in school, it is rarely connected with spectra. We usually think looking at a spectrum what we get is directly related to material properties. But as soon as we shine light on a sample, the electric fields polarize the matter and, thus, change it. This change is not noticeable for a diluted gas, but once the molecules are closer to each other, their polarization changes one of their neighbors that then again act back on the molecules in their vicinity. Many of us know the famous Clausius-Mossotti or Lorentz-Lorenz relation and use it to calculate the refractive index of mixtures. But polarization does not only affect the refractive index but also its twin, the absorption index (which is the decadic absorption coefficient divided by the wavenumber and $4\pi(\log_{10}e)$), so why do we not apply the Lorentz-Lorenz relation to evaluate IR spectra, but Beer's law instead, that denies the effect of light on matter?

In my opinion, a helpful analogy to the Bouguer-Beer-Lambert (BBL) law and the use of absorbance and their relation to wave optics and dispersion theory is a comparison of the geocentric and the heliocentric model/worldview. The geocentric model explains what most people experience in their daily life, i.e., the (seeming) movement of sun and moon around the earth in circles (and even the phases of moon). That the movements of the other planets make no sense in the geocentric worldview does not matter since those movements are harder to observe. Therefore, these strange movements are not an experience of daily life. The same is true for subtle peak shifts, variations in band shape, and intensity changes in the spectra of organic and biological compounds. As soon as you look closer and have the desire to understand all effects that are actually present in spectra, then you have to move on and apply wave optics and dispersion theory.

Still, the quantity absorbance can certainly be used. As discussed earlier, the absorption coefficient is proportional to the imaginary part of the complex index of refraction function. Quantitative evaluation of spectra therefore means in the end to determine the optical constants of the material investigated. Once these are evaluated, we can calculate the (true or corrected) absorbance. However, the way it is usually done in infrared spectroscopy is to set the negative decadic logarithm of the transmittance or the reflectance equal to (the so-called apparent) absorbance. In many cases, this is very approximative and sometimes lacks any sense as we will see later on when we calculate the transmittance and the reflectance based on Maxwell's equations.

At this point, the usual question that follows is, ok, but what about spectrophotometry? Indeed, it can be shown (and, to my best knowledge, we were the first to do this in 2016 [3]) that under certain circumstances, in addition to the very diluted gas, the BBL law is (nearly) compatible with Maxwell's equations. However, this is in general just because the measurement is not performed like suggested by Eq. (1.4), because it is not the negative decadic logarithm of the transmittance of a solution that is used, but the ratio of this transmittance and the transmittance of the pure solvent. Moreover, as we will also see, for infrared spectra of diluted solutions the same trick will **not** work.

I have come to the woeful conclusion that the use of absorbance as the main quantity in infrared spectroscopy as this became established since the 1980s is highly misleading and has strongly hindered the development of the field since then. One example, where the ignorance of wave optics has greatly impaired the quantitative evaluation of spectra in the last years, is the so-called electric field standing wave (EFSW) effect, which is nothing else but an interference phenomenon [4–6]. It can be fully understood with help of Eq. (1.2) as we will see in the course of this book. A very closely related effect is the occurrence of interference fringes in the spectra of thin films. For me, it is unsettling to see that the removal of such fringes is nowadays often sold as a mere baseline correction that is performed after improper conversion of transmittance or reflectance to absorbance. A proper wave optics-based correction, which is known since the beginning of the 1970s, additionally removes the sometimes dispersion artifact called influence of reflectance on transmittance spectra. As can easily be verified experimentally, reflectance is not even approximately constant around bands. This is caused by dispersion and cannot simply be removed by subtracting a constant baseline from absorbance spectra. As soon as we understand that not only interference effects are at play, it becomes clear that this issue will also persist for films of organic or biological matter on a nearly index-matched substrate like, e.g., CaF_2 crystals. From a wave optics-based perspective, it is not surprising that such films on CaF_2 can show up to 30% and more deviation between apparent absorbance and true or corrected absorbance. This example illustrates how crucial it is to have an adequate understanding of the physical phenomena occurring during light-matter interaction.

Next to deviations in intensity, also changes of the wavenumber positions of bands commonly occur. From the viewpoint of contemporary textbooks, it may be a surprise, but the wavenumber positions of the maxima in absorbance never reflect the oscillator positions [7]. The deviations may be small, but can be as large as $25\,\text{cm}^{-1}$ in transflection spectra. On top of that, it is even possible that additional bands occur far from oscillator positions due to wave optics-related effects. Overall, for materials that can be characterized by a scalar dielectric function, neglecting such effects may be possible to some extent, but to know when and if, a profound knowledge of this topic is essential. In a likewise manner, it is not only wave optics, which is treated as an orphan in literature, but also dispersion theory [8]. Have you ever seen a textbook of infrared spectroscopy, which explains where the Lorentz-profile is derived from? (If you don't know it, Lorentz-profiles are used, among other profiles, to perform band fitting of absorbance spectra. A serious spectrometer software includes a corresponding module, which lets you do this kind of spectrum evaluation.) Indeed, from dispersion theory! If this was known, it would be easy (one additional line of code in the aforementioned spectrometer software) to calculate the index of refraction function from absorbance spectra (but there is no spectrometer software available that is able to do that, at least not when I last checked this) [7]. Even worse, have you ever seen a textbook where the BBL law is derived from electromagnetic theory? Just to remind you, we are now talking of the concentration dependence of the absorbance [9].

$$A = \varepsilon^*(\tilde{\nu}) \cdot c \cdot d, \tag{1.5}$$

where $\varepsilon^*(\tilde{\nu})$ is the molar decadic absorption coefficient, c is the molar concentration, and d is the sample thickness (the oldest reference I found for this modern form of the BBL approximation is that of a paper of Fritz Weigert from 1916. He states that this form had been provided by Bunsen, but does not give a reference).

Now you may already wonder, what is wrong with that? If you go back to Beer's paper [10], you see that Eq. (1.5) is merely empirical. Is there any derivation of this law? Occasionally, one finds a kind of derivation that relies on the quantity absorption cross section of one molecule. It is argued that if one adds up the extinction cross section (which is logically the same if all molecules are chemically identical) of the molecules, one ends up with the absorbance linearly depending on the concentration. This so-called derivation is doubly flawed. First of all, one assumption of Beer's law is that the sample is homogenous. If one assumes molecules, a natural counterargument is that if there are molecules, then there is shadowing, if one molecule is directly behind another, and how will you account for that? Even better, the absorption cross section is defined by the ratio of absorption coefficient and number of molecules per unit volume. So, this is like chasing one's tail and going around in circles, but more about this is discussed later on in this chapter.

In fact, the derivation of Beer's law is simple, when you know the Clausius-Mossotti equation or, equivalently, the Lorentz-Lorenz equation and how these are derived, but it nevertheless took until 2019 before someone realized this [11]. Say, we have one molecular dipole moment $\mathbf{p}$ (which is defined by charge times distance), how do we calculate the macroscopic polarization $\mathbf{P}$ in case there is no interaction between the dipole moments? This is, indeed, very simple:

$$\mathbf{P} = N \cdot \mathbf{p}. \tag{1.6}$$

N is the number of dipole moments per unit volume. Furthermore, of $\mathbf{P}$ is known that there is the following proportionality between it and an applied electric field for not too high field strengths (this is the linear regime, in which we will stay within this book):

$$\mathbf{P} = \chi \cdot \varepsilon_0 \cdot \mathbf{E} = (\varepsilon_r - 1) \cdot \varepsilon_0 \cdot \mathbf{E}. \tag{1.7}$$

In this equation, χ is the electric susceptibility, ε_0 is the permittivity of free space, and ε_r is the relative dielectric constant. Accordingly, if we solve the equation for ε_r, we find

$$\varepsilon_r = 1 + \frac{\mathbf{P}}{\varepsilon_0 \cdot \mathbf{E}}. \tag{1.8}$$

Now, let's put the result for $\mathbf{P}$ from Eq. (1.6) into Eq. (1.8):

$$\varepsilon_r = 1 + \frac{N \cdot \mathbf{p}}{\varepsilon_0 \cdot \mathbf{E}}. \tag{1.9}$$

If we consider that the molecular dipole moment is the mean polarizability α times the electric field and that $N = N_A \cdot c$, with Avogadro's constant N_A and the molar concentration c, and that the dipole moments and the electric field are co-oriented (isotropic/scalar medium), we obtain

$$\varepsilon_r = 1 + c \frac{N_A \cdot \alpha}{\varepsilon_0}. \tag{1.10}$$

Note that, if we have absorption, both ε_r and α are complex quantities, as is the index of refraction $\hat{n}$, the imaginary part of which is the absorption index k. The relation between ε_r and $\hat{n}$ is introduced by Maxwell's wave equation (we come to that later):

$$\varepsilon_r = \hat{n}^2 = (n + ik)^2. \tag{1.11}$$

Therefore,

$$\hat{n}^2 = 1 + c \frac{N_A \cdot \alpha}{\varepsilon_0}, \tag{1.12}$$

or using that the imaginary part of ε_r is equal to $2 \cdot n \cdot k$:

$$k = c \frac{N_A \cdot \alpha''}{2n \cdot \varepsilon_0}, \tag{1.13}$$

where α'' is the imaginary part of the polarizability. Now, the absorbance depends on k in the following way:

$$A(\tilde{\nu}) = 4\pi (\log_{10} e) d\tilde{\nu} k(\tilde{\nu}). \tag{1.14}$$

d is the cuvette thickness. Putting Eqs. (1.13), (1.14) together, one arrives at (with the molar absorption coefficient $\varepsilon^*(\tilde{\nu})$)

$$A(\tilde{\nu}) = \underbrace{\frac{2\pi N_A (\log_{10} e)\tilde{\nu}\alpha''(\tilde{\nu})}{n(\tilde{\nu}) \cdot \varepsilon_0}}_{\varepsilon^*(\tilde{\nu})} \cdot c \cdot d. \tag{1.15}$$

There it is, Beer's law! On the first view, everything seems ok, but on the second view, what has the index of refraction to do with the absorbance? There it is once again, the assumption of having a diluted gas with large distances between individual molecules! This is an implicit assumption, meaning that it is generally not mentioned, not even in the most advanced textbook about this topic I came across so far [12]. It might be disturbing; not only the index of absorption but also the (real) index of refraction depends on the concentration of the molecules [13]. As long as it stays small, the index of refraction is very close to one and does not interfere. If $\alpha''(\tilde{\nu})$ is small, $n(\tilde{\nu})$ will stay constant even around bands and for larger concentrations close to those for (neat) liquids and solids (we will investigate this relationship more closely in Section 5.3). Once the index of refraction becomes important, however, we can readily predict that it will destroy the linear relationship between absorbance and concentration. Even worse, the peak shape will be altered and the peak will shift to higher wavenumbers, since the minimum of $n(\tilde{\nu})$ is located at higher wavenumbers relative to the oscillator position. Nevertheless, with the absorbance, all quantities that are derived from it, like said absorption cross section (which is, of course, also changing with concentration), can be meaningful within certain limits, but one must have profound knowledge of high-level theory in order not to overstep the mark. For those who still do believe (actually, scientists should never believe in anything science related, but see everything just as a hypothesis) in Beer's law, how about the Raman effect? As we all know, it is based on changes in the polarizability. But the existence of polarizabilities requires matter to be polarized. As already discussed, a result of this polarization is local electric fields different from externally applied ones. As a consequence, deviations from linearity are given. This means, being strict, the ones who use the Raman effect cannot believe at the same time in the correctness of Beer's law.

Note that, for the derivation of Beer's law, I had to assume that the applied electric field is equal to the local field around the molecules/atoms. Once again, the diluted gas approximation is key here to fulfill this condition. In the next step, one would have to consider the difference between applied and local fields and use the Clausius-Mossotti or, equivalently, the Lorentz-Lorenz relation for somewhat higher concentrations. Overall, however, this will not limit the linear regime of the concentration dependence of the absorbance much more, as I will demonstrate in Section 3.3. This is about as far as modern theory (which is about a hundred years old) goes to understand deviations from the linear concentration dependence that Beer's empiric law assumes [14]. This wording already suggests that there is more, and indeed there is. Before we discuss this in more detail, I will first emphasize something that probably already lingered in the back of your mind as it did in mine, before I was able to fully understand it. The Lorentz-Lorenz relation describes a merely physical interaction, in the sense that it only persists as long as you shine light on matter. This is in contrast to chemical interactions, which continue to be there in the dark. Usually, these chemical interactions are suspected to be the cause of every deviation from Beer's part of the BBL law. That this is not true can easily be shown with the help of IR spectra of (quasi)ideal mixtures like benzene-toluene, which obviously do also deviate from Beer's law [15]. However, again, this interaction exists only as long as light shines on the sample, because the polarization is caused by an externally applied electric field. This means that shining light on a sample does change this sample, and when we interpret a spectrum, we interpret something that reflects this changed state the sample is in. I have mentioned this already, but what I have not mentioned is the bad news that the Lorentz-Lorenz relation is itself approximative and we cannot use it to fully obtain the unchanged state of matter. The good news is that often we can safely neglect the error when we deal with organic or biological matter, except if we use such highly sensitive evaluation methods like 2D correlation spectroscopy or principal component analysis. For strong oscillators, however, the effect can be overwhelming. For example, when we reinvestigated solvatochromism in the visible spectral region using the Lorentz-Lorenz relation we found not only the position of bands can redshift by up to 50%, but also the molar absorption coefficient at the peak of this band can increase by up to 600% depending on the solvent [16]. Obviously, under such conditions, Beer's law cannot hold.

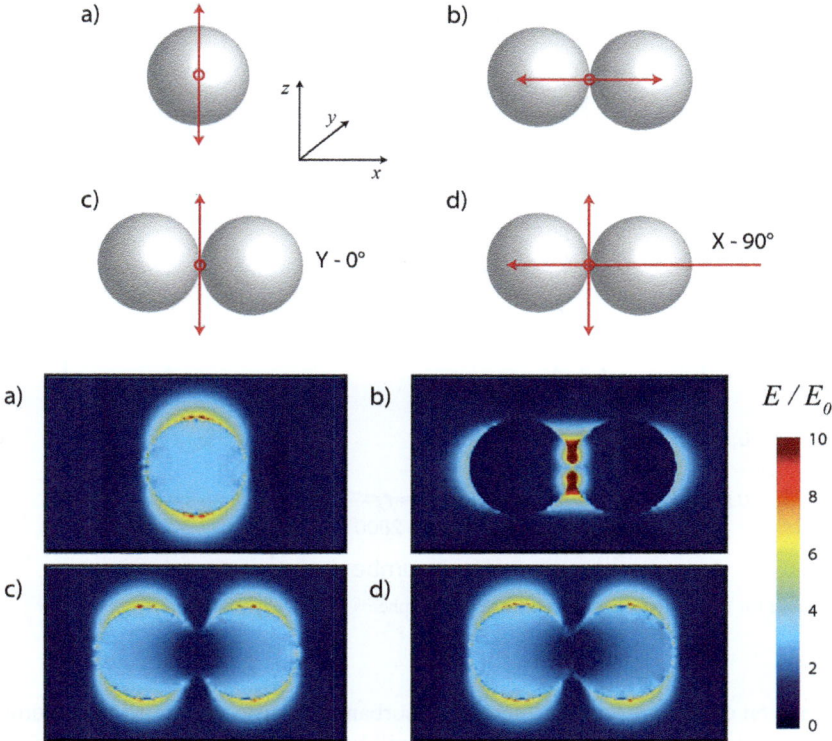

FIG. 1.1 Electric fields relative to the incoming electric field strengths around a single sphere (A) and touching spheres calculated by finite-difference time-domain simulations. (B) Light propagation direction along *y*, polarization direction along *z*. (C) Light propagation direction along *y*, polarization direction along *x*. (D) Light propagation direction along *x*, polarization direction along *z* [17].

Local field effects are derived under the assumption that the electric field is actually static or, at least, does not vary strongly over a molecule or a particle. While a molecule is comparably small, a particle or a crystallite is usually much larger and the precondition to have a nonvarying electric field is harder to obey. Mie theory is able to deal with varying fields. If you use Mie theory to calculate the electric field around a sphere, you literally can see that the field varies. What's more, if the spheres are much smaller than the wavelength, so that the corresponding sample would be micro-heterogenous, but homogenous for light, the static field assumption is still not true as can be seen in Fig. 1.1. We originally assumed that the local field is the same everywhere as the one applied, which also means that it is isotropic. Neither of these assumptions holds in the case of the two spheres (note that this is a situation that cannot be calculated by Mie theory, which only holds for a single sphere, or a bunch of spheres that are so far away from each other that they do not interact). The electric field intensity depends on the polarization of light relative to the orientation of the spheres and the incoming direction. Quite interesting is also case d) where one might expect that one sphere is *shadowed* by the other. Nevertheless, the electric field distribution is not much different from case c) where both spheres are illuminated by the incoming light. Keep in mind that we are in the wave and not in the ray regime, so that shadowing is a concept that does not make sense. Accordingly, assuming that one molecule could shadow another is like assuming that you would not get wet when you enter the sea directly behind another person.

Interestingly, even if we assume to have only a single sphere, so that nearfield or local field effects are absent, Beer's law can be shown to be a limiting law. Based on Mie theory, we can calculate separately the scattering, extinction, and absorption. To do that, for larger spheres a number of Bessel functions must be calculated, which are roughly proportional to $2\pi r n_m/\lambda$, where r is the radius of the sphere, n_m is the index of refraction of the surrounding medium, and λ is the wavelength. If r becomes small, then only a few terms are needed, and finally, the absorption cross section converges to the following relation:

$$C_{\text{abs}}(\tilde{\nu}) = 8\pi^2 r^3 \tilde{\nu} \text{Im}\left\{\frac{\varepsilon_r - 1}{\varepsilon_r + 2}\right\}. \tag{1.16}$$

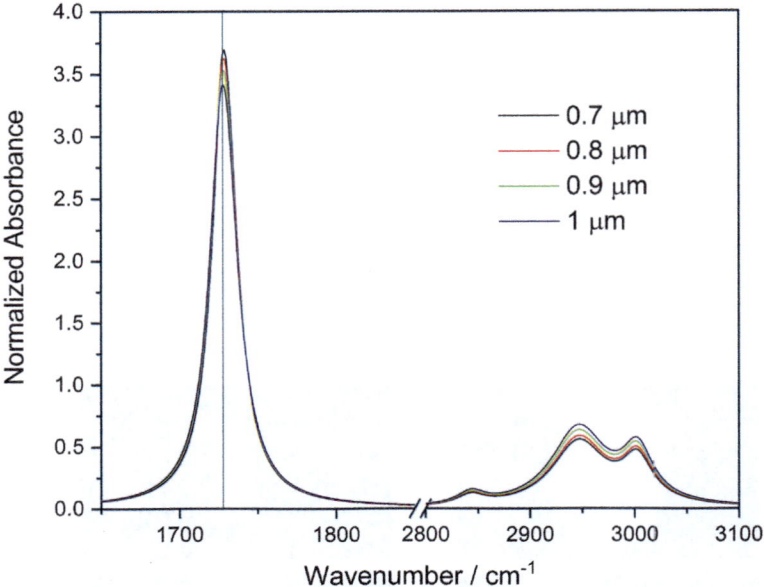

FIG. 1.2 Normalized absorbance (or absorption cross section) of PMMA spheres of different diameters.

This means that the absorption cross section, and, thereby, absorbance, is proportional to the volume of the sphere and we recover Beer's law, i.e., if we double the volume, absorbance will be doubled. However, Eq. (1.16) is again a limiting law, in full accordance with what we have derived so far. Accordingly, if we calculated the volume normalized absorption cross section following Mie theory for spheres of different radii, we can expect that the values will be a function of sphere size.

Let's test this assumption. In Fig. 1.2, you find corresponding results for poly(methyl methacrylate) (PMMA) spheres in vacuum and diameters ranging from 700 nm to 1 μm. Compared to the wavelength, these spheres are no longer small, so that one cannot assume a kind of effectively homogeneous medium. But even when we investigate the absorption cross sections of smaller spheres, smaller than 10 times the wavelength, say with a radius 200 nm, this absorption cross section is according to Mie theory not 8 times that of a sphere with a 100 nm radius as it would have to according to the BBL approximation.

In the previous paragraph, we could already guess that the BBL law is out of its limits once the medium can no longer be assumed as homogeneous. In fact, homogeneity is a very important requirement for the law to hold. Why? Because generally it is the intensities that are additive, i.e., reflectance or transmittance (absorbance is not an intensity in the sense that it is proportional to the electric field intensity, i.e., the electric field squared). This does not play a role as long as the sample is homogeneous, but otherwise additivity on the level of reflectance or transmittance is not compatible with the BBL law [18].

Consider a sample that is homogeneous for infrared light, which means that we would not see any heterogeneity if we investigated the sample with an IR microscope. Such a homogeneous sample can still be composed of various compounds, but these compounds must be intimately mixed on a level that is not resolvable for light and the composition must be the same everywhere. If this were the case, then every microscopic spectrum will be the same as the macroscopic average. Now assume that the compounds completely demix. In this case, the microscope will detect regions with only one compound and regions where there is more than one compound. In the latter case, the overall intensity spectrum will be a weighted average of the different intensity spectra, and this is also true for the macroscopic spectrum. Therefore, the macroscopic spectrum of a micro-heterogeneous sample will in general be different from that of a micro-homogeneous sample of the same composition. In other words, if you applied, e.g., a checkerboard pattern of two different colors on a window, a spectrometer behind the window would detect something else as if you simply mixed these colors and put the mixed color onto the window.

If you take a spectrum of the whole area or volume, then weak bands will still follow the BBL law to a good approximation. In contrast, stronger bands will increase nonlinearly with concentration if a larger area of the sample is recorded. In fact, spectra of heterogeneous samples change continuously with the structure size, e.g., with the edge length in case of said checkerboard structure. If this surprises you, then think of the wave function of a particle in a box. The wavefunction obviously changes with the size of the box, even if the potential outside the box would be finite. In case of light, the potential is represented by the dielectric function and one can easily imagine that the light intensity within structures depends on the structure size. This effect can have drastic consequences for spectral evaluation, but is commonly unknown [18].

That anisotropy-related effects, as in case of the two spheres, are a game changer, too, is also typically not fully realized on the level of the BBL. Usually in this case in the textbooks of infrared spectroscopy, a theory is invoked that has its origin sometime around the 50s of the last century [19]. At that time, it seemed that no full understanding of the optical properties of anisotropic materials existed (albeit, actually, the understanding was much more advanced among spectroscopists in earlier times [20] and based on Paul Drude PhD thesis [21] and the later experimental confirmation by Matossi and Dane [20], the theory from the 1950s can easily be falsified); therefore this (very) approximate theory was developed, which is often referred to as Linear Dichroism Theory. Unfortunately, it is fully based on absorbance and therefore inherited all aforementioned shortcomings right from the start. However, on top of that comes a number of additional weaknesses. The denial of the effects of interfaces (somehow it is very telling that Linear Dichroism Theory works strictly only for oriented gas molecules of a highly diluted gas) leads to the erroneous conclusion that it is only the angle of the transition moment relative to the polarization direction that dictates the intensity of a band in a spectrum. From basic electromagnetic theory, we know, however, that at interfaces only the tangential component of the electric field is continuous. This results in band shifts, if the transition moment is not oriented parallel to the surface as Drude's theory predicted and Matossi and Danes experiment and modeling efforts showed. The corresponding effect is related to the, for inorganic materials and layers, well-known Berreman effect, and both are transversal-optical longitudinal-optical shifts. These band shifts are also accompanied by changes of the band shapes. The shifts are dependent on the oscillator strengths of the vibrations and may not be recognizable for weaker transitions in organic and biological material. This is not true, however, for the intensities. Depending on the angle between transition moment and interface, absorbance can easily be up to 100% stronger for the same angle between transition moment and polarization direction, if it is not parallel to the interface [22]. While correction schemes for anisotropic materials may be extremely effortful (counterintuitively, the most complicated situation is a randomly oriented polycrystalline material where the crystallites are not small compared to the resolution limit of light!) [23], the spectra of homogenous and isotropic layers on substrates can, e.g., be corrected extremely fast and in an automatic manner, if we apply a little bit of wave optics. So it is puzzling to me, why so often obscure chemometrics is applied to apparent absorbance spectra, when wave optics can accomplish better results in shorter time.

If you want to get a more detailed overview of the shortcomings of the BBL law, which I will call in the following for the reasons above the BBL approximation, I suggest you to read Ref. [24].

Before we go into *medias res* (for an overview of wave optics and dispersion theory, cf. Fig. 1.3), I do not want to conclude this chapter without mentioning two further problems that we will solve in the course of this book. First of

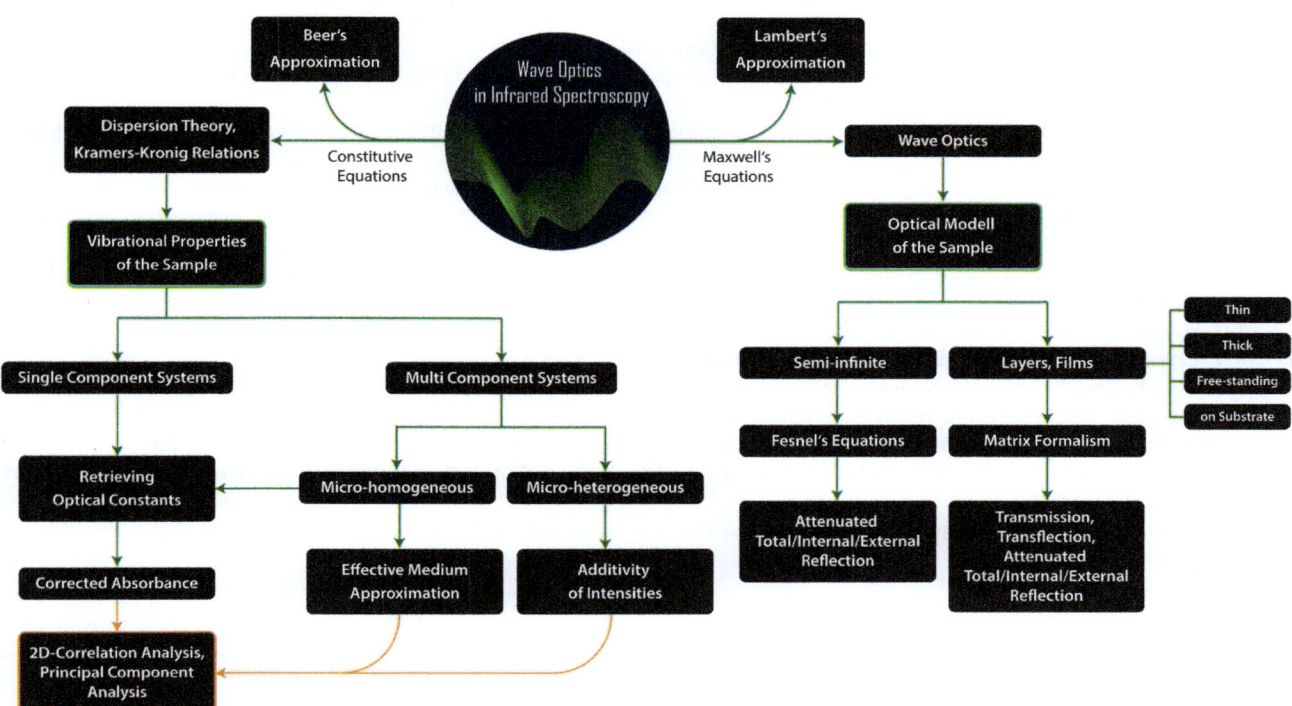

FIG. 1.3 Overview of wave optics and dispersion theory.

all, we do not stop—as books usually do—at materials with orthorhombic symmetry, i.e., we do not assume that the off-diagonal terms of the dielectric tensor function, which describes dispersion for anisotropic crystals, are zero. This means that monoclinic and triclinic crystals will be properly treated, so that their spectra can be understood and quantitatively evaluated. Furthermore, I will demonstrate that isotropy is just necessary, but not sufficient, for reducing a dielectric tensor to a scalar. In fact, as a consequence, at least three different types of optical isotropy exist. While this seems on the first view to be a philosophical insight with little practical relevance, I think, it is everything else, because usually textbooks not only refrain from treating materials with lower symmetry than orthorhombic, but quite often stick to materials that can be characterized by a scalar dielectric function (the special case that is assumed for the first part of this book). Even then it is first important to understand the more general principle before the simpler one can be appropriately appreciated.

References

[1] M. Born, E. Wolf, A.B. Bhatia, Principles of Optics: Electromagnetic Theory of Propagation, Interference and Diffraction of Light, Cambridge University Press, 1999.
[2] F. Matossi, Ergebnisse der Ultrarotforschung, in: F. Hund (Ed.), Ergebnisse der Exakten Naturwissenschaften: Siebzehnter Band, Springer, Berlin, Heidelberg, 1938, pp. 108–163.
[3] T.G. Mayerhöfer, H. Mutschke, J. Popp, Employing theories far beyond their limits—the case of the (Boguer-) Beer–Lambert law, ChemPhysChem 17 (2016) 1948–1955.
[4] T.G. Mayerhöfer, H. Mutschke, J. Popp, The electric field standing wave effect in infrared transmission spectroscopy, ChemPhysChem 18 (2017) 2916–2923.
[5] T.G. Mayerhöfer, J. Popp, The electric field standing wave effect in infrared transflection spectroscopy, Spectrochim. Acta A Mol. Biomol. Spectrosc. 191 (2018) 283–289.
[6] T.G. Mayerhöfer, S. Pahlow, U. Hübner, J. Popp, Removing interference-based effects from the infrared transflectance spectra of thin films on metallic substrates: a fast and wave optics conform solution, Analyst 143 (2018) 3164–3175.
[7] T.G. Mayerhöfer, J. Popp, Quantitative evaluation of infrared absorbance spectra—Lorentz profile versus Lorentz oscillator, ChemPhysChem 20 (2019) 31–36.
[8] T.G. Mayerhöfer, J. Popp, Beer's law—why absorbance depends (almost) linearly on concentration, ChemPhysChem 20 (2019) 511–515.
[9] F. Weigert, Über Absorptionsspektren und über eine einfache Methode zu ihrer quantitativen Bestimmung, Ber. Dtsch. Chem. Ges. 49 (1916) 1496–1532.
[10] Beer, Bestimmung der Absorption des rothen Lichts in farbigen Flüssigkeiten, Ann. Phys. 162 (1852) 78–88.
[11] T.G. Mayerhöfer, J. Popp, Beer's law derived from electromagnetic theory, Spectrochim. Acta A Mol. Biomol. Spectrosc. 215 (2019) 345–347.
[12] G. Kortüm, Kolorimetrie · Photometrie und Spektrometrie: Eine Anleitung zur Ausführung von Absorptions-, Emissions-, Fluorescenz-, Streuungs-, Trübungs- und Reflexionsmessungen, Springer, Berlin, Heidelberg, 1962.
[13] T.G. Mayerhöfer, A. Dabrowska, A. Schwaighofer, B. Lendl, J. Popp, Beyond Beer's law: why the index of refraction depends (almost) linearly on concentration, ChemPhysChem 21 (2020) 707–711.
[14] A. Aubret, M. Orrit, F. Kulzer, Understanding local-field correction factors in the framework of the Onsager–Böttcher model, ChemPhysChem 20 (2019) 345–355.
[15] T.G. Mayerhöfer, O. Ilchenko, A. Kutsyk, J. Popp, Beyond Beer's law: quasi-ideal binary liquid mixtures, Appl. Spectrosc. 76 (2022) 92–104.
[16] S. Spange, T.G. Mayerhöfer, The negative solvatochromism of Reichardt's dye B30—a complementary study, ChemPhysChem 23 (2022) e202200100.
[17] T.G. Mayerhöfer, S. Höfer, J. Popp, Deviations from Beer's law on the microscale—nonadditivity of absorption cross sections, Phys. Chem. Chem. Phys. 21 (2019) 9793–9801.
[18] T.G. Mayerhöfer, J. Popp, Beyond Beer's law: spectral mixing rules, Appl. Spectrosc. 74 (2020) 1287–1294.
[19] R. Zbinden, Infrared Spectroscopy of High Polymers, Academic Press, New York, London, 1964.
[20] F. Matossi, F. Dane, Reflexion, Dispersion und Absorption von Kalkspat im Absorptionsgebiet bei 7 μ, Z. Phys. 45 (1927) 501–507.
[21] P. Drude, Ueber die Gesetze der Reflexion und Brechung des Lichtes an der Grenze absorbirender Krystalle, Ann. Phys. 268 (1887) 584–625.
[22] T.G. Mayerhöfer, Employing theories far beyond their limits—linear dichroism theory, ChemPhysChem 19 (2018) 2123–2130.
[23] T. Mayerhöfer, Z. Shen, R. Keding, J. Musfeldt, Optical isotropy in polycrystalline Ba2TiSi2O8: testing the limits of a well established concept, Phys. Rev. B 71 (2005) 184116.
[24] T.G. Mayerhöfer, S. Pahlow, J. Popp, The Bouguer-Beer-Lambert law: shining light on the obscure, ChemPhysChem 21 (2020), https://doi.org/10.1002/cphc.202000464.

Chapter 2

Transition from the Bouguer-Beer-Lambert approximation to wave optics and dispersion theory

Originally, the first chapter of this book started with Maxwell's equations on the typical level for readers with physics background on bachelor level. This was not a good idea, since it is the primary goal of this book to help reconcile the different branches of infrared spectroscopy. Starting with Maxwell's equations may be too demanding for the typical molecular spectroscopists with a background in chemistry or biology. From my own background, I know that you are usually not able to immediately start with Maxwell's relations in vectorial and partial differential form and combine the results for plane electromagnetic waves with dispersion theory in full flavor. Instead, please see the following chapter as an intermediate and linkage between the BBL approximation on the one hand and wave optics and dispersion theory on the other. I will introduce some simplifications and approximations as well as explanations and formulas, which hopefully will arm you with the necessary knowledge to render access to a lot of information provided in the next chapters easier and motivate you to go on. Once you have mastered this chapter, you can try and see if this fundament suffices to understand what is provided in Chapters 6–10. But, as I already stated, the most important idea behind this chapter is, however, to make you fit for Chapters 3–5 and render a part of what you will learn there redundant so that you can focus more on the essentials. If you have a solid knowledge of electromagnetics and optics, you can simply skip this chapter.

2.1 The electric field and the electric displacement

There is no full understanding of infrared spectra without understanding the nature of light. A key quantity for this understanding is the electric field E. The source of an electric field is a charged particle or a distribution of charged particles that exert a force on other charged particles. This force can either be attractive or repellent depending on the sign of the charges. If we assume a distribution of charged particles as a source of the electric field, we can represent them with the help of the charge density ρ,

$$\rho = \frac{Q}{V}, \tag{2.1}$$

wherein Q represents the overall charge and V the volume. Usually, the charge density is defined in differential form by the number of charges per unit volume. As already stated, charges or charge density are the sources of the electric field that is explained by the force that it exerts on an infinitesimally small positive test charge q:

$$E = \frac{F}{q}. \tag{2.2}$$

This force exists at every point in space, and, in turn, it determines the electric field or the electric field strength at all points. It can be represented by electric field lines, which start in positive and end in negative charges. The number of field lines is usually chosen to be proportional to the electric field strength. A particular simple electric field is established between the plates of a plate capacitor, cf. Fig. 2.1. As long as the charges would only be in empty space, i.e., in a vacuum, the electric field would be all we need in the following, but since we want to understand spectra, we unavoidably require to understand how the electric field is changed when it enters matter, i.e., the samples. These changes are neglected when the BBL approximation is used, but one is often confronted with them. For the moment, the most important change is that the number of field lines inside matter is reduced. We can describe this reduction by the dielectric constant ε that relates the electric field in vacuum and in matter, E_{vac} and E_{mat}:

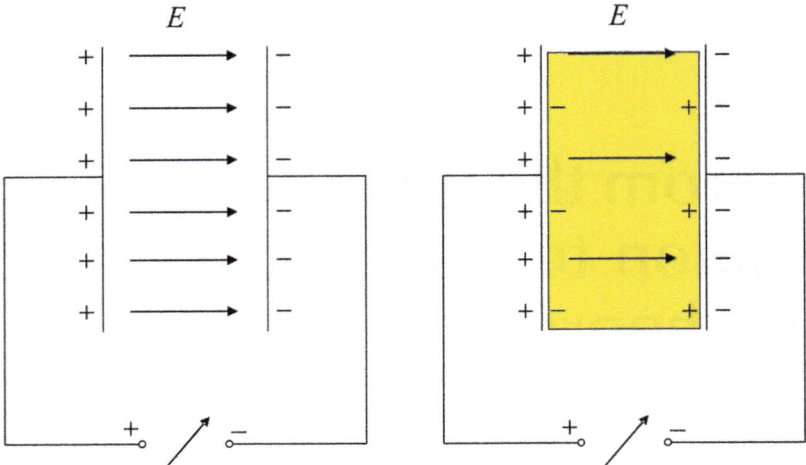

FIG. 2.1 Homogeneous electric field between the plates of a plate capacitor without *(left side)* and with matter *(right side)*.

$$E_{\text{vac}} = \varepsilon E_{\text{mat}}, \qquad (2.3)$$

The reason for this reduction is that the electric field polarizes matter and creates dipoles with dipole moment p,

$$p = \alpha E, \qquad (2.4)$$

which counteract the electric field so that it is reduced. In Eq. (2.4), α is the polarizability and we have assumed that the electric field is comparably weak. If it is not, the dependence between polarizability and the electric field will be nonlinear and contain higher order terms, but for the whole book we will assume that the relation between the dipole moment and the electric field is linear. The dipole moment itself is defined by the product of the distance l between a negative and positive charge $-q$ and q and the charge $p = q \cdot l$.

By defining the polarizability as a scalar, we have presumed that our medium is isotropic and homogeneous. Since we have not discussed a time dependence of the electric field, we consider it to be static for the moment, which means that its frequency is zero and its corresponding wavelength is infinite. In this case, the assumption of homogeneity is trivial in the mathematical sense, meaning that the static electric field in a plate capacitor will average over a potentially inhomogeneous medium in between the plates. Note that depending on the scale, every medium is inhomogeneous. The demand for homogeneity, which is intrinsic by assuming that the polarizability is no function of position within the medium, is much less easy to fulfill if we have to deal with a changing electric field like in the case of light, e.g., for the mid-IR spectral region with wavelengths between 2.5 and 25 µm. Please keep in mind right from the start a rule of thumb that says that for matter to be homogeneous for light, inhomogeneities should be smaller than about a tenth of the wavelength (in the UV/vis maybe even a hundredth of the wavelength due to much stronger scattering). This is well below the limit of resolution of a microscope, but even then, it can be shown that light, having wave properties, is already somewhat influenced by inhomogeneities, albeit this influence can be neglected in practice. The other extreme is the X-ray spectral regime, where the inhomogeneities, in this case the atoms, are so large compared to the wavelengths that the X-ray light is scattered by them.

Isotropic, on the other hand, means that there is no directionality with regard to the properties, and, as long as you can consider the medium of interest as homogeneous compared to the wavelength, isotropy is equivalent with a polarizability and a dielectric constant that are scalars. What does this mean? As you can see in Fig. 2.1, the field lines have a direction, the same along which the positive test charge is pushed by the electric field. In other words, the electric field is a vector field, i.e., to every point of space a direction is assigned. This direction does not change in vacuum, but in matter it is the usual case that in some directions it is easier to polarize matter than in others. This is not hard to understand if you think of a molecule. For example, a buckyball, i.e., a fullerene or C_{60} molecule, is nearly spherical and, therefore, its polarizability is isotropic. On the other hand, a benzene molecule will have a strongly anisotropic polarizability, which is generally very different in the direction along its sixfold rotation axis compared to the direction perpendicular to it. For such anisotropic molecules, the direction of the electric field and that of certain of their symmetry elements will not coincide. As a result, induced dipole moment and electric field will have different directions and the polarizability will become a tensor, which transforms one vector into the other. As a consequence, also the dielectric constant will become a tensor. For the moment, we will not discuss directional dependence and, in particular, anisotropy anymore until we come back to the latter starting from Chapter 11. We can do that because a liquid consisting of anisotropic molecules or a randomly oriented polycrystalline material can have a scalar macroscopic polarizability and dielectric function (the dielectric constant is frequency dependent), as long as it is homogeneous relative to the wavelengths of the incident light. Even if you do not intend to immerse yourself into the optics and spectroscopy of anisotropic materials, please do always keep these constraints in mind, since a lot of spectral features are caused or changed if these basic assumptions are violated.

Having introduced the polarization, we are able to introduce another field that is concomitant with the electric field, which is the electric displacement D. In vacuo, the electric field and the electric displacement only differ by a constant,

$$D = \varepsilon_0 E, \tag{2.5}$$

which is the permeability of free space, ε_0. The unit of the electric field is [V/m], while that of ε_0 is [C/(V m)]. This does not only tell us that the electric displacement has a unit different from the electric field, namely [C/m^2], but also that it is defined in a different way. Indeed, the electric displacement is not defined by the force on a test charge, but by an action, i.e., a displacement of the bound charges in matter. Since in a vacuum, there are no bound (or free) charges, the ratio of the number of field lines through a unit area perpendicular to these field lines is given by the permeability of free space. When going over to matter, some of the charges that would cause the electric field are compensated by induced dipoles (cf. Fig. 2.1). Therefore, the number of field lines per unit area perpendicular to the lines, that is the field line density, decreases, as field lines can only originate from uncompensated charges. This is different for the electric displacement that is in matter equal to

$$D = \varepsilon_0 E + Np = \varepsilon_0 E + P. \tag{2.6}$$

Here, N is the number of (induced) dipole moments per unit volume and P is the polarization. This assumes that the macroscopic quantity polarization is a mere volume average of the microscopic polarization and links macroscopic and microscopic quantities. This does not seem a big deal, but without that you know, we are already discussing important points with regard to Beer's approximation. In fact, we will discuss and see in later sections that a dipole certainly has some effect on neighboring dipoles, which, in turn, have again an impact on the original dipole and so on and so forth. We can certainly assume that the p in Eq. (2.6) is an average dipole moment at an instant in time after having switched on the electric field, where the interactions have already caused some changes that led to a new equilibrium. In this case, you have to keep in mind that this dipole moment is not the same as that which an isolated molecule in vacuum or in a highly diluted solution would have. In a first approximation, the connection between the dipole moment of an isolated molecule and the same molecule in a denser gas, a less diluted solution, or in a condensed phase can be established by the Lorentz-Lorenz relation (or, equivalently, the Clausius-Mossotti relation), which we will examine in detail in Chapter 5 and, in even more detail, in Chapter 10. Even though we do not discuss it now, I will already mention and you should keep in mind that the Lorentz-Lorenz relation only holds for a spherical molecule in an assumed homogeneous surrounding that nevertheless consists of the same spherical molecules (or the molecules of a solvent that are also assumed to be spherical). This shall already warn you that Lorentz-Lorenz theory is just the simplest theory that takes into account the effect of so-called local fields. In other words, polarization of matter leads to the fact that the electric field at the position of a molecule is not the same as the electric field that is applied from the exterior, except if you have an isolated molecule in a vacuum. Again, just as a hint, if you disregard this polarization effect and linearize a nonlinear relationship between absorption index and transition moment, you end up with Beer's approximation as will be demonstrated in detail in Section 5.3. You might find all these warnings about the applicability of certain relations and their limits somewhat strange, but I think that much too often simple relationships are presented as universal laws (at least in the mind of the reader), while they are nothing else in reality but approximations. In my mind, this is one of the reasons why IR spectroscopy does no longer evolve from a theoretical point of view (for instrumental development, this is definitely different, but sometimes I find asking myself if the product of the magnitude of theoretical knowledge and the quality of available instruments is a kind of fundamental constant).

Before we go on, please note that the unit of the polarization is [C/m^2], because the dipole moment is defined as charge multiplied by distance, and if we multiply this with the number of dipole moments per unit volume, N, we end up with [C/m^2], which is the same unit as the electric displacement. Since we assume scalar media, the polarization has the same direction as the electric displacement. Contrary to the electric field and the electric displacement, however, which originate in the positive charge and point to the negative charge, the dipole moment points from the negative charge to the positive charge. This seems to create a contradiction. However, if you look at the right side of Fig. 2.1, then you see that the charges inside the medium are opposed to the one on the plates. If you connect the charges inside the medium that are opposite to the plates, you get arrows in the same direction as those of the electric field. This means if you add the field lines of the electric field and those due to the dipoles inside the medium, you get the same number of field lines as for the electric field in

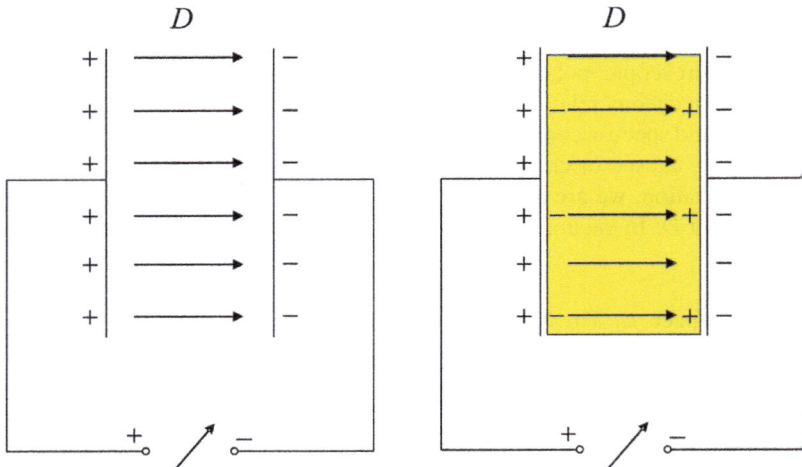

FIG. 2.2 Homogeneous electric displacement between the plates of a plate capacitor without *(left side)* and with matter *(right side)*.

vacuum (left side of Fig. 2.1). In other words, the number density of the field lines of the electric displacement in vacuo (Eq. 2.5) must be the same as that if you introduce matter (Eq. (2.6)). This is also shown in Fig. 2.2.

One could also say that the number (density) of field lines corresponds to the number (density) of charges on the plates. This number is not altered if matter is introduced within the plates as long as the switch is open and no current can flow; otherwise the number density of charges is increased and the field lines of the electric field stay constant. Anyway, as you can imagine, we only need either E or D as long as we talk about vacuum. This changes if we introduce matter, which we obviously have to do, since the sample the spectrum of which interests us is, of course, nothing else but this matter. In this case, we need both field quantities. The connection between both in matter can also be established in an alternative way:

$$D = \underbrace{\varepsilon_0 \varepsilon_r}_{\varepsilon} E. \tag{2.7}$$

In Eq. (2.7), ε_r represents the unitless relative dielectric constant. The product $\varepsilon_0 \varepsilon_r$ equals the dielectric constant or permittivity ε. Quite often in literature, however, the relative dielectric constant will be written as ε, and this is also what I will do in the following chapters if nothing else is mentioned. Eq. (2.6) and Eq. (2.7) put together give

$$\varepsilon_0 E + P = \varepsilon_0 \varepsilon_r E \rightarrow P = \varepsilon_0 (\varepsilon_r - 1) E = \varepsilon_0 \chi E, \tag{2.8}$$

with χ the electric susceptibility. It looks like that the electric susceptibility is just another way to express the dielectric constant, but for the derivation of Beer's approximation later on in Section 5.3 (cf. also Chapter 1), it makes sense to introduce it at this point.

I want to already alert you at this point about what happens in anisotropic media. In the general case, the direction of the electric field and the direction of polarization at the same point will not be the same, which means that also the E- and the D-field will have different directions. Again, we will end up with vector and tensor calculus. At the end of this chapter, I will introduce the vectorial form of the fields as you will need it for the next chapter, but until then we will stay with scalars.

2.2 The magnetic field and the magnetic induction

I do not want to spend too much words talking about the magnetic field and the magnetic induction, not because I do not consider them important, but because we more or less introduce them just because they are coupled via Maxwell's equations to the electric field and the electric displacement, but then we will do everything to get rid of them as soon as possible. The reason for this is that for the systems that we will focus on (actually for most of matter, but with the important exception of metamaterials) the interaction between them and the magnetic field is generally so weak that the related effects are not important (for the following, we also exclude circular birefringence and dichroism, but the formalism introduced in Chapter 12 allows to treat these two sides of the same coin and we will discuss vibrational circular dichroism in some detail in Chapter 16). In principle, the magnetic field B is defined analogously to the electric field by the force that a field of a certain strength exerts on a charged particle. It is a particle that is not stationary, but moving at a certain speed v. If the charged particle is traveling perpendicular to the magnetic field lines, and a force F is exerted on it, then B is given by

$$B = \frac{F}{qv}. \tag{2.9}$$

In practice, one would have to quantify first the force that is exerted on the test charge at rest to determine the electric field strength and subsequently the force at a speed v. The difference would then be the force due to the magnetic field alone and would allow to determine it. Based on its definition, the unit of the magnetic field is [N/(m A)].

Similar to the electric field and the electric displacement, in vacuo the magnetic field B and the magnetic field H are simply related by a constant μ_0, which is called the permeability of free space and has the unit [(V s)/(A m)]:

$$B = \mu_0 H. \tag{2.10}$$

Correspondingly, the magnetic field H has the unit [A/m] and becomes important in relation to matter. Accordingly, in matter the definition of the H-field is given by

$$H = \frac{B}{\mu_0} - M. \tag{2.11}$$

M is the magnetization, which is the counterpart of the polarization when we talk about magnetic fields and the changes these fields induce in matter. As Eq. (2.11) implies, it has the same unit as the magnetic field H. Another way of taking into account the changes inside matter is by using the permeability μ:

$$B = \underbrace{\mu_0 \mu_r}_{\mu} H. \tag{2.12}$$

The counterpart of the permeability μ for the electric field is the permittivity ε, and all I said about isotropy and homogeneity in the previous section, as well as about vectors and tensors, applies to the permeability as well, meaning that in the most general case μ is a tensor and becomes a scalar only if there is no directional dependence in a homogeneous medium. Even better, for all materials we will discuss, magnetization is so weak that we can generally consider μ as a scalar that is not frequency dependent and that is always equal to μ_0. In other words, for the magnetic field all kinds of matter that we discuss will be the same as vacuum. While we will touch the influence of metals as substrates or thin films, those will always be unstructured and large perpendicular to the direction of propagation of the light. A counterexample, where we could not neglect the influence of magnetic fields, because we actually use it to change the optical properties of matter, would be metamaterials with structured films in which charges would cause a μ_r different from unity—such materials will be excluded from discussion within this book. Before I close this section, if you ask yourself how we can get rid of the magnetic fields when they are the integral part of Maxwell's equations, the answer is that we will proficiently combine two of the four equations so that we obtain the so-called wave equation and by doing so, the magnetic fields will be removed from this equation (for fans of the magnetic fields, you can also turn the tables and remove the electric fields. This approach has in some instances also its justification, but those cases will not be discussed either). Nevertheless, the magnetic fields certainly will still be there and accompany the electric wave to form the electromagnetic wave that is light.

2.3 Maxwell's equations in simplified form

By now, you should be fit to make your first acquaintance with Maxwell's equations, and I will introduce you to them for the time being only in strongly simplified form.

2.3.1 1st equation—Gauss's law

The sources of the electric displacement D are charges. These charges, which are represented by the charge density $\rho = Q/V$, cause the electric field. The charges are the origins of the force that acts on other (test) charges and the electric displacement:

$$\frac{\partial D}{\partial z} = \rho. \tag{2.13}$$

In other words, if you integrate over the charge density in a volume, it is the same as if you count the number of field lines that pierce normally through a sphere that surrounds the volume.

Actually, in most practical cases that we will investigate, there are no free charges or, more precisely, no charges that would not be compensated by charges with the opposite sign (think of a neutral atom or molecule or of an ionic substance that is overall neutral, because all charges of the anions are compensated by those of the cations). This means that $\rho = 0$, and according to the first relation, the electric displacement does not change at all. $\rho = 0$ also means that the number of field lines does not change if we see it from a three-dimensional point of view, even if we stay with one dimension for the time being. You might wonder what we need such a relation for, and actually, in the one-dimensional case we cannot make use of it, but in the three-dimensional case it will lead to the so-called transversality conditions that will tell us that the E-field is transversal both to the direction of a wave and to the H-field in a scalar medium. We can also use them to derive the boundary conditions/continuity relations at an interface when we have nonnormal incidence. This is certainly something that cannot happen in the one-dimensional case, so for the time being the electric field strength (and also the magnetic field strength) will simply be continuous at an interface, which is just a line in this section, but will later be on a plane if we go 3D. Anyway, as already stated, for the moment we do not need this relation, pretty much as in the case where the BBL approximation would be applicable.

If you are unfamiliar with the notation $\partial/\partial t$, it indicates a partial derivative, which you have to use if you have to deal with a function that depends on more than one variable. Luckily all rules for conventional derivatives stay in place, as you treat all other variables as constants. Such partial derivatives you will also encounter in the following (Section 2.5) when you are introduced to harmonic waves, but we do not need them on a regular basis in this book. In fact, we will get rid of them very easily, because we will assume that we only deal with plane waves, but I will not reveal the explanation what plane waves are until we take a closer look onto the vectorial nature of the various fields we already have encountered in the previous sections.

2.3.2 2nd equation—Faraday's law of induction

This is the law of induction: Every change of the magnetic field in time leads to a response and a related change of the electric field intensity in space—it even can create an electric field where none existed before. In particular, the changes of the electric field strength tend to oppose the changes of the magnetic field strength (as indicated by the negative sign):

$$\frac{\partial E}{\partial z} = -\frac{\partial B}{\partial t}. \tag{2.14}$$

The usefulness of this relation becomes clear, when we combine it with the 4th equation.

2.3.3 3rd equation—Gauss's law for magnetism

Unlike for electric fields, there are no sources of the magnetic field—a magnetic monopole, the magnetic equivalent of a charge, does not exist (actually, one can argue that there are magnetic monopole quasi-particles, but they always occur pairwise). If a magnet is separated into its magnetic South and North pole, two magnets, each with its own South and North pole, result. Also, unlike electric fields, magnetic field lines are closed loops without beginning or end:

$$\frac{\partial B}{\partial z} = 0. \tag{2.15}$$

There is not much to say about this relation, except that it is not useful, like the 1st equation, in the 1D-case either.

2.3.4 4th equation—Ampere's circuital law

Magnetic fields can be generated in two ways, either by an electrical current or by a changing electromagnetic field. If we assume that there are no (net) free charges moving inside the medium, the following equation results:

$$\frac{\partial H}{\partial z} = \frac{\partial D}{\partial t}. \tag{2.16}$$

2.4 Deriving the wave equation

Maxwell's equations allow to predict the existence of electromagnetic waves. We start with the second and the fourth Maxwell's equations:

$$\frac{\partial E}{\partial z} + \frac{\partial B}{\partial t} = 0 \quad \text{(II)}$$
$$\frac{\partial H}{\partial z} - \frac{\partial D}{\partial t} = 0 \quad \text{(IV)}$$
(2.17)

Since we assumed that magnetization is zero, we can set $B = H$ in (II) and apply the operator $\partial/\partial z$ from the left:

$$\frac{\partial^2 E}{\partial z^2} + \frac{\partial}{\partial t}\frac{\partial H}{\partial z} = 0. \tag{2.18}$$

We now take (IV) and differentiate it with respect to time:

$$\frac{\partial}{\partial t}\frac{\partial H}{\partial z} - \frac{\partial^2 \varepsilon E}{\partial t^2} = 0. \tag{2.19}$$

Additionally, to obtain this result, we have used the relation $D = \varepsilon E$ (Eq. 2.7). If we combine (2.18) and (2.19), we get

$$\frac{\partial^2 E}{\partial z^2} + \varepsilon \frac{\partial^2 E}{\partial t^2} = 0. \tag{2.20}$$

Solutions to this equation, which is called the wave equation, are given by $E(z,t) = E_0 \sin(k_{sp} z - \omega t)$ and $E(z,t) = E_0 \cos(k_{sp} z - \omega t)$, where ω is the temporal angular frequency and k_{sp} is the spatial angular frequency with $\varepsilon = k_{sp}^2/\omega^2$, see next section, in which I focus on waves. These solutions are not the most general ones possible, but they are the most useful ones for our goals. If you are a chemist, the aforementioned equation should not shock you—you know it from quantum mechanics, which you see at the latest when you replace E by Ψ, and, actually, you are already used to it in its 3D form to which we come at the end of the chapter. In this context, have you ever thought about the strange course of conduct in spectroscopy based on the BBL approximation, where we allow matter to behave like waves, but deny the same to light?

2.5 One-dimensional and harmonic waves

In Sections 2.1 and 2.2, we have basically discussed electrostatics, but to understand the interaction of infrared light with matter we needed Maxwell's equations, which led us to the existence of waves. Therefore, we have to know and understand what waves are and how to describe them. For the moment, we leave things very simple and assume that we have only one-dimensional and harmonic waves that do not change shape. We will not be able to stay with assuming only one dimension any longer than until the end of this chapter. Otherwise, we would not be able to understand the important concepts like the polarization of waves (which you please do not confuse with the polarization of matter that we discussed in the previous sections; in case of waves, polarization is a directional property of this wave) and important consequences that interfaces have on the light properties and, thereby, on the spectra we measure. In contrast, we will keep the concept of harmonic waves from now on for the rest of this book. As a consequence, whatever we will discuss in the following, it will suffice to assume that the light waves have a sinusoidal form and can be described accordingly by sinus functions. This will be true in both, space and time. This means that if we would start to look at the value of the electric field at a certain point in space and from there check the values along a straight line, the values would change like a sine or a cosine function (a sine function can be seen as a cosine function shifted by $\pi/2$ to the left and vice versa). A typical picture that is used in this connection is that of a wave in water. However, there is an important difference, since such a wave dislocates the water particles transversal to its direction. While a light wave also is a transversal wave, nothing is dislocated; it is just the value of the electric field strength that changes along the direction of propagation in a sinusoidal way; see Fig. 2.3.

Therefore, at the starting point in space, the electric field strength is assumed to be zero, while after $\lambda/4$ it is at maximum. Here λ represents the wavelength, which is the distance between two consecutive maxima (or minima). Note that we come back to electrostatics if we suppose an infinite wavelength, which means that the electric field strength does not change (this certainly assumes that we look at the electric field inside of a plate capacitor; for a point charge, the electric field decreases with the inverse squared distance). As already mentioned, the harmonic wave has the same appearance in space as well as in time. The timely distance between two consecutive maxima is the period τ, which is the inverse of the frequency ν:

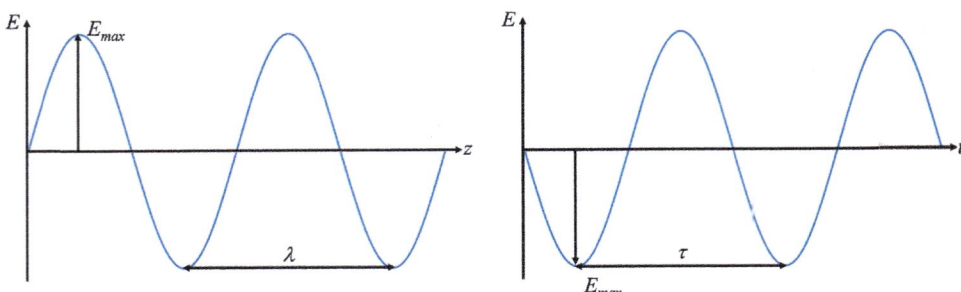

FIG. 2.3 A sinusoidal wave in space *(left panel)* and in time *(right panel).*

$$\nu \equiv \frac{1}{\tau}. \tag{2.21}$$

In other words, the frequency equals the number of maxima going through a certain fixed point in space per second. If we view the wave at two points in time that are a period apart, the wave will have traveled a distance equal to the wavelength, and it moves in free space with the speed of light c:

$$\lambda = c \cdot \tau. \tag{2.22}$$

To capture both the dependence of the wave in space and in time, we can describe the wave by [1],

$$E(z,t) = E_{\max} \sin(k_{sp} \cdot z - \omega \cdot t), \tag{2.23}$$

with k_{sp} the spatial angular frequency, which is in vacuum given by $k_{sp} = 2\pi/\lambda$ and $\omega = 2\pi\nu$ is the temporal angular frequency. k_{sp} is sometimes also termed the wavenumber. As a spectroscopist, you will have heard this term already, but then it is simply defined as the inverse of the wavelength: $\tilde{\nu} = 1/\lambda$. In fact, later on in the next chapters, k_{sp} and $\tilde{\nu}$ will be easier to distinguish, since k_{sp} will then be a vector pointing in the direction the wave is traveling and correspondingly called the wavevector, while the definition of $\tilde{\nu}$ will not change. At this point, I think it is important to already mention a source of confusion that I will in a more detailed way explain and try to remove in Section 2.8, when we talk about the way to describe waves in absorbing media. The point that is confusing is that a wave can also be described by [2],

$$E(z,t) = E_{\max} \sin(\omega \cdot t - k_{sp} \cdot z), \tag{2.24}$$

and the same sign switch of the argument is also possible and transferred to the exponential representation of a wave that I will introduce on the next page (if you do not believe this, please go back to Section 2.4. You can easily convince yourself that Eq. (2.24) is also a solution of the wave equation). If you see one or the other way of representing, just keep in mind that both are correct. Nevertheless, we have to make an important adaption later on when we discuss the damped harmonic oscillator, so this will be the latest point where one has to decide in which way the wave will be described. I will adapt the physics convention, which assumes a wave of the form of Eq. (2.23) (the other form is correspondingly called the electrical engineering convention, as it is usually used in this field, but do not rely on it).

The time dependence does not have a major importance for understanding the rest of the book, but the spatial dependence is extremely important. How can we get rid of the time dependence? Based on Eq. (2.23), it does not look like that there is mathematically an easy way for achieving this. To find this way, one has to perform a mathematical transformation, which is of eminent usefulness, actually not only for getting rid of the time dependence, which is why I will introduce it in the following.

I hope you are familiar with the concept of complex numbers. Just to remind you, they have been originally introduced to find solutions for situations where you have a negative argument of a square root like, in the simplest case, $\sqrt{-1}$. This problem is not solvable for real numbers, but if we define an imaginary number i for which $i^2 = -1$, then we have formally solved the problem. Our new complex number will then have a real part and an imaginary part like the complex number $1 + 2i$, which has the real part 1 and the imaginary part 2. In the same way, we can define a complex electric wave by

$$\begin{aligned}\widehat{E}(z,t) &= E_{\max}\left[\cos(k_{sp} \cdot z - \omega \cdot t) + i\sin(k_{sp} \cdot z - \omega \cdot t)\right], \\ &= E_{\max} \exp\left[i(k_{sp} \cdot z - \omega \cdot t)\right]\end{aligned} \tag{2.25}$$

where $E_{\max}$ is the amplitude. This is the so-called Euler's formula. Physically, $\widehat{E}(z,t)$ is not meaningful (at least, to my best knowledge, no meaning has been found so far), so that we will always retransform $\widehat{E}(z,t)$ by computing its real part $\mathrm{Re}\left[\widehat{E}(z,t)\right] = E(z,t)$.

If you look at Eq. (2.25), you may wonder how we arrived at the cosine function for the real part, while we started with the sine function. As already mentioned, you can convert every sine function to a cosine function by realizing that a cosine function is nothing but a sine function that is shifted by 90° or $\pi/2$: $\sin(x) = \cos(x - \pi/2)$—cf. Fig. 2.3. This shift is called a phase shift, because the phase φ is defined as $\varphi = k_{sp} \cdot z - \omega \cdot t$.

You may remember phase shifts from inductors and capacitors, if you have an alternating current. For an inductor, the current lacks behind the voltage, whereas for capacitors it is the other way round. In any way, you certainly can describe harmonic waves with the cosine function instead of the sine function as well.

Coming back to the description of a harmonic wave with an exponential function, we finally find that

$$E(z,t) = \text{Re}\left[E_{\max}\exp\left[i(k_{sp} \cdot z - \omega \cdot t)\right]\right] \\ = \text{Re}\left[E_{\max}\exp\left[i(k_{sp} \cdot z)\right]\exp\left[-i(\omega \cdot t)\right]\right], \quad (2.26)$$

which means that if only the spatial part of the wave changes, e.g., if a wave is transmitted through an interface, and one takes the ratio, then the temporal part $\exp[-i(\omega \cdot t)]$ can simply be canceled or not taken into account. The latter is also what we will do with the function $\text{Re}[x]$ and the exponential way to describe the wave, meaning that I will not add this function to any formula using the exponential notation of the wave for the rest of this book, but certainly a complex electric wave does not make sense, so please keep this in mind.

One property of the exponential notation that we will investigate briefly is differentiation. From differential calculus, you may remember that $(\sin[k_{sp}z])' = k_{sp}\cos[k_{sp}z]$ and $(\cos[k_{sp}x])' = -k_{sp}\sin[k_{sp}x]$ so that $(\sin[k_{sp}z])'' = -k_{sp}^2\sin[k_{sp}z]$. This means that the derivative of a harmonic wave stays a harmonic wave (and the same is certainly true if we use a cosine instead of the sine function). This is a property that is also handed down if we use the complex notation:

$$\frac{\partial E(zt)}{\partial z} = ik_{sp}E_{\max}\exp\left[i(k_{sp} \cdot z - \omega \cdot t)\right] = ik_{sp}E(zt)$$
$$\frac{\partial^2 E(zt)}{\partial z^2} = -k_{sp}^2 E_{\max}\exp\left[i(k_{sp} \cdot z - \omega \cdot t)\right] = -k_{sp}^2 E(zt)$$
$$\frac{\partial E(zt)}{\partial t} = -i\omega E_{\max}\exp\left[i(k_{sp} \cdot z - \omega \cdot t)\right] = -i\omega E(zt) \quad (2.27)$$
$$\frac{\partial^2 E(zt)}{\partial t^2} = -\omega^2 E_{\max}\exp\left[i(k_{sp} \cdot z - \omega \cdot t)\right] = -\omega^2 E(zt)$$

These relations allow us to write the second and the fourth Maxwell's equations in a somewhat simplified way, if we assume that the magnetic fields are also harmonic waves:

$$\frac{\partial E}{\partial z} + \frac{\partial B}{\partial t} = 0 \xrightarrow{M=0}$$
$$ik_{sp}E - i\omega H = 0 \to \quad \cdot \quad (2.28)$$
$$k_{sp}E - \omega H = 0$$

What we cannot see in this 1D-form, but what will come out when we are using the vectorial form, is that the electric field and the magnetic field are vectors perpendicular to each other. What also gets lost, as already mentioned earlier, is the understanding of polarization, which is also something the understanding of which I will convey at the end of this and in the next chapter. For those of you, who are already familiar with polarization, please see the electric, as well as the magnetic, field in this section as naturally polarized, which means that E-field and H-field are perpendicular to each other and parallel at the same time, because we superpose two linearly polarized and mutually perpendicular oriented electromagnetic waves with the same time dependencies. Note that we would have arrived also at Eq. (2.28), if I would have chosen the other convention for the representation of the (electromagnetic) wave.

The fourth Maxwell's equation simplifies congruently with the second to

$$\frac{\partial H}{\partial z} = \frac{\partial D}{\partial t} \to$$
$$k_{sp}H - \omega\varepsilon E = 0 \quad (2.29)$$

Consequently, we can also simplify the wave equation to

$$k_{sp}^2 + \omega^2\varepsilon_0\varepsilon_r = 0. \quad (2.30)$$

In this case, we obtain a very simple connection between the dielectric function ε and the index of refraction, which is for normal incidence $n = k_{sp}/\left(\omega\sqrt{\varepsilon_0}\right)$:

$$n^2 = \varepsilon_r. \tag{2.31}$$

This was one of the major outcomes of Maxwell's equations and used to check their correctness for decades after their publication, see, e.g., [3]. It also led to a shift of perception, and to the concept of a dielectric constant.

Note that ε is a material property that is connected with the induced dipole moments, i.e., polarizability, whereas k and ω and, as a consequence, n are wave properties. Eq. (2.31) is often abused in the sense that this important distinction is no longer made. Once you get to Chapter 12, you will easily distinguish ε and n, because the dielectric constant becomes a tensor that inherits its symmetry from the unit cell of a single crystal, while there are simply two different indices of refraction for a certain direction of the wave inside an anisotropic crystal. Please keep in mind that both the dielectric constant and the refractive index are frequency or wavenumber dependent. Therefore, they are better called dielectric function and index of refraction function (or, equivalently, refractive index function), respectively. Their frequency dependence is called dispersion, and we will deal with this dispersion in a simplified manner in the next section (Section 2.6).

It is also instructive to think about what happens if you use the other convention to specify the harmonic wave. In Eq. (2.27), the first derivatives certainly change signs, but the second derivatives stay the same. This is also true for the second and fourth of Maxwell's equations, and, finally the wave equation. This confirms the equality of both approaches and definitely makes it a matter of taste which convention to choose. I explain to you later why I prefer the taste of one convention over the other one.

Before we go on, I will discuss one last property of waves in general that is shared by electromagnetic waves and that is of fundamental importance. The property I am talking about is how the waves interfere with each other. For electromagnetic waves, the superposition principle, which you know from quantum mechanics, holds. In other words, the values of the individual waves simply add up at every point in space (or in time), since we deal or assume linearly correlated functions. The situation is particularly simple if two waves have the same wavelength or frequency, travel in the same direction, and have the same amplitude. In this case, we can distinguish two extreme cases. The first case assumes that both waves have the same phase. This means that the overall amplitude is $2E_{\max}$. The second case assumes that the two waves are shifted by exactly π, which implies that for any value $E(z,t)$ of the first wave, the second wave has the value $-E(z,t)$. In other words, the combined wave is a wave that has a value of zero at all points in space—it is extinguished.

If you find this concept strange, I am pretty sure that you experienced it yourself already and might enjoy it quite often—this is the way noise-canceling headphones work: For every sound wave that belongs to the noise, a second wave is generated with (ideally) the same amplitude, but phase-shifted by π. The only difference is that sound waves are longitudinal waves, i.e., compression and depressurization of air take place in the direction of the movement and not transversal like the movement of a volume element of water when it is affected by a wave. If you have problems to imagine the difference between longitudinal and transversal waves, have a look at Fig. 2.4. If you wonder why I keep on at that topic, when electromagnetic waves are always transversal, then be assured that they are not necessarily transversal or purely transversal under all circumstances. Again, the reason lies in the special effect of interfaces that are not perpendicular to the direction

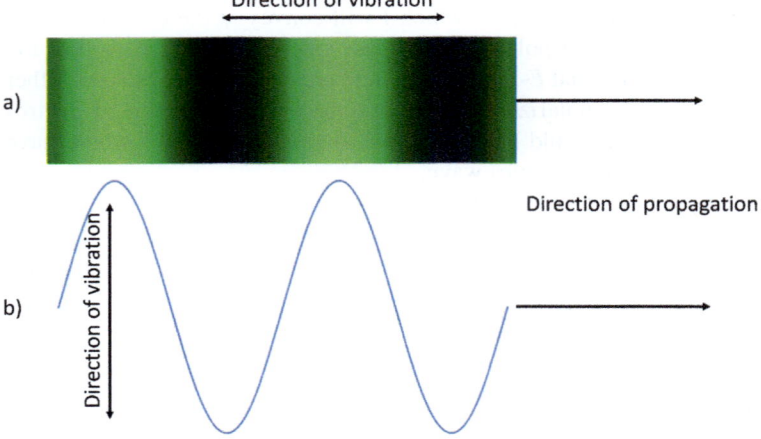

FIG. 2.4 (A) Longitudinal wave. (B) Transversal wave.

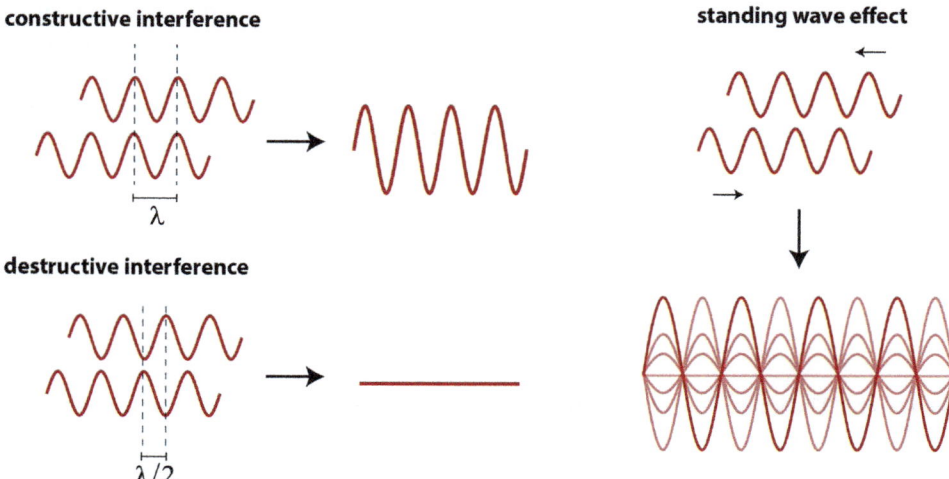

FIG. 2.5 Constructive and destructive interference as well as the standing wave effect.

of propagation and/or in media (samples!) that are anisotropic. So do not rely on the transversality and do not exclude automatically corresponding effects to take place.

Coming back to interference, so far, we have discussed only two forward traveling waves, but more important for optics and spectroscopy is the case when one wave is propagating forward and the other is propagating backward. Why is this case more important? Because a part of the wave that is traveling forward may become (partly) reflected at the next interface and then may interfere with the forward traveling wave. If the interface is highly or totally reflecting, a standing wave may arise, again with everything between perfectly constructive and destructive interference, cf. Fig. 2.5. Since the reflected part of a wave is always coherent with the original wave itself, interference can also take part with light that is as an ensemble of incoherent waves, meaning that it consists of many waves with phase relations that are in one point in time or in space different to those at later times or different locations. To give you some examples, the light of a bulb or of a globar is generally incoherent, whereas the light stemming from a laser is often coherent when it interacts with a sample.

As you can imagine, wave interference is of outstanding importance as you shine light on a sample. With light having an electric field strength E, you can get field strengths between the extreme cases 0 and $2E_{max}$ inside the sample or at an interface. Since absorption scales with the electric field intensity, i.e., E^2, absorption can be up to four times higher than you expect or lowered to the point where no light at all is absorbed in extreme cases.

For the moment, this shall be enough as a basis for the following, but as I already stated, we will return to waves and introduce them in vectorial form at the end of this chapter (Section 2.13).

2.6 Harmonic molecular vibrations and the dielectric function

In the last section, we have studied harmonic waves and their changes with regard to both, location and time. In the following, we focus only on harmonic changes in time, which means that the equations that I soon introduce have the harmonic wave as solution if we set the location as constant. We look at a very simple system that you know already by heart, which is a massless spring fixed on the ceiling with a mass m attached. We will slightly displace the mass by stretching the spring and then release the mass. The reason why we only slightly displace it is because we can then apply Hooke's law, which states that

$$F_s = k_{spring} \cdot x, \tag{2.32}$$

where F_s is the force that we need to stretch or compress the spring and x is the displacement of the weight from its equilibrium position. If this displacement stays small, as is assumed if you use Hooke's law, then displacement and force are linearly related with a proportionality constant k_{spring}, which is called the spring constant. Eq. (2.32) is the more common way to state Hooke's law, but molecular spectroscopists are usually more used to the following convention:

$$F_r = -k_{spring} \cdot x, \tag{2.33}$$

The negative sign is a consequence of the focus on the restoring force F_r and it and the displacement being antiparallel, meaning that this force tries to pull or push the weight back in the equilibrium position. If the weight is released after being displaced, it will move in the direction of the equilibrium position, but it will not stop at it. Instead, the movement will persist beyond this point, after which the spring is compressed, when it was stretched before and will, correspondingly, push the weight in the opposite direction and vice versa. In fact, by stretching or compressing the spring we introduce potential energy into the system that is given by

$$E_{\text{pot}} = \frac{1}{2} k_{\text{spring}} \cdot x^2, \tag{2.34}$$

and will be converted into kinetic energy,

$$E_{\text{kin}} = \frac{1}{2} m \cdot v^2, \tag{2.35}$$

once we release the weight. The kinetic energy will be at maximum in the equilibrium position, at which x and, thereby, the potential energy will be zero, as the sum of kinetic and potential energy stays constant. Beyond this position, the kinetic energy will decrease until the turning point is reached where the system again has only potential energy. If there is no energy dissipation, the system will repeat endlessly this circle, which is what we assume in the first place.

We proceed with Newton's second law, which states that the force acting on a constant mass experiences an acceleration proportional to this mass. The force is actually the one that the spring exerts according to Eq. (2.33),

$$F = -k_{\text{spring}} \cdot x = m \cdot a = m \frac{d^2 x}{dt^2}, \tag{2.36}$$

where we have made use of the fact that the acceleration a is the second derivative of the distance, in this case the displacement, with respect to time. Therefore, what we have to solve is the following equation:

$$m \frac{d^2 x}{dt^2} + k_{\text{spring}} \cdot x = 0. \tag{2.37}$$

What we need is a function the second derivative of which is the function itself multiplied by a constant. This means that we can use the functions that we have already discussed in the previous section:

$$f(x) = \begin{cases} A_{\max} \cos(-\omega_0 t) & \text{(I)} \\ A_{\max} \sin(-\omega_0 t) & \text{(II)} \\ A_{\max} \exp[-i(\omega_0 t)] & \text{(III)} \end{cases}. \tag{2.38}$$

Here, ω_0 is the eigenfrequency of the system (the frequency with which it would vibrate if there is neither agitation nor damping, as in this case). If we differentiate the functions twice with respect to time, we obtain

$$f''(x) = \begin{cases} -\omega_0^2 A_{\max} \cos(-\omega_0 t) & \text{(I)} \\ -\omega_0^2 A_{\max} \sin(-\omega_0 t) & \text{(II)} \\ -\omega_0^2 A_{\max} \exp[-i(\omega_0 t)] & \text{(III)} \end{cases}. \tag{2.39}$$

If we put any of these three functions into Eq. (2.37), the result is

$$\omega_0^2 m = k_{\text{spring}} \rightarrow \omega_0 = \sqrt{\frac{k_{\text{spring}}}{m}}, \tag{2.40}$$

which means that the eigenfrequency of the system is a function of the square root of the ratio of spring constant and mass.

Molecular spectroscopists quite often deal with vibrations in which a hydrogen atom is involved. In this case, the mass of the rest of the molecule is usually so high in comparison that in a good approximation only the hydrogen atom moves against the rest of the molecule, which means that the aforementioned simplification of having a single mass fixed to an unmovable object is justified to a large extent. Actually, since the bonding electron pair is usually a little bit displaced toward the next atom, e.g., C, O, or N, it is not an uncharged atom that moves, but an atom that carries a partial charge, which is what is relevant in the following.

Since this partial charge moves and the distance between it and the next atom/charge is varied, from the classical point of view it is a dipole moment that changes. As a consequence, this system can interact with electromagnetic radiation. To be more precise, it interacts with the electric field of a wave $E_{max}\cos(-\omega t)$, which changes periodically at the location of the molecule. As a consequence, this electric field acts on the partial charge q with the force F_E:

$$F_E = qE(t) = qE_{max}\cos(-\omega t). \tag{2.41}$$

If we again use Newton's second law and sum up all forces, we get

$$m\frac{d^2x(t)}{dt^2} + \omega_0^2 mx(t) = qE(t). \tag{2.42}$$

Since we have expressed the periodic change of the electric field with a cosine function, we also select this function for the change of the dipole moment and put it into Eq. (2.42) to obtain:

$$-\omega^2 mx(t) + \omega_0^2 mx(t) = qE(t). \tag{2.43}$$

If we solve for $x(t)$, the result is

$$x(t) = \frac{q}{m(\omega_0^2 - \omega^2)} E(t). \tag{2.44}$$

If we multiply $x(t)$ with the charge q, then we obtain the dipole moment $p(t)$ of the molecule. To continue, we need to remember some relations we have discussed in the beginning of this chapter. The first relation is Eq. (2.6), the essential part of which is that the macroscopic polarization $P(t) = Np(t)$ with N the number of molecules per unit volume, which is essentially a concentration. Therefore, we get

$$P(t) = \frac{Nq^2}{m(\omega_0^2 - \omega^2)} E(t). \tag{2.45}$$

In the next step, we need Eq. (2.8), $P = \varepsilon_0(\varepsilon_r - 1)E$, and equate both expressions for P, after which we can eliminate the electric field, providing

$$\left.\begin{array}{l} P = \dfrac{q^2 N/m}{\omega_0^2 - \omega^2} E \\ P = \varepsilon_0(\varepsilon_r - 1)E \end{array}\right\} \varepsilon_0(\varepsilon_r - 1) = \frac{q^2 N/m}{\omega_0^2 - \omega^2}. \tag{2.46}$$

We are one step away from the final result, for which we solve for the relative dielectric function ε_r:

$$\varepsilon_r = 1 + \underbrace{\frac{q^2 N}{\varepsilon_0 m}}_{S^2} \frac{1}{\omega_0^2 - \omega^2}. \tag{2.47}$$

S^2 is called the oscillator strength: $S^2 = q^2 N/(m\varepsilon_0)$ (note that sometimes you find different definitions of the classical oscillator strength; apart from the one I will use, e.g., based on S^2/ω_0^2). It is a very important quantity, not only because it is related to the transition moment but also because its square is proportional to N, which is the concentration of oscillators per unit volume and related to the molar concentration c by $N = N_A \cdot c$. This will sometimes open up a way to determine the concentration in a way that is multivariate, but much more reliable than chemometrics on the one hand. On the other hand, it will allow us to determine the dielectric function, and, with it, the refractive index function via Eq. (2.31). However, before this is possible, we need to understand that Eq. (2.47) is not realistic and works only well at frequencies at some distance from the oscillator's eigenfrequency ω_0. Why is this the case, what did we miss? Let's first have a look on the resulting function in the proximity of ω_0. Have a look at Fig. 2.6.

Now you can see what the problem is and why Eq. (2.47) only works at some distance from the eigenfrequency ω_0, namely because there is a pole, a singularity exactly at ω_0, where coming from lower frequencies, the value of ε_r approaches ∞, while coming from higher frequencies ε_r goes to $-\infty$. Physically, the related phenomenon is called a resonance disaster. In this scenario, the oscillator would take over energy indefinitely. As a consequence, its amplitude would grow without boundary, which is sometimes indeed a nearly realistic scenario, that has to be avoided in case of, e.g., bridges. One often stated example in this respect is the 1940 Tacoma Narrows Bridge collapse. Actually, this case was probably more complicated, and it happened despite of countermeasures, which are in case of bridges damping elements. How can damping elements help? To understand this, it is instructive to compare an undamped harmonic oscillation with one that is damped, as this is done in Fig. 2.7.

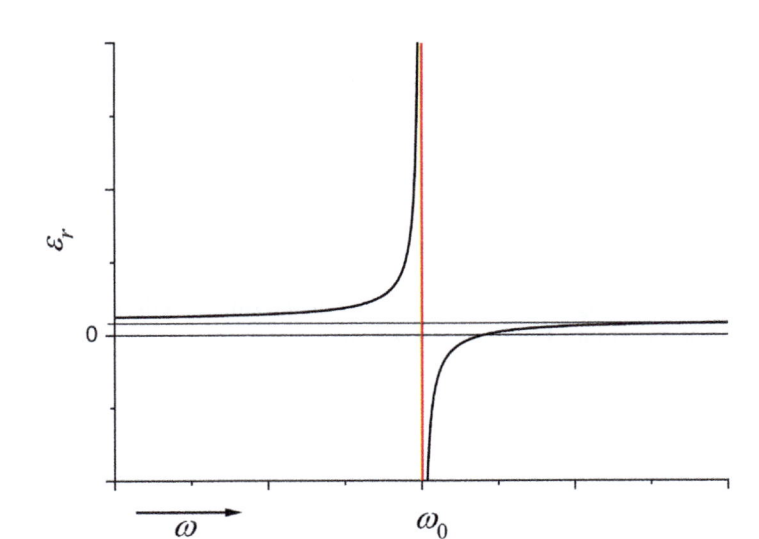

FIG. 2.6 Dispersion of the (relative) dielectric function around an eigenfrequency ω_0 under neglect of damping.

An undamped harmonic oscillation preserves the energy that is in the system. If you continue to add energy to the system, which happens if the eigenfrequency of the incoming light and the oscillating system agrees, the resonance disaster happens. Put in concrete terms, the dipole moments would grow without boundaries, which means that the distance between the (partial charges) would become infinite. In real matter, this cannot happen. We do not even have to invoke quantum mechanics and think about broken bonds between atoms. It suffices to think that we leave the realm of Hooke's law and harmonic vibrations, which means that higher order terms of the spring constant would kick in. These higher order terms are also responsible for the coupling of vibrations and the energy dissipation through this coupling, which is the reason why light energy can be converted into heat and spread.

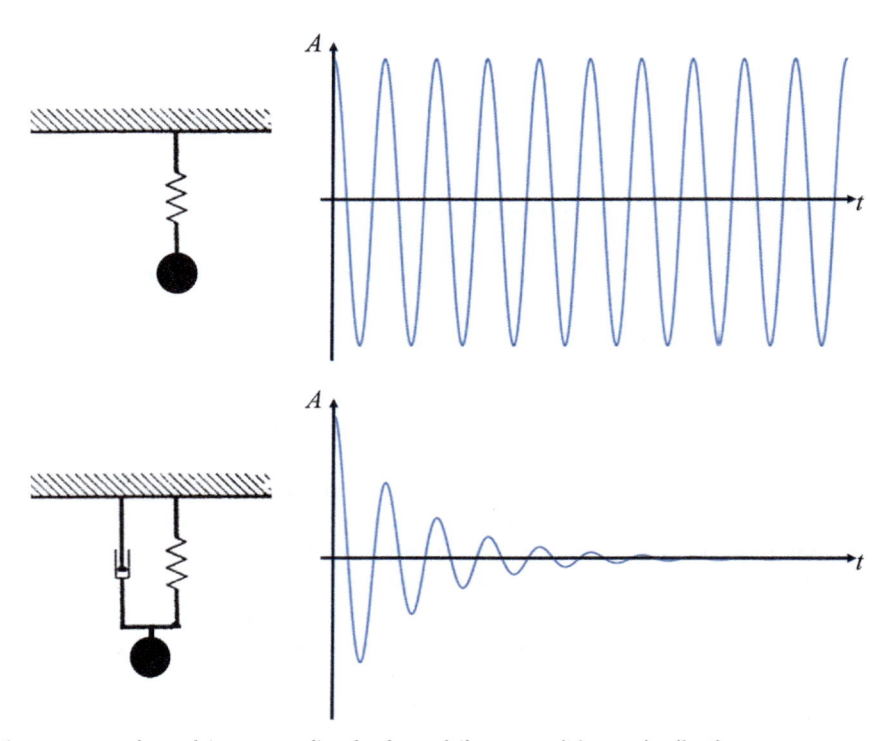

FIG. 2.7 Comparison between an undamped *(upper panel)* and a damped *(lower panel)* harmonic vibration.

While we will not examine the physical origin of this energy dissipation in detail (if you are interested, Max Born treated this problem in detail in [4]), we will simply introduce energy dissipation in our spring system by assuming that the spring is coupled to a honey pot in a cascade as shown in the lower panel of Fig. 2.7. This complicates somewhat our system and its mathematical description. If we go back to Eq. (2.42), we have to slightly modify it and introduce an extra term due to this damping. This extra term will, on empirical grounds, be proportional to the velocity of the vibration according to

$$m\frac{d^2x(t)}{dt^2} + m\gamma\frac{dx(t)}{dt} + \omega_0^2 mx(t) = qE(t), \tag{2.48}$$

with γ called the damping constant. It seems that Eq. (2.48) has increased strongly in complexity, but what happened is only that now neither the sine nor the cosine function can provide a solution, since they need to be derived twice before the original function appears again. What still remains is the exponential function (cf. Eq. 2.38), but with an important modification:

$$f'(x) = \begin{cases} -\omega_0 A_{\max} \sin(-\omega_0 t) & \text{(I)} \\ \omega_0 A_{\max} \cos(-\omega_0 t) & \text{(II)} \\ -i\omega_0 A_{\max} \exp[-i(\omega_0 t)] & \text{(III)} \end{cases}. \tag{2.49}$$

While the function $\exp[-i(\omega_0 t)]$ is regained already after one derivation, the overall function becomes imaginary due to the chain rule:

$$-\omega^2 mx(t) - i\omega\gamma mx(t) + \omega_0^2 mx(t) = qE(t). \tag{2.50}$$

If we, again, solve for $x(t)$, the result is

$$x(t) = \frac{q}{m(\omega_0^2 - \omega^2 - i\omega\gamma)} E(t). \tag{2.51}$$

The rest of the steps are the same as those employed to get from Eq. (2.45) to Eq. (2.47), and the final result is

$$\varepsilon_r = 1 + \frac{S^2}{\omega_0^2 - \omega^2 - i\omega\gamma}. \tag{2.52}$$

This means that our relative dielectric function is now a complex function with a real and an imaginary part, ε_r' and ε_r''. These parts and their dispersions are depicted in Fig. 2.8.

In some books, you may see Eq. (2.52) slightly altered with a positive sign in front of the term with the damping constant, which is a consequence of the sign conventions in the argument of the time-dependent part of the wave function as

FIG. 2.8 Dispersion of the (relative) dielectric function around an eigenfrequency ω_0 assuming damping in comparison with that without damping *(black curve)*.

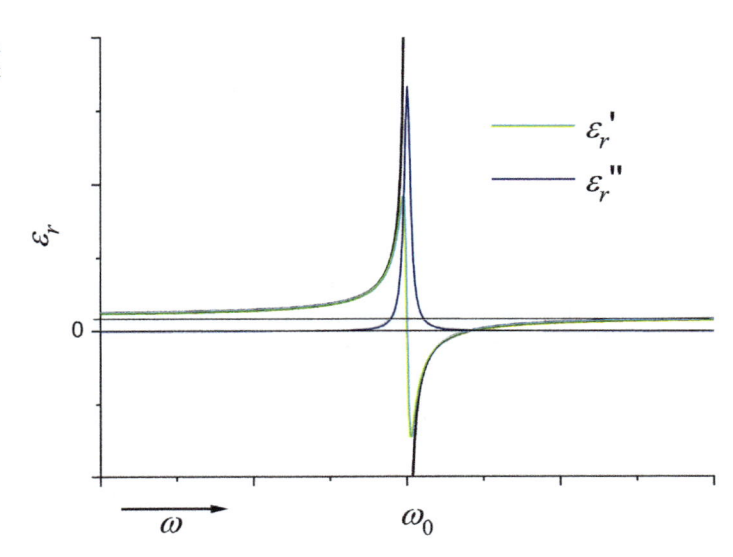

already discussed. Eq. (2.52) leads to a dielectric function of the form $\varepsilon_r = \varepsilon_r' + i\varepsilon_r''$, whereas the other convention leads to $\varepsilon_r = \varepsilon_r' - i\varepsilon_r''$. Please keep this in mind!

In Fig. 2.8, you also find the result for the case without dissipation of energy, i.e., the black curve. Obviously, this case correctly describes the real part of the dielectric function except in close proximity to the eigenfrequency (historically, it was not so easy to find an experimental prove—determining the refractive index close to an absorption was not an easy task as Ketteler or Ladenburg would tell you [5,6]). This is also why the early theories to describe the change of the dielectric function, or, historically, the square of the refractive index function, relied on Eq. (2.42) rather than on Eq. (2.48). The former equation was introduced by Sellmeier in 1872 [7] and the latter by Helmholtz in 1875 [8], but the form that is still used today is due to Ketteler and described in detail in his book from 1885 [5]. This was not long after Maxwell published the equations named after him, which he did in 1865 [9]. This is relevant, because of the dielectric constant or the dielectric function and its relation to the now complex refractive index function (I will indicate that the refractive index function features complex numbers by the symbol ˆ, but I will not do the same for the relative dielectric function, $\varepsilon_r = \hat{n}^2$, Eq. (2.31)), which was not known before and was one of the central results as I mentioned earlier. The first to factor in for dispersion theory that light consists of electromagnetic waves was Drude in 1887 [10]. Before Drude, all authors described the dispersion via $\hat{n}^2$ instead of ε_r and they did not consider electromagnetic theory to derive their dispersion relations, but thought of kinds of elastic vibrations and elastic waves in luminiferous ether, instead. Nevertheless, the formulas are the same (Fresnel certainly also thought of elastic waves in ether when he derived the equations named after him). Nevertheless, it is very important that dispersion can be explained well for $\hat{n}^2$, but not that well for $\hat{n}$—we will soon come back to this point, as it is very important to understand where *band fitting* comes from.

Back to Fig. 2.8, we see that the pole is superseded by a maximum and a minimum in ε_r'. ε_r'', on the other hand, indicates energy dissipation, or, in other words, absorption, which is relevant mainly in the vicinity of the eigen or resonance frequency, but can be neglected not far away from it. Nevertheless, if the oscillator strength is large, a sample may be opaque even far away from the eigenfrequency if it is thicker. As an example, vitreous silica has a strong oscillator at about $1100\,\mathrm{cm}^{-1}$ and it can be opaque up to $2000\,\mathrm{cm}^{-1}$ and beyond, which is why it cannot be employed as a substrate for transmission measurements if the fingerprint region is of interest.

Coming back to the question of a dispersion formula for $\hat{n}$ instead of $\hat{n}^2$, which is something that Lorentz thought of as an approximation in 1906 [11], we start with Eq. (2.52) and use $\varepsilon_r = \hat{n}^2$ (Eq. 2.31):

$$\hat{n}^2 = 1 + \frac{S^2}{\omega_0^2 - \omega^2 - i\omega\gamma}. \tag{2.53}$$

We apply the square root to both sides,

$$\hat{n} = \sqrt{1 + \frac{S^2}{\omega_0^2 - \omega^2 - i\omega\gamma}}, \tag{2.54}$$

and use that $\sqrt{1+x} \approx 1 + 1/2x$ if $x \ll 1$. This implies that S^2 or the oscillator strength must be small—please keep this condition in mind (in particular in the UV/Vis this is absolutely not the case, and it is very disturbing when spectroscopists expect bands to be Lorentzian!). If we assume that the oscillator strength is small, we find that

$$\hat{n} = n + ik = 1 + \frac{S^2/2}{\omega_0^2 - \omega^2 - i\omega\gamma}. \tag{2.55}$$

In Eq. (2.55), I have introduced the relation $\hat{n} = n + ik$. n is nothing but the real refractive index (function) that we used before. k is the index of absorption (function), which is the equivalent of ε_r'' and which will assure, as I will show you in Section 2.8, that the electromagnetic wave will experience energy dissipation through absorption by matter and, thereby, damping. The sign before k is certainly also dictated by the convention I chose to describe the waves.

What is not obvious from Eq. (2.55) is that this equation is equivalent with the assumption that Beer's approximation is valid. Let's first separate the real from the imaginary part. Since this is pure algebra (you actually just have to make the denominator real by multiplying it with its complex conjugate to arrive at the following equation), I just state the results:

$$\hat{n} = 1 + \underbrace{\frac{S^2(\omega_0^2 - \omega^2)/2}{(\omega_0^2 - \omega^2)^2 + \omega^2\gamma^2}}_{n} + i\underbrace{\frac{S^2\omega\gamma/2}{(\omega_0^2 - \omega^2)^2 + \omega^2\gamma^2}}_{k}. \tag{2.56}$$

For the infrared spectral region, it is common to employ wavenumbers instead of angular frequencies. If we use that $\lambda \cdot \nu = c$ and $1/\lambda = \tilde{\nu}$, as well as $\omega = 2\pi\nu$, we find that $\omega = 2\pi\tilde{\nu}c$ and that

$$k = \frac{S^2 \tilde{\nu} \gamma / 2}{\left(\tilde{\nu}_0^2 - \tilde{\nu}^2\right)^2 + \tilde{\nu}^2 \gamma^2}. \tag{2.57}$$

If we employ further the approximation that $\tilde{\nu}_0 \approx \tilde{\nu}$, as k is very different from zero only in the neighborhood of the eigen-wavenumber $\tilde{\nu}_0$, we arrive at

$$k = \frac{S^2}{2\gamma \tilde{\nu}_0} \frac{(\gamma/2)^2}{(\tilde{\nu}_0 - \tilde{\nu})^2 + (\gamma/2)^2}. \tag{2.58}$$

Eventually, we take into account that

$$A = k \frac{4\pi d \tilde{\nu}}{\ln 10}, \tag{2.59}$$

and multiply both sides of Eq. (2.58) by $4\pi d\tilde{\nu}/\ln 10$, where d is the sample thickness and $\ln 10$ is needed because absorbance is defined via the decadic logarithm instead of the natural logarithm (before the advent of calculators, it was much easier to work with decadic than with natural logarithms), which occurs in the Bouguer-Lambert approximation, to get absorbance:

$$A = S^2 \underbrace{\frac{\pi d}{\gamma \ln 10}}_{A_0} \frac{(\gamma/2)^2}{(\tilde{\nu}_0 - \tilde{\nu})^2 + (\gamma/2)^2}. \tag{2.60}$$

This is what is called a Lorentzian or Lorentz profile in honor of Lorentz as it was derived by him from the damped harmonic oscillator model (which is often called a Lorentz oscillator) [11]. This is why it is worrisome, if someone argues that a Lorentzian should not be called like that because there is no underlying physics. Instead, some colleagues opt for calling a Lorentzian a Cauchy function to emphasize the purely mathematical nature. It looks to me as this argument disregards the very basics of spectroscopy.

There is one important thing to note, which is that at every point the value of the imaginary part of a Lorentzian is linearly depending on S^2, which is itself linearly related to the concentration (this is something Ketteler already realized in 1885 by experiment, but since his derivations were not based on electromagnetic theory, he could not realize that the squared oscillator strength is intrinsically proportional to the number of dipole moments per unit volume). This is why assuming that a band in an absorbance spectrum is a Lorentzian is equivalent with assuming that Beer's approximation is strictly valid. In fact, for most oscillators in organic materials this is a good approximation, because for these oscillators, as a rule of thumb, $S < 200 \, \text{cm}^{-1}$ (the C=O band, e.g., in PMMA has about this strength, but actually consists of more than one oscillator). For a Lorentz oscillator, the peak maximum begins to blueshift for such strengths, as can be seen in Fig. 2.9.

Apart from the adherence to Beer's approximation, there are further peculiarities of the Lorentz-profile, e.g., that it is symmetric and its fullwidth at half height is equivalent to the damping constant γ as can be seen in Fig. 2.10, a figure similar to that which appeared first as far as I know in 1908 in Woldemar Voigt's book "Magneto- und Elektrooptik" [12]. These

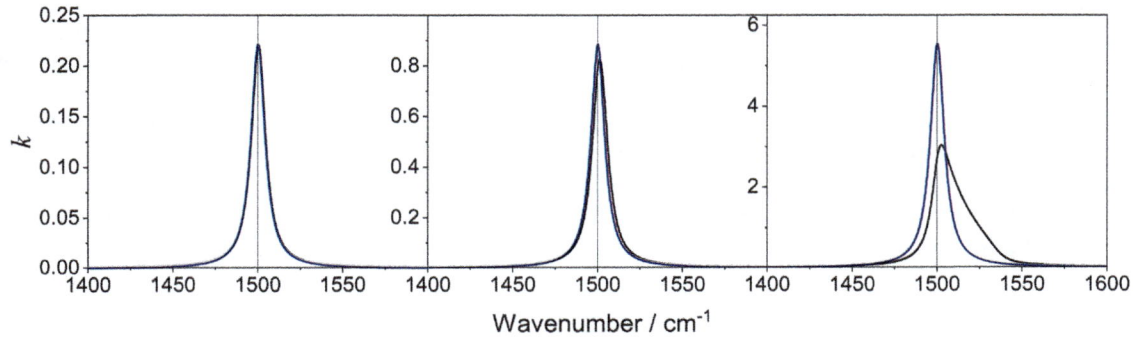

FIG. 2.9 Comparison between the absorption index functions for several oscillator strengths (from the left to the right: $S = 100/200/500 \, \text{cm}^{-1}$) based on the Lorentz oscillator *(black lines)* and on the Lorentz profile model *(blue lines)* based on Eq. (2.58).

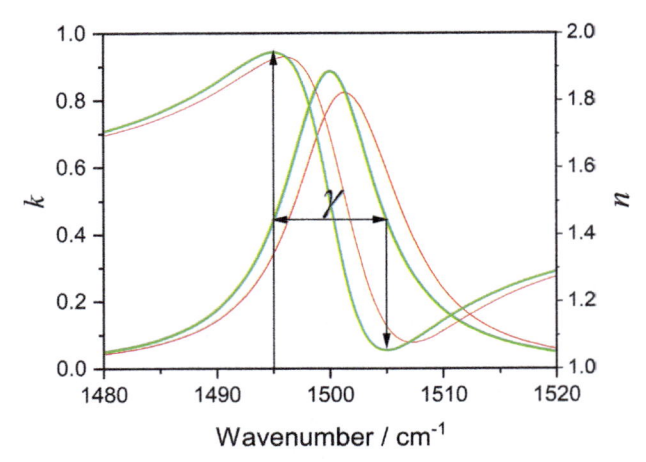

FIG. 2.10 Comparison between refractive index and absorption index functions for an oscillator with $S=200\,\text{cm}^{-1}$ based on the Lorentz oscillator *(red lines)* and on the Lorentz profile model *(green lines)*.

wavenumber points also indicate the maximum and the minimum of the corresponding refractive index function. Nevertheless, deviations from the Lorentz oscillator are already obvious at $S=200\,\text{cm}^{-1}$, and the Lorentz oscillator should be preferred. However, since it has been forgotten that band fitting is derived from dispersion theory, there is no option for using a Lorentz oscillator in the corresponding programs of the spectrometer companies. I also want to emphasize again that the seemingly nice properties of the Lorentz profile derive from the approximations that we introduced. The Lorentz oscillator, on the other hand, loses this symmetry, the property that the damping constant is the HWHH and that the peak indicates the oscillator position gradually with increasing oscillator strength, because $\varepsilon_r = \hat{n}^2$ (Eq. 2.31). If we would investigate its properties in ε_r'', we would actually see that it would have the same properties as the Lorentz profile in k (with regard to the symmetry, this can be seen in Fig. 2.9) or in absorbance spectra. But, in this context, you should never forget that the index of absorption is a wave property. Therefore, from this point of view, absorbance as a quantity defined via Eq. (2.59) is also a wave property and does only make sense for vanishing oscillator strength.

On a side note, Eq. (2.59) does also not agree with the definition of absorbance as $A = -\log_{10} T$, where T is the transmittance. This is the reason why we need to better understand what transmission as a process actually encompasses, which we will do in Section 2.9. But before that, I will introduce to you the Kramers-Kronig relations.

2.7 The Kramers-Kronig relations

If you have already read about the Kramers-Kronig relations and how they are derived, you are probably as traumatized as I was the first time, when I was confronted with them, mainly because I was missing corresponding mathematical knowledge. For a spectroscopist, in particular with a chemist's background, it seems you have to simply somehow accept that there are relations between the real and imaginary part of the complex refractive index function, as well as between those of the dielectric function. But what you do not have to do is to believe that those relations have been derived after deep thinking about causality and that a particular polarization state of matter can only be influenced by the current and past states or that there is no immediate effect—it takes some time before a steady state is established and transients have gone away, etc. Also, neither Hendrik Anthony Kramers nor Ralph Kronig thought about particular properties of a complex function that is analytic in the upper half-plane and what value an integral of a closed contour takes on according to Cauchy's theorem.

What Kramers and Kronig did was to use dispersion theory to derive the relations named after them. Maybe I am mistaken, but I do not think that there is something to learn for you by comprehending the derivation of the relations—I just want to show you the relations in this section and discuss some important points. But before I do this, I claim that you have already learnt that you can calculate the real from the imaginary part and vice versa. If you do not believe me, let me reproduce Eq. (2.56):

$$\hat{n} = 1 + \underbrace{\frac{S^2(\tilde{\nu}_0^2 - \tilde{\nu}^2)/2}{(\tilde{\nu}_0^2 - \tilde{\nu}^2)^2 + \tilde{\nu}^2\gamma^2}}_{n} + i\underbrace{\frac{S^2\tilde{\nu}^2\gamma/2}{(\tilde{\nu}_0^2 - \tilde{\nu}^2)^2 + \tilde{\nu}^2\gamma^2}}_{k}. \tag{2.61}$$

The denominators of $n-1$ and k are the same; therefore

$$\frac{n-1}{k} = \frac{S^2(\tilde{\nu}_0^2 - \tilde{\nu}^2)/2}{S^2 \tilde{\nu}\gamma/2} = \frac{\tilde{\nu}_0^2 - \tilde{\nu}^2}{\tilde{\nu}\gamma}. \tag{2.62}$$

As a consequence, we find the following relation to calculate k from n:

$$k = \gamma \frac{\tilde{\nu}(n-1)}{\tilde{\nu}_0^2 - \tilde{\nu}^2}. \tag{2.63}$$

This relation is exact (disregarding the approximation that we introduced to arrive at Eq. (2.56)), but useful only for spectra that consist of one band only (note that this does not necessarily mean that the material only has one transition from zero to infinite frequency; it rather means that the band must be separated from others by transparency regions).

For this band, you would have to know the damping constant γ and the oscillator position ω_0. Ketteler already extended the oscillator model to more than one oscillator and thought there might be situations where it would be necessary or thinkable that a summation over the contributions from different oscillators could be replaced by an integral. While I do not provide a strict derivation here, the corresponding Kramers-Kronig relation very much looks like as it would have been derived somehow by assuming an infinite number of oscillators whose damping constants become infinitely small, while actually they are not.

$$k(\tilde{\nu}) = -\frac{2\tilde{\nu}}{\pi} \wp \int_0^\infty \frac{n(\tilde{\nu}_0) - 1}{\tilde{\nu}_0^2 - \tilde{\nu}^2} d\tilde{\nu}_0. \tag{2.64}$$

The $\wp$ indicates that this is an improper integral (the function is not defined at $\tilde{\nu}_0 = \tilde{\nu}$), which is assigned a certain value, the Cauchy principal value. The second relation, which allows to calculate the refractive index function from absorbance spectra, cannot be derived in this way, but having the first relation, we can argue our way to the second very much the same way as Kramers did, namely via inversion [13]:

$$n(\tilde{\nu}) - 1 = \frac{2}{\pi} \wp \int_0^\infty \frac{\tilde{\nu}_0 k(\tilde{\nu}_0)}{\tilde{\nu}_0^2 - \tilde{\nu}^2} d\tilde{\nu}_0. \tag{2.65}$$

This relation looks very nice on the first view, but on closer inspection there seems to be a big flaw: The integration has to be carried out from zero to infinite wavenumber! This means you need to know the absorption index in the same infinite wavenumber range, which is impossible, simply because there is no spectrometer that can cover a (semi-)infinite spectral region (and even if, it would certainly be very expensive). Indeed, according to most textbooks, this is very problematic. It seems that somehow you have to extend the optical constant functions into the unknown based on physical grounds. But no worries! This needs to be done only if the material is not that well behaved unlike the one whose optical constant functions are schematically shown in Fig. 2.11 (please keep in mind that this illustration only makes sense for the scalar case, because $\varepsilon = \hat{n}^2$).

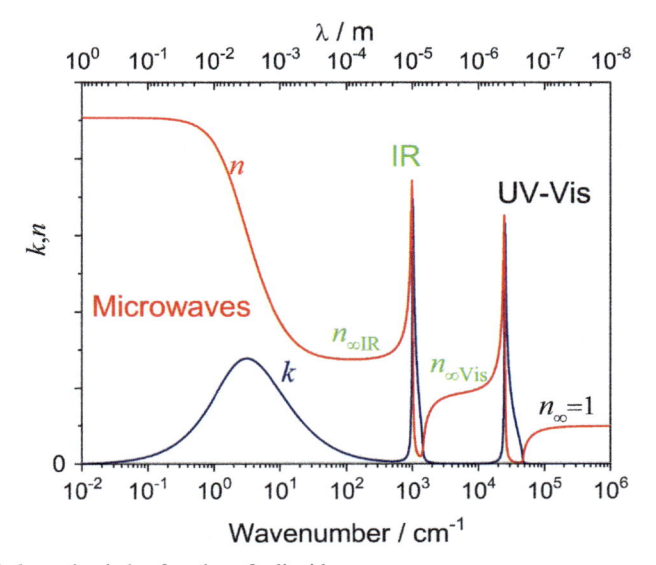

FIG. 2.11 Schematic refractive and absorption index function of a liquid.

Remember, we are dealing with integration! In particular, in case of the absorption index function the consequences are easy to visualize, since we are trying to determine areas. If the absorption index is zero, or close to zero, a transparency region results. If bands are circumscribed by transparency regions, the increase of the refractive index is proportional to the area of the absorption index within the opaque region. Certainly, the refractive index may change somewhat even in transparency regions, since the absorption index is just approximately zero, but these errors are negligible for spectroscopists (and this is even true if you use the optical constant functions for optical modeling, e.g., to find the potential structures for surface-enhanced IR spectroscopy).

In the case of the refractive index, it is a little bit more complicated, but you can condense what happened in spectral regions at higher wavenumbers to a good approximation in a constant, since dispersion caused in these higher wavenumber regions is small in lower ones and leads to inaccuracies that are much smaller than those caused by experimental errors. For most of the materials a molecular spectroscopists deals with, no problem will arise to use the Kramers-Kronig relations in the following form:

$$n(\tilde{\nu}) = n_\infty + \frac{2}{\pi} \wp \int_{\tilde{\nu}_e}^{\tilde{\nu}_f} \frac{\tilde{\nu}' k(\tilde{\nu}')}{\tilde{\nu}'^2 - \tilde{\nu}^2} d\tilde{\nu}', \tag{2.66}$$

where, e.g., $\tilde{\nu}_e = 400$ cm^{-1}, $\tilde{\nu}_f = 4000$ cm^{-1}, and n_∞ is the index of refraction at $\tilde{\nu}_f$ (strictly, it is not, but we will investigate this in Section 5.8 in detail). In case of inorganic materials, it may be necessary to measure also the far infrared spectral range, since due to the higher atom masses, the vibrations are usually located at lower wavenumbers. Problematic are only materials with free electrons, like conductors and doped semiconductors, since those stay opaque at low wavenumbers due to absorption caused by the free electrons.

Having discussed about the consequences of absorptions in matter, I will next introduce absorption into the harmonic electromagnetic waves.

2.8 The influence of absorption on the electromagnetic waves

Absorption in a material means that energy is drawn from the electromagnetic waves that traverse it. How is this introduced in the calculus? First of all, remember that we can use the exponential notation also for electromagnetic waves, cf. (2.26):

$$E(z,t) = \text{Re}[E_{\max}\exp[i(k_{sp} \cdot z)] \exp[-i(\omega \cdot t)]]. \tag{2.67}$$

In Eq. (2.67), we can again neglect the time dependence of the electric field for the following. But before we go on, I want to once more remind you that there are two different conventions with regard to the argument of the wave function.

Note that in literature quite often you find books where the authors switch back and forth between both conventions (in an early version of this book, I did the same), which is definitely not a good idea. Anyway, be aware of the fact that if you use different books, you have to check which convention is used. To remind you, the main consequence is that for the convention expressed in Eq. (2.67), $\hat{n} = n + ik$ must be used, whereas for the alternative convention $\hat{n} = n - ik$.

In section 0, I have introduced k_{sp} as the spatial angular frequency in vacuum: $k_{sp} = 2\pi/\lambda$. For waves in matter, k_{sp} needs to be modified:

$$k_{sp} = \hat{n}\frac{2\pi}{\lambda} = (n + ik)\frac{2\pi}{\lambda}. \tag{2.68}$$

If a light wave hits a medium, Eq. (2.68) only holds for normal incidence; for nonnormal incidence, it needs to be further altered as we will see in Chapter 3. How is the light wave modified in the medium? Let us first assume that the imaginary part equals zero.

As can be seen in Fig. 2.12, the wavelength is inversely proportional to the refractive index. Since the speed of the electromagnetic wave in matter v is given by $v = c \cdot n$, the frequency is not altered. This also follows from energy conservation and that the energy of a photon is given by $E = h \cdot \nu$, with h the Planck constant; for those who consider it strange that I change from the wave-view to the particle-view, an electromagnetic wave is certainly at the same time also a flow of photons. When we assume the medium to be absorbing, the wave will be damped; the stronger, the farther it progresses inside the medium. We will derive this in the following by replacing k_{sp} based on Eq. (2.68):

$$\begin{aligned} E(z) &= \text{Re}[E_{\max}\exp[i(2\pi(n+ik)\tilde{\nu} \cdot z)]] \\ &= \exp[-2\pi k\tilde{\nu} \cdot z]\,\text{Re}[E_{\max}\exp[i(2\pi n\tilde{\nu} \cdot z)]] \end{aligned}. \tag{2.69}$$

FIG. 2.12 From left to the right: An electromagnetic wave in vacuum, the same wave in a medium with refractive index $n = 2$ and in an absorbing medium with $n = 2 + 0.2i$.

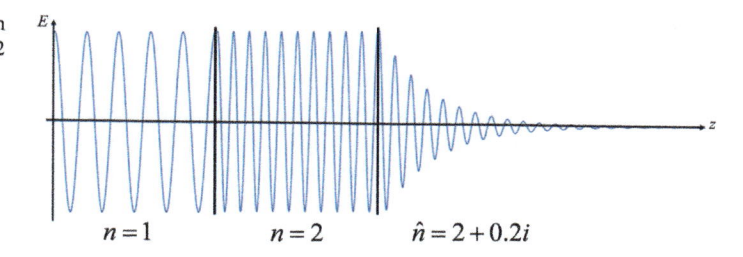

Therefore, the damping term introduced into the model for the dielectric function, i.e., to model absorption of light in the medium, leads correspondingly to an exponential decay of the light wave, which makes sense in terms of energy conservation.

From here, we could go back to what you are used to, i.e., the BBL approximation, by taking the square of $E(z)^2$, which is the electric field intensity I,

$$I(z) = E(z)^2 = E_{max}^2 \exp\left[-2\pi k \tilde{\nu} \cdot z\right]^2 |\exp[i(2\pi n \tilde{\nu} \cdot z)]|^2. \tag{2.70}$$

This is probably not what you would have expected, since there is still this imaginary exponential term. And why do we need the absolute value bars? Remember that we measure real quantities, but we introduced complex quantities, because they simplify the mathematical transformations. The bars remind us to get rid of the complex part, and we do this not by simply squaring the complex part, but by multiplying it with its complex conjugate (the same will be done in the next section for reflectance and transmittance):

$$|\exp[i(2\pi n \tilde{\nu} \cdot z)]|^2 = \exp[i(2\pi n \tilde{\nu} \cdot z)] \cdot \exp[-i(2\pi n \tilde{\nu} \cdot z)] = 1. \tag{2.71}$$

As a consequence, we obtain for the intensity (or, as we call it later on, the irradiance, which is the average energy per unit area per unit time):

$$I(z) = I_0 \exp[-4\pi k(z)\tilde{\nu} \cdot z]. \tag{2.72}$$

The quantity $\alpha(\tilde{\nu}) = 4\pi k(z)\tilde{\nu}$ is called the Napierian absorption coefficient (cf. Eq. 1.1). At this point, it makes sense to stop and consider what we got by Eq. (2.72). We started from an electromagnetic wave with its spatial angular frequency and obtained a material property, the absorption coefficient. Really? It looks like and on the level of this chapter, and even later on, you can keep this illusion if you like, although the spatial angular frequency will become the wavevector in Chapter 3. But as long as we will deal with a scalar dielectric function, it does not matter where inside such a material the wavevector is pointing to, the absorption coefficient will stay the same. You can imagine it as a sphere, where a vector pointing from the center to the surface will always have the same length. But for anisotropic materials, this will drastically change. The scalar dielectric function will become a tensor, which is accompanied by a tensor ellipsoid instead of a sphere. So, the absorption coefficient will also become a tensor and can be imagined as an ellipsoid? No, it will not (in spite of what can be read in certain textbooks)! What happens is that for a particular polarization of the wave you obtain two different values of $\alpha(\tilde{\nu})$ for a particular direction of the wavevector, since from an incident wave in general two different waves are generated inside the medium. The important point is that as long as we talk about materials with a scalar dielectric function, we can exchange it by the (complex) refractive index and see the latter as a material property.

In fact, however, the refractive index is a wave property like the absorption coefficient and the spatial angular frequency. Don't worry if you are not yet able to change your paradigms—you are in good company! Actually, if you step up beyond wave optics and dispersion theory (no worries, so far this is necessary only in very exceptionable cases that I will only touch on), you will have to see matter and light as being coupled in some cases, and the differences between the properties of light and matter become blurred anyway.

Coming back to Eq. (2.72), it looks like we have just proved the Bouguer-Lambert part of the BBL approximation by our simplified wave optics calculus if we identify z as the thickness of the sample. There is one thing to consider, though, which is that Eq. (2.72) is valid only inside a medium that is infinite, so that no part of the wave can be reflected and interfere with itself anywhere. If we have a sample, it usually has two interfaces, one at the front and one at its backside if it is a bulk sample. If it is a layer on a substrate, we have actually one interface more between layer and substrate. At each of these interfaces, the wave will split in a part that is transmitted through this interface and a part that is reflected. In a layer, part of the light is reflected multiple times, until it either leaves the layer or is absorbed. All of this is not taken care of in the BBL approximation, which is not very surprising, since Bouguer and Lambert were mainly interested in describing the absorption that light undergoes when it traverses through the atmosphere, where the changes of the refractive index are very small and also continuous, so there are no interfaces to consider. This can be very different for liquid and solid samples, and we need to inspect carefully the situation at hand to see if the use of the BBL approximation is actually justified. Or we simply use the more advanced theory that we develop in the following. At least for normal incidence and scalar media, this is not really much more complicated as you will see in the next section (actually, there are certainly other possibilities to avoid interference, which are, e.g., to roughen the surface of a sample or to use wedge-shaped samples. As a result, the interference fringes in spectra are indeed gone, but a quantitative evaluation of such spectra would be much more difficult and so complicated that the topic is out of scope of this book).

2.9 Reflection and transmission at an interface separating two scalar media under normal incidence

At first, we will inspect the simplest possible case of reflection from and transmission into a semi-infinite medium. Semi-infinite means that the medium has a starting point, or, better a starting surface perpendicular to the direction of the light beam, since we consider normal incidence, and then extends infinitely. As already stated, this assumes that a wave that is transmitted into this second medium will not experience any interface further on, so that no part of it will ever be reflected. The reason is that for this simplest case, we can only measure reflectance, since the detector can only be positioned in the incidence medium (actually, since we talk about normal incidence, it seems that it must be on the normal to the interface, like the light source, but this problem can be circumvented by a tricky optical configuration using semitransparent mirrors)—measuring in transmission mode would require that the detector can be located behind the sample, something which is certainly only possible if the sample is not semi-infinite.

Overall, this sounds like a case without practical value, but this changes if the sample is absorbing and thick enough that the light reflected from the backside cannot reach the frontside. This is the actual case we are interested in, since if the sample is not absorbing, we simply do not get a spectrum, neither in transmission nor in reflection. In practice, the sample will not be absorbing over the whole spectral regions, but only in certain regions. Remember, we already talked about vitreous silica and its spectrum. If you take a look at Fig. 2.13, you see reflectance spectra of two different samples of vitreous silica. The important difference between the two spectra is located in the spectral region above $2000\,cm^{-1}$. The red spectrum is from a sample thin enough, so that the measured reflectance is no longer only that from the first interface above about $2300\,cm^{-1}$. Instead, you see a contribution from the backside reflection. As long as you concentrate

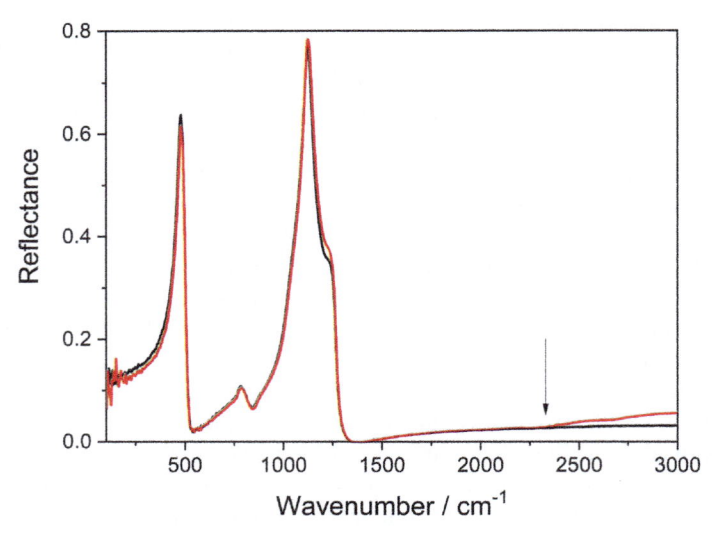

FIG. 2.13 Reflectance spectra of two different samples of vitreous silica.

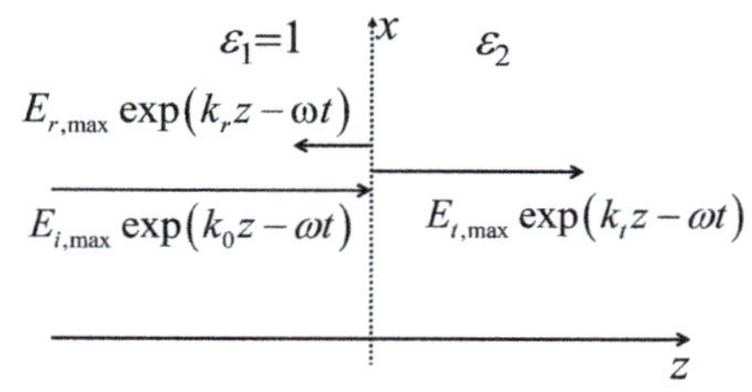

$$\varepsilon_1 = 1 \qquad \uparrow x \qquad \varepsilon_2$$

$$E_{r,\max}\exp(k_r z - \omega t) \leftarrow$$

$$\overline{E_{i,\max}\exp(k_0 z - \omega t)} \rightarrow \qquad E_{t,\max}\exp(k_t z - \omega t)$$

$$\longrightarrow z$$

SCHEME 2.1 A harmonic wave coming in from vacuum ($\varepsilon_1 = 1$) hits the surface of a sample (the surface of which is indicated by the dashed line along the x-axes) and is split into a transmitted and a reflected part.

on evaluating the lower spectral region, you are allowed to consider both spectra as originating from semi-infinite samples, but if you are interested in the reflectance in the nonabsorbing region above 2300 cm^{-1}, only the black spectrum could be evaluated by the theoretical considerations we discuss in the following (strictly, not even this one, because both spectra have been recorded at an angle of incidence of 20° with s-polarized radiation, and you have to digest the theoretical considerations in Chapter 3 to quantitatively evaluate it correctly).

The situation that we discuss in the following is shown in Scheme 2.1. A harmonic wave coming in from vacuum, which has a dielectric function that always equals unity independent from wavenumber ($\varepsilon_1 = 1$), hits the surface of a sample and is split into two parts. The first part is transmitted through the interface (t), and the second part is reflected (r) and traveling back through the incidence medium.

The question that we have to answer now is how large is the intensity of the reflected wave (actually, as already mentioned, the irradiance, but I allow us to be less strict in this chapter) in dependence of the wavenumber. We will also be able to calculate the intensity of the transmitted wave, but, as already discussed, this will not be helpful in practice as we cannot measure it.

How can we approach the problem? Actually, strictly speaking, we cannot, because by treating electric and magnetic fields as scalars, we could not make an important conclusion, which is that the tangential component of the electric field, i.e., the component parallel to the surface of the sample, is continuous at an interface. On the other hand, what helps us is that for normal incidence and scalar media, the electric field is always perpendicular to the direction of propagation, another conclusion that we could not make, so that for normal incidence and our problem at hand the fields only have tangential components. So, without knowing the details and challenges, our solution would have been the correct one anyway, but you are now warned that things get somewhat more complicated under different assumptions. Anyway, for the problem at hand, we can conclude that

$$\begin{aligned} E_{1,\tan} &= E_{2,\tan} \rightarrow \\ E_i + E_r &= E_t \end{aligned}. \tag{2.73}$$

In other words, the sum of the two electric field strengths, which stem from the incoming and the reflected wave, is equal to that of the transmitted electric field at the interface. This somehow seems odd on first view, since the incoming field should be the largest one, but by the change of direction by 180 degree, the amplitude of the reflected field becomes negative when the incoming field is positive and the other way round.

From the continuity of the tangential components of the magnetic field, we obtain a second equation:

$$\begin{aligned} H_{1,\tan} &= H_{2,\tan} \xrightarrow{k_{sp}E - \omega H = 0} \\ k_{sp,i}E_i + k_{sp,r}E_r &= k_{sp,t}E_t \xrightarrow{k = \omega\sqrt{\varepsilon}} \\ \sqrt{\varepsilon_1}E_i - \sqrt{\varepsilon_1}E_r &= \sqrt{\varepsilon_2}E_t \xrightarrow{\varepsilon_1 = 1} \\ E_i - E_r &= \sqrt{\varepsilon_2}E_t \end{aligned}. \tag{2.74}$$

Here, we have used Eq. (2.28) to replace the magnetic field by the electric field. Overall, we now have two equations in the electric field and three unknowns:

$$\begin{aligned} E_i + E_r &= E_t \quad \text{(I)} \\ E_i - E_r &= \sqrt{\varepsilon_2}E_t \quad \text{(II)} \end{aligned}. \tag{2.75}$$

This does not seem to be sufficient, but what we will do is express the reflected and the transmitted field in terms of the incident field, and you will see that this brings us directly to the quantities reflectance and transmittance, which we are after. So first, we replace E_t in the second equation in (2.75) by the left side of the first equation:

$$\begin{aligned} E_i - E_r &= \sqrt{\varepsilon_2}(E_i + E_r) \rightarrow \\ E_i - \sqrt{\varepsilon_2}E_i &= \sqrt{\varepsilon_2}E_r + E_r \rightarrow \\ (1 - \sqrt{\varepsilon_2})E_i &= (1 + \sqrt{\varepsilon_2})E_r \rightarrow, \\ r = \frac{E_r}{E_i} &= \frac{1 - \sqrt{\varepsilon_2}}{1 + \sqrt{\varepsilon_2}} = \frac{1 - \hat{n}_2}{1 + \hat{n}_2} \end{aligned} \qquad (2.76)$$

r is called the reflection coefficient. Next, we express the reflected field at the interface in terms of the transmitted and the incident field to obtain the ratio between these two fields:

$$\begin{aligned} E_i - (E_t - E_i) &= \sqrt{\varepsilon_2}E_t \rightarrow \\ 2E_i - E_t &= \sqrt{\varepsilon_2}E_t \rightarrow \\ 2E_i &= \sqrt{\varepsilon_2}E_t + E_t \rightarrow \\ 2E_i &= (1 + \sqrt{\varepsilon_2})E_t \\ t = \frac{E_t}{E_i} &= \frac{2}{1 + \sqrt{\varepsilon_2}} = \frac{2}{1 + \hat{n}_2} \end{aligned} \qquad (2.77)$$

t is called the transmission coefficient. Finally, we have to consider that we deal with intensities, which means that we have to square the reflection and the transmission coefficient to obtain the reflectance R and transmittance T. This actually works in case of the reflectance, but only since incoming and reflected wave are in the same medium. For the transmittance, we have to take into account that the transmitted wave is in a medium of different refractive index than the incident wave for reasons that are explained in detail in the Chapter 3 (cf. Poynting vector, Section 3.3). Overall, we find that (remember, the absolute value bars indicate that you have to multiply the reflection and transmission coefficients by their complex conjugates, since reflectance and transmittance are measurable quantities and, therefore, real valued):

$$\begin{aligned} r = \frac{E_r}{E_i} = \frac{1 - \hat{n}_2}{1 + \hat{n}_2} &\longrightarrow R = \left|\frac{E_r}{E_i}\right|^2 = |r|^2 \\ t = \frac{E_t}{E_i} = \frac{2}{1 + \hat{n}_2} &\longrightarrow T = \hat{n}_2 \left|\frac{E_t}{E_i}\right|^2 = \hat{n}_2|t|^2 \end{aligned} \qquad (2.78)$$

If $\hat{n}_2$ is real, $R+T=1$. If not, then $R+T+A=1$, where A is the absorptance (note that I will always use a nonitalic A to indicate the absorbance in contrast to the italic A for the absorptance). One must very carefully differentiate between both quantities, even if it can be shown that they become the more similar the weaker an oscillator, and, thereby, the absorption index, is. This is not that easy, because sometimes absorptance is called absorbance in physics literature. On the other hand, a molecular spectroscopist might not even have heard the term absorptance, although the (comparably) new methods like photothermal IR are actually able to measure absorptance (but **not** the absorbance as is often assumed!). Reflectance, on the other hand, seems to be not very useful for the molecular spectroscopist, since textbooks claim there is no easy way to interpret reflection spectra quantitatively. In 2021, however, IR refraction spectroscopy has been invented [14], which allows exactly that as long as the material is only weakly absorbing as most organic and biologic materials are (see Section 7.1). Nevertheless, next I will extend the formalism we just derived to include samples with finite thickness so that we can go back to absorbance if this is your dearest wish.

2.10 Transmission through a thick slab suspended in vacuum

We start with the transmission through a thick slab, because this case is more similar to what you are used to from the BBL approximation. Thick means in effect that we assume that the slab cannot be fabricated with good enough plane parallel surfaces, so that wave interference does not occur. It certainly looks to us by visual inspection of such substrates as if the front and rear faces would be plane parallel, but in reality, the slab will have thickness inhomogeneities larger than the wavelength of the light. This means that the light going through the slab will experience path differences, which lead to quasi-continuous distributions of phase differences between 0 and π, all with equal probabilities. Therefore, the photons that arrive at the detector will have experienced everything between constructive and destructive interference. Therefore, on average they will have experienced no interference at all. In textbooks, you can sometimes read that for materials with a thickness larger than the wavelength interference effects cannot occur anyway, since coherence is destroyed (coherence means, as explained earlier that the waves keep the same phase relation), but I can only repeat that this is nonsense,

SCHEME 2.2 Scheme to derive transmittance through a thick slab with thickness d embedded in vacuum.

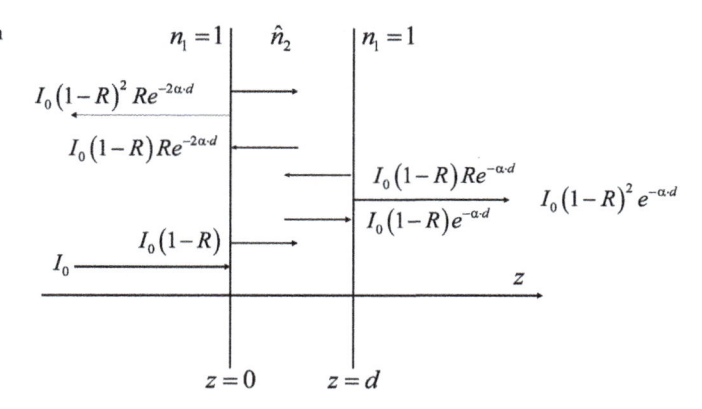

since if a part of a wave is reflected at some boundary, it will always be coherent to itself and, therefore, interfere with itself, even if the many different forward traveling waves are no longer coherent to their siblings. Have you ever measured transmittance of a silicon wafer of, say, about 0.5 mm thickness? Those can actually be fabricated with so low thickness tolerances that you will see interference fringes in a spectrum if you choose a high enough spectral resolution (if you do not believe me, take a look at Fig. 3 of [15]). But if we talk about glass or crystalline materials, e.g., CaF_2, and you measure the transmittance, interference fringes will be absent due to the thickness inhomogeneities. For this section, we assume this and the simplest possible case of a slab suspended in air with a thickness d and a complex index of refraction $\hat{n}_2$.

To understand what is going on, have a look at Scheme 2.2.

The beam (we are looking at intensities, so it is better to speak of beams, instead of waves) arrives with an intensity I_0 at the interface at $z=0$. At this interface, the beam is split into two parts. The intensity of the reflected beam is RI_0, and the intensity of the transmitted part equals $I_0(1-R)$. The latter part travels toward the second interface at $z=d$. Since the slab is absorbing with an absorption coefficient α, which relates to the index of absorption by (cf. Eq. (2.69)),

$$\alpha = 2\pi k \tilde{\nu}, \tag{2.79}$$

the part of the beam that is not absorbed has an intensity of $I_0(1-R)e^{-\alpha d}$. Due to the interface, this beam is again split into two. The first part is the reflected beam and has the intensity $I_0(1-R)Re^{-\alpha d}$, and the other part is finally transmitted, i.e., it leaves the slab, with the intensity $I_0(1-R)^2 e^{-\alpha d}$. The reflected beam travels back through the slab and, by absorption, loses again intensity, so that directly before the first interface its intensity amounts to $I_0(1-R)Re^{-2\alpha d}$, since the beam has traveled twice the distance d. Not surprisingly, this remaining beam is again split by the interface into two beams. The part transmitted through the interface has an intensity given by $I_0(1-R)^2 Re^{-2\alpha d}$, whereas the part traveling again forward to the slab has the intensity $I_0(1-R)R^2 e^{-2\alpha d}$. I guess you know by now how this will progress further, and you will be able to find out by yourself what will happen at the second interface and that this time the intensity $I_0(1-R)^2 R^2 e^{-3\alpha d}$ passes through the second interface. So, to determine the overall transmittance, you have to add $I_0(1-R)^2 e^{-\alpha d} + I_0(1-R)^2 R^2 e^{-3\alpha d} + \ldots$ $I_0(1-R)^2 R^{2m} e^{-(2m+1)\alpha d}$, where m is the number of times the beam that is transmitted has been reflected at the second interface.

All in all, we can also write the transmitted intensity I_t in this way:

$$I_t = I_0(1-R)^2 e^{-\alpha \cdot d}\left(1 + R^2 e^{-2\alpha \cdot d} + R^4 e^{-4\alpha \cdot d} + \ldots + R^{2m} e^{-2m\alpha \cdot d}\right). \tag{2.80}$$

This is an infinite series within the second pair of brackets according to $\sum_{k=0}^{\infty} x^k = 1/(1-x)$, with $x = R^2 e^{-2\alpha \cdot d}$, so that

$$T = \frac{I_t}{I_0} = \frac{(1-R)^2 \exp(-\alpha d)}{1 - R^2 \exp(-2\alpha d)}, \quad R = \left|\frac{1-\hat{n}_2}{1+\hat{n}_2}\right|^2. \tag{2.81}$$

If we assume that the oscillator strengths are not too high, which means that the absorption coefficients are assumed to be small, then R^2 will be small, as R is always smaller than unity, so that

$$T \approx (1-R)^2 \exp(-\alpha d). \tag{2.82}$$

This is actually the level of theory August Beer used in his seminal paper from 1852, i.e., he took into account the reflectance at each interface once, but neglected multiple reflections [16].

If we assume the oscillator strengths to be even smaller, then not only R^2 but also R will be small, i.e., $R \ll 1$. Then, Eq. (2.82) further simplifies to

$$T \approx \exp(-\alpha d), \tag{2.83}$$

and we regain the Bouguer-Lambert part of the BBL approximation.

Maybe you have heard that absorbance should not exceed 3 for most instruments, because then the detector noise overwhelms the signal. This is equivalent with stating that the transmittance should not be smaller than 0.1%, meaning that only one per mill of the initial intensity I_0 will reach the detector (using transmittance in this case actually makes much more sense than absorbance, because you deal with the quantity that is actually measured). Now, if I assume that $T = 0.001$ and d is 1 mm, how large can α be so that we can still apply the approximations? From Eq. (2.83), we get

$$\begin{aligned} 10^{-3} &< \exp(-\alpha \cdot 0.1 \text{ cm}) \to \\ \ln 10^{-3} &< -\alpha \cdot 0.1 \text{ cm} \to \\ 30 \cdot \ln 10 \text{ cm}^{-1} &> \alpha \to \\ 69.077 \text{ cm}^{-1} &> \alpha \end{aligned} \tag{2.84}$$

This is a fairly small absorption coefficient. If you are not so familiar with absorption coefficients, you can also view it from the perspective of the absorption index and assume a wavenumber of 1000 cm^{-1}. Based on Eq. (2.79), we divide α by 1000 and 2π and get that $k < 0.01$, which is definitely a very small value if you compare it with the values from Fig. 2.9. Just that you get a feeling for the numbers, for a typical band with a damping constant of 20 cm^{-1}, the oscillator strength would have to be smaller than 25 cm^{-1}. For those of you who know the spectrum of polyethylene, the oscillators would have to be still weaker than the wagging deformation vibration at 1367 cm^{-1}, which means that for most practical cases the sample thickness would be much too thick for enough light to reach the detector to show something else but noise. The thicknesses actually would have to be in the range of less than some 10 μm for organic materials and some 100 nm for inorganic materials. But this means that we are in the range of the wavelengths in the mid infrared (400–4000 cm^{-1} ≙ 25–2.5 μm), which tells us that we have to consider the wave properties of light.

2.11 Transmission through a thin slab suspended in vacuum

As discussed at the end of the previous section, if we want spectra without flatfoots (this is how we called bands with a transmittance <0.1 back in the days), then we have to use samples with layers having thicknesses in the range of the wavelength. If the surfaces are smooth, this would mean that there can be no thickness variations large enough to average out interference effects. Therefore, we will have to deal properly with them. Accordingly, we cannot use intensities anymore and must go one step back to the fields, which also means that we have to respect their location dependency. Luckily, the problem is not much more complicated as the one we treated in the last section. Instead of a beam, we have to see light as an electric wave, with an amplitude $E_{i,\text{max}}$ for the incoming wave, and see how this amplitude changes when it is reflected at or transmitted through an interface. When we have done this, we sum up all transmitted parts on the level of the fields, i.e., coherently (in the previous section we summed up on the level of the intensities, which is called incoherent summation). As we assume a thin layer suspended in vacuum, the first semi-infinite medium is vacuum, then comes the slab and, finally, again semi-infinite vacuum—have a look at Scheme 2.3.

Compared to the previous case of a thick slab, there is one additional issue that we need to factor in. In the previous section, we did not (and needed not to) distinguish between different directions of the light, which means the reflectance or transmittance was the same regardless if the beam penetrated from vacuum into the sample or the other way round. This symmetry is broken on the level of the electric field and the wave. If we derive the reflection and transmission coefficients along the lines we discussed in Section 2.9 for the case that light penetrates from medium 2 into medium 1 and compare the results with those for the opposite direction, we find that

$$\begin{aligned} t_{12} &= \frac{2}{1+\hat{n}_2}, & t_{21} &= \frac{2\hat{n}_2}{1+\hat{n}_2} = \hat{n}_2 t_{12}, \\ r_{12} &= \frac{1-\hat{n}_2}{1+\hat{n}_2}, & r_{21} &= \frac{\hat{n}_2-1}{1+\hat{n}_2} = -r_{12}. \end{aligned} \tag{2.85}$$

For the transmission coefficients, the asymmetry is removed when we go over to the transmittance by Eq. (2.78), whereas for the reflection coefficients it vanishes through squaring if we focus on what happens on a single interface. Since we have two interfaces, things are a little bit different for the case at hand. Anyway, this is just a small complication.

SCHEME 2.3 Scheme to derive transmittance through a thin slab embedded in vacuum.

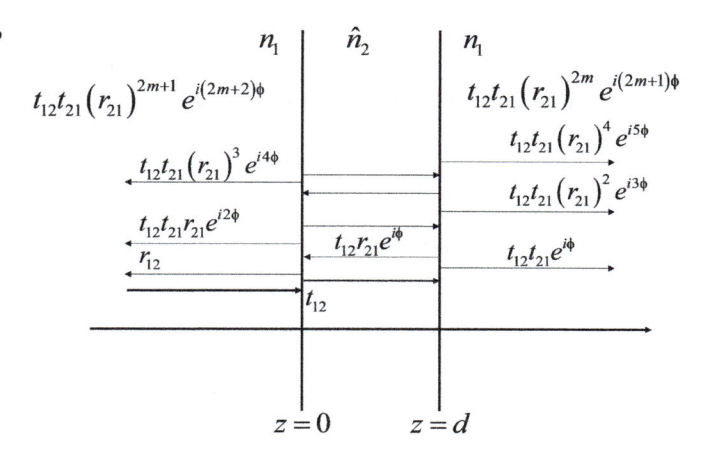

After the wave is transmitted through the first interface, the remaining electric field traveling in the forward direction is given by $t_{12}E_{i,max}$. While it travels through the slab, the amplitude is overall decreased by absorption, but it also changes periodically as this is shown in the right part of Fig. 2.12. When it reaches the second interface, its amplitude is given by $t_{12}E_{i,max}\exp(i\phi)$ with $\phi = 2\pi\hat{n}_2\tilde{\nu}d$. At this interface, the wave is again split into two parts, a transmitted part with $t_{12}t_{21}E_{i,max}\exp(i\phi)$ and a reflected part $t_{12}r_{21}E_{i,max}\exp(i\phi)$. When this reflected part arrived once more at the first interface, the electric field amounts to $t_{12}r_{21}E_{i,max}\exp(2i\phi)$. Again it is split into two waves, a transmitted part given by $t_{12}t_{21}r_{21}E_{i,max}\exp(2i\phi)$, which then interferes with the first reflected part $r_{12}E_{i,max}$, and a reflected part $t_{12}(r_{21})^2 E_{i,max}\exp(2i\phi)$, which travels again in forward direction and reaches with a field strength of $t_{12}(r_{21})^2 E_{i,max}\exp(3i\phi)$ the second interface. The transmitted part is therefore $t_{12}t_{21}(r_{21})^2 E_{i,max}\exp(3i\phi)$. If we follow the wave a couple of more cycles, we find that

$$E_t = E_{i,max} t_{12} t_{21} \exp(i\phi)\left(1 + (r_{21})^2 \exp(2i\phi) + (r_{21})^4 \exp(4i\phi) + \ldots + (r_{21})^{2m} \exp(2mi\phi)\right). \quad (2.86)$$

As in the previous section, this is an infinite series within the second pair of brackets according to $\sum_{k=0}^{\infty} x^k = 1/(1-x)$, with $x = (r_{21})^2 \exp(2i\phi)$, so that

$$T = |t|^2 = \left|\frac{t_{12}t_{21}\exp(i\phi)}{1 - r_{21}^2 \exp(2i\phi)}\right|^2. \quad (2.87)$$

Note that the incidence and the exit medium are one and the same; therefore $T = |t|^2$.

In order to understand what this result means, please forget for a moment absorption. For a phase equal to $\pi/2$, $\exp(i\phi)$ becomes minimal and $\exp(2i\phi)$ maximal, which means that transmittance becomes minimal (and reflectance becomes maximal). On the other hand, for π and any multiple of it, transmittance becomes unity (because reflectance becomes zero and $R + T = 1$). If you think about the definition of absorbance, which is in this case $A = -\log_{10}T$ ("transmittance absorbance"), you immediately see that something is wrong with this definition for layers or thin slabs, since except for $\phi = m\pi$, $A \neq 0$ even without absorption (and the baseline is not just a constant offset, but shows fringes).

How is it possible that for $\phi = m\pi$ reflectance becomes zero? Since $r_{21} = -r_{12}$ and the sample is nonabsorbing, the wave reflected from the first interface is canceled out by the waves reflected from the second interface and transmitted back through the first interface if they agree concerning the phase; therefore reflectance is zero.

In a similar way, we obtain for the reflectance

$$R = |r|^2 = \left|r_{12} + \frac{t_{12}t_{21}r_{21}\exp(2i\phi)}{1 - r_{21}^2 \exp(2i\phi)}\right|^2. \quad (2.88)$$

Let me emphasize again that simple measures in order to prevent interference effects from occurring, like roughening one side of the sample, are generally not a good idea. Indeed, this removes these effects, but you do not only lose information about the sample. What you get instead is actually much worse. In this case, diffuse reflection and/or transmission occurs, depending on if it is the first interface or the second or both that are rough. Microscopically this means that you get scattering either from holes or from tips. Unfortunately, the systems become immensely complex due to multiple scattering, so complex that an analytical solution of the problem is no longer possible. Therefore, it is impossible to solve the corresponding inverse problem, which is to evaluate the spectra of such samples quantitatively. In other words, it is impossible to determine the dielectric function and the oscillator parameter from such a sample, while it is not a big deal to do this in case of the spectra of free-standing films as long as they have plane-parallel surfaces, at least once you have mastered the content of the first part of this book. On top of that, you obtain the sample or layer thickness and/or the (nearly constant) refractive index or dielectric constant in the transparency region.

2.12 Transmission through a layer on a substrate suspended in vacuum

To solve the problem of calculating the transmission through a layer on a substrate suspended in vacuum assuming normal incidence is as far as we go concerning complexity in this chapter. Once you have finished this section, you can jump directly to Chapter 6 and you will have some basis to understand the rest of the first part of the book without immersing yourself to deep into theory. On the other hand, in this section we reach the point where we need a more structured approach to come to a result, which is provided in the following chapters, so why not simply proceed consecutively before looking at concrete problems?

Anyway, to understand how the transmittance through a layer on a substrate suspended in vacuum can be calculated for normal incidence, we need to make some amendments to what we discussed in Section 2.9. Remember, there we discussed normal incidence from vacuum onto the surface of a sample. This we need to extend first by assuming that the first medium is not vacuum, which means we need the corresponding relations if we cannot assume that $n_1 = 1$. Also, we will not assume that both media separated by the interface are semi-infinite.

How shall we proceed? In general, we may have no idea what happened to the wave before, i.e., we do not know the structure of the sample on the left (the direction from which the light is assumed to come from), and we have no idea what happens afterward, which means that we do not know the structure on the right side. No worries, this is actually not a problem! We just do not assume that the incoming wave from the left has the same amplitude as the wave that falls onto the layer stack out of vacuum. This means that only for the actual calculation of the transmittance through the whole layer stack we would have to know the field that arrives at the interface from the left. Instead, what we derive in the following gives us the ratio of whatever arrives to the parts that are reflected and transmitted. The ansatz is the same as in Section 2.8, except that we keep ε_1 instead of setting $\varepsilon_1 = 1$ in Eq. (2.74):

$$\begin{aligned} H_{1,\tan} &= H_{2,\tan} \xrightarrow{k_{sp}E - \omega H = 0} \\ k_{sp,i}E_i + k_{sp,r}E_r &= k_{sp,t}E_t \xrightarrow{k = \omega\sqrt{\varepsilon}} \\ \sqrt{\varepsilon_1}E_i - \sqrt{\varepsilon_1}E_r &= \sqrt{\varepsilon_2}E_t \longrightarrow \\ E_i - E_r &= \sqrt{\frac{\varepsilon_2}{\varepsilon_1}}E_t \end{aligned} \qquad (2.89)$$

The two equations in the electric field for the three unknowns then read:

$$\begin{aligned} E_i + E_r &= E_t \quad \text{(I)} \\ E_i - E_r &= \sqrt{\frac{\varepsilon_2}{\varepsilon_1}}E_t \quad \text{(II)} \end{aligned} \qquad (2.90)$$

Again, we replace E_t in the second equation in (2.90) by the left side of the first equation:

$$\begin{aligned} E_i - E_r &= \sqrt{\frac{\varepsilon_2}{\varepsilon_1}}(E_i + E_r) \rightarrow \\ E_i - \sqrt{\frac{\varepsilon_2}{\varepsilon_1}}E_i &= \sqrt{\frac{\varepsilon_2}{\varepsilon_1}}E_r + E_r \rightarrow \\ \left(1 - \sqrt{\frac{\varepsilon_2}{\varepsilon_1}}\right)E_i &= \left(1 + \sqrt{\frac{\varepsilon_2}{\varepsilon_1}}\right)E_r \rightarrow \\ (\sqrt{\varepsilon_1} - \sqrt{\varepsilon_2})E_i &= (\sqrt{\varepsilon_1} + \sqrt{\varepsilon_2})E_r \rightarrow \\ r_{12} = \frac{E_r}{E_i} &= \frac{\sqrt{\varepsilon_1} - \sqrt{\varepsilon_2}}{\sqrt{\varepsilon_1} + \sqrt{\varepsilon_2}} = \frac{\hat{n}_1 - \hat{n}_2}{\hat{n}_1 + \hat{n}_2} \end{aligned} \qquad (2.91)$$

Likewise, we obtain for the transmission coefficient:

$$t_{12} = \frac{E_t}{E_i} = \frac{2\sqrt{\varepsilon_1}}{\sqrt{\varepsilon_1} + \sqrt{\varepsilon_2}} = \frac{2\hat{n}_1}{1 + \hat{n}_2}. \quad (2.92)$$

Transmittance and reflectance are then calculated by

$$\begin{aligned} r_{12} &= \frac{\hat{n}_1 - \hat{n}_2}{\hat{n}_1 + \hat{n}_2} \rightarrow R_{12} = |r_{12}|^2 \\ t_{12} &= \frac{2\hat{n}_1}{\hat{n}_1 + \hat{n}_2} \rightarrow T_{12} = \frac{\hat{n}_2}{\hat{n}_1}|t_{12}|^2 \end{aligned}. \quad (2.93)$$

Since this time, it can, in principle, also happen that the wave does not come from the left side, but from the right side, we have to think about what would follow in such a situation. Actually, instead of assuming that the wave comes from the right side, we can simply flip the order of the two media and exchange the indices 1 and 2 in the aforementioned relations. The result is then

$$\begin{aligned} r_{21} &= \frac{\hat{n}_2 - \hat{n}_1}{\hat{n}_1 + \hat{n}_2} \rightarrow R_{21} = |r_{21}|^2 \\ t_{21} &= \frac{2\hat{n}_2}{\hat{n}_1 + \hat{n}_2} \rightarrow T_{21} = \frac{\hat{n}_1}{\hat{n}_2}|t_{21}|^2 \end{aligned}. \quad (2.94)$$

If we compare the results, i.e., Eqs. (2.93), (2.94), we see that

$$\begin{aligned} r_{21} &= -r_{12} \rightarrow R_{21} = R_{12} \\ \frac{1}{\hat{n}_1} t_{21} &= \frac{1}{\hat{n}_2} t_{12} \rightarrow T_{21} = T_{12} \end{aligned}. \quad (2.95)$$

Before we go on to derive how the transmittance through a layer on a substrate suspended in vacuum can be calculated, let us take a look at the situation that is illustrated in Scheme 2.4.

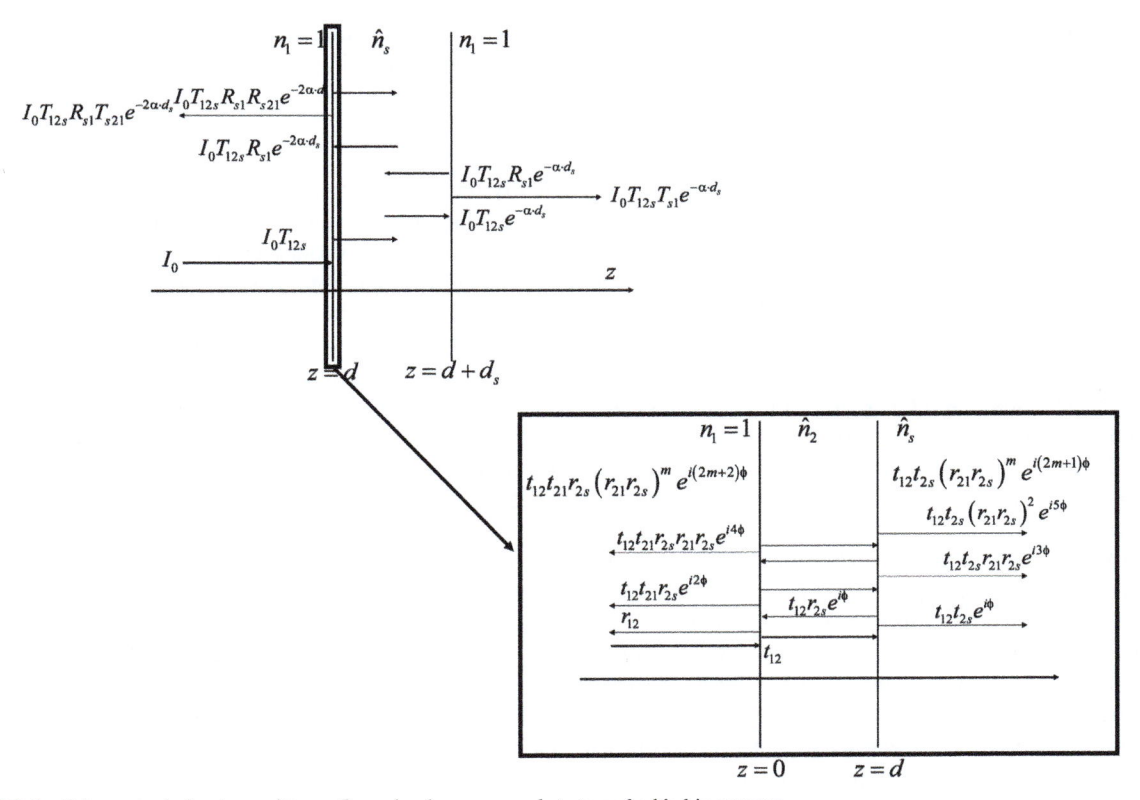

SCHEME 2.4 Scheme to derive transmittance through a layer on a substrate embedded in vacuum.

If we would try to derive the solution in the same way as for the thin and thick slab in the previous sections, we would be in trouble (at least, I would be), because this seems to be a rather complicated situation, also due to the different nature of the two layers, one leading to coherent superposition of waves, while the other shows incoherent multiple reflections.

Luckily, we can borrow from an idea that we discuss in Chapter 4 to solve such and much more complex situations systematically. We assume that the overall situation is similar to that of the thick slab, but that it has two different interfaces, $12s$ and $s1$ (cf. Scheme 2.4). Accordingly, the transmittance is given by

$$T = \frac{T_{12s}T_{s1}\exp(-\alpha d)}{1 - R_{s1}R_{s21}\exp(-2\alpha d)}. \tag{2.96}$$

In Eq. (2.96), $12s$ is a complex interface consisting of a so-called coherent packet. Compared to Eq. (2.81), we had to replace $(1-R)^2$ by $(1-R_{12s})(1-R_{s1}) = T_{12s}T_{s1}$ to account for the two different interfaces.

The complex interface or coherent packet consists itself of a coherent layer packet the transmittance T_{12s} and reflectance R_{s21} of which we have to determine separately, whereas T_{s1} and R_{s1} can be computed from Eq. (2.93) if medium 2 is considered as the substrate. For T_{12s} and R_{s21}, we find from Scheme 2.4 that (to determine R_{s21}, we have to switch 1 and s) [17]:

$$\begin{aligned} T_{12s} &= |t_{12s}|^2 = \left|\frac{t_{12}t_{2s}\exp(i\phi)}{1-r_{12}r_{2s}\exp(2i\phi)}\right|^2 \\ R_{s21} &= |r_{s21}|^2 = \left|r_{s2} + \frac{t_{s2}t_{2s}r_{21}\exp(2i\phi)}{1-r_{12}r_{2s}\exp(2i\phi)}\right|^2 \end{aligned}. \tag{2.97}$$

To derive Eq. (2.97) was definitely not an easy task, but if you could follow the derivations to this point, you should be able to skip the next two chapters and grasp the essence of the chapters behind in the scalar part if you do not have the leisure to deepen your understanding first. On the other hand, with some more patience and endurance and the content of the next section, you will not only be able to understand everything in this part more deeply, but you will also be fit for the tensorial part. In any way, I guess you have a pretty good feeling now about how complicated things can become as soon as you go over from the BBL approximation to the next level of theory, based on Maxwell's equations and dispersion theory. Is it worth the effort? This really depends. If you focus on analytical chemistry, you certainly can have the point of view that there will always be some error left, which you cannot explain, and you are certainly right. Even when you are at the end of this book, you won't be able to remove all sources of error, because nature is just too complex. On the other hand, if you see this as a kind of open game where every step is the destination, go on and develop, after you have worked through the book, some additional steps, which are not in this book—I am looking forward to reading about them!

2.13 Scalar and vector fields

So far, we have assumed that the electric field is a scalar. We actually started by even assuming that it is homogeneous, like that between the plates of a large plate capacitor. We then restricted this latter postulation by presuming that the electric field varies in a sinusoidal way, if there is no dissipation, and with a superposed exponential decay, if there is dissipation. What we cannot understand in this simplified picture is polarization and the fact that waves in scalar media are transversal or can be partly longitudinal in anisotropic media. Also, we cannot understand the interplay between electric and magnetic fields, as we could only introduce a severely curtailed form of Maxwell's equations. There is also no way of introducing anisotropy without vectors, since without direction the electric field and the electric displacement as well as the two different magnetic fields H and B can differ only by a constant. In addition, with the present assumptions the wave cannot have a direction in an anisotropic medium, which we cannot define as long as we do not know what a tensor is. We also cannot understand that the component of the electric and magnetic fields perpendicular to an interface must be treated differently from the two components that are parallel to the interface, which is why we generally assumed perpendicular incidence in this chapter until now.

There are certainly ways to cheat around or simply ignore the corresponding phenomena. One of these ways is of course to continue using the BBL approximation and adding to it its increment linear dichroism theory to treat optical anisotropy. This would mean to further ignore the wave properties of light, but if you like to progress, then it is time for you to get to the next level. You most probably actually know anyway already what I am going to show you in this section and the knowledge just lies dormant, since you did not need to relate and connect it to your knowledge about how spectroscopy works. After all, the treatment of particles as waves in quantum chemistry is also three-dimensional, so if you worked with Schrödinger's equation, then Maxwell's equations from which a corresponding wave equation results (certainly, this one was there first) should also pose no problem.

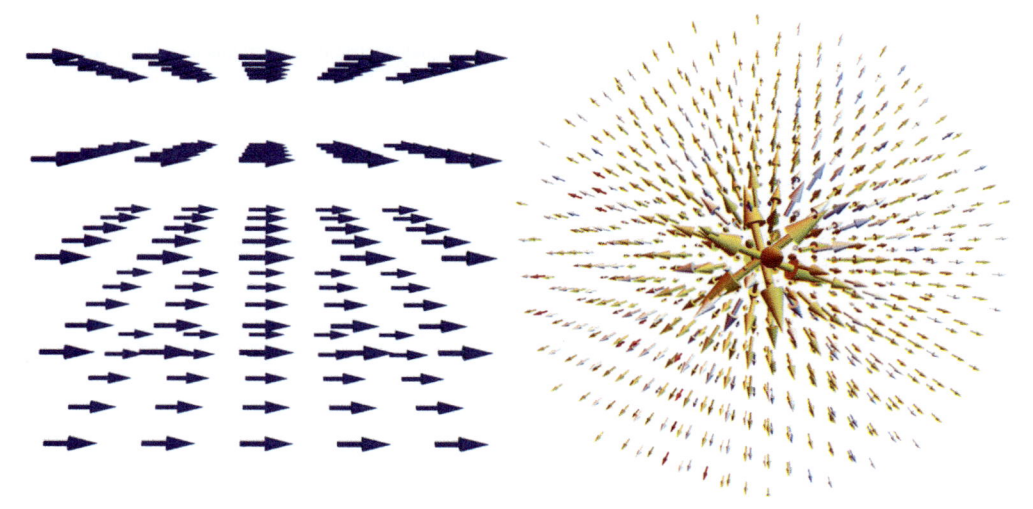

FIG. 2.14 Electric fields between a plate capacitor *(left side)* and of a point charge *(right side)*.

I start with a very simple problem you should also know already, which is a comparison between the electric fields within a plate capacitor and that of a point charge that is provided in Fig. 2.14.

The electric field between the plates of a plate capacitor is constant as long as we do not look at positions near the edges of the plates. However, the field certainly has a direction, because the field is explained as the force that acts on a test charge. This force has a direction, since the plates are oppositely charged. For the point charge, there would be a way to simplify the situation if we would use spherical coordinates, since the force always acts along the radius, but in this book, we stay with cartesian coordinates, because we focus on plane waves. The electric field of a point charge is given by

$$\mathbf{E} = \frac{1}{4\pi\varepsilon_0} \frac{q}{r^2} \frac{\mathbf{r}}{r} \quad \mathbf{r} = \begin{bmatrix} x \\ y \\ z \end{bmatrix} \quad r = \sqrt{x^2 + y^2 + z^2}. \tag{2.98}$$

Correspondingly, we will write plane waves generally in the following way:

$$\mathbf{E}(\mathbf{r},\omega) = \mathbf{E}_0 \exp(i(\mathbf{k}\cdot\mathbf{r} - \omega t)). \tag{2.99}$$

Since time is one-dimensional, there is no change compared to Section 2.5. What changes is the angular spatial frequency, which we will call from now on the *wavevector*. In particular, the direction of the wavevector is equivalent to the direction of the wave. We certainly could make things easier by assuming in general that our waves are traveling, say, in the z-direction. However, since we want to treat cases with nonnormal incidence, this would mean that our interfaces are not aligned parallel to the coordinate axes. This is something I will avoid in this book, because it drastically complicates the situations. Instead, we will generally keep the interfaces aligned with the coordinate axes and, in particular, assume that their normal is always aligned with the z-direction. Therefore, except for normal incidence, the wavevector will have components,

$$\mathbf{k} = k_0 [k_x, k_y, k_z]^T = k_0 [0, \sin\alpha_i, \cos\alpha_i]^T, \tag{2.100}$$

where α_i is the angle of incidence measured from the normal shared by all interfaces (cf. Scheme 2.5) and $k_0 = 2\pi\hat{n}/\lambda = 2\pi\tilde{\nu}\hat{n}$. I take the liberty and do not generalize the wavevector to the extreme of having three components different

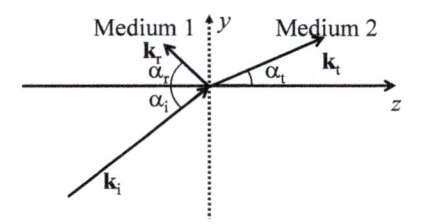

SCHEME 2.5 A plane wave traveling through two semi-infinite media nonnormally relative to their interface.

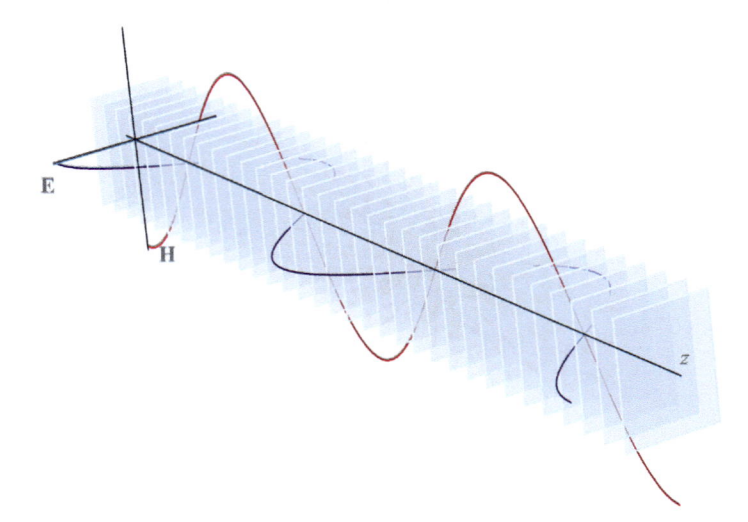

FIG. 2.15 Plane harmonic waves. The planes describe points of equal electric field strengths and directions.

from zero—I have never come across a case where this would have been necessary. Instead, we will assume through the further course of this book that the *x*-component of the wavevector is zero, which can be done "without any loss of generality."

If we would choose $\mathbf{k} = k_0[0, 0, 1]^T$, then the wave would travel in the *z*-direction. In this case the dot product $\mathbf{k} \cdot \mathbf{r}$ would result in

$$\mathbf{k} \cdot \mathbf{r} = k_0[0, 0, 1] \begin{bmatrix} r\sin\theta\cos\varphi \\ r\sin\theta\sin\varphi \\ r\cos\theta \end{bmatrix} = k_0 \cdot z, \tag{2.101}$$

which is the case that we have discussed throughout this chapter. Since there is no *x* or *y* dependence, all points in space have the same amplitude, i.e., we have the case of a plane harmonic wave as it is depicted in Fig. 2.15. The points lying in a plane all have the same electric field strength and the same direction of the electric field, and this also holds for the magnetic field. In particular, the direction of the electric field vectors in Fig. 2.15 is along the *x*-direction, $\mathbf{E}_0 = E_0[1, 0, 0]^T$, and that of the magnetic field in the *y*-direction $\mathbf{H}_0 = H_0[0, 1, 0]^T$, which means that the wave is linearly polarized. The form also stays constant in time, except if there is dissipation.

In this book, I will restrict myself mainly to linearly polarized light and combinations thereof without phase difference, which means that we will talk about naturally and elliptically polarized light. For naturally polarized light, we combine two harmonic waves with $\mathbf{E}_0 = E_0[1, 0, 0]^T$ and $\mathbf{E}_0 = E_0[0, 1, 0]^T$. In case of elliptically polarized light, the two waves are combined with different amplitudes, but we will discuss this case in more detail in Chapter 3.

To understand the short notation of Maxwell's equations, it is necessary to introduce a couple of other terms, namely the divergence of a vector field and its rotation. The divergence of $\mathbf{E}$ can be expressed as

$$\text{div}\,\mathbf{E} = \nabla \cdot \mathbf{E} = \frac{\partial E_x}{\partial x} + \frac{\partial E_y}{\partial y} + \frac{\partial E_z}{\partial z}, \tag{2.102}$$

where ∇ is the Nabla operator,

$$\nabla = \frac{\partial}{\partial x} + \frac{\partial}{\partial y} + \frac{\partial}{\partial z}. \tag{2.103}$$

The notation $\nabla \cdot \mathbf{E}$ tells you that you can treat the operation formally as a scalar product of the Nabla operator with a vector, in this case $\mathbf{E}$. As this implies, the result is a scalar field, i.e., at every point in space you get a scalar instead of a vector. The outcome would be the quantity of sources if you integrate over a volume containing each point, which would be in case of the electric field the number of free charges. If we go back to the electric field between the plates of an empty plate capacitor (cf. Fig. 2.14), then the electric field is constant in the direction from one plate to the other and zero in the normal directions. In any way, this means that the sum of the derivatives with regard to the coordinates is zero, and the same would certainly be true if we integrate over the volume within the plates. It is so, because there are no free charges in between the plates, which would otherwise render the field to be inhomogeneous.

In this book, we will only discuss situations where there are no free charges. So why would we care about the two Maxwell's equations, which feature the divergence? As we will see, those equations allow us to conclude that both, the electric and the magnetic field, are transversal to the direction of propagation of the wave and to each other in scalar media. In addition, these equations will allow us to conclude how the different components of the electric and magnetic field change at an interface between two media.

The rate of rotation of a vector field is determined by the curl operator:

$$\nabla \times \mathbf{E} = \begin{pmatrix} \frac{\partial E_z}{\partial y} - \frac{\partial E_y}{\partial z} \\ \frac{\partial E_x}{\partial z} - \frac{\partial E_z}{\partial x} \\ \frac{\partial E_y}{\partial x} - \frac{\partial E_x}{\partial y} \end{pmatrix}. \tag{2.104}$$

If we apply the curl operator to the field between the plates of a plate capacitor, the result is $\nabla \times \mathbf{E} = (0,0,0)^T$, since we only have a component of the electric field in z-direction, which is constant and two others perpendicular to it that are zero; therefore all derivatives are zero. Consequently, it makes no sense to show the field in a figure. Instead, we copy the field of a point source and compare it with its curl in Fig. 2.16.

As you can see, the resulting vector is always perpendicular to the original vector and the magnitude scales with the magnitude of the original, i.e., the electric field vector.

Formally, the cross-product between the curl operator and electric field can be written as follows:

$$\nabla \times \mathbf{E} = \begin{vmatrix} \hat{e}_x & \hat{e}_y & \hat{e}_z \\ \frac{\partial}{\partial x} & \frac{\partial}{\partial y} & \frac{\partial}{\partial z} \\ E_x & E_y & E_z \end{vmatrix} = \left(\frac{\partial E_z}{\partial y} - \frac{\partial E_y}{\partial z}\right)\hat{e}_x + \left(\frac{\partial E_x}{\partial z} - \frac{\partial E_z}{\partial x}\right)\hat{e}_y + \left(\frac{\partial E_y}{\partial x} - \frac{\partial E_x}{\partial y}\right)\hat{e}_z = \begin{pmatrix} \frac{\partial E_z}{\partial y} - \frac{\partial E_y}{\partial z} \\ \frac{\partial E_x}{\partial z} - \frac{\partial E_z}{\partial x} \\ \frac{\partial E_y}{\partial x} - \frac{\partial E_x}{\partial y} \end{pmatrix}. \tag{2.105}$$

This means that the relation for each component can be obtained by cyclically permuting the subscripts according to $x \to y$, $y \to z$, and $z \to x$. This operator is part of the other two Maxwell's equations and connects either the curl of the electric field with a timely change of the magnetic field or the curl of the magnetic field with a timely change of the electric displacement. In general, relatively complex relations can develop. Luckily, in the cases we are interested in, the relations stay comparably simple, in particular, if we align the direction and the polarization direction(s) of the plane wave with the axes of the cartesian coordinate system. For example, in case of a plane harmonic electromagnetic wave traveling in the z-direction in a scalar medium, which is x-polarized,

$$\mathbf{E} = E_0 \hat{e}_x \exp[i(k_z \cdot z - \omega t)], \tag{2.106}$$

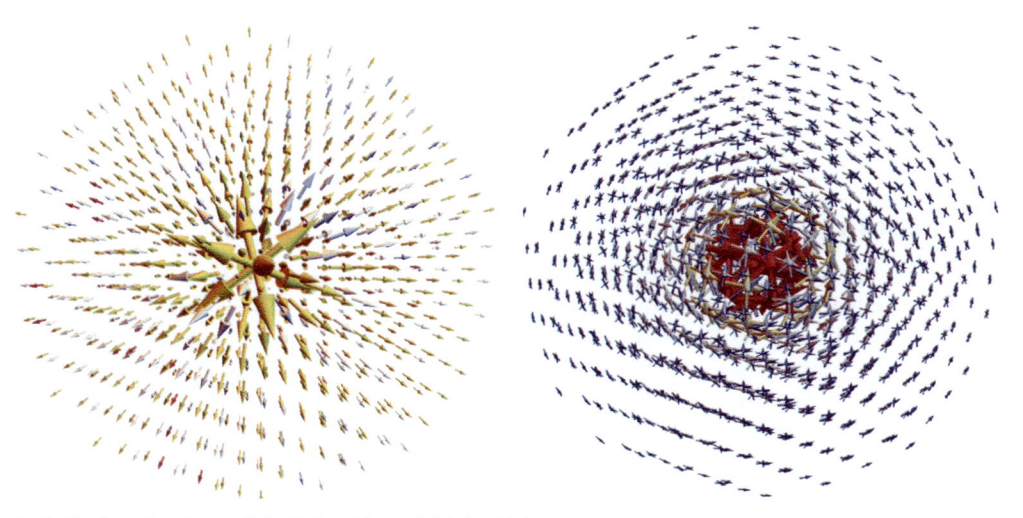

FIG. 2.16 Electric field of a point charge *(left side)* and its curl *(right side)*.

$\partial E_z = 0$, because $E_z = 0$ at all times. Furthermore, since we are dealing with plane waves, the direction and amplitude of the wave do not change in a plane perpendicular to z. This is the x-y-plane; therefore all derivatives with regard to x and y must be zero. Taking a look at Eq. (2.104), what remains is

$$\nabla \times \mathbf{E} = \begin{bmatrix} 0 \\ \frac{\partial E_x}{\partial z} \\ 0 \end{bmatrix} = i \begin{bmatrix} 0 \\ k_z \\ 0 \end{bmatrix} E_{0,x} \exp[i(k_z \cdot z - \omega t)], \tag{2.107}$$

which is much less complex than the entry point. In fact, thanks to the use of the exponential form of the plane harmonic waves, we will be able to display Maxwell's equations in a very simple form stripped of any derivatives in Section 3.4.

Finally, I introduce the vector product of the curl operator with itself. Say, we have an equation of the form,

$$\nabla \times \mathbf{E} + \frac{\partial}{\partial t} \mathbf{H} = 0, \tag{2.108}$$

and apply the curl operator from the left side:

$$\underbrace{\nabla \times \nabla \times \mathbf{E}}_{\nabla^2 \mathbf{E} = \Delta \mathbf{E}} + \nabla \times \frac{\partial}{\partial t} \mathbf{H} = 0. \tag{2.109}$$

Two consecutive applications of the Nabla operator can be replaced by the Laplace operator Δ.

Because of Schwarz's theorem, we can interchange the curl operator and the derivation with respect to time:

$$\Delta \mathbf{E} + \frac{\partial}{\partial t} \nabla \times \mathbf{H} = 0. \tag{2.110}$$

We can take a second equation, which states,

$$\nabla \times \mathbf{H} - \frac{\partial \varepsilon \mathbf{E}}{\partial t} = 0, \tag{2.111}$$

and derive it with regard to time:

$$\frac{\partial}{\partial t} \nabla \times \mathbf{H} - \frac{\partial^2 \varepsilon \mathbf{E}}{\partial t^2} = 0. \tag{2.112}$$

Then we combine Eqs. (2.110), (2.112) and get,

$$\Delta \mathbf{E} + \varepsilon \frac{\partial^2 \mathbf{E}}{\partial t^2} = 0, \tag{2.113}$$

which is obviously a wave equation (think again of Schrödinger's equation!). Finally, if we assume once more harmonic plane waves, the final form of Eq. (2.113) is:

$$\mathbf{k} \times (\mathbf{k} \times \mathbf{E}) + \omega^2 \varepsilon \mathbf{E} = 0. \tag{2.114}$$

If you have persevered to this point, you should be armed with enough knowledge to understand the rest of this book without larger difficulties. Enjoy!

2.14 Further reading

A very good introduction to waves in general is provided by the second chapter of the book "Optics" from Eugene Hecht [1], which I got recommended when I started my Ph.D. In this book, you also get a fairly good introduction to dispersion theory. I still think it is an extremely good overall introduction to optics and read. Nevertheless, be warned that it changes convention with regard to the representation of the waves between this chapter and Chapter 3. It uses in the latter chapter the convention, which leads to a minus sign before the imaginary part of the dielectric function and the absorption index, which is less common in the physics community, while it changes back and forth in Chapter 4.

References

[1] E. Hecht, Optics, fourth ed., 2002. Pearson Education.
[2] P. Yeh, Optical Waves in Layered Media, Wiley, 2005.
[3] C. Schaefer, F. Matossi, Das Ultrarote Spektrum, Verlag von Julius Springer, Berlin, 1930.
[4] M. Born, K. Huang, Dynamical Theory of Crystal Lattices, Clarendon Press, 1954.
[5] E. Ketteler, Theoretische Optik: gegründet auf das Bessel-Sellmeier'sche Princip. Zugleich mit den experimentellen Belegen, F. Vieweg, 1885.
[6] M.J. Taltavull, Rudolf Ladenburg and the first quantum interpretation of optical dispersion, Eur. Phys. J. H 45 (2020) 123–173.
[7] W. Sellmeier, Ueber die durch die Aetherschwingungen erregten Mitschwingungen der Koerpertheilchen und deren Rueckwirkung auf die ersteren, besonders zur Erklaerung der Dispersion und ihrer Anomalien, Ann. Phys. 223 (1872) 525–554.
[8] H. Helmholtz, Zur Theorie der anomalen Dispersion, Ann. Phys. 230 (1875) 582–596.
[9] J.C. Maxwell VIII, A dynamical theory of the electromagnetic field, Philos. Trans. R. Soc. Lond. 155 (1865) 459–512.
[10] P. Drude, Ueber die Beziehung der Dielectricitätsconstanten zum optischen Brechungsexponenten, Ann. Phys. 284 (1893) 536–545.
[11] H.A. Lorentz, The absoption and emission lines of gaseous bodies, Koninkl. Ned. Akad. Wetenschap. Proc. 8 (1906) 591–611.
[12] W. Voigt, Magneto- und Elektrooptik, B.G. Teubner, 1908.
[13] H.A. Kramers, Die Dispersion und Absorption von Röntgenstrahlen, Phys. Z. 30 (1929) 522–523.
[14] T.G. Mayerhöfer, V. Ivanovski, J. Popp, Infrared refraction spectroscopy, Appl. Spectrosc. 75 (2021) 1526–1531.
[15] S.W. King, M. Milosevic, A method to extract absorption coefficient of thin films from transmission spectra of the films on thick substrates, J. Appl. Phys. 111 (2012) 073109.
[16] Beer, Bestimmung der Absorption des rothen Lichts in farbigen Flüssigkeiten, Annalen der Physik 162 (1852) 78–88.
[17] E.D. Palik, N. Ginsburg, H.B. Rosenstock, R.T. Holm, Transmittance and reflectance of a thin absorbing film on a thick substrate, Appl. Opt. 17 (1978) 3345–3347.

Chapter 3

The electromagnetic field

3.1 Maxwell's relations

Spectroscopists usually do not feel comfortable with Maxwell's relations. During the 8th International Conference of Advanced Vibrational Spectroscopy in Vienna, I listened to a talk about an improved method to consider orientation in Raman spectroscopy. The talk also featured some infrared spectra for comparison purposes, and after the talk, I began a discussion with the PhD student who presented it. In the course of the discussion, I brought into play Maxwell's equations. As a reaction to that, the PhD student exclaimed something like "please consider, I am just a chemist!" I answered: "Yes, me too...."

As spectroscopists, we might not feel comfortable with Maxwell's equations, in particular with the magnetic field **H** and the magnetic induction **B**. Luckily, there is a way to get rid of both quantities so that we can focus on the electric field **E** and the electric displacement **D** when we deduce the existence of propagating waves from Maxwell's equations. We present them here in differential form, and since we will assume in the following that matter will be uncharged (volume charge density $\rho = 0$) and that no net currents will flow in it (current density $\mathbf{J} = \mathbf{0}$), they are simplified to

$$\begin{aligned} \nabla \cdot \mathbf{D} &= 0 \\ \nabla \times \mathbf{E} + \frac{\partial \mathbf{B}}{\partial t} &= 0 \\ \nabla \cdot \mathbf{B} &= 0 \\ \nabla \times \mathbf{H} - \frac{\partial \mathbf{D}}{\partial t} &= 0 \end{aligned} \tag{3.1}$$

As introduced in the preceding chapter, all fields are written as vectorial quantities (therefore in bold), in particular, because we will also treat anisotropic materials in the second part of this book. Overall, we actually see eight equations here, two that state that the divergence of the electric displacement and of the magnetic induction is zero,

$$\nabla \cdot \mathbf{W} = \frac{\partial W_x}{\partial x} + \frac{\partial W_y}{\partial y} + \frac{\partial W_z}{\partial z} = 0 \qquad \mathbf{W} = \mathbf{D}, \mathbf{B}, \tag{3.2}$$

which means that we assume that there are no sources or sinks, where the field lines of **D** and **B** start or end. Furthermore, two times three more, which state that from every temporal change of **D** and **B** a magnetic or electric field will be induced (or its shape will be altered if there was already one there before):

$$\nabla \times \mathbf{U} = \begin{pmatrix} \frac{\partial}{\partial x} \\ \frac{\partial}{\partial y} \\ \frac{\partial}{\partial z} \end{pmatrix} \times \begin{pmatrix} U_x \\ U_y \\ U_z \end{pmatrix} = \begin{pmatrix} \frac{\partial U_z}{\partial y} - \frac{\partial U_y}{\partial z} \\ \frac{\partial U_x}{\partial z} - \frac{\partial U_z}{\partial x} \\ \frac{\partial U_y}{\partial x} - \frac{\partial U_x}{\partial y} \end{pmatrix} = \begin{pmatrix} \frac{\partial V_x}{\partial t} \\ \frac{\partial V_y}{\partial t} \\ \frac{\partial V_z}{\partial t} \end{pmatrix} \qquad \mathbf{U} = \mathbf{E}, \mathbf{H}; \quad \mathbf{V} = \mathbf{B}, -\mathbf{D}. \tag{3.3}$$

A very important point to consider is that we overall have eight equations, but there are 12 unknowns (three components per field). Accordingly, we cannot solve Maxwell's equations unambiguously, if we do not evoke further relations. These further relations manifest themselves in the form of the so-called constitutive relations or material equations, which are at the same time the defining equations for **D** and **B**:

$$\begin{aligned} \mathbf{D} &= \varepsilon_0 \mathbf{E} + \mathbf{P} \\ \mathbf{H} &= \frac{\mathbf{B}}{\mu_0} - \mathbf{M} \end{aligned} \tag{3.4}$$

In Eq. (3.4), **P** is the macroscopic polarization that incorporates the reaction of matter to the external field **E** and that is, since we also want to allow anisotropic media, not necessarily co-linear to the generating electric field. For the magnetization **M**, we will assume that it is zero, which is the case for all materials that we will usually encounter in this book (but, e.g., not for metamaterials, etc.). ε_0 is the permittivity and μ_0 is the permeability of free space.

3.2 Boundary conditions

The boundary conditions seemed to me, when I encountered them first, as not very important. I was completely wrong. Nowadays, these conditions always come to my mind first, when I think about the failure of BBL-based infrared spectroscopy. As a chemist, it is for me the fundamental difference between a gas, which takes on every volume that is offered to it, and liquids and solids, both of which have a very well-defined volume and, by that, also a surface. These two terms, volume and surface, are also the key terms to understand the derivation of the first boundary condition. In general, the Gauss divergence theorem may be easy to understand as it just states that everything that goes into a volume or leaves it has to pass the surface that encloses it. Somewhat more scientifically formulated, if we have a vectorial property **F**, then the relation between the following volume integral and the surface integral is given by

$$\int \nabla \cdot \mathbf{F} dV = \int \mathbf{F} dS. \tag{3.5}$$

The value of the Gauss divergence theorem becomes immediately clear, if we assume that the volume we are interested in is separated by an interface like in Scheme 3.1. We now reduce the height of the cylinder in a way that the interface is still contained within the cylinder until it reaches zero. The surface of this cylinder consists then only of the two circles normal to the interface. Accordingly, Eq. (3.5) is reduced to

$$\mathbf{n} \cdot (\mathbf{F}_1 - \mathbf{F}_2) = 0, \tag{3.6}$$

Since the volume becomes zero in the limit of zero height. If we now replace the left part of Eq. (3.5) with the first and the third Maxwell equation, the result is

$$\begin{aligned} \mathbf{n} \cdot (\mathbf{B}_1 - \mathbf{B}_2) &= 0 \rightarrow \\ B_{1,\perp} &= B_{2,\perp} \xrightarrow{\mathbf{M}_1 = \mathbf{M}_2 = 0} \\ H_{1,\perp} &= H_{2,\perp} \\ \mathbf{n} \cdot (\mathbf{D}_1 - \mathbf{D}_2) &= 0 \\ D_{1,\perp} &= D_{2,\perp} \rightarrow \\ \varepsilon_{1,\perp} E_{1,\perp} &= \varepsilon_{2,\perp} E_{2,\perp} \end{aligned} \tag{3.7}$$

This means that the normal components of the **B** and the **D** fields are continuous. Since we assumed that the magnetization is zero, also the normal components of the **H** field are continuous. This is different for the electric field. Since the polarization is nonzero, the electric field is **not** continuous. We obtain the relation between the electric field before (1) and after (2) the interface by the following equation:

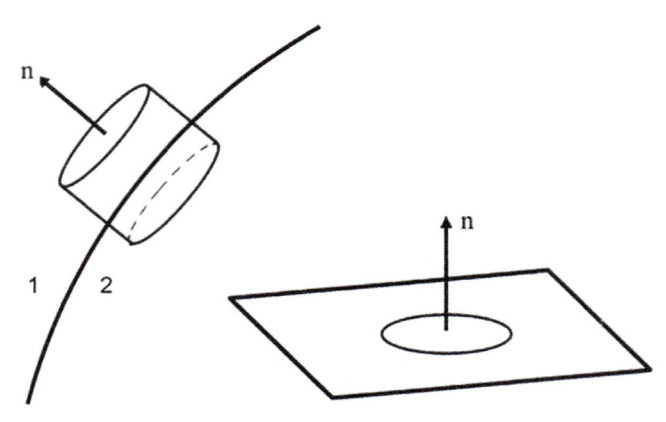

SCHEME 3.1 A cylindrical volume that is intersected by an interface (left side) and the same volume after its height is reduced to zero (right side).

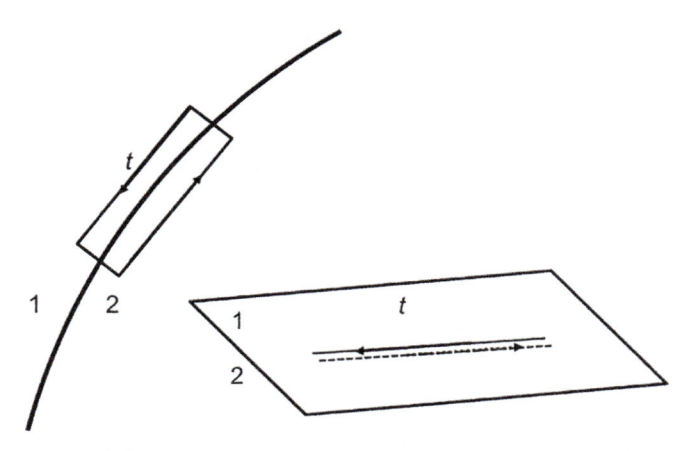

SCHEME 3.2 Scheme to illustrate an area (left side) that is intersected by an interface and the same area after its height is reduced to zero (right side).

$$D = \varepsilon_0 E + P = \underbrace{\varepsilon_0 \varepsilon_r}_{\varepsilon} E, \tag{3.8}$$

wherein ε and ε_r are the dielectric tensor and the relative dielectric tensor, which are functions of frequency/wavelength/wavenumber (in Eq. (3.7), we have tacitly assumed that the dielectric tensor can be diagonalized and has its principal component normal to the interface designated with the symbol $\perp$).

What happens with the tangential components of the fields? We can derive a further boundary condition with help of Stoke's theorem:

$$\int \nabla \times F dA = \int F \cdot dl. \tag{3.9}$$

Again, we do not try to find a mathematically strict proof and sacrifice rigor for clarity. The statement of Stoke's theorem is immediately clear, in particular, if we let the height of a rectangle approach zero as it is illustrated in Scheme 3.2: Obviously the tangential components of the vector field must be equal to compensate each other to zero:

$$n \times (F_1 - F_2) = 0, \tag{3.10}$$

Therefore, if we insert the second or the fourth Maxwell's equations into Eq. (3.10), we obtain the following boundary conditions:

$$\begin{aligned} E_{1,t} &= E_{2,t} \\ H_{1,t} &= H_{2,t} \end{aligned}. \tag{3.11}$$

These boundary conditions are not only pivotal for the calculation of reflectance and transmittance, but also immediately allow us to conclude that it is not only the orientation relative to the polarization direction that is important but also the orientation relative to the interface.

3.3 Energy density and flux

In this section, I introduce the energy density and energy flux. We will need the energy flux, respectively, Poynting's vector, since it is required to calculate transmittance at an interface from one medium to another, if the angle of incidence is different from zero. On the other hand, for spectroscopists who are interested in transmittance, the latter point is of no practical relevance. The reason is that in this case it is not the transmittance from one medium to another, but the transmittance through a stack of media, e.g., a layer on a substrate. In such cases, the detector is usually located in the same medium as the source. Therefore, the fact that the squared amplitude of the electric field in the last medium is not always the same as the energy flux inside the medium could be ignored. On the other hand, even if I wanted to keep wave optics to the minimum, necessary to understand how strongly infrared spectra are affected by it, the appearance of an otherwise unmotivated factor in the calculations of transmittance could be unsettling. The work performed by an electric field, which equals the energy dissipation, is the dot-product between charge density J and the electric field E: $J \cdot E$. I withhold the charge density in Maxwell's fourth equation (3.1), which actually reads if the charge density is considered:

$$\nabla \times \mathbf{H} - \frac{\partial \mathbf{D}}{\partial t} = \mathbf{J}. \tag{3.12}$$

For the moment, we use it in this form and replace the charge density with the left part to yield:

$$\mathbf{J} \cdot \mathbf{E} = \mathbf{E} \cdot (\nabla \times \mathbf{H}) - \mathbf{E} \cdot \frac{\partial \mathbf{D}}{\partial t}. \tag{3.13}$$

It is certainly not immediately clear how this could help. The same is true for the next step where we invoke a purely mathematical vector identity:

$$\mathbf{J} \cdot \mathbf{E} = \mathbf{E} \cdot (\nabla \times \mathbf{H}) - \mathbf{E} \cdot \frac{\partial \mathbf{D}}{\partial t} \xrightarrow{\nabla \cdot (\mathbf{E} \times \mathbf{H}) = \mathbf{H} \cdot (\nabla \times \mathbf{E}) + \mathbf{E} \cdot (\nabla \times \mathbf{H})}$$
$$\mathbf{J} \cdot \mathbf{E} = -\nabla \cdot (\mathbf{E} \times \mathbf{H}) + \mathbf{H} \cdot (\nabla \times \mathbf{E}) - \mathbf{E} \cdot \frac{\partial \mathbf{D}}{\partial t} \tag{3.14}$$

While this looks even more complex than before, we can use Maxwell's second equation to gain two terms (actually three, but the last two belong together):

$$\mathbf{J} \cdot \mathbf{E} = \underbrace{-\nabla \cdot (\mathbf{E} \times \mathbf{H})}_{\text{Energy flow}} \underbrace{-\mathbf{H} \cdot \frac{\partial \mathbf{B}}{\partial t} - \mathbf{E} \cdot \frac{\partial \mathbf{D}}{\partial t}}_{\text{Decrease of energy density}}. \tag{3.15}$$

If we assume that our medium is linear (which means that the refractive indices are virtually independent of the electric field strengths, which is the case if the latter are not too strong), we can write:

$$-\mathbf{J} \cdot \mathbf{E} = \nabla \cdot (\mathbf{E} \times \mathbf{H}) + \frac{\partial \frac{1}{2}(\mathbf{E} \cdot \mathbf{D} + \mathbf{H} \cdot \mathbf{B})}{\partial t} \rightarrow$$
$$-\mathbf{J} \cdot \mathbf{E} = \nabla \cdot \mathbf{S} + \frac{\partial U}{\partial t} \tag{3.16}$$

In Eq. (3.16), $\mathbf{S}$ is Poynting's vector that gives the energy flux. Its amount is the power per unit area transported in the direction of the vector, and U is the energy density. Both quantities are usually formulated slightly differently:

$$\mathbf{S} = \frac{1}{2} \operatorname{Re}[\mathbf{E} \times \mathbf{H}^*]$$
$$U = \frac{1}{4} \operatorname{Re}[\mathbf{E} \cdot \mathbf{D}^* + \mathbf{H} \cdot \mathbf{B}^*] \tag{3.17}$$

In Eq. (3.17), the asterisk denotes the complex conjugate. The latter representation is in particular important, when the complex number representation is used for the electric field.

3.4 The wave equation

For all the problems that we are going to tackle in the framework of this book, we will assume that the light beams we are dealing with can be represented by plane waves. Based on my experience, this is no serious restriction—I never experienced deviations between experiment and simulation that were caused by this assumption. In the following, we will see how Maxwell's equations automatically lead to the existence of such waves. We start with the second and the fourth Maxwell's equations:

$$\nabla \times \mathbf{E} + \frac{\partial \mathbf{B}}{\partial t} = 0 \quad \text{(I)}$$
$$\nabla \times \mathbf{H} - \frac{\partial \mathbf{D}}{\partial t} = 0 \quad \text{(II)} \tag{3.18}$$

Since we assumed that magnetization is zero, we can set $\mathbf{B} = \mathbf{H}$ in (I) and apply the curl operator. The order of the differentiation with respect to time and the curl operator can be changed, therefore

$$\nabla \times \nabla \times \mathbf{E} + \frac{\partial}{\partial t} \nabla \times \mathbf{H} = 0. \tag{3.19}$$

We now take (II) and differentiate it with respect to time:

$$\frac{\partial}{\partial t} \nabla \times \mathbf{H} - \frac{\partial^2 \varepsilon \mathbf{E}}{\partial t^2} = 0. \tag{3.20}$$

Additionally, we have used the relation $\mathbf{D} = \varepsilon \mathbf{E}$. If we combine Eqs. (3.19), (3.20), the result is

$$\underbrace{\nabla \times \nabla \times \mathbf{E}}_{\nabla^2 \mathbf{E}} + \varepsilon \frac{\partial^2 \mathbf{E}}{\partial t^2} = 0. \tag{3.21}$$

Solutions to this equation, which is called the wave equation, are given by $\mathbf{E}(\mathbf{r},t) = \mathbf{E}_0 \sin(\mathbf{k} \cdot \mathbf{r} - \omega t)$ and $\mathbf{E}(\mathbf{r},t) = \mathbf{E}_0 \cos(\mathbf{k} \cdot \mathbf{r} - \omega t)$. Here, ω is the angular frequency given by $\omega = 2\pi\nu$, where ν is the frequency and $\mathbf{k}$ is the wave vector given by $\mathbf{k} = \omega/cn\mathbf{s} = 2\pi\tilde{\nu}\mathbf{s}$, with $\mathbf{s}$ being a unit vector in the direction of propagation and n is the refractive index, whose properties we will investigate further down.

If we keep in mind that electric fields are always real, we can, for convenience, also construct a solution to the wave equation from a linear combination of both solutions according to the following:

$$\mathbf{E}(\mathbf{r},t) = \mathbf{E}_0 \cos(\mathbf{k} \cdot \mathbf{r} - \omega t) + i\mathbf{E}_0 \sin(\mathbf{k} \cdot \mathbf{r} - \omega t) = \mathbf{E}_0 \exp[i(\mathbf{k} \cdot \mathbf{r} - \omega t)]. \tag{3.22}$$

However, since we multiplied the second solution with the imaginary number i, only the real part has, strictly speaking, any physical relevance, so that the physically meaningful form of Eq. (3.22) is the following:

$$\mathbf{E}(\mathbf{r}t) = \mathrm{Re}[\mathbf{E}_0 \exp[i(\mathbf{k} \cdot \mathbf{r} - \omega t)]]. \tag{3.23}$$

When I first encountered Eq. (3.23), I had a very hard time to imagine how such a plane wave would look like (even though I additionally used the book "Optics" by Eugene Hecht, which really gives very basic and illustrative examples for the changes of the wave with time and position [1]). To simplify things, we investigate how the polarization direction of $\mathbf{E}_0$ (the amplitude) and the change of position of locations of the same amplitude $\mathbf{r}(t)$ are connected. To that end, let's use that $\nabla \cdot \mathbf{D} = 0$:

$$\nabla \cdot \mathbf{D}(\mathbf{r},t) = i\mathbf{k} \cdot \mathbf{D}(\mathbf{r},t) = 0 \rightarrow \mathbf{k} \cdot \mathbf{D}(\mathbf{r},t) = 0. \tag{3.24}$$

As a consequence, the direction of $\mathbf{D}(\mathbf{r},t)$ is perpendicular to the direction of the wave vector, which is the direction of propagation. In general, this is not valid for $\mathbf{E}(\mathbf{r},t)$, since for an anisotropic material the direction of $\mathbf{D}(\mathbf{r},t)$ and $\mathbf{E}(\mathbf{r},t)$ do not coincide. However, for the first part of the book we restrict ourselves to the case where the dielectric tensor is a scalar.

In this case, in order to find out the relative orientation between the electric field and the magnetic field in the plane wave model, we investigate the rotation of $\mathbf{E}(\mathbf{r},t)$:

$$\nabla \times \mathbf{E} \xrightarrow{\mathbf{E} = \mathbf{E}_0 \exp[i(\mathbf{k} \cdot \mathbf{r} - \omega t)]} \begin{pmatrix} \frac{\partial E_z}{\partial y} - \frac{\partial E_y}{\partial z} \\ \frac{\partial E_x}{\partial z} - \frac{\partial E_z}{\partial x} \\ \frac{\partial E_y}{\partial x} - \frac{\partial E_y}{\partial y} \end{pmatrix} = i \begin{pmatrix} E_z k_y - E_y k_z \\ E_x k_z - E_z k_x \\ E_y k_x - E_x k_y \end{pmatrix} = i\mathbf{k} \times \mathbf{E}. \tag{3.25}$$

In Eqs. (3.24), (3.25), we see one big advantage of writing a plane wave in the exponential form, which is we can easily evaluate derivatives and are able to replace calculating the derivative by multiplying it with either the negative wave vector (if it is a derivative with respect to location) or the angular frequency (if the derivation is with respect to time) multiplied with the imaginary number. If we use this in the second equation of Eq. (3.1),

$$\begin{array}{c} \nabla \times \mathbf{E} + \frac{\partial \mathbf{B}}{\partial t} = 0 \xrightarrow{\mathbf{M}=0} \\ i\mathbf{k} \times \mathbf{E} - i\omega \mathbf{H} = 0 \rightarrow \mathbf{k} \times \mathbf{E} = \omega \mathbf{H} \end{array}. \tag{3.26}$$

Therefore, the magnetic field is perpendicular to both $\mathbf{E}$ and $\mathbf{k}$ (for a material where the magnetization is nonzero and not co-oriented to the magnetic field, it is $\mathbf{B}$ that is perpendicular to the direction of propagation).

It may be easier to grasp Eq. (3.26), if we simplify things somewhat. Throughout the book, we will assume that the different media are stacked and we will assume the stack direction to be along the z-coordinate. Furthermore, in addition to the idealization of plane waves, we introduce a further idealization and assume that the properties of the media do not vary perpendicular to the stacking direction, i.e., along the x-y-plane (the dimensions of samples in the x-y-plane are usually very large compared to the wavelength; in addition, the comparison between an experiment and simulations justifies our assumptions). While the plane waves do not necessarily travel along the z-direction (we discuss nonnormal incidence later on), we will assume this simplification to illustrate Eq. (3.24):

$$\begin{pmatrix} 0 \\ 0 \\ k_z \end{pmatrix} \times \mathbf{E}(\mathbf{r}, t) = 0 \rightarrow \mathbf{E}_0 = \begin{pmatrix} E_{0,x} \\ E_{0,y} \\ 0 \end{pmatrix}. \tag{3.27}$$

We even simplify Eq. (3.27) further by assuming that $E_{0,y} = 0$. As we will see later, this means that we assume that the plane wave does not only travel along z, but is additionally x-polarized. Therefore

$$\nabla \times \mathbf{E} = -\frac{\partial \mathbf{B}}{\partial t} \xrightarrow{\mathbf{E} = \mathbf{E}_0 \exp[i(k_z \cdot z - \omega t)]} \nabla \times \mathbf{E} = \begin{pmatrix} \frac{\partial E_z}{\partial y} - \frac{\partial E_y}{\partial z} \\ \frac{\partial E_x}{\partial z} - \frac{\partial E_z}{\partial x} \\ \frac{\partial E_y}{\partial x} - \frac{\partial E_x}{\partial y} \end{pmatrix} = i \begin{pmatrix} 0 \\ k_z E_x \\ 0 \end{pmatrix} = i \begin{pmatrix} 0 \\ k_z \\ 0 \end{pmatrix} E_{0,x} \exp[i(k_z \cdot z - \omega t)] \rightarrow$$

$$\begin{pmatrix} 0 \\ k_z \\ 0 \end{pmatrix} E_{0,x} \exp[i(k_z \cdot z - \omega t)] = \omega H_{0,y} \exp[i(k_z \cdot z - \omega t)]$$
. (3.28)

Accordingly, the magnetic field is y-polarized. It may seem to some readers that I am too explicit at this point. Since we need these calculations again when we derive formulas for the reflectance and transmittance for layer stacks, I'll take the risk of discussing seemingly trivial steps to make things easier when we have to concentrate on other problems.

To summarize this section, we can now formulate Maxwell's equations in an equivalent form, but thanks to the plane wave assumption fully based on simple vector calculations:

$$\begin{aligned} \mathbf{k} \cdot \mathbf{D} &= 0 \\ \mathbf{k} \times \mathbf{E} - \omega \mathbf{B} &= 0 \\ \mathbf{k} \cdot \mathbf{B} &= 0 \\ \mathbf{k} \times \mathbf{H} + \omega \mathbf{D} &= 0 \end{aligned} \tag{3.29}$$

Furthermore, we can provide a simpler form of the wave equation:

$$\mathbf{k} \times (\mathbf{k} \times \mathbf{E}) + \omega^2 \varepsilon \mathbf{E} = 0. \tag{3.30}$$

And, last, but not least, Poynting's vector can be simplified into the following form:

$$\mathbf{S} = \frac{1}{2} \mathrm{Re}[\mathbf{E} \times \mathbf{H}^*] \xrightarrow{\mathbf{k} \times \mathbf{E} = \omega \mathbf{H}} \mathbf{S} = \frac{\mathbf{E} \times \mathbf{k} \times \mathbf{E}}{2\omega} = \frac{\mathbf{k}}{2\omega} |E_0|^2. \tag{3.31}$$

We will use this equation to calculate the reflectance R and transmittance T in the next section.

3.5 Polarized waves

We have already worked with polarized waves in the last section without explicitly mentioning this. Since we plan to focus on anisotropic materials in the second part of this book (to be more precise, only on materials of dielectric anisotropy), to develop a feeling for polarization is very important. But not only that: The calculations of reflectance and transmittance can become very complex for general anisotropy and nonnormal incidence; therefore, it is advantageous to separate two different cases where the light is polarized either perpendicular to the plane of incidence or parallel to the plane of incidence. In fact, the separation of these cases generally makes sense and simplifies the math considerably. The reason for this simplification is based on the fact that perpendicular polarized light only has a component tangential to the interface between the incidence medium and sample, whereas parallel polarized light consists of a tangential as well as a normal component.

In addition to so-called linear polarized light, we also have to introduce elliptically polarized light, since media with dielectric anisotropy generally convert linear polarized light into elliptically polarized light. For the incidence medium, we will usually assume that this medium can be described by a scalar dielectric function (like vacuum, air, or materials that are used as crystals for attenuated total reflection like Ge, Si, or ZnSe). Accordingly, **E** and **H** and **k** are mutually perpendicular in such a material and it is sufficient to focus on the direction of **E** to describe the polarization. For linear polarized light, the direction of **E** is perpendicular to **k** and confined to a line. In contrast, elliptically polarized light can be seen to consist of a superposition of two linear polarized waves along two mutually perpendicular axes that have a certain phase difference. In fact, this is exactly into what linear polarized light is transformed when it is transmitted through an anisotropic material. Assume a plane wave propagating in the z-direction:

$$\mathbf{E}(z,t) = \mathrm{Re}[\mathbf{E}_0 \exp[i(\mathbf{k} \cdot \mathbf{z} - \omega t)]]. \tag{3.32}$$

Said superposition of two plane waves traveling along z and being polarized along x and y can then be formulated as follows:

$$\underbrace{E_x(z,t) = E_{0,x}\cos(k_z z - \omega t) \quad E_y(z,t) = E_{0,y}\cos(k_z z - \omega t + \delta)}_{\mathbf{E}_0 = \mathbf{E}_{0,x} + \mathbf{E}_{0,y}\exp[i\delta_y]}. \tag{3.33}$$

The ellipsoidal nature of the polarization is easily verified. To do that, we normalize the x-polarized wave and square it:

$$\begin{aligned} E_x &= E_{0,x}\cos(k_z z - \omega t) \\ \rightarrow \frac{E_x}{E_{0,x}}\cos\delta &= \cos(k_z z - \omega t)\cos\delta \\ \rightarrow \left(\frac{E_x}{E_{0,x}}\right)^2 &= \cos^2(k_z z - \omega t) = 1 - \sin^2(k_z z - \omega t) \end{aligned} \tag{3.34}$$

Furthermore, we also normalize the y-polarized wave:

$$\begin{aligned} E_y &= E_{0,y}\cos(k_z z - \omega t + \delta) \\ \frac{E_y}{E_{0,y}} &= \cos(k_z z - \omega t + \delta) = \cos(k_z z - \omega t)\cos\delta - \sin(k_z z - \omega t)\sin\delta \end{aligned}. \tag{3.35}$$

We then subtract Eq. (3.34) from Eq. (3.35) and obtain an ellipse with semiaxes of length $E_{0,x}$ and $E_{0,y}$ that are at an angle δ relative to the axes of the coordinate system (cf. also Scheme 3.3):

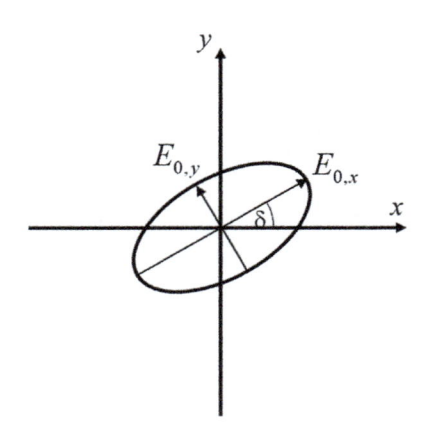

SCHEME 3.3 Polarization ellipse.

$$\frac{E_y}{E_{0,y}} - \frac{E_x}{E_{0,x}}\cos\delta = -\sin(k_z z - \omega t)\sin\delta \xrightarrow{\sqrt{1-\left(\frac{E_x}{E_{0,x}}\right)^2}=\sin(k_z z-\omega t)}$$
$$\left(\frac{E_y}{E_{0,y}} - \frac{E_x}{E_{0,x}}\cos\delta\right)^2 = \left(1-\left(\frac{E_x}{E_{0,x}}\right)^2\right)\sin^2\delta \rightarrow \quad (3.36)$$
$$\left(\frac{E_x}{E_{0,x}}\right)^2 + \left(\frac{E_y}{E_{0,y}}\right)^2 - \frac{2\cos\delta}{E_{0,x}E_{0,y}}E_x E_y = \sin^2\delta$$

From Eq. (3.36), we regain linear polarization for the case that the phase difference is a multiple of π:

$$\delta = \delta_y - \delta_x = m\pi \qquad m = 0, 1. \quad (3.37)$$

Accordingly, it follows that

$$\frac{E_x}{E_y} = (-1)^m \frac{E_{0,x}}{E_{0,y}}. \quad (3.38)$$

This means that the ratio of the two vectors on the left side is the ratio of the two amplitudes, which is itself a constant. Therefore, the light is linearly polarized. For most of the book, with the exception of Chapter 16, this is the important polarization state. For Chapter 16, in which vibrational circular dichroism plays the main role, another polarization state is important, namely circularly polarized light. For such light, the conditions are that the phase difference is 90 degrees and that the amplitudes are equal:

$$\delta = \delta_y - \delta_x = \pm\frac{1}{2}\pi$$
$$\frac{E_{0,x}}{E_{0,y}} = 1 \quad (3.39)$$

Concerning right and left circularly polarized light, two different conventions are possible, according to which a wavefront either falls onto the observer or starts at the observer. In vibrational circular dichroism, the convention is used that the wavefront starts at the observer. Therefore, if the vector rotates counterclockwise when the observer looks along the axis of propagation, the light is left circularly polarized (LCP), cf. Fig. 3.1. As shown in the figure, circularly polarized light results from properly combined linearly polarized light. Properly combining left and right circularly polarized light, on the other hand, results in linearly polarized light.

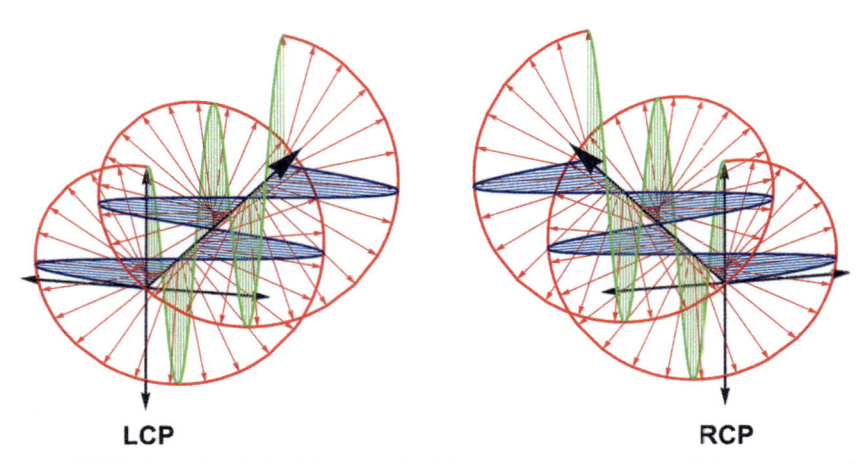

FIG. 3.1 Left (LCP) and right (RCP) circularly polarized light combined from two linearly polarized light waves according to the convention used in vibrational circular dichroism.

3.6 Further reading

I suggest in particular the book of Pochi Yeh "Optical waves in layered media," which adds many aspects that I thought were less important for the course of the book (but I might be in error, and knowing more is always a good strategy!) [2]. Furthermore, I suggest the Born and Wolf "Principles of Optics," which might be no surprise at all [3]. Also, at a somewhat more introductory level, Hecht's "Optics" is a good supplement to the other two books, but certainly also an excellent read on its own [1].

References

[1] E. Hecht, Optics, fourth ed., 2002. Pearson Education.
[2] P. Yeh, Optical Waves in Layered Media, Wiley, 2005.
[3] M. Born, E. Wolf, A.B. Bhatia, Principles of Optics: Electromagnetic Theory of Propagation, Interference and Diffraction of Light, Cambridge University Press, 1999.

Chapter 4

Reflection and transmission of plane waves

4.1 Reflection and transmission at an interface separating two scalar media under normal incidence

We start with the simplest case which is given by a plane wave reaching an interface that separates two semiinfinite scalar media under normal incidence. By a scalar medium, we understand a medium that is homogenous and can be characterized in the interesting spectral range by a scalar dielectric function. In most cases, we are not interested in the incidence medium (the medium where the plane wave has its origin) that consists of vacuum or air, but we will also have to understand what changes if the incidence medium has a dielectric function different from unity, since then attenuated total reflection (ATR) can occur (although not for normal incidence). The so-called exit medium usually is the sample. Why should these media be semiinfinite? Actually, we are simplifying the situation, if we assume that both media are semiinfinite, because this means that the part of the wave that is reflected at the interface and is traveling backward in the incidence medium will never hit another interface. Therefore, an again-reflected wave that is traveling back to the interface will never exist. Accordingly, we do not have to take care of such multiple reflections and the consequences that result from a superposition of the forward and backward traveling waves. Actually, this is not completely true, because even for an semiinfinite incidence and exit medium, the original wave and the part that is reflected will be superposed. Since, as we will see, the original wave and its reflected alter ego have the same phase at the interface and they are traveling in the same medium, their phases will have a fixed relationship over the whole incidence medium. Accordingly, they are said to be coherent. For the exit medium, if it is also assumed to be semiinfinite, the transmitted part of the wave is presumed to travel forward endlessly. Because in this medium therefore only a forward traveling wave exists, a superposition will not take place. In practice, a sample will certainly never be semiinfinite. However, since we want to do spectroscopy, it most probably has absorption bands. Even somewhat away from the maximum of such a band, absorption is still not zero, so a finite thickness will be sufficient to lead to pseudo-semiinfiniteness. Absorption needs to be just high enough that light from the backside of the sample does not reach the first interface again. If this happens, like, e.g., for inorganic glass samples some $1000 \, \text{cm}^{-1}$ away from the highest-wavenumber absorption, then a step of the reflectance can be seen, which is an indication that the optical model that is assumed would have to be changed to a more complex one with a layer in-between semiinfinite media. For the moment, however, we will presume that our media are nonabsorbing and in fact semiinfinite. As a last assumption, we will suppose that the interface between the two media is plane parallel and smooth.

We will in the following focus on quantity I which is called the spectral irradiance (or, for a spectroscopist, the intensity), which is the irradiance of a surface per unit frequency, wavelength, or wavenumber. In particular, irradiance is the radiant flux (power) that is received/reflected or transmitted by a surface per unit area. We will call

- I_0 the received radiant flux (r.f.)
- I_R the reflected r.f.
- I_T the transmitted r.f.

Furthermore, we will define the reflectance R as the ratio of the reflected and the received r.f.,

$$R = \frac{I_R}{I_0}, \tag{4.1}$$

and the transmittance T as the ratio of the transmitted and received r.f.,

$$T = \frac{I_T}{I_0}. \tag{4.2}$$

How do we obtain the reflected and the transmitted irradiances? This is what we introduced Poynting's vector for in the preceding chapter. Remember the definition,

$$\mathbf{S}_j = \frac{\mathbf{k}}{2\omega} |E_j|^2 \quad j = i, t, r, \tag{4.3}$$

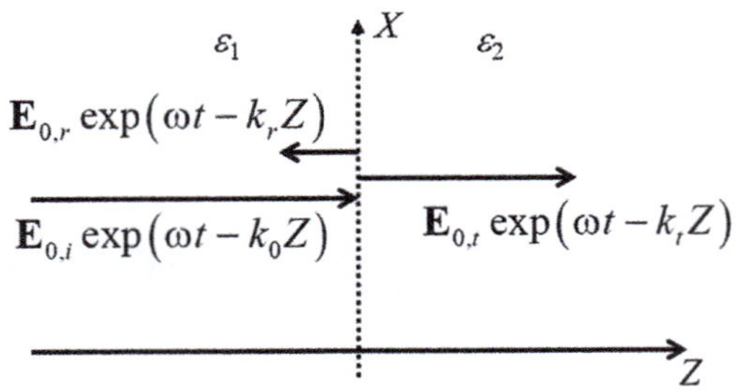

58 PART | I Scalar theory

$$\mathbf{E}_{0,r}\exp(\omega t - k_r Z)$$
$$\mathbf{E}_{0,i}\exp(\omega t - k_0 Z) \quad \mathbf{E}_{0,t}\exp(\omega t - k_t Z)$$

SCHEME 4.1 A plane wave traveling along Z through two semiinfinite media with dielectric constants ε_1 and ε_2 normal to their interface.

where i,t,r stands for "incoming," "transmitted," and "reflected", respectively. Eq. (4.3) enables us to calculate the flux when we know the corresponding electric field strengths. Therefore, what we need to do is to calculate the electric fields at the interface keeping in mind the continuity conditions that we derived in the last chapter. Scheme 4.1 illustrates the situation.

The fluxes have the same direction as the waves. From Maxwell's equations we know, as discussed in the last chapter, that the direction of **E** is tangential to the direction of propagation and, therefore, tangential to the interface. The recipe to calculate the reflectance and transmittance includes three steps:

(1) Obtain two equations from the continuity of the tangential electric and magnetic fields. Convert **H** to **E** using Maxwell's equations.
(2) Use the two equations to express
  (a) $\mathbf{E}_t$ in terms of $\mathbf{E}_i$ and $\mathbf{E}_r$ to obtain the ratio between $\mathbf{E}_r$ and $\mathbf{E}_i$.
  (b) $\mathbf{E}_r$ in terms of $\mathbf{E}_i$ and $\mathbf{E}_t$ to obtain the ratio between $\mathbf{E}_t$ and $\mathbf{E}_i$.
(3) Calculate the flux S in the direction of Z to obtain I_R and I_T. From the flux calculate R and T.

Since these steps are the basic building blocks for the calculation of R and T up to layered media of arbitrary dielectric anisotropy, I will guide you step by step through it.

(1) Obtain two equations from the continuity of the tangential electric and magnetic fields. Convert **H** to **E** using Maxwell equations.

The first continuity relation tells us that the tangential components of the electric fields are continuous. Accordingly:

$$\mathbf{E}_{1,\tan} = \mathbf{E}_{2,\tan} \rightarrow$$
$$\mathbf{E}_i + \mathbf{E}_r = \mathbf{E}_t. \tag{4.4}$$

On the left side, i.e., in medium 1, we have a superposition of the incoming and the reflected field. This superposition must be equal to the transmitted electric field in medium 2. The form of Eq. (4.4) is particularly simple, because there are only tangential components of the electric field. The same is also true for the magnetic fields at the interface:

$$\mathbf{H}_{1,\tan} = \mathbf{H}_{2,\tan} \xrightarrow{\mathbf{k}\times\mathbf{E}-\omega\mathbf{H}=0}$$
$$\mathbf{k}_i \times \mathbf{E}_i + \mathbf{k}_r \times \mathbf{E}_r = \mathbf{k}_t \times \mathbf{E}_t \longrightarrow$$
$$\begin{pmatrix} 0 \\ 0 \\ k_{i,Z} \end{pmatrix} \times \mathbf{E}_i + \begin{pmatrix} 0 \\ 0 \\ k_{r,Z} \end{pmatrix} \times \mathbf{E}_r = \begin{pmatrix} 0 \\ 0 \\ k_{t,Z} \end{pmatrix} \times \mathbf{E}_t \xrightarrow{|\mathbf{k}|=\omega\sqrt{\varepsilon}}$$
$$\sqrt{\varepsilon_1}\mathbf{E}_i - \sqrt{\varepsilon_1}\mathbf{E}_r = \sqrt{\varepsilon_2}\mathbf{E}_t \longrightarrow$$
$$\mathbf{E}_i - \mathbf{E}_r = \sqrt{\frac{\varepsilon_2}{\varepsilon_1}}\mathbf{E}_t. \tag{4.5}$$

The first transformation of Eq. (4.5) uses the second equation of Eq. (3.30) (noting that $\mathbf{H} = \mathbf{B}$, since $\mu = 1$). For the second equation, we take into account that the wave vector has only one component in the Z-direction. For the third equation, we have to keep in mind that the direction of the reflected wave is reversed (and therefore the sign changes). As a result, we have two relations between the incident, the transmitted, and the reflected electric field:

$$\mathbf{E}_i + \mathbf{E}_r = \mathbf{E}_t \quad (\text{I})$$
$$\mathbf{E}_i - \mathbf{E}_r = \sqrt{\frac{\varepsilon_2}{\varepsilon_1}}\mathbf{E}_t \quad (\text{II}). \tag{4.6}$$

(2) Use the two equations to express

(a) $\mathbf{E}_t$ in terms of $\mathbf{E}_i$ and $\mathbf{E}_r$ to obtain the ratio between $\mathbf{E}_r$ and $\mathbf{E}_i$.

To that end, we replace $\mathbf{E}_t$ in the second equation in (4.6) by the left side of the first equation:

$$\mathbf{E}_i - \mathbf{E}_r = \sqrt{\frac{\varepsilon_2}{\varepsilon_1}}(\mathbf{E}_i + \mathbf{E}_r) \rightarrow$$

$$\mathbf{E}_i - \sqrt{\frac{\varepsilon_2}{\varepsilon_1}}\mathbf{E}_i = \sqrt{\frac{\varepsilon_2}{\varepsilon_1}}\mathbf{E}_r + \mathbf{E}_r \rightarrow$$

$$\left(1 - \sqrt{\frac{\varepsilon_2}{\varepsilon_1}}\right)\mathbf{E}_i = \left(1 + \sqrt{\frac{\varepsilon_2}{\varepsilon_1}}\right)\mathbf{E}_r \rightarrow \qquad (4.7)$$

$$r = \frac{\mathbf{E}_r}{\mathbf{E}_i} = \frac{1 - \sqrt{\frac{\varepsilon_2}{\varepsilon_1}}}{1 + \sqrt{\frac{\varepsilon_2}{\varepsilon_1}}}.$$

By that, we obtain the ratio between the reflected and the incoming electric field which is called the reflection coefficient r.

(b) $\mathbf{E}_r$ in terms of $\mathbf{E}_i$ and $\mathbf{E}_t$ to obtain the ratio between $\mathbf{E}_t$ and $\mathbf{E}_i$.

From the first equation of (4.6) we note that $\mathbf{E}_r = \mathbf{E}_t - \mathbf{E}_i$. We insert this result into the second equation and get the transmission coefficient t:

$$\mathbf{E}_i - (\mathbf{E}_t - \mathbf{E}_i) = \sqrt{\frac{\varepsilon_2}{\varepsilon_1}}\mathbf{E}_t \rightarrow$$

$$2\mathbf{E}_i - \mathbf{E}_t = \sqrt{\frac{\varepsilon_2}{\varepsilon_1}}\mathbf{E}_t \rightarrow$$

$$2\mathbf{E}_i = \sqrt{\frac{\varepsilon_2}{\varepsilon_1}}\mathbf{E}_t + \mathbf{E}_t \rightarrow \qquad (4.8)$$

$$2\mathbf{E}_i = \left(1 + \sqrt{\frac{\varepsilon_2}{\varepsilon_1}}\right)\mathbf{E}_t$$

$$t = \frac{\mathbf{E}_t}{\mathbf{E}_i} = \frac{2}{1 + \sqrt{\frac{\varepsilon_2}{\varepsilon_1}}}.$$

Now we are ready for the final step:

(3) Calculate flux $\mathbf{S}$ in the direction of Z to obtain I_R and I_T. From that calculate R and T.

Of general relevance is only the energy flow perpendicular to, i.e., through, the interface. This is a, in this case, trivial condition, since the wave is moving perpendicular to the interface:

$$r = \frac{\mathbf{E}_r}{\mathbf{E}_i} = \frac{1 - \sqrt{\frac{\varepsilon_2}{\varepsilon_1}}}{1 + \sqrt{\frac{\varepsilon_2}{\varepsilon_1}}}$$

$$t = \frac{\mathbf{E}_t}{\mathbf{E}_i} = \frac{2}{1 + \sqrt{\frac{\varepsilon_2}{\varepsilon_1}}} \qquad (4.9)$$

$$R = \frac{S_r}{S_i} \xrightarrow{S_j = \frac{k_j}{2\omega}|E_j|^2} R = \frac{k_{z,r}}{k_{z,i}}\left|\frac{\mathbf{E}_r}{\mathbf{E}_i}\right|^2 = |r|^2$$

$$T = \frac{S_t}{S_i} \xrightarrow{S_j = \frac{k_j}{2\omega}|E_j|^2} T = \frac{k_{z,t}}{k_{z,i}}\left|\frac{\mathbf{E}_t}{\mathbf{E}_i}\right|^2 = \sqrt{\frac{\varepsilon_2}{\varepsilon_1}}|t|^2 = \frac{n_2}{n_1}|t|^2.$$

Here, n_j is the index of refraction of the medium j. If the ε_j are real numbers, then $R + T = 1$, which can easily be verified using the results of Eq. (4.9). Certainly, both quantities add up to unity not by accident. The result follows the law of energy conservation, since what is not reflected, must be transmitted (so far, we have excluded absorption from the discussion by assuming that our dielectric function is real).

4.2 Reflection and transmission at an interface separating two scalar semiinfinite media under nonnormal incidence

Assuming normal incidence, as in the preceding section, polarization does not play a role, since the media are not anisotropic. In case of light that is nonnormally incident, it makes sense to separate two particular cases, even for scalar media, which differ with regard to polarization. To understand the difference, we first have to define the so-called plane of incidence. The definition is absolutely straightforward, as the plane of incidence is defined on the one hand by the direction of the incoming plane wave, and, on the other hand, by the direction of the reflected wave (for normal incidence, both are colinear, and a plane of incidence cannot be defined). The polarization directions which we will now define, are *s*-polarized ("*s*" stands for the German *senkrecht*, which means perpendicular) and *p*-polarized ("*p*" is short for parallel, which works in more than one language). Sometimes, you might also experience an alternative nomenclature, which I want to introduce, even if it is much less common in spectroscopy. In this nomenclature, *s*-polarized is called *transverse electric*, or, in short, TE. This means in both cases the same, namely, that the electric field vector is perpendicular to the plane of incidence. Somewhat in contrast, but at the same time along with TE, the opposite is called TM, which stands for *transverse magnetic* and is in this case referring to the direction of the magnetic field vector. As long as we are talking about scalar media, which are homogenous in the sense that the inhomogeneities are smaller than the resolution limit, both definitions are fully synonymous. In any way, the decisive difference between normal and nonnormal incidence is that *p*-polarized light has not only a component of the electric field tangential to the interface but also one normal to it, and the latter component increases with the angle of incidence (which is the angle between the normal to the interface and the direction of the incoming light). I have already mentioned that this normal component makes a big difference for the spectroscopist, since the larger this component is, the more band shapes and peak positions will deviate from those resulting from *s*-polarized light.

Let us investigate the situation for nonnormal incidence in some more detail (cf. Scheme 4.2). We assume that Z is the direction perpendicular to the interface (note that we are using capital letters for the Cartesian coordinates, since later on, when we are discussing anisotropic materials, we will need small letters to denote the axes of coordinate systems fixed inside the materials. Accordingly, capital letters belong to laboratory coordinate systems throughout the book). The incoming, reflected, and transmitted wave vectors all lie in the plane of incidence, which we denote as the Y-Z plane, in this case without loss of generality. Therefore, for *s*-polarized or TE light, the polarization direction is parallel to the X-axis. For *p*-polarized or TM light, the polarization direction is located somewhere between the Z- and the Y-axis. Using the angle of incidence α_i, we can state that the component along Z is proportional to $\sin \alpha_i$.

How can we determine the angle of reflection α_r and the angle of refraction α_t? Actually, there is even more that we need to focus on. If we take again a look at our plane wave, $\mathbf{E}(\mathbf{r}, t) = \operatorname{Re}[\mathbf{E}_0 \exp(\mathbf{k} \cdot \mathbf{r} - \omega t)]$, and assume that at $t = 0$ it hits the interface ($Z = 0$), then the continuity relations require that the phases are spatially equal:

$$(\mathbf{k}_i \cdot \mathbf{r})_{Z=0} = (\mathbf{k}_r \cdot \mathbf{r})_{Z=0} = (\mathbf{k}_t \cdot \mathbf{r})_{Z=0}. \tag{4.10}$$

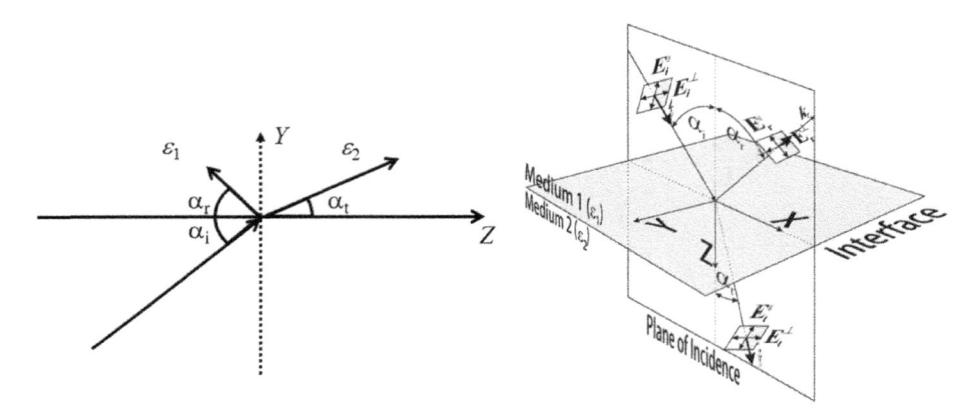

SCHEME 4.2 A plane wave traveling through two semiinfinite media nonnormal to their interface.

From Eq. (4.10), we can deduce that the following relations must hold (remember, $k_X = 0$!):

$$\rightarrow Y k_{i,Y} = Y k_{r,Y} = Y k_{t,Y}$$
$$\rightarrow k_{i,Y} = k_{r,Y} = k_{t,Y}$$
$$k_{i,Y} = n_1 \sin \alpha_i \qquad (4.11)$$
$$k_{r,Y} = n_1 \sin \alpha_r$$
$$k_{t,Y} = n_2 \sin \alpha_t.$$

From that, we can deduce Fresnel's law:

$$k_{i,Y} = k_{r,Y} = k_{t,Y},$$
$$n_1 \sin \alpha_i = n_1 \sin \alpha_r = n_2 \sin \alpha_t. \qquad (4.12)$$

This immediately allows us to conclude that the angle of incidence and the angle of reflectance must be equal. Furthermore, if we assume that our incidence medium has a lower index of refraction than the exit medium, then

$$\arcsin\left(\frac{n_1}{n_2} \sin \alpha_i\right) = \alpha_t \rightarrow \alpha_i > \alpha_t. \qquad (4.13)$$

Note that in order to match the Y-components of the wave vector at the interface, it is automatically required that the Z-components in the different media must differ. To summarize, all wave vectors, $\mathbf{k}_i$, $\mathbf{k}_r$, $\mathbf{k}_t$ must lie in a plane, the plane of incidence. Furthermore, the tangential components k_Y must be equal. Our recipe to calculate reflectance and transmittance contains again three steps:

(1) Calculate $\mathbf{H}$ from $\mathbf{E}$ using Maxwell's equations.
(2) Obtain two equations from the continuity of the tangential components of the electric and magnetic fields. Calculate from these r_s and t_s or r_p and t_p.
(3) Calculate the flux $\mathbf{S}$ in the direction of Z to obtain I_R and I_T. From that calculate R_s and T_s or R_p and T_p.

The plane wave can be written in the following way:

$$\mathbf{E}_j = \mathbf{E}_{0,j} \exp[i(\mathbf{k}_i \cdot \mathbf{r} - \omega t)] = \mathbf{E}_{0,j} \exp\left[i(Y k_{Y,j} + Z k_{Z,j} - \omega t)\right] = \mathbf{E}_{0,j} \exp\left[i\omega \underbrace{\left(\frac{n_j}{c}(Y \sin \alpha_j + Z \cos \alpha_j) - t\right)}\right] \qquad (4.14)$$

$$\mathbf{E}_j = \mathbf{E}_{0,j} \exp[i\varphi] \quad j = i, r, t.$$

Once more, i, r, t denote the incoming, the reflected, and the transmitted wave. The Maxwell equation that we need for the first step is again $\mathbf{k}_i \times \mathbf{E} = \omega \mu_0 \mathbf{H}$ (Eq. 3.27).

4.2.1 s-Polarized light

To follow the above recipe, we first have to calculate $\mathbf{H}$ from $\mathbf{E}$. Remember, for s-polarization the electric field is polarized parallel to the X-axis. According to Scheme 4.3, which illustrates the situation, the polarization direction as indicated is actually antiparallel to X (assuming a right-handed coordinate system, X points into the page's (or the screen's)

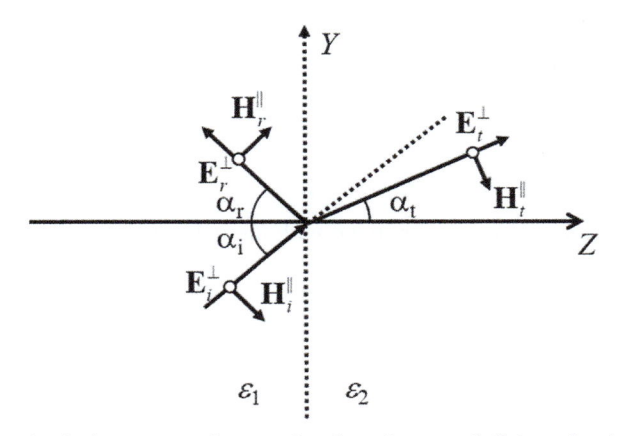

SCHEME 4.3 A s-polarized plane wave impinging nonnormally on an interface of two semiinfinite and scalar media.

plane, whereas the electric field vector points toward you, the reader. Accordingly, the incident electric field can be written as:

$$E_{i,X} = -E_i^{\perp} \exp[i\varphi]. \tag{4.15}$$

Applying Eq. (3.27) leads to:

$$\mathbf{k}_i \times \mathbf{E} = \omega\mu_0 \mathbf{H} \rightarrow \mathbf{H} = \mathbf{k} \times \mathbf{E}/\omega\mu_0$$

$$\mathbf{k}_i \times \mathbf{E} = \begin{pmatrix} 0 \\ E_X k_{Z,i} \\ -E_X k_{Y,i} \end{pmatrix}, \quad \mathbf{k}_i = \sqrt{\varepsilon_1}\frac{\omega}{c}\begin{pmatrix} 0 \\ \sin\alpha_i \\ \cos\alpha_i \end{pmatrix}$$

$$H_{i,Y} = -\sqrt{\varepsilon_1}\frac{\cos\alpha_i}{\mu_0 c}E_i^{\perp}\exp[i\varphi]$$

$$H_{i,Z} = \sqrt{\varepsilon_1}\frac{\sin\alpha_i}{\mu_0 c}E_i^{\perp}\exp[i\varphi]. \tag{4.16}$$

Accordingly, the incident magnetic field has two components, one that is antiparallel to the Y-axis and a second component in the Z-direction. Since the electric field is continuous at the interface, its direction is preserved upon reflection and transmission. Therefore, a change of direction of the magnetic field would be the sole consequence of a directional change of the wave vector. Since the Z-component of the wave changes upon reflection, the wave vector of the reflected wave indeed needs to be antiparallel to the Z-axis. Consequently, the Y-component of the magnetic field of the reflected wave $H_{r,Y}$ is positive:

$$H_{r,Y} = \sqrt{\varepsilon_1}\frac{\cos\alpha_i}{\mu_0 c}E_i^{\perp}\exp[i\varphi]. \tag{4.17}$$

The form of the Z-component $H_{r,Z}$ is similar to the one of $H_{i,Z}$ (Eq. 4.16). For the transmitted waves, the directions of both components of $\mathbf{H}_t^{\parallel}$ do not change. However, even when the signs are not varied, since the value of the index of refraction is altered in the second medium, so are the values of the components of the magnetic field:

$$H_{t,Y} = -\sqrt{\varepsilon_2}\frac{\cos\alpha_t}{\mu_0 c}E_i^{\perp}\exp[i\varphi]$$

$$H_{t,Z} = \sqrt{\varepsilon_2}\frac{\sin\alpha_t}{\mu_0 c}E_i^{\perp}\exp[i\varphi]. \tag{4.18}$$

Now we have everything that we need to calculate the reflection and the transmission coefficient. First, we note that the exponential term in the Y-components of the magnetic fields is the same in Eqs. (4.15)–(4.18); therefore, we can drop this term. From the continuity relation of the tangential components of the electric field, we obtain the first equation:

$$E_X(medium\,1) = E_X(medium\,2) \rightarrow$$
$$E_{i,X} + E_{r,X} = E_{t,X}. \tag{4.19}$$

In the same way, as for normal incidence, the second equation is obtained from the continuity of the tangential components of the magnetic fields,

$$H_Y(medium\,1) = H_Y(medium\,2) \rightarrow$$
$$H_{i,Y} + H_{r,Y} = H_{t,Y}. \tag{4.20}$$

By replacing the magnetic fields with the results obtained with the help of Eqs. (4.16)–(4.18):

$$-E_{i,X}\sqrt{\varepsilon_1}\cos\alpha_i + E_{r,X}\sqrt{\varepsilon_1}\cos\alpha_r = -E_{t,X}\sqrt{\varepsilon_2}\cos\alpha_t \xrightarrow{\alpha_i=\alpha_r}$$
$$(E_{i,X} - E_{r,Y})\sqrt{\varepsilon_1}\cos\alpha_i = E_{t,X}\sqrt{\varepsilon_2}\cos\alpha_t. \tag{4.21}$$

Here, we have taken advantage of the fact that the angle of incidence and the angle of reflectance are equal. To get the reflection coefficient r_s, we replace $E_{t,X}$ in Eq. (4.21) with the left side of the result from Eq. (4.19):

$$(E_{i,X} - E_{r,X})\sqrt{\varepsilon_1}\cos\alpha_i = (E_{i,X} + E_{r,X})\sqrt{\varepsilon_2}\cos\alpha_t \rightarrow$$
$$E_{i,X}(\sqrt{\varepsilon_1}\cos\alpha_i - \sqrt{\varepsilon_2}\cos\alpha_t) = E_{r,X}(\sqrt{\varepsilon_1}\cos\alpha_i + \sqrt{\varepsilon_2}\cos\alpha_t) \rightarrow$$
$$r_s = \frac{E_{r,X}}{E_{i,X}} = \frac{\sqrt{\varepsilon_1}\cos\alpha_i - \sqrt{\varepsilon_2}\cos\alpha_t}{\sqrt{\varepsilon_1}\cos\alpha_i + \sqrt{\varepsilon_2}\cos\alpha_t}.$$

(4.22)

To derive the solution for the transmission coefficient, we replace $E_{r,X}$ from Eq. (4.21) by $E_{t,X} - E_{i,X}$:

$$(E_{i,X} - E_{t,X} + E_{i,X})\sqrt{\varepsilon_1}\cos\alpha_i = E_{t,X}\sqrt{\varepsilon_2}\cos\alpha_t \rightarrow$$
$$2E_{i,X}\sqrt{\varepsilon_1}\cos\alpha_i = E_{t,X}(\sqrt{\varepsilon_1}\cos\alpha_i + \sqrt{\varepsilon_2}\cos\alpha_t) \rightarrow$$
$$t_s = \frac{E_{t,X}}{E_{i,X}} = \frac{2\sqrt{\varepsilon_1}\cos\alpha_i}{\sqrt{\varepsilon_1}\cos\alpha_i + \sqrt{\varepsilon_2}\cos\alpha_t}.$$

(4.23)

I remember that, after having studied the corresponding equations for some time, I asked myself where I should get the angle of transmittance from. Maybe you do not share my problem in this respect, but for me, the following derivation would have been helpful. To solve this problem, we first replace the cosine function with the sine function and then use Fresnel's law to replace $\sin\alpha_t$:

$$\sin^2\alpha_t + \cos^2\alpha_t = 1 \longrightarrow \cos\alpha_t = \sqrt{1 - \sin^2\alpha_t} \xrightarrow{n_1\sin\alpha_i = n_2\sin\alpha_t}$$
$$\cos\alpha_t = \sqrt{1 - \left(\frac{n_1}{n_2}\sin\alpha_i\right)^2} \rightarrow$$
$$\cos\alpha_t = \sqrt{1 - \frac{\varepsilon_1}{\varepsilon_2}\sin^2\alpha_i}.$$

(4.24)

Employing Eq. (4.24) we arrive at the final expressions for the transmission and reflection coefficient which contain only known quantities:

$$r_s = \frac{E_{r,X}}{E_{i,X}} = \frac{\sqrt{\varepsilon_1}\cos\alpha_i - \sqrt{\varepsilon_2 - \varepsilon_1\sin^2\alpha_i}}{\sqrt{\varepsilon_1}\cos\alpha_i + \sqrt{\varepsilon_2 - \varepsilon_1\sin^2\alpha_i}}$$
$$t_s = \frac{E_{t,X}}{E_{i,X}} = \frac{2\sqrt{\varepsilon_1}\cos\alpha_i}{\sqrt{\varepsilon_1}\cos\alpha_i + \sqrt{\varepsilon_2 - \varepsilon_1\sin^2\alpha_i}}.$$

(4.25)

4.2.2 *p*-Polarized light

We again follow the same recipe, but this time we have to consider that for *p*-polarization the electric field is polarized along both, the *Y*- as well as the *Z*-axis. According to Scheme 4.4, the *Y*-component is positive for the incident, the reflected,

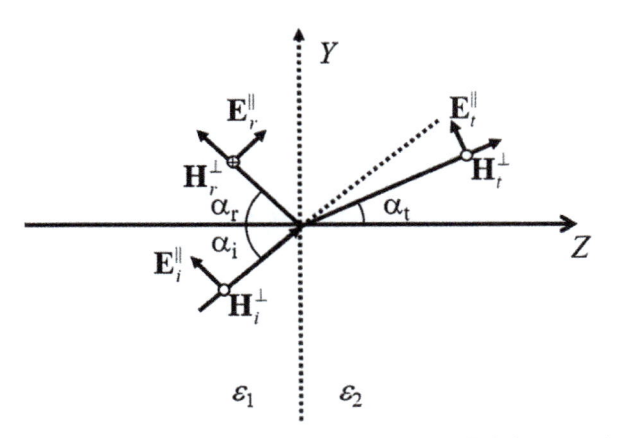

SCHEME 4.4 A *p*-polarized plane wave impinging nonnormally on an interface of two semiinfinite and scalar media.

and the transmitted wave, while the Z-component is positive only for the reflected wave. Since the incident and transmitted electric field can be written as,

$$E_{j,Y} = E_j^{\parallel} \cos \alpha_j \exp[i\varphi]$$
$$E_{j,Z} = -E_j^{\parallel} \sin \alpha_j \exp[i\varphi] \quad j = i, t \tag{4.26}$$

the corresponding electric field of the reflected wave is given by:

$$E_{r,Y} = E_r^{\parallel} \cos \alpha_i \exp[i\varphi]$$
$$E_{r,Z} = E_r^{\parallel} \sin \alpha_i \exp[i\varphi]. \tag{4.27}$$

Applying Eq. (3.27) to the incident wave leads to:

$$\mathbf{k} \times \mathbf{E} = \begin{pmatrix} E_Z k_Y - E_Y k_Z \\ -E_Z k_X \\ E_Y k_X \end{pmatrix}, \quad \mathbf{k}_i = \sqrt{\varepsilon_1} \frac{\omega}{c} \begin{pmatrix} 0 \\ \sin \alpha_i \\ \cos \alpha_i \end{pmatrix}$$

$$H_{i,X} = \sqrt{\varepsilon_1} \frac{1}{\mu_0 c} \left[\sin \alpha_i \left(-E_i^{\parallel} \sin \alpha_i \exp[i\varphi] \right) - \cos \alpha_i E_i^{\parallel} \cos \alpha_i \exp[i\varphi] \right] = -\sqrt{\varepsilon_1} \frac{1}{\mu_0 c} E_i^{\parallel} \exp[i\varphi]. \tag{4.28}$$

Therefore, the incident magnetic field has only one component, which is antiparallel to the X-axis. The transmitted wave keeps this direction for the magnetic field, while the direction is reversed for the reflected wave:

$$\mathbf{k} \times \mathbf{E} = \begin{pmatrix} E_Z k_Y - E_Y k_Z \\ -E_Z k_X \\ E_Y k_X \end{pmatrix}, \quad \mathbf{k}_r = \sqrt{\varepsilon_1} \frac{\omega}{c} \begin{pmatrix} 0 \\ \sin \alpha_i \\ -\cos \alpha_i \end{pmatrix}$$

$$H_{r,X} = \sqrt{\varepsilon_1} \frac{1}{\mu_0 c} \left[\sin \alpha_i \left(E_r^{\parallel} \sin \alpha_i \exp[i\varphi] \right) + \cos \alpha_i E_r^{\parallel} \cos \alpha_i \exp[i\varphi] \right] = \sqrt{\varepsilon_1} \frac{1}{\mu_0 c} E_r^{\parallel} \exp[i\varphi]. \tag{4.29}$$

The magnetic field for the transmitted wave is correspondingly given by:

$$H_{t,X} = -\sqrt{\varepsilon_2} \frac{1}{\mu_0 c} E_t^{\parallel} \exp[i\varphi]. \tag{4.30}$$

Again, we have everything what we need to calculate the reflection and the transmission coefficient. For a second time, we drop the exponential term, which is the same in Eqs. (4.26)–(4.30). From the continuity relation of the tangential components of the electric field we obtain the first equation:

$$E_Y(medium\,1) = E_Y(medium\,2) \rightarrow$$
$$E_i^{\parallel} \cos \alpha_i + E_r^{\parallel} \cos \alpha_r = E_t^{\parallel} \cos \alpha_t. \tag{4.31}$$

Once more, the second equation is obtained from the continuity of the tangential components of the magnetic fields,

$$H_X(medium\,1) = H_X(medium\,2)$$
$$H_{i,X} - H_{r,X} = H_{t,X}, \tag{4.32}$$

by replacing the magnetic fields with the results obtained with the help of Eqs. (4.28)–(4.30):

$$\sqrt{\varepsilon_1} \left(E_i^{\parallel} - E_r^{\parallel} \right) = \sqrt{\varepsilon_2} E_t^{\parallel}. \tag{4.33}$$

To calculate the reflection coefficient r_p, we replace $E_t^{\|}$ in Eq. (4.33) by employing Eq. (4.31):

$$\sqrt{\varepsilon_1}\left(E_i^{\|} - E_r^{\|}\right) = \sqrt{\varepsilon_2}\frac{E_i^{\|}\cos\alpha_i + E_r^{\|}\cos\alpha_r}{\cos\alpha_t} \rightarrow$$

$$E_i^{\|}\left(\sqrt{\varepsilon_1}\cos\alpha_t - \sqrt{\varepsilon_2}\cos\alpha_i\right) = E_r^{\|}\left(\sqrt{\varepsilon_1}\cos\alpha_t + \sqrt{\varepsilon_2}\cos\alpha_i\right) \rightarrow \quad (4.34)$$

$$r_p = \frac{E_r^{\|}}{E_i^{\|}} = \frac{\sqrt{\varepsilon_1}\cos\alpha_t - \sqrt{\varepsilon_2}\cos\alpha_i}{\sqrt{\varepsilon_1}\cos\alpha_t + \sqrt{\varepsilon_2}\cos\alpha_i}.$$

To derive the solution for the transmission coefficient, we replace $E_r^{\|}$ with the help of Eq. (4.31):

$$\sqrt{\varepsilon_1}E_i^{\|} - \sqrt{\varepsilon_2}E_t^{\|} = \sqrt{\varepsilon_1}\frac{E_t^{\|}\cos\alpha_t - E_i^{\|}\cos\alpha_i}{\cos\alpha_i} \rightarrow$$

$$\sqrt{\varepsilon_1}\cos\alpha_i E_i^{\|} - \sqrt{\varepsilon_2}\cos\alpha_i E_t^{\|} = \sqrt{\varepsilon_1}\cos\alpha_t E_t^{\|} - \sqrt{\varepsilon_1}\cos\alpha_i E_i^{\|} \rightarrow$$

$$2\sqrt{\varepsilon_1}\cos\alpha_i E_i^{\|} = E_t^{\|}\left(\sqrt{\varepsilon_1}\cos\alpha_t + \sqrt{\varepsilon_2}\cos\alpha_i\right) \rightarrow \quad (4.35)$$

$$t_p = \frac{E_t^{\|}}{E_i^{\|}} = \frac{2\sqrt{\varepsilon_1}\cos\alpha_i}{\sqrt{\varepsilon_1}\cos\alpha_t + \sqrt{\varepsilon_2}\cos\alpha_i}.$$

Employing Eq. (4.24) we arrive at the final expressions for the transmission and reflection coefficient which contain only known quantities:

$$r_p = \frac{\sqrt{\varepsilon_1}\sqrt{1 - \frac{\varepsilon_1}{\varepsilon_2}\sin^2\alpha_i} - \sqrt{\varepsilon_2}\cos\alpha_i}{\sqrt{\varepsilon_1}\sqrt{1 - \frac{\varepsilon_1}{\varepsilon_2}\sin^2\alpha_i} + \sqrt{\varepsilon_2}\cos\alpha_i}$$

$$t_p = \frac{2\sqrt{\varepsilon_1}\cos\alpha_i}{\sqrt{\varepsilon_1}\sqrt{1 - \frac{\varepsilon_1}{\varepsilon_2}\sin^2\alpha_i} + \sqrt{\varepsilon_2}\cos\alpha_i}.$$

(4.36)

4.2.3 Calculation of reflectance and transmittance

Finally, to obtain the reflectance and the transmittance for both cases, s- as well as p-polarized incident light, we have to calculate the flux $\mathbf{S}$ in the direction of Z to obtain I_R and I_T. From that we can calculate R_s and T_s as well as R_p and T_p:

$$R_s = \frac{|\mathbf{Z}\cdot\mathbf{S}_{r,s}|}{|\mathbf{Z}\cdot\mathbf{S}_{i,s}|} = \frac{k_{Z,r}}{k_{Z,i}}|r_s|^2, \quad R_p = \frac{|\mathbf{Z}\cdot\mathbf{S}_{r,p}|}{|\mathbf{Z}\cdot\mathbf{S}_{i,p}|} = \frac{k_{Z,r}}{k_{Z,i}}|r_p|^2$$

$$T_s = \frac{|\mathbf{Z}\cdot\mathbf{S}_{t,s}|}{|\mathbf{Z}\cdot\mathbf{S}_{i,s}|} = \frac{k_{Z,t}}{k_{Z,i}}|t_s|^2, \quad T_p = \frac{|\mathbf{Z}\cdot\mathbf{S}_{t,p}|}{|\mathbf{Z}\cdot\mathbf{S}_{i,p}|} = \frac{k_{Z,t}}{k_{Z,i}}|t_p|^2.$$

(4.37)

To calculate the reflectance, we have to evaluate the ratio of the Z-component of the wave vector of the incident and the reflected wave. Since incidence and reflection take part in the same medium and under the same angle, this ratio is simply unity:

$$\frac{k_{Z,r}}{k_{Z,i}} = \frac{\sqrt{\varepsilon_1}\cos\alpha_r}{\sqrt{\varepsilon_1}\cos\alpha_i} \xrightarrow{|\alpha_i|=|\alpha_r|} \frac{k_{Z,r}}{k_{Z,i}} = 1. \quad (4.38)$$

The corresponding ratio for the transmittance, in contrast, is not unity, since the incidence medium and the medium in which the light is transmitted are different:

$$\frac{k_{Z,t}}{k_{Z,i}} = \frac{\sqrt{\varepsilon_2}\cos\alpha_t}{\sqrt{\varepsilon_1}\cos\alpha_i} = \frac{\sqrt{\varepsilon_2}\sqrt{1 - \frac{\varepsilon_1}{\varepsilon_2}\sin^2\alpha_i}}{\sqrt{\varepsilon_1}\cos\alpha_i}. \quad (4.39)$$

Accordingly, as for normal incidence, the reflectance is simply given by:

$$R_j = |r_j|^2 \quad j = s, p, \tag{4.40}$$

whereas for the transmittance the ratio in Eq. (4.39) has to be accounted for:

$$T_j = \frac{\sqrt{\varepsilon_2}\sqrt{1 - \frac{\varepsilon_1}{\varepsilon_2}\sin^2\alpha_i}}{\sqrt{\varepsilon_1}\cos\alpha_i}|t_j|^2 \quad j = s, p. \tag{4.41}$$

4.2.4 Example: Dependence of the reflectance from the angle of incidence

For the following example, we assume that the incidence medium is a vacuum, i.e., $\varepsilon_1 = 1$. The exit medium is characterized by $\varepsilon_2 = 4$ (accordingly, the indices of refraction are $n_1 = 1$ and $n_2 = 2$). For the angle of incidence $\alpha_i = 0°$ (normal incidence), the polarization leads to the fact that R_p (better called R_Y in this case, since the plane of incidence is not defined for $\alpha_i = 0°$) has only a component parallel to the interface as has R_s (better called, for the same reason, R_X). Accordingly, in this case, $R_Y = R_X$ and therefore R_s and R_p have the same starting point, cf. Fig. 4.1. In contrast to R_p, R_s is a monotonically increasing function which ends at $R_s = 1$ at $\alpha = 90°$, since then the plane wave is traveling perpendicular to the interface and, accordingly, is no longer reflected. R_p first decreases until it reaches zero at $\alpha \approx 63°$ in this particular example. The angle at which R_p becomes zero is called Brewster's angle. Since at this angle, the parallel polarized component is not reflected, naturally polarized light becomes s-polarized upon reflection, which can be (and has been) exploited to construct polarizers (e.g., using Se [1]).

If we focus on the transmitted light, we find that a part of the s-polarized light is transmitted. Therefore, the transmitted light is made up of both, p-, and also s-polarized light, the ratio of the amplitudes depends on the index of refraction of the second medium. At least in our example, this has no practical relevance, as the assumption of an semiinfinite exit medium in any way excludes in principle any use of the transmitted light. This is also the reason why we do not show the dependence of the transmittance on the angle of incidence as there is no way to check the correctness of the calculation (to verify the theory, it would be, strictly speaking, necessary to implement a detector into the second medium with the same optical properties as the second medium. On the other hand, since $R + T = 1$, such a test would not be necessary, since if the reflectance adheres to the derived relations, so would transmittance). For angles larger than Brewster's angle, R_p is also monotonically increasing and reaches for symmetry reasons unity at the same angle as R_s, namely at 90°. What is remarkable and what you should keep in mind is that for natural polarized light, the reflectance nearly does not change for small angles of incidence <20°.

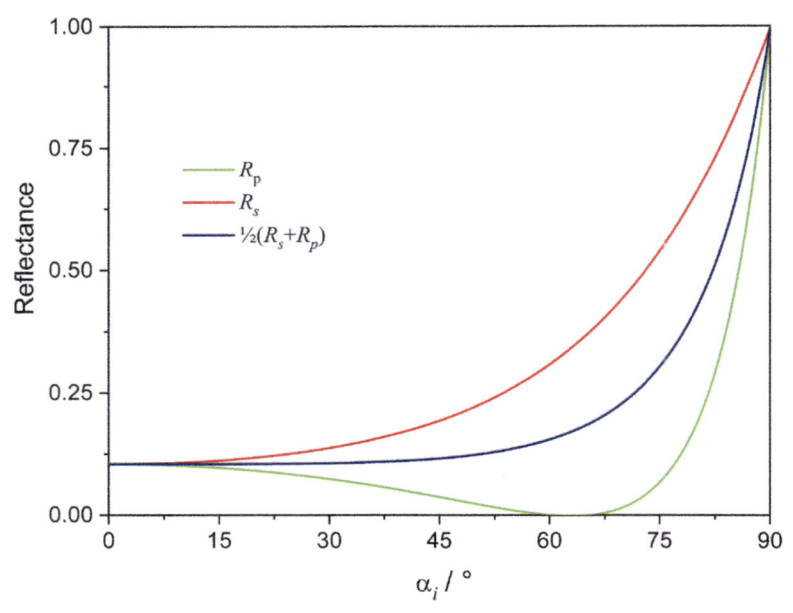

FIG. 4.1 Dependence of R_s, R_p, and naturally polarized light from the angle of incidence.

4.3 Reflection and transmission at an interface separating two scalar media under nonnormal incidence-absorbing media

On the first view, it seems that the relations that we derived in the preceding chapter are of little practical relevance. Infrared spectroscopy relies on the fact that materials absorb. Certainly, there are spectral regions where there are no bands, but this does not mean that absorption is zero (as already mentioned in Chapter 2, SiO_2 bulk glass samples become, depending on their thickness, nontransparent below about $2000\,\text{cm}^{-1}$, even when the highest-wavenumber band is located between about 1100 and $1200\,\text{cm}^{-1}$). In any way, including absorption into the formulas is not a problem anymore, since computer programs can easily handle imaginary numbers properly. Therefore, the good news is that all relations we discussed so far keep their form (and this will also hold for all relations introduced in upcoming parts of this book). For example, if we deal with the refractive index, we just have to replace the real index of refraction in the formulas with its complex counterpart (the same holds for the dielectric function):

$$\hat{n} = n + ik. \tag{4.42}$$

Herein, n is the (real) index of refraction as we have discussed it so far. Analogously, k is the index of absorption (not to be confused with the wave vector), which will be a very important quantity as it is the bridge between wave optics-based and Beer-Lambert-based quantitative infrared spectroscopy. A first insight into what happens as a consequence of the introduction of the index of absorption can be gained, if we assume a plane wave traveling in the Z-direction being (for no particular reason, except to simplify things) X-polarized:

$$E_X(Z,t) = E_{0,X} \exp(i(k_Z \cdot Z - \omega t)). \tag{4.43}$$

Since $k_Z = \hat{n}\frac{\omega}{c} = (n + ik)\frac{\omega}{c}$, the wave can be written as:

$$E_X(Z,t) = E_{0,X} \exp\left(i\left((n+ik)\frac{\omega}{c} \cdot Z - \omega t\right)\right). \tag{4.44}$$

And if we now separate real and imaginary part of the wave,

$$E_X(Z,t) = E_{0,X} \underbrace{\exp\left(-\frac{\omega}{c}k \cdot Z\right)}_{\text{Exponential decay}} \exp\left(i\left(\frac{\omega}{c}n \cdot Z - \omega t\right)\right), \tag{4.45}$$

we see that the electric field strength, and therefore also the light's intensity, decays exponentially. So, this certainly must mean that the use of absorbance is justified and the Beer-Lambert approximation has rightfully its merits. If there were no interfaces to cross and the wave would be traveling in a homogenous medium forever, certainly, because then we can indeed assume that the electric field intensity at a certain Z is fully determined by the irradiance after exiting the light source:

$$I(Z) = I_0 \exp\left(-\frac{2\omega k}{c} \cdot Z\right) = I_0 \exp(-\alpha \cdot Z) \rightarrow A = -\log_{10}\left(\frac{I(Z)}{I_0}\right) = \alpha \cdot Z. \tag{4.46}$$

This means that under these exceptional conditions using the absorbance A definitely makes sense. Under all other conditions, it is more meaningful to use the absorptance A, which we obtain from energy conservation:

$$\frac{I_R}{I_0} + \frac{I_T}{I_0} + \frac{I_A}{I_0} = R + T + A = 1. \tag{4.47}$$

When thinking about Eq. (4.47), keep in mind that the interfaces are assumed to be smooth and the media to be homogenous, which means that scattering does not take place. Note that we have defined absorbance and absorptance along the IUPAC standards [2]. In literature, these terms are often used differently or even synonymously and, sometimes, what definition is used becomes only clear through the context.

For R_s, absorption does not alter the curves substantially as can be seen in Fig. 4.2. R_s remains to be a monotonically increasing function of the angle of incidence, just the starting value for $\alpha = 0°$ increases. In case of R_p, the changes are also, on the first view, not drastic. The function still possesses a minimum, which is, however, no longer a zero. In addition, this minimum is shifted to higher angles of incidence. How does absorption alter the reflection from the interface between two media? In general, absorption increases reflection. This seems to be a paradox, but indeed, the strongest reflecting material in the infrared is gold due to having the highest index of absorption. As a consequence of this high reflection, not much radiation is actually absorbed (this completely changes in powdered form, which is why metal powders are usually black). For not-too-high indices of absorption, the dependence of reflection on the angle of incidence is depicted in Fig. 4.2.

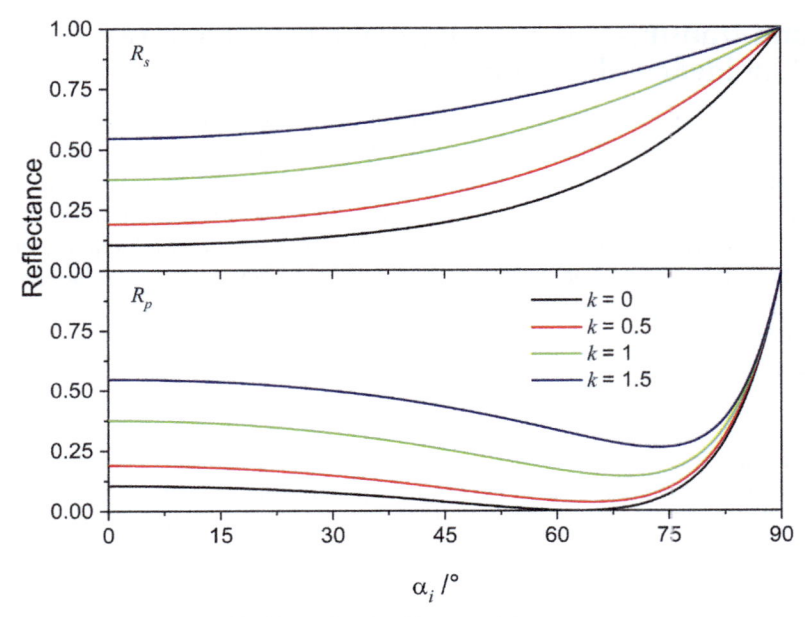

FIG. 4.2 Dependence of R_s and R_p from the angle of incidence for absorbing materials.

4.4 Reflection and transmission at an interface separating two scalar media under nonnormal incidence—Total/internal reflection

In the preceding examples, we somewhat automatically assumed incidence from vacuum or air, but what if the incidence medium is the optically denser medium, meaning that it has a higher index of refraction? This seems to be a situation that gets us into trouble, since from Fresnel's law we have deduced that for $n_1 < n_2$, $\alpha_i > \alpha_t$. Consequently, if we reverse the situation, then the angle of refraction must be larger than the angle of incidence. However, we cannot arbitrarily increase the angle of refraction—once it equals 90°, the wave is actually no longer penetrating into the second medium, but travels parallel to the interface. To derive this, we start from Eq. (4.14) and focus on the transmitted wave:

$$\mathbf{E}_t = \mathbf{E}_{0,t} \exp\left[i\omega\left(\frac{n_2}{c}(Y\sin\alpha_t + Z\cos\alpha_t) - t\right)\right]. \tag{4.48}$$

Due to Fresnel's law, Eq. (4.12), the Y-component of the wave vector stays the same and we can transform (4.48) with the help of Eq. (4.24) into:

$$\mathbf{E}_t = \mathbf{E}_{0,t} \exp\left[i\omega\left(\frac{n_1}{c}Y\sin\alpha_i - \frac{n_1}{c}Z\sqrt{n_2^2 - n_1^2\sin^2\alpha_i} - t\right)\right]. \tag{4.49}$$

For $\alpha_t = 90°$, $\sin\alpha_t = 1$, accordingly, the corresponding angle of incidence, which is called the critical angle α_c, becomes:

$$\alpha_c = \arcsin\left(\frac{n_2}{n_1}\right). \tag{4.50}$$

Above α_c, $n_2^2 - n_1^2\sin^2\alpha_i < 0$. Nevertheless, Eq. (4.49) is still meaningful. To see this, we replace $n_2^2 - n_1^2\sin^2\alpha_i$ by $i^2(n_1^2\sin^2\alpha_i - n_2^2)$ and obtain

$$\mathbf{E}_t = \mathbf{E}_{0,t} \exp\left[i\omega\left(\frac{n_1}{c}Y\sin\alpha_i + i\frac{n_1}{c}Z\sqrt{n_1^2\sin^2\alpha_i - n_2^2} - t\right)\right] \tag{4.51}$$

which we can then transform into:

$$\mathbf{E}_t = \mathbf{E}_{0,t} \exp\left[\omega\frac{n_1}{c}Z\sqrt{n_1^2\sin^2\alpha_i - n_2^2}\right]\exp\left[i\omega\left(\frac{n_1}{c}Y\sin\alpha_i - t\right)\right]. \tag{4.52}$$

This result obviously is the cause of trouble, because our wave would grow exponentially with Z. Remembering that we actually deal with real waves and that the real part remains the same, even if we multiply the phase in the exponential function by -1 we have to repair our unphysical results by introducing $-i^2(n_1^2\sin^2\alpha_i - n_2^2)$ instead of $i^2(n_1^2\sin^2\alpha_i - n_2^2)$ to assure that the field decays exponentially with increasing Z:

$$\mathbf{E}_t = \mathbf{E}_{0,t} \exp\left[-Z\frac{\omega}{c}n_1\sqrt{n_1^2\sin^2\alpha_i - n_2^2}\right] \exp\left[i\omega\left(t - \frac{n_1}{c}Y\sin\alpha_i\right)\right]. \quad (4.53)$$

One or the other reader might think that changing the sign is an act of serious arbitrariness. To convince those, it might be enough to point out that there are two ways to introduce a complex index of refraction as discussed in Chapter 2, namely, in the form $\hat{n} = n + ik$ and $\hat{n} = n - ik$. The use of one or the other form depends on the definition of the complex form of the wave and the fact that the form of the complex index of refraction must lead to exponential decay for nonzero k. Eq. (4.53) describes an electric wave, which has just a Y-component in the phase part, meaning that it travels along the interface between medium 1 and 2. Accordingly, there is no energy flux through the interface as long as n_2 is real. The electric field intensity as a function of distance from the interface and wavenumber is illustrated in Fig. 4.3. The exponential decay of the field intensity with distance from the interface (black line in Fig. 4.3) is clearly visible, as is the scaling with the wavenumber (remember that $\omega/c = 2\pi\tilde{\nu}$). The penetration depth is further influenced by the refractive indices of both media and the angle of incidence according to Eq. (4.53).

Eq. (4.50) implies that the higher the index of refraction of the incidence medium is, the smaller becomes the critical angle α_c. The dependence of the reflectance on the angle of incidence still shows the same characteristics as discussed in the preceding section, but the reflectance already reaches unity at the critical angle and stays at this value. This can be seen in Fig. 4.4.

Why do we actually care about this particular situation? Since the wave cannot really enter the second medium, it cannot probe it, so this situation seems to be useless for the spectroscopist, right? On the contrary, this situation is highly interesting! Since we are interested in absorbing media, it is instructive to see how the situation changes, if we allow the second medium to be absorbing. If we assume a complex index of refraction, two changes can be realized in Eq. (4.53) concerning the first exponential function:

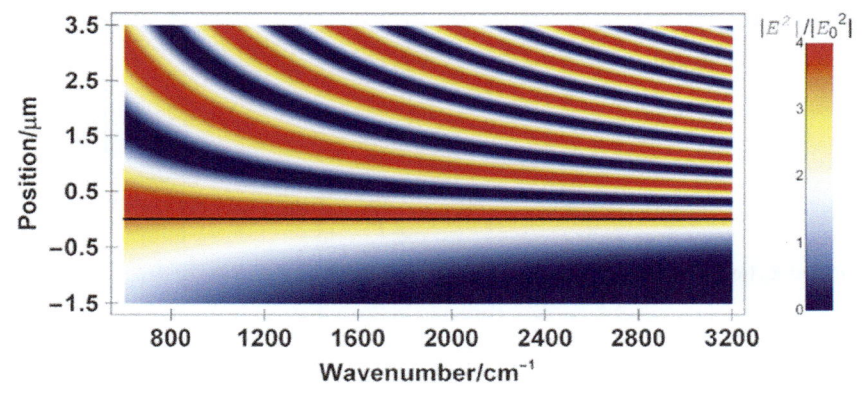

FIG. 4.3 Electric field intensity at the interface for s-polarized incident light ($\alpha_i = 30°$, $n_1 = 3.4$, $n_2 = 1.4$). The incidence medium extends from above down to the black line, which denotes the interface.

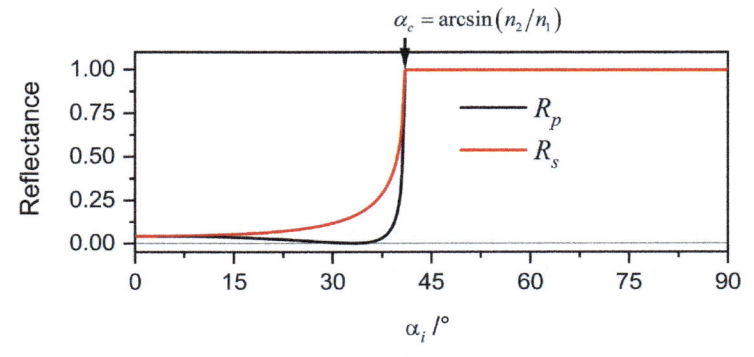

FIG. 4.4 Dependence of R_s and R_p from the angle of incidence in case of total (internal) reflection.

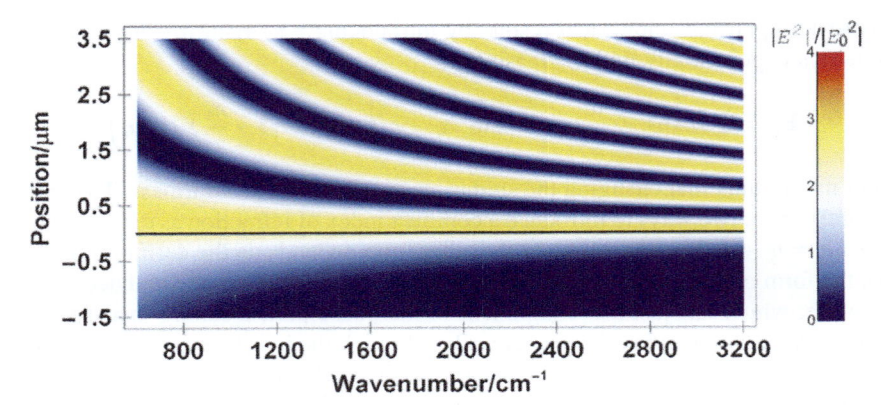

FIG. 4.5 Electric field intensity at the interface for s-polarized incident light ($\alpha_i = 30°$, $n_1 = 3.4$, $\hat{n}_2 = 1.4 + i\, 0.54$).

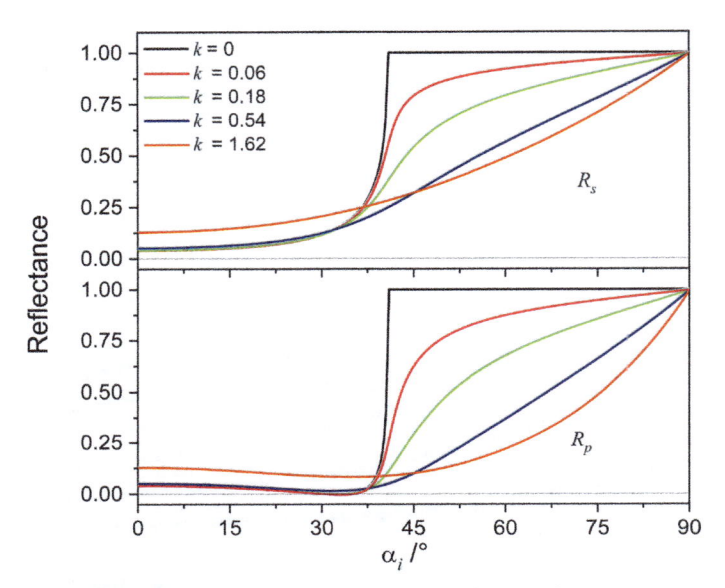

FIG. 4.6 Dependence of R_s and R_p from the angle of incidence for $n_1 > n_2$ and absorbing materials.

$$-Z\frac{\omega}{c}n_1\sqrt{n_1^2\sin^2\alpha_i - n_2^2} \xrightarrow{\hat{n}_2 = n_2 + ik_2} -Z\frac{\omega}{c}n_1\sqrt{n_1^2\sin^2\alpha_i - (n_2^2 - k_2^2) + 2ink}. \quad (4.54)$$

Without evaluating (4.54) further, it is obvious that the wave would still be evanescent, but with faster decay as can also be seen in Fig. 4.5.

From this figure, it is also obvious that absorption attenuates the reflectance, and it is no longer total; therefore, this method is usually called ATR (attenuated total reflection) spectroscopy. In fact, for larger absorption indices, any resemblance with total reflection gets lost as is illustrated in Fig. 4.6.

The second change is that with increasing absorption index, an increasing part of the Z-component becomes real, which means the wave is no longer tangential to the interface, and an increasing flux of energy into the second medium does occur where it is absorbed.

4.5 Reflection and transmission at an interface separating two scalar media under nonnormal incidence—Matrix formalism

The formalisms that I presented in the last sections seemed to be of limited use for infrared spectroscopy, since, while measurements of reflection and total internal reflection play an important role, an understanding of the most important technique, which is the transmission technique, cannot be advanced (at the time this book was printed, it may be that the ATR technique is more often employed than transmittance measurements. But if we integrate over all times, transmittance should still be in the lead. Anyway, one should aim at understanding all existing measurement techniques). Indeed, to understand transmission spectroscopy from a wave optical point of view, we need to introduce three media at least: The incidence medium, the sample, and the exit medium, where the first and the last are considered semiinfinite and are usually air or vacuum.

A typical example is a pressed tablet of an IR-transparent material containing a small amount of sample of finely grained consistency to form a medium that is homogenous to IR-radiation to avoid scattering and a further complication due to micro-heterogeneous mixing, which is rather unknown and will be discussed in Chapters 6 and 10. Further examples would be freestanding films (e.g., polymers) or sections of bulk samples. In case of inorganic materials, it is, however, very hard to impossible to prepare sections thin enough so that they could be used for transmission measurements, at least not in the spectral range around their main bands (fundamentals) as thicknesses of the order of 100 nm would be required. In such cases, layers could be prepared, e.g., by vapor deposition methods, on IR-transparent substrates. As a result, we have already two media of limited thickness that are transmitted by radiation!

Furthermore, if liquids are to be investigated, we actually need one medium more, as liquids are usually contained in cuvettes, so that the cuvette material is transmitted twice, when light passes through the cuvette. Even if this is the same material with usually the same thickness before and after the sample, this adds another degree of complexity.

Why would that be the case? Because in the infrared spectral range, absorption, even that of organic liquids, is strong enough to suggest that the thickness of a cuvette should be in the range of some 10 μm (note that the use of solvents is usually not an option in infrared spectroscopy, as it is hard to find a proper solvent that is not chemically similar and does not have absorption bands in the same spectral regions as the liquid to be investigated). In this case, as for freestanding films and films on a substrate, we have not only to consider that we have multiple reflections but also that we have interference. As a consequence, the measurement cannot be referenced to the measurement of an empty cuvette, since the index of refraction differences between cuvette material and content is usually much higher than if filled. Accordingly, interference effects will be enhanced, as we will see later.

In thick substrates on the other hand, interference effects are often absent (according to the textbooks, interference is absent, if the thickness of the medium is no longer in the same range as the wavelength of light and/or if the light comes from an incoherent light source. Actually, coherence has nothing to do with the ratio of the wavelength to the thickness of the substrate, since we are confronted with interference fringes even in the spectra of a 1 mm thick Si substrate with light from an incoherent light source. This is because every light wave interferes with its reflected self, which is of course perfectly coherent to the original wave. A better explanation for the absence of interference fringes in the case of thick substrates is therefore that missing plane planarity destroys the interference effects). Overall, this makes things somewhat complicated, as we would need to derive a formula for the freestanding film, another one for the layer on a substrate, and another one for the liquid in the cuvette. Moreover, what do we do, if we have more than one layer on the substrate? Therefore, instead of deriving all these particular formulas, we go the inverse way and provide a general formalism, which contains all these special cases (and much more). Then, when we discuss particular cases, we go backward and derive or state the explicit formulas. We can then regain the vividness that we have to give up for the moment in order to encompass all possible cases.

We first assume that we have exclusively a coherent superposition of waves. In addition to the freestanding film or section, this would be the case for a layer stack of thin films with smooth and plane parallel interfaces. However, it could also be a film on a substrate as long as the substrate surfaces are plane parallel to each other and the spectral resolution is high enough (e.g., $<4\,\text{cm}^{-1}$ for a 1-mm-thick Si substrate). In the layer, or in each layer of the layer stack, there exist in general forward and backward traveling waves, since reflection occurs at each of the interfaces, which leads to multiple reflections of the incoming wave. It seems that this is a very confusing situation, but there is a recipe to find a consistent and general solution in a very concise way.

4.5.1 Matrix formulation for *s*-polarized waves at a single interface

First, we extend the situation at an interface of the layer stack by assuming that we have an incoming and a reflected wave from both sides of the interface as it is depicted in Scheme 4.5.

Accordingly, the continuity of the transversal components of the electric and magnetic fields leads to the following two equations:

$$\begin{aligned} E_{1,X} + E'_{1,X} &= E_{2,X} + E'_{2,X} \\ E_{1,X} n_1 \cos\alpha_1 - E'_{1,X} n_1 \cos\alpha_1 &= E_{2,X} n_2 \cos\alpha_2 - E'_{2,X} n_2 \cos\alpha_2. \end{aligned} \qquad (4.55)$$

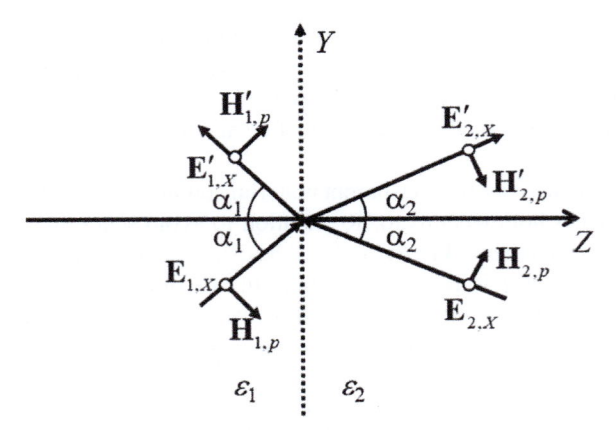

SCHEME 4.5 Forward and backward traveling s-polarized plane waves impinging nonnormally on an interface of two semiinfinite and scalar media.

These can be expressed in matrix form as:

$$\underbrace{\begin{pmatrix} 1 & 1 \\ n_1 \cos \alpha_1 & -n_1 \cos \alpha_1 \end{pmatrix}}_{\mathbf{D}_s(1)} \begin{pmatrix} E_{1,X} \\ E'_{1,X} \end{pmatrix} = \underbrace{\begin{pmatrix} 1 & 1 \\ n_2 \cos \alpha_2 & -n_2 \cos \alpha_2 \end{pmatrix}}_{\mathbf{D}_s(2)} \begin{pmatrix} E_{2,X} \\ E'_{2,X} \end{pmatrix}. \tag{4.56}$$

Here, $\mathbf{D}_s(j)$ denotes the so-called dynamical matrix for s-polarized light and the medium j. For a single interface, we obtain the reflection and transmission coefficients r_s and t_s in the following way:

$$\mathbf{D}_s(1) \begin{pmatrix} E_{1,X} \\ E'_{1,X} \end{pmatrix} = \mathbf{D}_s(2) \begin{pmatrix} E_{2,X} \\ E'_{2,X} \end{pmatrix} \rightarrow$$

$$\begin{pmatrix} E_{1,X} \\ E'_{1,X} \end{pmatrix} = \underbrace{\mathbf{D}_s(1)^{-1} \mathbf{D}_s(2)}_{\mathbf{M}_s} \begin{pmatrix} E_{2,X} \\ E'_{2,X} \end{pmatrix} \rightarrow \tag{4.57}$$

$$\begin{pmatrix} E_{1,X} \\ E'_{1,X} \end{pmatrix} = \mathbf{M}_s \begin{pmatrix} E_{2,X} \\ E'_{2,X} \end{pmatrix} \rightarrow r_s = \left(\frac{M_{21,s}}{M_{11,s}}\right)_{E'_{2,X}=0}, \quad t_s = \left(\frac{1}{M_{11,s}}\right)_{E'_{2,X}=0}.$$

Certainly, Eq. (4.57) yields the same result as Eq. (4.25), when we assume that there is no incoming wave traveling in the $-Z$ direction ($E'_{2,X}=0$).

4.5.2 Matrix formulation for *p*-polarized waves at a single interface

Equivalent to Section 4.5.1 we now assume two p-polarized waves incoming from both sides of an interface. The situation is depicted in Scheme 4.6.

The continuity of the transversal components of the electric and magnetic fields again leads to two equations,

$$E_{1,p} \cos \alpha_1 + E'_{1,p} \cos \alpha_1 = E_{2,p} \cos \alpha_2 + E'_{2,p} \cos \alpha_2$$
$$E_{1,p} n_1 - E'_{1,p} n_1 = E_{2,p} n_2 - E'_{2,p} n_2, \tag{4.58}$$

which expressed in matrix form take on the following form:

$$\underbrace{\begin{pmatrix} \cos \alpha_1 & \cos \alpha_1 \\ n_1 & -n_1 \end{pmatrix}}_{\mathbf{D}_p(1)} \begin{pmatrix} E_{1,p} \\ E'_{1,p} \end{pmatrix} = \underbrace{\begin{pmatrix} \cos \alpha_2 & \cos \alpha_2 \\ n_2 & -n_2 \end{pmatrix}}_{\mathbf{D}_p(2)} \begin{pmatrix} E_{2,p} \\ E'_{2,p} \end{pmatrix}. \tag{4.59}$$

In equivalence to its counterpart $\mathbf{D}_s(j)$, $\mathbf{D}_p(j)$ denotes the dynamical matrix for p-polarized light and the medium j. Assuming that there is no incoming wave from the negative Z-direction, we recover the result from Eq. (4.36):

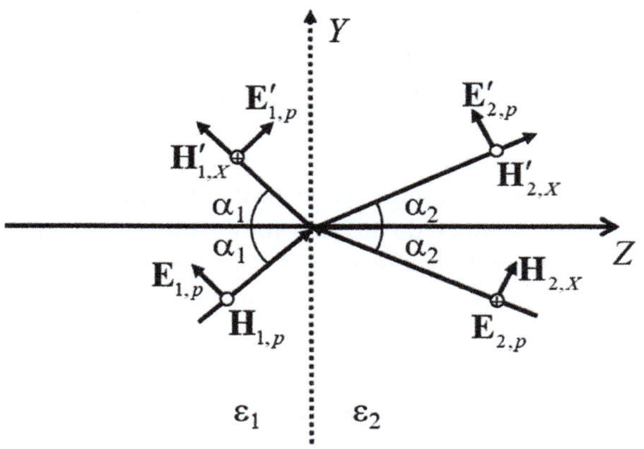

SCHEME 4.6 Forward and backward traveling p-polarized plane waves impinging nonnormally on an interface of two semiinfinite and scalar media.

$$\mathbf{D}_p(1)\begin{pmatrix} E_{1,p} \\ E'_{1,p} \end{pmatrix} = \mathbf{D}_p(2)\begin{pmatrix} E_{2,p} \\ E'_{2,p} \end{pmatrix} \rightarrow$$

$$\begin{pmatrix} E_{1,p} \\ E'_{1,p} \end{pmatrix} = \underbrace{\mathbf{D}_p(1)^{-1}\mathbf{D}_p(2)}_{M_p}\begin{pmatrix} E_{2,p} \\ E'_{2,p} \end{pmatrix} \rightarrow \qquad (4.60)$$

$$\begin{pmatrix} E_{1,p} \\ E'_{1,p} \end{pmatrix} = \mathbf{M}_p \begin{pmatrix} E_{2,p} \\ E'_{2,p} \end{pmatrix} \rightarrow r_p = \left(\frac{M_{21,p}}{M_{11,p}}\right)_{E'_{2,p}=0}, \quad t_p = \left(\frac{1}{M_{11,p}}\right)_{E'_{2,p}=0}.$$

4.5.3 Combined matrix formulation for waves at a single interface

Contrary to the trend to reduce dimensionality as far as possible, we here go the opposite way and increase it. To that end, we define a 4×4 matrix from the 2 2×2 matrices for s- and p-polarized light by placing the matrix for s-polarized light in the top-left quadrant, while we put the matrix for p-polarized light in the bottom-right quadrant:

$$\underbrace{\begin{pmatrix} 1 & 1 & 0 & 0 \\ n_1 \cos\alpha_1 & -n_1\cos\alpha_1 & 0 & 0 \\ 0 & 0 & \cos\alpha_1 & \cos\alpha_1 \\ 0 & 0 & n_1 & -n_1 \end{pmatrix}}_{D_1}\begin{pmatrix} E_{1,s} \\ E'_{1,s} \\ E_{1,p} \\ E'_{1,p} \end{pmatrix} = \underbrace{\begin{pmatrix} 1 & 1 & 0 & 0 \\ n_2 \cos\alpha_2 & -n_2\cos\alpha_2 & 0 & 0 \\ 0 & 0 & \cos\alpha_2 & \cos\alpha_2 \\ 0 & 0 & n_2 & -n_2 \end{pmatrix}}_{D_2}\begin{pmatrix} E_{2,s} \\ E'_{2,s} \\ E_{2,p} \\ E'_{2,p} \end{pmatrix}.$$

(4.61)

For the moment, this looks somewhat odd, in particular, since the remaining quadrants are filled with zeros. On the other hand, it is not too hard to imagine that by using a 4×4 matrix formalism we obtain the results for the reflection and transmission coefficients for s-and p-polarized light in one go:

$$\begin{pmatrix} E_{1,s} \\ E'_{1,s} \\ E_{1,p} \\ E'_{1,p} \end{pmatrix} = \underbrace{\mathbf{D}_1^{-1}\mathbf{D}_2}_{M}\begin{pmatrix} E_{2,s} \\ E'_{2,s} \\ E_{2,p} \\ E'_{2,p} \end{pmatrix} \rightarrow \begin{pmatrix} E_{1,s} \\ E'_{1,s} \\ E_{1,p} \\ E'_{1,p} \end{pmatrix} = \mathbf{M}\begin{pmatrix} E_{2,s} \\ E'_{2,s} \\ E_{2,p} \\ E'_{2,p} \end{pmatrix} \qquad (4.62)$$

$$r_s = \left(\frac{M_{21}}{M_{11}}\right)_{E'_{2,s},E_{2,p},E'_{2,p}=0}, \quad t_s = \left(\frac{1}{M_{11}}\right)_{E'_{2,s},E_{2,p},E'_{2,p}=0}, \quad r_p = \left(\frac{M_{43}}{M_{33}}\right)_{E_{2,s},E'_{2,s},E'_{2,p}=0}, \quad t_p = \left(\frac{1}{M_{33}}\right)_{E_{2,s},E'_{2,s},E'_{2,p}=0}.$$

Here, $\mathbf{D}_1$ and $\mathbf{D}_2$ are the dynamical 4×4 matrices of medium 1 and 2. Since nowadays neither memory nor speed of computers poses a problem for the calculations at hand, the use of a sparse matrix of large dimensionality should be ok.

However, there is a further reason why I wanted to introduce Eq. (4.61), which is that we generally need 4×4 matrices when we discuss anisotropic media in the second part of this book. For those media, the components, which are zero now, are a measure of s-polarized light that is converted to p-polarized light and vice versa. Since the 4×4 matrix formalism for anisotropic media becomes singular when it is used for scalar media (we will discuss this in Chapter 12 in detail), it is good to know that we can directly use the matrices that we just derived instead.

Having introduced the 4×4 matrices, we take a step back to the 2×2 matrices to shorten the notation. We can do this, as we examine some properties of the matrices, which do not alter for the different polarization directions. We therefore suppress the subscript, which would indicate the polarization direction keeping in mind that the difference between s-polarized matrices and p-polarized matrices just surfaces once we have to specify the components of the dynamical matrices.

4.5.4 A layer sandwiched by two semiinfinite media

You are now ready for the next level. Actually, once you have mastered this one, you will be able to compile formulas for arbitrary layer stacks provided that they consist of thin, homogenous, scalar media with smooth interfaces (thin in this context always means that interference effects do exist). To simplify things, we paradoxically need to make them first a little bit more complex by introducing a new nomenclature. So far, we have just differentiated between fields on the two sides of an interface, but now we have to differentiate between fields in a layer directly on the right side of the left interface (the one with the lower value of Z) and those on the left side of the right interface. Those will, in general, be different, not only because of absorption, but also, as we will see later, because the field strength is not constant over the layer and might decrease in positive Z-direction, even in the absence of absorption! We therefore define E_{jL}^+ as the field strength of the wave that is traveling in the positive Z-direction (therefore the "+") immediately after the wave entered the medium j, i.e., directly behind the left interface of the jth medium (the interface between the $(j-1)$th and the jth medium). Likewise, E_{jR}^+ is the field strength of the same wave immediately before it exits the jth medium, i.e., directly at the left side of the right interface (between the jth and the $(j+1)$th medium). For the wave traveling antiparallel to the Z-direction everything is analogous, the only change is that "+" is replaced by "−" to indicate the change in the traveling direction. The situation just described is illustrated in Scheme 4.7 for the case that $j = 2$.

The total field strength in medium j at the coordinate Z is then found by adding up the two fields belonging to the forward and the backward traveling wave:

$$E_j(Z) = R_j \exp(ik_{jZ}Z) + L_j \exp(-ik_{jZ}Z) = E_j^+(Z) + E_j^-(Z). \tag{4.63}$$

R_j and L_j are the amplitudes in the medium j. What has not been explained so far are the matrix products in Scheme 4.7. As we know from Eq. (4.62), the matrix product $\mathbf{D}_{j-1}^{-1}\mathbf{D}_j$ generally links the fields before and after an interface, but what is the function of the matrix $\mathbf{P}_2$? Obviously, this matrix, which we call the propagation matrix further on, describes the alteration of the electric fields when the wave propagates from the left interface to the right interface. Therefore, we can write,

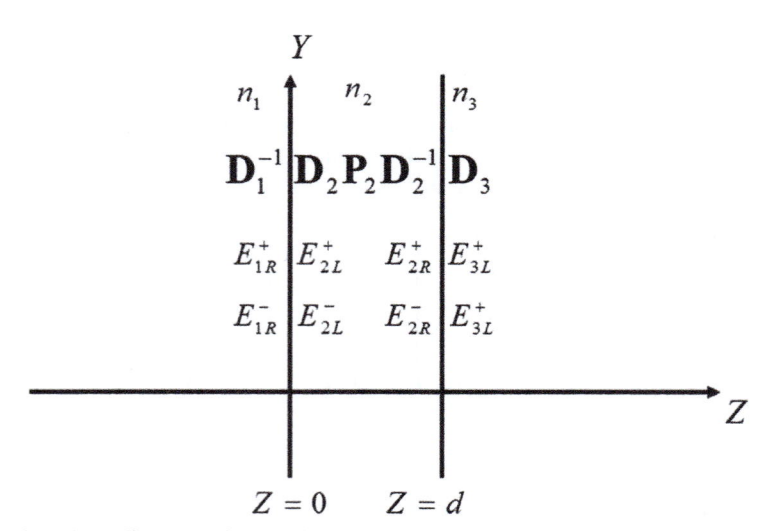

SCHEME 4.7 Forward and backward traveling waves in a one-layer system with incidence medium (medium 1)/layer/exit medium (medium 3).

$$\begin{pmatrix} E_{2L}^+ \\ E_{2L}^- \end{pmatrix} = \mathbf{P}_2 \begin{pmatrix} E_{2R}^+ \\ E_{2R}^- \end{pmatrix}, \tag{4.64}$$

wherein the matrix $\mathbf{P}_2$ is given by:

$$\mathbf{P}_2 = \begin{pmatrix} \exp(i\phi_2) & 0 \\ 0 & \exp(-i\phi_2) \end{pmatrix}. \tag{4.65}$$

Herein, ϕ_2 is the phase of the wave in medium 2:

$$\phi_2 = k_{2Z}d = \frac{2\pi}{\lambda} n_2 d \cos\alpha_2. \tag{4.66}$$

Who thinks that Eq. (4.66) appears somewhat "out of the blue" is requested to go back to Eq. (4.43), where Eq. (4.66) originates from (we have just given up the time dependence, because it is not relevant for the calculation of transmittance and reflectance). The difference of the sign in the arguments of the exponentials in Eq. (4.65) simply results from the fact that the element (1,1) belongs to the forward traveling wave (therefore, as a memory hook, the real part must show an exponential decay when the wave propagates in the positive Z-direction), whereas element (2,2) belongs to the backward traveling wave (and decays when propagating in the negative Z-direction). The only difference is that we have to consider also nonnormal incidence and we do that by the factor $\cos\alpha_2$ (be reminded that you can easily calculate this factor, cf. Eq. 4.24). It may be an unnecessary remark, but for scalar media ϕ_2 does not have a polarization dependence, so that $\mathbf{P}_2$ has the same form regardless of polarization. Accordingly, in the 4×4 matrix form, a propagation matrix for a scalar medium is of diagonal form with the same elements at (1,1) and (3,3) positions as well as at (2,2) and (4,4), respectively.

If we put everything together, we can calculate the fields in medium 3 directly behind the second interface from the fields in medium 1 directly before the first interface according to,

$$\begin{pmatrix} E_{1R}^+ \\ E_{1R}^- \end{pmatrix} = \underbrace{\mathbf{D}_1^{-1} \mathbf{D}_2 \mathbf{P}_2 \mathbf{D}_2^{-1} \mathbf{D}_3}_{\mathbf{M}} \begin{pmatrix} E_{3L}^+ \\ E_{3L}^- \end{pmatrix}, \tag{4.67}$$

and, from the matrix $\mathbf{M}$, reflection and transmission coefficient, as well as reflectance and transmittance using

$$r_l = \left(\frac{M_{21,l}}{M_{11,l}}\right)_{E_{3L,l}^-=0}, \quad t_l = \left(\frac{1}{M_{11,l}}\right)_{E_{3L,l}^-=0}, \quad l = s, p. \tag{4.68}$$

Formally, we can also state the matrix product in the following form:

$$\mathbf{D}_{j-1}^{-1} \mathbf{D}_j = \mathbf{D}_{(j-1)j}. \tag{4.69}$$

The matrix $\mathbf{D}_{(j-1)j}$ can be written as:

$$\mathbf{D}_{(j-1)j} = \frac{1}{t_{(j-1)j}} \begin{pmatrix} 1 & r_{(j-1)j} \\ r_{(j-1)j} & 1 \end{pmatrix}. \tag{4.70}$$

Here, we have to distinguish between the different polarizations of the wave. For s-polarization, we have

$$r_{(j-1)j,s} = \frac{k_{(j-1)Z} - k_{jZ}}{k_{(j-1)Z} + k_{jZ}}, \quad t_{(j-1)j,s} = \frac{2k_{(j-1)Z}}{k_{(j-1)Z} + k_{jZ}}. \tag{4.71}$$

For p-polarization, we find

$$r_{(j-1)j,p} = \frac{n_{j-1}^2 k_{jZ} - n_j^2 k_{(j-1)Z}}{n_{j-1}^2 k_{jZ} + n_j^2 k_{(j-1)Z}}, \quad t_{(j-1)j,p} = \frac{2n_{j-1} n_j k_{(j-1)Z}}{n_{j-1}^2 k_{jZ} + n_j^2 k_{(j-1)Z}}. \tag{4.72}$$

Eq. (4.68) can easily be generalized for an arbitrary number of layers. However, before we do this, let us first put some flesh on the bones and take a look at a simple example. This example shall consist of the calculation of the reflection and transmission of a slab that is suspended in vacuum $n_1 = n_3 = 1$. Furthermore, we assume normal incidence so that $k_{Zj} = n_j$. Under the latter condition, the problem degenerates and the 4×4 matrix contains two identical 2×2 matrices $\mathbf{M}$ since there is no difference between s- and p-polarization (cf. Section 4.1):

$$\mathbf{M} = \underbrace{\begin{pmatrix} \frac{1}{2} & \frac{1}{2} \\ \frac{1}{2} & -\frac{1}{2} \end{pmatrix}}_{\mathbf{D}_1^{-1}} \underbrace{\begin{pmatrix} 1 & 1 \\ n_2 & -n_2 \end{pmatrix}}_{\mathbf{D}_2} \underbrace{\begin{pmatrix} \exp(i\phi_2) & 0 \\ 0 & \exp(-i\phi_2) \end{pmatrix}}_{\mathbf{P}_2} \underbrace{\begin{pmatrix} \frac{1}{2} & \frac{1}{2n_2} \\ \frac{1}{2} & -\frac{1}{2n_2} \end{pmatrix}}_{\mathbf{D}_2^{-1}} \underbrace{\begin{pmatrix} 1 & 1 \\ 1 & -1 \end{pmatrix}}_{\mathbf{D}_1}. \qquad (4.73)$$

With the help of the matrix **M**, we can calculate the reflectance and transmittance according to:

$$t = \frac{1}{M_{11}} = \frac{t_{12}t_{21}\exp(i\phi)}{1 - r_{21}^2 \exp(2i\phi)} \quad t_{12} = \frac{2}{1+n_2}, t_{21} = \frac{2n_2}{1+n_2},$$

$$r = \frac{M_{21}}{M_{11}} = r_{12} + \frac{t_{12}t_{21}r_{21}\exp(2i\phi)}{1 - r_{21}^2 \exp(2i\phi)} \quad r_{12} = \frac{1-n_2}{1+n_2}, r_{21} = \frac{n_2-1}{1+n_2} = -r_{12},$$

$$T = |t|^2,$$

$$R = |r|^2,$$

$$\frac{k_{3Z}}{k_{1Z}} = 1.$$

(4.74)

When I look at the results, I have to admit that it would not be immediately clear to me that Eq. (4.74) indeed accounts for multiple reflections of light inside the slab that coherently interfere. As we will discuss this problem at length in Chapter 6, since it is one of the effects that can cause large deviations from the BBL approximation, I will not go into the details. Instead, I will for the moment just provide a hint for those of you who cannot wait: Just calculate at every interface for every pass of the wave the electric fields of the waves being reflected and refracted and add them all up as this was done by G.B. Airy in 1833 to explain the color of Newton's rings (cf. Section 2.11). [3].

4.5.5 Arbitrary number of layers

So far, we followed to a large extent pathways that have been influenced by [4]. Yeh certainly also provided a formalism to calculate the reflectance, transmittance, and absorptance for an arbitrary number of layers. We are, however, not only interested in the calculation of these quantities but also in the calculation of the field strengths and the field intensities at an arbitrary Z, while, so far, we have calculated these only at interfaces. The interest in these calculations derives from the possibility to calculate field and intensity maps, which I introduced already in Section 4.4. These maps allow a very intuitive access to changes of absorption with layer thicknesses and wavenumber. They will be essential, or at least illustrative, for understanding the related interference effects and are worth the extra effort. To be able to perform the corresponding calculations, we will intermix the approaches from [4,5] in the following.

The situation for an arbitrary number of layers is depicted in Scheme 4.8. The calculation of the fields at the interface between layer m and exit medium (medium $m+1$) is straightforward. Looking at Eq. (4.67), we see that the product of matrices begins from the left side with the inverse of the dynamical matrix of the incidence medium (in this case medium 1) $\mathbf{D}_1^{-1}$. At the right side, the product is completed by the dynamical matrix of the exit medium $\mathbf{D}_3$. In between we find the matrix product $\mathbf{D}_2\mathbf{P}_2\mathbf{D}_2^{-1}$. Accordingly, if we would have a second layer, its contribution to the matrix **M** would be

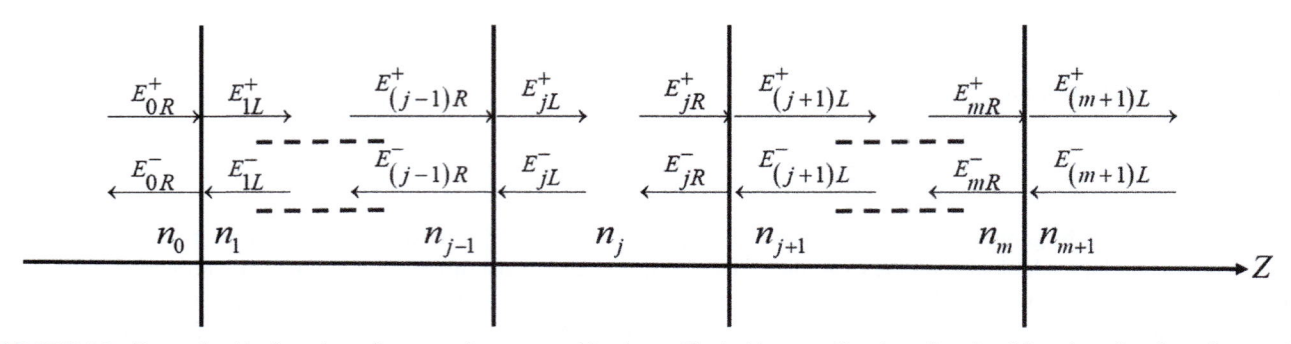

SCHEME 4.8 Forward and backward traveling waves in a system with m layers. The incidence medium is medium 0 and the exit medium is medium $m+1$.

$\mathbf{D}_3\mathbf{P}_3\mathbf{D}_3^{-1}$. Overall, if we rename the incidence medium as medium 0 and the exit medium as medium $m+1$, we can describe the forward and backward traveling waves in this exit medium in terms of the forward and backward traveling waves in medium 0 and the different dynamical and propagation matrices as:

$$\begin{pmatrix} E_{0R}^+ \\ E_{0R}^- \end{pmatrix} = \underbrace{\mathbf{D}_0^{-1} \left(\prod_{i=1}^m \mathbf{D}_i \mathbf{P}_i \mathbf{D}_i^{-1} \right) \mathbf{D}_{m+1}}_{\mathbf{M}} \begin{pmatrix} E_{(m+1)L}^+ \\ E_{(m+1)L}^- \end{pmatrix}. \tag{4.75}$$

Again, we use one 2×2 matrix for s-polarization and one 2×2 matrix for p-polarization. t and r as well as T and R are calculated in exactly the same way from the matrix $\mathbf{M}$ as in Eq. (4.68):

$$\begin{aligned}
r &= \left(\frac{E_{0R}^-}{E_{0R}^+} \right)_{E_{(m+1)L}^- = 0} = \frac{M_{21}}{M_{11}} \\
t &= \left(\frac{E_{(m+1)L}^+}{E_{0R}^+} \right)_{E_{(m+1)L}^- = 0} = \frac{1}{M_{11}} \\
r' &= \left(\frac{E_{(m+1)L}^+}{E_{(m+1)L}^-} \right)_{E_{0R}^+ = 0} = -\frac{M_{12}}{M_{11}} \\
t' &= \left(\frac{E_{0R}^-}{E_{(m+1)L}^-} \right)_{E_{0R}^+ = 0} = \frac{\det \mathbf{M}}{M_{11}}.
\end{aligned} \tag{4.76}$$

We nevertheless repeat the formulas here, because we also want to introduce the reflectance from the backside of the layer stack, r', which is calculated under the assumption that light is incident from the backside (accordingly, medium 0 is then the exit medium and $E_{0R}^+ = 0$) and the corresponding transmittance t').

To make you acquainted with the notation of [5] I provide again the formulas for the calculation of the reflectance (which is trivially the same) and the transmittance,

$$\begin{aligned}
R_l &= |r_l|^2, \quad l = s, p \\
T_s &= \frac{|t_s|^2 \operatorname{Re}(n_{m+1} \cos \alpha_{m+1})}{n_0 \cos \alpha_0} \\
T_p &= \frac{|t_p|^2 \operatorname{Re}(n_{m+1}^* \cos \alpha_{m+1})}{n_0 \cos \alpha_0},
\end{aligned} \tag{4.77}$$

where n_{m+1}^* is the complex conjugate of n_{m+1} (or, simply, n_{m+1}, if it is real). Note that if we deal with absorbing media, we assume here that the incidence medium is nonabsorbing. For a spectroscopist, certainly, nothing else would actually make sense, because otherwise light would usually not reach sample, let alone the detector.

4.5.6 Calculating the electric field strengths of a layered medium—Coherent layers

If you are not interested in being able to calculate field maps, this section can be skipped. If you just care about calculating field maps for completely coherent systems, commercial software is available. On the other hand, this excludes to my best knowledge, e.g., calculating the field map of a thin layer on a substrate. So, if you are interested in calculating this field map, you have to either ask Emanuele Centurioni or me to calculate the field map for you, or, better, work through this and the following sections. To use Centurioni's formalism, [5] we first have to subdivide the layer according to Scheme 4.8 into two parts ranging from medium 0 to medium j and from medium j to medium $m+1$. More precisely, we make the cut at the interface $j/j+1$. If we want to calculate the fields directly on the left side of this interface, i.e., on the right end of the layer, from the fields left from the first interface 0/1, we have in analogy to Eq. (4.75):

$$\begin{pmatrix} E_{0R}^+ \\ E_{0R}^- \end{pmatrix} = \mathbf{D}_0^{-1} \left(\prod_{i=1}^{j-1} \mathbf{D}_i \mathbf{P}_i \mathbf{D}_i^{-1} \right) \mathbf{D}_j \mathbf{P}_j \begin{pmatrix} E_{jR}^+ \\ E_{jR}^- \end{pmatrix}. \tag{4.78}$$

There is one difference, though, which is that in Eq. (4.75), the final fields are those right from the last interface. This time, however, we are interested in those left from the next interface. This is the reason why we also need to include the propagation matrix $\mathbf{P}_j$. The second part, conversely, agrees formally completely with Eq. (4.75):

$$\begin{pmatrix} E_{jR}^+ \\ E_{jR}^- \end{pmatrix} = \underbrace{\mathbf{D}_j^{-1} \left(\prod_{i=j+1}^{m} \mathbf{D}_i \mathbf{P}_i \mathbf{D}_i^{-1} \right) \mathbf{D}_{m+1}}_{\mathbf{M}_{j(m+1)}} \begin{pmatrix} E_{(m+1)L}^+ \\ E_{(m+1)L}^- \end{pmatrix}. \tag{4.79}$$

On the other hand, instead of performing this calculation in two steps, we can do this in one step according to Eq. (4.75), and if we use the result for the transmission coefficient (Eq. 4.76),

$$\left(\frac{E_{(m+1)L}^+}{E_{0R}^+} \right)_{E_{0R}^- = 0} = \frac{1}{M_{11}} \longrightarrow \begin{pmatrix} \frac{1}{M_{11}} \\ 0 \end{pmatrix} E_{0R}^+ = E_{(m+1)L}^+, \tag{4.80}$$

we can reformulate Eq. (4.79) in the following way (note that we assume in the derivation of t in Eq. (4.76) that $E_{(m+1)L}^- = 0$ since the exit medium is semiinfinite):

$$\begin{pmatrix} E_{jR}^+ \\ E_{jR}^- \end{pmatrix} = \mathbf{M}_{j(m+1)} \begin{pmatrix} E_{(m+1)L}^+ \\ E_{(m+1)L}^- \end{pmatrix} \longrightarrow \begin{pmatrix} E_{jR}^+ \\ E_{jR}^- \end{pmatrix} = \mathbf{M}_{j(m+1)} \begin{pmatrix} \frac{1}{M_{11}} \\ 0 \end{pmatrix} E_{0R}^+. \tag{4.81}$$

This does not seem to be a big deal, and when I first saw this equation, I thought to myself, ok, what is this equation good for and where do I get E_{0R}^+ from? It is good, that I kept this question to myself. For field maps, the calculated fields are always relative to E_{0R}^+, so its explicit value is not needed and we can set it as unity. If we now take into account, that we can calculate the fields $E_j^+(Z)$ and $E_j^-(Z)$ in the jth layer at a position Z by propagating backward, i.e., by

$$\begin{pmatrix} E_j^+(Z) \\ E_j^-(Z) \end{pmatrix} = \mathbf{P}\left(\phi_j \frac{d_j - Z}{d_j} \right) \begin{pmatrix} E_{jR}^+ \\ E_{jR}^- \end{pmatrix} = \mathbf{P}\left(\phi_j \frac{d_j - Z}{d_j} \right) \mathbf{M}_{j(m+1)} \begin{pmatrix} \frac{1}{M_{11}} \\ 0 \end{pmatrix} E_{0R}^+, \tag{4.82}$$

we have everything we need to calculate the field maps.

Note that what stands in parentheses after $\mathbf{P}$ is in this case the argument, i.e., ϕ_j in the exponential functions has to be replaced by this argument. The total field that we are interested in is then simply the sum of $E_j^+(Z)$ and $E_j^-(Z)$.

4.5.7 Incoherent layers

We have already talked about cases when layers or layer stacks can be considered incoherent (which actually has nothing to do with using a noncoherent light source as you sometimes read in literature). If the thickness of the medium is homogenous, then it will nevertheless seem to be incoherent when either the thickness is increased (provided certainly, that the spectral resolution is limited) or the spectral resolution is correspondingly decreased, provided that the incidence is normal. In any case, typical examples of incoherent layers are pressed pellets (KBr or CsI technique), substrates, or the walls of a cuvette, when, as discussed above, no interference fringes are visible (still, interference fringes from layers on a substrate or in the inside of the cuvette can appear; those cases will be subject of the next section). How can we treat these cases? This is actually simple: We treat the waves as beams, i.e., we remove interference through superposition by squaring the fields (actually multiplying the fields with their complex conjugates) and looking at intensities instead. This does not mean that we now assume the BBL approximation as valid (cf. Eqs. (1.3), (1.4)), since we can still have multiple reflections inside our layers. But, as we will see later, the BBL approximation is a much better estimate in such cases. Another good news is that our matrix formalism for the treatment of coherent layers can (nearly) be taken over without changes, the only difference is that we are no longer dealing with fields, but with intensities, which we will denote with the capital letter U ($U = |E|^2$). The corresponding situation for an arbitrary number of incoherent layers is depicted in Scheme 4.9.

Due to the correspondence of the formalisms, we can formulate the relations between an incoming beam with intensity $U_{0'R}^+$ and the intensity of the reflected beam $U_{0'R}^-$ and the transmitted beam $U_{(m'+1)L}^+$ as:

$$\begin{pmatrix} U_{0'R}^+ \\ U_{0'R}^- \end{pmatrix} = \underbrace{\overline{\mathbf{D}}_{0'}^{-1} \left(\prod_{i'=1}^{m'} \overline{\mathbf{D}}_{i'} \overline{\mathbf{P}}_{i'} \overline{\mathbf{D}}_{i'}^{-1} \right) \overline{\mathbf{D}}_{m'+1}}_{\overline{\mathbf{M}}} \begin{pmatrix} U_{(m'+1)L}^+ \\ U_{(m'+1)L}^- \end{pmatrix}. \tag{4.83}$$

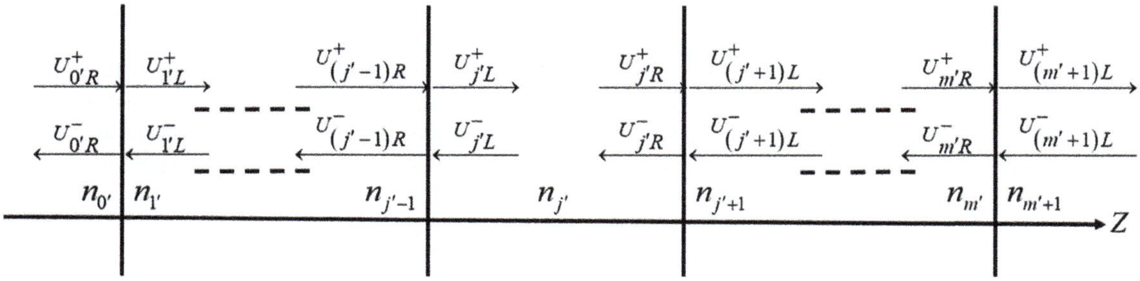

SCHEME 4.9 Forward and backward traveling beams in a system with m' layers. The incidence medium is medium $0'$ and the exit medium is medium $m'+1$.

To indicate that we now deal with intensities rather than with electric fields and to distinguish the matrices from those for the fields, we mark the intensity matrices by horizontal lines above the letters. The propagation matrices $\overline{\mathbf{P}}_{j'}$ are now given by (in perfect analogy to the use of intensities):

$$\overline{\mathbf{P}}_{j'} = \begin{pmatrix} |\exp(i\phi_{j'})|^2 & 0 \\ 0 & |\exp(-i\phi_{j'})|^2 \end{pmatrix}. \tag{4.84}$$

As a consequence, only the imaginary part remains. If it is zero, then the beam intensity does not change at all in this medium.

Deviations can though be found in the dynamical matrices $\overline{\mathbf{D}}_{(j'-1)j'}$, which are less symmetric than the corresponding ones in coherent layers (at least if absorption plays a role):

$$\overline{\mathbf{D}}_{(j'-1)j'} = \frac{1}{|t_{(j'-1)j'}|^2} \begin{pmatrix} 1 & -|r'_{(j-1)'j'}|^2 \\ |r_{(j'-1)j'}|^2 & |t_{(j'-1)j'} t'_{(j'-1)j'}|^2 - |r_{(j'-1)j'} r'_{(j'-1)j'}|^2 \end{pmatrix}. \tag{4.85}$$

In Eq. (4.85), the reflection and transmission coefficients are those for the coherent layers and can therefore be calculated in the usual way by Eqs. (4.71) for s-polarization and (4.72) for p-polarization. From the matrix, $\overline{\mathbf{M}}$ we can also obtain reflection and transmission coefficients for the whole incoherent layer stack according to:

$$\begin{aligned} \overline{r} &= \left(\frac{U^-_{0'R}}{U^+_{0'R}}\right)_{U^-_{(m'+1)L}=0} = \frac{\overline{M}_{21}}{\overline{M}_{11}} \\ \overline{t} &= \left(\frac{U^+_{(m'+1)L}}{U^+_{0'R}}\right)_{U^-_{(m'+1)L}=0} = \frac{1}{\overline{M}_{11}} \\ \overline{r}' &= \left(\frac{U^+_{(m'+1)L}}{U^-_{(m'+1)L}}\right)_{U^+_{0'R}=0} = -\frac{\overline{M}_{12}}{\overline{M}_{11}} \\ \overline{t}' &= \left(\frac{U^-_{0'R}}{U^-_{(m'+1)L}}\right)_{U^+_{0'R}=0} = \frac{\det \overline{\mathbf{M}}}{\overline{M}_{11}}. \end{aligned} \tag{4.86}$$

These reflection and transmission coefficients, however, are already intensities. Accordingly, the reflection coefficient equals the reflectance, while the transmission coefficients need to be corrected to obtain the transmittance:

$$\begin{aligned} R &= \overline{r} \\ T_s &= \frac{\overline{t}_s \operatorname{Re}(n_{m'+1} \cos \alpha_{m'+1})}{n_{0'} \cos \alpha_{0'}} \\ T_p &= \frac{\overline{t}_p \operatorname{Re}(n^*_{m'+1} \cos \alpha_{m'+1})}{n_{0'} \cos \alpha_{0'}}. \end{aligned} \tag{4.87}$$

This completes the calculation of reflectance and transmittance for an incoherent layer stack. As an observant reader, you may have noticed that the numbering of the media was slightly different compared to the incoherent case. Not that the order of the numbers would have been changed, but the numbers were given an apostrophe. The rationale behind this change is that in the next section we intermix coherent layer and incoherent layer packages and need to distinguish between the different layers. On the first view, the way it is done is highly confusing, at least it was for me. I hope I am able to convey the formalism in a manner, which is less unsettling for you (if not, you were at least warned what to expect in contrast to me).

4.5.8 Mixed coherent and incoherent layers

The best way of thinking of such a mixed coherent and incoherent layer stack is to keep first only the incoherent layers and the corresponding formalism from the preceding section. If we take the examples that we already talked about, e.g., a coherent layer stack on a thick substrate, you would first think of incidence medium/substrate/exit medium and use the mathematical framework of the preceding section. Depending on whether the light first hits the coherent layer stack and then the substrate or vice versa, you would then replace either the first or the second interface with the coherent layer stack as this is shown in Scheme 4.10.

Let us assume it is the first interface incidence medium/substrate where the coherent layer stack is situated. Then, to calculate the transmittance and reflectance of the whole coherent/incoherent layer stack, the dynamical matrix $\overline{\mathbf{D}}_{0'1'}$ would be given by,

$$\overline{\mathbf{D}}_{i'(i'+1)} = \begin{pmatrix} |M_{11}|^2 & -|M_{12}|^2 \\ |M_{21}|^2 & \dfrac{|\det \mathbf{M}|^2 - |M_{12}M_{21}|^2}{|M_{11}|^2} \end{pmatrix}, \qquad (4.88)$$

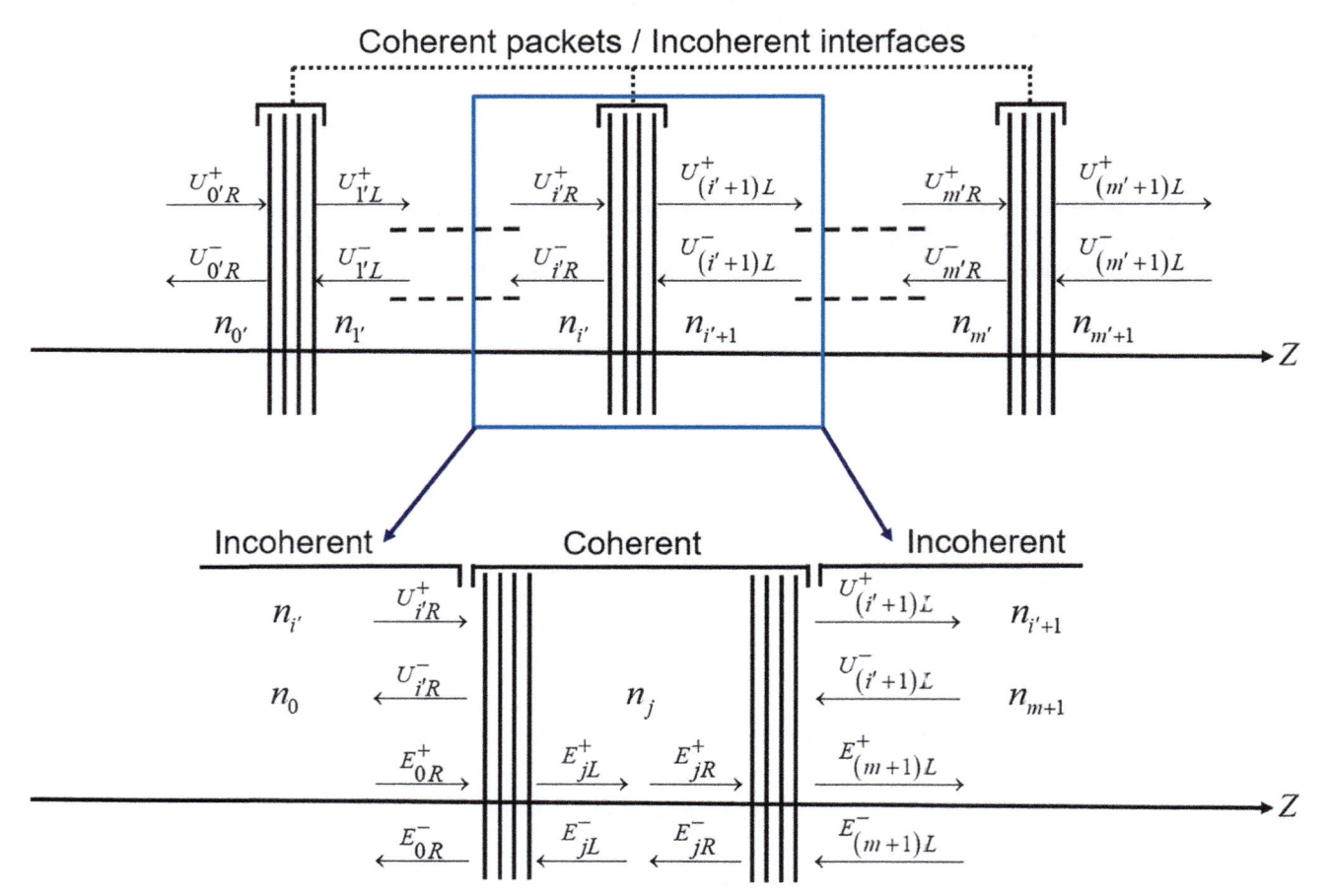

SCHEME 4.10 Forward and backward traveling beams in a mixed coherent/incoherent system with m' incoherent layers. The incidence medium is medium $0'$ and the exit medium is medium $m'+1$. The enlarged coherent system (lower part), which builds the interface between incoherent layer i' and $(i'+1)$ consists of m coherent layers whose incidence medium is medium i' and whose exit medium is medium $(i'+1)$.

if we set $i' = 0'$. The matrix **M** in Eq. (4.88) you would calculate according to Section 4.5.5. Your incidence medium for the coherent layer stack would be the incoherent medium on the left side (in our example medium $0'$) and the exit medium the incoherent medium on the right side (the substrate). In general, the recipe for the calculation would be the following:

- Divide the multilayer packet into incoherent layers and packets of coherent layers.
- Use for every coherent layer packet the formalism for coherent layers.
- Assume that for a number of m coherent layers, the layers 0 and $m+1$ correspond to the incoherent layers $i'-1$ and i'.

By that, every simple interface or packet of coherent layers can be converted into the corresponding incoherent interface and the formalism for the incoherent layers can be used to calculate the transmittance and reflectance.

Note that, if we would like to, we certainly could also construct 4×4 matrices in the same way as already explained for a coherent layer stack to perform the calculations for s- and p-polarization in one go.

4.5.9 Calculating the electric field strengths of a layered medium—Mixed coherent-incoherent multilayers

The first steps are very similar to the one that we need to perform in order to calculate the electric field strength in a pure coherent layer stack. Therefore, we divide in Eq. (4.83), which we repeat here for convenience,

$$\begin{pmatrix} U^+_{0'R} \\ U^-_{0'R} \end{pmatrix} = \underbrace{\overline{\mathbf{D}}_{0'}^{-1} \left(\prod_{j'=1}^{m'} \overline{\mathbf{D}}_{j'} \overline{\mathbf{P}}_{j'} \overline{\mathbf{D}}_{j'}^{-1} \right) \overline{\mathbf{D}}_{m'+1}}_{\overline{\mathbf{M}}} \begin{pmatrix} U^+_{(m'+1)L} \\ U^-_{(m'+1)L} \end{pmatrix}, \tag{4.89}$$

the matrix $\overline{\mathbf{M}}$ into the following two parts,

$$\begin{pmatrix} U^+_{0'R} \\ U^-_{0'R} \end{pmatrix} = \overline{\mathbf{D}}_{0'}^{-1} \left(\prod_{j'=1}^{i'-1} \overline{\mathbf{D}}_{j'} \overline{\mathbf{P}}_{j'} \overline{\mathbf{D}}_{j'}^{-1} \right) \overline{\mathbf{D}}_{i'} \overline{\mathbf{P}}_{i'} \begin{pmatrix} U^+_{i'R} \\ U^-_{i'R} \end{pmatrix}, \tag{4.90}$$

and

$$\begin{pmatrix} U^+_{i'R} \\ U^-_{i'R} \end{pmatrix} = \underbrace{\overline{\mathbf{D}}_{i'}^{-1} \left(\prod_{j'=i'+1}^{m'} \overline{\mathbf{D}}_{j'} \overline{\mathbf{P}}_{j'} \overline{\mathbf{D}}_{j'}^{-1} \right) \overline{\mathbf{D}}_{m'+1}}_{\overline{\mathbf{M}}_{i'(m'+1)}} \begin{pmatrix} U^+_{(m'+1)L} \\ U^-_{(m'+1)L} \end{pmatrix}. \tag{4.91}$$

Dividing the matrix $\overline{\mathbf{M}}$ at the right interface of the ith layer is equivalent to dividing the layer stack into two stacks. It is, not by accident, an incoherent layer that is adjacent to a coherent layer stack (cf. Scheme 4.10) where we divide the stack. However, this division does not alter the calculation of the field inside the incoherent layer i', it is just of importance because we will afterward calculate the fields inside the coherent layer packet.

Eq. (4.91) can be evaluated starting at the intensity in the nonabsorbing incidence medium $U^+_{0'R}$ in the following way:

$$\begin{pmatrix} U^+_{i'R} \\ U^-_{i'R} \end{pmatrix} = \overline{\mathbf{M}}_{i'(m'+1)} \begin{pmatrix} \frac{1}{\overline{M}_{11}} \\ 0 \end{pmatrix} U^+_{0'R}. \tag{4.92}$$

The intensities of the forward traveling beam $U^+_{i'}(Z)$ and the backward traveling beam $U^-_{i'}(Z)$ inside layer i' can then be obtained by:

$$\begin{pmatrix} U^+_{i'}(Z) \\ U^-_{i'}(Z) \end{pmatrix} = \overline{\mathbf{P}} \left(\phi_{i'} \frac{d_{i'} - Z}{d_{i'}} \right) \begin{pmatrix} U^+_{i'R} \\ U^-_{i'R} \end{pmatrix} = \overline{\mathbf{P}} \left(\phi_{i'} \frac{d_{i'} - Z}{d_{i'}} \right) \overline{\mathbf{M}}_{i'(m'+1)} \begin{pmatrix} \frac{1}{\overline{M}_{11}} \\ 0 \end{pmatrix} U^+_{0'R}. \tag{4.93}$$

The overall intensity $U_{i'}(Z)$ is then the sum of $U^+_{i'}(Z)$ and $U^-_{i'}(Z)$,

$$U_{i'}(Z) = U^+_{i'}(Z) + U^-_{i'}(Z), \tag{4.94}$$

where Z is measured from the interface $(i'-1)i'$.

How do we now evaluate the electric field strength or intensity in the coherent layer stack that is sandwiched by the incoherent layers i' and $i'+1$? Partly, we can revert to Section 4.5.6. Why just partly? Because we need to identify E_{0R}^+ which is now definitely not equal to unity since the beam/wave has already passed to a number of incoherent layers or mixed layers before arriving at the left side of the interface between i' and $i'+1$, therefore it has definitely been weakened by reflection and, potentially, absorption. However, this is not our only problem! In Section 4.5.6, we considered the last medium $m+1$ an semiinfinite exit medium, which means that $E_{(m+1)L}^-$ equals zero, since there is no backward traveling wave in such a medium. We do not need to inspect Scheme 4.10 more closely, to see that medium $i'+1$ is not semiinfinite and therefore $E_{(m+1)L}^-$ not zero. But does this mean we cannot use the formalism presented in Section 4.5.6? Yes, we can! We simply split the calculation and first drive E_{0R}^+ from the left side through the coherent layer stack assuming $E_{(m+1)L}^-$ to be zero. Afterward we start a second calculation where we drive $E_{(m+1)L}^-$ backward through the coherent layer stack where we presume that E_{0R}^+ is zero. Then, we calculate the overall field strengths/intensities simply by superposition.

We start by assuming that the wave is coming from the left (green color)-hand side. To do that we need, as mentioned above, E_{0R}^+, which is not a problem at all. We obtain it simply from $U_{i'R}^+$ in the following way:

$$\frac{E_{0R}^+}{E_{0'R}^+} = \left(\frac{U_{i'R}^+}{U_{0'R}^+}\right)^{\frac{1}{2}}. \tag{4.95}$$

As usual we can simply set $U_{0'R}^+$, the intensity of the beam that hits the first interface on the left side, unity and obtain from Eq. (4.92):

$$E_{0R}^+ = \left\{(1,0)\overline{\mathbf{M}}_{i'(m'+1)}\left(\frac{\frac{1}{\overline{M}_{11}}}{0}\right)\right\}^{\frac{1}{2}} E_{0'R}^+. \tag{4.96}$$

Based on E_{0R}^+, we can then calculate the electric field of the forward and backward traveling waves in layer j of the coherent layer packet based on Eq. (4.81):

$$\begin{pmatrix} E_j^+(Z) \\ E_j^-(Z) \end{pmatrix} = \mathbf{P}\left(\phi_j \frac{d_j - Z}{d_j}\right)\begin{pmatrix} E_{jR}^+ \\ E_{jR}^- \end{pmatrix} = \mathbf{P}\left(\phi_j \frac{d_j - Z}{d_j}\right)\mathbf{M}_{j(m+1)}\left(\frac{\frac{1}{\overline{M}_{11}}}{0}\right)E_{0R}^+. \tag{4.97}$$

Finally, we find the intensity in the jth layer of the coherent layer stack due to light coming from the left-hand side at Z by adding the fields due to the forward and backward traveling waves and squaring the sum:

$$U_{jL}(Z) = |E_j(Z)|^2 = |E_j^+(Z) + E_j^-(Z)|^2. \tag{4.98}$$

Next, we determine the intensity due to the beam coming from the right (red color) side of the jth layer at position Z in the coherent packet $U_{jR}(Z)$. To do this we need the intensity of the light beam that travels in a negative direction in the incoherent layer $i'+1$:

$$\frac{E_{(m+1)L}^-}{E_{0'R}^+} = \left(\frac{U_{(i'+1)L}^-}{U_{0'R}^+}\right)^{\frac{1}{2}}. \tag{4.99}$$

To take into account that $E_{(m+1)L}^- \neq 0$ and that $E_{0R}^+ = 0$, we have to alter Eq. (4.81) accordingly:

$$\begin{pmatrix} E_{jR}^+ \\ E_{jR}^- \end{pmatrix} = \mathbf{M}_{j(m+1)}\begin{pmatrix} E_{(m+1)L}^+ \\ E_{(m+1)L}^- \end{pmatrix} = \mathbf{M}_{j(m+1)}\begin{pmatrix} -\frac{M_{12}}{M_{11}} \\ 1 \end{pmatrix} E_{(m+1)L}^-. \tag{4.100}$$

Based on Eqs. (4.100), (4.99), we obtain for $E_{(m+1)L}^-$ in terms of $E_{0'R}^+$:

$$E_{(m+1)L}^- = \left\{(0,1)\overline{\mathbf{P}}_{(i'+1)}\overline{\mathbf{M}}_{(i'+1)(m'+1)}\left(\frac{\frac{1}{\overline{M}_{11}}}{0}\right)\right\}^{\frac{1}{2}} E_{0'R}^+. \tag{4.101}$$

Now we are able to calculate the electric field of the forward and backward traveling waves in layer j of the coherent layer packet:

$$\begin{pmatrix} E_j^+(Z) \\ E_j^-(Z) \end{pmatrix} = \mathbf{P}\left(\phi_j \frac{d_j - Z}{d_j}\right) \begin{pmatrix} E_{jR}^+ \\ E_{jR}^- \end{pmatrix} = \mathbf{P}\left(\phi_j \frac{d_j - Z}{d_j}\right) \mathbf{M}_{j(m+1)} \begin{pmatrix} -\dfrac{M_{12}}{M_{11}} \\ 1 \end{pmatrix} E_{(m+1)L}^-. \qquad (4.102)$$

The calculation of $U_{jR}(Z)$ is analogous to Eq. (4.98):

$$U_{jR}(Z) = |E_j(Z)|^2 = |E_j^+(Z) + E_j^-(Z)|^2. \qquad (4.103)$$

Now that we know $U_{jL}(Z)$ and $U_{jR}(Z)$ at every point and for every wavenumber of interest (generally all calculations in this chapter remain the same and have to be executed for every wavenumber point in the spectral range of interest), we simply sum up both intensities according to:

$$U_j(Z) = U_{jL}(Z) + U_{jR}(Z). \qquad (4.104)$$

For the simple case of a nonabsorbing coherent layer on top of an incoherent substrate with light normally inciding and first hitting the layer and then the substrate, the effect due to $U_{jR}(Z)$ certainly depends on the index of refraction of the substrate. For highly reflecting materials like Si, the effect can be considerable so that Si can certainly not be seen as an semi-infinite medium as is shown in Fig. 4.7.

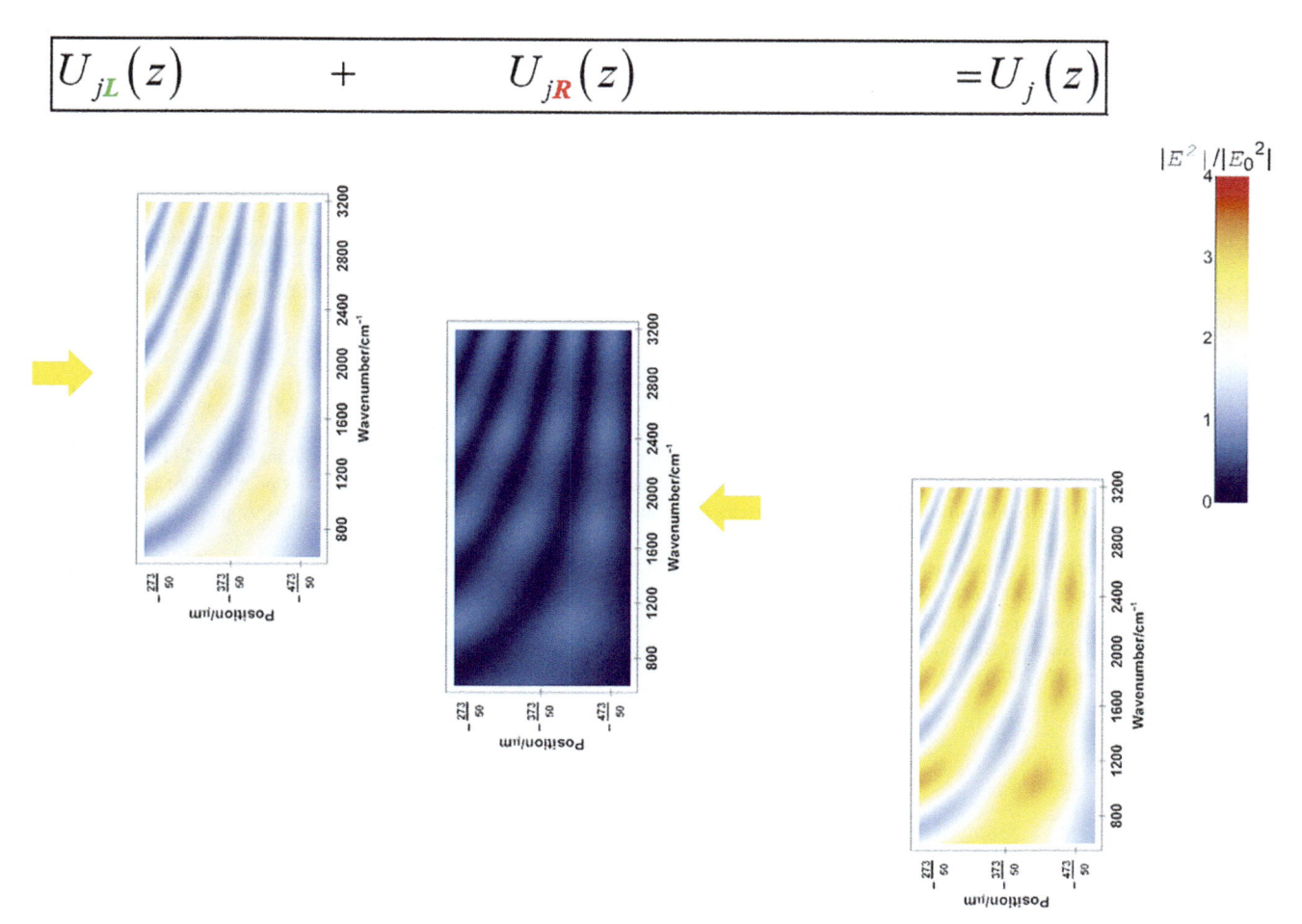

FIG. 4.7 Electric field intensity of a coherent layer on top of an incoherent substrate for normal incidence due to the incident light (left panel), the light that is reflected from the backside of the substrate (center panel) and the sum effect (right panel). The coherent layer is assumed to be nonabsorbing with an index of refraction $n = 1.5$, and the substrate is assumed to be Si.

4.6 Further reading

As for the preceding chapter, I highly recommend the book of Pocchi Yeh "Optical waves in layered media," which I have used for large parts of this chapter as basis and inspiration [4]. In addition, Born and Wolf's "Principles of Optics" is an excellent read [6] and also, Hecht's "Optics" [7]. The best book about infrared spectroscopy in combination with the description of optics and optical models is Light and Matter 1a of the series Handbuch der Physik edited by Ludwig Genzel, with articles authored by Ely Eugene Bell and William Cochran, which deals with Optical Constants and their Measurement and Phonons in Perfect Crystals [8]. To understand phenomena related to total internal reflection, I recommend "Internal Reflection and ATR Spectroscopy" [9] by M. Milosevic, a book which is (necessarily for ATR spectroscopy) fully based on wave optics. Furthermore, it is obvious that this chapter cannot cover the whole range of electromagnetic phenomena, which can be observed in infrared spectra. A very important part that is missing are scattering effects. As long as we restrict ourselves to scalar media, Bohren and Huffman's book "Absorption and scattering of light by small particles" [10] is highly recommended (this, however, does not extend to the treatment of anisotropy, where the book would definitely require a revision and update). In addition to the paper of Centurioni [5], it might be also worthwhile to look at different papers which introduced and updated the (transfer) matrix formalism over the years. The first who introduced the matrix formalism for coherent layer stacks to calculate reflectance and transmittance was Abelès at the end of the 40ies of the last century [11]. The first who extended the formalism to mixed coherent—incoherent layers seems to be Harbecke in 1986 [12]. Worth mentioning are also two papers by Ohta and Ishida, which seem to be the first who occupied themselves with the calculation of electric field strength and intensity at general positions within the layer [13,14]. Finally, also partial coherence was introduced [15,16].

References

[1] C. Schaefer, F. Matossi, Das Ultrarote Spektrum, Verlag von Julius Springer, Berlin, 1930.
[2] https://goldbook.iupac.org/.
[3] G.B. Airy, VI. On the phænomena of Newton's rings when formed between two transparent substances of different refractive powers, Lond. Edinb. Dublin Philos. Mag. J. Sci 2 (1833) 20–30.
[4] P. Yeh, Optical Waves in Layered Media, Wiley, 2005.
[5] E. Centurioni, Generalized matrix method for calculation of internal light energy flux in mixed coherent and incoherent multilayers, Appl. Opt. 44 (2005) 7532–7539.
[6] M. Born, E. Wolf, A.B. Bhatia, Principles of Optics: Electromagnetic Theory of Propagation, Interference and Diffraction of Light, Cambridge University Press, 1999.
[7] E. Hecht, Optics, fourth ed., Pearson Education, 2002.
[8] L. Genzel, S. Flügge, Licht und Materie 1a, Springer-Verlag, 1967.
[9] M. Milosevic, Internal Reflection and ATR Spectroscopy, Wiley, 2012.
[10] C.F. Bohren, D.R. Huffman, Absorption and Scattering of Light by Small Particles, Wiley, 1983.
[11] F. Abelès, Recherches sur la propagation des ondes électromagnétiques sinusoïdales dans les milieux stratifiés, Ann. Phys. 12 (1950) 596–640.
[12] B. Harbecke, Coherent and incoherent reflection and transmission of multilayer structures, Appl. Phys. B Lasers Opt. 39 (1986) 165–170.
[13] K. Ohta, H. Ishida, Matrix formalism for calculation of electric field intensity of light in stratified multilayered films, Appl. Opt. 29 (1990) 1952–1959.
[14] K. Ohta, H. Ishida, Matrix formalism for calculation of the light beam intensity in stratified multilayered films, and its use in the analysis of emission spectra, Appl. Opt. 29 (1990) 2466–2473.
[15] J.S.C. Prentice, Coherent, partially coherent and incoherent light absorption in thin-film multilayer structures, J. Phys. D Appl. Phys. 33 (2000) 3139.
[16] C.C. Katsidis, D.I. Siapkas, General transfer-matrix method for optical multilayer systems with coherent, partially coherent, and incoherent interference, Appl. Opt. 41 (2002) 3978–3987.

Chapter 5

Dispersion relations

As spectroscopists, our life is much more complicated than if we were just occupied with wave optics, because then it would often be sufficient to solve a certain system of equations only for a particular dielectric constant or an index of refraction. For us, the dielectric constant is not really constant, and it can considerably change when we vary frequency. Therefore, from here on, we call it dielectric function to reflect this variance. At the same time, we should better call the index of refraction also a function. However, the index of refraction function is not at the center of attention in this chapter, because it is, as we will see, only indirectly a material property (this is why historically it was not the dispersion, i.e., the frequency dependence of the index of refraction that was investigated, but its square, also because the dielectric function did not get much attention before Maxwell's theory [1]). Unfortunately, this also means automatically that the index of absorption function, as its imaginary part, is neither a material property. Therefore, it also does not automatically make sense to try to analyze an absorbance spectrum (which is the index of absorption function times the wavenumber and a constant) by a band fit, but we will see later on that there are important exceptions.

Schematically, the change of the real and imaginary part of the dielectric function is illustrated in Fig. 5.1. In the regime of very high frequencies/wavenumbers where the energies of the waves are too high to excite the transitions between different energy levels of electrons of an atom, molecule, or solid structure resonantly, the dielectric function has the constant value of unity, which means that there is essentially no difference to vacuum. Once the wavenumber is lowered and the light interacts with the gas, liquid, or solid, this changes dramatically, since absorption sets in and the imaginary part of the dielectric function begins to have values >0. At the same time, the real part changes correspondingly, so there seems to be some relation between both (indeed, as we will see later, there is a relation that connects the real with the imaginary part, which requires that changes in one part are reflected in the other, see Section 5.8). The shape of the bands in the UV/Vis can be very complicated depending on, whether we are looking at an isolated atom (line spectrum), a molecule (might have bands with shapes like in Fig. 5.1), or ionic or metallic solids (comparably complicated band structures).

Those are (mostly!) not the subject of this book and the seemingly only effect those excitations have, is that between the visible spectral range and the onset of excitable vibrations, the dielectric function stays real with values greater than unity (the deviation from unity is the higher the more intense the absorptions in the UV/Vis spectral range are; the reason for this we will also explore in the course of this chapter). In this spectral region, the near-infrared region (NIR), which extends from the visible region down to about $4000\,\mathrm{cm}^{-1}$, higher harmonics and combinations of the fundamental vibrations can occur. These are visible due to anharmonicity, i.e., deviations from Hooke's law, but they usually have low to very low intensities so we will neglect them in our discussion. Below $4000\,\mathrm{cm}^{-1}$, the fundamental vibrations occur. For gases, we will not discuss the dielectric functions explicitly. One reason is that deviations from the BBL approximation only set in at higher pressure—at lower pressure, the concentrations of the molecules are also low which, as we will see later, decreases oscillator strengths and, with it, also the deviations. The vibrations of molecules in the gaseous state are accompanied by rotational transitions, which cause the bands to have a characteristic fine structure or a characteristic band shape with a local minimum in the center of the band, depending on the spectral resolution. For liquids, this fine structure usually does not occur (very much like in gases under higher pressure due to line broadening) as in solids, but for the latter because rotations are usually not possible (instead they are transformed into lattice vibrations). The number of fundamental vibrations depends on the number of atoms in the molecule or in the unit cell and the symmetry of the vibrations (for details, we refer to the older textbooks of vibrational spectroscopy, see further reading at the end of this chapter). I just want to point out the selection rule, which helps to determine, if a certain vibration is infrared active:

$$\left(\frac{\partial \mu}{\partial r}\right)_{r_0} = M_{ij}. \tag{5.1}$$

Accordingly, with a change of the distance r between atoms, the dipole moment μ must change to make the transition moment M_{ij} non-zero (The dipole moment can also change from zero to some finite value. In other words, a permanent dipole moment is not needed, just a change of it). A change in the dipole moment on a microscopic level is connected with a

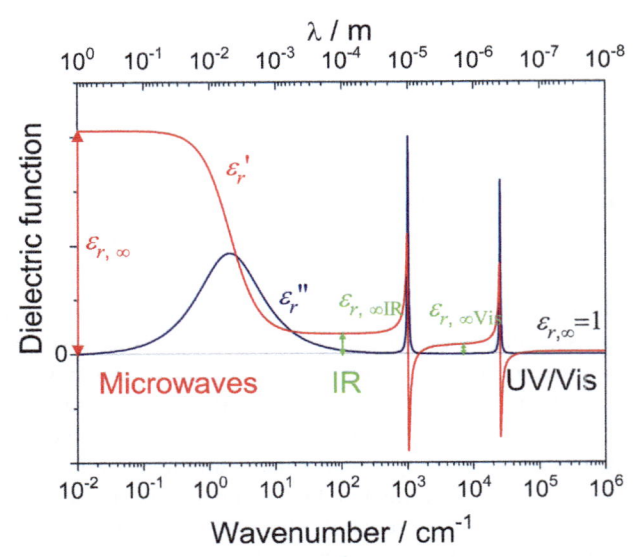

FIG. 5.1 Schematic dispersion of the dielectric function of a liquid. For a solid, the broad absorption with the lowest wavenumber position would vanish, since it does not have rotational degrees of freedom.

change of polarization on the macroscopic level, and this will help us to connect the microscopic changes to the macroscopic theory that is represented by Maxwell's equations.

In general, the resonance wavenumber $\tilde{\nu}_{0,j}$ shifts to lower values the larger the masses of the vibrating atoms and the smaller the force constants of the bonds between the atoms are. Therefore, for many salts, e.g., the halides, the resonance wavenumbers (the only one, some, or all) shift below $400\,\text{cm}^{-1}$. This wavenumber is usually considered as the boundary between the far infrared region (FIR, Newfangled: Terahertz spectral range) and the mid-infrared region (MIR). If we went further down in the spectrum, something that we will not undertake, but for completeness, I will bring it up, the frequency would be so low that permanent dipoles in liquids can follow the electromagnetic radiation. As a consequence, liquids can reach very high values of the real part of the dielectric function at low wavenumbers, in contrast to corresponding solids, wherein a free or semi-free rotation is not possible.

Basically, we are coming back to the materials equations (Eq. 3.5), which are necessary to complete Maxwell's equations and provide a unique solution for electromagnetic fields in matter:

$$\mathbf{D} = \varepsilon_0 \mathbf{E} + \mathbf{P} = \underbrace{\varepsilon_0 \varepsilon_r}_{\varepsilon} \mathbf{E} \cdot \tag{5.2}$$

In general, polarization induced by an electric field will counteract the electric field and reduce it as illustrated in Fig. 5.2. The reduction in the field lines is proportional to ε_r, the relative dielectric constant. In contrast, the **D** field does not change. It stays constant, e.g., in a plate capacitor, if the matter is brought into the field. For the static case, if a direct current voltage is applied to the plate capacitor, the situation will not change after a very short induction phase. When we now switch to alternating current voltage, for low frequencies ε will not vary strongly until the frequency is so high that the permanent dipoles are no longer able to follow the electric field (provided they could, as in liquids). This will lower ε, which will afterward stay approximately constant until the frequency is so high that dipole changes due to vibrations will no longer be possible. In the following, we will convert this semi-quantitative model to a quantitative one which will be able to describe spectra in the infrared spectral region to an astonishing high degree, as you will soon see in this chapter.

5.1 Dispersion relation—Uncoupled oscillator model

The usual reason why one would study the infrared properties of molecules consisting of two atoms first is because this is the easiest possible case. This is why we also start with this type of molecule to derive the dispersion relation. A molecule consisting of two atoms can only be infrared active if the two atoms are different, because an isolated homonuclear molecule could not fulfill the condition of Eq. (5.1), since its dipole moment cannot be changed by a change of the distance between the atoms (This would be different, if the homonuclear molecule is not isolated. In particular in the condensed phase, also homonuclear materials can have an IR-spectrum due to Van der Waals interactions.). Therefore, we need two different atoms with the masses m_1 and m_2. The chemical bond between the two masses is responsible for an attractive

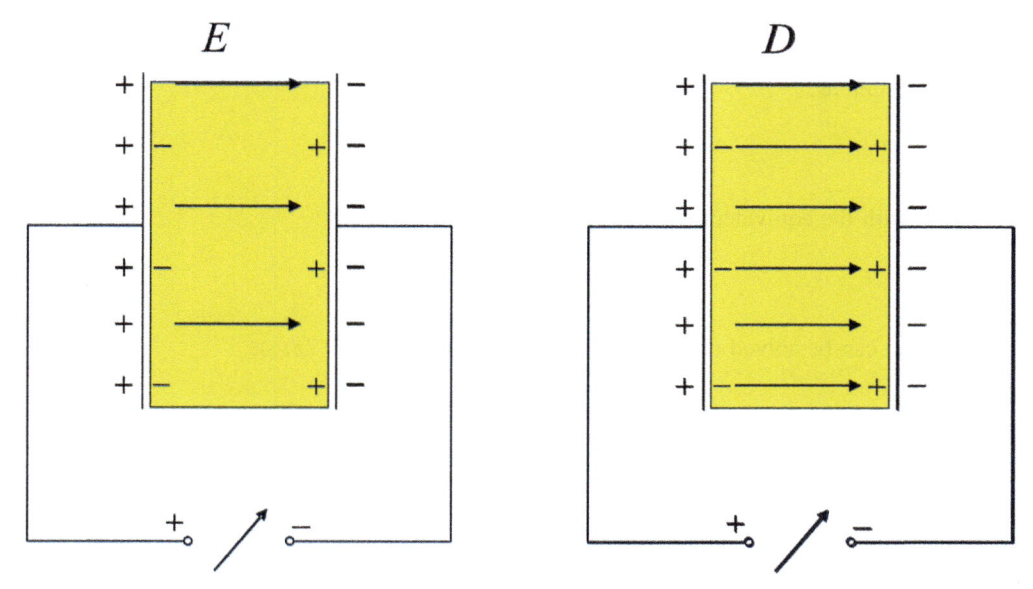

FIG. 5.2 Situation in a plate capacitor if matter is introduced: An electric field induces and/or orients dipoles in matter, which are antiparallel to the applied field. This will lower the number of field lines of the electric field, whereas that of the electric displacement stays the same.

force, if the distance r between them is larger than a certain equilibrium distance r_0 and a repulsive force, if $r < r_0$. We will assume that the deviation Δr between r and r_0 ($r - r_0 = \Delta r$) is small so that we can assume that the force F is proportional to this deviation (Certainly, the force F as well as the deviation Δr and, later on, the electric field E and the polarization P are vectorial quantities. Since the quantities are either parallel or antiparallel, I decided to omit indicating this in the formulas in this section):

$$F = -k \cdot \Delta r. \tag{5.3}$$

If the molecule is in its equilibrium position, the two atoms may have the coordinates x_1 and x_2 and if they are out of equilibrium then their positions change by Δx_1 and Δx_2. Therefore, $\Delta r = \Delta x_1 - \Delta x_2$ and Eq. (5.3) can be rewritten as:

$$F = -k \cdot (\Delta x_1 - \Delta x_2). \tag{5.4}$$

At the same time, the force is equal to the mass multiplied by the acceleration according to Newton's second law:

$$F = m_1 \frac{d^2 \Delta x_1}{dt^2} = m_2 \frac{d^2 \Delta x_2}{dt^2} = -k \cdot (\Delta x_1 - \Delta x_2). \tag{5.5}$$

A vibration is characterized by the fact that the center of gravity of the molecule does not change position. Therefore,

$$m_1 \Delta x_1 + m_2 \Delta x_2 = 0, \tag{5.6}$$

from which we gain an expression for Δx_2 in terms of Δx_1 and the two masses:

$$\Delta x_2 = -\frac{m_1}{m_2} \Delta x_1. \tag{5.7}$$

We replace Δx_2 in Eq. (5.5) by Eq. (5.7):

$$m_1 \frac{d^2 \Delta x_1}{dt^2} = -k \cdot \left(\Delta x_1 + \frac{m_1}{m_2} \Delta x_1 \right) \rightarrow$$
$$-\frac{d^2 \Delta x_1}{dt^2} = k \cdot \Delta x_1 \left(\frac{1}{m_1} + \frac{1}{m_2} \right) \tag{5.8}$$

The term in parentheses is the inverse of the so-called reduced mass μ given by:

$$\mu = \left(\frac{1}{m_1} + \frac{1}{m_2} \right)^{-1} = \frac{m_1 m_2}{m_1 + m_2}. \tag{5.9}$$

From Eq. (5.9), it is obvious that, if there is one atom much heavier than the other (say $m_1 \gg m_2$), then the reduced mass is approximately equal to the lighter atom, which means that practically only this atom moves. In any case, using the reduced mass, Eq. (5.8) can be written as:

$$-\frac{d^2 \Delta x_1}{dt^2} = \frac{k}{\mu} \Delta x_1. \tag{5.10}$$

We combine Eq. (5.10) with the equivalent equation for Δx_2 and obtain:

$$-\frac{d^2 \Delta r}{dt^2} = \frac{k}{\mu} \Delta r. \tag{5.11}$$

The differential Eq. (5.11) can be solved using the ansatz $\Delta r = A \cos \omega_0 t$ ($\omega_0 = 2\pi\nu_0$):

$$\begin{aligned}\frac{d^2 \Delta r}{dt^2} &= -4\pi^2 \nu_0^2 \cos \omega_0 t \rightarrow \\ \frac{k}{\mu} &= \omega_0^2 \rightarrow \\ \omega_0 &= \sqrt{\frac{k}{\mu}}\end{aligned} \tag{5.12}$$

ν_0 is the so-called eigenfrequency.

This is the derivation for the unforced vibration. If we shine a light on the molecule, we can drive this vibration and, accordingly, we have to add to Eq. (5.11) a term for the additional/driving force $F_E = qE(r,t)$ exerted by the electric field, wherein q is the so-called effective charge. We will additionally consider that the molecule is small compared to the wavelength of infrared light. Therefore, we can neglect the dependence of the electric field from the location and assume it to be homogenous at the location of the molecule $F_E = qE(t) = qE_0 \cos(-\omega t)$:

$$\mu \frac{d^2 \Delta r}{dt^2} + \omega_0^2 \mu \Delta r = qE_0 \cos(-\omega t). \tag{5.13}$$

This time the ansatz is that the solution must have the same frequency and time dependence as the exciting electric field, $\Delta r(t) = \Delta r_0 \cos(-\omega t)$. Accordingly, the solution is

$$\Delta r(t) = \frac{q}{\mu(\omega_0^2 - \omega^2)} E(t). \tag{5.14}$$

This is microscopically, if multiplied by the charge, the dipole moment p. When we multiply the dipole moment with the number of dipole moments per unit volume N, we obtain the polarization:

$$P = \underbrace{q\Delta r}_{p} N. \tag{5.15}$$

We can now compare Eq. (5.15) with the macroscopic equation for the polarization $\mathbf{P} = \varepsilon_0(\varepsilon_r - 1)\mathbf{E}$ (cf. Eq. 5.2):

$$\left.\begin{aligned}\mathbf{P} &= \frac{q^2 N/\mu}{\omega_0^2 - \omega^2} \mathbf{E} \\ \mathbf{P} &= \varepsilon_0(\varepsilon_r - 1)\mathbf{E}\end{aligned}\right\} \varepsilon_0(\varepsilon_r - 1) = \frac{q^2 N/\mu}{\omega_0^2 - \omega^2}. \tag{5.16}$$

Therefore, we obtain the following relation for ε_r,

$$\varepsilon_r = 1 + \frac{S^2}{\omega_0^2 - \omega^2}, \tag{5.17}$$

where we have set $S^2 = q^2 N/(\mu \varepsilon_0)$. S^2 is called the oscillator strength. The change of ε_r frequency is displayed in Fig. 5.3.

If we look at Eq. (5.17) and/or at Fig. 5.3, it becomes evident that this dispersion formula can only be used away from the eigenfrequency, because the function has a pole at ω_0 at which a resonance disaster would happen. We also know already from the fact that Eq. (5.17) is a real function, that it cannot be used to describe the dielectric function of a real material because it does not have an imaginary part, which would introduce absorption. How can we solve this problem? In fact,

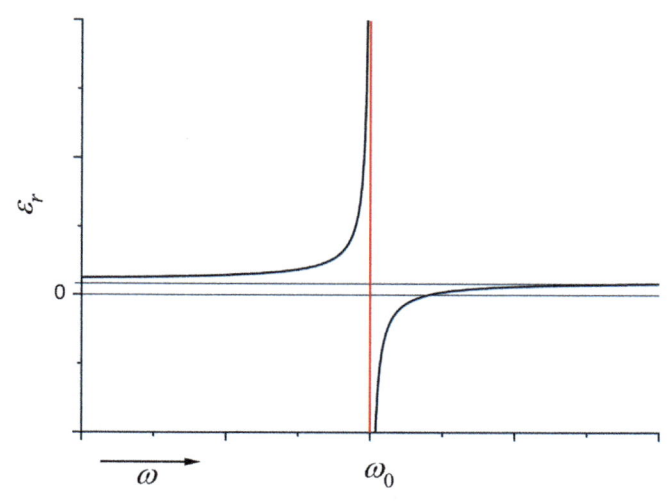

FIG. 5.3 Dispersion of the (relative) dielectric function around an eigenfrequency ω_0 according to Eq. (5.17).

there must be some kind of damping, on the one hand, to prevent the resonance disaster, which we do not observe, and, on the other hand, to account for absorption, which we do observe. One potential reason for this absorption would be that a changing dipole radiates and loses energy, but it seems that the corresponding loss would not be enough to account fully for the magnitude of the damping experimentally observed. Therefore, some part of the energy loss needs to be due to non-radiative mechanisms like it is introduced by anharmonicity. In any way, introducing damping more or less empirically in a way that is proportional to the velocity of the vibration might be the proper way to correct Eq. (5.17). If we introduce a corresponding damping term into Eq. (5.13), it then reads,

$$\mu \frac{d^2 \Delta r}{dt^2} + \mu \gamma \frac{d \Delta r}{dt} + \omega_0^2 \mu \Delta r = q E_0 \exp(-i \omega t), \tag{5.18}$$

wherein γ is the damping constant. Correspondingly, we changed over to the complex form for the exciting electric field. Accordingly, the new ansatz $\Delta r(t)$ is now $\Delta r(t) = \Delta r_0 \exp(-i\omega t)$. The corresponding solution is now found to be complex:

$$\varepsilon_r = 1 + \frac{S^2}{\omega_0^2 - \omega^2 - i \omega \gamma}. \tag{5.19}$$

You might come across a slightly different-looking solution where the damping term $i\omega\gamma$ has a positive sign. The sign depends on the way you have defined the complex form of the electromagnetic wave. If the time dependence is positive, then the damping term has a positive sign, but in the complex form of the dielectric function and the index of refraction function the imaginary part must have a negative sign. It all depends on the used convention, which must be stringent; otherwise, the amplitude of the waves might increase despite absorption.

The new curve shape(s) is/are displayed in Fig. 5.4. ε_r', the real part of the relative dielectric function, is now finite at the eigenfrequency. Coming from higher frequencies, it becomes negative, has its minimum at a higher frequency than the eigenfrequency, before it changes its sign again, and has a maximum at a lower frequency than the eigenfrequency. Between maximum and minimum, the real part shows so-called anomalous dispersion ($d\varepsilon_r'/d\omega < 0$) which changes to normal dispersion ($d\varepsilon_r'/d\omega > 0$) after the maximum. ε_r'' is the imaginary part, which becomes small away from the eigenfrequency and has its maximum exactly at ω_0.

What about the real and the imaginary part of the index of refraction function? Actually, to answer this question, we need a more elaborate discussion, which gives us a very valuable insight. This insight was well-known nearly a hundred years ago in the 1930s, but got somewhat lost in the conventional/contemporary books of vibrational spectroscopy. However, before we come to this insight, let us first slightly transform Eq. (5.19). First, we agree upon the usual convention, that from now on, when we speak of the dielectric function, or, later on, of the dielectric function tensor, we always mean the relative dielectric function (tensor) and can suppress the subscript "r." Furthermore, as we as vibrational spectroscopists are more acquainted with wavenumbers, we convert the angular frequency to wavenumber by expanding the numerator and

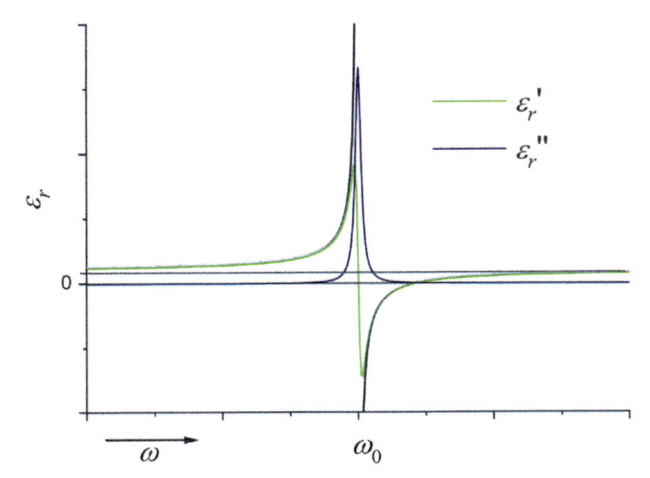

FIG. 5.4 Dispersion of the (relative) dielectric function around an eigenfrequency ω_0 assuming damping according to Eq. (5.19) (real part of the dielectric function: *green curve*, imaginary part: *blue curve*) in comparison with that without damping *(black curve)*.

the denominator by $(2\pi c)^{-1}$. The unit of the corresponding square root of the oscillator strength S, the damping constant γ, and the resonance wavenumber $\tilde{v}_0$ is then cm^{-1}:

$$\varepsilon = 1 + \frac{S^2}{\tilde{v}_0^2 - \tilde{v}^2 - i\tilde{v}\gamma}. \tag{5.20}$$

Let us for a moment further investigate this solution. Assume to that end, that we are not dealing with an isolated molecule, but instead with condensed matter formed by molecules. We further assume that the form of Eq. (5.20) stays the same and only the number of oscillators per unit volume N changes. How do the values of ε change with increasing oscillator strength (due to increasing transition moments)? For reasons that become obvious afterward, we will extend this question also to the index of refraction function. The square root of the oscillator strengths shall have the values $S = 50$, 200, and 500 cm^{-1}. $S = 50$ cm^{-1} is a value for an average weak vibration for organic/biologic matter. If we inspect Fig. 5.5, we see that the real part shows only a very weak change around the resonance position (1500 cm^{-1}). The same is true for the real part of the index of refraction function. Also, in unison are the two imaginary parts, which show symmetric bands around the resonance position. This drastically changes for the oscillator with $S = 200$ cm^{-1}, which could be considered strong for organic (a typical example would be the C=O vibration in ketones), but still weak for inorganic materials. Here, the real parts show asymmetric behavior around the resonance position and the changes are already substantial and can no longer be neglected (which is proved later, i.e., in the next chapter, when we show spectra of organic layers on a CaF$_2$ substrate). In addition, while the imaginary part of the dielectric function shows a curve shape symmetric to the resonance position independent of S, this has changed for the index of absorption function. Obviously, the part on the low-wavenumber side is lowered compared to the high-wavenumber side and the maximum is no longer at the resonance position (why this is a not well-known fact for the molecular spectroscopist we will discuss in Section 5.2).

Now it is time to remember that absorbance is proportional to the index of absorption. Therefore, the absorbance maximum indicates only the resonance position to a good approximation, if the oscillator is weak. For stronger oscillators, the maxima begin to shift to higher wavenumbers. If you are experienced in infrared spectroscopy, you might have heard of the practice to separate overlapping bands by a fit of absorbance spectra which is called *band fitting* or *band deconvolution* (a practice that is promoted by the instrument manufacturers, as accompanying software is often able to perform such fits). You certainly can conduct such fits, but you must be aware of the errors introduced right from the start. Not only the band positions may not be accurate but also the increasingly asymmetric band shapes might cause problems (cf. Section 5.2). Moreover, we have not talked about further problems introduced by dispersion and, potentially, interference.

What about condensed matter consisting of unit cells rather than molecules? You might have heard that in these cases vibrations are called phonons and have very different properties from those of molecules. We will come to those later on in more detail, but there are no obvious principal differences between Eq. (5.20) and the equation we will derive for phonons. As you will learn from other textbooks, what we just concluded is completely valid for both, liquid and solid matter (there are less than a handful of textbooks which actually tell you "yes and no," among them two authored by Max Born [2,3], but we will come to this later). Before we turn our attention to this derivation, let me point out that infrared spectroscopists

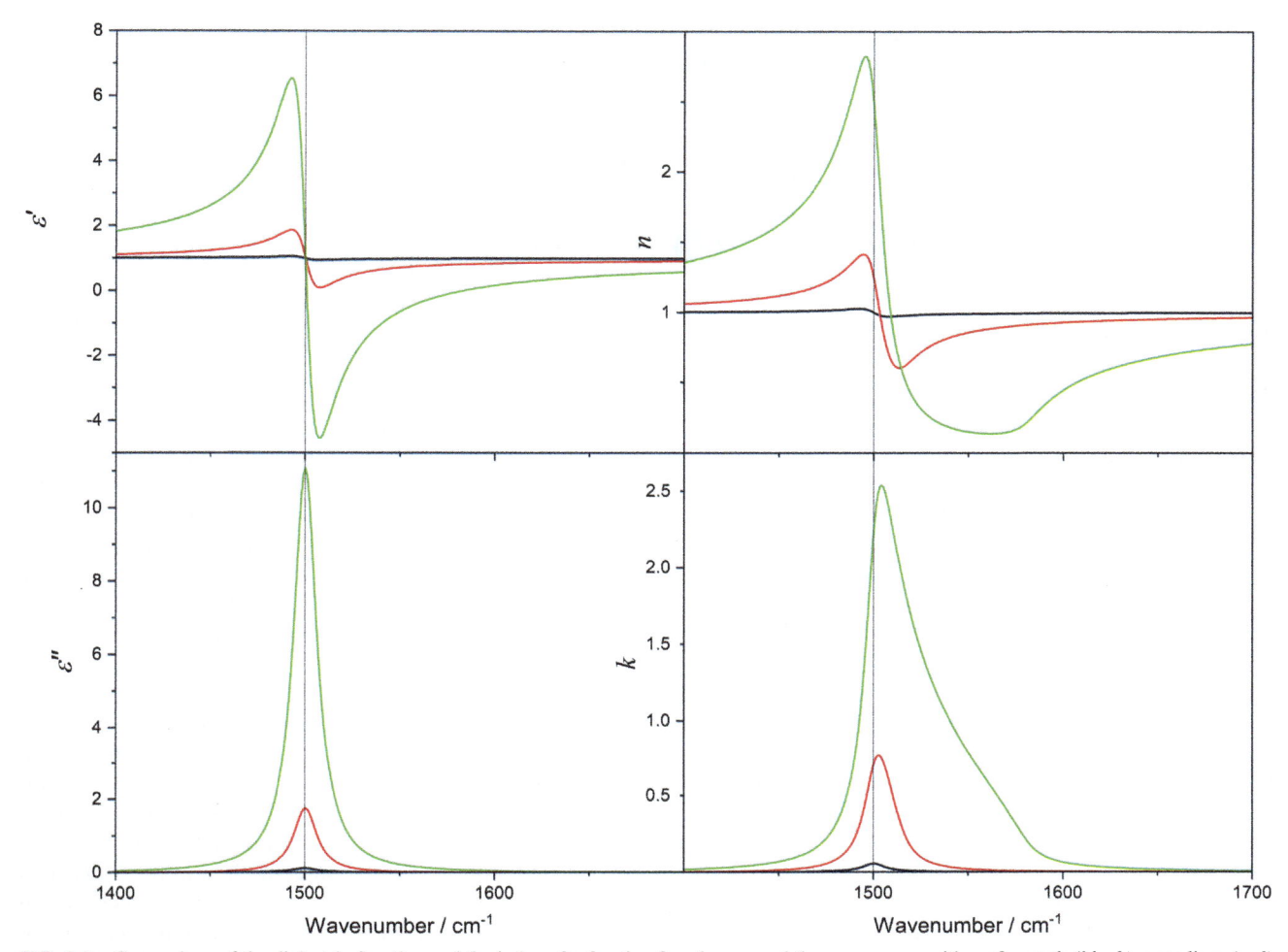

FIG. 5.5 Comparison of the dielectric function and the index of refraction function around the resonance position of a weak *(black)*, a medium *(red)*, and a strong *(green)* vibration. The square root of the oscillator strengths has the values $S = 50$, 200, and 500 cm^{-1}.

dealing with inorganic matter never would employ absorbance spectra just for the reasons above, and you can easily imagine why, if you look at the curves for $S=500\,\text{cm}^{-1}$. Dispersion and the asymmetric band shape of the imaginary part of the index of refraction function are just much too strong to be ignored. Therefore, it is necessary to apply *dispersion analysis*, if quantitative spectra evaluation of inorganic materials is the goal (this certainly includes resonance positions), a method that we will pay extensive attention to in this book.

Before we turn our focus to condensed matter, we introduce a first change to Eq. (5.20) by considering that there exist UV/Vis absorptions with much higher frequency than the infrared bands. Accordingly, there is usually a negligible influence of these absorptions in the IR-spectral range apart from introducing a constant offset of the real part: Usually, the absorption index far away from the resonance position is practically zero. Hence, what cannot be seen in Fig. 5.5, as the range is not extended enough, is that dispersion is very small far away from resonances. The only thing that needs to be covered (we will scrutinize this in greater detail in Section 5.8.4) is that each absorption increases both, the real part of the dielectric function and the real part of the index of refraction function at wavenumbers lower than the resonance. This offset is the larger the grander the oscillator strength is (cf. Fig. 5.5) and is called in Fig. 5.1 $\varepsilon_{r,\infty Vis}$. Out of convention, we call this offset dielectric background ε_∞ (we will meet again the dielectric background when we introduce the Kramers-Kronig relations and discuss it in more detail in Section 5.8):

$$\varepsilon = \varepsilon_\infty + \frac{S^2}{\tilde{v}_0^2 - \tilde{v}^2 - i\tilde{v}\gamma}. \tag{5.21}$$

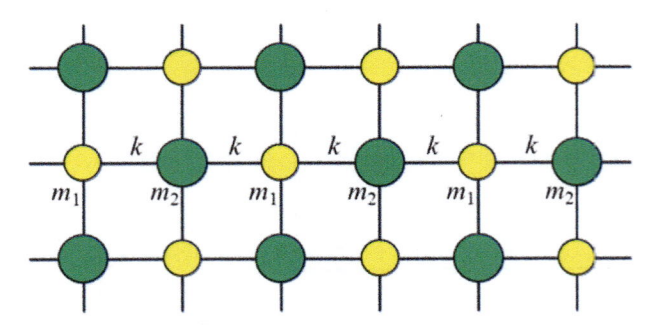

FIG. 5.6 Plane in a hypothetic crystal with NaCl structure consisting of two atoms in a unit cell.

Now that we have described dispersion only for a single molecule consisting of two atoms, it is certainly necessary to extend this description to larger molecules and liquids and solids. In liquids, as in glassy or amorphous solids, the situation can be very complicated. The reason is that those bond lengths and angles have a distribution and, at this point, it is unclear how to factor this in and treat this situation properly. Therefore, the fewest complications should occur, if we do not think about this problem first, but choose instead to understand dispersion for a well-ordered crystal. To simplify things further, we investigate the kind of crystal that is illustrated in Fig. 5.6. Such crystals actually are realized in nature, since thanks to the two different ion sizes good space-filling can be achieved, which would be definitely suboptimal for a crystal consisting of only one kind of atoms or ions. An appropriate structure could, e.g., be the NaCl-structure depending on the relative ion diameters (consider that Fig. 5.6 is just the 2-dimensional representation of a three-dimensional structure). I also have chosen this structure because it can be seen as a somewhat further developed model based on the molecule we discussed previously with two different masses m_1 and m_2 and a bond with the force constant k between the atoms.

Although the situation is already considerably simplified compared to the actual situation in some more complex crystals, I want to simplify it further and assume that we have just a chain (and not a 3D structure) of alternating atoms. The point I want to emphasize can actually be made also with this very simple model. Without going too much into detail, we can say that the potential vibrations in such a chain can be distinguished by the wavelength λ_k of the vibration and how the two atoms in one unit cell move. The first possibility is that in principle both atoms in a unit cell move in the same direction. This possibility is depicted in Fig. 5.7. How does the wavelength of the vibration come into play? It depends on the phases of the different unit cells relative to each other. The shortest possible wavelength is realized when in two neighboring unit cells the heavier atoms move in opposite directions (in this limiting case all the lighter atoms stay fixed). If the wavelength equals four unit cells, then in the second unit cell the two atoms would have, e.g., moved to the right and reached the maximum displacement, while in the next unit cell to the right or to the left, both atoms are in an equilibrium

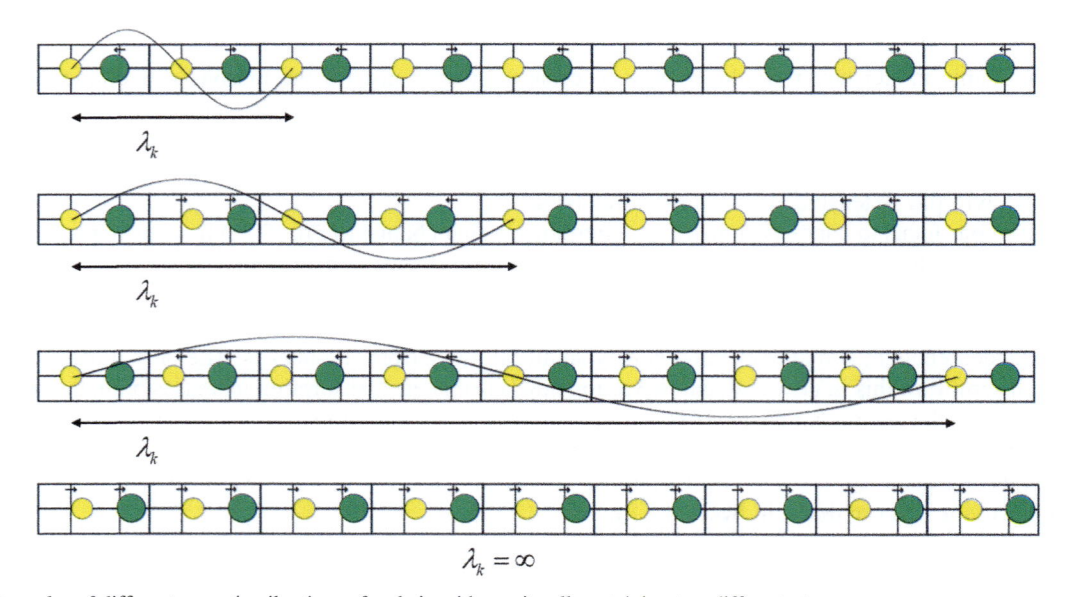

FIG. 5.7 Examples of different acoustic vibrations of a chain with a unit cell containing two different atoms.

position. In the fourth unit cell, both atoms moved to the left and have their maximum displacement, while in the fifth unit cell, both atoms are again in equilibrium positions (this is certainly just a particular snap-shot only to illustrate the principle). The larger the wavelength of the vibrations, the more unit cells are involved until the series repeats itself. In the extreme, the direction of movement and the velocity of the two atoms in the unit cell are the same and in every unit cell, the movements are in phase. This would mean that the wavelength of the vibration is infinite. This, however, is nothing else but a translation of the chain as a whole. Overall, this special kind of vibration is infrared inactive (as are the others of this type), as the dipole moment does not change (only the wavelength).

The situation is fundamentally different for the second kind of vibrations. For these vibrations, the two atoms within a unit cell generally move in opposite directions. For the shortest possible wavelength, now the atoms with the higher mass stay fixed and the lighter atoms move antiparallel in adjacent unit cells but with the same velocity. Again, for infinite wavelength, every unit cell shows the same movement of the two atoms, i.e., every unit cell (and every atom of the same kind) is in phase (cf. Fig. 5.8). Since for such vibrations, the dipole moment continuously changes, and they constitute the so-called optical branch.

I do not repeat the calculation here, which is based in principle on an ansatz analogous to that of Eq. (5.5), for reasons you will understand in a minute. Instead, we just present the result in Fig. 5.9 (the calculation can be found, e.g., in detail in the book "The infrared spectra of minerals" in Chapter 3 [4]).

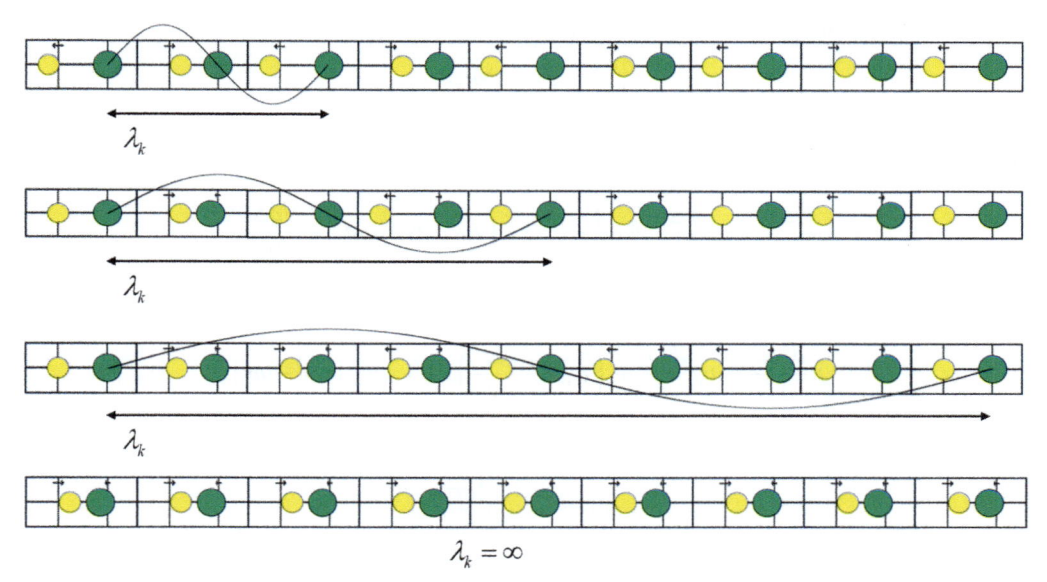

FIG. 5.8 Different optical vibrations of a chain with a unit cell containing two different atoms.

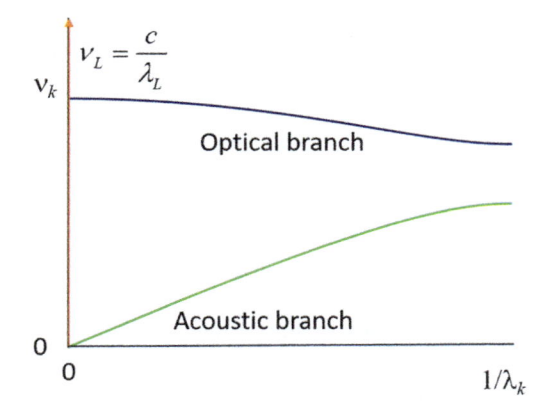

FIG. 5.9 Dispersion of the lattice. In comparison, the dispersion of the photon proceeds parallel to the y-axis *(orange line)*.

Fig. 5.9 shows the dispersion of the vibrations of the chain. On the abscissa, we find the inverse of the wavelength, which means that for zero wavenumber the wavelength is infinite, i.e., the movement of the atoms in every unit cell is the same. In contrast, at the end of the two branches, the acoustic and the optical branch, the atoms of the same kind in neighbored unit cells perform antiparallel movements (to be more precise, only the heavier atoms move for the end of the acoustic branch and only the lighter move for the optical branch as already mentioned).

However, the most important point is, that due to the need for the conservation of momentum, the frequency/wavenumber of the photon must be the same as that of the vibration of the chain. Since the proportionality constant for the relation between frequency and the inverse of the wavelength is the speed of light, the dispersion curve for the phonon, which is actually a line, is practically parallel to the ordinate. Accordingly, the intersection between the light line and the optical branch (the intersection with the acoustic branch is not relevant; since the dipole moment does not change for the acoustic branch at infinite wavelength, the corresponding movement cannot be excited by light) is practically identical to the vibration with infinite wavelength. At this wavelength, the frequency is finite and we find it to be similar to the frequency for the isolated molecule we discussed above.

There is, however, one complication that we have not taken into account so far. When Drude derived the dispersion relation for solids, he used relations that were identical to those that we have employed for the free and isolated molecule, i.e., Eq. (5.18). As was soon criticized (see, e.g., [5]), he assumed that the electric field at the location of the vibrating entity is the same as the one that has been applied externally. In fact, for condensed matter and dense gases, this is not the case as had been shown independently by Lorenz [6] and Lorentz [7] (Lorentz-Lorenz relation). For an isotropic material, the extra field at the location of the vibrating entity may be of the order of,

$$E_L = \frac{P}{3\varepsilon_0} = \frac{Nq\Delta r}{3\varepsilon_0}, \tag{5.22}$$

if the vibrating entity is in an environment of cubic symmetry. Accordingly, the equation of motion has to be modified,

$$\mu\frac{d^2\Delta r}{dt^2} + \mu\gamma\frac{d\Delta r}{dt} + \omega_0^2\mu\Delta r = q\left(E_0 \exp(-i\omega t) + \frac{Nq\Delta r}{3\varepsilon_0}\right), \tag{5.23}$$

which can be reformulated as:

$$\mu\frac{d^2\Delta r}{dt^2} + \mu\gamma\frac{d\Delta r}{dt} + \left(\omega_0^2\mu - \frac{Nq^2}{3\varepsilon_0}\right)\Delta r = qE_0 \exp(-i\omega t). \tag{5.24}$$

If we now set,

$$\omega_0'^2 = \omega_0^2 - \frac{Nq^2}{3\varepsilon_0\mu}, \tag{5.25}$$

we formally regain Eq. (5.18), so that also the dispersion relation, Eq. (5.21), keeps the same form. We just have to keep in mind that the resonance frequency is obviously reduced under the effect of the local field. It is possible to reckon the reduction by remembering that $S^2 = q^2 N/(\mu\varepsilon_0)$,

$$\tilde{\nu}_0'^2 = \tilde{\nu}_0^2 - \frac{1}{3}S^2, \tag{5.26}$$

which means that for stronger oscillators at lower resonance wavenumbers, the effect could be tremendous, whereas for medium strength oscillators, like C=O, it would be about $4\,\text{cm}^{-1}$, which explains why it is usually not considered for organic materials. Nevertheless, even if the shift could be easily $50\,\text{cm}^{-1}$ for inorganic materials and more, it is also usually disregarded in dispersion analysis of inorganic materials. At least, any mentioning of the local field effect in the dispersion equation has been missing right from the start [8,9]. It seems that the only one who had been considering and treating it in the context of infrared spectroscopy had been Hadni [10]. In fact, the problem is more serious for more than one type of absorption, because then those couple, the more the stronger the absorptions are. We come back to this problem later in this section.

Overall, our original problem has the same solution in three dimensions and for more than two atoms in the unit cell as long as the crystal is of cubic symmetry. This means to understand what is vibrating how in cubic crystals we just have to take a look at one unit cell. The number of vibrations is then $3N$, where N is in this case the number of atoms in the unit cell. Certainly, not all will be infrared active and, in cubic crystals, every infrared active vibration is threefold degenerated (for more details concerning symmetry analysis, we refer, e.g., again to ref. [4]). There is just a slight adjustment that we have to make in Eq. (5.21) in order to allow for more than one vibration:

$$\varepsilon = \varepsilon_\infty + \sum_{j=1}^{N} \frac{S_j^2}{\tilde{v}_{0,j}^2 - \tilde{v}^2 - i\tilde{v}\gamma_j}. \tag{5.27}$$

In Eq. (5.27), N is, deviating from the definition above, the number of infrared active vibrations, including combinations and higher harmonics.

You might ask yourself how Eq. (5.27) should be named. Indeed, this is a question that cannot be answered unambiguously. Quite often one finds the name Lorentz assigned to it, but also Drude is mentioned frequently, so it could be named the Drude-Lorentz model (even if this name in this context is usually used to describe the dispersion of metals). However, dispersion mattered already before Maxwell's equations, and Eq. (5.27) was earlier derived just assuming mere mechanical vibrations, probably by Helmholtz who seems to have been the first to introduce damping [11]. In this context, also Ketteler should be mentioned, who also earned merits, e.g., by extending the dispersion formula for more than one excitation [12]. At this point, in particular, the interpretation of what is actually vibrating and how damping comes into play was very important and many names appear, e.g. Planck [13], Koláček [14], Goldhammer [15], etc. For this reason, we refer in the following to this oscillator model by the name classical damped harmonic oscillator (CDHO) model.

There might remain some skepticism, in particular, due to the derivation of Eq. (5.27), that this equation is indeed able to properly reflect and represent the infrared spectral properties of real materials. For your reassurance, I will show you a comparison of experimental and theoretical spectra based on Eq. (5.27) in Fig. 5.10.

From the very good agreement, you could think that there is no more room for improvement. However, if you look closer at the band with the highest wavenumber, then there is some deviation, in particular, close to the band maximum or at the adjacent band minima. For this band, S is around $500\,\text{cm}^{-1}$. For bands with even higher oscillator strength, the deviations begin to increase. Typical examples are the principal spectra of $NdGaO_3$ of which we use one in some of the following sections to exemplify these deviations. For $NdGaO_3$, the deviations are particularly strong for the very intense band located between about 300 and $500\,\text{cm}^{-1}$. While the small minima and maxima on top of the broad band are most probably caused by surface roughness and corresponding scattering effects [16], we focus here, as in the previous example, on the deviations at the right wing of the band and the deviations in the adjacent minima. A potential explanation for these deviations is that for such strong bands, the coupling of vibrations plays a strong role. Another one makes deviations from Hooke's law responsible. We will come back to these explanations and related improvements of the oscillator model in later sections of this chapter.

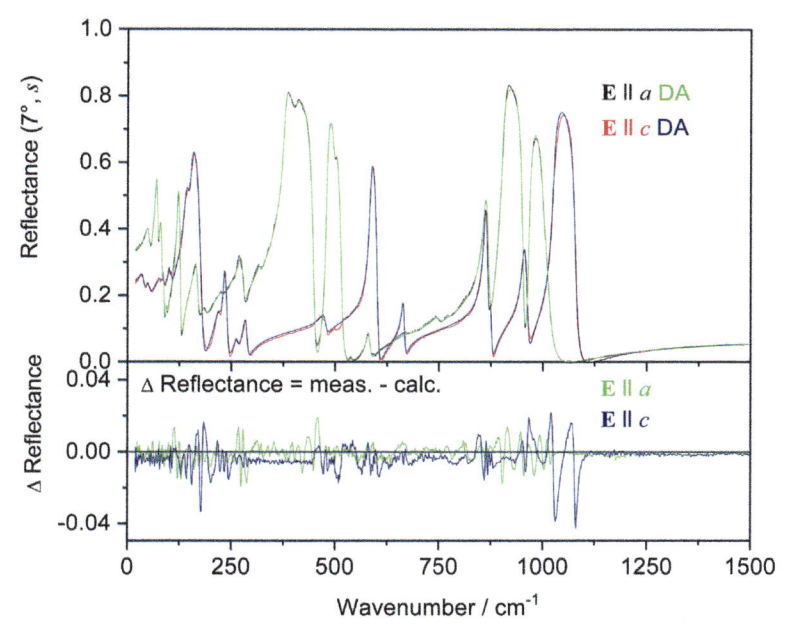

FIG. 5.10 Comparison between experimental reflectance spectra *(red and black lines)* and simulated spectra *(green and blue lines)* based on Eq. (5.27). The experimental spectra are those of single-crystal Fresnoite taken with polarized light and polarization direction along the crystallographic *a*- and *c*-axis.

A part of me wishes this section would be at its end at this point, and actually originally it was. Above I told you, that the dispersion relation does not change if we go over from individual molecules, i.e., gases, to condensed matter. Unfortunately, this is both true and incorrect at the same time. If you want to understand this statement, there are only two books that provide an explanation for it, namely [2,3], (there is actually a third book where the corresponding effect is mentioned, but the author more or less mentions it and then turns away from this subject telling his readers that it is not important [17]. I can fully understand why because I also wished the following would not be of importance). In fact, when we talked about the Lorentz-Lorenz relation above, I just used the usual argument that can be found in many textbooks that the local field correction just means that the oscillator position that we get from dispersion analysis is redshifted compared to that without local field. But this is only true for materials that have just one oscillator (there certainly may be materials with only one active IR-vibration, but here I am talking about all possible excitations including those in the UV and VIS spectral regions). To understand what is going on if there is more than one oscillator, we have to attack the problem from another direction and go back to Eq. (1.9):

$$\varepsilon = 1 + \frac{N \cdot \mathbf{p}}{\varepsilon_0 \cdot \mathbf{E}}. \tag{5.28}$$

The electric field $\mathbf{E}$ in this equation is the applied electric field. The field that is effective at the location of a molecule or in the unit cell is different. According to Eq. (5.22), we have to add to the applied electric field $\mathbf{E}_L$ to obtain the local field. Under the assumption of cubic crystal symmetry, we obtain:

$$\mathbf{E}_{loc} = \mathbf{E} + \mathbf{E}_L = \mathbf{E} + \frac{1}{3\varepsilon_0} \mathbf{P}. \tag{5.29}$$

Therefore, the actual macroscopic polarization $\mathbf{P} = N \cdot \mathbf{p} = N \cdot \alpha \cdot \mathbf{E}_{loc}$ is given by:

$$\mathbf{P} = N \cdot \alpha \cdot \left(\mathbf{E} + \frac{1}{3\varepsilon_0} \mathbf{P} \right) = \frac{N \cdot \alpha}{1 - \frac{N \cdot \alpha}{3\varepsilon_0}} \cdot \mathbf{E}. \tag{5.30}$$

Consequently, for the dielectric function, the following result is obtained:

$$\varepsilon = 1 + \frac{N \cdot \alpha}{\varepsilon_0 - \frac{1}{3} N \cdot \alpha} = \frac{1 + \frac{2N \cdot \alpha}{3 \varepsilon_0}}{1 - \frac{1}{3} \frac{N \cdot \alpha}{\varepsilon_0}}. \tag{5.31}$$

If we express α for one oscillator, we get

$$\varepsilon - 1 = 3 \left(\frac{1}{1 - \frac{S^2}{\tilde{v}_0^2 - \tilde{v}^2 - i\tilde{v}\gamma}} - 1 \right) = \frac{3 \frac{S^2}{\tilde{v}_0^2 - \tilde{v}^2 - i\tilde{v}\gamma}}{1 - \frac{S^2}{\tilde{v}_0^2 - \tilde{v}^2 - i\tilde{v}\gamma}} \rightarrow \varepsilon_r - 1 = 3 \frac{S^2}{(\tilde{v}_0^2 - S^2) - \tilde{v}^2 - i\tilde{v}\gamma}, \tag{5.32}$$

which can be brought into the following form [18].

$$\varepsilon = 1 + \frac{S^2}{\left(\tilde{v}_0^2 - \frac{1}{3} S^2 \right) - \tilde{v}^2 - i\tilde{v}\gamma} \rightarrow \varepsilon = 1 + \frac{S^2}{\tilde{v}_0'^2 - \tilde{v}^2 - i\tilde{v}\gamma}, \tag{5.33}$$

and, as I have shown above, this just means a shift of the oscillator position with increasing oscillator strength. But for more than one oscillator, this is no longer true! If we consider the polarizability as,

$$\alpha(\tilde{v}) = \sum_i \frac{q_i^2/\mu_i}{\tilde{v}_{0,i}^2 - \tilde{v}^2 - i\tilde{v}\gamma_i}, \tag{5.34}$$

it was shown by placing this into Eq. (5.31) that the dielectric function can still be expressed by [18].

$$\varepsilon = 1 + \sum_i \frac{S_i'^2}{\tilde{v}_{0,i}'^2 - \tilde{v}^2 - i\tilde{v}\gamma_i}, \tag{5.35}$$

but now not only the oscillator positions but also the oscillator strengths are different. This seems to be still good news, as the form of the dispersion relation remains unchanged. But there is also a downside, which is that from determining $S_i'^2$ you do not get immediate information about charges and masses, since $S_i'^2 \neq Nq_i^2/(\varepsilon_0\mu_i)$. Even worse for many applications is that the oscillator strength is no longer proportional to the number of oscillators per unit volume N, but we will come to investigate the consequences in more detail later. Another downside is that we cannot simply test if local field effects play a role by using Eq. (5.31) and comparing the results with those from employing Eq. (5.35), since both will work equally well. Well, then we can simply use Eq. (5.31) instead of Eq. (5.35), right? Actually, it should be possible to express Eq. (5.34) as,

$$\alpha(\tilde{v}) = \frac{\varepsilon_0}{N}(\varepsilon_\infty - 1) + \sum_i \frac{q_i^2/\mu_i}{\tilde{v}_{0,i}^2 - \tilde{v}^2 - i\tilde{v}\gamma_i}. \tag{5.36}$$

At least for the case of a medium with cubic symmetry, this should solve our problem (but unfortunately not in general for scalar media!).

On the other hand, it is also possible to model the dielectric function in the IR with small modifications of the classical model. If we assume that

$$\varepsilon = \varepsilon_\infty + \frac{N\cdot\alpha}{\varepsilon_0 - \frac{1}{3}N\cdot\alpha}, \tag{5.37}$$

we can reformulate this as:

$$\varepsilon = \frac{\varepsilon_\infty\left(1 - \frac{1}{3}\frac{N}{\varepsilon_0}\cdot\alpha\right) + \frac{N}{\varepsilon_0}\cdot\alpha}{1 - \frac{1}{3}\frac{N}{\varepsilon_0}\cdot\alpha} = \frac{\varepsilon_\infty + \left(1 - \frac{1}{3}\varepsilon_\infty\right)\frac{N}{\varepsilon_0}\cdot\alpha}{1 - \frac{1}{3}\frac{N}{\varepsilon_0}\cdot\alpha}. \tag{5.38}$$

Therefore, we finally get:

$$\varepsilon = \frac{\varepsilon_\infty + \left(1 - \frac{1}{3}\varepsilon_\infty\right)\sum_{j=1}^{N}\frac{S_j^2}{\tilde{v}_{0,j}^2 - \tilde{v}^2 - i\tilde{v}\gamma_j}}{1 - \frac{1}{3}\sum_{j=1}^{N}\frac{S_j^2}{\tilde{v}_{0,j}^2 - \tilde{v}^2 - i\tilde{v}\gamma_j}}. \tag{5.39}$$

This equation allows to obtain the oscillator parameters for the situation when the local field effect is not present. In particular, S_j^2 should then still be proportional to the concentration of the transition moments. On the other hand, it certainly only works when the Lorentz-Lorenz model is strictly valid and this is, unfortunately, not generally the case which you will also see in Chapter 10. This is also probably the reason that nowadays the local field is often omitted, because historically, when it was realized that the Lorentz-Lorenz model is not commonly applicable, the local field was often indicated with an asterisk instead of explicit corrections. After some time, the asterisk was omitted and the local field was no longer mentioned. In the realm of organic and biological samples, this might be justified due to the low oscillator strengths. But, as we will see later on in the course of this book, even then it may play a role, e.g., for 2D-correlation analysis (Chapter 8) or for principal component analysis (Chapter 9).

At this point, you may also wonder why I did not discuss the relation to quantum mechanics—do we need to know anything that we do not already know? You have seen for yourself in Fig. 5.10 how well you can fit an experimental spectrum with the classical dispersion formula. The quantities that would need a different interpretation when you use the classical dispersion formula in the infrared spectral region are the oscillator position and the oscillator strength. As you already know from the conventional textbooks about infrared spectroscopy, the position is equivalent to an energy difference between the ground and the first excited vibrational state. For vibrations at higher wavenumbers, it is to a good approximation only the ground state which is occupied. This is not necessarily the case for vibrations in the FIR. For such vibrations, the oscillator strength is proportional to the difference in the number of molecules in the ground and in the excited state(s) (the BBL approximation would need to be modified accordingly) [19]. This also brings us to the second point that needs to be discussed. As we assume harmonic vibrations, the energy difference between two neighboring vibrational states is always the same.

In fact, in the spectra we often also see higher harmonics, i.e., starting from the ground state, the excited state is not the next one, but (in most cases for higher harmonics) the second excited state. Those oscillators have often much lower oscillator strength, but from the point of view of dispersion theory, one cannot distinguish a fundamental from a higher harmonic. As a last comment, we usually assume that a transition starts from the ground state. The reason is that, at least for the mid-infrared spectral region, even the first excited state is not very populated according to the Boltzmann distribution, also because the vibrational states are non-degenerated (e.g., for room temperature only about one of 10,000 molecules is in the first excited state). This ratio is higher in the far infrared spectral region, but even then, this usually needs not to be considered. If you want to do that nevertheless, Kramers and Heisenberg developed a form of the dispersion theory which takes into account the Boltzmann distribution [20], but I will not discuss this form in this book. Overall, it must be said that in the form of quantum mechanics developed by Heisenberg, dispersion theory played an important role, but this is a different story. Maybe, as a final note, Ladenburg should also be mentioned in this context, because he helped to understand why dispersion theory and the parameters used only had to be slightly reinterpreted after the advent of quantum theory [21].

5.2 Excursus: Lorentz profile vs. Lorentz oscillator

In the last section, I said concerning absorbance band fitting "You certainly can perform such fits, but you must be aware of the errors introduced right from the start." I have to admit that I also was not really aware of the extent of these errors when I originally wrote this part, but I took writing this book as a good reason to inspect them in detail [22]. It is actually hard to believe what I discovered, so this section might be a little bit strange for a textbook, but I want you to share my thoughts on this topic.

Band fitting/deconvolution is something that is not specific to infrared spectroscopy. It is performed also for bands in the UV/Vis and even in the X-ray spectral range using the same kind of mathematical functions, namely, Lorentz, Gauß, and Voigt profiles, where the latter is a convolution of the Lorentz and the Gauß profile (sometimes also sums and products of Lorentz and Gauß profiles are used, which might be convenient to fit spectra, but lack any physical background or justification). By the way, calling them mathematical functions, which is often done in literature, completely denies their physical origin and indeed, literature often states there is none, in particular, the typical contemporary textbook of infrared spectroscopy ("note that the Lorentz profile should actually be called Cauchy function...").

In the last section, we derived the model of the damped harmonic oscillator and the corresponding dispersion relations. As stated, Hendrik Antoon Lorentz was one of the scientists whose merits must be acknowledged in this regard and some do this by terming a damped harmonic oscillator Lorentz oscillator. May it be that there is actually some connection between a Lorentz profile and a Lorentz oscillator? If you read through the first paper about band fitting in the infrared (at least to my knowledge this is ref. [23]), then you learn nothing about this connection and also in later reviews about band shapes and band fitting, a connection is not established (see, e.g., [24,25]; sometimes you at least can read that the naming of the Lorentz profile is in honor of his merits in deriving the dispersion relations without further explanation). This is not very glorious for the field, and it does not become better, if you can read in certain textbooks that using the damped harmonic oscillator model is certainly much more advanced than using band fitting with Lorentz profiles [26], without somebody having ever investigated this in detail. Even better, the following derivation shows that for the comparably weak oscillators present in organic and biological matter the Lorentz oscillator is more useful than the Lorentz oscillator. Do you think in the last sentence I confused something? Actually, sometimes it makes sense to go back to the original literature. It may be no surprise, but Lorentz himself established the connection between the classical damped harmonic (Lorentz) oscillator and the Lorentz profile [27]. He did this to my best knowledge only for one oscillator system, which is why in the following I show you the extension for more than one oscillator that came to my mind when reading [27].

Let me remind you first of the formula we derived for the classical damped harmonic oscillator model in the preceding section:

$$\varepsilon = \varepsilon_\infty + \sum_{j=1}^{N} \frac{S_j^2}{\tilde{v}_{0,j}^2 - \tilde{v}^2 - i\tilde{v}\gamma_j}. \tag{5.40}$$

For cubic crystals and along the principal axes for optically uniaxial, orthorhombic, and monoclinic crystals, $\hat{n} = \sqrt{\varepsilon}$, therefore

$$\hat{n} = \sqrt{\varepsilon_\infty + \underbrace{\sum_{j=1}^{N} \frac{S_j^2}{\tilde{v}_{0,j}^2 - \tilde{v}^2 - i\tilde{v}\gamma_j}}_{x}}. \tag{5.41}$$

In the next step, we extract $\sqrt{\varepsilon_\infty}$,

$$\hat{n} = \sqrt{\varepsilon_\infty}\sqrt{1 + x/\varepsilon_\infty}, \tag{5.42}$$

and realize that $\sqrt{\varepsilon_\infty} = n_\infty$:

$$\widehat{n} = n_\infty \sqrt{1 + \underbrace{x/\varepsilon_\infty}_{y}}. \tag{5.43}$$

Now we use the series expansion,

$$\widehat{n} = (1 + y)^{\frac{1}{2}} = 1 + y/2 - y^2/8 + \ldots, \tag{5.44}$$

and for small x and y (an assumption which we have yet to justify), we obtain:

$$\widehat{n} = n_\infty (1 + y/2). \tag{5.45}$$

This is the step, which, in the end, renders the contribution of the individual oscillators to the absorbance additive, something which is now a basic assumption in parts of the community, that is rarely questioned.

Next, we find that:

$$\widehat{n} = n_\infty + \frac{x}{2n_\infty}. \tag{5.46}$$

And, if we resubstitute x:

$$\widehat{n} = n_\infty + \sum_{j=1}^{N} \frac{S_j^2}{2n_\infty} \frac{1}{\tilde{v}_{0,j}^2 - \tilde{v}^2 - i\tilde{v}\gamma_j}. \tag{5.47}$$

Now we separate real and imaginary part and obtain:

$$\widehat{n}(\tilde{v}) = n_\infty + \underbrace{\sum_{j=1}^{N} \frac{S_j^2\left(\tilde{v}_j^2 - \tilde{v}^2\right)/(2n_\infty)}{\left(\tilde{v}_{0,j}^2 - \tilde{v}^2\right)^2 + \tilde{v}^2\gamma_j^2}}_{n(\tilde{v})} + i\underbrace{\sum_{j=1}^{N} \frac{S_j^2 \tilde{v}\gamma_j/(2n_\infty)}{\left(\tilde{v}_{0,j}^2 - \tilde{v}^2\right)^2 + \tilde{v}^2\gamma_j^2}}_{k(\tilde{v})}. \tag{5.48}$$

If you did not go through Section 2.6, you might not be familiar with the benefit of the above simplification. Therefore, I reproduce Fig. 2.10 in the following:

As you can see in Fig. 5.11, the above simplification preserves the properties of the classical damped harmonic oscillator also for the complex refractive index function. Accordingly, the imaginary part is centered at the oscillator position and symmetric around it if only one band is present. At half height, it has the width of the damping constant γ. Between the corresponding wavenumbers, the real part shows anomalous dispersion. Correspondingly, its two extrema are situated at these wavenumbers. In other words, the basic assumption is that the refractive index function is a disguised dielectric function, an approximation that is only valid, be reminded, in the limit of vanishing oscillator strength.

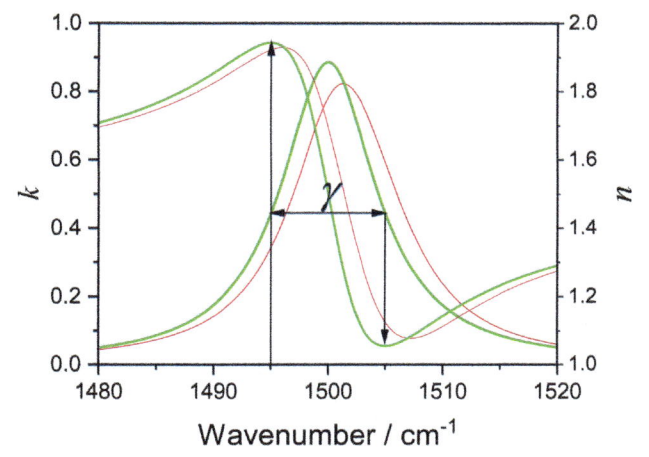

FIG. 5.11 Comparison between refractive index and absorption index functions for an oscillator with a strength of $200^2\,\text{cm}^{-2}$ based on the Lorentz oscillator *(red lines)* and on the Lorentz profile model *(green lines)*.

In the next step, we use that around bands $\tilde{v}_{0,j} \approx \tilde{v}$ and obtain the index of absorption function (for the real part this is not a good and an also unnecessary approximation):

$$k(\tilde{v}) = \sum_{j=1}^{N} \frac{S_j^2}{2n_\infty \gamma_j} \frac{(\gamma_j/2)^2}{\tilde{v}_{0,j}\left[(\tilde{v}_{0,j} - \tilde{v})^2 + (\gamma_j/2)^2\right]}. \tag{5.49}$$

Eventually, we multiply Eq. (5.49) by $4\pi d\tilde{v}/\ln 10$ from both sides to get absorbance:

$$A(\tilde{v}) = \frac{2\pi d}{n_\infty \ln 10} \sum_{j=1}^{N} \frac{S_j^2}{\gamma_j} \frac{\tilde{v}(\gamma_j/2)^2}{\tilde{v}_{0,j}\left[(\tilde{v}_{0,j} - \tilde{v})^2 + (\gamma_j/2)^2\right]}. \tag{5.50}$$

Finally, we again employ that $\tilde{v}_{0,j} \approx \tilde{v}$:

$$A(\tilde{v}) = \frac{2\pi d}{n_\infty \ln 10} \sum_{j=1}^{N} \frac{S_j^2}{\gamma_j} \frac{(\gamma_j/2)^2}{(\tilde{v}_{0,j} - \tilde{v})^2 + (\gamma_j/2)^2}. \tag{5.51}$$

Therefore, if the approximations hold, then we have a very good physical reason to use Lorentz profiles. Before we do that, I want to spend a couple of words on the parameters. From Eq. (5.51), we realize that the peak position is supposed to be the oscillator position. With what we know from the preceding section, this cannot be correct. Indeed, it is not, but for weak oscillators this deviation is small. Let's have a look at a corresponding comparison in Fig. 5.12.

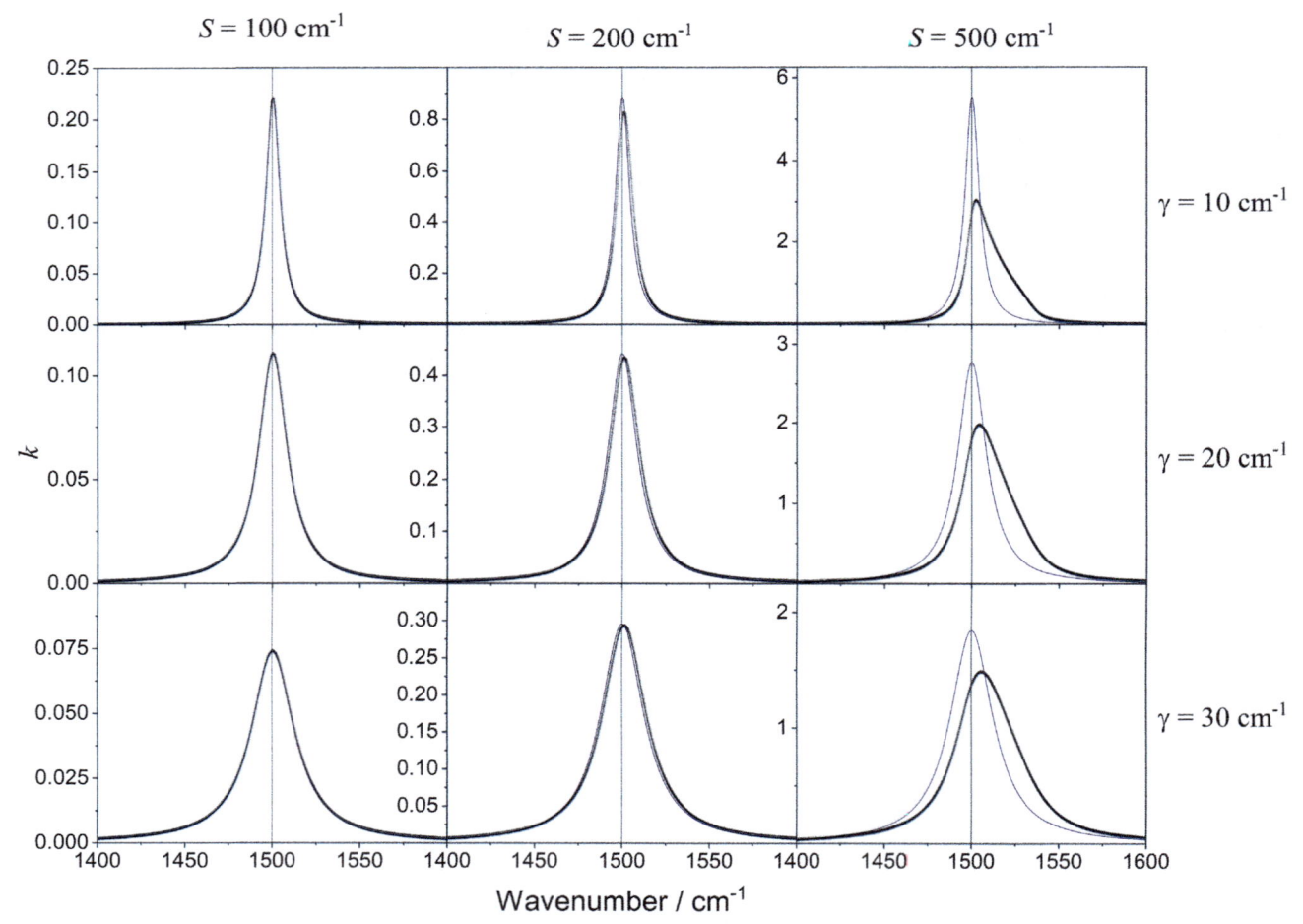

FIG. 5.12 Comparison between absorption index functions for several oscillator strengths and damping constants based on the CDHO model (*thick black lines*) and on the Lorentz profile model (*thin blue lines*) according to Eq. (5.49). (Credit: Reproduced with permission from T.G. Mayerhöfer, J. Popp, Quantitative evaluation of infrared absorbance spectra – Lorentz profile versus Lorentz oscillator, ChemPhysChem, 20 (2019) 31–36.)

The Lorentz profile generally is symmetric around the oscillator position, whereas the CDHO leads to blue shifts and increasingly asymmetric shapes with increasing oscillator strengths. Up to about $S = 200 \text{cm}^{-1}$, which includes most organic and biological oscillators; however, the effects stay small. On the other hand, it is possible, even in the range of beginning deviations between the shape of the CDHO bands and the Lorentz profile, to model the former with the latter. Therefore, even when the oscillator parameters might deviate somewhat, the description of the index of absorption function is still accurate, and so is the index of refraction function, if the oscillator parameters are used in Eq. (5.48). This really advances band fitting or band deconvolution from a simple tool towards a qualified instrument to quantitatively analyze infrared absorbance spectra in a way nearly adequate to dispersion analysis. This is why we attributed a new name to it, namely Poor Man's Dispersion Analysis, in analogy to the Poor Man's Kramers-Kronig Analysis, which we will introduce in the last part of this chapter. An advantage of Poor Man's Dispersion Analysis is that the reparameterization can easily be introduced in existing spectrometer software. With an additional line of code, it is then even possible to gain the full set of optical constants in the fitted spectral range.

In retrospective, I must admit that even 4 years after [28] was published, still no spectrometer software existed that is based on Eq. (5.48). What I also did not take into account is that Eq. (5.48) might be used and simplified further (or that Eq. (5.51) might be applied for the dispersion of the complex refractive index). Note that Lorentz used simplifications, because Lorentz thought that the original dispersion relation derived for the damped harmonic oscillator model is too complex to discuss and understand (see p. 154 of his book "The Theory of Electrons and its applications to the phenomena of Light and radiant Heat" [29]). For the calculation of the dielectric function or the complex index of refraction function, however, there is nowadays no benefit to work with simplifications—the calculations are in no way faster if you use simplified versions. In contrast, further simplifications can be very detrimental under certain circumstances, which is why we are discussing an example in the following.

We start from,

$$\widehat{n} = n_\infty + \sum_{j=1}^{N} \frac{S_j^2}{2n_\infty} \frac{1}{\tilde{v}_{0,j}^2 - \tilde{v}^2 - i\tilde{v}\gamma_j}, \tag{5.52}$$

and consider that $\tilde{v}_{0,j} \approx \tilde{v}$ and, thus, $\tilde{v}_{0,j}^2 - \tilde{v}^2 = (\tilde{v}_{0,j} + \tilde{v})(\tilde{v}_{0,j} - \tilde{v}) \approx 2\tilde{v}_{0,j}(\tilde{v}_{0,j} - \tilde{v})$ to arrive at:

$$\widehat{n}(\tilde{v}) \approx n_\infty + \sum_{j=1}^{N} \frac{S_j^2}{4\tilde{v}_{0,j}n_\infty} \frac{1}{\tilde{v}_{0,j} - \tilde{v} - i\gamma_j/2}. \tag{5.53}$$

The constants in $S_j^2/(4\tilde{v}_{0,j}n_\infty)$ can be combined into a new constant describing the amplitude, while the oscillator position and the damping constant keep their respective meanings.

How would Eq. (5.53) perform in comparison with the classical damped harmonic oscillator model? Absorption index functions for the same oscillator as employed to generate Fig. 5.12 can be found in Fig. 5.13. The differences are marginal—the values computed by Eq. (5.53) are somewhat higher, but the differences can be noticed only by closer inspection of the third column featuring the strong oscillator with $S = 500 \text{cm}^{-1}$. Much more interesting is the comparison of the calculated real parts of the complex refractive index functions displayed in Fig. 5.14. In their case, for the strong oscillator ($S = 500 \text{cm}^{-1}$) definitely a red line is crossed, at least for the typical damping constants $\gamma = 10 \text{cm}^{-1}$ and $\gamma = 20 \text{cm}^{-1}$, where the refractive index becomes negative according to Eq. (5.53) (there are materials with negative refractive indices, but only among the metamaterials and not among conventional dielectrics). Therefore, while the model according to Eq. (5.53) could in principle be used for organic and biological materials, it is not useful for inorganic materials with their large oscillator strengths, not only because the model leads to unphysical results like negative refractive indices but also because it cannot reproduce the bandshapes of inorganic materials (I tried it and attempts to employ Eq. (5.53) are definitely bound to failure). The Lorentz profiles have further disadvantages based on the fact that they are not Kramers-Kronig compatible—this means that if you apply the Kramers-Kronig relations to calculate the real from the imaginary part or vice versa, the result will deviate from what you obtain by the relations introduced in this section (the Kramers-Kronig relations is something that we will discuss in more details in Section 5.8). Another disadvantage that we will scrutinize later on is how individual Lorentz profiles interact and couple through their real part in comparison with damped harmonic oscillators with and without the local field of Lorentz (cf. Section 7.3.1).

5.3 Excursus: Dispersion relations and Beer's approximation

What has the dispersion relation to do with Beer's approximation (and here, I speak indeed not about the Bouguer-Lambert part, but just about Beer's part)? And if there is a connection, why has it not been found a hundred years ago? Most probably it has, but the corresponding paper (or papers) might have been forgotten, and/or might have been published in some

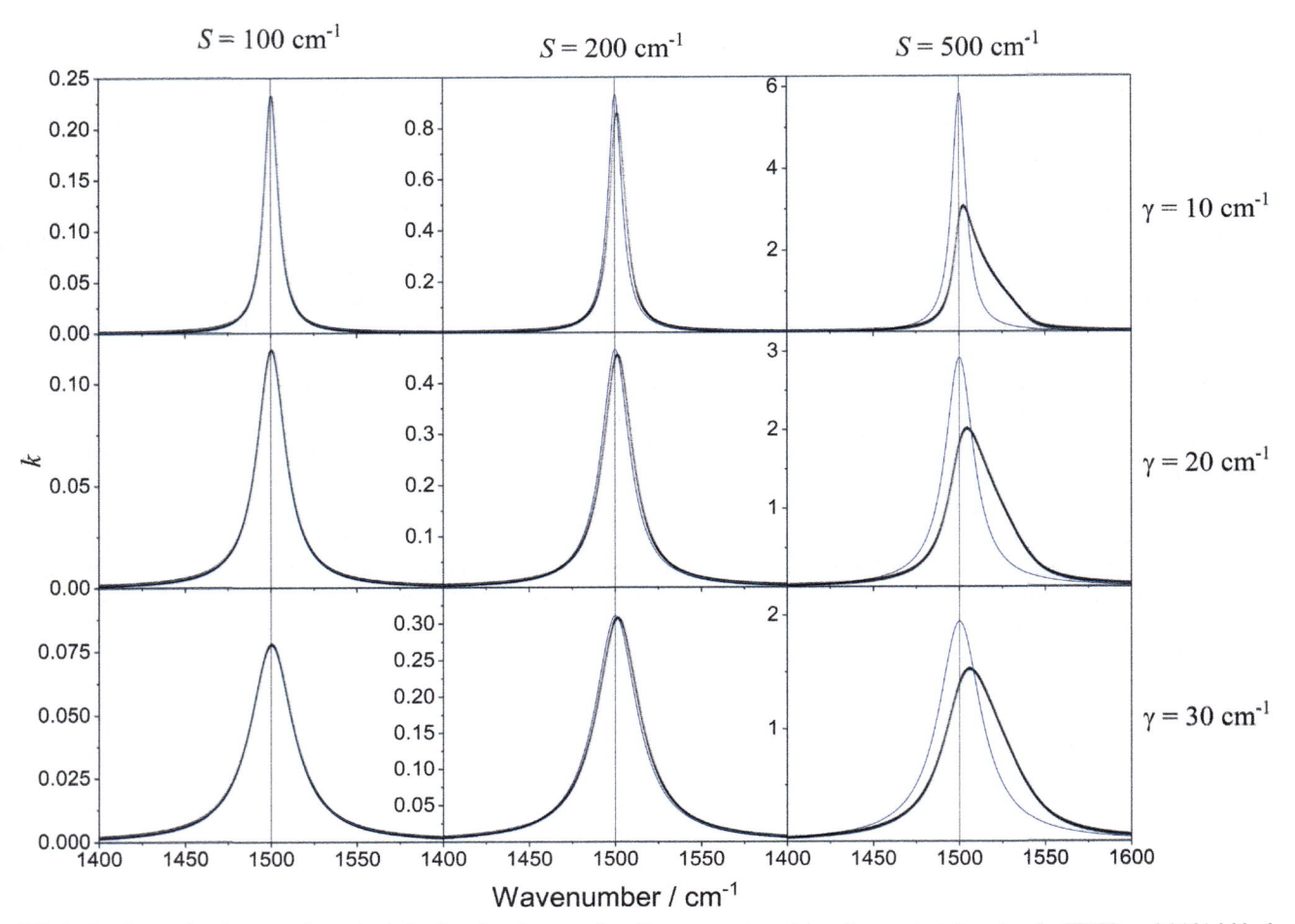

FIG. 5.13 Comparison between absorption index functions for several oscillator strengths and damping constants based on the CDHO model *(thick black lines)* and on the Lorentz profile model *(thin blue lines)* according to Eq. (5.53).

language, which is not common (on the other hand, I knew for 20 years that concentration is somehow a parameter in the dielectric function, before this bothered me enough to establish the connection). Generally, we have the problem that there seems to be a deep chasm between chemistry and physics, not only with respect to optical spectroscopy, which might also be a potential explanation why you cannot find the following in any related textbook.

Let us go back to Eq. (5.15):

$$\mathbf{P} = \underbrace{q \Delta \mathbf{r}}_{\mathbf{p}} N. \tag{5.54}$$

I told you in Section 5.1 that N is the number of oscillators per unit volume. As every chemist knows, the molar concentration c is defined as the number of molecules or ions in a unit volume divided by the Avogadro constant:

$$c = \frac{N}{N_A}. \tag{5.55}$$

If we use this fundamental relation and plug it into Eq. (5.54), we find that the macroscopic polarization is proportional to the molar concentration of the oscillators:

$$\mathbf{P} = q \cdot \Delta \mathbf{r} \cdot N_A \cdot c. \tag{5.56}$$

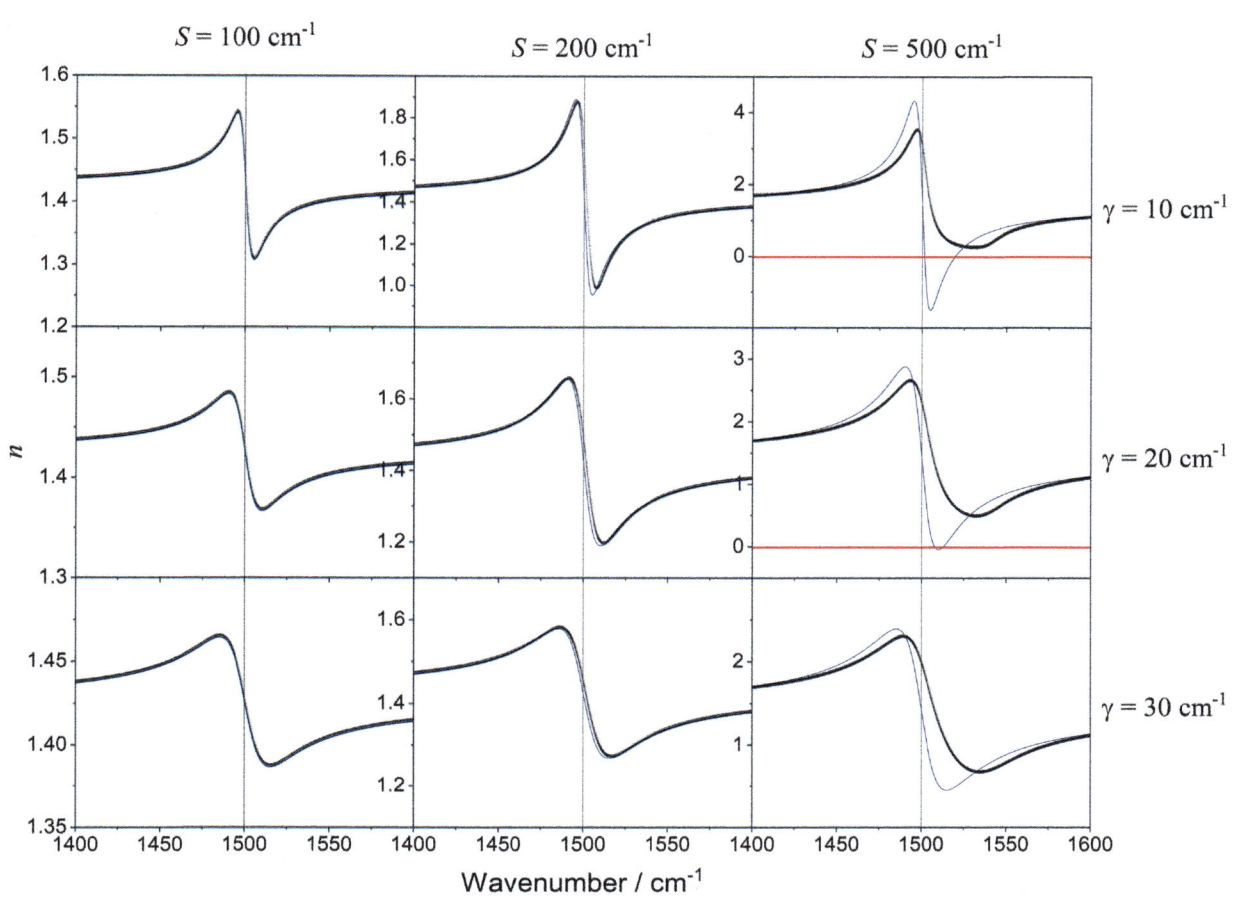

FIG. 5.14 Comparison between refractive index functions for several oscillator strengths and damping constants based on the CDHO model *(thick black lines)* and on the Lorentz profile model *(thin blue lines)* according to Eq. (5.53).

At this point, let me remind you, that we assume that there are no interactions between the individual dipole moments/oscillators, which becomes very important later on, to arrive at Eqs. (5.54), (5.56). For simplicity, we assume that we just have one oscillator, so that

$$\Delta \mathbf{r}(t) = \frac{q}{\mu(\tilde{v}_0^2 - \tilde{v}^2 - i\tilde{v}\gamma)} \mathbf{E}(t). \tag{5.57}$$

Accordingly, the relative dielectric function will be given by:

$$\varepsilon = 1 + \frac{N_A c}{\mu \varepsilon_0} \frac{q^2}{\tilde{v}_0^2 - \tilde{v}^2 - i\tilde{v}\gamma}. \tag{5.58}$$

At this point, we define a molar oscillator strength S^{*2},

$$S^{*2} = q^2 N_A / (\mu \varepsilon_0), \tag{5.59}$$

with which Eq. (5.58) transforms into:

$$\varepsilon = 1 + c \frac{S^{*2}}{\tilde{v}_0^2 - \tilde{v}^2 - i\tilde{v}\gamma}, \tag{5.60}$$

Let me discuss the result—what we found seems to be nothing special. Obviously, the change of both, the real and the imaginary part of the dielectric function is proportional to the concentration of the oscillator, which seems to be completely reasonable. Obviously, this is nothing else but Beer's approximation in somewhat different form; therefore, it should be a small step to derive it from here, would there not be a small problem. As we know, according to Maxwell's relations, the dielectric constant is equal to the square of the complex index of refraction (this can be derived from Eq. (3.31)):

$$\varepsilon = \widehat{n}^2. \tag{5.61}$$

As a consequence, we find the index of the refraction function:

$$\widehat{n}(\tilde{v}) = \sqrt{1 + c\frac{S^{*2}}{\tilde{v}_0{}^2 - \tilde{v}^2 - i\tilde{v}\gamma}}. \tag{5.62}$$

This still does not look like Beer's approximation. We just used Lorentz's approximations in the last section. Accordingly, we can employ Eq. (5.44) and assume that we can retain the linear term and neglect all that are of higher order:

$$\widehat{n}(\tilde{v}) \approx 1 + c\frac{S^{*2}/2}{\tilde{v}_0{}^2 - \tilde{v}^2 - i\tilde{v}\gamma}. \tag{5.63}$$

As we know already well, the index of absorption function is then given by:

$$k(\tilde{v}) \approx c\frac{S^{*2}\tilde{v}\gamma/2}{\left(\tilde{v}_0^2 - \tilde{v}^2\right)^2 + \tilde{v}^2\gamma^2}. \tag{5.64}$$

Since absorbance is given as,

$$A(\tilde{v}) = \frac{4\pi d\tilde{v}}{\ln 10}k(\tilde{v}), \tag{5.65}$$

we can combine this and the previous equation to yield:

$$A(\tilde{v}) = \frac{4\pi d\tilde{v}}{\ln 10} \cdot c \cdot \frac{S^{*2}\tilde{v}\gamma/2}{\left(\tilde{v}_0^2 - \tilde{v}^2\right)^2 + \tilde{v}^2\gamma^2}, \tag{5.66}$$

Beer's approximation states that absorbance is the product of molar absorption coefficient $\varepsilon^*(\tilde{v})$, concentration, and thickness d of a sample:

$$A(\tilde{v}) = \varepsilon^*(\tilde{v})cd, \tag{5.67}$$

If we finally compare Eq. (5.66) with Eq. (5.67), we find that

$$\varepsilon^*(\tilde{v}) = \frac{2\pi}{\ln 10}\frac{S^{*2}\tilde{v}^2\gamma}{\left(\tilde{v}_0^2 - \tilde{v}^2\right)^2 + \tilde{v}^2\gamma^2}. \tag{5.68}$$

As a résumé, we see that Beer's approximation is nothing else but a rewritten dispersion relation. Note that it is not even necessary to assume a certain line shape (i.e., that of a damped harmonic oscillator in this case), to arrive at this result (cf. Chapter 1). What remains to be scrutinized if Lorentz's approximation holds, and what happens if not. Before we do this, let us first think of the highest possible concentrations of oscillators: Obviously those are reached for neat substances. If we go back to Fig. 5.12, we see that we will not run into problems with organic and biological matter, because oscillators (with the exception of the O—H and, possibly, the C=O vibrations) are usually weak and the product of S^{2*} and c (which is the oscillator strength), will never exceed values that lead to strong deviations from linearity. On the other hand, for somewhat stronger oscillators, the higher order terms in Eq. (5.44) become important and can cause considerable spectral changes and deviations from linearity, if we look at plots of the absorbance against the concentration as illustrated in Fig. 5.15.

For a medium strong oscillator ($S^* = 70$ (L cm^{-1})/mol, $\varepsilon^* \approx 10^3$ L/(mol cm^{-1}), neat at $c = 50$ mol/L), and comparably low concentrations, it is hard to see any changes (the black curve in the left panel of Fig. 5.15 is completely hidden under the red curve). At 5 mol/L, the deviations begin to set in, and at higher concentrations the changes are everything else but subtle.

Note that I am definitely not the first using dispersion theory to connect to Beer's approximation. Max Planck had, as I already stated, derived its own flavor of dispersion theory in 1902 [13]. One year later, he published a paper in which he investigated the consequences of his theory for bands of the absorption index semi-quantitatively. In fact, he could already derive in this way that Beer's approximation will only hold good for weak oscillators. While he did not perform the final step to derive Beer's approximation in a strict way from electromagnetic theory, I cannot believe that his conclusions were actually totally ignored in the literature. I could not even find the smallest reference in any book about spectroscopy, not

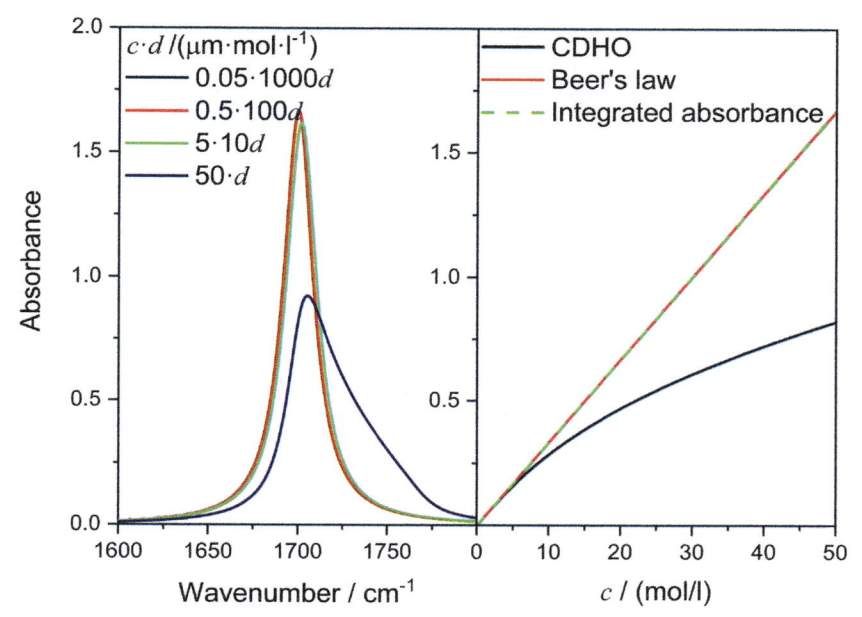

FIG. 5.15 *Left panel*: Spectrum of a medium strong oscillator for different concentrations and thicknesses. For clarity, the product of concentration and thickness is constant for all four spectra. *Left panel*: Concentration dependence of the absorbance at the peak maximum of the band and for the integrated absorbance. (Credit: Reproduced with permission and in an altered form from T.G. Mayerhöfer, J. Popp, Beer's law – why absorbance depends (almost) linearly on concentration, ChemPhysChem, 20 (2019) 511–515.)

even in the book of Kortüm [30], which is otherwise way ahead of all modern books dealing with spectroscopy (which are, accordingly, perfect examples of demodernization in some parts of science). Again, an example that makes me wonder how many treasures remain hidden in the old literature waiting to be discovered or reinvented, but a very drastic one, which shows that not even getting the Nobel prize saves all your findings from being ignored.

What Planck, however, did not know is that there is a possibility to solve the problem. So far, we have shown that Beer's approximation is no longer correct pointwise, irrespective if we pick the oscillator position or any other point of a spectrum, for high concentrations. Later, in Section 5.8, we will derive so-called Kramers-Kronig sum rules. These sum rules link, e.g., the area under the absorption index, to be more precise, the absorption index times the wavenumber, to the oscillator strength [31]:

$$\int_0^\infty \tilde{v} k(\tilde{v}) d\tilde{v} = \frac{\pi}{4} S^2 = \frac{\pi}{4} c S^{*2}. \tag{5.69}$$

The sum rules were not known around 1920—they had been developed after dispersion theory had been transformed into quantum mechanics, and it seems that this was already too late to influence the simplified theory around absorbance. However, Rudolf Ladenburg derived Eq. (5.69) based on Kirchhoff's law already in 1914 [32] and Eq. (5.69) is known as line strength in the field of gas spectroscopy. It is strange that Eq. (5.69) is nowadays mostly unknown in the field of IR spectroscopy.

What has Eq. (5.69) to do with absorbance? The integral contains the product of wavenumber and absorption index, exactly like absorbance, so that if integrated absorbance is divided by $4\pi d/\ln 10$ the oscillator strength results which is linearly depending on the concentration if local field effects can be neglected. Still, the usefulness of Eq. (5.69) seems to be very limited, since one has to integrate the absorbance over the whole spectral range. On the other hand, spectral ranges with $k \approx 0$ do virtually not contribute. Therefore, it is possible to dissect the spectral range, and, as long as bands do not overlap, the area of an absorbance band is equal to the oscillator strength of the underlying oscillator. Therefore, even without applying dispersion analysis, it is possible to "regain" Beer's approximation (cf. Fig. 5.15) as long as the applied electric field is approximately the same as the one that is effective at the location of the absorbing species.

One more comment on the Lorentz profiles used to fit absorbance before we move on. It seems that in plasma physics it is never assumed that Beer's approximation would work pointwise—absorbance bands are usually fitted with Lorentz-profiles and the band area/oscillator strength is used instead. Can you find the hidden paradox? Indeed, the derivation of the Lorentz-profile is analogous to that of Beer's approximation. Therefore, absorbance Lorentz profiles follow Beer's approximation automatically at each point.

In practice, however, as already pointed out in Section 5.1, the merits of Eq. (5.69) may be smaller than expected, in particular in the infrared spectral range, because of the local field effect. So far, in this section, we have assumed that the local electric field is that which is externally applied. If we assume that the Lorentz-Lorenz relation holds and if there would be only one oscillator, it would be sufficient to replace $\tilde{v}_0^2$ by $\tilde{v}'^2_0$ (cf. Eq. 5.26) in Eq. (5.62):

$$\hat{n} = \sqrt{1 + c\frac{S^{*2}}{\tilde{v}'^2_0 - \tilde{v}^2 - i\tilde{v}\gamma}} = \sqrt{1 + c\frac{S^{*2}}{\tilde{v}_0^2 - \frac{c}{3}S^{*2} - \tilde{v}^2 - i\tilde{v}\gamma}}. \quad (5.70)$$

Without the local field, the maximum of the absorbance shifted to higher wavenumbers with increasing concentration. Thanks to the correction term, this behavior is compensated up to about 5 mol/L for the particular example we are looking at (cf. Fig. 5.17). Accordingly, the deviations from Beer's approximation are smaller up to concentrations of about 20 mol/L, before the stronger shift to lower wavenumbers due to the local field leading to stronger deviations from Beer's approximation. Still, Eq. (5.69) would allow us to determine the oscillator strength from the integrated absorbance, and, thereby, the concentration.

In reality, we always have more than one oscillator. Therefore, our starting point is Eq. (5.31) with the small modification that we replace N by $N_A \cdot c$:

$$\varepsilon = \frac{1 + \frac{2}{3}c\frac{N_A \cdot \alpha}{\varepsilon_0}}{1 - \frac{1}{3}c\frac{N_A \cdot \alpha}{\varepsilon_0}}. \quad (5.71)$$

From Eq. (5.71), it is really hard to see how this relation should transform into one with a linear dependence of the absorbance from the concentration for lower concentrations. Therefore, let me rewrite it to the original form that was derived by both Lorenz and Lorentz (with the difference that we keep N replaced by $N_A \cdot c$):

$$\frac{\hat{n}^2 - 1}{\hat{n}^2 + 2} = \frac{1}{3}c\frac{N_A \cdot \alpha}{\varepsilon_0}. \quad (5.72)$$

In some other optics textbooks, simplifications of this formula are presented for diluted gases. For those, n is not very different from 1 in non-absorbing regions. Therefore, $\hat{n}^2 + 2 \approx 3$ and,

$$\hat{n}^2 - 1 = c\frac{N_A \cdot \alpha}{\varepsilon_0}. \quad (5.73)$$

Furthermore, it is usually argued that under the same constraint $\hat{n}^2 - 1 \approx 2(\hat{n} - 1)$ and, hence,

$$\hat{n} - 1 = c\frac{N_A \cdot \alpha}{2\varepsilon_0}. \quad (5.74)$$

What is not realized in literature is that Eq. (5.74) is nothing else but the derivation of Beer's empiric law [33]! Certainly, neither Lorenz nor Lorentz replaced N by $N_A \cdot c$, but Lorenz was actually trained as a chemist and he used the relation derived by him in 1865 to determine Loschmidt's constant which is directly related to N_A [34]. In its original form, the Lorentz-Lorenz equation considered only the real index of refraction, but when we use it with a complex index of refraction and only keep the imaginary part, we obtain,

$$k(\tilde{v}) = c\frac{N_A \cdot \alpha''(\tilde{v})}{2\varepsilon_0}, \quad (5.75)$$

which is Beer's approximation, when combined with Eq. (5.65). Furthermore, Eq. (5.73) certainly requires to assume that the local field is the same as the applied field. Note that Beer's approximation is not the only approximation that can be gained from Eq. (5.74). If we focus on the real part, we obtain,

$$n - 1 = c\frac{N_A \cdot \alpha'}{2\varepsilon_0}, \quad (5.76)$$

which means that the increase of the refractive index due to an absorption is also approximately proportional to the molar concentration. That this relation works well in many cases is known from the vast field of refractive index sensing, but also, e.g., from the determination of the sugar content in grapes, which is conveniently performed by a refractometer. Recently, the group of Bernhard Lendl in Vienna has built a quantum cascade laser-based Mach-Zehnder interferometer, with which the change of the index of refraction due to a vibrational band can be determined. With this instrument, the nearly linearly increase of the refractive index, and, with it, Eq. (5.76) could be proved [35,36].

The important question is, how much diluted is sufficient for Beer's approximation to hold? Usually, it is assumed that it holds for gases at atmospheric pressure. This assumes that the concentration is approximately a thousandth of the concentration of a neat substance. We will therefore in the following investigate the concentration dependence of the absorbance when we change the concentration from a thousandth part to that of a neat substance. Before we do this, I need to add, that the Lorentz-Lorenz equation can certainly also be applied to mixtures. When for one substance the polarizability can be expressed as a sum of contributions of independent oscillators,

$$\alpha(\tilde{v}) = \sum_i \frac{q_i^2/\mu_i}{\tilde{v}_{0,i}^2 - \tilde{v}^2 - i\tilde{v}\gamma_i}, \tag{5.77}$$

we can assume that for a mixture of j substances, all oscillators are still independent so that the polarizability of the mixture is given by [33]:

$$\alpha(\tilde{v}) = \sum_j c_j \sum_i \frac{q_{ji}^2/\mu_{ji}}{\tilde{v}_{0,ji}^2 - \tilde{v}^2 - i\tilde{v}\gamma_{ji}}, \tag{5.78}$$

In the following, we will assume for simplification that we have a hypothetic substance with one oscillator in the UV/Vis spectral region and one additional oscillator in the IR. Furthermore, in the mixture, or, better, in the solution, the first hypothetic substance is the solute and we have a second hypothetic substance with only a UV oscillator which does not overlap with that of the first hypothetic substance. Both substances shall have the same concentration $c = 50$ mol/L when they are neat. The corresponding absorbance spectra are shown in Fig. 5.16.

With the exception of (D), right panel, all spectra are normalized so that a constant product of concentration and thickness, $c \cdot d$, was assumed. Accordingly, if Beer's approximation would generally hold, all spectra would be congruent. In fact, this is practically the case for the spectra with $c = 0.05$ and 0.5 mol/L. Accordingly, the black spectra are hidden behind the red spectra. If we first focus on the spectra of the solute in vacuum, the changes are still small from 0.5 to 5 mol/L. If the local field is neglected as for (A), the IR band is slightly blueshifted, whereas it is redshifted if it is assumed that the local field only causes a decrease in the oscillator position. In both cases, the intensity is slightly lowered in contrast to the

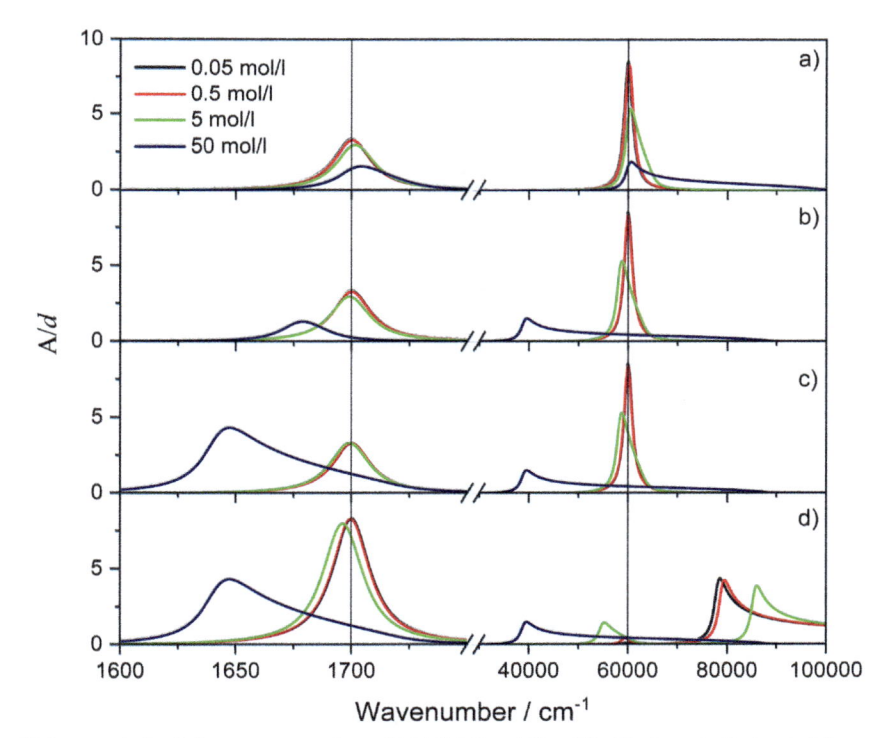

FIG. 5.16 Absorbance-thickness ratio for different concentrations of 0.05, 0.5, 5, and 50 mol/L of the hypothetical material assuming $c \cdot d = 5 \cdot 10^3$ cm·mol/L (1600–1750 cm^{-1}) and 50 cm·mol/L (30,000–100,000 cm^{-1}). (A) Conventional dispersion formula. (B) Conventional dispersion formula assuming a redshift according to Eq. (5.70). (C) Lorentz-Lorenz formula Eq. (5.71) in combination with Eq. (5.77). (D) Same as (C) but with hypothetical solvent (from 30,000 to 100,000 cm^{-1} a constant $d = 10^{-6}$ cm was chosen instead of a constant $c \cdot d$) [33].

coupled case (C). For 50 mol/L, these changes are, which is more than apparent, much stronger. Responsible for the increase in absorbance for the coupled case is obviously the increased field strength and, therefore, the higher field intensity to which absorption (but not absorbance, cf. Chapter 1) is proportional. In the UV/Vis spectral range, in contrast, the cases (B) and (C) are practically not distinguishable. The reason why these cases are so similar is that the oscillator strength of the UV/Vis band is much stronger, so the local electric field is only marginally increased by the comparably weak IR oscillator. The absorbance-thickness ratio must decrease, because, as we will see later when we discuss the Kramers-Kronig sum rules (Chapter 5.8), in this particular case the area of the band remains constant.

If we no longer assume vacuum as a solvent, the spectral behavior changes dramatically due to the coupling of all oscillators. While there is still not much difference between $c = 0.05$ and 0.5 mol/L for constant $c \cdot d$, the intensity is strongly increased due to the coupling with the oscillator of the solvent, since for the Lorentz-Lorenz theory, oscillators belonging to one component cannot be distinguished from the other. This coupling also leads to a stronger redshift than without the solvent. This increased intensity must decay somewhat for increasing concentration of the solute, since this means a decreasing concentration of the solvent. Accordingly, for increasing the concentration of the solute, the band of the solvent blueshifts. Overall, without solvent, we expect Beer's approximation to hold at least for densities of 1% of that of the condensed phase. In fact, linear behavior is predicted practically for densities of up to 10% ($c = 5$ mol/L) as can be seen from Fig. 5.17. Employing integrated absorbance instead of peak absorbance is not very useful in the infrared spectral range due to the strong coupling, the linear range is only slightly enhanced. In the UV spectral range using integrated absorbance works, because there is only one very strong oscillator. As soon as it couples to the oscillator of the solvent, linearity is also lost for higher concentrations, but it is kept much longer than for the peak absorbance.

Overall, it seems that for the infrared spectral range, the deviations from Beer's approximation should not be experimentally noticeable, probably with the exception of the case when inorganic materials would be investigated by the pellet method. Certainly, higher concentrations lead to very small transmittances, but much brighter light sources like quantum cascade lasers and more sensitive detectors might shift the experimental limits in the future. On the other hand, the attenuated total reflection technique can be used for higher concentrations without any problems (except that the spectra need to be corrected before they are useful, see Section 6.5). To prove the existence of local field effects and to exclude changes in chemical interactions as causes for spectral changes, there is only one possibility, which is to investigate the mixing behavior of a system that is as close to ideal as possible. Such a system is a mixture of Benzene and Toluene. Benzene is nonpolar and Toluene only has a very weak polarity. Their molecular structure is so similar that Raoult's law is obeyed, which means that the volumes are additive over the whole mole fraction range. X-ray and neutron diffraction indicate that the short-range order changes are very small in the mixtures—in general the molecules have 12 neighbors and are

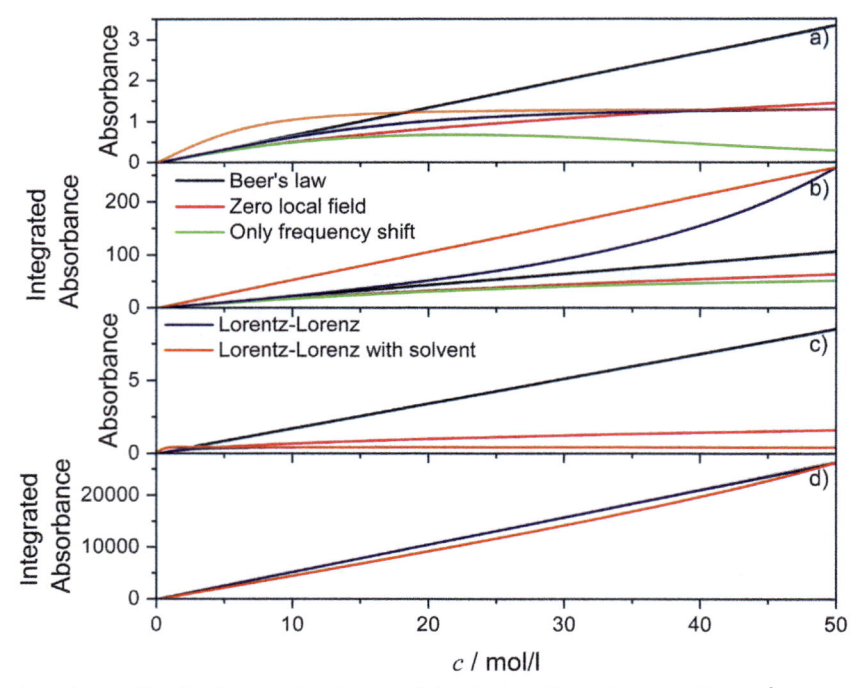

FIG. 5.17 Concentration dependence of the absorbance and the integrated absorbance. (A) absorbance at $1700\,\text{cm}^{-1}$. (B) absorbance integrated from 100 to $3000\,\text{cm}^{-1}$. (C) absorbance at $60,000\,\text{cm}^{-1}$. (D) absorbance integrated between $30,000$ and $100,000\,\text{cm}^{-1}$. $d = 1\,\mu\text{m}$ for (A) and (B) and $d = 0.01\,\mu\text{m}$ for (C) and (D) [33].

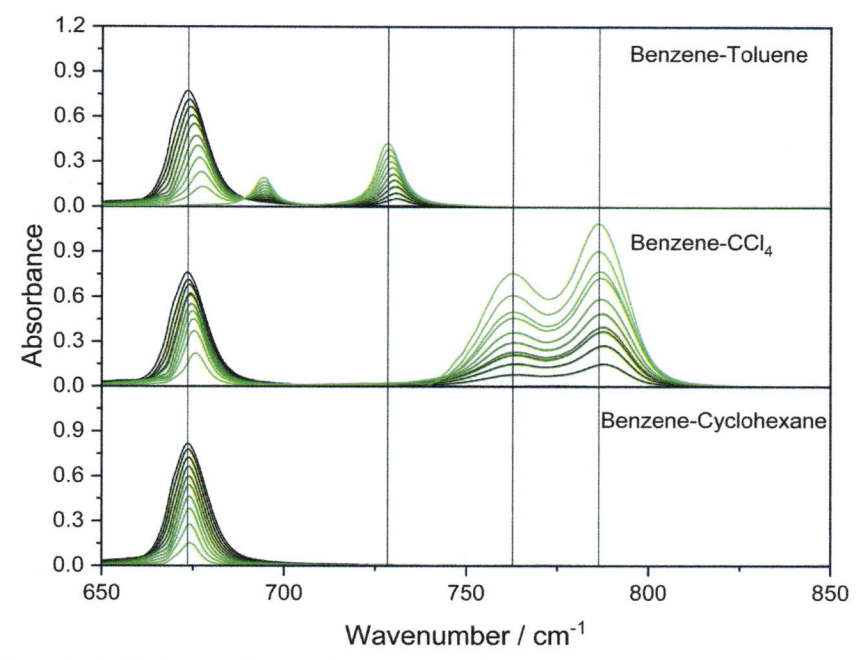

FIG. 5.18 Spectra of mixtures in the ideal systems Benzene-Toluene, Benzene-Carbon tetrachloride, and Benzene-Cyclohexane.

well-ordered, pretty much as in the crystalline state. What surprises me is that if you scan the literature, this system has not been examined by infrared spectroscopy previously. More or less by accident I learnt that two colleagues, Oleksii Ilchenko and Andrii Kutsyk, had investigated not only the system Benzene-Toluene, but also the systems Benzene-Carbon tetrachloride and Benzene-Cyclohexane as well. I was very excited when I saw these at that time yet unpublished spectra, because they proved the existence of local field effects [37]. What they also prove is that the situation is much more complex as the one assumed to derive the Lorentz-Lorenz relation. The spectra are depicted in Fig. 5.18. The spectral range that is shown is the one with the strongest oscillators (except for Cyclohexane, where the C—H stretching vibrations below $3000\,cm^{-1}$ also lead to strong bands). Obviously, the bands are blueshifted with decreasing volume fraction, something which could be expected from local field effects. However, this blue shift depends on the system as can be seen for the Benzene band at $774\,cm^{-1}$. Interestingly, the shift is strongest for the system which should be the one closest to the ideal state, namely, the system Benzene-Toluene. In this system, the strongest Toluene band is nearly as much shifted as the Benzene band, even if its oscillator strength is lower. In contrast, the C—Cl bands in Carbon tetrachloride blueshift only by a maximum $1\,cm^{-1}$, albeit they are clearly stronger than the Benzene band. Obviously, in contrast to the Lorentz-Lorenz equation, the local fields for the two different molecules can be dissimilar. Also, the Benzene and Toluene molecules are not spherical as assumed by the Lorentz-Lorenz relation and they possess an anisotropic polarizability. Furthermore, since there is a short-range order, the surrounding of the molecule cannot be described as a continuum, the more since this short-range order will change strongly when the volume fraction is varied, maybe except for the system Benzene-Toluene. It is therefore clear that the Lorentz-Lorenz equation is nothing else but an approximation, albeit a much better one than Beer's approximation as can be seen in Fig. 5.19 [37].

With regard to more advanced relations than Lorentz-Lorenz, it must be said that for 100 years a lot of effort had been put into their derivation, see, e.g. the books of Carl Böttcher [38,39] and Chapter 10. Unfortunately, while the theory is extremely complex, I did not see anything sophisticated enough to treat the problem at hand. In addition, it seems that many of the more sophisticated alterations could not be proved by experiment. In the years that followed it seemed to me that another direction was taken to model the systems under a nonstatic electric field, which was to apply molecular dynamics simulations. It is worth to make some words about the results. In fact, it was possible to model the far infrared spectrum of Benzene. Indeed, liquid Benzene has a far infrared spectrum, despite the fact that it has no permanent dipole moment! However, the induced dipole moments are sufficient that radiation exerts a force to orient them along the electric field direction, which leads to a broad band in the range 0–$200\,cm^{-1}$ peaking at about $75\,cm^{-1}$, which can be explained well by molecular dynamics simulations assuming point quadrupoles and the thereby generated local fields [40,41]. It is really amazing that the knowledge that infrared spectroscopy is sensitive to near-range order was present in this community, but did not make it to the community of spectroscopists interested in intramolecular vibrations.

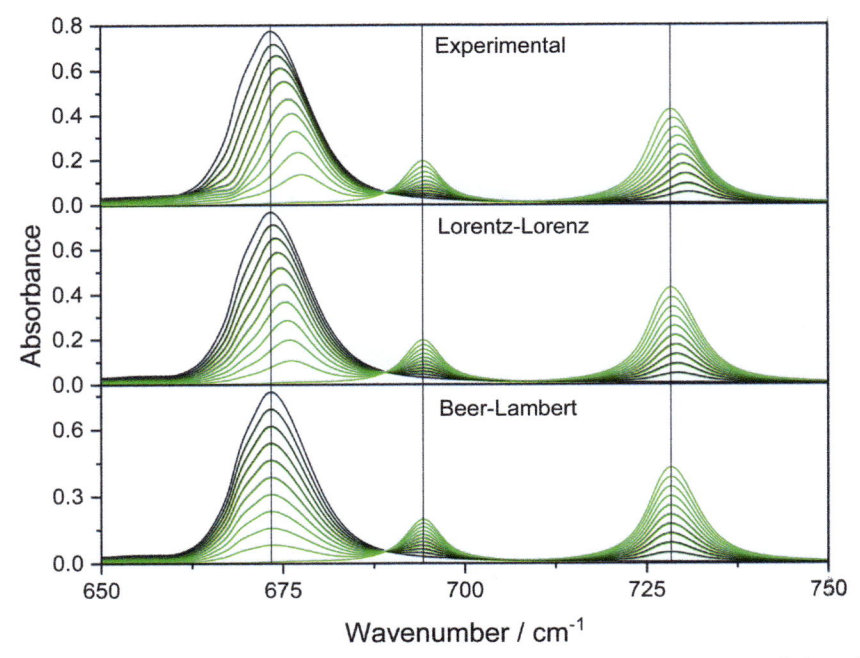

FIG. 5.19 Comparison of experimental and forward calculated spectra of mixtures in the ideal systems Benzene-Toluene [37].

5.4 Dispersion relation—Coupled oscillator model

If there is more than one vibrational oscillator, quantum mechanical calculations showed already at the beginning of the 1960ies that "the damping of the different modes is not necessarily independent" [42]. This kind of coupling is not due to a local field effect, which couples oscillators independent of the difference between their eigenfrequencies, but due to mechanical coupling which increases with the proximity of the eigenfrequencies of the oscillators. It also needs to be emphasized that it is not anharmonicity that underlies this model, but the coupling of two harmonic oscillators. Accordingly, the classical equations of motion need to be modified and can be written as:

$$\begin{aligned} \frac{d^2 x_1}{dt^2} + (\gamma_1 + \gamma_{12})\frac{dx_1}{dt} + \gamma_{12}\frac{dx_2}{dt} + \omega_1^2 x_1 &= q_1 E_x \\ \frac{d^2 x_2}{dt^2} + (\gamma_2 + \gamma_{12})\frac{dx_2}{dt} + \gamma_{12}\frac{dx_1}{dt} + \omega_2^2 x_2 &= q_2 E_x \end{aligned} \quad (5.79)$$

The corresponding model concept is illustrated in Scheme 5.1, where damping is implemented by the honeypots. The honeypot between the two charges q_1 and q_2 introduces the interaction between the oscillators. This interaction reflects itself by the fact that the two equations in Eq. (5.79) are no longer independent and linked through the interaction damping constant γ_{12} (if the interaction damping constant is zero, then two independent equations of motions result, which conforms to the standard model where such an interaction is not taken into account). Nevertheless, the solutions are found in the same way as for Eq. (5.18) and it is assumed that both displacements x_1 and x_2 as well as the exciting electric field have the same time dependence, namely $\exp(-i\omega t)$.

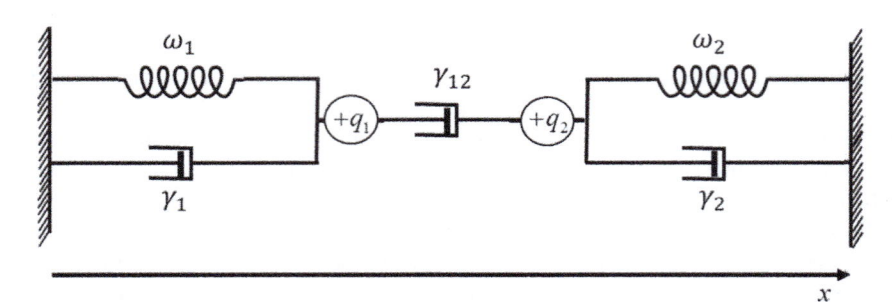

SCHEME 5.1 Mechanical model of two coupled oscillators with interaction damping.

As can be expected, the solution becomes a little bit more complex than in the case of independent oscillators. For the displacement of the first effective charge q_1 we find:

$$x_1 = \frac{q_1 E_x + \dfrac{i\omega \gamma_{12} q_2 E_x}{\omega_{0,2}^2 - \omega^2 - i\omega(\gamma_2 + \gamma_{12})}}{\omega_{0,1}^2 - \omega^2 - i\omega(\gamma_1 + \gamma_{12}) - \dfrac{\omega^2 \gamma_{12}^2}{\omega_{0,2}^2 - \omega^2 - i\omega(\gamma_2 + \gamma_{12})}} \exp(-i\omega t). \tag{5.80}$$

The solution for the displacement of the second effective charge q_2 is obtained by an interchange of the subscripts 1 and 2 according to:

$$x_2 = \left(\text{same with } \begin{array}{c} 1 \to 2 \\ 2 \to 1 \end{array}\right). \tag{5.81}$$

The polarization is again obtained in the same way as for uncoupled oscillators and the result for the contribution of the first oscillator to the dielectric function ε_1 is found to be:

$$\varepsilon_1 = \frac{S_1^2 + \dfrac{i S_1 S_2 \tilde{v} \gamma_{12}}{\tilde{v}_{0,2}^2 - \tilde{v}^2 - i\tilde{v}(\gamma_2 + \gamma_{12})}}{\tilde{v}_{0,1}^2 - \tilde{v}^2 - i\tilde{v}(\gamma_1 + \gamma_{12}) + \dfrac{\tilde{v}^2 \gamma_{12}^2}{\tilde{v}_{0,2}^2 - \tilde{v}^2 - i\tilde{v}(\gamma_2 + \gamma_{12})}}. \tag{5.82}$$

The contribution of the second oscillator ε_2 is again obtained by interchanging the subscripts. The dielectric function for this two-oscillator system is then $\varepsilon(\tilde{v}) = \varepsilon_\infty + \varepsilon_1(\tilde{v}) + \varepsilon_2(\tilde{v})$.

From Eq. (5.82), we can draw two conclusions. First, in the additional term in the numerator, we see a product of the square roots of two oscillator strengths. For negative effective charges, the S_j becomes negative (remember: $S_j = q\sqrt{N/(\mu\varepsilon_0)}$), which has no consequences for uncoupled systems as then only the magnitude of the square root of the oscillator strength can be obtained by the fit. In contrast, for coupled systems also the charge and its sign could in principle be attained. Second, even non-IR active oscillators can influence IR-active bands and thereby change band shapes. The oscillators that are non-IR active can also be called *silent modes*. If the second oscillator is silent, then its oscillator strength $S_2^2 = 0$. Consequently, $\varepsilon_2(\tilde{v}) = 0$ and $\varepsilon_1(\tilde{v})$ is given by [43]:

$$\varepsilon_1 = \frac{S_1^2}{\tilde{v}_{0,1}^2 - \tilde{v}^2 - i\tilde{v}(\gamma_1 + \gamma_{12}) + \dfrac{\tilde{v}^2 \gamma_{12}^2}{\tilde{v}_{0,2}^2 - \tilde{v}^2 - i\tilde{v}(\gamma_2 + \gamma_{12})}}. \tag{5.83}$$

Obviously, the result differs from the one obtained for uncoupled oscillators. To understand what is going on, it makes sense to start with the spectral changes introduced by the silent mode/oscillator.

As can be seen in Fig. 5.20, the silent mode becomes visible in the spectrum at its position ($\tilde{v}_{0,2} = 600\,\text{cm}^{-1}$) as a dip in the imaginary part of the dielectric function. The driven mode, which is characterized by a symmetric Lorentz-profile in ε'' in the uncoupled state, becomes asymmetric through coupling and its peak maximum is shifted toward the second mode. This shift is also seen in ε', but at lower wavenumbers away from the driven mode. Overall, the coupling to the silent mode does not seem to affect the values of ε' very strongly. Based on the Kramers-Kronig sum rules (cf. Section 5.8.3), the area of ε'' due to the active mode should not be changed by the silent mode since its oscillator strength is zero.

If also the second mode is driven, the situation becomes much more complex. Let us first assume that for both modes the charges have the same sign. This is the usual situation discussed in the literature (but it remains unclear if this is also the usual situation experienced in reality). The resulting dielectric functions are displayed in Fig. 5.21.

If both charges have the same sign, the modes repel each other, but the peak maxima in the imaginary part of the dielectric function do virtually not shift. As a consequence, the peak shapes in the imaginary part become asymmetric. In addition, this part of the dielectric function becomes smaller in between both peaks compared to the uncoupled oscillators. At the location of the respective other mode, the contributions show a dispersion-like shape and become negative in between the modes. This contribution is responsible for the decrease of the imaginary part between the peaks relative to the uncoupled modes (the sum of both contributions, however, always stays positive, so that no unphysical behavior arises).

For the case that the effective charges have different signs, the behavior is the opposite, see Fig. 5.22. In the resulting dielectric function, the modes are attracted to each other. Therefore, each one shifts towards the other and the values in between are higher than for the uncoupled oscillators. Again, at certain spectral regions, the individual contributions to the imaginary parts become negative.

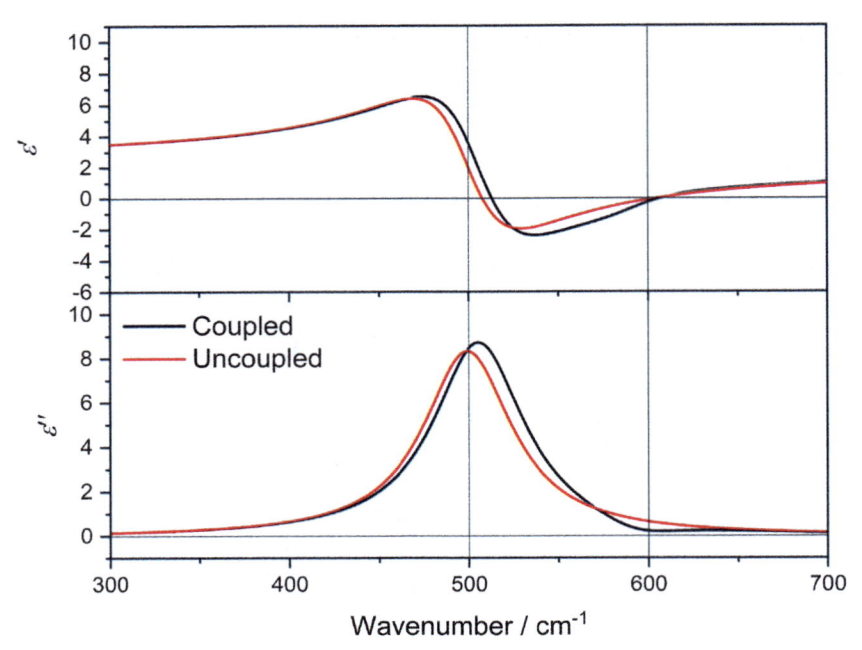

FIG. 5.20 Comparison between the real and imaginary parts of the dielectric function due to two coupled oscillators, one of which, located at 600 cm^{-1}, is not driven, and due to a single oscillator at 500 cm^{-1}.

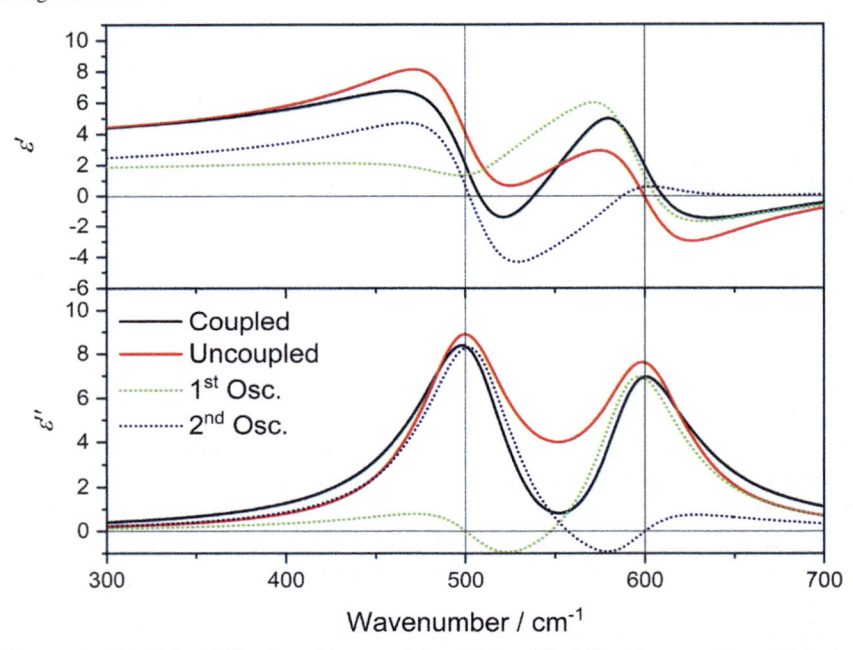

FIG. 5.21 Real and imaginary part of the dielectric function of two coupled oscillators *(black lines)* in comparison with that of two uncoupled oscillators *(red lines)* in each case located at 500 and 600 cm^{-1}. S_1 and S_2 have the same sign. The contributions of the individual oscillators are also shown *(blue and green dotted lines)*.

Note that it is also possible to combine $\varepsilon_1(\tilde{\nu})$ and $\varepsilon_2(\tilde{\nu})$:

$$\varepsilon_1(\tilde{\nu}) + \varepsilon_2(\tilde{\nu}) = \frac{S_1^2\left(\tilde{\nu}_{0,2}^2 - \tilde{\nu}^2 - i\tilde{\nu}(\gamma_2 + \gamma_{12})\right) + 2iS_1S_2\tilde{\nu}\gamma_{12} + S_2^2\left(\tilde{\nu}_{0,1}^2 - \tilde{\nu}^2 - i\tilde{\nu}(\gamma_1 + \gamma_{12})\right)}{\left(\tilde{\nu}_{0,1}^2 - \tilde{\nu}^2 - i\tilde{\nu}(\gamma_1 + \gamma_{12})\right)\left(\tilde{\nu}_{0,2}^2 - \tilde{\nu}^2 - i\tilde{\nu}(\gamma_2 + \gamma_{12})\right) + \tilde{\nu}^2\gamma_{12}^2}. \quad (5.84)$$

In practice, it is possible to mix uncoupled and coupled oscillators simply by adding the term(s) for $\varepsilon_1(\tilde{\nu}) + \varepsilon_2(\tilde{\nu})$ to Eq. (5.27):

$$\varepsilon = \varepsilon_\infty + \varepsilon_1(\tilde{\nu}) + \varepsilon_2(\tilde{\nu}) + \sum_{j=3}^{N} \frac{S_j^2}{\tilde{\nu}_{0,j}^2 - \tilde{\nu}^2 - i\tilde{\nu}\gamma_j}. \quad (5.85)$$

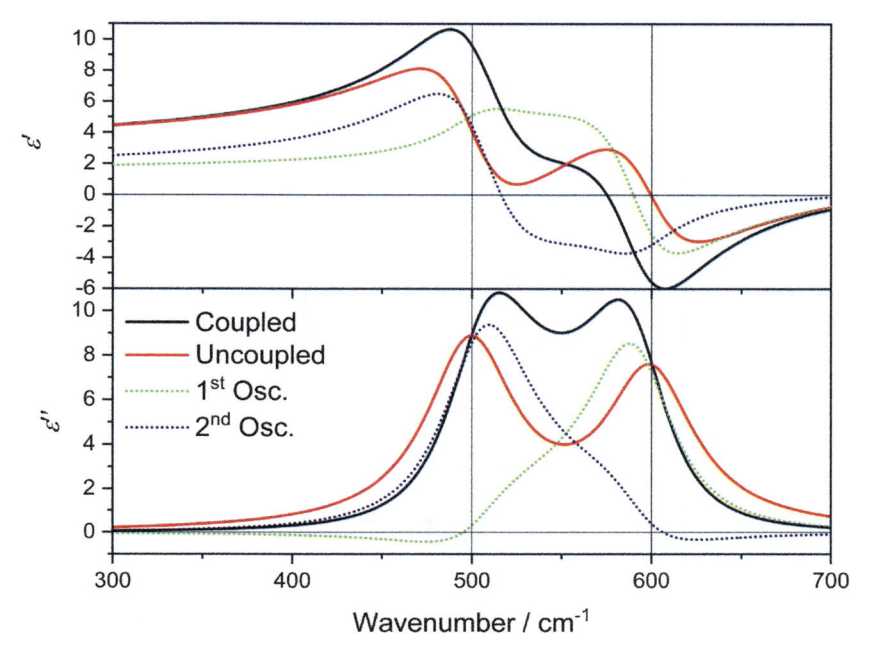

FIG. 5.22 Real and imaginary part of the dielectric function of two coupled oscillators *(black lines)* in comparison with that of two uncoupled oscillators *(red lines)* in each case located at 500 and 600 cm^{-1}. S_1 and S_2 have different signs. The contributions of the individual oscillators are also shown *(blue and green dotted lines)*.

Generally, neighboring oscillators (neighboring in the spectral sense, i.e., related to their wavenumber position) show stronger coupling effects than those whose oscillator positions $\tilde{v}_{0,j}$ are more different. In this sense, let us assume that we have a system of five oscillators, like in NdGaO$_3$, and that only directly neighboring oscillators couple (cf. Scheme 5.2). We can then write the resulting equations of motions in matrix form as:

$$\begin{pmatrix} A_1 & i\tilde{v}\gamma_{12} & 0 & 0 & 0 \\ i\tilde{v}\gamma_{12} & A_2 & i\tilde{v}\gamma_{23} & 0 & 0 \\ 0 & i\tilde{v}\gamma_{23} & A_3 & i\tilde{v}\gamma_{34} & 0 \\ 0 & 0 & i\tilde{v}\gamma_{34} & A_4 & i\tilde{v}\gamma_{45} \\ 0 & 0 & 0 & i\tilde{v}\gamma_{45} & A_5 \end{pmatrix} \begin{pmatrix} x_1 \\ x_2 \\ x_3 \\ x_4 \\ x_5 \end{pmatrix} = \begin{pmatrix} q_1 \\ q_2 \\ q_3 \\ q_4 \\ q_5 \end{pmatrix} E$$

$$A_1 = \tilde{v}_{0,1}^2 - \tilde{v}^2 - i\tilde{v}(\gamma_1 + \gamma_{12})$$
$$A_2 = \tilde{v}_{0,2}^2 - \tilde{v}^2 - i\tilde{v}(\gamma_2 + \gamma_{12} + \gamma_{23})$$
$$A_3 = \tilde{v}_{0,3}^2 - \tilde{v}^2 - i\tilde{v}(\gamma_3 + \gamma_{23} + \gamma_{34})$$
$$A_4 = \tilde{v}_{0,4}^2 - \tilde{v}^2 - i\tilde{v}(\gamma_4 + \gamma_{34} + \gamma_{45})$$
$$A_5 = \tilde{v}_{0,5}^2 - \tilde{v}^2 - i\tilde{v}(\gamma_5 + \gamma_{45})$$

(5.86)

The matrix in Eq. (5.86) is relatively sparse. Nevertheless, it is convenient to use a computer algebra system like Mathematica in order to solve the corresponding equations and to formulate the dispersion relations. Since the complexity increases nonlinearly, so does the effort to calculate the dielectric function even when there is only one (to be more precise, $(N-1)/N$) additional parameters per oscillator. Nowadays, it is possible to afford the computational extra effort, but is the result worth this effort? The improvement with respect to the conventional dispersion relation according to Eq. (5.27) is depicted in Fig. 5.23.

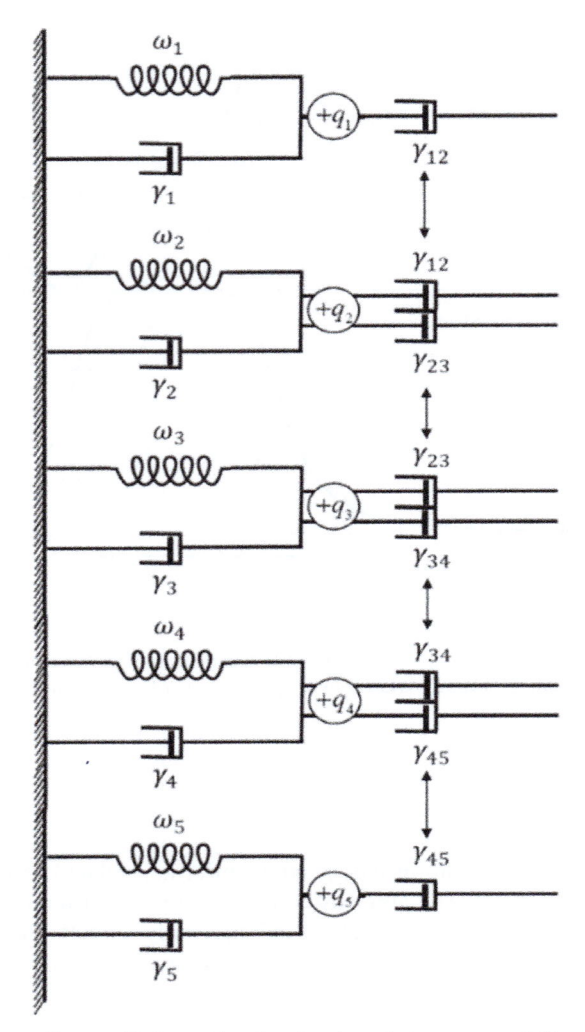

SCHEME 5.2 Mechanical model of five oscillators with interaction damping through the nearest neighbors.

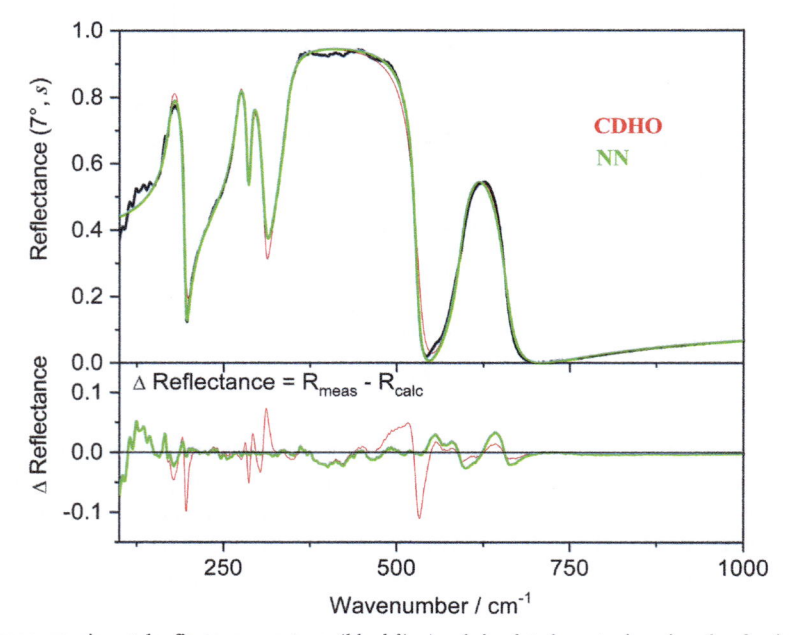

FIG. 5.23 Comparison between experimental reflectance spectrum *(black line)* and simulated spectra based on the classical damped harmonic oscillator model *(red line)* and assuming coupling of the nearest neighbors *(green line)*.

Obviously, the correspondence between measured and modeled spectrum based on the nearest neighbor model is in general much better, in particular at the wings of the broad band between about 300 and 500 cm^{-1} and the average error decreased in our particular example by nearly one-third.

Nevertheless, there is also some regression, e.g., the agreement becomes worse around the high-wavenumber band.

If the coupling between the nearest neighbors can improve the correspondence between model and experiment so drastically, it seems natural to check if further improvement is possible by incorporating also the coupling to the next nearest neighbors into the model:

$$\begin{pmatrix} A_1 & i\tilde{v}\gamma_{12} & i\tilde{v}\gamma_{13} & 0 & 0 \\ i\tilde{v}\gamma_{12} & A_2 & i\tilde{v}\gamma_{23} & i\tilde{v}\gamma_{24} & 0 \\ i\tilde{v}\gamma_{13} & i\tilde{v}\gamma_{23} & A_3 & i\tilde{v}\gamma_{34} & i\tilde{v}\gamma_{35} \\ 0 & i\tilde{v}\gamma_{24} & i\tilde{v}\gamma_{34} & A_4 & i\tilde{v}\gamma_{45} \\ 0 & 0 & i\tilde{v}\gamma_{35} & i\tilde{v}\gamma_{45} & A_5 \end{pmatrix} \begin{pmatrix} x_1 \\ x_2 \\ x_3 \\ x_4 \\ x_5 \end{pmatrix} = \begin{pmatrix} z_1 \\ z_2 \\ z_3 \\ z_4 \\ z_5 \end{pmatrix} E$$

$$A_1 = \tilde{v}_{0,1}^2 - \tilde{v}^2 - i\tilde{v}(\gamma_1 + \gamma_{12} + \gamma_{13})$$
$$A_2 = \tilde{v}_{0,2}^2 - \tilde{v}^2 - i\tilde{v}(\gamma_2 + \gamma_{12} + \gamma_{23} + \gamma_{24})$$
$$A_3 = \tilde{v}_{0,3}^2 - \tilde{v}^2 - i\tilde{v}(\gamma_3 + \gamma_{13} + \gamma_{23} + \gamma_{34} + \gamma_{35})$$
$$A_4 = \tilde{v}_{0,4}^2 - \tilde{v}^2 - i\tilde{v}(\gamma_4 + \gamma_{24} + \gamma_{34} + \gamma_{45})$$
$$A_5 = \tilde{v}_{0,5}^2 - \tilde{v}^2 - i\tilde{v}(\gamma_5 + \gamma_{35} + \gamma_{45})$$

(5.87)

Needless to say, that in this case the corresponding dispersion relations become again much more complex and the corresponding fit is now considerably slowed down. The result, which may be somewhat surprising even though I already stated that coupling effects depend on the spectral distance, is depicted in Fig. 5.24. It is hard to make out any improvement, only the further regression of the correspondence between experiment and model for the band with the highest wavenumber is obvious. Indeed, the decrease in the average error is only 1%! Even less encouraging is that nearly arbitrary values for the next-nearest neighbor damping constants lead to the same overall fit quality. This leads me to the conclusion that the model becomes overdetermined and the results lose physical significance. We could certainly go on and investigate the full model where each oscillator couples with each other, but I did this and it was not worth the effort. For further improvement, a fundamentally improved model would be required.

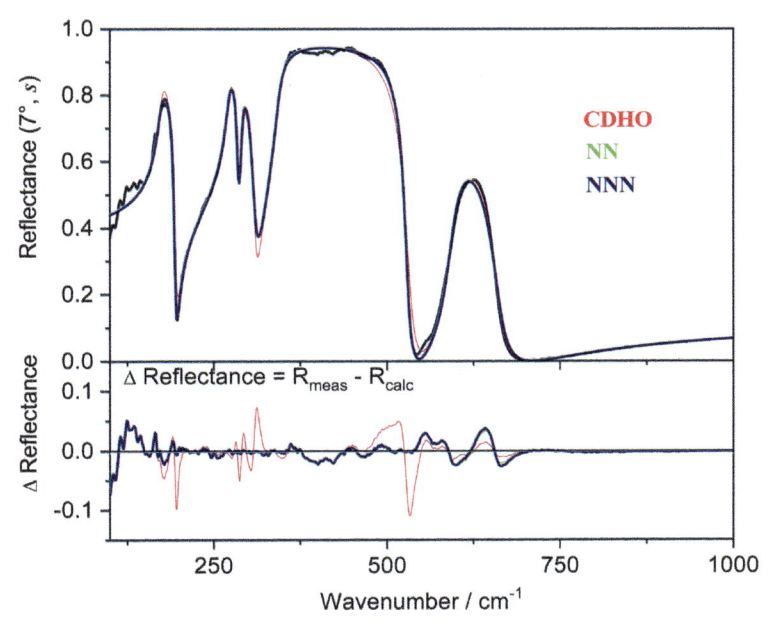

FIG. 5.24 Comparison between experimental reflectance spectrum *(black line)* and simulated spectra based on the classical damped harmonic oscillator model *(red line)*, assuming coupling of the nearest neighbors *(green line)* and on the dispersion relation assuming coupling of the next nearest neighbors *(blue line)*.

5.5 Dispersion relation—Semi-empirical four-parameter models

With four parameters I can fit an elephant, and with five I can make him wiggle his trunk.

Johannes von Neumann

5.5.1 Berreman-Unterwal model

To gain the CDHO dispersion model, we have assumed that the force F that drives the atoms back to their equilibrium position is proportional to the deviation from this position. Therefore, one derives a harmonic potential and, by applying quantum mechanics, that the vibration absorbs only one photon of equivalent energy. Accordingly, in simple cubic crystals like NaCl with only one (but threefold degenerated) infrared active vibration, there should be only one band in the spectrum. In contrast, the reststrahlen bands of such simple cubic crystals always show a structure, a kind of shoulder (but actually a dent), which could only be ascribed to a second active vibration. Since impurities could eventually be excluded as an explanation for these second bands, anharmonicity was considered. Indeed, it could be shown that for the anharmonic case the acoustic and the optical branch couple and that those frequencies with a horizontal tangent need to be added to explain the secondary band structures [44]. While the dispersion relation in principle keeps its form, the damping constant γ becomes a complicated function of the wavenumber, which cannot easily be adapted. Therefore, a couple of semiempirical approaches have been developed. From those I want to introduce three to you, the first of which was developed by Berreman and Unterwal [45]. Berreman and Unterwal developed their model just from the observation that there is one pole and one zero per oscillator, cf. Fig. 5.25.

Hence, it is possible to describe the dielectric function with more than one oscillator with a function of the following form:

$$\varepsilon' = \varepsilon_\infty \prod_{j=1}^{N} \frac{\tilde{v} - Z_j}{\tilde{v} - P_j}. \tag{5.88}$$

The poles are easily identified with the resonance positions of the oscillators, whereas we have not talked so far about what the zeros could represent.

When we discussed the vibrations in a chain (cf. Fig. 5.9), we did not consider how the direction of the light and the electric field vector must be oriented relative to the chain. Usually, light waves are transversal, which means that the direction of the incoming light must be perpendicular to the chain. In contrast, for Raman spectroscopy, the situation is somewhat different. In noncentrosymmetric crystals infrared active vibrations are also Raman active as the principle of mutual exclusion does not apply to the corresponding symmetries. As a consequence, we can compare the frequencies that we obtain from infrared spectra with those from Raman spectra. Surprisingly, in Raman spectra, we find in principle two bands for each infrared active vibration, in dependence of the direction of the incoming and the scattered light. In particular, if the incoming light and the scattered light are parallel and/or antiparallel (the latter is the case for a microscope!) to the

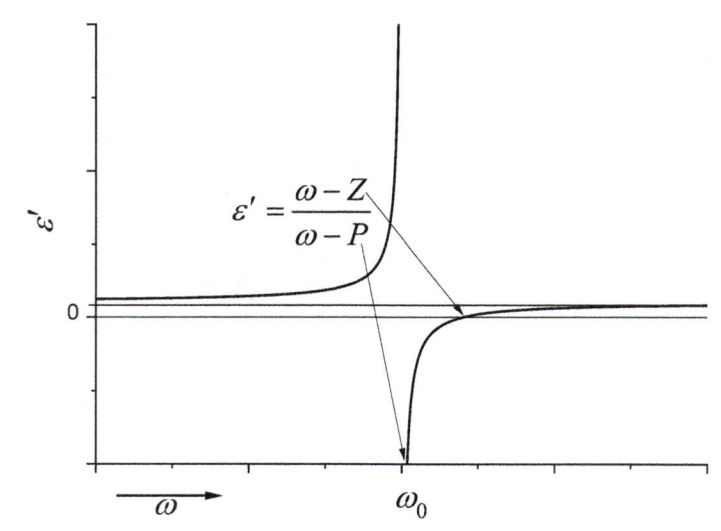

FIG. 5.25 Pole (P) and zero (Z) in a real dielectric function.

transition moment, only the Raman band at a higher wavenumber occurs. The shift between the two bands depends on the strength of the oscillator. How can we understand this?

Since the transition moment and electric field are parallel, $\nabla \cdot \mathbf{E} \neq 0$. However, as we deal with materials with no free charges, $\nabla \cdot \mathbf{D} = 0$ (Eq. 3.2). Because $\mathbf{D} = \varepsilon \mathbf{E}$, it follows that $\varepsilon \nabla \cdot \mathbf{E} = 0$. Both can only be fulfilled if $\varepsilon = 0$, which means that $\mathbf{D} = 0$. At the same time, $\mathbf{D} = \varepsilon_0 \mathbf{E} + \mathbf{P}$, therefore, with, $\mathbf{D} = 0$ it follows that the polarization is antiparallel to the electric field: $\varepsilon_0 \mathbf{E} = -\mathbf{P}$ [46].

This is also the physical reason why a peak is often not located at the oscillator position, but can be found somewhere between the corresponding pole and the zero of the dielectric function (or between the maximum of the imaginary part of the dielectric function and the maximum of the imaginary part of the inverse dielectric function, if absorption is taken into account). Since the polarization is oriented antiparallel relative to the electric field at $\varepsilon = 0$ and longitudinal relative to the direction of the wave, this case is called the longitudinal optical or LO mode, whereas the transversal optical mode is called the TO mode.

As I already stated, it is not possible to excite the LO mode directly by one-phonon processes like they are common in IR spectroscopy. Nevertheless, we will see that the LO mode position will be in many instances, e.g., for parallel polarized light at higher angles of incidence and, in particular, also for anisotropic media, relevant. How is the LO position related to the band strength? First of all, it is important to note that for $\varepsilon' < 0$, that is for stronger oscillators and wavenumbers between the pole and the zero in Fig. 5.25, the index of refraction becomes purely imaginary and the wave cannot longer penetrate into the material. As a consequence, reflectance becomes unity between TO and LO positions, which drift apart further with increasing oscillator strength. As already mentioned above, for the more physically relevant situation that we have damping and, correspondingly, the dielectric function is complex, it is no longer the zero of the real part of the dielectric function that determines the LO position, but instead the maximum of the negative imaginary part of the dielectric loss function $-\text{Im}(1/\varepsilon)$. Considering damping explicitly, the common form of the semiempirical four-parameter model is the following:

$$\varepsilon = \varepsilon_\infty \prod_{j=1}^{N} \frac{\tilde{v}_{LO,j}^2 - \tilde{v}^2 - i\tilde{v}\gamma_{LO,j}}{\tilde{v}_{TO,j}^2 - \tilde{v}^2 - i\tilde{v}\gamma_{TO,j}}. \tag{5.89}$$

Herein, the TO and LO mode positions of the jth mode are $\tilde{v}_{TO,j}$ and $\tilde{v}_{LO,j}$ and $\gamma_{TO,j}$ and $\gamma_{LO,j}$ their respective damping constants. Accordingly, a concrete functional dependence of the damping constant from the wavenumber is not being assumed. Instead, two particular values, one at the TO and one at the LO position, are employed. To remain on physical grounds, we have to assume that $\gamma_{LO,j} - \gamma_{TO,j} > 0$ otherwise we obtain the unphysical result that the imaginary part of the dielectric function can attain negative values which would imply emission. This is a serious restriction which also limits the applicability of Eq. (5.89). It has been shown that this condition can be somewhat relaxed by not requiring that the above condition must apply to each oscillator individually, but to demand instead that the sum of all differences is positive [47]:

$$\sum_{j=1}^{N} (\gamma_{LO,j} - \gamma_{TO,j}) > 0. \tag{5.90}$$

This is certainly a much less demanding condition, which however does not work as reliable as one would assume [48] and even when it is obeyed, unphysical results can be obtained. If the latter does not happen, one problem persists, which is that weaker and less important oscillators take on unrealistic values to compensate for stronger oscillators, for which $\gamma_{LO,j} - \gamma_{TO,j} < 0$. We will come back to this issue later in Section 5.6 and investigate it in more detail. For the moment, it may suffice to state that the condition in Eq. (5.90) needs to be implanted in some way as a penalty function if Eq. (5.89) is used for dispersion analysis. Nevertheless, dispersion analysis may require much more supervision and user intervention if Eq. (5.89) is employed, than if the CDHO model according to Eq. (5.27) is used, in particular if there are many and/or overlapping bands in a spectrum. Nevertheless, when successfully applied, the average error substantially decreases, in our example in Fig. 5.26 by more than 25%, which is less than what is attainable with the coupled oscillator model, but, given the much smaller computational effort, not bad at all.

5.5.2 Kim oscillator

This is a model I was made aware of by Wolfgang Theiss (R.I.P.) who authored the software SCOUT for dispersion analysis. The Kim oscillator was originally suggested to model the optical dielectric function of zinc-blende semiconductors in the visible part of the spectrum [49], but it can also be integrated into the classical CDHO model [50] and used to some advantage for the infrared spectral range. The authors claim that "it is found to be more generally valid than the

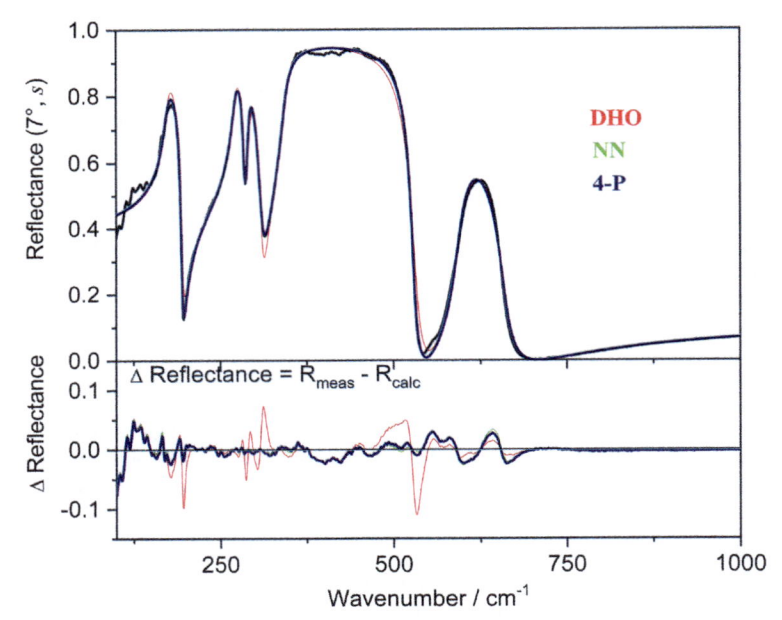

FIG. 5.26 Comparison between experimental reflectance spectrum *(black line)* and simulated spectra based on the classical damped harmonic oscillator model *(red line)*, on the dispersion relation assuming coupling of the nearest neighbors *(green line)* and on the dispersion relation based on the semi-empirical 4 parameter model *(blue line)*.

harmonic-oscillator model…" and that it is Kramers-Kronig conform. Apart from some advantages for the description of the dielectric functions of semiconductors in the visible spectral range, it also assumes that the damping constant is actually no constant but varies, something which also motivates the introduction of some of the other models in this section. A big advantage of the Kim oscillator is that it allows to switch seamlessly from a Lorentzian to a Gaussian bandshape. This is possible thanks to,

$$\gamma_j = \gamma_{0,j} \exp\left(-\frac{1}{1+\sigma_j^2}\left(\frac{\tilde{\nu}-\tilde{\nu}_{0,j}}{\gamma_0}\right)^2\right), \tag{5.91}$$

in which σ_j is the so-called Gauss-Lorentz switch. For $\sigma_j = 0$, the *j*th oscillator shows a Gaussian profile with regard to the imaginary part of the dielectric function. This profile becomes a Lorentz profile if $\sigma_j \geq 10$, cf. Fig. 5.27. The advantages compared to the Brendel oscillator, which I introduce and discuss in Section 5.4, are that it is mathematically less

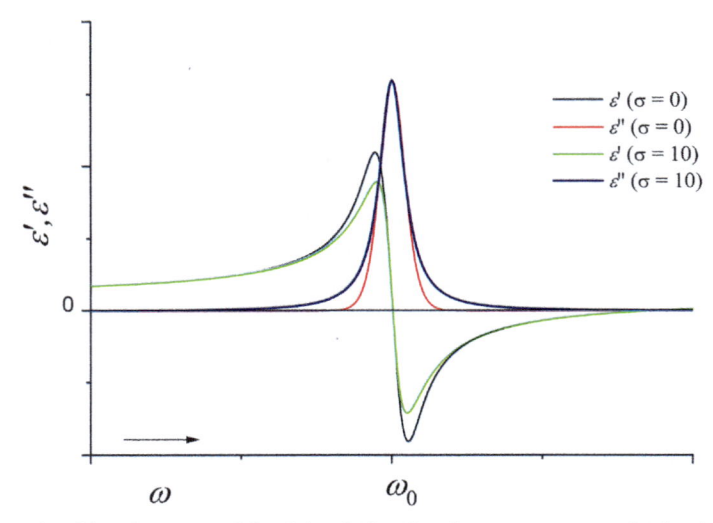

FIG. 5.27 Comparison between real and imaginary part of the dielectric function close to a resonance for the classical **damped harmonic oscillator** model *(black and red curves)* and the corresponding curves for a Kim oscillator *(green and blue)*.

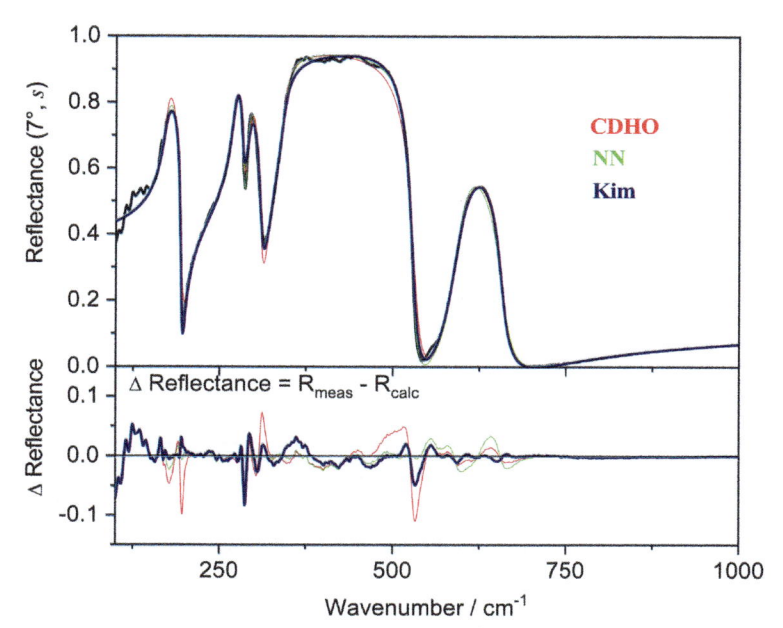

FIG. 5.28 Comparison between experimental reflectance spectrum *(black line)* and simulated spectra based on the classical damped harmonic oscillator model *(red line)*, on the dispersion relation assuming coupling of the nearest neighbors *(green line)*, and on the Kim oscillator *(blue line)*.

demanding and that if σ_j is fixed at zero, a Gaussian profile with only three parameters is obtained that is physically based on the damped harmonic oscillator.

At this point, I do not want to go into the discussion about lineshapes, Lorentzian- and Gaussian broadening, etc. Suffice to say for the moment that I think that this discussion is mainly needed for gases and that for liquids and solids in most cases the classical damped harmonic oscillator with its Lorentzian lineshape is adequate. On the other hand, there are cases where only oscillators with a Gaussian lineshape can fit certain IR bands of condensed matter, e.g., that of the HO-stretching vibrations in (liquid) water. In comparison with the damped harmonic oscillator model and the nearest neighbors model, the Kim oscillator performs favorably concerning the first and only very slightly better than the second. Apart from all discussions about the physical meaningfulness, it definitely allows to capture the optical constant functions well (cf. Fig. 5.28) with one important drawback, which is that the more the switch forces the lineshape towards Gaussian, the less the real and imaginary parts are Kramers-Kronig conform (cf. Section 5.8.1). This means that the real part derived from the imaginary part by the Kramers-Kronig relations deviates from the one calculated by the dispersion relation despite the authors' claim that their model is Kramers-Kronig conform.

5.5.3 Classical model with frequency-dependent damping constant

As already stated in the preceding sections, directly coupled and/or anharmonic vibrations lead to a frequency-dependent damping constant. Born and Huang treated this situation in their famous book "Dynamical theory of Crystal Lattices" and derived that the damping constant should be frequency dependent [51].

If one does not know the particular form of a function (and, in fact, this is a formidable problem which was not solved satisfactorily in the middle of the 1980s [52], after which, it seems, any endeavor towards a solution was discontinued), it is always a possibility to develop the unknown relation in an infinite series and just use the nth partial sum until the remaining error falls under a certain limit. For the damping constant, the following series was suggested [53]:

$$\gamma = \gamma_0 \left(1 + a\tilde{\nu}^2 + b\tilde{\nu}^4 + c\tilde{\nu}^6 \ldots \right). \tag{5.92}$$

Since with enough parameters it is always possible to obtain a better fit, we limit our discussion right from the start to the case $a \neq 0$; $b, c, \ldots = 0$, which means that we have another four-parameter model based on the CDHO model, for which we replace γ by $\gamma_0(1 + a\tilde{\nu}^2)$. The resulting real and imaginary parts of the dielectric function are somewhat altered, but not too much (in particular no shift of the extrema, only a small change of band shapes and no alteration away from the extrema), so that this seems to be a very promising approach (cf. Fig. 5.29).

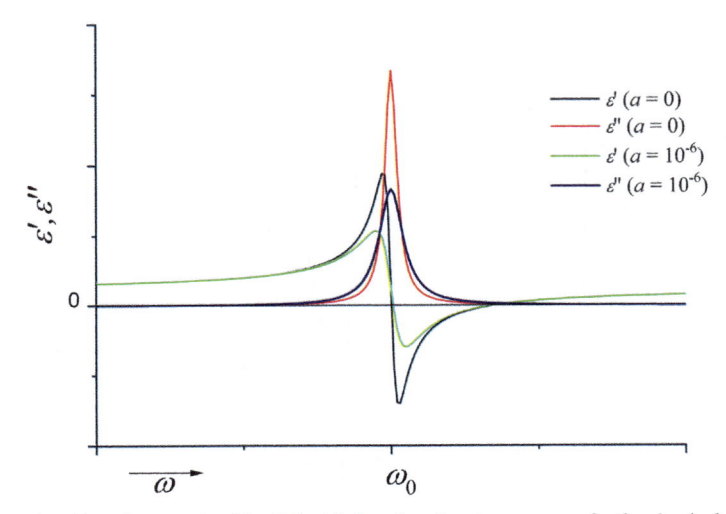

FIG. 5.29 Comparison between real and imaginary parts of the dielectric function close to resonance for the classical damped harmonic oscillator model with a frequency-independent damping constant *(black and red curves)* and the corresponding curves for a frequency-dependent damping constant according to $\gamma = \gamma_0(1 + a\tilde{\nu}^2)$ *(green and blue)*.

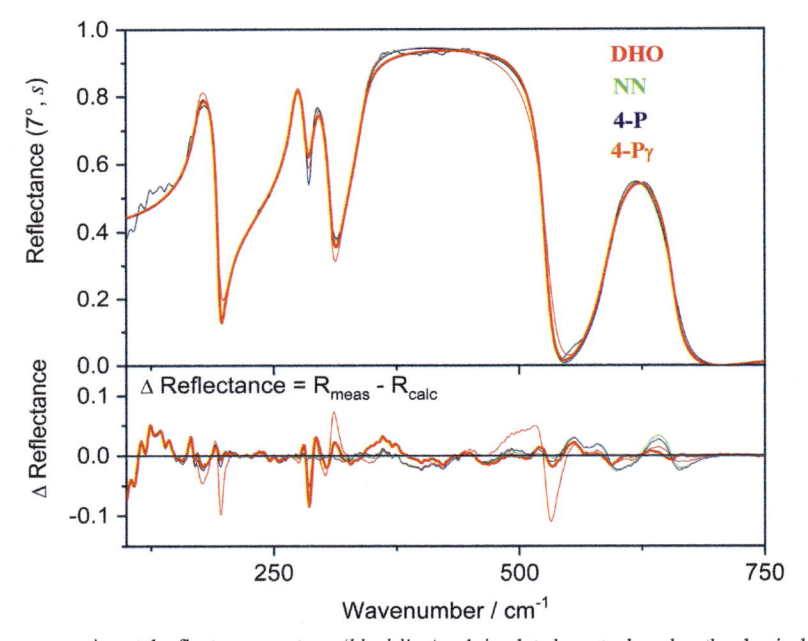

FIG. 5.30 Comparison between experimental reflectance spectrum *(black line)* and simulated spectra based on the classical dispersion relation *(red line)*, on the dispersion relation assuming coupling of the nearest neighbors *(green line)*, on the dispersion relation based on the semiempirical 4 parameter model *(blue line)* and on the classical dispersion relation with frequency-dependent damping constant *(orange line)*.

How does it perform in terms of deviation from the experimental spectra? Looking at Fig. 5.30, it seems that the frequency-dependent damping constant leads to a much better fit of the right wing of a very strong band at the expense of the fit of the left wing. Overall, the average error is somewhat smaller than that of the nearest neighbor model. Given the fact that this model combines the advantages of the classical model, i.e., nearly the same small computational effort and less intervention and supervision with a clearly reduced average error this model also seems to be a good choice.

5.5.4 Classical model with complex oscillator strength

Despite the improvements possible due to the models already introduced in this section, there might still remain a wish to improve the agreement between experiment and model further without sacrificing too much physical meaningfulness of the oscillator parameters (Of course, a better agreement can always be achieved by simply using more oscillators or introducing combinations of the different parameters, and you will be made acquainted in Section 5.8 with a method where any significance is sacrificed for a good agreement even in case of countless spectral features). In the case of the coupled oscillator scheme introduced in Section 5.4, the dispersion relations can get very complex fast, even if one considers only the next neighbors. Maybe, there would be a way to approximate this model? Indeed, Humlíček, Henn, and Cardona found such an approximation, which is given by [54]:

$$\varepsilon = \varepsilon_\infty + \sum_{j=1}^{N} \frac{S_j^2 - i\tilde{v}\sigma_j}{\tilde{v}_{\Omega,j}^2 - \tilde{v}^2 - i\tilde{v}\Gamma_j}. \tag{5.93}$$

This leads to a kind of complex oscillator strength. The idea behind this model is to bundle all interactions due to coupling in the interaction parameter σ_j. The need for multiplying the imaginary part of the oscillator strength by the wavenumber is to obtain the complex conjugate if $\tilde{v}$ is replaced by $-\tilde{v}$ in Eq. (5.93) to ensure Kramers-Kronig conformity (cf. Section 5.8.1). If we write Eq. (5.79) in a generalized form,

$$\frac{d^2 x_j}{dt^2} + \gamma_j \frac{dx_j}{dt} + \sum_k \gamma_{jk} \frac{d(x_k - x_j)}{dt} + \tilde{v}_{0,j}^2 x_j = q_j E_x, \tag{5.94}$$

it is easy to see that the algebra behind would be indeed cumbersome (cf. Section 5.4). To simplify the situation, we assume that for the two modes j and k the damping constants γ_j and γ_k, as well as their interaction damping constant γ_{jk} are small compared with the resonance wavenumbers. In this case, the above equation reduces to:

$$\begin{aligned}(\tilde{v}_{0,j}^2 - \tilde{v}^2 - i\tilde{v}\gamma_j)x_j - i\tilde{v}\gamma_{jk}(x_j - x_k) = q_j E_x \\ (\tilde{v}_{0,k}^2 - \tilde{v}^2 - i\tilde{v}\gamma_k)x_k - i\tilde{v}\gamma_{jk}(x_k - x_j) = q_k E_x\end{aligned}. \tag{5.95}$$

Furthermore, we assume that the difference between $\tilde{v}_{0,j}$ and $\tilde{v}_{0,k}$ is large compared to γ_j and γ_k, as well as γ_{jk}. In a spectral region in the vicinity around $\tilde{v}_{0,j}$, solutions to Eq. (5.95) can be approximated by:

$$\begin{aligned}x_j &\approx \frac{q_j - i\tilde{v}\gamma_{jk}q_k/(\tilde{v}_{0,k}^2 - \tilde{v}_{0,j}^2)}{\tilde{v}_{0,j}^2[1 + 2i\tilde{v}\gamma_{jk}q_k/(\tilde{v}_{0,k}^2 - \tilde{v}_{0,j}^2)] - \tilde{v}^2 - i\tilde{v}(\gamma_j + \gamma_{jk})} E_x \\ x_k &\approx \frac{-i\tilde{v}\gamma_{jk}q_k/(\tilde{v}_{0,k}^2 - \tilde{v}_{0,j}^2)}{\tilde{v}_{0,j}^2[1 + 2i\tilde{v}\gamma_{jk}q_k/(\tilde{v}_{0,k}^2 - \tilde{v}_{0,j}^2)] - \tilde{v}^2 - i\tilde{v}(\gamma_j + \gamma_{jk})} E_x\end{aligned}. \tag{5.96}$$

The first part of Eq. (5.96) is similar to Eq. (5.93) for a particular j in case that,

$$\begin{aligned}\tilde{v}_{\Omega,j} &= \sqrt{\tilde{v}_{0,j}^2[1 + 2i\tilde{v}\gamma_{jk}q_k(\tilde{v}_{0,k}^2 - \tilde{v}_{0,j}^2)]}, \Gamma_j = (\gamma_j + \gamma_{jk}) \\ \sigma_j &= \gamma_{jk}q_j q_k N/[(\mu\varepsilon_0)(\tilde{v}_{0,k}^2 - \tilde{v}_{0,j}^2)] = \gamma_{jk} S_j S_k/(\tilde{v}_{0,k}^2 - \tilde{v}_{0,j}^2)\end{aligned}. \tag{5.97}$$

Note that from the necessity that the dielectric function has to decrease faster than $1/\tilde{v}$ if $|\tilde{v}| \to \infty$ (see Section 5.8.3), a sum rule can be derived according to which:

$$\sum_{j=1}^{N} \sigma_j = 0. \tag{5.98}$$

Despite some of you might have the feeling that this derivation is a little far-fetched, the correspondence of the model with the experimental values is excellent as is demonstrated by Fig. 5.31. It is not only of the same order of magnitude as the nearest neighbor model but even slightly better.

The particular charm of this model is not only its usefulness with regard to capturing the dielectric function particularly accurately. Additionally, it also can be directly linked to so-called Fano-like profiles, very much like this is possible for the coupled oscillator model. These profiles are very important to comprehend the various coupling effects in IR spectra and their importance for understanding surface-enhanced IR spectroscopy and strong coupling (cf. Sections 7.2 and 7.3). The model presented in Eq. (5.93) can be linked to the classical Fano line shape $(\epsilon + Q)^2/(\epsilon^2 + 1)$ in the limit of weak coupling $Q \gg 1$ with $Q = -2\tilde{v}_{\Omega,j}/\sigma_j$ if we introduce the reduced wavenumber $\epsilon = (\tilde{v}_{\Omega,j} - \tilde{v})/(\Gamma/2)$ and approximate the imaginary part around the resonance wavenumber by:

$$\varepsilon'' = const. - \frac{\sigma_j}{\Gamma_j} \frac{\epsilon - \tilde{v}_{\Omega,j}/\sigma_j}{\epsilon^2 + 1}. \tag{5.99}$$

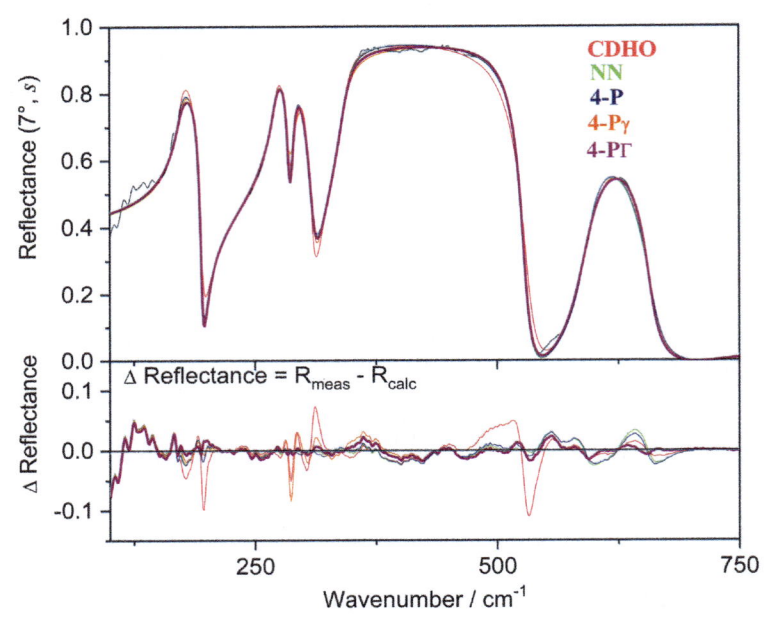

FIG. 5.31 Comparison between experimental reflectance spectrum *(black line)* and simulated spectra based on the classical dispersion relation *(red line)*, on the dispersion relation assuming coupling of the nearest neighbors *(green line)*, on the dispersion relation based on the semi-empirical four-parameter model *(blue line)*, on the classical dispersion relation with frequency-dependent damping constant *(orange line)* and on the classical dispersion relation with interaction parameter *(magenta line)*.

Personally, I prefer a slightly different representation of Eq. (5.93) which better illustrates the switch between the typical Lorentz profile and the Fano-like band shape. It is loosely based on an idea from Terry, who included a phase factor in the CDHO [55].

$$\varepsilon = \varepsilon_\infty + \sum_{j=1}^{N} \frac{S_j^2 \exp(i\phi_j)}{\tilde{v}_{0,j}^2 - \tilde{v}^2 - i\tilde{v}\gamma_j}, \tag{5.100}$$

In contrast to Terry's oscillator model my suggestion respects the Kramers-Kronig conformity:

$$\varepsilon = \varepsilon_\infty + \sum_{j=1}^{N} \frac{S_j^2 \left(\cos\phi_j + i\tilde{v} \sin\phi_j / \tilde{v}_{0,j} \right)}{\tilde{v}_{0,j}^2 - \tilde{v}^2 - i\tilde{v}\gamma_j}. \tag{5.101}$$

It is easy to see that for $\phi_j = 2n\pi$, $n = \pm 1, 2, 3\ldots$ the CDHO model is recovered. How do the oscillators look like for other phases in-between? Since $\tilde{v} \approx \tilde{v}_{0,j}$ around a band, the imaginary part is ½ π ahead of the real part (cf. Fig. 5.32). Depending on what is more important for you, exactly the same values for the real and the phase-shifted imaginary part or Kramers-Kronig conformity, you can either choose Eq. (5.100) or Eq. (5.101). For dispersion analysis, the differences are small with advantages for Eq. (5.101) due to the Kramers-Kronig conformity. Compared to Eq. (5.93), Eq. (5.101) leads to the same fit quality, superior to all other models.

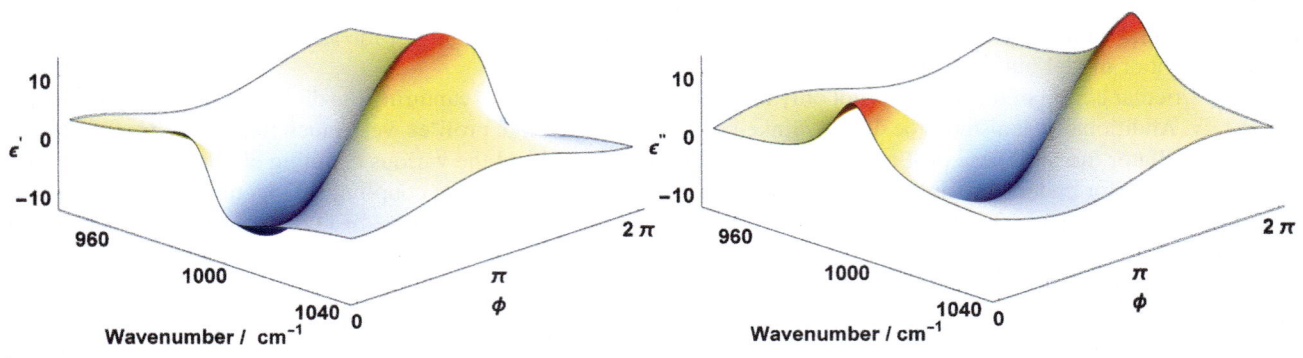

FIG. 5.32 Real and imaginary part of the oscillator model with complex oscillator strength according to Eq. (5.101) in dependence of the phase.

5.5.5 Convolution model

For glasses and amorphous solids, the classical oscillator model seems to require an extension. From the experimental point of view, this is due to the experience that for glasses/amorphous materials, for which also polycrystalline forms exist (glasses/amorphous solids with the same composition as a crystal), the glasses often show bands at the same positions which are, however, broadened. This broadening can have several causes. In the Random Network Theory, it is basically assumed that glass is completely disordered; therefore, there would be a distribution of different bond angles and strengths and, accordingly, a distribution of different oscillator frequencies. In contrast, Crystallite Theory assumes that glasses consist of crystallites, which are so small that they are X-ray amorphous. As a consequence, there would be no distribution of resonance frequencies, but very high damping constants instead, since the damping constant is the inverse lifetime of a phonon, which obviously decreases with the size of the crystallite. In practice, both theories have their drawbacks and it can be assumed that both factors, resonance broadening and damping constant increase, play a role. Therefore, Efimov and Makarova suggested the following dispersion relation which they termed "convolution model" [56]:

$$\varepsilon(\tilde{v}) = \varepsilon_\infty + \sum_{i=1}^{N} \frac{S_i^2}{\sqrt{2\pi}\sigma_i} \int_{-\infty}^{+\infty} \frac{\exp\left(-(x-\tilde{v}_i)^2/2\sigma_i\right)}{x^2 - \tilde{v}^2 - i\tilde{v}\gamma_i} \, dx. \tag{5.102}$$

The convolution model has σ_i as a fourth parameter, which describes the standard deviation of the resonance wavenumber from its mean value $\tilde{v}_{0,i}$. It is instructive to investigate the spectral changes that a band undergoes with increasing σ_i in comparison with those that occur to a band when the damping constant is increased (cf. Fig. 5.33, left panel). Indeed, increasing the standard deviation leads to a marked blue shift in the resonance wavenumber which is not observed when spectra from glasses and crystalline samples are compared (indeed, if there are blue shifts, then it is usually the polycrystalline sample that shows blue shifts due to an optical crystallite size effect causing a TO-LO shift, see Chapter 15). Taken together with the fact that unrealistically large standard deviations result and the fits become highly arbitrary (cf. Fig. 5.33, right side, where different starting conditions for the high wavenumber bands lead to similar average errors within 1%, but also to drastically altered values of the oscillator parameters). While the oscillator parameters lose their physical significance, still a somewhat better fit compared to using the classical model is possible, which means that the optical constants

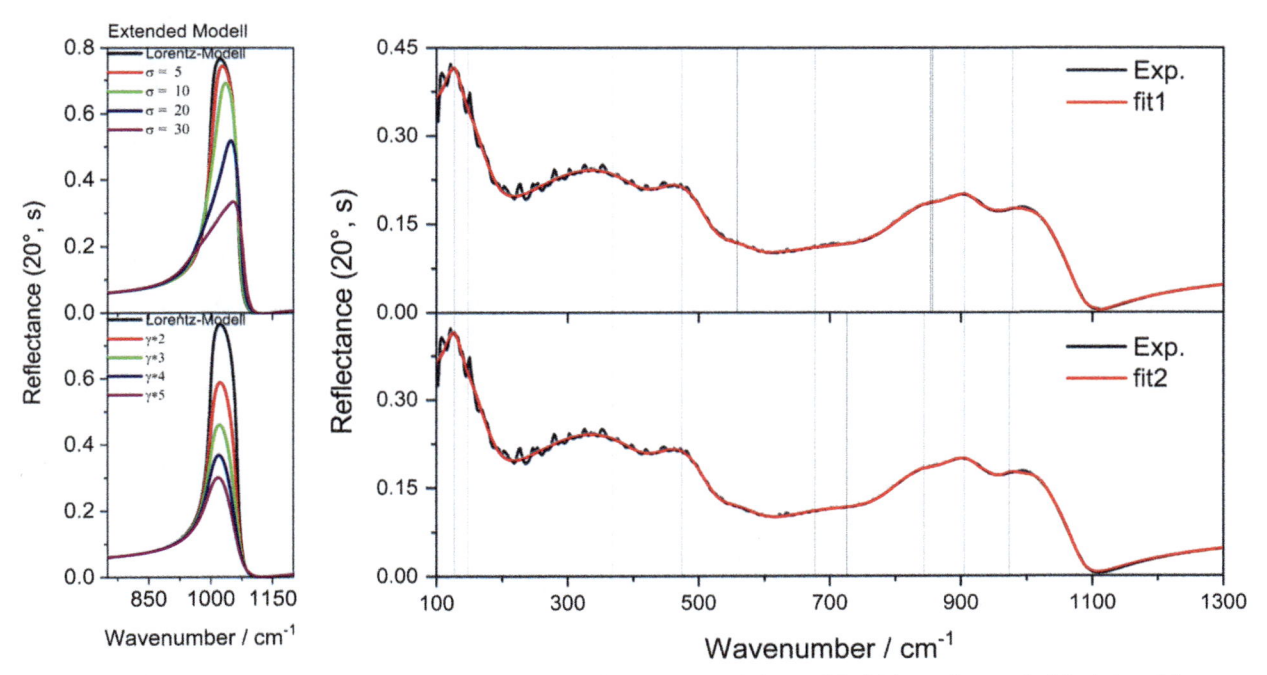

FIG. 5.33 *Left panel*: Comparison of spectral changes of a band according to the convolution model with increasing standard deviation of the resonance wavenumber and of spectral changes of a band due to the CDHO model with increasing damping constant. *Right panel*: Comparison of two oscillator fits of a Fresnoite glass with two very different results concerning the oscillator parameters but similar standard errors (*dashed lines* indicate oscillator positions) [57].

might be more accurately represented. However, since the convolution model is much more computationally demanding, increasing the number of oscillators and using the classical model might also do the trick (or using one of the other four-parameter models). On a side note, the convolution model is often termed Brendel oscillator, after the first author of ref. [58], where the same model has been suggested somewhat later as in ref. [56].

In my opinion, the convolution model is intrinsically flawed because it assumes that glasses consist of elementary cells that are of cubic symmetry. Indeed, glasses are optically isotropic like a crystal of this symmetry, but in contrast to single crystals, glasses show regions of medium order, which are randomly oriented and these regions are in general anisotropic. Therefore, orientational averaging plays a role, which explains to a good degree the weaker intensities in glasses, so that this averaging and an increase of the damping constants can explain the spectral changes (cf. also Chapter 15) [57]. Certainly, there will also exist a distribution of resonance frequencies, but, realistic distributions will lead to minor spectral changes according to Eq. (5.102) that need not be considered.

5.6 Dispersion relation—Inverse dielectric function model

While band shapes and peak positions of bands in reflectance and transmittance spectra of samples with scalar dielectric functions recorded with *s*-polarized light change comparably weakly with increasing angle of incidence, the changes are much more prominent for *p*-polarized light. Interestingly, band maxima shift and band shapes change characteristically. This is, for reasons explained in the second part of this book, in particular, obvious when the material under investigation is polycrystalline for strongly anisotropic materials and crystallites, which are small compared to the wavelength. We have already introduced Fresnoite, which can easily be prepared in this form by annealing a Fresnoite glass sample. The reflectance spectra of correspondingly prepared polycrystalline Fresnoite with *p*-polarized light and comparably low and high angles of incidence are shown in Fig. 5.34. Obviously, the spectrum with a low angle of incidence shows features

$$R_p = \left| \frac{\left(\varepsilon\cos^2\alpha_i\right)^{\frac{1}{2}} - \left(1 - \frac{1}{\varepsilon}\sin^2\alpha_i\right)^{\frac{1}{2}}}{\left(\varepsilon\cos^2\alpha_i\right)^{\frac{1}{2}} + \left(1 - \frac{1}{\varepsilon}\sin^2\alpha_i\right)^{\frac{1}{2}}} \right|^2$$

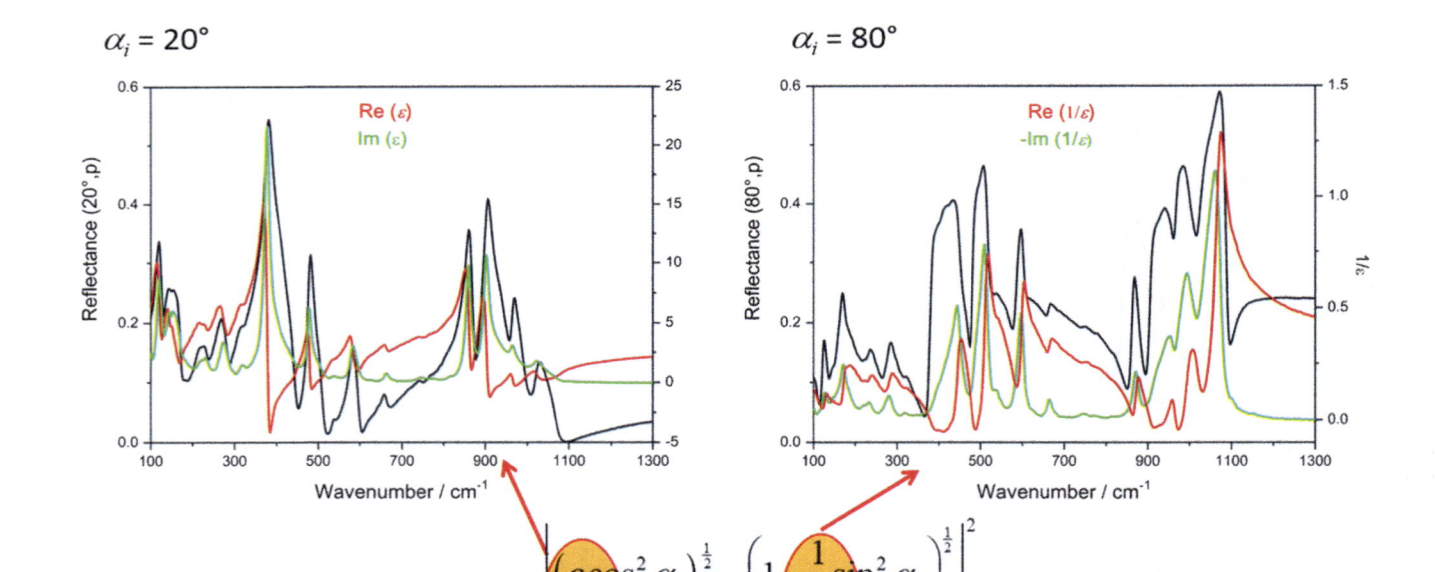

FIG. 5.34 *Left panel*: Comparison of the reflectance spectrum of polycrystalline Fresnoite with small crystallites compared to the resolution limit recorded with an angle of incidence of 20° and *p*-polarized light with the real and the imaginary part of the dielectric function. *Right panel*: Comparison of the reflectance spectrum of polycrystalline Fresnoite with small crystallites recorded with an angle of incidence of 80° and *p*-polarized light with the real and the negative imaginary part of the inverse dielectric function.

comparable to those of the dielectric function, which is not very surprising, as we derived in Section 4.2.2. Accordingly, the reflection coefficient is given by:

$$r_p = \frac{\sqrt{1 - \frac{1}{\varepsilon}\sin^2\alpha_i} - \sqrt{\varepsilon}\cos\alpha_i}{\sqrt{1 - \frac{1}{\varepsilon}\sin^2\alpha_i} + \sqrt{\varepsilon}\cos\alpha_i}, \qquad (5.103)$$

where we have assumed that the incidence medium is vacuum. If we write the relation for r_p in slightly different form and multiply it with its conjugate complex, we obtain:

$$R_p = \left| \frac{(\varepsilon\cos^2\alpha_i)^{\frac{1}{2}} - \left(1 - \frac{1}{\varepsilon}\sin^2\alpha_i\right)^{\frac{1}{2}}}{(\varepsilon\cos^2\alpha_i)^{\frac{1}{2}} + \left(1 - \frac{1}{\varepsilon}\sin^2\alpha_i\right)^{\frac{1}{2}}} \right|^2. \qquad (5.104)$$

For low angles of incidence α_i, $\cos^2\alpha_i$ is close to unity and the dielectric function dominates, whereas for higher angles of incidence, its inverse takes over and governs the shape of the bands and their position in the spectrum (this has been discussed in the literature already some time ago [59], but it is everything else but well known). Accordingly, it might be of advantage to model a corresponding spectrum using the inverse dielectric function, which can also be described in a form very similar to the classical oscillator model:

$$\varepsilon^{-1} = \varepsilon_\infty^{-1} - \sum_{j=1}^{N} \frac{S_{j,LO}^2}{\tilde{v}_{j,LO}^2 - \tilde{v}^2 - i\tilde{v}\gamma_{j,LO}}. \qquad (5.105)$$

It seems that Humlíček was the first to introduce and use Eq. (5.105) [60], but it is not clear how this equation was derived (probably it was introduced empirically; another way to introduce Eq. (5.105) would be to use the Kramers-Kronig sum rules, Section 5.8.3, and conclude that the inverse of the dielectric function has a similar form as the dielectric function itself); therefore, we will attempt to derive this equation in the following, also because the physical meaning of the LO oscillator strength does not become clear in contrast to the other parameters. To do this, we assume a 3D lattice of cubic symmetry consisting of two different atoms (ions) per unit cell exactly like in the case of the CDHO. We start with the equation of the (transversal) damped motion of two atoms under the assumption of a forced oscillation, which we know already (Eq. (5.18), where Δr is replaced by x, E stands for the amplitude of $\mathbf{E}$ (the same holds for P) and the subscript 0 by TO indicate transversal optical vibrations):

$$\mu\frac{d^2x}{dt^2} + \mu\gamma_{TO}\frac{dx}{dt} + \mu\omega_{TO}^2 x = qE. \qquad (5.106)$$

Characteristic for transversal optical vibrations is that the vibrations are perpendicular to the wave propagation like the changes of the $\mathbf{E}$-field and the $\mathbf{H}$-field in scalar media. For longitudinal optical (LO) vibrations, however, the vibrations would occur parallel to the wave propagation [61]. Such vibrations cannot be excited in cubic crystal by infrared radiation, but if the crystal is noncentrosymmetric, vibrations exist which are both, infrared and Raman active; in this case, LO vibrations/phonons can be excited by Raman spectroscopy [62], e.g., by using a 180° geometry like in a microscope, where backscattered Raman-shifted photons are registered. Longitudinal vibrations in an isotropic crystal are accompanied by an additional net polarization introduced by the electric field, qP/ε_0 [61], which increases the resonance frequency from ω_{TO} to ω_{LO}:

$$\mu\frac{d^2x}{dt^2} + \mu\gamma_{LO}\frac{dx}{dt} + \mu\omega_{LO}^2 x = qE + qP/\varepsilon_0. \qquad (5.107)$$

Here, γ_{LO} is the LO damping constant. This damping constant is for harmonic vibrations equal to γ_{TO}, but it can, for modeling purposes, also be seen as a free parameter. There are two different paths we can follow to solve Eq. (5.107). The first path is analogous to the derivation of the influence of the local field of Lorentz according to Eqs. (5.23)–(5.26):

$$\mu\frac{d^2x}{dt^2} + \mu\gamma_{LO}\frac{dx}{dt} + \mu\omega_{LO}^2 x = qE + qP/\varepsilon_0 = qE + q^2Nx/\varepsilon_0 \rightarrow$$

$$\mu\frac{d^2x}{dt^2} + \mu\gamma_{LO'}\frac{dx}{dt} + \mu(\omega_{LO}^2 - q^2N/(\mu\varepsilon_0))x = qE \rightarrow \qquad (5.108)$$

$$\mu\frac{d^2x}{dt^2} + \mu\gamma_{LO'}\frac{dx}{dt} + \mu\omega_{LO'}^2 x = qE$$

Like in the case of the local field of Lorentz, the resonance frequency is reduced, in this case ω_{LO} will be decreased to $\omega_{LO'}$. If we set the same time dependence for the electric field and the displacement x and use that $P = Np = NxqE$, we obtain,

$$P = \frac{Nq^2 E}{\mu\left(\omega_{LO'}^2 - \omega^2 - i\gamma_{LO'}\omega\right)}, \tag{5.109}$$

and, finally, with $S^2 = Nq^2/\mu\varepsilon_0$ for the dielectric function:

$$\varepsilon(\omega) = 1 + \frac{S^2}{\omega_{LO'}^2 - \omega^2 - i\gamma_{LO'}\omega}. \tag{5.110}$$

To determine the difference between ω_{LO} and $\omega_{LO'}$, we use our definition of $\omega_{LO'}$,

$$\omega_{LO'}^2 = \omega_{LO}^2 - S^2, \tag{5.111}$$

together with the formula for the transversal optical dielectric function,

$$\varepsilon(\omega) = 1 + \frac{S^2}{\omega_{TO}^2 - \omega^2 - i\gamma_{TO}\omega}, \tag{5.112}$$

in which, we set $\omega = 0$ to obtain:

$$\varepsilon(0) = 1 + \frac{S^2}{\omega_{TO}^2} \rightarrow \omega_{TO}^2(\varepsilon(0) - 1) = S^2. \tag{5.113}$$

In addition, we invoke the Lyddane-Sachs-Teller (LST) relation [63].

$$\frac{\omega_{LO}^2}{\omega_{TO}^2} = \frac{\varepsilon(0)}{\varepsilon_\infty}. \tag{5.114}$$

In this relation, we set $\varepsilon_\infty = 1$ and with the resulting relation $\omega_{LO}^2 = \omega_{TO}^2 \varepsilon(0)$, we get:

$$\omega_{LO}^2 - \omega_{TO}^2 = S^2. \tag{5.115}$$

If we employ this result for S^2 in Eq. (5.111), the result is

$$\omega_{LO'}^2 = \omega_{TO}^2. \tag{5.116}$$

Obviously, there is no difference between the longitudinal dielectric function and its transversal counterpart, which is certainly what would be expected of a cubic crystal for long wavelength vibrations. Now let us pursue the second path.

We know that $E + P/\varepsilon_0 = \varepsilon E$. Therefore

$$\mu \frac{d^2 x}{dt^2} + \mu\gamma_{LO} \frac{dx}{dt} + \mu\omega_{LO}^2 x = q\varepsilon E. \tag{5.117}$$

If we now assume, as usual, that the fields E as well as x have the same time harmonic dependence $\exp(-i\omega t)$, we find that

$$-\mu\omega^2 x - \mu i\gamma_{LO}\omega x + \mu\omega_{LO}^2 x = q\varepsilon E, \tag{5.118}$$

with the solution:

$$x = \frac{q\varepsilon E}{\mu\left(\omega_{LO}^2 - \omega^2 - i\gamma_{LO}\omega\right)}. \tag{5.119}$$

If we multiply the solution with the number of oscillators per unit volume N and the effective charge q, the result is

$$P = \frac{Nq^2 \varepsilon E}{\mu\left(\omega_{LO}^2 - \omega^2 - i\gamma_{LO}\omega\right)}. \tag{5.120}$$

Using $P = \varepsilon_0(\varepsilon - 1)E$, getting rid of E and rearranging gives,

$$\frac{\varepsilon - 1}{\varepsilon} = 1 - \frac{1}{\varepsilon} = \frac{Nq^2}{\mu\varepsilon_0\left(\omega_{LO}^2 - \omega^2 - i\gamma_{LO}\omega\right)}, \tag{5.121}$$

from which we get:

$$\varepsilon(\omega)^{-1} = 1 - \frac{S_{LO}^2}{\omega_{LO}^2 - \omega^2 - i\gamma_{LO}\omega}. \tag{5.122}$$

Accordingly, it follows that $S_{LO}^2 = S_{TO}^2 = S^2 = Nq^2/\mu\varepsilon_0$. Therefore, for a dispersion relation with only one oscillator, there is no difference between S_{LO} and S_{TO}. Converting Eq. (5.122) to wavenumbers and extending it to multiple oscillators seems to lead directly to Eq. (5.105), but there are some more pitfalls to consider. Before discussing them, let me point out that the negative sign in front of the fraction is particularly important. In the empirically used relations, this sign was usually positive, which is obviously incorrect. From one of the so-called Kramers-Kronig sum rules (cf. Section 5.8), it is immediately clear that the sign must be negative:

$$\int_0^\infty \tilde{v}\,\text{Im}(-1/\varepsilon(\tilde{v}))d\tilde{v} = \frac{\pi}{2}S^2. \tag{5.123}$$

As a first step toward a multioscillator version of Eq. (5.122) we want to include the dielectric background ε_∞. In a naïve way, we could simply replace unity by ε_∞, like this is done in the empiric version, but is this correct? We start by writing:

$$\begin{aligned}\varepsilon(\tilde{v}) &= \varepsilon_\infty + \frac{S^2}{\tilde{v}_{TO}^2 - \tilde{v}^2 - i\gamma_{TO}\tilde{v}} \quad \text{(I)} \\ \varepsilon(\tilde{v})^{-1} &= \varepsilon_\infty^{-1} - \frac{S^2}{\tilde{v}_{LO}^2 - \tilde{v}^2 - i\gamma_{LO}\tilde{v}} \quad \text{(II)}\end{aligned} \tag{5.124}$$

We assume that we want to evaluate Eq. (5.124) for a wavenumber $\tilde{v}_f$ which is larger than both $\tilde{v}_{TO}$ and $\tilde{v}_{LO}$, $\tilde{v}_f > \tilde{v}_{TO}, \tilde{v}_{LO}$. If we square $\tilde{v}_f$ than $\tilde{v}_f^2 \gg \tilde{v}_{TO}^2, \tilde{v}_{LO}^2$. In addition, since damping constants are usually small compared to the wavenumber, the imaginary parts will be very small if $\tilde{v}_f > \tilde{v}_{TO}, \tilde{v}_{LO}$. If we choose a $\tilde{v}_f$, which is in the transparency range between the mid-infrared and UV/Vis spectral range, we obtain:

$$\begin{aligned}\varepsilon(\tilde{v}) &= \varepsilon_\infty - \frac{S_{TO}^2}{\tilde{v}_f^2} \quad \text{(I)} \\ \varepsilon(\tilde{v})^{-1} &= \varepsilon_\infty^{-1} + \frac{S_{LO}^2}{\tilde{v}_f^2} \quad \text{(II)}\end{aligned} \tag{5.125}$$

By inverting both sides of Eq. (5.125), (II), we get:

$$\varepsilon(\tilde{v}) = \frac{1}{\varepsilon_\infty^{-1} - \frac{S_{LO}^2}{-\tilde{v}_f^2}} = \varepsilon_\infty \frac{1}{1 + \frac{\varepsilon_\infty S_{LO}^2}{\tilde{v}_f^2}}. \tag{5.126}$$

If we employ the approximation $1/(1+x) = \sum_{k=0}^\infty (-1)^k x^k \approx 1 - x$ with $x = \varepsilon_\infty S_{LO}^2/\tilde{v}_f^2$, which is exact for $x \to 0$, and which is valid except for exceptionally strong oscillators, since $S^2 \ll \tilde{v}_f^2$, we obtain:

$$\varepsilon(\tilde{v}) = \varepsilon_\infty - \frac{\varepsilon_\infty^2 S_{LO}^2}{\tilde{v}_f^2}. \tag{5.127}$$

If we compare Eq. (5.127) with Eq. (5.125), (II) the result is

$$S^2 = S_{TO}^2 = \varepsilon_\infty^2 S_{LO}^2. \tag{5.128}$$

For realistic dielectric backgrounds and oscillator strengths, the differences between the results of the inverted Eq. (5.124), I and the inverse dielectric function Eq. (5.124), II are on the order of 0.005% [48]. Therefore, if the oscillators are well-separated, it is sufficient to model the TO oscillator strength. The LO oscillator strength can then be calculated from Eq. (5.128). In case that the oscillators are not separated by a transparency region, they are coupled, the stronger the smaller their wavenumber difference is (cf. Section 7.3.1). Nevertheless, the LO oscillator strength is still comparable to the uncoupled case, which helps to assign LO modes. This assignment seems to be trivial, since when the wavenumber is increased, a TO oscillator is always followed by a LO oscillator. Therefore, often in literature, the LO mode is naïvely assigned to this TO mode. This is a typical case of the ignorance of older literature, since Gervais has already shown

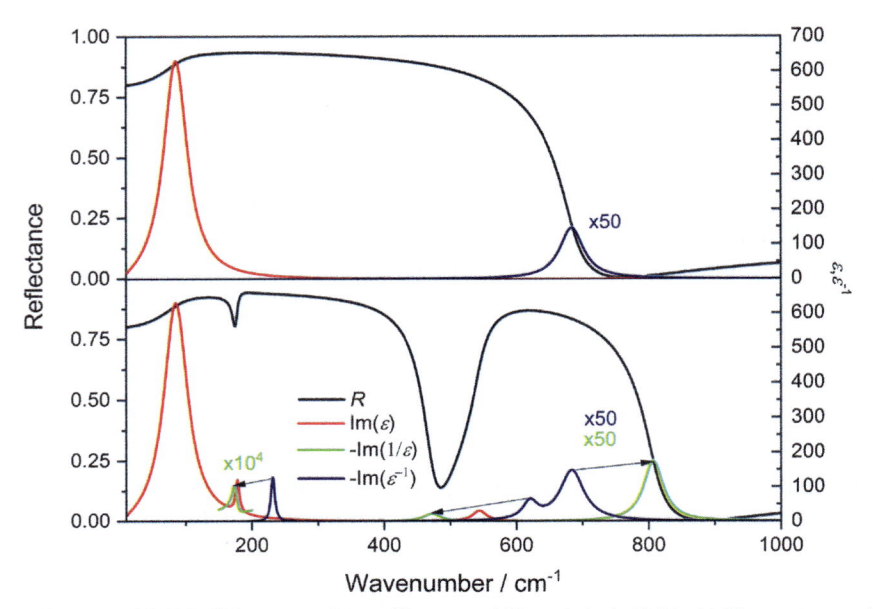

FIG. 5.35 *Upper panel*: Reflectance of SrTiO$_3$ if the two weaker oscillators would be switched off *(black)*. The maximum of the imaginary part of the dielectric function indicates the TO position, the maxima of the imaginary part of the inverse dielectric functions determine the LO position (*green*: inverse of the dielectric function, blue: inverse dielectric function modeled according to Eq. (5.124)) *Lower panel*: The same as for the upper panel, but with the two weaker oscillators switched on. The *arrows* indicate the shift of the LO positions from the uncoupled model to the positions for the inverted dielectric function. Note that the minima of $1/\varepsilon$ are now at lower wavenumbers than their counterparts and that $1/\varepsilon \neq \varepsilon^{-1}$ due to mode coupling in case of the former. Reflectance, dielectric, and loss functions have been modeled using the data provided in [65].

in 1977 that an LO mode can also have a lower wavenumber than its TO counterpart [64]. How is it possible then that nevertheless TO and LO always alternate with one another? Imagine you have a very strong mode, e.g., in SrTiO$_3$ and two weaker modes (cf. Fig. 5.35). If we switch off the two weaker modes, then reflectance is particularly high between the maxima of the dielectric function ε and its inverse $1/\varepsilon$ (since there is only one oscillator, the inverse is the same as calculating ε^{-1} by Eq. (5.124) together with Eq. (5.128)). If we now switch back on the two weaker oscillators, the maximum of ε^{-1} stays, but the corresponding one of $1/\varepsilon$ clearly blueshifts due to mode coupling (certainly, ε^{-1} and $1/\varepsilon$ are mathematically equivalent, but here we use the terms to differentiate two ways of computing the inverse dielectric function) [66]. The peak value has, however, only slightly changed, so that it is absolutely clear that this is the LO position which belongs to the strong mode. If we investigate closer the two weaker modes, then we see that those maxima of $1/\varepsilon$, which belong to the two weaker modes, are both redshifted. If we assign the number 1 to the strong mode, then the consecutive order of the modes is TO$_1$/LO$_2$/TO$_2$/LO$_3$/TO$_3$/LO$_1$, so that the TO-LO rule is obeyed, while we have an outer pair TO$_1$/..../LO$_1$, which encloses two consecutive inner pairs LO$_2$/TO$_2$/LO$_3$/TO$_3$. The first mode is so strong, that the real part of the dielectric function becomes negative, like in metals below the plasma frequency (cf. Section 5.7), which is why such absorptions have been called in former times metal-like. In fact, we could call indeed the first mode a plasmon-like mode, because it leads to a reflectance spectrum which very much resembles that of metals, except that the mode position is not at zero wavenumbers like for a free electron in a plasmon. On the other hand, it is possible to describe with the same theory that we just applied also the coupling of plasmons with LO modes to a much higher degree than what was possible before [48]. This is of great interest, because the concentration of free electrons can be varied by doping in semiconductors, and thereby, the plasmon oscillator strength ("plasma frequency"). This increases the LO wavenumber position. If a semiconductor possesses also phonon modes, their LO modes will begin to interact and couple with the plasmon, and, once the plasmon's LO wavenumber position is larger, the TO and LO mode of the phonon will be inverted.

Before this happens, the plasmon's LO position and mode strength can be well-approximated by capturing the influence of the phonons by a constant ε_0, which would be the dielectric constant at zero wavenumber (for more details, see Section 7.3.1) [48]. Once the LO position passes the LO phonon with the lowest wavenumber, ε_0 needs to be reduced by the contribution of this phonon, etc. If we start with $\varepsilon_{0,1}$ as the dielectric constant at zero wavenumber and $\varepsilon_{0,j}$ is the static dielectric constant without the contribution of the $j-1$ phonon, an approximate description of the properties of the plasmon is possible. If the phonons are well separated, so that there are transparency regions in between, the dielectric function in these transparency regions can be described by:

$$\varepsilon(\tilde{v}) = 1 - \frac{\sum_{j=1}^{N} S_{TO,j}^2}{\tilde{v}_f^2} \quad \text{(I)}$$

$$\varepsilon(\tilde{v})^{-1} = 1 + \frac{\sum_{j=1}^{N} \frac{S_{TO,j}^2}{\varepsilon_{0,j}^2}}{\tilde{v}_f^2} \quad \text{(II)}$$

(5.129)

Overall, even if, like in the case of the local field of Lorentz, the inverse dielectric functions with and without the additional field, $1/\varepsilon$ and ε^{-1}, are no longer equal due to coupling, the comparison between the uncoupled and the coupled case can still be valuable for mode assignment and for the understanding of the coupling.

Coming back to the merits of inverse dielectric function modeling, in particular for higher angles of incidence and p-polarization, Eq. (5.105) allows describing experimental spectra with an accuracy of up to two times better than the classical model [67]. Interestingly, even for experimental spectra recorded with s-polarized light and low angle of incidence, improved accuracy is possible by using inverse dielectric function modeling as can be seen in Fig. 5.36, where the residual sum of squares using the inverse dielectric function model to fit the experimental data is smaller than the classical model, but higher than the Berreman-Unterwal model.

With inverse dielectric function modeling, it is now possible to directly determine $\tilde{v}_{LO,j}$ and $\gamma_{LO,j}$. Using the values of $\tilde{v}_{TO,j}$ and $\gamma_{TO,j}$ gained by the CDHO model, we can compare all values with those obtained from the Berreman-Unterwal model. While $\tilde{v}_{TO,j}$ as well as $\tilde{v}_{LO,j}$ usually agree within the resolution of the spectrum, for the damping constants the agreement is less satisfying. In particular, it can be observed, that the necessary condition for the Berreman-Unterwal model to produce positive values for the imaginary part of the dielectric function over the whole spectral range, $\sum_{j=1}^{N} (\gamma_{LO,j} - \gamma_{TO,j}) > 0$ (Eq. 5.90) is not obeyed for the damping constants derived from the classical and the inverse model. Accordingly, a fit of experimental data with the Berreman-Unterwal model using a penalty function to force compliance to Eq. (5.90), leads to the effect that the damping constants of weaker and less important oscillators compensate for the unphysical behavior of stronger oscillators as already stated in Section 5.5.1.

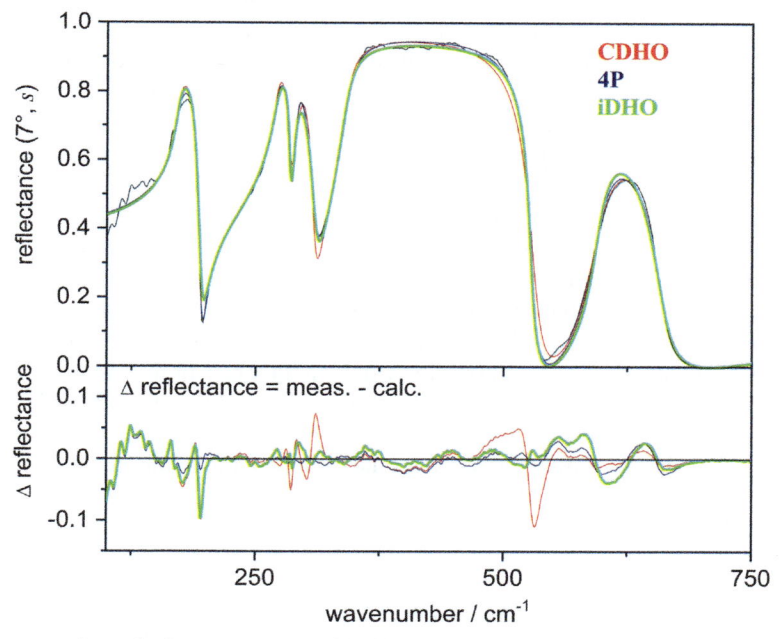

FIG. 5.36 Comparison between experimental reflectance spectrum *(black line)* and simulated spectra based on the classical dispersion relation *(red line)*, on the dispersion relation based on the semiempirical four-parameter model *(blue line)*, and on the inverse damped harmonic oscillator-based relation *(green line)*.

Overall, the strength of the inverse dielectric function oscillator model is emphasized, if it is employed to p-polarized spectra recorded at high angles of incidence, and there exists a particularly important application for this model in connection with so-called perpendicular modes, which will be detailed in the second part of this book when anisotropic media and their peculiarities will be discussed.

5.7 Dispersion relation—Drude model

Metals are special kinds of materials, especially when we talk about the dielectric function. This is because not the lattice vibrations dominate their infrared spectrum, but the behavior of the (quasi-)free electrons. If we characterize the electrons as free, then because we can use the classical damped harmonic oscillator model and modify it accordingly by assuming there is no restoring force (m_e is the electron mass and $-e$ the corresponding charge):

$$m_e \frac{d^2 \Delta r}{dt^2} + m_e \gamma \frac{d \Delta r}{dt} = -eE_0 \exp(-i\omega t). \tag{5.130}$$

The solution, which can also be found by setting $\tilde{\nu}_0 = 0$ in Eq. (5.21), is,

$$\varepsilon = \varepsilon_\infty - \frac{\tilde{\nu}_p^2}{\tilde{\nu}^2 + i\tilde{\nu}\gamma}, \tag{5.131}$$

$$\frac{Ne^2}{\varepsilon_0 m_e} = S^2 = \tilde{\nu}_p^2$$

if we redefine the oscillator strength as the square of the plasma frequency $\tilde{\nu}_p$. This frequency or wavenumber would be the turning point above which the metal becomes a conventional dielectric material if there would be no damping and $\varepsilon_\infty = 1$. Regardless of the latter conditions, this point is always characterized by the fact that the real part of the dielectric function becomes negative below it. The physical reason for the damping is that the mean free path of the electrons is limited and they interact with the atoms. In any way, assuming $\gamma = 0$, the dielectric function gets negative for wavenumbers lower than the plasma frequency. Accordingly, the square root becomes imaginary and changes from being real above the plasma frequency to being purely imaginary below it (this is indicated by the color change in the center panel of Fig. 5.37 from green to blue). Accordingly, for zero damping, reflectance becomes total ($R = 1$) at the plasma frequency, as can be seen in the upper part of Fig. 5.37, since the waves inside the metal are then evanescent (the metal is still assumed not to be a perfect conductor; therefore, the waves penetrate the interface. However, with decreasing wavenumber the imaginary part increases and the penetration depth becomes smaller). If we allow damping (for Fig. 5.37, we used the plasma frequency and the damping constant of gold), the imaginary part at low wavenumbers is about one order of magnitude lower than without damping (for Cu and Al it would be even lower). For higher wavenumbers, however, it begins to approach the imaginary part computed without damping (accordingly the red curve begins to approach the blue curve). Above the plasma frequency, k approaches zero, but not immediately, so that a real metal does not instantaneously become transparent above $\tilde{\nu}_p$.

The reflectance nevertheless differs only by minute amounts above $\tilde{\nu}_p$ from that calculated under the assumption of zero damping, while immediately below $\tilde{\nu}_p$ the increase of the reflectance is less steep and unity will not be reached. Especially, in the infrared range $R \approx 0.98$ and, in particular when the transflection technique is used (layers on highly reflecting substrates), we found it necessary to correct for the wavenumber dependence of the reflectance of gold (it might be advisable in general for external reflectance measurements, where usually gold mirrors are used as reference, to correct for the fact that the reflectance of gold does not equal 100%. This correction would neither be a problem nor a big effort). In this context, it is also important to point out two further observations. First of all, once $\varepsilon_\infty > 1$, $\tilde{\nu}_p$ from Eq. (5.131) is strictly speaking no longer the turning point. For the example of gold, we have $\varepsilon_\infty = 1.54$, which shifts the turning point from $\tilde{\nu}_p = 69,930 \text{ cm}^{-1}$ [68,69] to $56,347 \text{ cm}^{-1}$. Secondly, it is not only ε_∞ that moves the turning point. In the case of gold, we also have to consider comparably low-lying d-d transitions. To take those into proper consideration, Eq. (5.131) has to be augmented by contributions from two so-called critical point transitions,

$$\varepsilon(\tilde{\nu}) = \varepsilon_\infty - \frac{\tilde{\nu}_p^2}{\tilde{\nu}^2 + i\tilde{\nu}\gamma} + \sum_{j=1}^{2} \frac{S_j^2}{\tilde{\nu}_{0,j}} \left[\frac{e^{i\phi_j}}{\tilde{\nu}_{0,j} - \tilde{\nu} - i\gamma_j} + \frac{e^{-i\phi_j}}{\tilde{\nu}_{0,j} + \tilde{\nu} + i\gamma_j} \right], \tag{5.132}$$

where $S_j^2 (= A_j)$ represents the oscillator strengths (the critical point amplitudes), the $\tilde{\nu}_{0,j}$ stand for the interband transition wavenumbers, and the γ_j and ϕ_j are the corresponding damping constants and phases, respectively (the corresponding

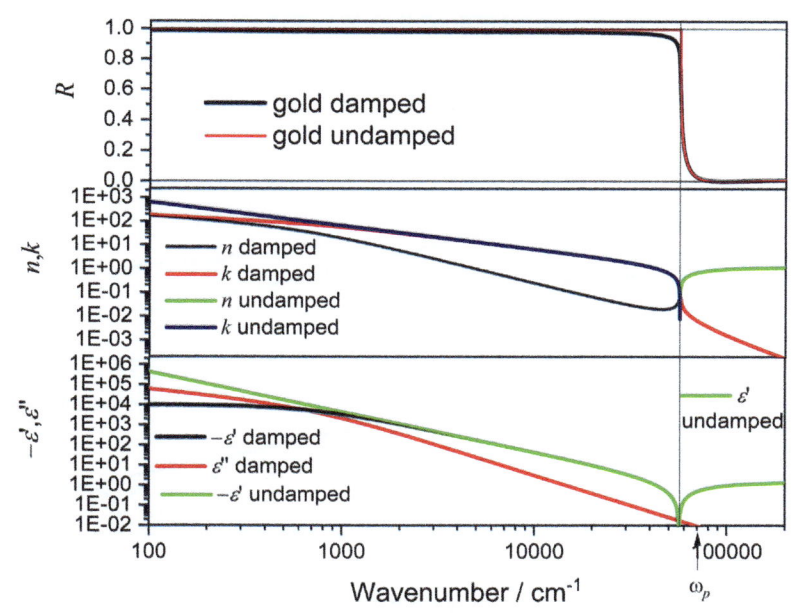

FIG. 5.37 *Upper panel*: Reflectance of gold with and without damping in the absence of interband transitions. *Center panel*: Corresponding real and imaginary part of the index of refraction functions. *Lower panel*: Corresponding negative real and positive imaginary part of the dielectric function. Note that the imaginary part is zero everywhere if there is no damping. After the turning point, indicated by the vertical line, the real part of the dielectric function becomes positive. Therefore, it is then depicted as positive real part.

oscillator model is comparable with that of Eq. (5.93) in that it assumes complex oscillator strengths; however, using it for oscillators in the infrared turned out to be not useful, at least for me) [68,69]. The resulting reflectance spectrum and dielectric function are compared to that without interband transitions in Fig. 5.38. The most obvious change is that the steep decrease of the reflectance already starts at about $15,000\,\text{cm}^{-1}$ and that afterward the spectrum is characterized by the two d-d features. The reason for the early start is, in contrast to the situation without d-d transitions, not the real part of the dielectric function becoming positive, because this point is actually increased in wavenumber to about $64,150\,\text{cm}^{-1}$. Instead, the reason can unanimously be attributed to the deviations between the dielectric functions with and without d-d transitions and the corresponding changes in the index of refraction function. In the regions of the FIR and MIR, and even in the low wavenumber parts of the NIR, the differences concerning the reflectance between the two models (with and without considering the d-d transitions) are actually very small. Therefore, it would be possible to simply omit the d-d transitions and use ε_∞, the plasmon frequency (wavenumber) and damping constant provided in [68,69] directly in Eq. (5.131).

5.8 Kramers-Kronig relations and sum rules

5.8.1 The basics

Hendrik Anthony Kramers and Ralph (de Laer) Kronig developed eponymous relations independently of each other in 1926 [70] and 1927 [71]. To be more precise, Kronig was first, but derived only one of the relations. Only one of the relations?! Those of you with prior knowledge (but with no more than I had, before I read the very revealing essay of Bohren about this topic [72]) may become a little wary because of this, as both relations belong together in the same way as conjoined twins:

$$\varepsilon'(\tilde{\nu}) - 1 = \frac{2}{\pi} \wp \int_0^\infty \frac{\varepsilon''(\tilde{\nu}')\tilde{\nu}'}{\tilde{\nu}'^2 - \tilde{\nu}^2} d\tilde{\nu}'$$

$$\varepsilon''(\tilde{\nu}) = -\frac{2\tilde{\nu}}{\pi} \wp \int_0^\infty \frac{\varepsilon'(\tilde{\nu}') - 1}{\tilde{\nu}'^2 - \tilde{\nu}^2} d\tilde{\nu}'$$

(5.133)

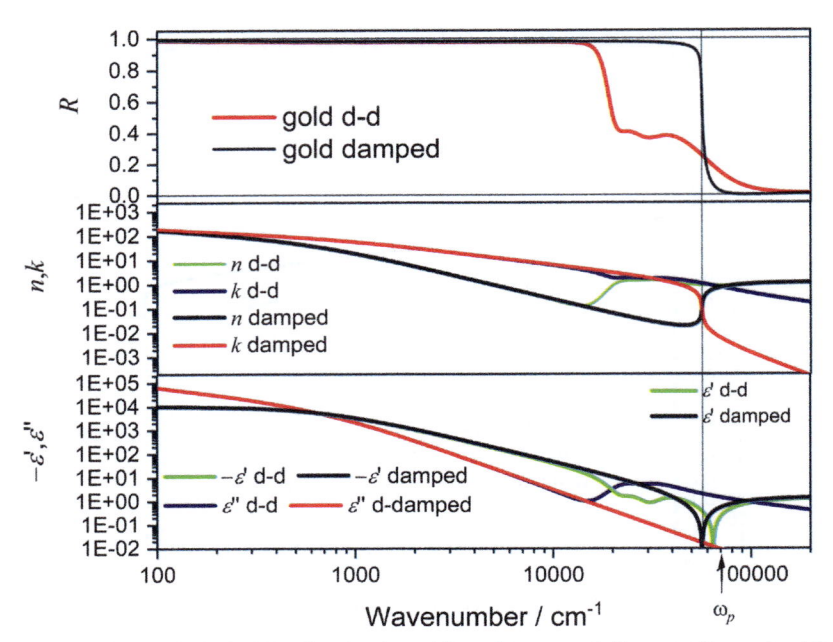

FIG. 5.38 *Upper panel*: Reflectance of gold with and without interband transitions. *Center panel*: Corresponding real and imaginary parts of the index of refraction functions. *Lower panel*: Corresponding negative real and positive imaginary parts of the dielectric function. After the turning point, indicated by the vertical line, the real parts of the dielectric function become positive. Therefore, they are then depicted as the positive real parts.

In Eq. (5.133), $\wp$ stands for the Cauchy principal value, which actually means that the integrals essentially consist of two parts with a gap at the point of discontinuity $\tilde{v}' = \tilde{v}$:

$$\varepsilon'(\tilde{v}) - 1 = \frac{2}{\pi} \wp \int_0^\infty \frac{\varepsilon''(\tilde{v}')\tilde{v}'}{\tilde{v}'^2 - \tilde{v}^2} d\tilde{v}' = \frac{2}{\pi} \lim_{\delta \to 0} \left(\int_0^{\tilde{v}-\delta} \frac{\varepsilon''(\tilde{v}')\tilde{v}'}{\tilde{v}'^2 - \tilde{v}^2} d\tilde{v}' + \int_{\tilde{v}+\delta}^\infty \frac{\varepsilon''(\tilde{v}')\tilde{v}'}{\tilde{v}'^2 - \tilde{v}^2} d\tilde{v}' \right)$$

$$\varepsilon''(\tilde{v}) = -\frac{2\tilde{v}}{\pi} \wp \int_0^\infty \frac{\varepsilon'(\tilde{v}') - 1}{\tilde{v}'^2 - \tilde{v}^2} d\tilde{v}' = -\frac{2\tilde{v}}{\pi} \lim_{\delta \to 0} \left(\int_0^{\tilde{v}-\delta} \frac{\varepsilon'(\tilde{v}') - 1}{\tilde{v}'^2 - \tilde{v}^2} d\tilde{v}' + \int_{\tilde{v}+\delta}^\infty \frac{\varepsilon'(\tilde{v}') - 1}{\tilde{v}'^2 - \tilde{v}^2} d\tilde{v}' \right)$$

(5.134)

In fact, the integrals are special cases of Hilbert transform pairs and it is possible to derive them from the assumption of causality, i.e., that a signal at present can only have been influenced by the past and be influenced by the present, but not by the future. Nevertheless, originally the Kramers-Kronig relations (KKR) have been derived simply from dispersion theory.

Be it as it may, there are a number of important conclusions that can be drawn from Eqs. (5.133), (5.134). First of all, real and imaginary parts of the dielectric function are not independent of each other. In fact, the following consequences can be derived:

$$\begin{aligned} &\text{(I)} \quad \varepsilon'' = 0 \to \varepsilon' = 1 \\ &\text{(II)} \quad \Delta\varepsilon'(\tilde{v}) \neq 0 \leftrightarrow \Delta\varepsilon''(\tilde{v}) \neq 0 \\ &\text{(III)} \quad \lim_{\tilde{v} \to \infty} \varepsilon'(\tilde{v}) = 1 \end{aligned}$$

(5.135)

According to the first consequence, if there is no absorption over the whole wavenumber range then the real part stays at unity, which is even for absorbing matter the limit for high wavenumbers according to the third conclusion. Furthermore, if the real part shows changes with wavenumber, so must the imaginary part (and, actually, vice versa).

Essentially, this offers the new view on Fig. 5.1 that I promised at the beginning of this chapter. I have copied this figure for convenience as Fig. 5.39.

Accordingly, we start with a relative dielectric constant of unity at very high wavenumbers. From there, while decreasing the wavenumber, each absorption will be somehow summed up and added to the real part. The contribution seems to increase with increasing oscillator strength. This can be seen in the transparent regions where the dielectric

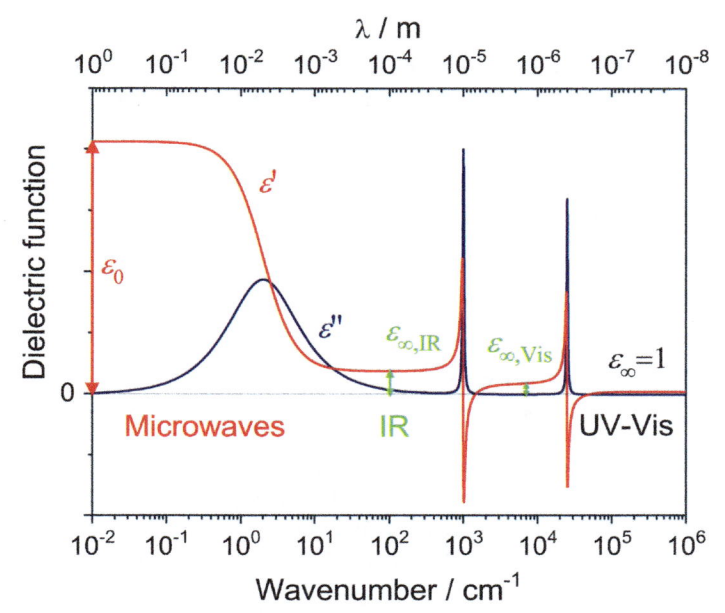

FIG. 5.39 Schematic dispersion of the dielectric function of a liquid.

function is nearly wavenumber independent (this is actually the conclusion which Kronig drew for the relatively limited spectral region of the X-rays). That the inversion is also correct was then stated by Kramers. In fact, Eqs. (5.133), (5.134) were first formulated for the real and the imaginary part of the complex index of refraction function and not for the (relative) dielectric function. Still, at that time, the dielectric function led a shadowy existence compared to the index of refraction function; e.g., it was mentioned just once by Planck [13] and not a single time by Lorentz [27] in their papers about their forms/flavors of dispersion theory. Furthermore, in 1930, Schaefer and Matossi thought it still necessary in their book about IR spectroscopy to check the result of Maxwell's wave equations that $\varepsilon = n^2$ for isotropic and perfectly homogenous materials and dedicated a whole section to the answer to this question. In addition, they still did not think of a dielectric function as we know it today. Instead, they looked at a dielectric constant at zero or infinite frequencies. The KKR relations stay nevertheless valid, even if the real part of the dielectric function is replaced by the real part of the index of refraction and the imaginary part by the index of absorption and they work as well for the real and imaginary part of the (mean) polarizability and even for the real and imaginary part of the inverse dielectric function!

Before we can use Eqs. (5.133), (5.134) in practice, we first have to take a number of practical hurdles. First of all, it is not possible to record a spectrum over the whole spectral region (or, to put some limit to it, over the whole spectral region where absorption occurs). This seems to be a thought-terminating cliché or, more figurative, a knockout argument, but this argument is overall harmless except for materials with free electrons with no low-frequency transparency regions. The challenge, that we do not know the spectrum at wavenumber regions above the infrared, is the usual case for an infrared spectroscopist, and we have solved the problem by introducing ε_∞ (we will discuss this quantity in more detail in Section 5.8.4). Indeed, the following modified form is valid to a very high degree for the infrared spectral range (actually for the spectral regions up to and including the infrared) [73]:

$$\varepsilon'(\tilde{v}) = \varepsilon_\infty + \frac{2}{\pi} \wp \int_0^{\tilde{v}(\varepsilon_\infty)} \frac{\varepsilon''(\tilde{v}')\tilde{v}'}{\tilde{v}'^2 - \tilde{v}^2} d\tilde{v}'. \tag{5.136}$$

As long as we terminate the integral at the lower wavenumber in a transparency region, the problem of not knowing what is going on below is not a big problem, since the integration is actually carried out from high wavenumbers down. Even if I cannot offer here strict proof, it is easy to generate a dielectric model function assuming harmonic oscillators, which are fully Kramers-Kronig compatible, with two transparency regions above and below the infrared and carry out some tests to see that lower-lying oscillators do not generate complications, at least if they are spectrally well-separated and not too strong (Drude contributions from free electrons—see Section 5.7, however, must not be present as already mentioned; such tests have been carried out amply in [73]). Actually, for the first transform provided in Eq. (5.136), which is in fact the more important for the molecular spectroscopist due to the focus on absorbance, we see that the integration can be terminated once ε'' falls below a certain threshold. In practice, this threshold is often reached after a few tens of wavenumbers below the first and above the last infrared absorption (in particular, given the comparably weak oscillators in organic and biological matter). In this sense, the integral is well-behaving and its evaluation usually not a problem, even if we do not start at zero.

The only remaining obstacle to perform these tests is efficient ways to evaluate Eq. (5.136). A first naïve solution, following Eq. (5.134), would be the following:

$$\varepsilon'(\tilde{v}) = \varepsilon_\infty + \frac{2}{\pi} \lim_{\delta \to 0} \left(\int_0^{\tilde{v}-\delta} \frac{\varepsilon''(\tilde{v}')\tilde{v}'}{\tilde{v}'^2 - \tilde{v}^2} d\tilde{v}' + \int_{\tilde{v}+\delta}^{\infty} \frac{\varepsilon''(\tilde{v}')\tilde{v}'}{\tilde{v}'^2 - \tilde{v}^2} d\tilde{v}' \right)$$

$$\approx \varepsilon_\infty + \frac{2}{\pi} \left[\sum_{j=1}^{n-1} \frac{\varepsilon'(\tilde{v}_j')\tilde{v}_j'}{\tilde{v}_j'^2 - \tilde{v}^2} + \sum_{n+1}^{N} \frac{\varepsilon'(\tilde{v}_j')\tilde{v}_j'}{\tilde{v}_j'^2 - \tilde{v}^2} \right], \quad \tilde{v}_n'^2 = \tilde{v}^2$$

(5.137)

Eq. (5.137) is known to have a very slow convergence as it omits to calculate the value at the point of discontinuity. Methods have been developed to calculate this missing value at least approximately, but Maclaurin's formula is not only much more accurate [74], but also more efficient, since the number of wavenumber points is halved (and computing time increases with the square of the number of points). Maclaurin's formula is given by,

$$\varepsilon'(\tilde{v}) = \varepsilon_\infty + 2d \frac{2}{\pi} \left[\sum_j \frac{\varepsilon''(\tilde{v}_j)\tilde{v}_j}{\tilde{v}_j^2 - \tilde{v}^2} \right],$$

(5.138)

where d is the wavenumber difference between two points. The starting value depends on the parity of j and the step size $d = 2j$: When k is odd, $j = 2,4,6\ldots$ and when it is even, $j = 1,3,5\ldots$

The fast convergence of Eq. (5.138) is illustrated in Fig. 5.40, where already for $d = 2\,\text{cm}^{-1}$ the result practically agrees with the one derived from the harmonic oscillator model.

While Maclaurin's formula is fast and for all practical problems accurate enough, it lacks some flexibility, as it requires equal-spaced wavenumber points. This problem can be removed, but only if higher-precision arithmetic software is employed [75]. A more flexible method is based on the piecewise dielectric function model introduced by Kuzmenko

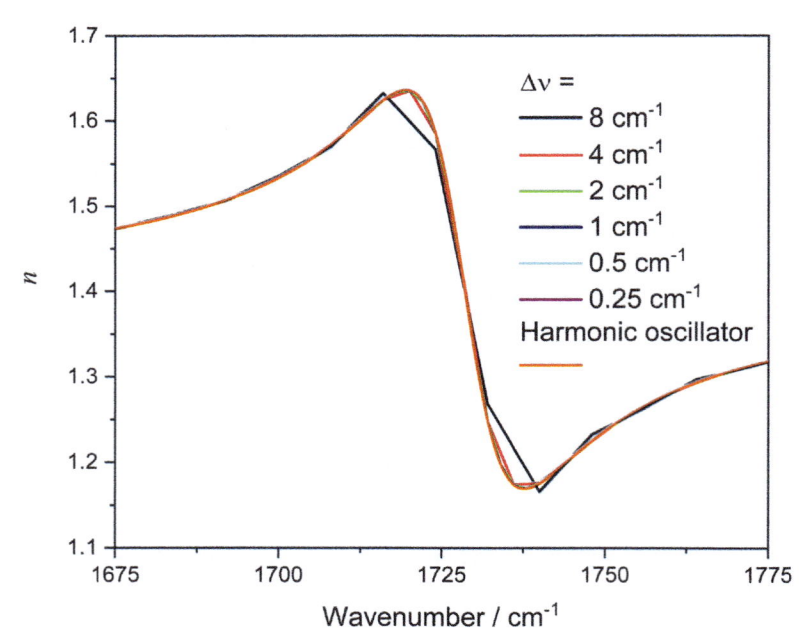

FIG. 5.40 Convergence of Maclaurin's formula for the index of refraction around the C=O band of PMMA.

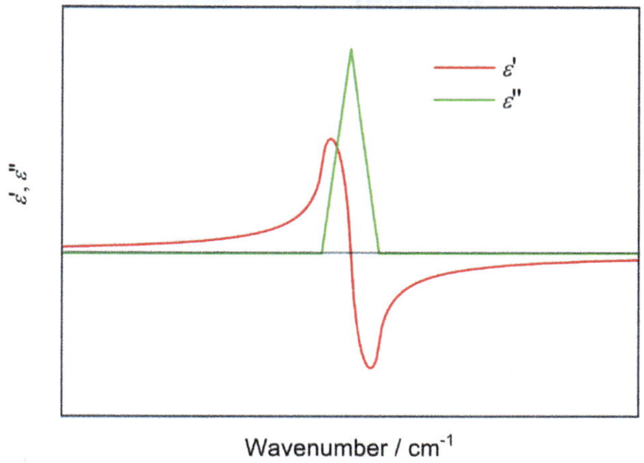

FIG. 5.41 Triangular function for representing the imaginary part of the dielectric function and corresponding real part.

[76]. Accordingly, the imaginary part of the dielectric function (or, alternatively, the index of absorption function) is represented by triangular functions as displayed in Fig. 5.41:

$$\varepsilon_i''(\tilde{v}_i) = \begin{cases} \dfrac{\tilde{v} - \tilde{v}_{i-1}}{\tilde{v}_i - \tilde{v}_{i-1}}, & \tilde{v}_{i-1} < \tilde{v} \leq \tilde{v}_i \\ \dfrac{\tilde{v}_{i+1} - \tilde{v}}{\tilde{v}_{i+1} - \tilde{v}_i}, & \tilde{v}_i < \tilde{v} \leq \tilde{v}_{i+1} \\ 0 & \text{otherwise} \end{cases}, \qquad (5.139)$$

For each wavenumber point, one triangular function is introduced and multiplied with its amplitude A_i. This amplitude is simply given by the value of the imaginary part of the dielectric function or index of absorption function at this point. For the real part, an analytical solution can be provided,

$$\varepsilon_i'(\tilde{v}_i) = \frac{1}{\pi} \wp \int_{-\infty}^{\infty} \frac{x \varepsilon_i''(x)}{x^2 - \tilde{v}^2} dx = -\frac{1}{\pi} \left[\frac{g(\tilde{v}\tilde{v}_{i-1})}{\tilde{v}_i - \tilde{v}_{i-1}} - \frac{(\tilde{v}_{i+1} - \tilde{v}_{i-1}) g(\tilde{v}\tilde{v}_i)}{(\tilde{v}_i - \tilde{v}_{i-1})(\tilde{v}_{i+1} - \tilde{v}_i)} + \frac{g(\tilde{v}\tilde{v}_{i+1})}{\tilde{v}_{i+1} - \tilde{v}_i} \right], \qquad (5.140)$$

$$g(xy) = (x + y) \ln|x + y| + (x - y) \ln|x - y|$$

so that the overall dielectric function (or index of refraction function) is then given by:

$$\varepsilon(\tilde{v}_i) = \varepsilon_\infty + \sum_{i=2}^{N-1} A_i [\varepsilon_i'(\tilde{v}_i) + i\varepsilon_i''(\tilde{v}_i)]. \qquad (5.141)$$

We called this procedure Poor Man's Kramers-Kronig analysis (PMKKA) [77], but, as was brought to our attention later, we were not the first who had this idea [78]. Note that I have changed the sign in Eq. (5.140) compared to the solution originally provided in Kuzmenko's paper, because it did not work as expected. But still there was a problem, which I thought was due to a missing Kramers-Kronig conformity and which I could not pinpoint. If we again use PMMA as a model, we can use the imaginary part from the classical damped harmonic oscillator model as input, calculate the real part, and compare it with the one directly gained from CDHO. This comparison is shown in Fig. 5.42. Obviously, the PMKKA underestimates the index of refraction increasingly with decreasing wavenumber. The same trend is found for other materials (the error increases with increasing amplitudes of the triangular functions), so that it seems that Eq. (5.140) is plagued by a systematic inaccuracy. Since I needed PMKKA to perform better, I came, more by accident, across an empiric improvement (in green) that decreases the error by about three orders of magnitude:

$$\varepsilon_i'(\tilde{v}_i) = -\frac{1}{\pi} \frac{\tilde{v}_i}{\tilde{v}} \left[\frac{g(\tilde{v}, \tilde{v}_{i-1})}{\tilde{v}_i - \tilde{v}_{i-1}} - \frac{(\tilde{v}_{i+1} - \tilde{v}_{i-1}) g(\tilde{v}, \tilde{v}_i)}{(\tilde{v}_i - \tilde{v}_{i-1})(\tilde{v}_{i+1} - \tilde{v}_i)} + \frac{g(\tilde{v}, \tilde{v}_{i+1})}{\tilde{v}_{i+1} - \tilde{v}_i} \right]. \qquad (5.142)$$

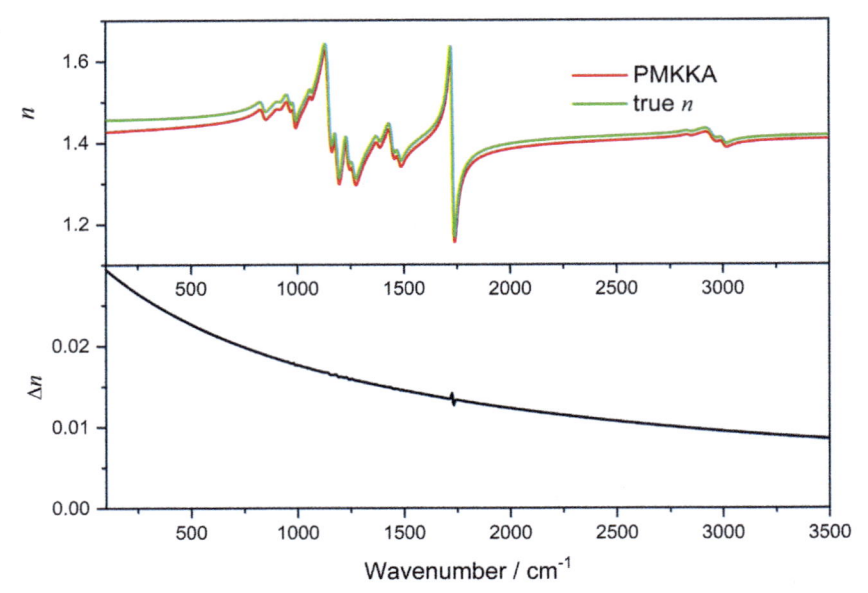

FIG. 5.42 Comparison of the real part of the refractive index function as gained from the CDHO model *(green curve)* with the one obtained from PMKKA.

The real part according to Eq. (5.142) is compared with that gained from Eq. (5.140) and from Maclaurin's formula in Fig. 5.43. The improvement is clearly visible in Fig. 5.43, but I was always aware of the fact that this is just a provisional solution. In fact, much after I thought up Eq. (5.142), I realized, also thanks to some fruitful quarrel with an unnamed reviewer during the review process of the corresponding paper [79] that a different approach that can be brought in a form similar to the one of Kuzmenko, did not suffer from the problem [80]. Before I could revisit the problem, Rousseau et al. did it, motivated by a copy&paste error in ref. [79] that found its way unfortunately also into refs. [77, 81, 82], which made them think that I had solved a problem but introduced another. Anyway, the correct form of Eq. (5.140) is the following [83,84]:

$$\varepsilon_i'(\tilde{\nu}_i) = \frac{1}{\pi} \left[\frac{g(\tilde{\nu}\tilde{\nu}_{i-1})}{\tilde{\nu}_i - \tilde{\nu}_{i-1}} - \frac{(\tilde{\nu}_{i+1} - \tilde{\nu}_{i-1})g(\tilde{\nu}\tilde{\nu}_i)}{(\tilde{\nu}_i - \tilde{\nu}_{i-1})(\tilde{\nu}_{i+1} - \tilde{\nu}_i)} + \frac{g(\tilde{\nu}\tilde{\nu}_{i+1})}{\tilde{\nu}_{i+1} - \tilde{\nu}_i} \right]$$
$$g(xy) = (x+y)\ln|x+y| - (x-y)\ln|x-y| \quad \text{for } x \neq y$$
$$g(xy) = (x+y)\ln|x+y| \quad \text{for } x = y$$

(5.143)

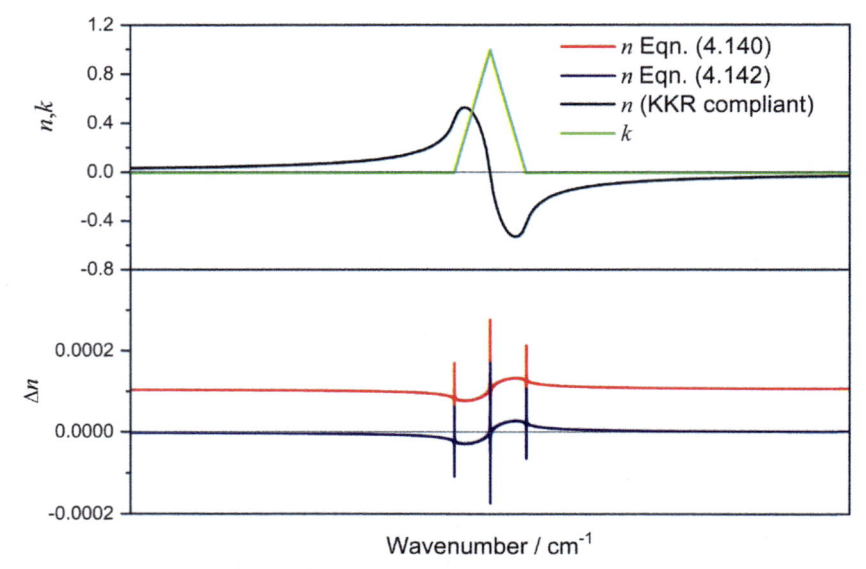

FIG. 5.43 *Upper panel*: Triangular function for representing the imaginary part of the dielectric function and corresponding real parts calculated by Eqs. (5.140), (5.142) and Mclaurin's formula (Eq. 5.138). *Lower panel*: Difference between Eq. (5.138) and Eqs. (5.140), (5.142).

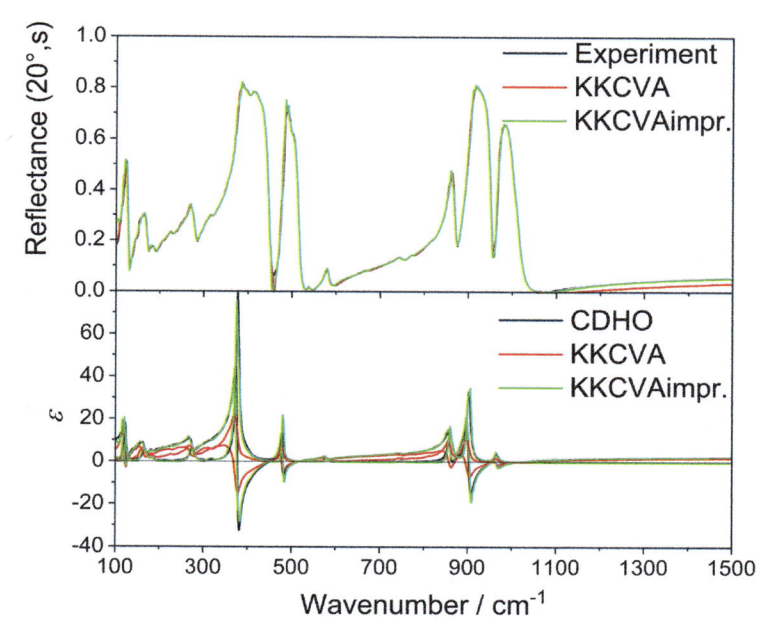

FIG. 5.44 *Upper panel*: Experimental principal reflectance spectrum of the a-axis of Fresnoite and modeled spectra using KKCVA following Eqs. (5.140) *(red line)* and (5.142) *(green line)*. *Lower panel*: Real and imaginary parts of the dielectric functions determined by dispersion analysis with damped harmonic oscillators *(black line)* and KKCVA following Eqs. (5.140) *(red line)* and (5.142) *(green line)*.

With this analytical solution, which also corrects the one provided in [85], the numerical error decreases compared to my provisionally introduced ad hoc solution (Eq. 5.142) by some additional orders of magnitude. Nevertheless, at the same time, the additional improvement is marginal, since it is much smaller than the line thickness. Anyway, both Eqs. (5.140), (5.142) are now obsolete thanks to Eq. (5.143).

Based on Eq. (5.143), it is now possible to fully exploit the advantages of Kuzmenko's idea, which originally was to use the triangular functions for dispersion analysis. Certainly, this idea has a drawback, namely the possibility to gain physically meaningful parameters like the oscillator position and strength, which give insight into the properties of the material, is lost. Another drawback is that the possibility to drastically reduce the amount of data also has to be given up. For Kramers-Kronig constrained variational analysis (KKCVA), no reduction is achieved, if it is performed as I suggest it, namely, with one triangular function for each spectral point. Corresponding fits of the *a*-axis reflectance spectrum of Fresnoite ($Ba_2TiSi_2O_8$) are shown in Fig. 5.44, where for the first fit Eq. (5.140) is employed and for the second Eq. (5.142).

Interestingly, the mean residual sum of squares of both fits is comparably small if only the range up to $1100\,cm^{-1}$ is considered. At $1100\,cm^{-1}$ and above, there are no more triangular functions and the errors in the real part of the dielectric function can no longer be compensated by introducing errors in the imaginary part. Correspondingly, the improved KKCVA yields a real and an imaginary part which are very comparable to their counterparts gained by the classical CDHO model (but also fits very small features that would require a lot of additional effort to be captured properly by the CDHO model). In contrast, the original KKCVA gives a real and an imaginary part which strongly deviates from the other two. Accordingly, the original recommendation was to use it only in combination with the CDHO model, a severe limitation which is fully lifted if the improved KKCVA is employed.

Obviously, the answer to the question if a certain lineshape/dielectric function model is conforming to the Kramers-Kronig relation is not only an academic one but also important for the question if this model is successfully applicable in practice. In fact, it has been shown by Keefe that Lorentz profiles as introduced in Section 5.2, are not Kramers-Kronig conform (and the same is true also for Gaussian profiles) [86]. The simple reason for this problem is that the Kramers-Kronig relations require the real part of the resulting functions to be an even function and the imaginary part to be an odd function, a condition that is fulfilled by the classical damped harmonic oscillator model (this is actually the same problem that Kuzmenko's original formulation of the real part of the triangular function base above suffered from). In practice, however, it seems according to Keefe's findings that noteworthy deviations only exist for oscillators in the FIR range below $100\,cm^{-1}$. So, if we are interested only in the MIR spectral range, we should be on the safe side?! In fact, yes, but only if the material of interest has just a few bands which do not overlap. Otherwise, the nonconformity can lead to considerable errors as I will demonstrate in the following, again on the example of PMMA (Fig. 5.45) and the *a*-axis spectrum of Fresnoite (Fig. 5.46).

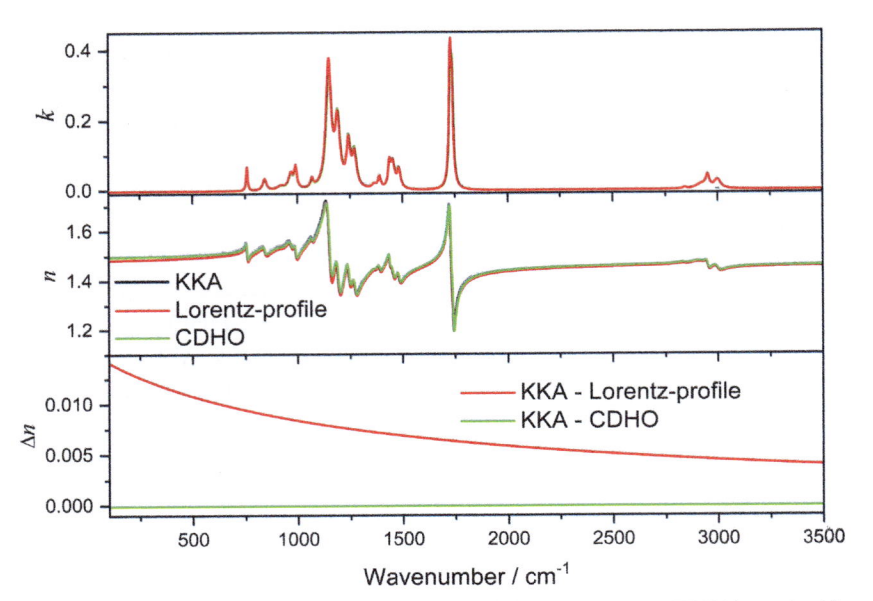

FIG. 5.45 *Upper panel*: Comparison of the imaginary part of the complex index of refraction function of PMMA as gained from the CDHO model *(green curve)* with the one obtained from the Lorentz profiles following Eq. (5.53). *Center panel*: Comparison of the real part of the complex index of refraction function of PMMA as gained from the CDHO model *(green curve)* with the one obtained from the Lorentz profiles following Eq. (5.53) and from a KKA of the imaginary part of the Lorentz profiles. *Lower part*: Differences between the real parts due to the dispersion relations and the real parts calculated from the imaginary parts by KKA.

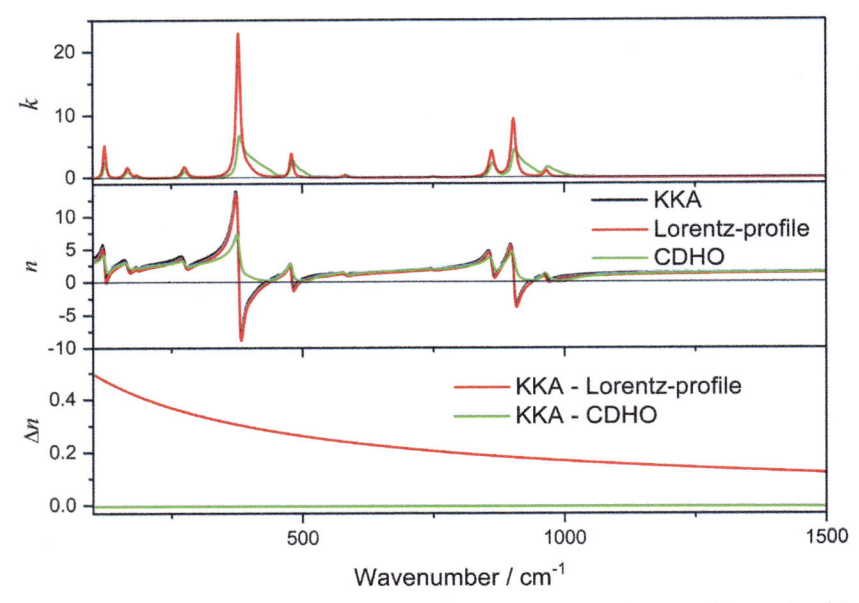

FIG. 5.46 *Upper panel*: Comparison of the imaginary part of the complex index of refraction function of the a-axis of Fresnoite as gained from the CDHO model *(green curve)* with the one obtained from the Lorentz profiles following Eq. (5.53). *Center panel*: Comparison of the real part of the complex index of refraction function of a-axis of Fresnoite as gained from the CDHO model *(green curve)* with the one obtained from the Lorentz profiles following Eq. (5.53) and from a KKA of the imaginary part of the Lorentz profiles. *Lower part*: Differences between the real parts due to the dispersion relations and the real parts calculated from the imaginary parts by KKA.

If Eq. (5.53) is used to generate the complex refractive index function of PMMA, which I reproduce here for convenience,

$$\widehat{n}(\tilde{v}) \approx n_\infty + \sum_{j=1}^{N} \frac{S_j^2}{2\tilde{v}_{0,j} n_\infty} \frac{1}{\tilde{v}_{0,j} - \tilde{v} - i\gamma_j/2}, \tag{5.144}$$

the bands agree with those gained from the CDHO model within line thickness, which is in line with our conclusion that the Lorentz profiles can be used to model absorbance spectra of materials with not too large oscillator strengths. The corresponding real part, however, suffers obviously from the missing KK-conformity, although about 1% error would still be tolerable, could we not get a perfect KK-conformity by simply using the CDHO model.

The situation clearly changes when I apply Eq. (5.53) for inorganic materials like Fresnoite. Fig. 5.46 shows what we already know from Section 5.2, namely, that the complex refractive index function generated by a sum of Lorentz profiles is a disguised dielectric function with spectral regions where the real part becomes negative and far too intense, and symmetric bands in the imaginary part. As a consequence, the missing KK conformity leads to errors of up to 10%, which is definitely intolerable. Therefore, the use of Lorentz profiles is severely hampered and should be abandoned for such materials.

While the determination of the real part of the dielectric function or the index of refraction function from the imaginary part seems to be a comparably easy task, the opposite seems to be true for determining the imaginary part from the real part. Before I tried it myself, I relied on the opinions of others, but I did not really understand what the problem actually is. Therefore, it may be of interest to demonstrate what happens if you integrate the refractive index function in a limited spectral range. To do this, we can actually use again Maclaurin's formula, but in a correspondingly modified form:

$$\varepsilon''(\tilde{v}) = -\tilde{v}d\frac{4}{\pi}\left[\sum_j \frac{\varepsilon'(\tilde{v}_j) - 1}{\tilde{v}_j^2 - \tilde{v}^2}\right]. \quad (5.145)$$

Just to remind you, that all equations remain valid if you replace the dielectric function with the refractive index function, I have carried out the corresponding calculations for the corresponding function of PMMA in the range of 700–4000 cm^{-1}. The result is depicted in Fig. 5.47.

Obviously, the red absorption index spectrum, which was obtained for the integration range 700–4000 cm^{-1} is not a very useful result. It seems that we somehow need to extrapolate the determined refractive index (actually, for organic and biological matter, you can more or less measure it to a good approximation, see Section 7.1) in the region where we do not know it. We certainly could speculate, but actually this is not necessary, and with just using what we know from the previous sections, we can obtain the blue curve in Fig. 5.47 with which most practicians would be satisfied. Let's have again a look at Eq. (5.133). It seems there is something that no one has seen so far, at least I could not find it in literature and, e.g., you can also not find it in the book "Kramers-Kronig Relations in Optical Materials Research," which seems to be comprehensive with respect to the KKR [87]. So what is there to see? The term $\varepsilon'(\tilde{v}') - 1$, where the starting value at high frequency, i.e., unity, is subtracted from the actual value of the real part, which means that the integration of the first band at high frequency contains areas above and below zero. We know, however, that the value of the real parts of the dielectric function and the index of the refraction

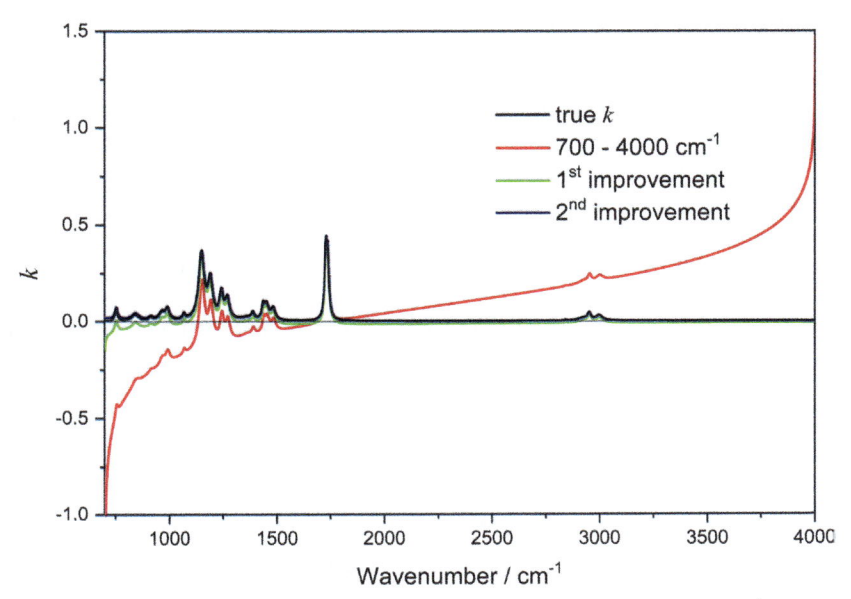

FIG. 5.47 Comparison of the original/true k-spectrum of PMMA with those obtained in the range 700–4000 cm^{-1} by integrating only over the spectral region presented in the figure *(red curve)*, and taking into account the region above *(green curve)* and below *(blue curve)*.

function increase with every absorption. Therefore, we must not disregard this increase and actually, we know the increase: It is simply $\varepsilon_\infty - 1$ or $n_\infty - 1$ (actually, this is a simplification, but one that works very well, which we will discuss at the end of this section). Therefore, as in Eq. (5.136), we replace unity with either ε_∞ or n_∞ [73].

$$\varepsilon''(\tilde{v}) = -\frac{2\tilde{v}}{\pi} \wp \int_0^{\tilde{v}_f} \frac{\varepsilon'(\tilde{v}') - \varepsilon_\infty}{\tilde{v}'^2 - \tilde{v}^2} d\tilde{v}',$$

$$\varepsilon''(\tilde{v}) = -\tilde{v} d \frac{4}{\pi} \left[\sum_j \frac{\varepsilon'(\tilde{v}_j) - \varepsilon_\infty}{\tilde{v}_j^2 - \tilde{v}^2} \right]$$

(5.146)

And change the upper limit from ∞ to $\tilde{v}_f$, a wavenumber point in the transparency region between MIR and Vis, e.g., $4000 \, \text{cm}^{-1}$.

The result is the green spectrum, which is obviously much more acceptable than the red one. Only in the low wavenumber range of the spectrum we still have a problem, which is due to the influence of the unknown progression in the range down to zero wavenumber. We know that if there would be no more absorption in this range, the real part of the dielectric function or the index of refraction would decrease with decreasing wavenumber (normal dispersion); therefore, we could model this with a Lorentz oscillator or a number or Lorentz oscillators. On the other hand, there still could be other oscillators located in the unknown range (and for PMMA, there actually are). Therefore, let us for simplicity assume that the real functions keep the first known value for the whole wavenumber range. The result is a slight offset in the 700–1200 cm^{-1} range of the blue curve, which is a sign that we overestimate the oscillator strength cumulated in low wavenumber region, but given the fact that we do not have to do any modeling of the unknown ranges, I would consider this a vast improvement. Even better, we can recycle these ideas in the next section, where they are even more helpful.

5.8.2 Determination of the optical constants directly from transmittance or reflectance

From a practical point of view, it seems that the above paragraphs contain everything what you need to know to determine optical constants from experimental spectra. If you have some experience in recording spectra and if you know the available spectrometer software, you may wonder why I have not yet discussed methods that allow to extract the optical constants from reflectance spectra (and, actually, also from transmittance spectra) via the Kramers-Kronig relations, when those are obviously so important that they are part of these spectrometer software packages. Before I added the following paragraphs, I would have said that their practical relevance is very low for infrared spectroscopists. Not only is the corresponding method a kind of black box, but every spectrometer manufacturer does it their own way, which means that if you use the different available programs with the same spectrum you end up not knowing what the optical constants of your material are because of the strong differences you find between the results [88,89]. The reason for the deviations is the same as described above for the calculation of the imaginary from the real parts. The difference in this case is, however, that reflectance and transmittance do not become zero at zero wavenumber, and the necessary extrapolations at lower wavenumbers than the experimental ones may differ between different vendors. The same problem affects the extrapolation of the reflectance and transmittance for higher wavenumbers. These reasons and the fact that better alternatives exist made me for a long time think that it is not worth to use the method in practice.

The starting point of the method is that it is possible to write the reflection coefficient and the transmission coefficient in the following form,

$$r = \frac{1 - \sqrt{\frac{\varepsilon_2}{\varepsilon_1}}}{1 + \sqrt{\frac{\varepsilon_2}{\varepsilon_1}}} = \rho \cdot \exp(i\phi)$$

$$t = \frac{2}{1 + \sqrt{\frac{\varepsilon_2}{\varepsilon_1}}} = \tau \cdot \exp(i\phi)$$

(5.147)

which takes into account that the sample (medium 2) has, in contrast to the incidence medium (medium 1), a complex dielectric function to account for absorption. When I first saw Eq. (5.147), I really had trouble understanding how the right forms could be equivalent to the left relations, but the former is in the first place just the polar forms of complex numbers with ρ and τ being the length of vectors in the complex plane and ϕ the angle between the vectors and the real axis. The right forms in Eq. (5.147) can certainly be also written as

$$\begin{aligned} \ln r &= \ln \rho + i\phi \\ \ln t &= \ln \tau + i\phi \end{aligned} \quad (5.148)$$

The next step is to argue (or even to omit this argumentation) that in principle $\ln\rho$ and $\ln\tau$ must be linked to the phase very much in the same way as the real to the imaginary part of the index of refraction. According to T. S. Robinson, who had the idea first [90], the corresponding equations read:

$$\begin{aligned} \phi(\tilde{\nu}) &= -\frac{2\tilde{\nu}}{\pi} \wp \int_0^\infty \frac{\ln \rho(\tilde{\nu}')}{\tilde{\nu}'^2 - \tilde{\nu}^2} d\tilde{\nu}' \\ \phi(\tilde{\nu}) &= -\frac{2\tilde{\nu}}{\pi} \wp \int_0^\infty \frac{\ln \tau(\tilde{\nu}')}{\tilde{\nu}'^2 - \tilde{\nu}^2} d\tilde{\nu}' \end{aligned} \quad (5.149)$$

Eq. (5.149) seems to be unusable, but once you realize that,

$$\begin{aligned} \ln \rho(\tilde{\nu}') &= \ln|r(\tilde{\nu}')| = \frac{1}{2}\ln|r(\tilde{\nu}')|^2 = \frac{1}{2}\ln R(\tilde{\nu}') \\ \ln \tau(\tilde{\nu}') &= \ln|t(\tilde{\nu}')| = \frac{1}{2}\ln|t(\tilde{\nu}')|^2 = \frac{1}{2}\ln T(\tilde{\nu}') \end{aligned}, \quad (5.150)$$

Eq. (5.149) takes on the following form:

$$\begin{aligned} \phi(\tilde{\nu}) &= -\frac{\tilde{\nu}}{\pi} \wp \int_0^\infty \frac{\ln R(\tilde{\nu}')}{\tilde{\nu}'^2 - \tilde{\nu}^2} d\tilde{\nu}' \\ \phi(\tilde{\nu}) &= -\frac{\tilde{\nu}}{\pi} \wp \int_0^\infty \frac{\ln T(\tilde{\nu}')}{\tilde{\nu}'^2 - \tilde{\nu}^2} d\tilde{\nu}' \end{aligned}. \quad (5.151)$$

I suggest not to use Eq. (5.151). Instead, I want to introduce you to a transformation of Eq. (5.151), that ensures that one big practical problem can be removed in a simple way. This transformation has been suggested by Berreman, who also extended the use of reflectance to the case of arbitrary angle of incidence [91]. The suggestion of Berreman can also be found in the book of David Tanner [92], who states that on the one hand Berreman's suggestion removes possible calibration errors and, on the other, that "both numerator and denominator of the integrand are zero when" $\tilde{\nu}' = \tilde{\nu}$. It is based on the following integral,

$$\wp \int_0^\infty \frac{1}{\tilde{\nu}'^2 - \tilde{\nu}^2} d\tilde{\nu}' = 0, \quad (5.152)$$

which is zero, because the negative area for $\tilde{\nu}' < \tilde{\nu}$ compensates the positive area for $\tilde{\nu}' > \tilde{\nu}$. Thus, it is possible to add an arbitrary multiple of said integral to Eq. (5.151),

$$\begin{aligned} \phi(\tilde{\nu}) &= -\frac{\tilde{\nu}}{\pi} \wp \int_0^\infty \frac{\ln R(\tilde{\nu}')}{\tilde{\nu}'^2 - \tilde{\nu}^2} d\tilde{\nu}' + \frac{\tilde{\nu}}{\pi} \ln R(\tilde{\nu}) \wp \int_0^\infty \frac{1}{\tilde{\nu}'^2 - \tilde{\nu}^2} d\tilde{\nu}' \\ \phi(\tilde{\nu}) &= -\frac{\tilde{\nu}}{\pi} \wp \int_0^\infty \frac{\ln T(\tilde{\nu}')}{\tilde{\nu}'^2 - \tilde{\nu}^2} d\tilde{\nu}' + \frac{\tilde{\nu}}{\pi} \ln T(\tilde{\nu}) \wp \int_0^\infty \frac{1}{\tilde{\nu}'^2 - \tilde{\nu}^2} d\tilde{\nu}' \end{aligned}, \quad (5.153)$$

to yield:

$$\phi(\tilde{v}) = -\frac{\tilde{v}}{\pi}\wp \int_0^\infty \frac{\ln(R(\tilde{v}')/R(\tilde{v}))}{\tilde{v}'^2 - \tilde{v}^2}d\tilde{v}'$$
$$\phi(\tilde{v}) = -\frac{\tilde{v}}{\pi}\wp \int_0^\infty \frac{\ln(T(\tilde{v}')/T(\tilde{v}))}{\tilde{v}'^2 - \tilde{v}^2}d\tilde{v}'$$
(5.154)

In David Tanners' book, you can also find helpful hints on how to extend your spectra beyond the spectral region of your measurement, if you want to do this. If you deal with molecular materials and the usual inorganics excluding semiconductors and conductors, I actually would advise against doing it in this way, and I come to the reason in a minute. First, let us discuss what to do if you have successfully determined the phase. Then you can invert the equation for the calculation of the reflectance to yield

$$\widehat{n}(\tilde{v}) = \frac{1 + \sqrt{R(\tilde{v})}\exp[i\phi(\tilde{v})]}{1 - \sqrt{R(\tilde{v})}\exp[i\phi(\tilde{v})]},$$
(5.155)

not only for a zero angle of incidence but also for the general case with the proper extension provided by Berreman for non-zero angles of incidence [91].

I consider this an excellent method if you like to do a lot of fiddling until you have properly extended your measured spectrum. Unfortunately, this usually takes a lot of time and requires experience. Even worse, what happens if you try to automatize this is that you no longer know if you can trust the result, except if you know exactly what you do and invest some brain into the method. From the previous section, you already know that there might be a way to remove the problems. Indeed, if you take a moment to think of the idea behind ε_∞ or n_∞, you might get the idea that there must be a corresponding R_∞. You do not even have to calculate it, in most cases, it is the reflectance value at the upper end of your spectrum, say at $4000\,\text{cm}^{-1}$ (in case that your reflectance shows a step at higher wavenumbers, this is due to the fact that your sample is no longer an infinite medium after the step and the part reflected at the backside of your sample increases the reflectance; then take the reflectance at lower wavenumbers than that at which the step occurs).

Then Eq. (5.154) transforms into [73]:

$$\phi(\tilde{v}) = -\frac{\tilde{v}}{\pi}\wp \int_0^{\tilde{v}_f} \frac{\ln(R(\tilde{v}')/R_\infty)}{\tilde{v}'^2 - \tilde{v}^2}d\tilde{v}'$$
$$\phi(\tilde{v}) = -\frac{\tilde{v}}{\pi}\wp \int_0^{\tilde{v}_f} \frac{\ln(T(\tilde{v}')/T_\infty)}{\tilde{v}'^2 - \tilde{v}^2}d\tilde{v}'$$
(5.156)

Note that based on the derivation above (Eq. 5.154), this is no problem at all, since we subtract zero anyway, i.e., the value of the integral does not change. On the other hand, this removes the offset of the logarithm of the reflectance and makes the integration behave nicely. As a second improvement, we extend our measurement to the low wavenumber region by taking the first measured reflectance value from the low wavenumber side. This should work even better than in the case above, since the logarithm should even change less than the reflectance itself. If you want to check this hypothesis by yourself, use again Maclaurin's formula to calculate the phase:

$$\phi(\tilde{v}) = -\tilde{v}d\frac{2}{\pi}\left[\sum_j \frac{\ln\left[R(\tilde{v}_j)/R_\infty\right]}{\tilde{v}_j^2 - \tilde{v}^2}\right].$$
(5.157)

If you do not want to test it, take a look at Fig. 5.48. The situation is pretty much comparable to the situation we had in the last section when we wanted to calculate the absorption index function from the refractive index function. Most of the improvement is based on the use of R_∞, the second idea does a good job at improving the result at lower wavenumbers. But with the phase, we cannot do anything except to realize how similar it is already to the index of absorption spectrum. Anyway, how large are the errors if we look at the optical constants derived from the reflectance and the phase?

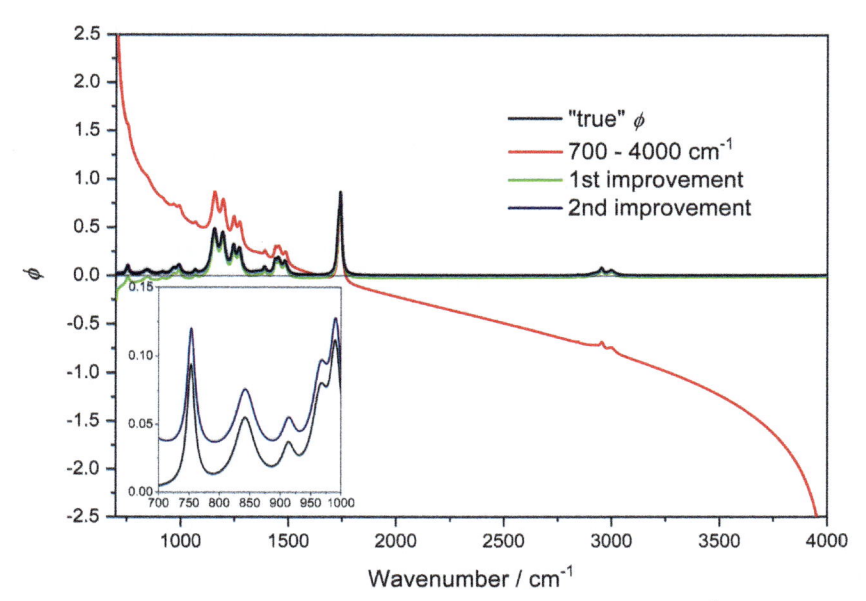

FIG. 5.48 Comparison of the true phase spectrum of PMMA with those obtained in the range 700–4000 cm^{-1} by integrating only over the spectral region presented in the figure *(red curve)*, and taking into account the region above *(green curve)* and below *(blue curve)*.

As Fig. 5.49 shows, nearly the complete error accumulates in the absorption index function as long as we restrict ourself to the region we actually know, i.e., from 700 cm^{-1} upwards, and the error is comparably small. In practice, if you have a noisy spectrum, I would not use one particular reflectance value for R_∞, but instead fit a line through the values, e.g., from 3500 to 4000 cm^{-1}. Apart from this, the above implementation is robust and so fast that it certainly can also be executed for hyperspectral imaging.

Nevertheless, there is still room for improvement. In the following, I will introduce two additional methods to decrease the errors in the absorption index function, at least, if there are no systematic errors in the experimental spectrum.

The first idea (method II) is based on our observation that the error accumulates in the absorption index function. If this is the case, then we could apply a two-step process, the first step of which also consists of applying the above process to calculate the optical constants, but we do not extend the spectra to the low-wavenumber range, i.e., we just use the first improvement. In the second step, we use the so-gained refractive index function and calculate from it in a consecutive

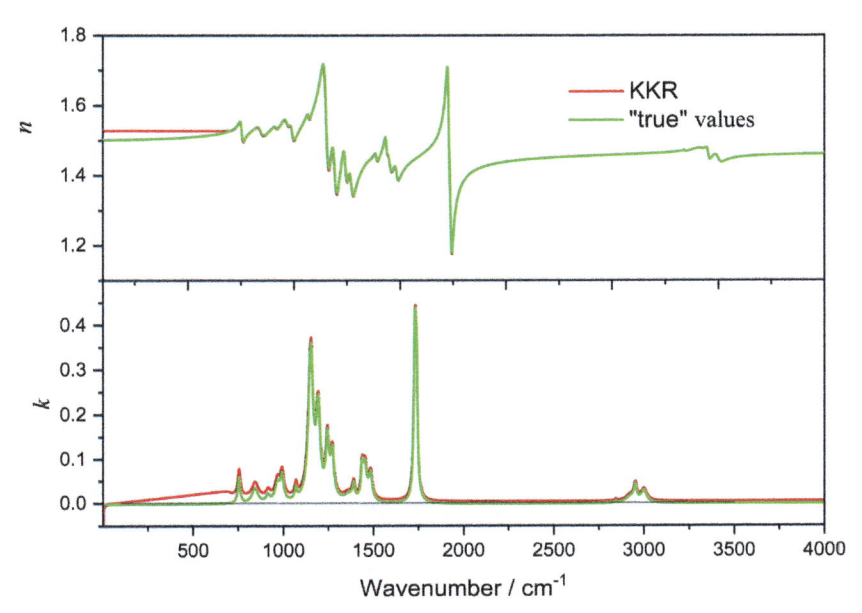

FIG. 5.49 Comparison of the true and the via KKR regained refractive index functions of PMMA.

Kramers-Kronig analysis the absorption index function. Note that it is really important to use only the first improvement and not the second, because if you start from the second, the resulting absorption index function is left unchanged, i.e., it is the same as after the first step.

The second idea (method III) is based on an iterative procedure. Instead of assuming a physics-based model to speculate how the spectra of a certain material can be extrapolated to low wavenumbers, we use a very practical approach that is applicable to any material with the exception of materials that feature a Drude term (semiconductors and conductors). When we assume that there are no absorptions in the lower wavenumber region, then the reflectance must decrease. In a first approximation, this would be a linear decrease. To assure that at the limit of the experimental spectral range, spectrum, and approximation are seamless, we use the following function [73].

$$R(\tilde{v}) = R(0)\frac{\tilde{v} - \tilde{v}_e}{\tilde{v}_e} + R(\tilde{v}_e)\frac{\tilde{v}}{\tilde{v}_e}, \tag{5.158}$$

in which the fit parameter is $R(0)$. Since there is no absorption in the low wavenumber range, the resulting phase function, which strongly resembles the absorption index function, is supposed to be zero, too. Therefore, the fit criterium is

$$\sum_{j=0}^{e-1} \left(\phi(\tilde{v}_j)\right)^2 \stackrel{!}{=} 0. \tag{5.159}$$

The sum starts at zero wavenumber and extends to the wavenumber point before $\tilde{v}_e$. Certainly, a linear change of the reflectance in the low wavenumber region is in general not a very good approximation, but, as you will see, it works.

Aside from the above criteria, we could also demand that the optical constants that result from the fit should provide a perfect match when they are used to calculate the reflectance. Since, as can be seen from Fig. 5.49, the errors of $k(\tilde{v})$ are more pronounced at low wavenumbers, the sum over the squared difference between original $R(\tilde{v})$ and the calculated one is taken only over the first hundred wavenumbers starting at $\tilde{v}_e$.

Now we are ready to compare the results for all four methods (cf. Fig. 5.50). Note that for the comparison I have, by intent, chosen a comparably unfavorable situation where the known range starts at $700\,\text{cm}^{-1}$. In practice, when the spectrum is known down to $400\,\text{cm}^{-1}$, the errors should be even smaller. On the other hand, usually, the empty channel spectrum is already comparable weak at $400\,\text{cm}^{-1}$ for standard equipment, which means that systematic errors may be larger, and it might be necessary to cut the spectrum.

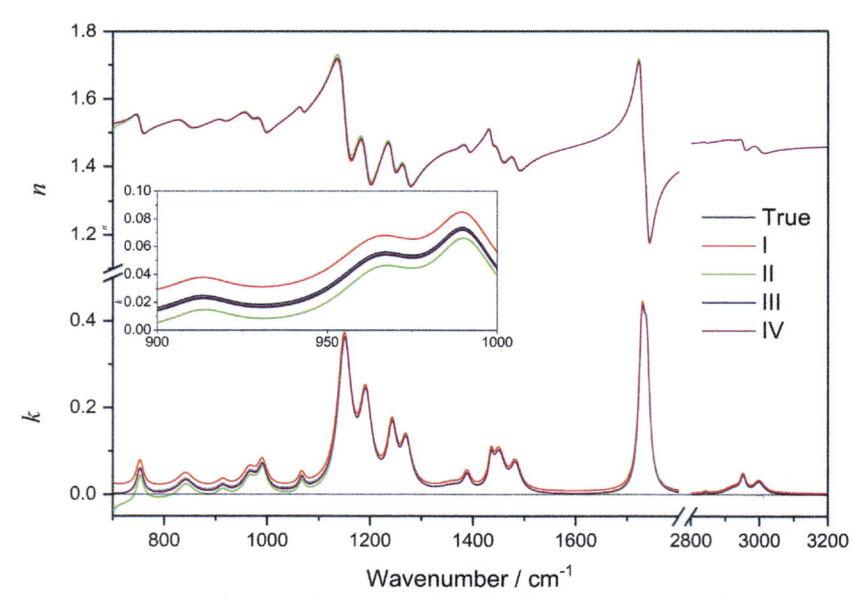

FIG. 5.50 Comparison of the results of the diverse processing methods for Kramers-Kronig analysis. The black curves are the true optical constants that have been used to generate the reflectance in the range 700–4000 cm^{-1}. I is the standard method where a constant reflectance is assumed below 700 cm^{-1} with the same value as at 700 cm^{-1}. II is the two-step procedure and for III a linear reflectance decrease with decreasing wavenumber is fitted. In case of IV, the fit is carried out by minimizing the errors between 700 and 800 cm^{-1}.

With regard to the refractive index function, the performance is excellent with the exception of method II, where some deviations are visible at stronger bands. This is not surprising, because the idea of method II is eventually to distribute the errors more equally, but this is important to know for refraction spectroscopy (cf. Section 7.1). Let me emphasize that you can use refraction spectroscopy of thick weakly absorbing materials like organic substances also for quantitative investigations, even if you rely just on the refractive index values, because of the twin of Beer's approximation discussed in Section 5.3, which states that changes of the refractive index are approximately proportional to the concentration [35]. An immediate application that comes to my mind is the concentration determination of the ingredients in tablets. With regard to the absorption index function, methods III and IV seem to be nearly equally performant, with small advantages for method IV, but for a final conclusion, more experiments have yet to be carried out.

Apart from the practical use of the above equations, I want to provide in the following some more fundamental insights. It seems that so far most colleagues took a look onto these equations who were not part of the molecular spectroscopy-oriented community. Since $-\ln T$ is different from $-\log_{10} T$ just by a multiplicative constant and, thus yields also a kind of absorbance. Therefore, when $-\ln T$ is divided by $4\pi d \tilde{\nu}$ the absorption index results. Accordingly, in the limiting case of weak absorptions, Eq. (5.154) becomes quasi-identical to the Kramers-Kronig relations for the complex refractive index. Actually, I forgot that this holds only for $-\ln T$ and the absorption index from which the refractive index can be calculated. However, $-\ln R$ does not contribute, or does it? Actually, it does, and on a more direct route than $-\ln T$ does. How is this possible? Imagine, we are carrying out an attenuated total reflection experiment. Assume that we have PMMA in contact with a Germanium ATR crystal and measure the reflectance at an angle of incidence of 60° with the polarizer at 45° (1:1 mixture of s- and p-polarized light). What would be the result we get? We certainly do not have to carry out this experiment, we can compute the result using Fresnel's equations (cf. Section 4.2). The result is provided in Fig. 5.51.

Obviously, apart from a scaling factor, $-\ln R$ agrees with the index of absorption (by the way, the scaling factor becomes unity if we would use p-polarized light and 45° for the incidence angle). Actually, for lower angles of incidence and ATR crystals with lower indices of refraction, the deviations are stronger, but it is obvious that these deviations scale with oscillator strength (we will come back to this problem in Section 6.5). This result is not really surprising. This is why ATR spectroscopy works comparably well for molecular spectroscopists. Coming back to our starting point, since $-\ln R$ is nothing but the scaled absorption index, the negative phase will, of course, be nothing else but the (scaled) refractive index function. This means that also for this case, Eq. (5.151) is nothing special, but reduces to the usual Kramers-Kronig relation. Therefore, this can be used as a test case to see if the spectrometer software works, if you really want to use it.

If you are one of the few nowadays, who own an ATR accessory which allows you to set the angle of incidence below the critical angle, you are heartily invited to do so and perform some measurements. If you cannot wait for the result, here is what happens if you measure at an angle of incidence close to zero (cf. Fig. 5.52).

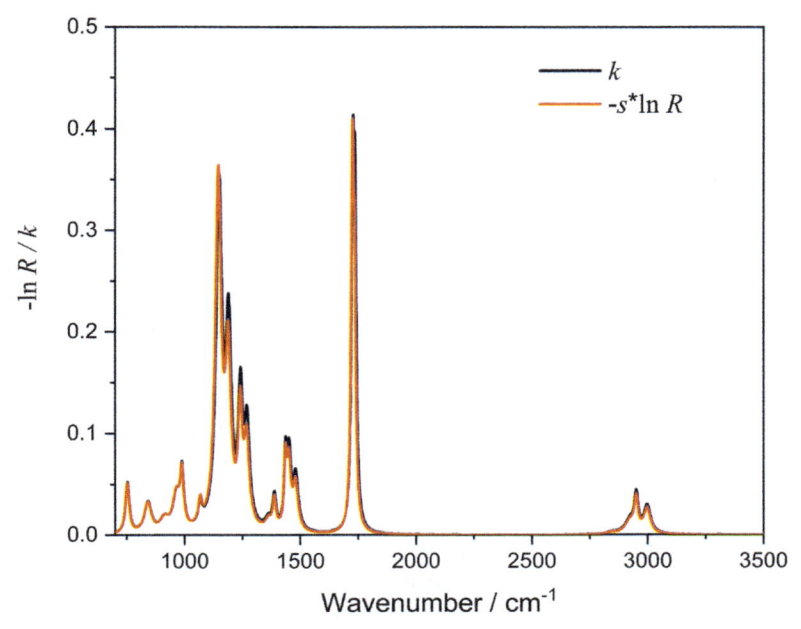

FIG. 5.51 Comparison of the scaled value of $-\ln R$ for an angle of incidence equal to 60° and the imaginary part of the complex index of refraction.

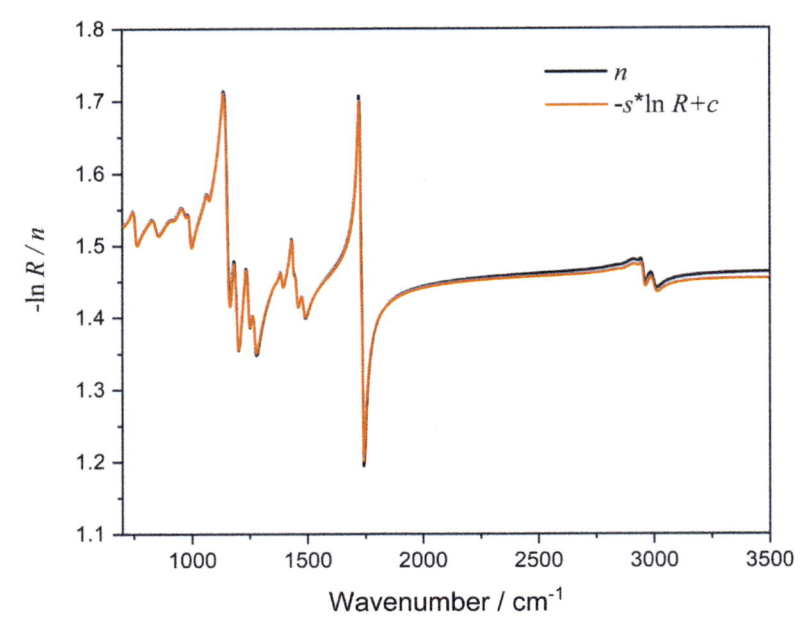

FIG. 5.52 Comparison of the scaled value of $-\ln R$ for zero angle of incidence and the real part of the complex index of refraction.

This means that for weakly absorbing materials, an ATR accessory gives you the chance to determine the phase, very much like in spectroscopic ellipsometry, but at much lower costs, just by switching from below the critical angle to above the critical angle. If you use a Ge crystal and incidence angles close to zero, you can use the reflectance formula for normal incidence, make a series expansion, and neglect higher order terms:

$$-\ln R = -\ln \left| \frac{1 - \sqrt{\frac{\varepsilon_2}{16}}}{1 + \sqrt{\frac{\varepsilon_2}{16}}} \right|^2 \approx n. \tag{5.160}$$

This explains why the natural logarithm of the reflectance is closely related to the real part of the index of refraction. On the other hand, for incidence from air, it is $-\ln R$ which resembles the index of refraction apart from a factor and a constant offset.

Unfortunately, while the process of reconstructing the phase is relatively straightforward for normal incidence, this becomes more complicated for nonnormal incidence or when the incidence medium has a refractive index different from unity. In these cases, certain singularities can arise, leading to additional terms in the phase computation [93]. If the refractive index of the incidence medium n_1 is different from unity, it can be at some wavenumber be equal to that of the sample n_2. As a consequence, reflectance becomes zero and a singularity in $\ln R$ results (the absorption indices of both media are assumed to be zero). For non-perpendicular incidence and s-polarization, an additional singularity is caused in case that $n_2 = n_1 \sin^2 \alpha_i$, i.e., on the onset of total reflection. For p-polarization, an additional singularity occurs if $n_2 = n_1 \tan^2 \alpha_i$, i.e., at Brewster's angle. When performing a contour integration in the complex plane, these singularities introduce additional terms known as phase corrections or Blaschke factors that need to be taken into account when computing the phase [93,94]. Unfortunately, these terms contain unknown quantities, making it challenging to determine the phase accurately. However, approximations have been developed if samples obey the total internal reflection condition at the low and high wavenumber end of the investigated spectral range, allowing for the determination of optical constants with satisfactory accuracy for most organic materials using s-polarized light [95–98].

For p-polarized light, the situation is slightly different. In theory, a correction term proposed by Bertie and Lan could be applied to this polarization [98]. However, calculating the real and imaginary parts of the complex refractive index requires solving a quadratic equation, the solutions of which were already provided by Berreman [91]. Berreman also discussed the circumstances under which each solution should be applied, but these circumstances depend on the wavenumber and cannot be specifically formulated. This has likely prevented previous authors from using p-polarized light in their investigations [95–98], except for Yamamoto and Ishida, who developed an iterative method to select the correct solution of the quadratic equation [99]. Unfortunately, this method significantly increases the complexity and time required to analyze a spectrum, making it unattractive. This is, in particular, unfortunate, since p-polarized light usually does have a much larger penetration depth than s-polarized light. Therefore, I provide in the following also a simple calculation method applicable for p-polarization.

As discussed above, for total internal reflection or attenuated total reflection $\ln R$ resembles the absorption index. Therefore we employ the formula suggested by Bertie and Lan [98]:

$$\phi(\tilde{\nu}) = -\frac{1}{\pi} \wp \int_0^\infty \frac{\tilde{\nu}' \ln R_j(\tilde{\nu}')}{\tilde{\nu}'^2 - \tilde{\nu}^2} d\tilde{\nu}'. \tag{5.161}$$

Very much like its pendant for the absorption index, the formula is insensitive to its use for a limited spectral range if the limits are located in transparency regions of the sample [73]. In particular for organic materials with their comparably weak oscillator strengths, errors due to the limited spectral range should be very small.

In Eq. (5.161), the correction terms due to the singularities are missing. If those are added, the phase is for s-polarized light computed by [100],

$$\phi(\tilde{\nu}) = -\frac{1}{\pi} \wp \int_0^\infty \frac{\tilde{\nu}' \ln R_s(\tilde{\nu}')}{\tilde{\nu}'^2 - \tilde{\nu}^2} d\tilde{\nu}'$$
$$+ \pi - 2 \arctan \left[\sqrt{n_1(\tilde{\nu})^2 \sin^2 \alpha_i - n_\infty^2} / (n_1(\tilde{\nu}) \cos \alpha_i) \right], \tag{5.162}$$
$$- \frac{S^2}{\tilde{\nu}^2} \frac{1}{\pi} \wp \int_0^\infty \frac{\tilde{\nu}' \ln R_s(\tilde{\nu}')}{\tilde{\nu}'^2 - \tilde{\nu}_\infty^2} d\tilde{\nu}'$$

and for p-polarized light by:

$$\phi(\tilde{\nu}) = -\frac{1}{\pi} \wp \int_0^\infty \frac{\tilde{\nu}' \ln R_p(\tilde{\nu}')}{\tilde{\nu}'^2 - \tilde{\nu}^2} d\tilde{\nu}'$$
$$+ \pi - 2 \arctan \left[\left(n_1(\tilde{\nu}) \sqrt{n_1(\tilde{\nu})^2 \sin^2 \alpha_i - n_\infty^2} \right) / \left(n_\infty^2 \cos \alpha_i \right) \right], \tag{5.163}$$
$$- \frac{S^2}{\tilde{\nu}^2} \frac{1}{\pi} \wp \int_0^\infty \frac{\tilde{\nu}' \ln R_p(\tilde{\nu}')}{\tilde{\nu}'^2 - \tilde{\nu}_\infty^2} d\tilde{\nu}'$$

I have incorporated a semi-empirical correction factor into the formula given by Bertie and Lan. This correction factor, $S^2/\tilde{\nu}^2$, is based on dispersion theory. Accordingly, S^2 represents a kind of oscillator strength [100], but exhibits only slight variations between different materials. Additionally, it increases slightly with increasing strength of the oscillators of the material under consideration. Although it is generally unaffected by the polarization state, it does decrease with both the angle of incidence and the refractive index of the ATR crystal. For germanium specifically, the optimal value for the correction factor is $S = 0$, regardless of the angle of incidence. This means that the last term in Eqs. (5.162), (5.163) becomes unnecessary. Anyway, due to the a priori unknown nature of the correct optical constants functions, the correction factors can never be exact, particularly when dealing with singularities.

Once the phase is known, the complex refractive index function for s-polarized light can be computed by:

$$\hat{n}_2(\tilde{\nu}) = n_1(\tilde{\nu}) \sqrt{\sin^2 \alpha_i + \cos^2 \alpha_i \left(\frac{1 + \sqrt{R_s(\tilde{\nu})} \exp[i\phi(\tilde{\nu})]}{1 - \sqrt{R_s(\tilde{\nu})} \exp[i\phi(\tilde{\nu})]} \right)^2 }, \tag{5.164}$$

In contrast to existing literature, I can provide a straightforward criterion for determining the correct sign in calculating optical constants for p-polarized light. This criterion states that the square root in the numerator should be subtracted if the phase is positive, and vice versa. With this criterion, the calculation of optical constants becomes simple and direct, even for *p*-polarized light:

$$\hat{n}_2(\tilde{v}) = n_1(\tilde{v}) \sqrt{\frac{1 - (\mathrm{sgn}\,\phi(\tilde{v}))\sqrt{1 - 4\sin^2\alpha_i \cos^2\alpha_i \left(\frac{1 + \sqrt{R_p(\tilde{v})}\exp[i\phi(\tilde{v})]}{1 - \sqrt{R_p(\tilde{v})}\exp[i\phi(\tilde{v})]}\right)^2}}{2\cos^2\alpha_i \left(\frac{1 + \sqrt{R_p(\tilde{v})}\exp[i\phi(\tilde{v})]}{1 - \sqrt{R_p(\tilde{v})}\exp[i\phi(\tilde{v})]}\right)^2}}, \quad (5.165)$$

How good do Eqs. (5.162), (5.163) in combination with Eqs. (5.164), (5.165) perform? I tested some both on artificial and experimental spectra and found they perform very well. If I generate the ATR spectra by Fresnel's equations using the optical constants of PMMA and the refractive index function of the ZnSe according to Tatian [101], and apply the KKA, the results are close to perfect as can be seen in Fig. 5.53. Accordingly, the calculated phases, refractive, and absorption index functions agree within line thickness with the corresponding original curves. For the phases, I have investigated this in some more detail and the results are presented in Fig. 5.54. Accordingly, except at the low wavenumber end, the deviations are generally less than half a permille. If the constant $K = S^2/\tilde{v}^2$ is set to unity, the deviations increase to about 1% for *p*-polarization. If the third term, which is actually a constant, in Eq. (5.162) for *s*-polarized and Eq. (5.163) for *p*-polarized light is omitted, the errors decrease with increasing wavenumber. Finally, the orange and green curves are obtained without correction terms.

5.8.3 The sum rules

Originally, I thought that the preceding sections provide all of the KKR that could be interesting for the practitioner, but then I stumbled across the so-called Kramers-Kronig sum rules. When first confronted with them, I still did not see a practical application of these rules. Now, after having used them to show that integral absorbance could to some extent save Beer's approximation (in the absence of local field effects) [31] and after having employed their way of derivation to understand the connection between TO and LO oscillator strength [48], I think I leave it to the reader to decide if these relations are useful. I will not introduce all of them. If you are interested, further sum rules can be found, e.g., in ref. [102].

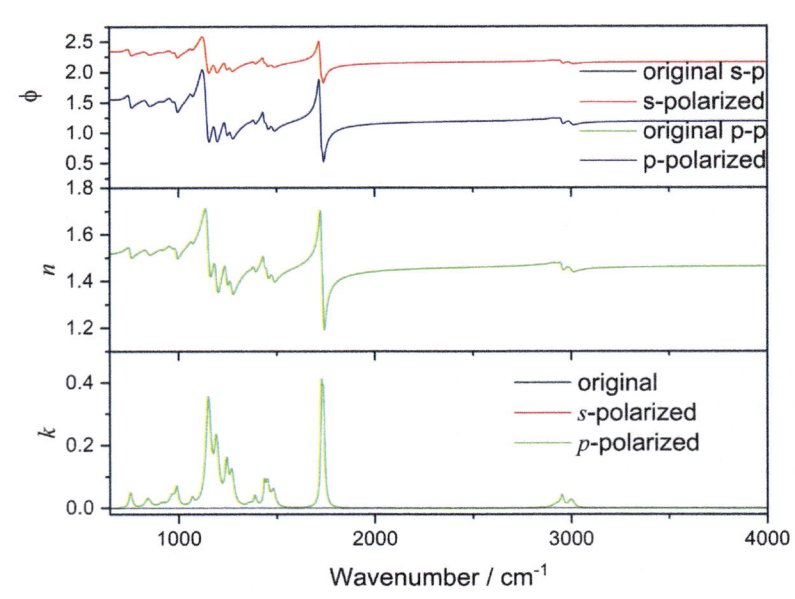

FIG. 5.53 Phase *(upper panel)*, refractive index *(center panel)*, and absorption index *(lower panel)* determined from a KKR analysis of ATR spectra of PMMA generated by Fresnel's equations and employing Eqs. (5.162), (5.164) for *s*-polarized and Eqs. (5.163), (5.165) for *p*-polarized light. In case of the phases, the original curves are virtually indifferentiable from the computed ones. The same is true for the original and the determined refractive and absorption index functions.

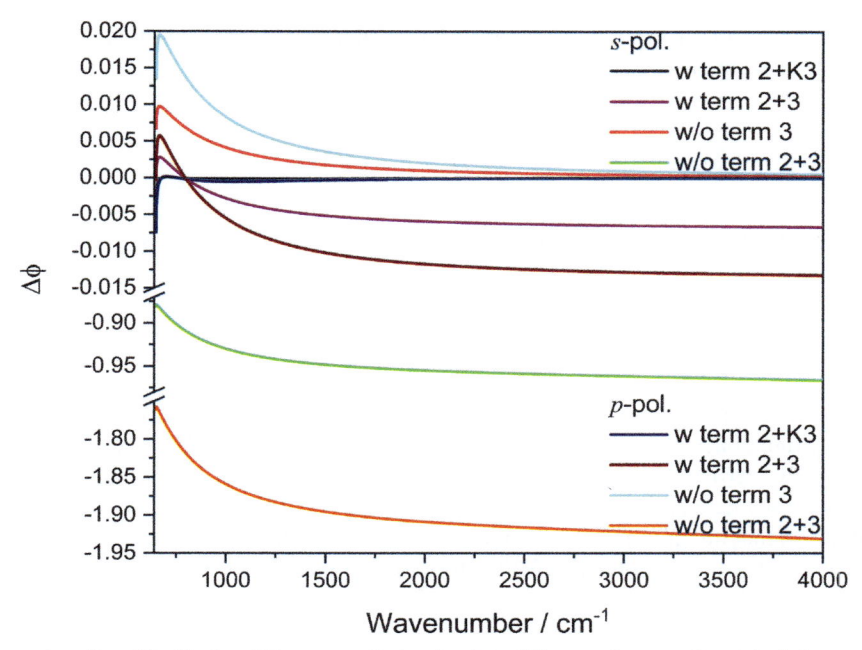

FIG. 5.54 Phase differences in radians. The *black and blue curves* display the phase differences between the original phases and the ones determined by Eqs. (5.162) for *s*-polarized and Eq. (5.163) for *p*-polarized light. If the constant $K = S^2/\tilde{\nu}^2$ is set to unity (i.e., the formula provided by Bertie and Lan [98] is employed), the pink and brown curves result. Omitting term 3, the *red* and turquoise curves are obtained. Finally, without correction terms, the green and orange curves were computed.

To derive the first sum rule, we start with Eq. (5.20):

$$\varepsilon = 1 + \frac{S^2}{\tilde{\nu}_{TO}^2 - \tilde{\nu}^2 - i\gamma_{TO}\tilde{\nu}}. \tag{5.166}$$

The real part ε' of Eq. (5.166) is given by:

$$\varepsilon' = 1 + \frac{S^2(\tilde{\nu}_{TO}^2 - \tilde{\nu}^2)}{(\tilde{\nu}_{TO}^2 - \tilde{\nu}^2)^2 + \tilde{\nu}^2\gamma^2}.$$

If we eliminate in the numerator the factor $\tilde{\nu}_{TO}^2 - \tilde{\nu}^2$, we obtain:

$$\varepsilon' = 1 + \frac{S^2}{\tilde{\nu}_{TO}^2 - \tilde{\nu}^2 + \frac{\tilde{\nu}^2\gamma^2}{\tilde{\nu}_{TO}^2 - \tilde{\nu}^2}}. \tag{5.167}$$

Increasing the wavenumber $\tilde{\nu}$ to a value (much) higher than the oscillator position, so that $\tilde{\nu}_{TO}^2 \ll \tilde{\nu}^2$ and, $\gamma^2 \ll \tilde{\nu}^2$, the value of the real relative dielectric function at this wavenumber is found to be:

$$\varepsilon' = 1 - \frac{S^2}{\tilde{\nu}^2}. \tag{5.168}$$

This can be compared to the real part determined by the KKRs, keeping in mind that $\tilde{\nu}$ is very large (Eq. 5.133):

$$1 - \frac{S^2}{\tilde{\nu}^2} = 1 + \frac{2}{\pi}\wp \int_0^\infty \frac{\varepsilon''(\tilde{\nu}')\tilde{\nu}'}{\tilde{\nu}'^2 - \tilde{\nu}^2} d\tilde{\nu}'. \tag{5.169}$$

In Eq. (5.169), we split the integral into two parts, one from 0 to $\tilde{\nu}_f$ and a second one from $\tilde{\nu}_f$ to ∞, with $\tilde{\nu}_f < \tilde{\nu}$ and the wavenumber starting from which $\varepsilon''(\tilde{\nu})$ is practically zero:

$$1 - \frac{S^2}{\tilde{\nu}^2} = 1 + \frac{2}{\pi}\wp \int_0^{\tilde{\nu}_f} \frac{\varepsilon''(\tilde{\nu}')\tilde{\nu}'}{\tilde{\nu}'^2 - \tilde{\nu}^2} d\tilde{\nu}' + \frac{2}{\pi}\wp \int_{\tilde{\nu}_f}^\infty \frac{\varepsilon''(\tilde{\nu}')\tilde{\nu}'}{\tilde{\nu}'^2 - \tilde{\nu}^2} d\tilde{\nu}'. \tag{5.170}$$

If $\varepsilon''(\tilde{v})$ is virtually zero, the second integral is also zero, so that (note that there is no longer a pole in the denominator; therefore, a regular integral is obtained),

$$\frac{S^2}{\tilde{v}^2} = \frac{2}{\pi} \int_0^{\tilde{v}_f} \frac{\varepsilon''(\tilde{v}')\tilde{v}'}{\tilde{v}'^2} d\tilde{v}', \qquad (5.171)$$

and, finally ($\tilde{v}_f$ can be replaced by ∞ since the value of the integral does not change by this replacement):

$$\int_0^\infty \tilde{v} \varepsilon''(\tilde{v}) d\tilde{v} = \frac{\pi}{2} S^2. \qquad (5.172)$$

This rule (Eq. 5.172) therefore tells us, that if we multiply the imaginary part of the dielectric function with the wavenumber and integrate the resulting function, we obtain the squared oscillator strength times $\pi/2$. We can go one step further, since we know that $\varepsilon''(\tilde{v})$ is practically zero once the wavenumber is somewhat smaller or larger than the resonance wavenumber. What you understand by somewhat certainly depends on the damping constant, but outside of $\tilde{v}_{TO} \pm 3\gamma$, the contribution to the integral will be small, so that integration between these limits may be sufficient, except when local field effects cannot be neglected. This means that we can apply Eq. (5.172) for isolated bands to obtain the oscillator strength of the corresponding oscillator:

$$\int_{\tilde{v}_{TO,j}-3\gamma_j}^{\tilde{v}_{TO,j}+3\gamma_j} \tilde{v} \varepsilon''(\tilde{v}) d\tilde{v} \approx \frac{\pi}{2} S_j^2. \qquad (5.173)$$

Note that it does not matter, if we replace unity by ε_∞ in Eq. (5.166), since we then would have to take Eq. (5.136) and ε_∞ would be canceled out like unity in Eq. (5.170). If we consider that

$$n(\tilde{v}) = \sqrt{\varepsilon(\tilde{v})} = \sqrt{1 - \frac{S^2}{\tilde{v}^2}} \approx 1 - \frac{S^2}{2\tilde{v}^2}, \qquad (5.174)$$

and,

$$n(\tilde{v}) - 1 = \frac{2}{\pi} \wp \int_0^\infty \frac{k(\tilde{v}')\tilde{v}'}{\tilde{v}'^2 - \tilde{v}^2} d\tilde{v}', \qquad (5.175)$$

we arrive at,

$$1 - \frac{S^2}{2\tilde{v}^2} = 1 + \frac{2}{\pi} \int_0^\infty \frac{k(\tilde{v}')\tilde{v}'}{-\tilde{v}^2} d\tilde{v}', \qquad (5.176)$$

so that we obtain,

$$\frac{\pi}{4} S^2 = \int_0^\infty k(\tilde{v}')\tilde{v}' d\tilde{v}'. \qquad (5.177)$$

Since absorbance $A = 4\pi(\log_{10} e) \cdot \tilde{v} \cdot k(\tilde{v})$, it is possible by determining the area of an absorbance band also the oscillator strength of the oscillator that causes the band, and, thereby, its concentration if the molar oscillator strength is known [31].

Interestingly, Eq. (5.177) is older than the Kramers-Kronig relations themselves. How is this possible? Dispersion, and, in particular, anomalous dispersion was of great interest for Rudolf Ladenburg since about 1908 [21]. In particular, Ladenburg was concerned with understanding oscillator strength in terms of the number and density of electrons that contributed to electronic transitions. In this context, in 1914, he took Planck's famous law of blackbody radiation together with Kirchhoff's law and arrived after some fiddling at Eq. (5.177) [32]. To ensure its correctness, he also derived it from dispersion theory in the flavor of Woldemar Voigt [103]. This explains why you can find Eq. (5.177) to be already firmly established in literature at the same time when the Kramers-Kronig relations were published. In fact, Eq. (5.177) was also derived in somewhat different form from simplified dispersion relations [104]:

$$\frac{\pi}{8 n_\infty \tilde{v}_{TO}} S^2 = \int_0^\infty k(\tilde{v}')d\tilde{v}'. \tag{5.178}$$

In a fashion similar to the one that led to Eq. (5.173), we obtain:

$$\int_{\tilde{v}_{TO,j}-3\gamma_j}^{\tilde{v}_{TO,j}+3\gamma_j} \tilde{v}\varepsilon''(\tilde{v})d\tilde{v} \approx \frac{\pi}{4} S_j^2. \tag{5.179}$$

Thanks to having established the link between TO and LO oscillator strength in Section 5.6, it is now very simple to derive the sum rule for the loss function. We start from,

$$\varepsilon^{-1} = 1 - \frac{S^2}{\tilde{v}_{LO}^2 - \tilde{v}^2 - i\gamma_{LO}\tilde{v}}, \tag{5.180}$$

the real part of which is given by:

$$\varepsilon'^{-1} = 1 - \frac{S^2(\tilde{v}_{LO}^2 - \tilde{v}^2)}{(\tilde{v}_{LO}^2 - \tilde{v}^2)^2 + \tilde{v}^2\gamma_{LO}^2}. \tag{5.181}$$

Again, increasing the wavenumber $\tilde{v}$ to a value (much) higher than the oscillator position, so that $\tilde{v}_{LO}^2 \ll \tilde{v}^2$ and $\gamma^2 \ll \tilde{v}^2$, the value of the inverse real relative dielectric function at this wavenumber is found to be:

$$\varepsilon'^{-1} = 1 + \frac{S^2}{\tilde{v}^2}. \tag{5.182}$$

Like the real and the imaginary parts of the dielectric function, the real and the imaginary parts of the inverse dielectric function are connected via the Kramers-Kronig relations. Therefore:

$$\int_0^\infty \tilde{v}\varepsilon''^{-1}(\tilde{v})d\tilde{v} = -\frac{\pi}{2}S^2. \tag{5.183}$$

The negative sign on the right side is a consequence of the fact that the imaginary part of the inverse dielectric function is negative. If we introduce ε_∞, we need to modify Eq. (5.183), with $\tilde{v}_f$ being the wavenumber at which $\varepsilon'(\tilde{v}_f) = \varepsilon_\infty$:

$$\int_0^{\tilde{v}_f} \tilde{v}\varepsilon''^{-1}(\tilde{v})d\tilde{v} = -\frac{\pi}{2}\frac{S^2}{\varepsilon_\infty^2}. \tag{5.184}$$

Finally, there is a further sum rule, which concerns the refractive index:

$$\int_0^\infty (n(\tilde{v}) - 1)d\tilde{v} = 0. \tag{5.185}$$

This means that the refractive index function has the same area above and below a line drawn through unity, the value of the vacuum refractive index. This sum rule can be easily derived from Eq. (5.133) if we assume a very high wavenumber $\tilde{v}_f$ well above the resonance wavenumbers at which absorption takes place:

$$0 = k(\tilde{v}_f) = -\frac{2\tilde{v}_f}{\pi} \wp \int_0^\infty \frac{n(\tilde{v}') - 1}{\tilde{v}'^2 - \tilde{v}_f^2} d\tilde{v}'. \tag{5.186}$$

Since $\tilde{v}' \ll \tilde{v}_f$,

$$0 = \int_0^\infty \frac{n(\tilde{v}') - 1}{-\tilde{v}_f^2} d\tilde{v}', \tag{5.187}$$

from which Eq. (5.185) follows. Similar relations are valid for the real part of the dielectric function as well as for the real part of the inverse dielectric function. In addition, we can also formulate partial sum rules, where $\tilde{v}_f$ is, e.g., a wavenumber in the NIR:

$$\int_0^{\tilde{v}_f} (n(\tilde{v}) - n_\infty) d\tilde{v} = 0. \tag{5.188}$$

5.8.4 The dielectric and the refractive index background

This is a less important section in the sense that you are not enabled to do anything new from what you learn. On the other hand, I was told twice on two different occasions by reviewers that there is strictly speaking no such thing as a dielectric background. Why? Because the use of ε_∞ or n_∞ implies that the absorptions in the infrared spectral range do not influence the UV/Vis spectral region and vice versa, which is indeed not strictly correct. The reason is that dispersion can become small, but never stops except at zero frequency. In this sense, it is correct to say that the use of ε_∞ or n_∞ produces dispersion relations that are no longer Kramers-Kronig conform. This seems to be unfortunate. However, the above statements are correct only in the mathematical sense, but nonsense from a practical point of view as everyone knows who used dispersion theory more than a few times. In fact, the effects are so small that they can be safely disregarded, but this has already been known well before English became the dominant language in science, which means that for most it is hard to find access to the corresponding papers where this finding had been established a long time ago. For this reason, it might be interesting to give a historical sketch to the point where presumably the dielectric background made its first appearance.

As I have already stated at the beginning of this chapter, Kramers and Kronig derived the relations named after them from dispersion theory [70,72,105], and not from considerations of causality, as it is usually done nowadays. Therefore, Kramers-Kronig analysis and dispersion analysis, have the same roots grounded in dispersion theory. Modern dispersion theory started in the second half of the 19th century. Before, Cauchy had provided first dispersion formulas which worked well in spectral regions where materials are transparent [106]. From the discovery of anomalous dispersion, a need for better dispersion formulas arose. Successful attempts were undertaken by Sellmeier [107,108], Helmholtz [11], and Ketteler [12,109]. In particular, the latter two advanced dispersion theory close to its contemporary form by introducing absorption, although their derivation was not yet based on the latest findings, i.e., Maxwell's equations. The first to employ the latter equations for the derivation was Drude, who compared the result with those obtained by Helmholtz and Ketteler and found all forms to be equivalent [110]. Many derivations, including those of Planck [13] and Lorentz [27], relied on the assumption of a single oscillator. Under this assumption, the dielectric function or the squared refractive index function is a term consisting of unity, the value for infinite frequency, plus one frequency-dependent term (cf. Eq. 5.19). As already detailed at the beginning of this chapter, in real materials, multiple oscillators exist, which can be grouped due to their origin as electronic excitations situated in the UV/vis and vibrational excitations in the infrared spectral region as well as orientation excitations in the microwave region (cf. Section 5.1). Do, e.g., the dispersive terms located in the IR have an influence on the dispersion in the UV/Vis spectral region and vice-versa? A dielectric or refractive index background assumes that such contributions are virtually zero. Is this justified? Actually, already Cauchy's formulas contained a constant difference from unity which is such a background. The first to introduce n_∞ seemed to be Ketteler [12,109]. Goldhammer also used it [5] and thought that its value is proportional to the square of the oscillator strength $S^2_{UV/Vis}$, which is equivalent to assume that

$$\frac{\pi}{4} S^2_{UV/Vis} = \int_{\tilde{v}_f}^\infty k(\tilde{v}') \tilde{v}' d\tilde{v}', \tag{5.189}$$

which agrees with the results of the previous section ($\tilde{v}_f$ is a wavenumber in the transparency range between IR and UV/vis). For those who are not yet convinced, there is another way to look at the problem. If we go back to the most general form of the dispersion relation and separate the sum into two,

$$\varepsilon(\tilde{v}) = 1 + \sum_{IR} \frac{S_j^2}{\tilde{v}_{0,j}^2 - \tilde{v}^2 - i\tilde{v}\gamma_j} + \sum_{UV/Vis} \frac{S_k^2}{\tilde{v}_{0,k}^2 - \tilde{v}^2 - i\tilde{v}\gamma_k}, \qquad (5.190)$$

one describing the contributions in the UV/vis spectral range and the other in the IR spectral range, and if we further assume that we are in the infrared spectral range with $\tilde{v} \approx \tilde{v}_{0,j} < \tilde{v}_{0,k}$, we can conclude that $\tilde{v}_k^2 \gg \tilde{v}^2 > i\tilde{v}\gamma_k$. Accordingly, we find that,

$$\varepsilon(\tilde{v}) \approx 1 + \sum_{IR} \frac{S_j^2}{\tilde{v}_{0,j}^2 - \tilde{v}^2 - i\tilde{v}\gamma_j} + \sum_{UV/Vis} \frac{S_k^2}{\tilde{v}_{0,k}^2}, \qquad (5.191)$$

and the third term does no longer depend on the wavenumber so that:

$$\varepsilon_\infty = 1 + \sum_{UV/Vis} \frac{S_k^2}{\tilde{v}_{0,k}^2}. \qquad (5.192)$$

A somewhat different way of deriving is chosen in the Born and Wolf (Section 2.3, 7th ed.) [3], where the constant $A = \varepsilon_\infty - 1$. Note that strictly speaking,

$$\varepsilon_\infty = \lim_{\tilde{v} \to 0} \left(1 + \sum_{UV/Vis} \frac{S_k^2}{\tilde{v}_{0,k}^2 - \tilde{v}^2 - i\tilde{v}\gamma_k} \right), \qquad (5.193)$$

but as has been already realized by Ketteler [12] and as I will show below the differences are actually small.

Another assumption the origin of which is hidden somewhere in the beginnings of modern dispersion theory is that the dispersion caused by absorptions in the UV/Vis in the transparency region can be modeled by employing a single oscillator ([3], also in Section 2.3, 7th ed.). For PMMA this can definitely be shown to be correct. For the proof, which is provided in Fig. 5.55, I assumed the following parameters for this oscillator: $S = 109,000 \, \text{cm}^{-1}$, $\gamma = 8000 \, \text{cm}^{-1}$ and $\tilde{v}_0 = 100,000 \, \text{cm}^{-1}$. These parameters generate an index of refraction of 1.491 at 589 nm and of 1.479 for $\lambda = \infty / \tilde{v} = 0$ and lead to a strong resemblance of the dispersion in the transparent region with that of real PMMA within the fluctuation margin of literature values [111–114]. On a side note, it is interesting to scrutinize the comparably strong variation of the literature values at

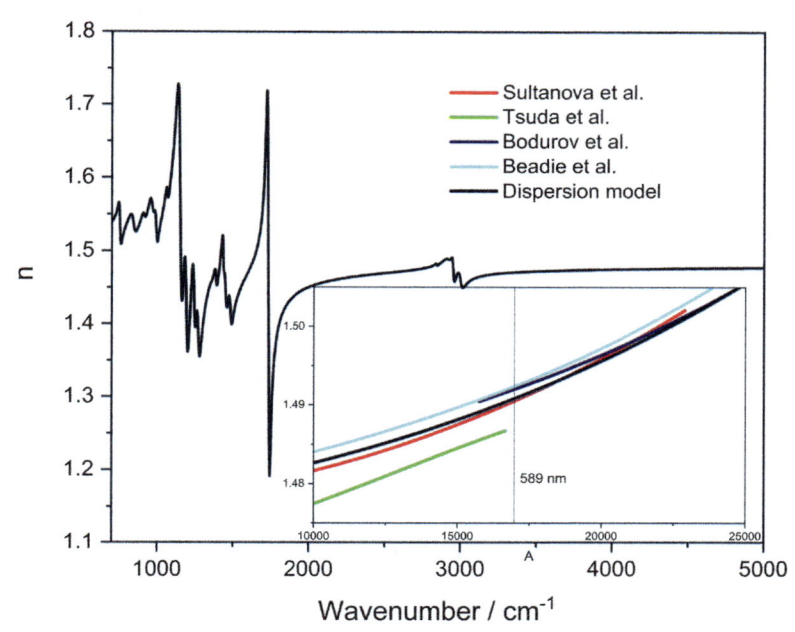

FIG. 5.55 Index of refraction of PMMA as determined by dispersion analysis employing an additional oscillator in the UV to model the dispersion in the transparency region as shown in the inset [73].

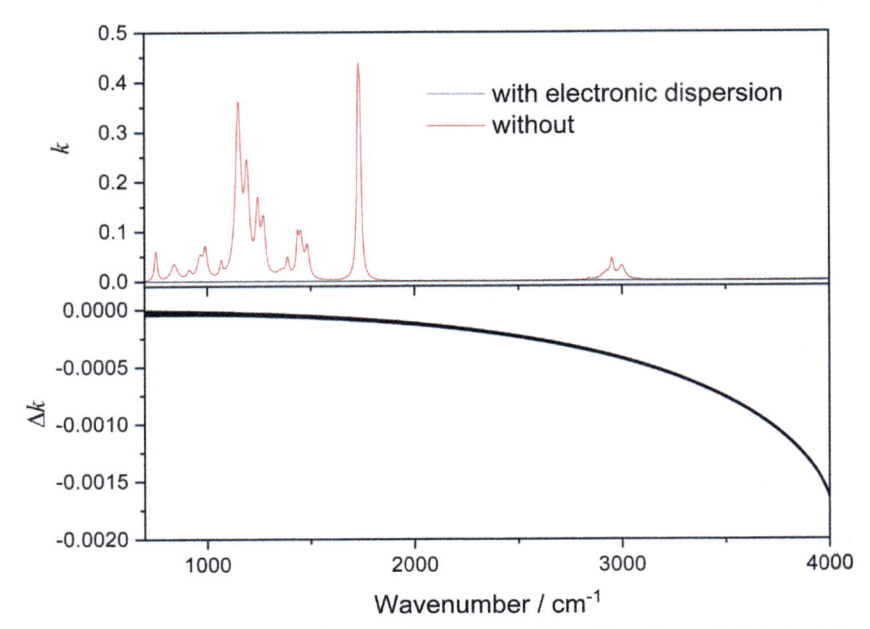

FIG. 5.56 *Upper panel*: Comparison of the absorption index function of PMMA according to dispersion analysis including the UV/Vis oscillator and the regained absorption index function from a KKA of the refractive index function assuming $n_\infty = 1.476$. *Lower panel*: Differences of the spectra in the upper panel.

$\lambda = 589$ nm (16,977 cm^{-1}), which is the wavelength at which refractive indices in the visible spectral region are usually measured (indicated by a vertical line in the inset of Fig. 5.55), which I mention, because these values often have been used as n_∞ in the past [115,116].

Due to the dispersion of the refractive index in the transparency region (cf. inset of Fig. 5.55), Bertie and Lan suggested to introduce Cauchy terms to take the dispersion into account for KKA, although they determined the influence of dispersion on the results of KKA in the IR spectral region to be marginal [117], completely in line with Ketteler. I have checked their finding in the case of PMMA by performing KKA of the refractive index function in the range 0–4000 cm^{-1} assuming an $n_\infty = 1.476$ (why I did take this particular value is explained below) and compared it to the original absorption index function in Fig. 5.56. Obviously, the spectra agree within line thickness and have maximal relative deviations in the range of absorptions of less than 0.1%. Overall, I conclude that the use of a constant to represent the contribution of the electronic excitations situated in the UV/Vis in the infrared spectral region is generally justified and should not lead to detectable errors under usual measurement conditions in agreement with the previous finding of Bertie and Lan [117].

According to above findings, we can write,

$$k(\tilde{v}) = -\frac{2\tilde{v}}{\pi} \wp \int_0^{\tilde{v}_f} \frac{n(\tilde{v}') - n_\infty}{\tilde{v}'^2 - \tilde{v}^2} d\tilde{v}', \qquad (5.194)$$

In practice, the best value for n_∞ or ε_∞ cannot be easily determined except by trial and error. Instead, one possibility is to take the value of the refractive index in the NIR at the point of minimum dispersion as n_∞ (or of the real part of the dielectric function as ε_∞), before dispersion due to the vibrational transitions sets in markedly. In case of PMMA, this point would be actually a region between 8415 and 8427 cm^{-1} (there is a plateau with minimal dispersion within the numerical accuracy, cf. Fig. 5.57), where the refractive index has the value 1.481 (cf. Fig. 5.57, at zero wavenumber n_∞ is not much different, namely 1.479).

Based on newer findings, however, there is a better way to estimate n_∞. This better way consists of finding the value for which Eq. (5.194) produces a zero slope at $\tilde{v}_f$, which results in the value $n_\infty = 1.476$ for PMMA (cf. Fig. 5.58).

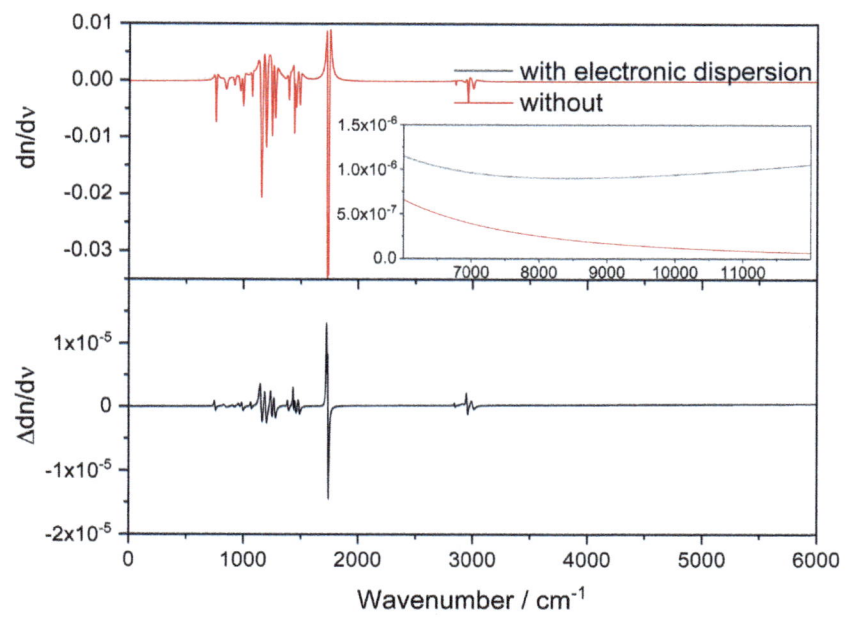

FIG. 5.57 *Upper panel*: Comparison of the dispersion of the refractive index function of PMMA with and without the UV/vis oscillator. *Lower panel*: Difference of the dispersion with and without the electronic part.

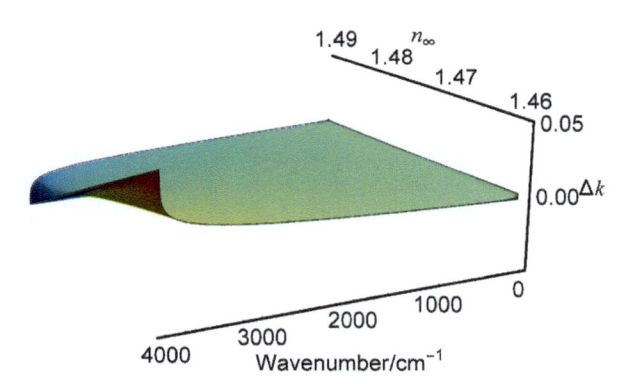

FIG. 5.58 Dependence of the result for Eq. (5.194) from the value used for n_∞ with $\tilde{\nu}_f = 4000\,\text{cm}^{-1}$.

5.9 Further reading

I could just scratch the surface concerning the description of phonons and their interaction with photons. If you want to learn more, I certainly recommend Born's and Huang's book Dynamical Theory of Crystal Lattices [51]. If you want to immerse yourself in more recent insights, I would suggest the book Light and Matter 1d of the series Handbuch der Physik, authored by Heinz Bilz, Dieter Strauch, Roland Wehner, which deals with Vibrational Infrared and Raman Spectra of Non-Metals based on (inharmonic) lattice dynamics [52]. After this book from the middle of the 1980s, it seems the interest in this topic has strongly flattened out. More on the student level, I suggest "Introduction to Solid State Physics" by Charles Kittel [61].

Optics books usually just introduce the classical damped harmonic oscillator model, but if you are more interested in the Kramers-Kronig relations, again the Book of Yeh [118] might be of interest to you, in addition to "Kramers-Kronig Relations in Optical Materials Research" [87]. Oscillator models aside from the classical model are generally beyond the scope of books about optics, but you also will not find anything about this in conventional infrared spectroscopy books, which even fail to make connections between band fitting and dispersion analysis. Some more models, but mostly linked to describing the band gap in the optical regions, can be found in books about spectroscopic ellipsometry like, e.g., in [119]. Furthermore, with regard to the local field of Lorentz, the book by Hadni [120] may be of interest to you (but it does not go beyond the redshift approximation—coupling through local fields is not included). A really good choice, based also on lecture notes, would be the book from one of the reviewers of my habilitation thesis, namely, David Tanner.

It is a valuable complement of this one, as it is less focused on optics, but more on the material aspects (but it also does not go beyond the redshift approximation) [92]. And then there is certainly the possibility to immerse yourself into the original literature which I cited therefore in a more detailed way than in the preceding chapters. If you just want to learn more about symmetry and selection rules, the standard textbooks of infrared spectroscopy are what you are looking for, e.g., refs. [26, 121, 122].

References

[1] J.C. Maxwell, VIII., A dynamical theory of the electromagnetic field, Philos. Trans. R. Soc. Lond. 155 (1865) 459–512.
[2] M. Born, Optik: Ein Lehrbuch der elektromagnetischen Lichttheorie, Julius Springer, 1933.
[3] M. Born, E. Wolf, A.B. Bhatia, Principles of Optics: Electromagnetic Theory of Propagation, Interference and Diffraction of Light, Cambridge University Press, 1999.
[4] V.C. Farmer, The Infrared Spectra of Minerals, Mineralogical Society, 1974.
[5] D.A. Goldhammer, Dispersion und Absorption des Lichtes in ruhenden isotropen Körpern: Theorie und ihre Folgerungen, Teubner, 1913.
[6] L. Lorenz, Ueber die Refractionsconstante, Ann. Phys. 247 (1880) 70–103.
[7] H.A. Lorentz, Ueber die Beziehung zwischen der Fortpflanzungsgeschwindigkeit des Lichtes und der Körperdichte, Ann. Phys. 245 (1880) 641–665.
[8] M. Czerny, Messungen am Steinsalz im Ultraroten zur Prüfung der Dispersionstheorie, Z. Phys. 65 (1930) 600–631.
[9] W. Spitzer, D. Kleinman, Infrared lattice bands of quartz, Phys. Rev. 121 (1961) 1324–1335.
[10] A. Hadni, Essentials of Modern Physics Applied to the Study of the Infrared, Pergamon Press, 1967.
[11] H. Helmholtz, Zur Theorie der anomalen dispersion, Ann. Phys. 230 (1875) 582–596.
[12] E. Ketteler, Zur Handhabung der Dispersionsformeln, Ann. Phys. 266 (1887) 299–316.
[13] M. Planck, Zur elektromagnetischen Theorie der Dispersion in isotropen Nichtleitern, Sitzungsberichte der Königlich Preussischen Akademie der Wissenschaften I (1902) 470–494.
[14] F. Koláček, Versuch einer Dispersionserklärung vom Standpunkte der elektromagnetischen Lichttheorie, Ann. Phys. 268 (1887) 224–255.
[15] D.A. Goldhammer, Die Dispersion und Absorption des Lichtes nach der electrischen Lichttheorie, Ann. Phys. 283 (1892) 93–106.
[16] D. Berreman, Anomalous Reststrahl structure from slight surface roughness, Phys. Rev. 163 (1967) 855–864.
[17] A.M. Efimov, Optical Constants of Inorganic Glasses, Taylor & Francis, 1995.
[18] K.F. Herzfeld, K.L. Wolf, Die Dispersion von Kaliumchlorid und Natriumchlorid, Ann. Phys. 383 (1925) 35–56.
[19] K.F. Renk, Basics of Laser Physics: For Students of Science and Engineering, Springer International Publishing, 2017.
[20] H.A. Kramers, W. Heisenberg, Über die Streuung von Strahlung durch Atome, Z. Phys. 31 (1925) 681–708.
[21] M.J. Taltavull, Rudolf Ladenburg and the first quantum interpretation of optical dispersion, Eur. Phys. J. H 45 (2020) 123–173.
[22] T.G. Mayerhöfer, J. Popp, Beer's law – why absorbance depends (almost) linearly on concentration, ChemPhysChem 20 (2019) 511–515.
[23] J.J. Fox, A.E. Martin, Absorption of the CH_2 group in the region of 3µ, Proc. R. Soc. Lond. Ser. A. Math. Phys. Sci. 167 (1938) 257–281.
[24] K.S. Seshadri, R.N. Jones, The shapes and intensities of infrared absorption bands—a review, Spectrochim. Acta 19 (1963) 1013–1085.
[25] W.F. Maddams, The scope and limitations of curve fitting, Appl. Spectrosc. 34 (1980) 245–267.
[26] P.R. Griffiths, J.A. De Haseth, Fourier Transform Infrared Spectrometry, Wiley, 2007.
[27] H.A. Lorentz, The absoption and emission lines of gaseous bodies, Koninkl. Ned. Akad. Wetenschap. Proc. 8 (1906) 591–611.
[28] T.G. Mayerhöfer, J. Popp, Quantitative evaluation of infrared absorbance spectra – Lorentz profile versus Lorentz oscillator, ChemPhysChem 20 (2019) 31–36.
[29] H.A. Lorentz, The Theory of Electrons and its Applications to the Phenomena of Light and Radiant Heat, G.E. Stechert & Company, 1916.
[30] G. Kortüm, Kolorimetrie · Photometrie und Spektrometrie: Eine Anleitung zur Ausführung von Absorptions-, Emissions-, Fluorescenz-, Streuungs-, Trübungs- und Reflexionsmessungen, Springer, Berlin, Heidelberg, 1962.
[31] T.G. Mayerhöfer, A.V. Pipa, J. Popp, Beer's law-why integrated absorbance depends linearly on concentration, ChemPhysChem 20 (2019) 2748–2753.
[32] R. Ladenburg, Über die Zahl der an der Emission von Spektrallinien beteiligten Atome, Verh. Dtsch. Phys. Ges. 16 (1914) 765–779.
[33] T.G. Mayerhöfer, J. Popp, Beyond Beer's law: revisiting the Lorentz-Lorenz equation, ChemPhysChem 21 (2020) 1218–1223.
[34] H. Kragh, The Lorenz-Lorentz formula: origin and early history, Substantia 2 (2018) 7–18.
[35] T.G. Mayerhöfer, A. Dabrowska, A. Schwaighofer, B. Lendl, J. Popp, Beyond Beer's law: why the index of refraction depends (almost) linearly on concentration, ChemPhysChem 21 (2020) 707–711.
[36] A. Dabrowska, S. Lindner, A. Schwaighofer, B. Lendl, Mid-IR dispersion spectroscopy – a new avenue for liquid phase analysis, Spectrochim. Acta A Mol. Biomol. Spectrosc. 286 (2023) 122014.
[37] T.G. Mayerhöfer, O. Ilchenko, A. Kutsyk, J. Popp, Beyond Beer's law: quasi-ideal binary liquid mixtures, Appl. Spectrosc. 76 (2022) 92–104.
[38] C.J.F. Böttcher, Theory of Electric Polarisation, Elsevier, 1973.
[39] C.J.F. Böttcher, O.C. van Belle, P. Bordewijk, A. Rip, Theory of Electric Polarization, Elsevier Scientific Publishing Company, 1973.
[40] B. Guillot, Line shapes in dense fluids: the problem, some answers, future directions, AIP Conf. Proc. 216 (1990) 453–472.
[41] M. Besnard, N. del Campo, J. Yarwood, B. Catlow, Far-infrared spectroscopy of liquid benzene - long ranged and short ranged dynamics in the neat liquid and in solution, J. Mol. Liq. 62 (1994) 33–54.

[42] A. Barker, J. Hopfield, Coupled-optical-phonon-mode theory of the infrared dispersion in $BaTiO_3$, $SrTiO_3$, and $KTaO_3$, Phys. Rev. 135 (1964) A1732–A1737.
[43] T. Möller, P. Becker, L. Bohaty, J. Hemberger, M. Gruninger, Infrared-active phonon modes in monoclinic multiferroic $MnWO_4$, Phys. Rev. B 90 (2014).
[44] M. Born, M. Blackman, Über die Feinstruktur der Reststrahlen, Z. Phys. 82 (1933) 551–558.
[45] D.W. Berreman, F.C. Unterwal, Adjusting poles and zeros of dielectric dispersion to fit reststrahlen of $PrCl_3$ and $LaCl_3$, Phys. Rev. 174 (1968) 791.
[46] A. Barker, Transverse and longitudinal optic mode study in MgF_2 and ZnF_2, Phys. Rev. 136 (1964) A1290–A1295.
[47] M. Schubert, T.E. Tiwald, C.M. Herzinger, Infrared dielectric anisotropy and phonon modes of sapphire, Phys. Rev. B 61 (2000) 8187–8201.
[48] T.G. Mayerhöfer, S. Höfer, V. Ivanovski, J. Popp, Understanding longitudinal optical oscillator strengths and mode order, Phys. B Condens. Matter 597 (2020) 412398.
[49] C.C. Kim, J.W. Garland, H. Abad, P.M. Raccah, Modeling the optical dielectric function of semiconductors: extension of the critical-point parabolic-band approximation, Phys. Rev. B 45 (1992) 11749–11767.
[50] A.B. Djurišić, E.H. Li, Modeling the index of refraction of insulating solids with a modified Lorentz oscillator model, Appl. Opt. 37 (1998) 5291–5297.
[51] M. Born, K. Huang, Dynamical Theory of Crystal Lattices, Clarendon Press, 1954.
[52] L. Genzel, H. Bilz, in: L. Genzel, H. Bilz (Eds.), Handbuch der Physik: Licht und Materie, 1 ed., Vol. 25, 2d, Springer, 1984.
[53] A. Chaves, R.S. Katiyar, S.P.S. Porto, Coupled modes with A_1 symmetry in tetragonal $BaTiO_3$, Phys. Rev. B 10 (1974) 3522–3533.
[54] J. Humlicek, R. Henn, M. Cardona, Infrared vibrations in $LaSrGaO_4$ and $LaSrAlO_4$, Phys. Rev. B 61 (2000) 14554–14563.
[55] F.L.J. Terry, A modified harmonic oscillator approximation scheme for the dielectric constants of $Al_xGa_{1-x}As$, J. Appl. Phys. 70 (1991) 409–417.
[56] A.M. Efimov, Quantitative IR spectroscopy: applications to studying glass structure and properties, J. Non-Cryst. Solids 203 (1996) 1–11.
[57] T.G. Mayerhöfer, H.H. Dunken, R. Keding, C. Rüssel, Interpretation and modeling of IR-reflectance spectra of glasses considering medium range order, J. Non-Cryst. Solids 333 (2004) 172–181.
[58] R. Brendel, D. Bormann, An infrared dielectric function model for amorphous solids, J. Appl. Phys. 71 (1992) 1.
[59] O.E. Piro, Optical properties, reflectance, and transmittance of anisotropic absorbing crystal plates, Phys. Rev. B 36 (1987) 3427–3435.
[60] J. Humlíček, Transverse and longitudinal vibration modes in α-quartz, Philos. Magaz. Part B 70 (1994) 699–710.
[61] C. Kittel, Introduction to Solid State Physics, Wiley, 2004.
[62] P.M.A. Sherwood, Vibrational Spectroscopy of Solids, Cambridge University Press, 1972.
[63] R.H. Lyddane, R.G. Sachs, E. Teller, On the polar vibrations of alkali halides, Phys. Rev. 59 (1941) 673–676.
[64] F. Gervais, Infrared dispersion in several polar-mode crystals, Opt. Commun. 22 (1977) 116–118.
[65] W. Spitzer, R. Miller, D. Kleinman, L. Howarth, Far infrared dielectric dispersion in $BaTiO_3$, $SrTiO_3$, and TiO_2, Phys. Rev. 126 (1962) 1710–1721.
[66] T.G. Mayerhöfer, J. Popp, Revisiting longitudinal optical modes in materials with plasmon and plasmon-like absorptions – $SrTiO_3$ and β-Ga_2O_3, Phys. B Condens. Matter 590 (2020) 412229.
[67] T.G. Mayerhöfer, V. Ivanovski, J. Popp, Dispersion analysis with inverse dielectric function modelling, Spectrochim. Acta A Mol. Biomol. Spectrosc. 168 (2016) 212–217.
[68] P.G. Etchegoin, E.C. Le Ru, M. Meyer, An analytic model for the optical properties of gold, J. Chem. Phys. 125 (2006) 164705.
[69] P.G. Etchegoin, E.C. Le Ru, M. Meyer, Erratum: "An analytic model for the optical properties of gold" [J. Chem. Phys. 125, 164705 (2006)], J. Chem. Phys. 127 (2007) 189901.
[70] R.D.L. Kronig, On the theory of dispersion of X-rays, J. Opt. Soc. Am. 12 (1926) 547–556.
[71] H.A. Kramers, La diffusion de la lumiere par les atomes, in: Atti del Congresso Internazionale dei Fisici, Como-Pavia-Roma, 2, 1927, pp. 545–557.
[72] C.F. Bohren, What did Kramers and Kronig do and how did they do it? Eur. J. Phys. 31 (2010) 573.
[73] T.G. Mayerhöfer, V. Ivanovski, J. Popp, Infrared refraction spectroscopy – Kramers-Kronig analysis revisited, Spectrochim. Acta A Mol. Biomol. Spectrosc. 270 (2022) 120799.
[74] K. Ohta, H. Ishida, Comparison among several numerical integration methods for Kramers-Kronig transformation, Appl. Spectrosc. 42 (1988) 952–957.
[75] F.W. King, Efficient numerical approach to the evaluation of Kramers–Kronig transforms, J. Opt. Soc. Am. B 19 (2002) 2427–2436.
[76] A.B. Kuzmenko, Kramers-Kronig constrained variational analysis of optical spectra, Rev. Sci. Instrum. 76 (2005).
[77] T.G. Mayerhöfer, S. Pahlow, U. Hübner, J. Popp, Removing interference-based effects from the infrared transflectance spectra of thin films on metallic substrates: a fast and wave optics conform solution, Analyst 143 (2018) 3164–3175.
[78] T.V. Dijk, D. Mayerich, P.S. Carney, R. Bhargava, Recovery of absorption spectra from Fourier transform infrared (FT-IR) microspectroscopic measurements of intact spheres, Appl. Spectrosc. 67 (2013) 546–552.
[79] T.G. Mayerhöfer, J. Popp, Improving poor Man's Kramers-Kronig analysis and Kramers-Kronig constrained variational analysis, Spectrochim. Acta A Mol. Biomol. Spectrosc. 213 (2019) 391–396.
[80] B. Johs, J.S. Hale, Dielectric function representation by B-splines, Phys. Status Solidi A 205 (2008) 715–719.
[81] T.G. Mayerhöfer, S. Pahlow, U. Hübner, J. Popp, Removing interference-based effects from infrared spectra – interference fringes re-revisited, Analyst 145 (2020) 3385–3394.
[82] T.G. Mayerhöfer, S. Pahlow, U. Hübner, J. Popp, CaF_2: an ideal substrate material for infrared spectroscopy? Anal. Chem. 92 (2020) 9024–9031.
[83] E. Rousseau, N. Izard, J.-L. Bantignies, D. Felbacq, Comment on the paper "Improving Poor Man's Kramers-Kronig analysis and Kramers-Kronig constrained variational analysis", Spectrochim. Acta Part A: Mol. Biomol. Spectrosc. 259 (2021) 119849.
[84] S. Nakov, E. Sobakinskaya, F. Müh, A unified framework for the numerical evaluation of the Q-subtractive Kramers–Kronig relations and application to the reconstruction of optical constants of quartz, Spectrochim. Acta A Mol. Biomol. Spectrosc. 288 (2023) 122157.

[85] J. Levallois, I.O. Nedoliuk, I. Crassee, A.B. Kuzmenko, Magneto-optical Kramers-Kronig analysis, Rev. Sci. Instrum. 86 (2015) 033906.
[86] C.D. Keefe, Curvefitting imaginary components of optical properties: restrictions on the lineshape due to causality, J. Mol. Spectrosc. 205 (2001) 261–268.
[87] V. Lucarini, J.J. Saarinen, K.E. Peiponen, E.M. Vartiainen, Kramers-Kronig Relations in Optical Materials Research, Springer, Berlin, Heidelberg, 2006.
[88] P. Lichvar, M. Liska, D. Galusek, What is the true Kramers-Kronig transform? Ceramics-Silikáty 46 (2002) 25–27.
[89] A. Kocak, S.L. Berets, V. Milosevic, M. Milosevic, Using the Kramers—Kronig method to determine optical constants and evaluating its suitability as a linear transform for near-normal front-surface reflectance spectra, Appl. Spectrosc. 60 (2006) 1004–1007.
[90] T.S. Robinson, Optical constants by reflection, Proc. Phys. Soc. Sect. B 65 (1952) 910–911.
[91] D.W. Berreman, Kramers-Kronig analysis of reflectance measured at oblique incidence, Appl. Opt. 6 (1967) 1519–1521.
[92] D.B. Tanner, Optical Effects in Solids, Cambridge University Press, Cambridge, 2019.
[93] J.S. Plaskett, P.N. Schatz, On the Robinson and Price (Kramers—Kronig) method of interpreting reflection data taken through a transparent window, J. Chem. Phys. 38 (1963) 612–617.
[94] V. Hopfe, P. Bussemer, E. Richter, P. Klobes, P- and s-polarized FTIR reflectance spectroscopy at oblique incidence by Kramers-Kronig transformation, J. Phys. D. Appl. Phys. 25 (1992) 288.
[95] J.A. Bardwell, M.J. Dignam, Extensions of the Kramers–Kronig transformation that cover a wide range of practical spectroscopic applications, J. Chem. Phys. 83 (1985) 5468–5478.
[96] J.B. Huang, M.W. Urban, Evaluation and analysis of attenuated total reflectance FT-IR spectra using Kramers-Kronig transforms, Appl. Spectrosc. 46 (1992) 1666–1672.
[97] K. Yamamoto, A. Masui, H. Ishida, Kramers–Kronig analysis of infrared reflection spectra with perpendicular polarization, Appl. Opt. 33 (1994) 6285–6293.
[98] J.E. Bertie, Z. Lan, An accurate modified Kramers–Kronig transformation from reflectance to phase shift on attenuated total reflection, J. Chem. Phys. 105 (1996) 8502–8514.
[99] K. Yamamoto, H. Ishida, Kramers-Kronig analysis of infrared reflection spectra with parallel polarization for isotropic materials, Spectrochim. Acta Part A Mol. Biomol. Spectrosc. 50 (1994) 2079–2090.
[100] T.G. Mayerhöfer, W.D.P. Costa, J. Popp, Sophisticated attenuated total reflection correction within seconds for unpolarized incident light at 45°, Appl. Spectrosc. 78 (3) (2024) 321–328.
[101] B. Tatian, Fitting refractive-index data with the Sellmeier dispersion formula, Appl. Opt. 23 (1984) 4477–4485.
[102] M. Altarelli, D.L. Dexter, H.M. Nussenzveig, D.Y. Smith, Superconvergence and sum rules for the optical constants, Phys. Rev. B 6 (1972) 4502–4509.
[103] W. Voigt, Magneto- und Elektrooptik, B.G. Teubner, 1908.
[104] W. Schütz, Die Gesamtabsorption als Maß für die Anzahl der Dispersionselektronen. Mit 6 Abbildungen, Zeitschrift fur Astrophysik 1 (1930) 300.
[105] H.A. Kramers, Die Dispersion und Absorption von Röntgenstrahlen, Phys. Z. 30 (1929) 522–523.
[106] H. Kayser, Handbuch der Spektroskopie, Vol. 4, Verlag von S. Hirzel, 1908.
[107] W. Sellmeier, Ueber die durch die Aetherschwingungen erregten Mitschwingungen der Körpertheilchen und deren Rückwirkung auf die ersteren, besonders zur Erklärung der Dispersion und ihrer Anomalien, Ann. Phys. 223 (1872) 386–403.
[108] W. Sellmeier, Ueber die durch die Aetherschwingungen erregten Mitschwingungen der Koerpertheilchen und deren Rueckwirkung auf die ersteren, besonders zur Erklaerung der Dispersion und ihrer Anomalien, Ann. Phys. 223 (1872) 525–554.
[109] E. Ketteler, Theoretische Optik: gegründet auf das Bessel-Sellmeier'sche Princip. Zugleich mit den experimentellen Belegen, F. Vieweg, 1885.
[110] P. Drude, Ueber die Beziehung der Dielectricitätsconstanten zum optischen Brechungsexponenten, Ann. Phys. 284 (1893) 536–545.
[111] N. Sultanova, S. Kasarova, I. Nikolov, Dispersion properties of optical polymers, Acta Phys. Pol. A 116 (2009) 585–587.
[112] S. Tsuda, S. Yamaguchi, Y. Kanamori, H. Yugami, Spectral and angular shaping of infrared radiation in a polymer resonator with molecular vibrational modes, Opt. Express 26 (2018) 6899.
[113] I. Bodurov, I. Vlaeva, A. Viraneva, T. Yovcheva, S. Sainov, Modified design of a laser refractometer, Nanosci. Nanotechnol. 16 (2016) 31–33.
[114] G. Beadie, M. Brindza, R.A. Flynn, A. Rosenberg, J.S. Shirk, Refractive index measurements of poly(methyl methacrylate) (PMMA) from 0.4-1.6 μm, Appl. Opt. 54 (2015) F139–F143.
[115] J.P. Hawranek, P. Neelakantan, R.P. Young, R.N. Jones, The control of errors in i.r. spectrophotometry—IV. Corrections for dispersion distortion and the evaluation of both optical constants, Spectrochim. Acta A: Mol. Spectrosc. 32 (1976) 85–98.
[116] J.E. Bertie, S.L. Zhang, C. Dale Keefe, Measurement and use of absolute infrared absorption intensities of neat liquids, Vib. Spectrosc. 8 (1995) 215–229.
[117] J.E. Bertie, Z. Lan, The refractive index of colorless liquids in the visible and infrared: contributions from the absorption of infrared and ultraviolet radiation and the electronic molar polarizability below 20 500 cm^{-1}, J. Chem. Phys. 103 (1995) 10152–10161.
[118] P. Yeh, Optical Waves in Layered Media, Wiley, 2005.
[119] H.G. Tompkins, E.A. Irene, Handbook of Ellipsometry, Springer, Berlin, Heidelberg, 2005.
[120] A. Hadni, Essentials of Modern Physics Applied to the Study of the Infrared: International Series of Monographs in Infrared Science and Technology, Pergamon Press Ltd., Oxford, 1967.
[121] J.M. Chalmers, P.R. Griffiths, Handbook of Vibrational Spectroscopy, J. Wiley, Chichester, 2002.
[122] M. Tasumi, Introduction to Experimental Infrared Spectroscopy: Fundamentals and Practical Methods, Wiley, 2014.

Chapter 6

Deviations from the (Bouguer-) Beer-Lambert approximation

In Chapter 1, I gave you a lot of arguments why absorbance is a quantity, which, in practice, is sometimes of very limited usefulness. One of the main reasons for this limited usefulness was that its experimental determination via the negative decadic logarithm of the transmittance, transflectance, or reflectance is usually not correct or accurate. For attenuated total reflection (ATR) measurements, the deviations between $A(\tilde{\nu}) = 4\pi(\log_{10}e)d\tilde{\nu}k(\tilde{\nu})$ and $A(\tilde{\nu}) = -\log_{10}(R/R_0)$ have been determined quantitatively already a long time ago [1], and for this special kind of spectroscopy, indeed sometimes a wave optics-conform spectral evaluation is performed. But if this is not the case, and this happens quite often, the deviations can be considerable or, at least, significant. For other infrared specialties, systematic, quantitative, and comprehensive comparisons between the results for the low-level and the high-level theory revealing fundamental errors of the (Bouguer-) Beer-Lambert (BBL) approximation seem to be missing, at least before we started to investigate them. I will provide such comparisons in the following, together with wave optics-conform formulas gained from the general formalism provided in Chapter 3 to calculate transmittance and/or reflectance. The general approach is always the same and depicted in Scheme 6.1.

Accordingly, we will start with at a complex index of refraction function $\hat{n}(\tilde{\nu}) = n(\tilde{\nu}) + ik(\tilde{\nu})$, which we generate, e.g., from oscillator parameters. This function will be put into said analytical formulas, which are derived by the wave optics-conform formalism developed in Chapter 4. With the analytical formulas, we calculate the transmittance, transflectance, or reflectance. From these quantities, we compute the negative decadic logarithm and divide it by the thickness, the wavenumber, and $4\pi(\log_{10}e)$ to obtain the index of absorption function $k(\tilde{\nu})$. The latter can then be compared to the original one to determine the errors.

Interestingly, as we will see, these errors include shifts of the maxima, which means that the errors are not proportional to the signal, and also depend on which method is used to obtain the spectrum. An overview of these methods, restricted to the most popular ones, can be found in Scheme 6.2. Accordingly, for transmission, the simplest samples are the free-standing film and the thick film, e.g., a pellet. The difference between the film (a simple layer) and the thick film is that in case of the former interference, effects play a role, while these are excluded for the thick films.

For a thick film to show detectable transmission, absorption must be comparably small like, e.g., in case of higher harmonics or in pellets, i.e., if a strongly absorbing material is embedded in a nonabsorbing matrix, so that the resulting mixture is homogenous to the probing light (ideally, the grain size must be much smaller than the wavelength, otherwise errors due to scattering and, in particular, due to the fact that transmittances are averaged for microheterogeneous samples take over, see Section 6.6).

An alternative to employing a free-standing film is to place or deposit the film onto a transparent substrate. In this case, the natural question arises, if it does make a difference, if first the film is illuminated, or if the incoming light reaches initially the substrate before it hits the film.

Furthermore, the film could be placed between two thick layers. A typical example is the measurement of a liquid in the infrared, since there is practically no solvent, which does not have its own infrared bands, and, since *similia similibus solvuntur*, the solvent is likely to possess its bands at similar positions as the solute, much unlike to the situation in UV/Vis spectroscopy. Again, only overtones and combination bands may be weak enough to use optically thick cuvettes.

For reflection measurements, the use of the BBL approximation generally does not make sense (but its twin regarding the nearly linear change of the refractive index function with molar concentration can make sense, cf. Sections 5.3 and 7.1), except if there is a highly reflecting substrate, such as a metal (then, the method is often called transflection spectroscopy, since the light is transmitted through the film and then reflected by the substrate before it is again transmitted through the film). Another exception is the situation where the incidence medium has a higher index of refraction than the sample, at least in nonabsorbing regions (but keep in mind that the resulting apparent absorbance given by $-\log_{10}(R/R_0)$ is, depending on oscillator strength and some more parameters, only an approximation for the true absorbance in the sense that dividing it

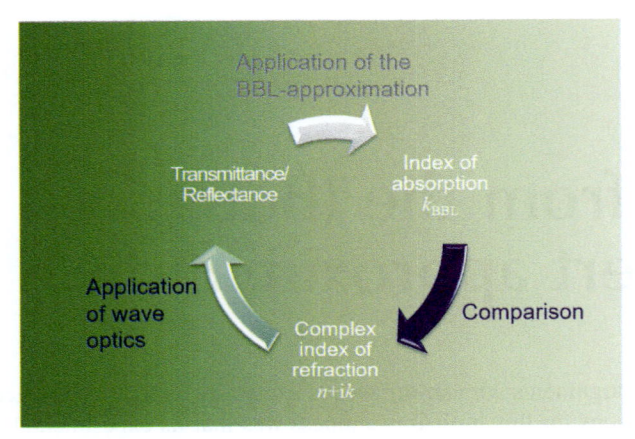

SCHEME 6.1 Determination of the errors of the (Bouguer-)Beer-Lambert approximation.

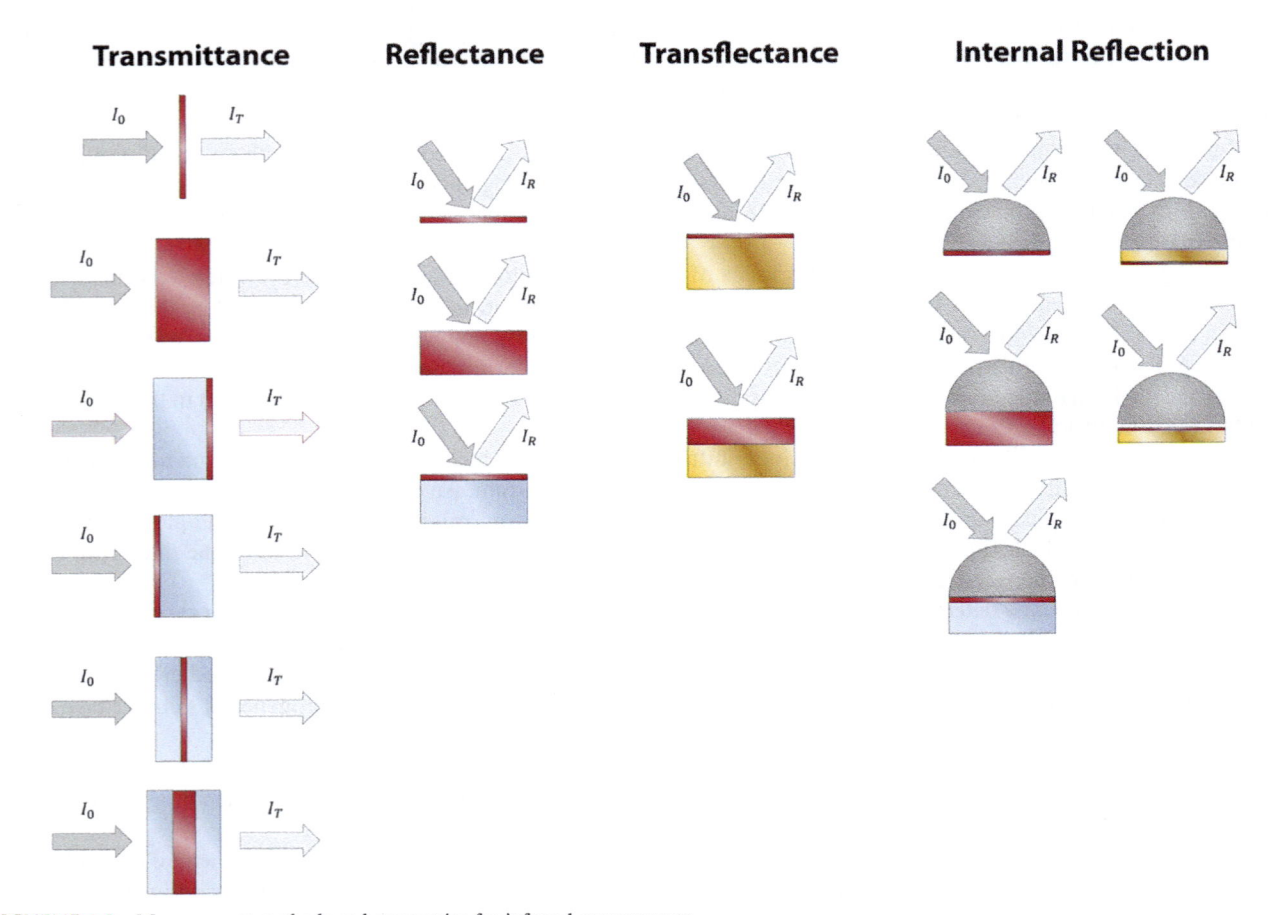

SCHEME 6.2 Measurement methods and geometries for infrared spectroscopy.

by $4\pi(\log_{10} e) d\tilde{\nu}$ does not give the absorption index). Sometimes and under certain conditions, it may make sense to apply/deposit a nonabsorbing layer or a thin gold layer onto the ATR crystal (= the incidence medium) or to perform a kind of transflection experiment in the ATR geometry with an additional air gap between crystal and sample layer, but we will not have a closer look on these geometries before the next chapter.

Since, as you will see, the errors by neglecting wave optics can be formidable, I will also introduce a correction scheme, which we recently developed further and which we combined with dispersion analysis to generate a one-click solution to calculate the corrected absorbance (which is equivalent in most cases to a determination of the optical constants). A corresponding program performs all necessary corrections within less a minute on a conventional office PC. Of course, this solution will be combinable with Poor man's dispersion analysis (cf. Section 5.2) to a full solution, which includes the determination of the dispersion parameters. Alternatively, I will introduce conventional dispersion analysis at the end of this chapter, which is also able to perform the necessary corrections, but needs somewhat more and more profound operator interventions. For inorganic samples in form of layers or thick samples, the latter method is, however, practically without alternative (see, e.g. Ref. [2]).

6.1 Transmittance of a slab embedded in vacuum/air

The simplest form of a potential sample for infrared spectroscopy is that of a slab embedded in vacuum/air, i.e., a pressed pellet of an absorbing material in a nonabsorbing host material like KBr or CsI. The absorbing material has to be strongly diluted (for reasons, we will discuss next) and the powder must consist of grains much smaller than the wavelength (for reasons already discussed previously), which must be homogeneously distributed to form a microhomogeneous sample. A microhomogeneous sample always displays the same spectrum when recorded by an IR microscope, regardless of the position of the beam on the sample; this is a condition not easy to fulfill, because it not only requires submicron-sized powder, but also that this powder is not concentrated between the growing crystallites of the embedding material, when the sample is pressed [3]. For pressed pellets, we can then apply the following mixing rule (which is a mixing rule nearly on the simplistic level of the BBL approximation, i.e., without local-field effects and higher-order terms) [4]:

$$\langle \hat{n} \rangle = \varphi \hat{n}_i + (1 - \varphi) n_m. \tag{6.1}$$

Here, the averaged index of refraction is given by a sum where the first summand is the complex index of refraction of the inclusions n_i multiplied by the volume fraction φ and the second summand is the rest of the volume occupied by the matrix with real refractive index n_m.

For such a slab, the transmittance can be calculated by Eqs. (4.83)–(4.87), which strongly simplify under the assumptions previously. First of all, since there is only one slab, Eq. (4.83) reads

$$\begin{pmatrix} U^+_{0'R} \\ U^-_{0'R} \end{pmatrix} = \underbrace{\overline{D}_{0'}^{-1} \overline{D}_{1'} \overline{P}_{1'} \overline{D}_{1'}^{-1} \overline{D}_{0'}}_{\overline{M}} \begin{pmatrix} U^+_{0'L} \\ U^-_{0'L} \end{pmatrix}, \tag{6.2}$$

where $\overline{P}_{1'}$ is given by:

$$\overline{P}_{j'} = \begin{pmatrix} \exp(-\alpha(\tilde{\nu})d) & 0 \\ 0 & \exp(\alpha(\tilde{\nu})d) \end{pmatrix}. \tag{6.3}$$

Here, $\alpha(\tilde{\nu})$ is the Napierian absorption coefficient as defined in Chapter 1. If we assume the usual transmission technique with incident light perpendicular to the surface (incidence angle $\alpha_i = 0$), the two matrices $\overline{D}_{0'1'}$ and $\overline{D}_{1'0'}$ can be written as follows:

$$\overline{D}_{0'1'} = \frac{1}{\left|\frac{2}{1+\hat{n}_1}\right|^2} \begin{pmatrix} 1 & -\left|\frac{\hat{n}_1-1}{1+\hat{n}_1}\right|^2 \\ \left|\frac{1-\hat{n}_1}{1+\hat{n}_1}\right|^2 & \left|\frac{2}{1+\hat{n}_1} \cdot \frac{2\hat{n}_1}{1+\hat{n}_1}\right|^2 - \left|\frac{1-\hat{n}_1}{1+\hat{n}_1} \cdot \frac{\hat{n}_1-1}{1+\hat{n}_1}\right|^2 \end{pmatrix}$$

$$\overline{D}_{1'0'} = \frac{1}{\left|\frac{2\hat{n}_1}{1+\hat{n}_1}\right|^2} \begin{pmatrix} 1 & -\left|\frac{1-\hat{n}_1}{1+\hat{n}_1}\right|^2 \\ \left|\frac{\hat{n}_1-1}{1+\hat{n}_1}\right|^2 & \left|\frac{2}{1+\hat{n}_1} \cdot \frac{2\hat{n}_1}{1+\hat{n}_1}\right|^2 - \left|\frac{1-\hat{n}_1}{1+\hat{n}_1} \cdot \frac{\hat{n}_1-1}{1+\hat{n}_1}\right|^2 \end{pmatrix}. \tag{6.4}$$

From the matrix $\overline{M}$, we just need the component $\overline{M}_{11}$:

$$T = \overline{t} = \frac{1}{\overline{M}_{11}}. \tag{6.5}$$

The calculation still seems to be tedious, but the result is comparably simple:

$$T = \frac{(1 - R_{01})^2 \exp(-\alpha d)}{1 - R_{01}^2 \exp(-2\alpha d)}, \qquad R_{01} = \left|\frac{1-\hat{n}_1}{1+\hat{n}_1}\right|^2. \tag{6.6}$$

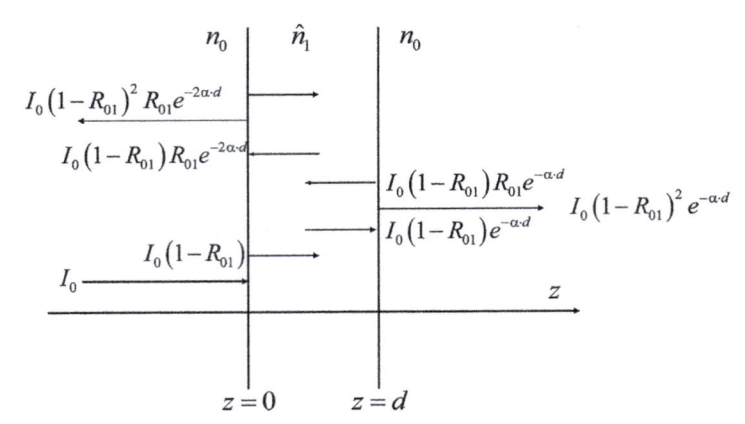

SCHEME 6.3 Scheme to derive Eq. (6.6) by an approach similar to the one of Airy.

Of course, this result could also have been derived by an approach similar to the one of Airy [5] from Eq. (6.7) (Scheme 6.3),

$$I_t = I_0(1 - R_{01})^2 e^{-\alpha \cdot d}(1 + R_{01}^2 e^{-2\alpha \cdot d} + R_{01}^4 e^{-4\alpha \cdot d} + \cdots), \tag{6.7}$$

realizing that the infinite series $\sum_{k=0}^{\infty} x^k = 1/(1-x)$.

Let us for this exercise assume that we have a pellet, which consists of KBr and diopside with different volume fractions ranging from 0.005 to 0.05. We use the different index of refraction functions [4] and calculate the average index of refraction function according to Eq. (6.1). Then, we put the result into Eq. (6.6) and compute the transmittance. Following the procedure depicted in Scheme 6.1, we take the negative decadic logarithm of the transmittance, which is supposed to be the absorbance A. Following the BBL approximation, we just need to divide A by $4\pi(\log_{10} e) d\tilde{\nu} \varphi$ to obtain the index of absorption function $k(\tilde{\nu})$, which should be independent of the volume fraction.

This is, however, not the case, as can easily be seen in Fig. 6.1. Obviously, with increasing φ, $k(\tilde{\nu})$ decreases, but not only that, also the features level out for higher volume fractions in dependence of the oscillator strength. Quite often, it is assumed that this is an effect of shadowing. However, shadowing is an effect that does not exist in homogenous media in the realm of wave optics (a wave bends around small objects, cf. also Fig. 1.1D); it is an effect related to the fact that transmittance gets close to zero. Nevertheless, the baseline has to be corrected for reflection, which we have done here properly. Accordingly, the features get more reflection-like, instead to become completely smoothed. In any way, it seems that the BBL approximation already fails in the simplest case where we do not even have interference!.

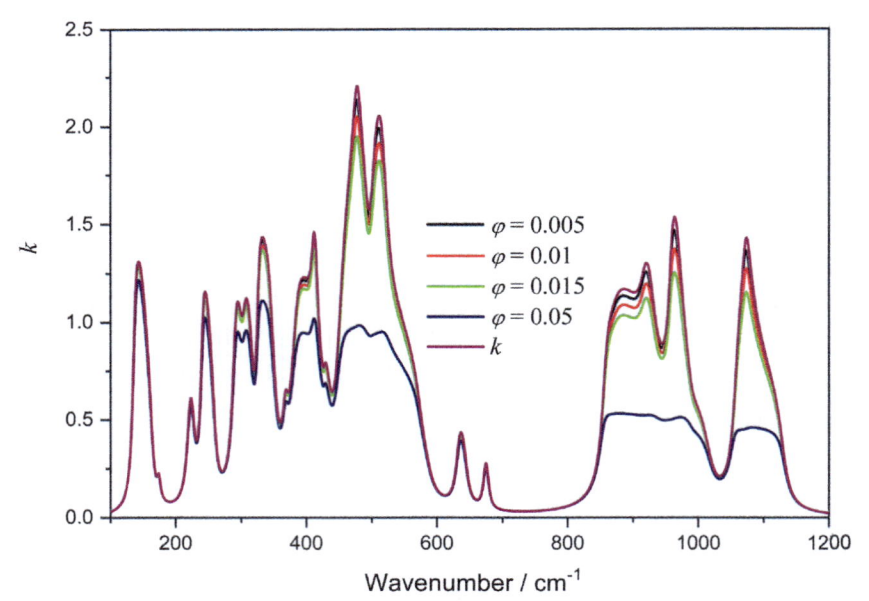

FIG. 6.1 Original index of absorption function $k(\tilde{\nu})$ of diopside in comparison with those regained from the application of the BBL approximation following Eq. (6.6) for different volume fractions φ in a material with $n_m = 1.55$ [4].

How is this possible? In spectrophotometry in the UV/Vis spectral region, the situation is very comparable and there the BBL approximation works (certainly apart from other potential problems like the principal nonlinearity of absorbance, stray light, associations of the molecules, etc.). Indeed, there the BBL works, but actually in a modified form. Remember that I stated that in optics, the BBL approximation describes the propagation inside a medium and not the transmission through it. Does that mean that in spectrophotometry we deal with propagation? Certainly not, but the formula applied in spectrophotometry is slightly different:

$$-\lg\frac{T_\varphi}{T_{\varphi=0}} = -\lg\frac{I_\varphi}{I_{\varphi=0}} = A. \tag{6.8}$$

According to Eq. (6.8), we do not take the negative decadic logarithm of the transmittance of the solution itself (or of the pellet with analyte), but the ratio of this transmittance and the one of the pure solute. We do this not only to remove the influence of the light source. In fact, using this particular ratio has a very physical background. Obviously, this was not clear before I derived it, at least this is what we learnt from Ref. [6], a paper that was approved by Peter Griffiths, the author of the two most influential books on infrared spectroscopy these days [7,8]. Here comes the proof. The abovementioned transmittance ratio can be written based on Eq. (6.6) as

$$\frac{T_\varphi}{T_{\varphi=0}} = \frac{(1-R_{01,\varphi})^2 \exp(-\alpha d)}{1-R_{01,\varphi}^2 \exp(-2\alpha d)} \cdot \frac{1-R_{01,\varphi=0}^2}{(1-R_{01,\varphi=0})^2}, \tag{6.9}$$

where $T_{\varphi=0}$ is the transmittance of the pellet made out of pure matrix material. Note that if we assume that there is no scattering, then the thickness of this pellet does not play any role, and it is just needed to correct for multiple reflections inside the pellet with $\varphi \neq 0$. To see why the BBL approximation works in this case, we first assume that $\varphi \ll 0$. As a consequence, $\langle \hat{n} \rangle \approx n_m$ and we can set $(1-R_{01,\varphi})^2 \approx (1-R_{01,\varphi=0})^2$:

$$\frac{T_\varphi}{T_{\varphi=0}} = \frac{(1-R_{01,\varphi=0}^2)\exp(-\alpha d)}{1-R_{01,\varphi=0}^2 \exp(-2\alpha d)}, \tag{6.10}$$

$R_{01,\varphi=0} \ll 1$, so if it is squared, it can be neglected,

$$\frac{T_f}{T_{f=0}} = \exp(-\alpha \cdot d) = 10^{-a \cdot d}, \tag{6.11}$$

and we obtain the BBL approximation (for KBr, e.g., the index of refraction in the infrared spectral region is about 1.53, which means that the reflectance is about 4.4%, which, if squared, results in about 2‰. Even for CsI ($n_{CsI} \approx 1.74$), we find that $R_{01,\varphi=0}^2 \approx 5‰$). Does Eq. (6.9) indeed perform better? The corresponding results are depicted in Fig. 6.2.

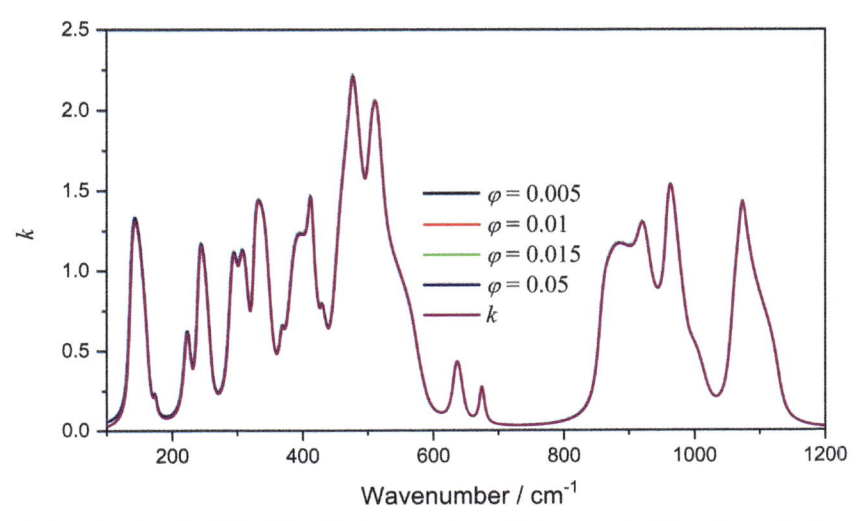

FIG. 6.2 Original index of absorption function $k(\tilde{\nu})$ of diopside in comparison with those regained from the application of the BBL approximation following Eq. (6.9) for different volume fractions φ in a material with $n_m = 1.55$ [4].

The difference is convincing. Obviously, the ratioing of the transmittance of a solution to that of the pure solvent is able to bare the BBL approximation and render it possible to really describe transmittance! Nevertheless, we always have to consider that the (normal) reflectance of the matrix material must be small. As soon as the index of refraction of the matrix material is larger than about 2, even the ratioing does no longer help and the BBL approximation will break down again [4]. A comprehensive analysis of the situation and potential errors can be found in Ref. [9].

6.2 Transmittance of a free-standing film embedded in vacuum/air

For quite some time, it is known (but not well-known!) that multiple reflections at interfaces may cause problems in connection with the quantitative interpretation of spectra using the BBL approximation, as the BBL approximation does not account for this phenomenon. What has this to do with a free-standing film? We have already considered multiple reflections for the (thick) slab in the preceding example. In fact, we do not only get multiple reflections for a thin film, but interference effects additionally. Marianus Czerny had already known this in 1930 (this is not really surprising; his doctorate supervisor had been Heinrich Rubens, who was the direct successor of Paul Drude as Professor for Physics and Director of the Physical Institute of the Friedrich Wilhelm University in Berlin; furthermore, Czerny played the Cello. This would be nothing special, if it had not been Albert Einstein and Max Planck, he played together with during his time in Berlin. Max Planck had, e.g., explicitly pointed out in 1903 [10] that "die Bildung stehender Wellen," i.e., an electric field standing wave effect, is obstructive if absorption is to be quantified. I take it for granted that Czerny not only played music with Einstein and Planck, but also read their papers and discussed those with them).

Unfortunately, the knowledge of the detrimental effects of interference on the BBL approximation got somehow lost during World War II. Out of the blue, it was rediscovered in 1955 [11], but no physical explanation for the deviations from the BBL approximation was given. At least, the phenomenon seemed important enough to be included in a textbook, even when again the effect was explained by multiple reflections [12]. After being forgotten for about 20 years, the effect was again described and a method for its correction was developed [13,14], before it went into oblivion once more. Since then, it was rediscovered a couple of times [15–17], but it never again found its entry into a contemporary textbook of infrared spectroscopy. Even worse, the fundamental significance of the effect has been forgotten, which is in my eyes a consequence of using absorbance and denying light wave properties. A symptom of this misery is that numerous approaches exist to remove interference fringes from absorbance spectra with pure mathematical formalisms (I refrain from giving any references. They are easy to find and I do not want to give those credit who do not use wave optics-based approaches to "correct" only the baseline effect), not understanding the origin of the fringes. Certainly, it is possible to remove the symptom and get smooth baselines by applying these methods, but it is not possible to cure the disease, which requires to understand what interference really means and causes in absorbing systems. I hope I can lay the foundation for this understanding in this section, in particular because the underlying motives will be our companions for some more sections.

First of all, what is the difference between a layer and a thick slab as long as the interfaces are plane parallel? In both cases, as I have reminded you already a couple of times, multiple reflections lead to deviations from the BBL approximation. For the thick slab, it is assumed that interferences are averaged out due to thickness inhomogeneities, when the thickness is much larger than the wavelength [18]. However, e.g., silica substrates are fabricated so accurately nowadays, that even for substrates with thicknesses in the mm range [19] you quite often see interference fringes (the fact that an incoherent source might be in use, like the usual Globar, cannot prevent interference effects, as each wave interacts with and is coherent to itself). These can simply be removed by decreasing the resolution to 2 or $4\,\text{cm}^{-1}$, as the thicknesses are comparably large, and this without consequences, as the substrate is nonabsorbing. Why does this make a difference? Because absorption is proportional to the (electric field) intensity, and, if interference takes place, the (electric field) intensity does not decrease exponentially as the BBL approximation assumes, but can even locally increase [20]. Furthermore, as we will see, the average electric field intensity in case of a standing wave is a function of the wavenumber (to be more precise, of the ratio of film thickness and wavelength), i.e., the same index of absorption will lead at one wavenumber to a stronger absorption than at another assuming a wavenumber independent absorption index. But let us investigate the related phenomena in some more detail. Luckily, I have already provided you the mathematical framework earlier, so I will just repeat here the result of the calculations, which were carried out in Section 4.5.4 and led to Eq. (4.74):

$$t = \frac{1}{M_{11}} = \frac{t_{01}t_{10}\exp(i\phi)}{1 - r_{10}^2 \exp(2i\phi)} \quad t_{01} = \frac{2}{1+n_1}, t_{10} = \frac{2n_1}{1+n_1},$$

$$r = \frac{M_{21}}{M_{11}} = r_{01} + \frac{t_{01}t_{10}r_{10}\exp(2i\phi)}{1 - r_{10}^2 \exp(2i\phi)} \quad r_{01} = \frac{1-n_1}{1+n_1}, r_{10} = \frac{n_1-1}{1+n_1} = -r_{01}.$$

(6.12)

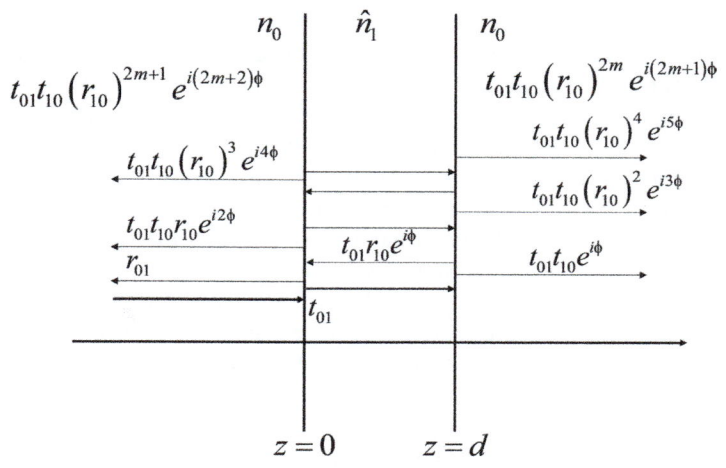

SCHEME 6.4 Scheme to derive Eq. (6.12) by an approach similar to the one of Airy.

In the same section, I promised you to provide a shortcut to the solution, which is along the same lines as the one offered to you in the previous section, i.e., based on Airy's formalism. The shortcut can be found in Scheme 6.4.

To make you understand the essence of the interference effect, we need more than the previous formula, namely, the formulas to calculate the electric field strength and/or intensity for a coherent layer provided in Section 4.5.6. Let us assume that we have a comparably weakly absorbing material with an absorption of the same strength every $200\,\text{cm}^{-1}$. The decadic absorption coefficient of such a material is depicted in the left panel of Fig. 6.3.

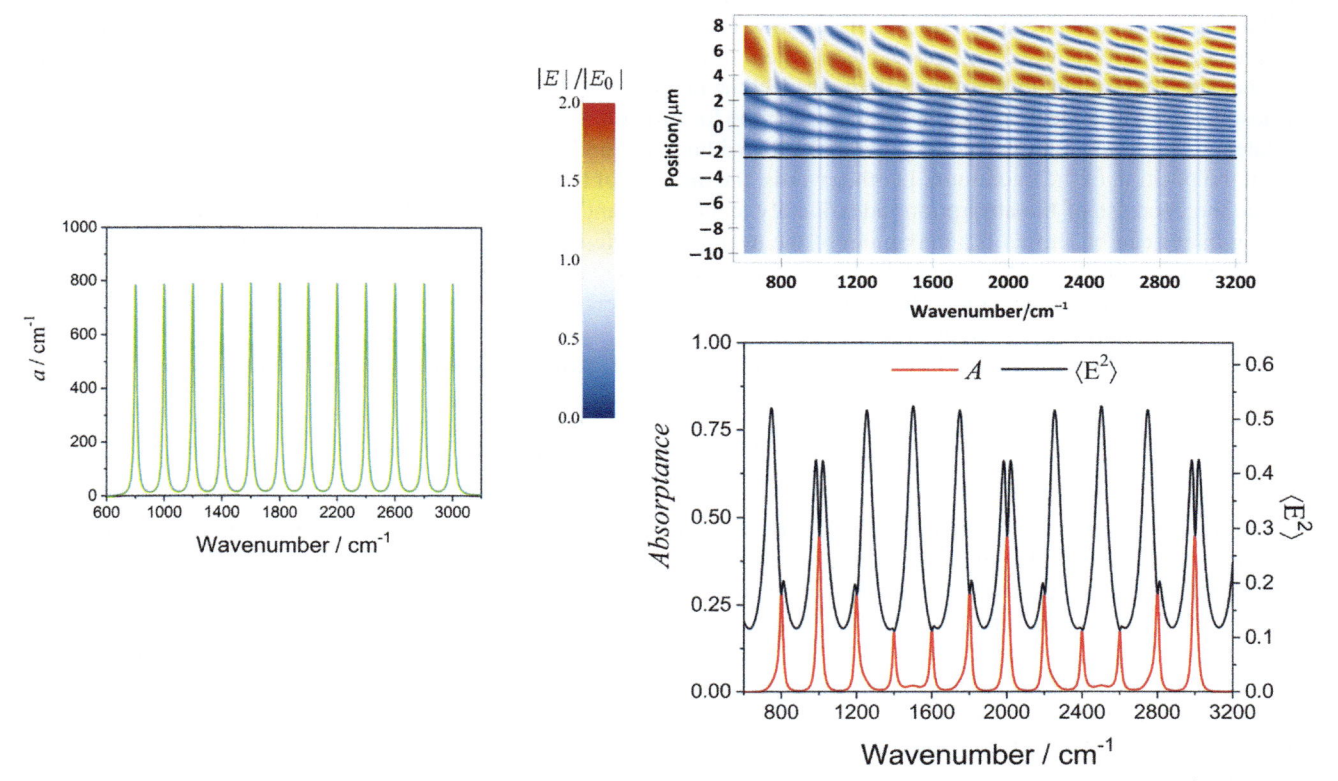

FIG. 6.3 Interference effect in transmission. Left panel: Artificial material with an oscillator with the same strengths every $200\,\text{cm}^{-1}$ starting from $800\,\text{cm}^{-1}$. Right side, upper panel: Electric field strength in a free-standing layer of 5 μm thickness limited by the *black lines*. Right side, lower panel: Absorptance and average electric field intensity within the layer [20].

To better visualize the effects, we have assumed that $n_\infty = 4$, which means that they are unrealistically strong for organic and biological matter. But for the moment, let us concentrate on the effect. Later, I will provide you with more realistic numbers for materials with lower refractive index.

On the right side, upper panel, you see the electric field strengths. Remember, laterally the electric field does not change, therefore, what you see is only the change along the z-direction (which is also the incidence direction of the light wave) in dependence of the wavenumber. The layer itself is assumed to be 5 μm thick, starts at $z = 2.5$ μm, and extends to -2.5 μm. The region below illustrates the electric field strength after the passage of the light through the layer. Before the light propagates within the layer, it has to pass through the first interface. Since the difference of the indices of refraction between incidence medium (=vacuum) is strong, so is reflectance. However, reflectance depends on wavenumber, and while it is, e.g., strong at 1400 cm^{-1}, it is zero at 1500 cm^{-1}, because of the standing wave present in the layer. The wavenumber position of zero reflectance can easily be calculated, because reflectance is zero, when the reflection coefficient r becomes zero (reflectance at the first and the second interface leads to waves that cancel each other in front of the sample) [20]:

$$\rho = \frac{n-1}{n+1} = r_{10} = -r_{01}$$

$$r = \frac{\rho(e^{2i\phi}-1)}{1-\rho^2 e^{2i\phi}} \overset{!}{=} 0 \rightarrow e^{2i\phi} = 1 \rightarrow \phi = m\pi, m = 0, 1, 2\ldots. \quad (6.13)$$

$$\xrightarrow{\phi = 2\pi n_1 \tilde{\nu} d} 2n_1 \tilde{\nu}_m d = m \rightarrow \tilde{\nu}_m = \frac{m}{2n_1 d}$$

Due to this zero reflectance, the electric field, thereby the light intensity, is stronger inside the layer than at other wavenumbers as a higher portion (actually 100%) enters it. As you can see in the right lower panel, at wavenumbers with zero reflectance, the average electric field intensity (= light intensity) is about 4 times higher than at the minima. Assuming that absorption is comparably weak, the absorptance (not the absorbance!) is proportional to the average electric field intensity. Even though absorption lowers the average electric field intensity, the fraction of intensity that is absorbed (= the absorptance) if the oscillator is located at a reflectance minimum is much higher than at a reflectance maximum.

But this is not all that can be noticed! If you look at the bands at 800 and 1200 cm^{-1}, then you can see that the bands become asymmetric: Toward the nearest electric field intensity maximum, the intensity falls off noticeably slower than on the other side of the band. Furthermore, if you inspect the range between the two bands at 1400 and 1600 cm^{-1}, you see that the absorptance has a local maximum in between at the location of an electric field intensity maximum. In fact, as we will see later, such a maximum is able to generate bands without any underlying oscillator, just because the absorption is not zero at the position of the maximum. In other words, the increase of the average electric field intensity is able to overcompensate the decrease of the absorption index locally and evoke bands by that. In fact, not only new bands can occur, but existing bands can be shifted if a maximum of the average electric field intensity is close to a maximum of the index of absorption. Overall, it is not just the BBL approximation that the spectrum is no longer conform with, but a number of surprising new features can come to life, which could not be understood by a spectroscopist who is not aware of the fact that wave optics plays an important role in spectra.

Before I go on, here is how you have to calculate absorptance if the oscillator is not weak as assumed for our discussion:

$$dI_A/I_0 \equiv dA = \alpha(\tilde{\nu})E^2 dl \rightarrow A = \int_0^d \alpha(\tilde{\nu})E^2 dl. \quad (6.14)$$

Accordingly, you have to integrate the product of the Napierian absorption coefficient and the electric field intensity over the z-coordinate of the layer (do you remember? The first part of Eq. (6.14) is the same as in Eq. (1.3)!). Let me emphasize again, that absorbance is in this case (and, e.g., also in surface- or interference-enhanced infrared spectroscopies), strictly speaking, a nonsense quantity, as it does no longer depend on the electric field intensity by its definition (in $A = \varepsilon \cdot c \cdot d$, neither the molar decadic absorption coefficient nor the molar concentration or the thickness depend on the electric field strength). On the other hand, however, it is assumed, when the absorbance is calculated as $A = -\log_{10}T$, to depend on the electric field intensity. This is because the transmittance does certainly depend on E^2, but because of $A = -\log_{10}T$, in a logarithmic way. So, obviously, two things here are connected by force, which do not fit together.

I promised you to employ realistic values for the index of refraction and the oscillator strengths. To perform corresponding calculations, I chose to use the infrared-optical parameters of polyethylene. The results are depicted in Fig. 6.4.

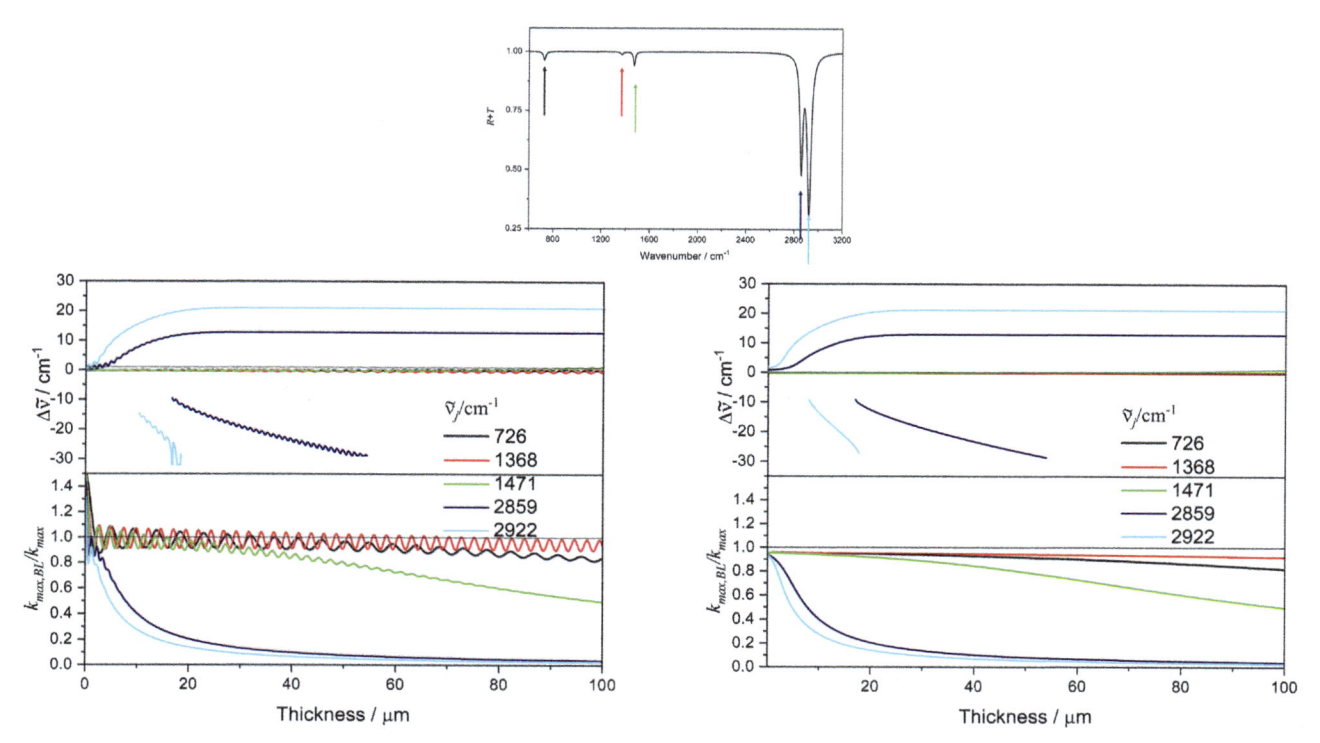

FIG. 6.4 Interference effect in a freestanding polyethylene film. The upper panel shows the sum of the reflectance and transmittance spectrum of a freestanding polyethylene layer and the *arrows* indicate the individual peaks. The lower panels on the right side show the by the BBL approximation calculated changes of the peak positions and the by the BBL approximation calculated absorption indices relative to the original ones in dependence of the thickness of the freestanding layer [20].

To remove the interference fringes and the baseline, I calculated the sum $R + T$. One corresponding spectrum is shown in the upper panel of Fig. 6.4. Despite of having obviously removed the fringes by the addition, $R + T$ and with it, the fringes in $-\log_{10}(R+T)$ the calculated absorption index still oscillates as can be seen in the lower left panel, where the index of absorption at the maximum of the different bands is depicted (It is obvious that $R + T$ must oscillate, since $1 - (R + T)$ does, which is the absorptance). The absorption indices shown are the ones calculated from the BBL approximation relative to the original ones. Accordingly, the values should be unity, independent of thickness, if the BBL approximation would hold strictly. For comparison, the right lower panel shows the same for the incoherent case, i.e., without interference effects. From the latter, we see that just due to reflection, the deviations from the BBL approximation increase, on the one hand with increasing oscillator strength (at least approximately) and on the other hand with increasing thickness. With interference effects, the values oscillate around the values for the incoherent case. In particular for the stronger oscillators at 2859 and 2922 cm^{-1}, it must be stated that it makes actually no sense to measure layers thicker than 5–10 μm, except if very bright light sources are used.

Quite interesting, in particular for comparably thin layers up to about two microns, the deviation from the BBL approximation can reach up to 50%, but only if there is interference. The reason for these deviations is that here the reflection from the first interface plays an important role, and, as we saw, this reflection can be totally (in the presence of absorption, nearly totally) suppressed by interference at a certain wavenumber. The closer a band is to $m = 0$ (cf. Eq. (6.13), i.e., the smaller its wavenumber), the more it gets enhanced for very thin layers. Once the layer thickness grows, the higher wavenumber bands decrease faster in intensity, since the position of the first reflectance maximum decreases in wavenumber and comes spectrally closer. For a thickness of a little more than 1.1 μm (cf. Fig. 6.5), the second reflectance minimum $m = 1$ reaches the higher wavenumber bands and, consequently, they show a maximum. However, since their maximum absorption indices are comparably high, only the weaker band reaches a value of 1. Also, the stronger a band, the faster the undulation ceases, because with increasing thickness, the portion of light that is reflected from the second interface and reaches again the first decreases.

The undulations of the intensities due to the interference effects also affect the positions of the band maxima. To understand the change in the incoherent case, it is important to recall that it is actually not $-\log_{10}T$, but $-\log_{10}(R+T)$ that was plotted in Fig. 6.4. Accordingly, for larger thickness, T reaches eventually zero and R is what remains. This causes a strong

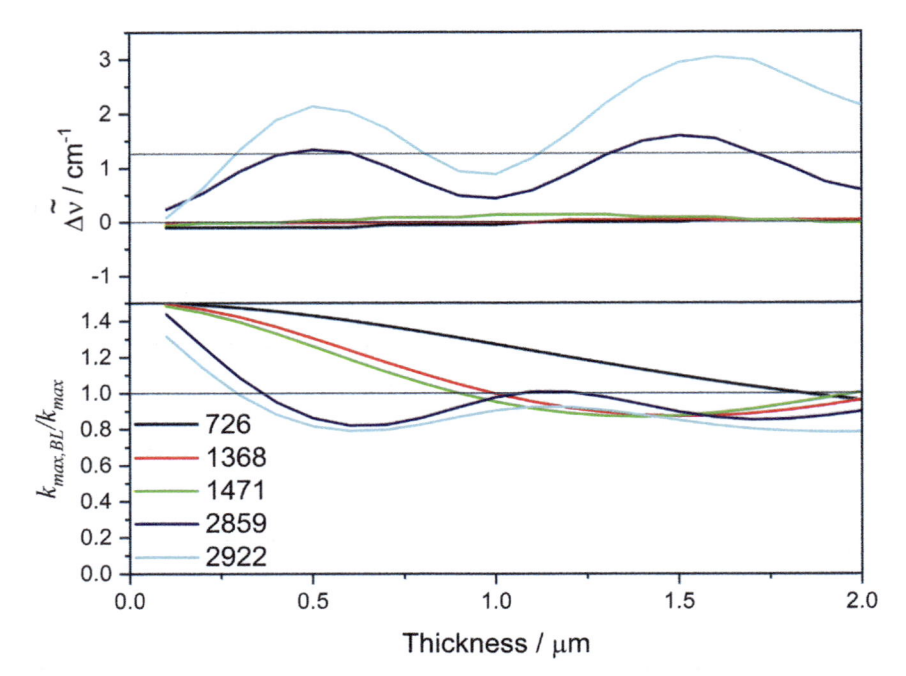

FIG. 6.5 Interference effect in a freestanding polyethylene film—zoom of the lower left panel of Fig. 6.4.

shift for stronger bands, since for the comparably weak oscillators in organic and biological materials, reflectance is mostly influenced by dispersion and the maximum of the real part of the dielectric function is redshifted compared to the oscillator position. The superposition of R and T also leads to the phenomenon that intermediately a second band maximum appears.

Is this a problem of the baseline correction, and do the effects go away if we would perform the baseline correction in another way? Yes and no. Yes, for larger thicknesses, this will help, except for the incoherent case. Why? Because reflectance from the first interface is independent of thickness. It just gets suppressed in the case of interference, since for $d \to 0$, $m = 0$, and the corresponding canceling of the reflectance from the first and the second interface extends over the whole spectral range. This is different for the incoherent case, since there is nothing that could cancel reflectance. On the other hand, in case of thin layers, thickness variations are extremely small and, consequently, the incoherent case loses significance.

Therefore, reflectance does not influence the band positions in transmittance for small thicknesses, even not for larger oscillator strength. However, the fluctuations of the band positions due to interferences still occur, regardless of how the baseline correction is carried out. For a freestanding film consisting of organic or biological material, these changes may be negligible. Quite the contrary is true for inorganic oscillators, see, e.g. Refs. [21,22]. But do not think you are on the safe side, if you do not deal with those, because in the next section, you will be shown that by employing a quite popular sampling technique, the fluctuations of band positions can become extremely large, even for organic and biological matter, and the same is true for the variations of intensity ratios.

6.3 Reflection of a layer on a highly reflecting substrate—Transflection

Recording an infrared spectrum of a layer on a highly reflecting substrate like a highly conducting metal, e.g., aluminum, silver or gold, and evaluating it, seems to be very simple: The metal is assumed to prevent the occurrence of interference fringes [8], the light goes through the layer twice, thereby enhancing sensitivity and the spectra seem to look like transmittance spectra. That things are actually not that simple, could have known again early, but first warnings by Dannenberg et al. were generally ignored [23]. Later on, as in case of free-standing films, further evidence went unnoticed [16,24], before the transflection technique was seen as perfect for biological samples, namely, for sliced tissue on low-e glass slides. In addition to the seeming advantages already mentioned, these glass slides are much cheaper than the usual IR transparent substrates like CaF_2 or ZnSe, less brittle, and fitting to the workflow in pathology. Then, in 2012, seemingly strange variations of the relative intensities of bands with layer thickness were found [25]. The so-called standing wave artifacts were made responsible for this effect. These standing waves should represent the electric field nodes on the surface of the metal, until it was shown that these electric field standing waves are not able to cause the effects observed. As a consequence, it was assumed that the other interface, the one between layer and incidence medium, must be responsible [26]. Nearly simultaneously, we showed that there is indeed an interference effect (electric field standing wave effect), which is, however, the same as that for free-standing films [27]. The only difference is that reflection at the interface between metal and layer is very large and, therefore, the effects are much stronger than in case of free-standing films, at least if the dielectric function of the layer is not very high. In the same paper, we first investigated the case of a nonabsorbing film on metals. In contrast to what is written in the conventional textbooks of infrared spectroscopy, interference fringes are not completely absent in transflection spectra. Their amplitude is a function of the (optical) conductivity and this amplitude is only zero for a so-called perfect conductor, which has a reflectivity of 100%. Accordingly, the amplitudes of the interference fringes increase from gold to silver and to aluminum. In this respect, it becomes important to ponder on the origin of the fringes. For a metal, the transmittance is zero. Therefore, $R + A = 1$. Since the layer is assumed to be nonabsorbing, this means that the electric field intensities must be strong enough to force the metal to absorb rather than reflect at certain wavenumbers. To understand this, we need to investigate the corresponding formulas.

The optical model that we require is actually the same as for the free-standing film, and the only difference is that the exit medium 2 is not identical with the incidence medium 0. Accordingly,

$$r = \frac{r_{01} + r_{12}\exp(2i\phi)}{1 + r_{01}r_{12}\exp(2i\phi)} \qquad r_{01} = \frac{1 - n_1}{1 + n_1}, r_{12} = \frac{n_1 - n_2}{n_1 + n_2}. \tag{6.15}$$

For a perfect conductor, $n_2 = \infty(1+i)$ and $r_{12} = -1$. As a consequence, if n_1 is real, $R = 1$. This is consistent with the fact that a wave cannot penetrate into a perfect conductor, while waves are evanescent in real metals. Therefore, light can be absorbed in this case. For real metals, the condition for minimum reflection has been derived by Park in 1964 [28]:

$$d = \frac{\lambda}{2}\left[\frac{2m+1}{2n_1} - \frac{1}{2\pi n_1}\tan^{-1}\left\{\frac{2n_1 k_2}{\left(n_1^2 - n_2^2 - k_2^2\right)^2}\right\}\right] \qquad m = 0, 1, 2.... \tag{6.16}$$

We have used the optical constants of gold and plugged them into Eq. (6.15) to calculate absorptance and reflectance under the assumption that $n_1 = 4$. Furthermore, we have calculated the electric field strengths for a 5-μm-thick layer on gold. The results are displayed in Fig. 6.6. Obviously, the wavenumber position of the minima of the reflectance correlates with the wavenumber positions of the standing wave, which causes absorption in the metal. The corresponding absorptance is proportional to the average electric field intensity as could be expected. If the index of refraction of the layer is significantly smaller, like for organic and biologic matter, the period of the fringes becomes larger and their intensity decreases substantially, which may make it more understandable that their existence went unnoticed or was denied. For example, for a 5-μm-thick nonabsorbing layer with $n_1 = 1.5$, I have calculated a peak height of slightly under 2% in absorptance (the peak height decreases slightly with decreasing wavenumber). For silver, one finds peak heights between 4.3% and 5.2% and for aluminum already heights between 6.5% and 9.4%. The overall increasing peak heights concur with the order of increasing reflectance of the metal in the MIR ($R(Au) > R(Ag) > R(Al)$), while the decrease of the peak height with decreasing wavenumber seems to be connected to the increase of the reflectance with decreasing wavenumber. This increase is much stronger for Al than for Ag and in particular for Au.

If we assume the layer to be absorbing, the changes are much more drastic than for the free-standing film. Obviously, responsible for these changes is the interface sample/gold and its extremely strong reflectance. Accordingly, there is a much stronger interaction between the incoming wave, which is only weakly attenuated by reflection from the first interface air/sample, and the reflected wave in comparison with the situation for a free-standing film. Therefore, even for organic and biologic materials, the deviations from Beer's approximation are strong as can be seen in Fig. 6.7, left side. The interference fringes can be detected by the fact that in-between peaks the baseline does not reach zero and shows undulations. If we increase the background dielectric function of the layer to $\varepsilon_\infty = 16$, then we see that those undulations are changed into additional peaks. This seems to be a strange concept for vibrational spectroscopists (at least as far as I remember). On the level of the BBL approximation, every peak indicates a transition from one quantum mechanical state to another and is exactly at the position, which corresponds to its transition energy. Fig. 6.7, however, clearly demonstrates that these additional peaks are caused by the spectrally stronger concentrated higher electric field intensities, since the peak maxima of the additional peaks agree with the maxima of $\langle E^2 \rangle$. If the peak maximum of the dielectric function gets closer to the peak maximum of $\langle E^2 \rangle$, then the two peaks combine to a single one, the wavenumber position of which is shifted away from the peak maximum of $\langle E^2 \rangle$ the farther away both are in terms of wavenumbers. On the other hand, if we change the thickness, then the positions of the peak maxima of $\langle E^2 \rangle$ change their wavenumber positions. This helps us to understand the next figure, where for one fixed molecular peak (fixed in the sense that the oscillator parameters, and, in particular, the

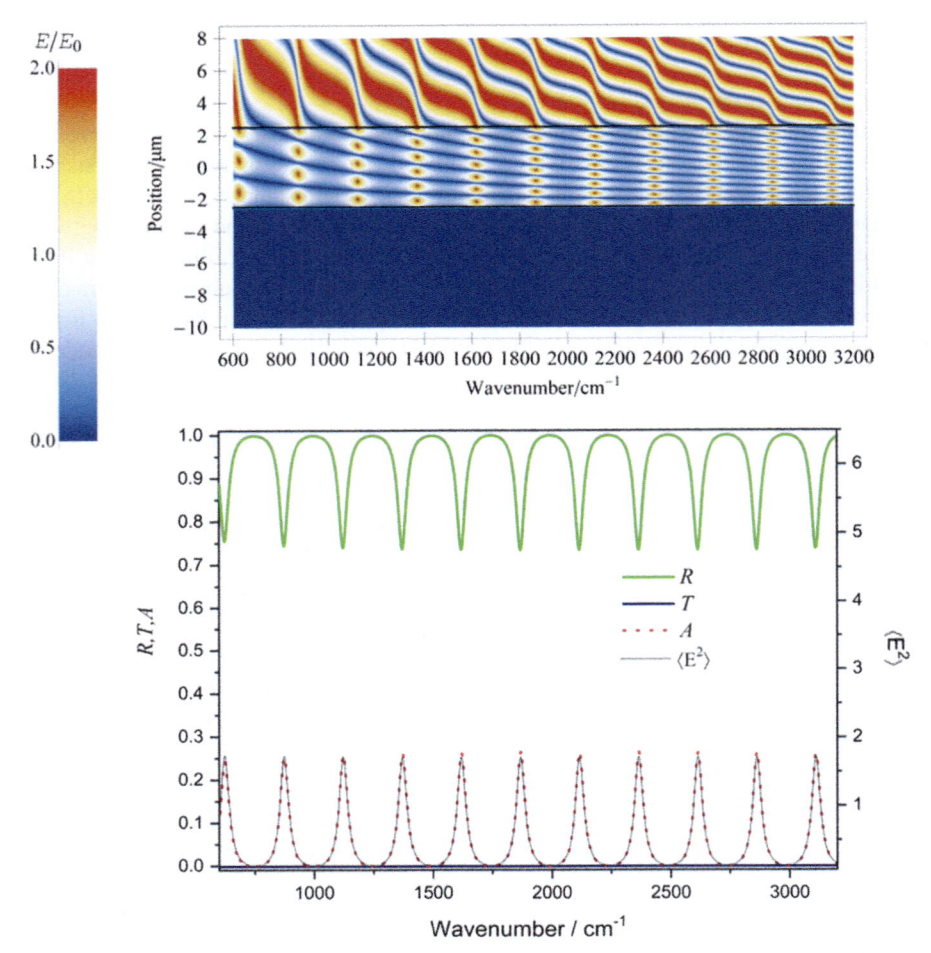

FIG. 6.6 Interference effect in a 5-μm-thick (situated between 2.5 and −2.5 μm) nonabsorbing layer on gold. The upper panel illustrates the electric field strengths in dependence of position and wavenumber, and the lower panel reflectance *(green line)*, transmittance *(blue line)*, absorptance *(red dots)*, and electric field intensity *(black line)*.

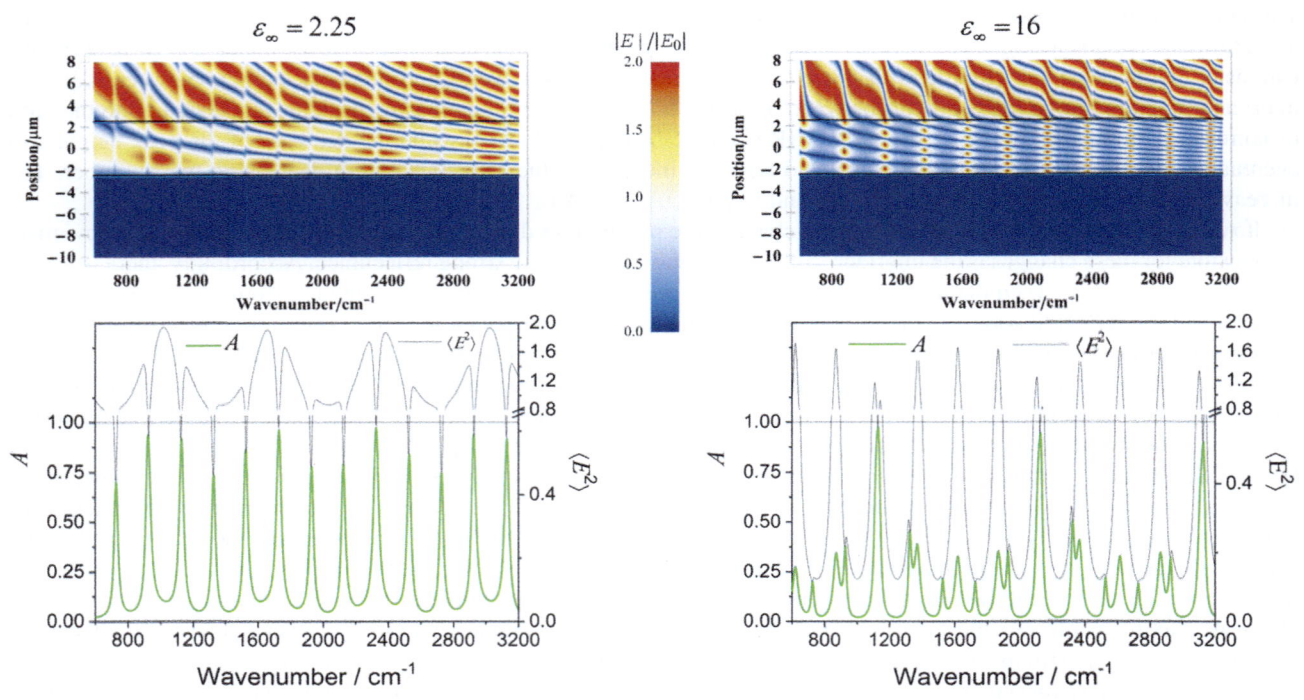

FIG. 6.7 Interference effect in a 5-μm-thick (situated between 2.5 and −2.5 μm) weakly absorbing layer on gold. The upper panels illustrate the electric field strengths in dependence of position and wavenumber, and the lower panel absorptance *(green line)*, and electric field intensity *(black line)*. The oscillators have the same oscillator strengths as those displayed in Fig. 6.3.

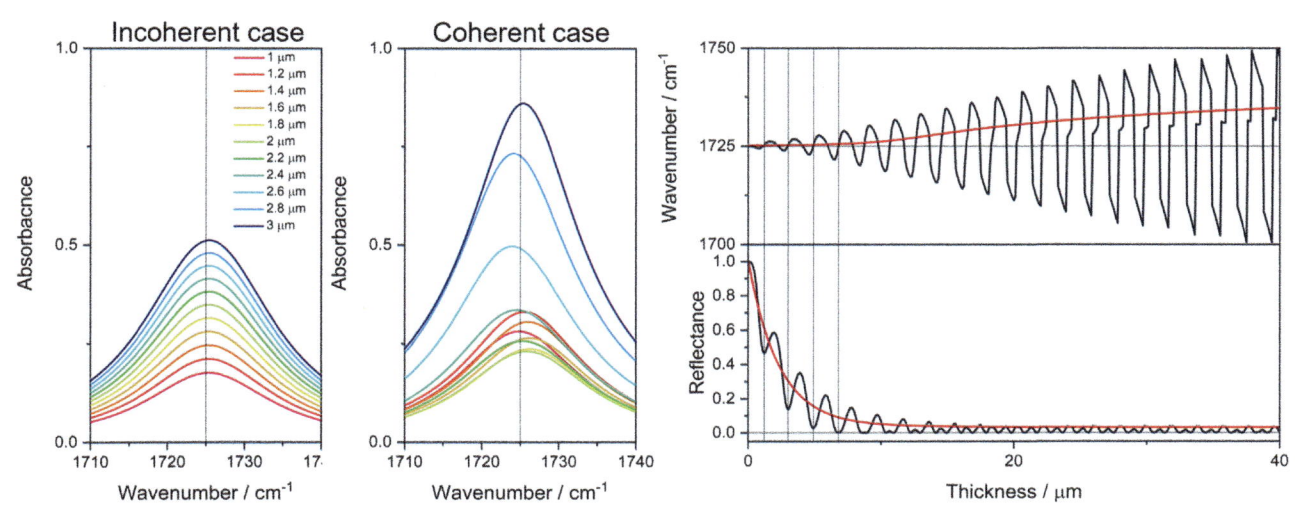

FIG. 6.8 Calculated peak position change in dependence of the thickness of the layer (PMMA) on gold for the incoherent case (left panel) and the coherent case (center panel) for certain thicknesses. Right upper panel: Dependence of the peak position on the layer thickness in the coherent case *(in black)*. The *red curve* shows the same for an incoherent superposition of waves in the layer. Lower right panel: Variation of the reflectance assuming coherent *(black)* and incoherent *(red)* superposition of the waves in the layer.

oscillator position, do not change), the thickness is increased. It can be seen in Fig. 6.8 that for the coherent case, the peak position begins to oscillate around the oscillator position. The comparison with the reflectance reveals that one value equals the other when the reflectance at the oscillator position is either at minimum or maximum, which coincides with an electric field intensity minimum or maximum. With increasing thickness, the oscillations increase, and due to the formation of standing waves, the reflectance can get even smaller than the reflectance of the first interface between layer and air. This becomes obvious through the comparison with the incoherent case where, with increasing thickness, the reflectance of the layer approaches that of the first interface. Since absorption is comparably strong at the oscillator position, at least for an organic material, a thickness of about 10 μm is sufficient that no light back reflected by the gold layer is able to reach and pass through the layer/air interface and arrive at the detector. It is instructive to examine how the reflectance absorbance, i.e., $-\log_{10}(R/R_0)$, where R is the reflectance with layer and R_0 is the reflectance without, changes with the layer thickness. This dependence is shown in Fig. 6.9 at the oscillator position of the C=O vibration of PMMA. Obviously, the reflectance absorbance does not change linearly with the thickness. It is even possible that the reflectance absorbance decreases with increasing thickness and that the amplitude of the changes increases with increasing thickness. But even in the absence of interference effects, i.e., in the incoherent case, reflectance absorbance is not depending linearly on the thickness. The reason is that, as mentioned previously, the reflectance does not become zero with increasing thickness, but approaches eventually the reflectance of the interface layer/air. All in all, while on the first view the transflection technique seems to be very advantageous and simple, its use needs profound knowledge about how to quantitatively evaluate spectra due to possible occurrence of considerable deviations from the BBL approximation.

6.4 Transmission of a layer on a transparent substrate

There are not many IR-transparent materials, since such materials must have their own vibrations at very low frequency, or these vibrations need to be infrared inactive without the material being a metal. Among the latter materials are Ge and Si, both of which are not that much valued because of their large dielectric constant/index of refraction function (both materials belong to the cubic crystal system, so we can use both for stating the material properties, even when the index of refraction is actually a wave property). Those lead not only to high reflectivity, but also to considerable interference effects. Among the former materials, with a low wavenumber vibrational mode, are TlBr-TlI (KRS-5), ZnSe, ZnS, BaF_2, CaF_2, NaCl, etc. Some of those substrate materials have the disadvantage that they are mechanically not very stable, others are not biocompatible [29] or the material is even toxic like in case of KRS-5. For the following, we focus on Si, ZnSe, and CaF_2, because they encompass all the important physics/optics. As can be seen in Fig. 6.10, Si is characterized by an only slowly decreasing index of refraction with decreasing wavenumber, since it has no infrared-active vibrational mode. ZnSe has a lower index of refraction, but starting from about 1500 cm^{-1}, the change of the index of refraction becomes stronger than that of Si because of its phonon mode. This means that in the fingerprint region of many organic and biologic compounds,

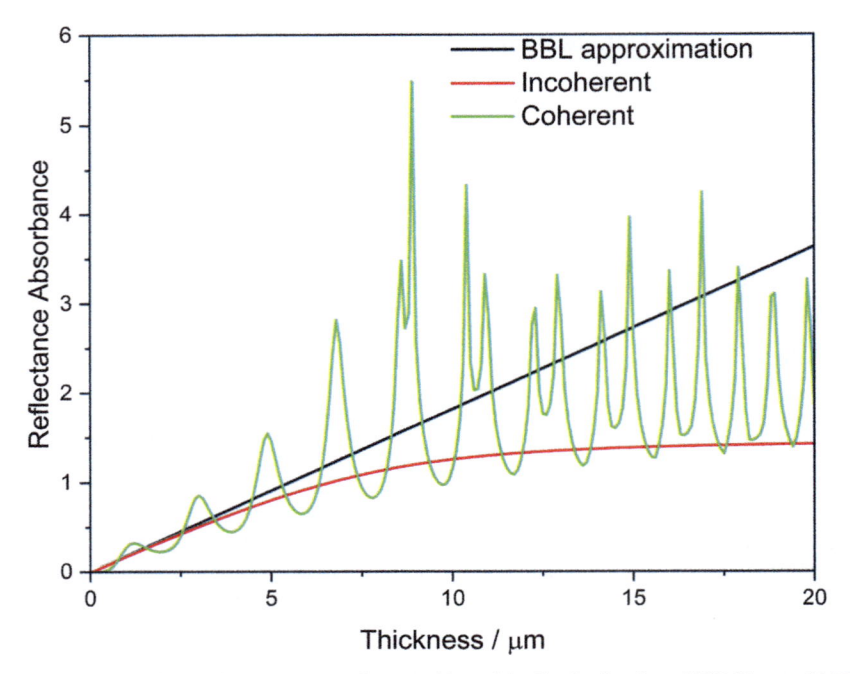

FIG. 6.9 Reflectance absorbance change with thickness at the oscillator position of the C=O vibration of PMMA on gold. The *black line* represents the BBL approximation, whereas the *green and the red lines* show the calculated thickness dependence for the coherent and the incoherent case, respectively.

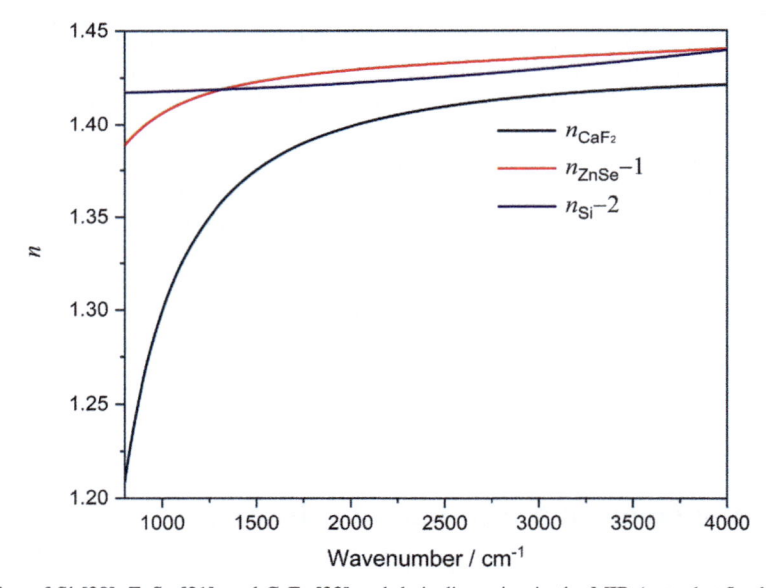

FIG. 6.10 Indices of refraction of Si [30], ZnSe [31], and CaF$_2$ [32] and their dispersion in the MIR (note that I subtracted unity from the index of refraction of ZnSe and twice unity from that of CaF$_2$).

the index of refraction begins to vary strongly. This is even truer for CaF$_2$. So, even if CaF$_2$ has the lowest index of refraction mismatch in comparison with these compounds, the change of the index in the fingerprint region is considerable. Also, as we will see, even for such compounds with their comparably weak oscillator strengths, the dispersion that accompanies an absorption band has also a significant influence on band shape and intensity.

The optics of films on transparent substrates is somewhat more complicated since in comparison, e.g., with the free-standing film, the symmetry with regard to incidence and exit medium is broken. One interesting question in this regard is, does it make a difference, if light is first falling onto the layer and exiting the substrate or if the situation is vice versa? The answer is yes and no at the same time, but we will investigate this in detail.

In the following, we assume a thin film on a thick substrate, which means that we have a coherent layer packet of air/film/substrate, which shares an incoherent interface with the incoherent layer given by the substrate, so that overall, we have an incoherent multilayer in the sense of Section 4.5.8. We will first calculate the reflection and transmission coefficients for the coherent packet where we denote air (or vacuum) as medium 0 as usual, the medium of the film as medium 1, and, newly, the substrate as medium 2. Accordingly, we can write the packet in short form as 0/1/2 and the main difference to Section 6.2 is that the packet is no longer symmetric, which complicates the formulas somewhat. Also, we will allow nonnormal incidence, which also increases the complexity slightly. First of all, we will write down the wavevectors k_i in the different media:

$$k_0 = n_0 \cdot \cos\alpha_i = \cos\alpha_i$$
$$k_j = \sqrt{n_j^2 - n_0^2 \sin^2\alpha_i} = \sqrt{n_j^2 - \sin^2\alpha_i} \quad j = 1, 2. \tag{6.17}$$

Here, we have assumed that our incidence medium has the dielectric constant $\varepsilon_0 = n_0^2 = 1$ independent of frequency or wavenumber. For the reflection and transmission coefficients, we find the following relations for s-polarized light:

$$r_{ij,s} = \frac{k_i - k_j}{k_i + k_j}, \quad r_{ji,s} = \frac{k_j - k_i}{k_i + k_j} = -r_{ij,s}, \quad t_{ij,s} = \frac{2k_i}{k_i + k_j}, \quad t_{ji,s} = \frac{2k_j}{k_i + k_j}. \tag{6.18}$$

For p-polarized light, we find the reflection and transmission coefficients when we replace k_i by $n_i^2 k_j$ and, correspondingly, k_j by $n_j^2 k_i$:

$$r_{ij,p} = \frac{n_i^2 k_j - n_j^2 k_i}{n_i^2 k_j + n_j^2 k_i}, \quad r_{ji,p} = \frac{n_j^2 k_i - n_i^2 k_j}{n_i^2 k_j + n_j^2 k_i} = -r_{ij,p}, \quad t_{ij,p} = \frac{2n_i^2 k_j}{n_i^2 k_j + n_j^2 k_i}, \quad t_{ji,p} = \frac{2n_j^2 k_i}{n_i^2 k_j + n_j^2 k_i}. \tag{6.19}$$

For the calculation of the reflection and transmission coefficients of our coherent layer packet, we have to consider that our packet is, as already stated, asymmetric, and, since multiple reflection also occur in incoherent layer stacks, light waves are passing through not only in the direction $0 \to 1 \to 2$ (012) but also in reversed order (210):

$$r_{012} = \frac{r_{01} + r_{12}\exp(2i\phi_1)}{1 + r_{01}r_{12}\exp(2i\phi_1)}, \quad r_{210} = \frac{r_{21} + r_{10}\exp(2i\phi_1)}{1 + r_{21}r_{10}\exp(2i\phi_1)},$$
$$t_{012} = \frac{t_{01}t_{12}\exp(i\phi_1)}{1 + r_{01}r_{12}\exp(2i\phi_1)}, \quad t_{210} = \frac{t_{21}t_{10}\exp(i\phi_1)}{1 + r_{21}r_{10}\exp(2i\phi_1)}. \tag{6.20}$$

The coefficients for reflection and transmission are the same independent of the polarization. The phase of the wave is given by $\phi_1 = 2\pi \cdot \tilde{\nu} \cdot d_1 \cdot n_1$, wherein d_1 is the thickness of the layer. The overall transmittance T and reflectance R can then be calculated according to [33].

$$R = R_{012} + \frac{T_{012}R_{20}T_{210}|\exp(2i\phi_2)|^2}{1 - R_{210}R_{20}|\exp(2i\phi_2)|^2},$$
$$T = \frac{T_{012}T_{20}|\exp(i\phi_2)|^2}{1 - R_{210}R_{20}|\exp(2i\phi_2)|^2}, \tag{6.21}$$

where it is assumed that the exit medium is the same as the incidence medium, $\phi_2 = 2\pi \cdot \tilde{\nu} \cdot d_2 \cdot n_2$ analogously to ϕ_1, and the capital letters R_m and T_m represent the products of r_m and t_m, respectively, with their complex conjugates (cf. also, e.g., Harbecke [34]).

A very important result for such asymmetric layer stacks is that the transmittance does, in contrast to the reflectance, not depend on from which side the layer stack is illuminated. As a consequence, it does not matter for the transmittance absorbance if first the layer or the substrate is illuminated, the result will always be identical. If the same is not valid in general for the reflectance, then we can already conclude that the absorptance, i.e., $1 - R - T$, must be dependent on the illumination direction. I will show this in the following first by calculating the absorbance and the absorptance for a 2 μm thick layer of PMMA on the different substrates. The results of these calculations are shown in Fig. 6.11.

Note that the absorptance does not show interference fringes, very much like in the case of the freestanding film, because reflectance maxima in nonabsorbing regions are accompanied by transmittance minima and conservation of energy requires then that $R+T=1$. In case of the absorbance, we removed the fringes naïvely by calculating the transmittance assuming a nonabsorbing layer with a nonvarying dielectric function being equal to ε_∞ of PMMA and the same thickness (this method is still much more advanced than many of those which are usually employed). The transmittance spectra of the

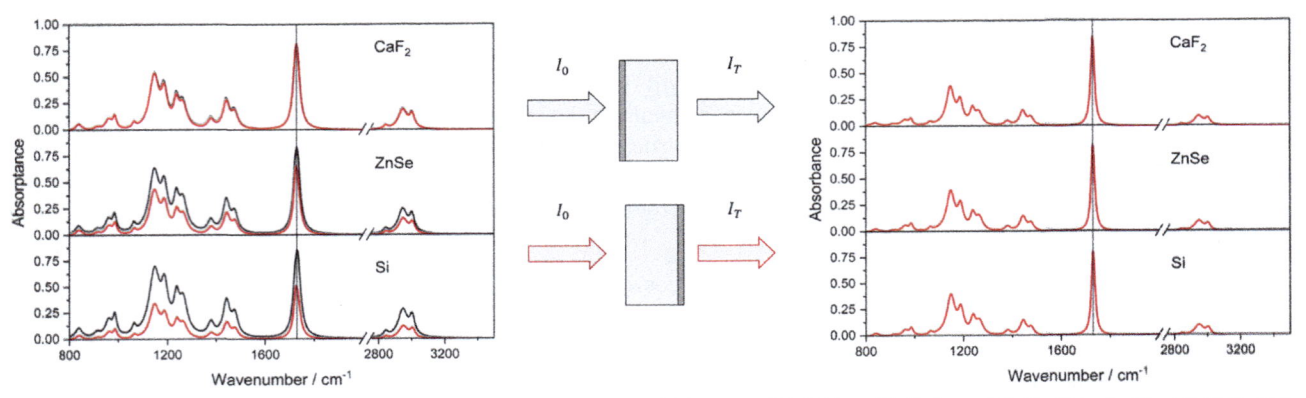

FIG. 6.11 Calculated absorptance (left) and absorbance (right) of a 2 μm thick layer of PMMA on Si, ZnSe, and CaF$_2$. If the incident light first hits the layer, the *black curves* result, and for the reversed situation, the *red curves* have been computed.

system PMMA/substrate were then ratioed to the result of the previous calculation. On the first view, this seems to work quite well. But before discussing this finding further, we first note that the absorbance spectra are identical as expected, irrespective if the light first hits the layer or the substrate. This is different in case of the absorptance. Here, the difference between the absorptances increases with increasing index of refraction of the substrate. This can easily be understood, because when the index of refraction of the substrate is high and the incoming light illuminates first the substrate, then the reflectance from this first interface is also high and the intensity of the light is already strongly decreased before it can be absorbed by the layer. On the other hand, when the light first hits the layer, the comparably low index of refraction of the PMMA leads only to a comparably weak reflectance. This also explains why the absorptance difference is small in case of the CaF$_2$ substrate, because its index of refraction is over wide spectral ranges not very different from that of the PMMA. This effect can be visualized very effectively by the field maps in Fig. 6.12.

The field maps for the nonabsorbing films on Si and ZnSe clearly show that the interference effects are strongest for the case when the layer is first illuminated, whereas it is actually the other way round for the CaF$_2$ substrate in the region below 800 cm^{-1}, because of the strong decrease of the index of refraction of CaF$_2$ in this region.

Now, let us focus on the interference effects. To inspect them in detail, we have calculated the spectra for PMMA with various thicknesses on Si and picked one of peaks of the C—H vibrations as indicated in Fig. 6.13 [35].

It is certainly not surprising to see that the absorbance also shows fringes as does the transmittance. Since many purely mathematical approaches exist to remove these fringes, which do not take notice of the physics and neglect the effect on the absorption due to the changing of the electric field intensity, it may be instructive to spend some more words on, e.g., the naïve approach mentioned previously, which at least has some physical background compared to other methods of fringe

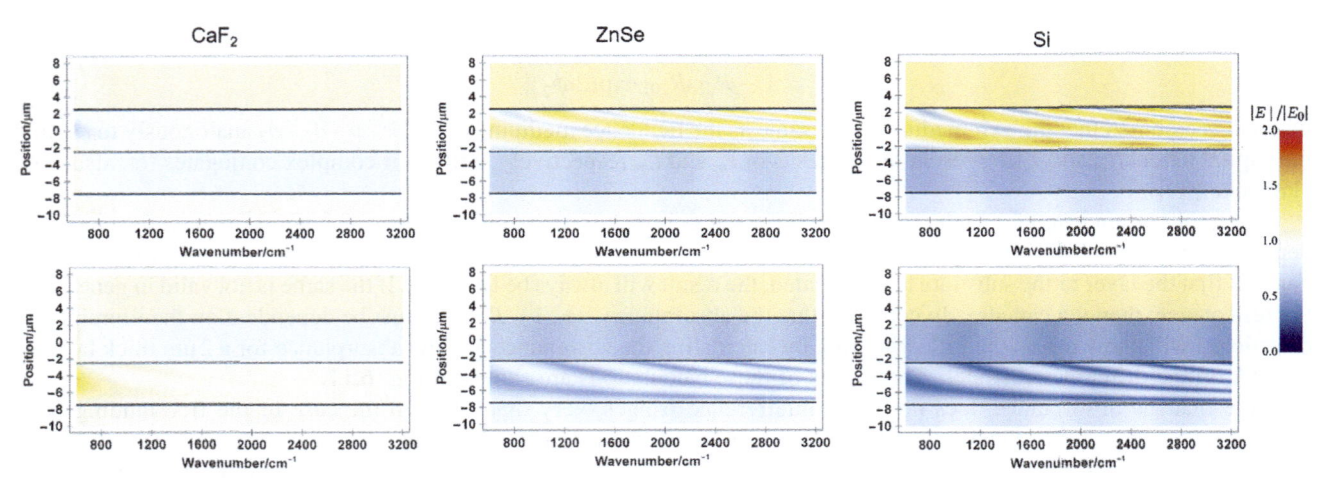

FIG. 6.12 Field maps of Si, ZnSe, and CaF$_2$ substrates with a nonabsorbing 5 μm thick layer on top (first row) and beneath the substrate (second row) relative to the illumination direction. For better illustration, the thickness of the layer and of the substrate is the same. Since the substrates have been assumed to be nonabsorbing and the light waves inside them to be incoherent, their thickness does not matter for the calculation.

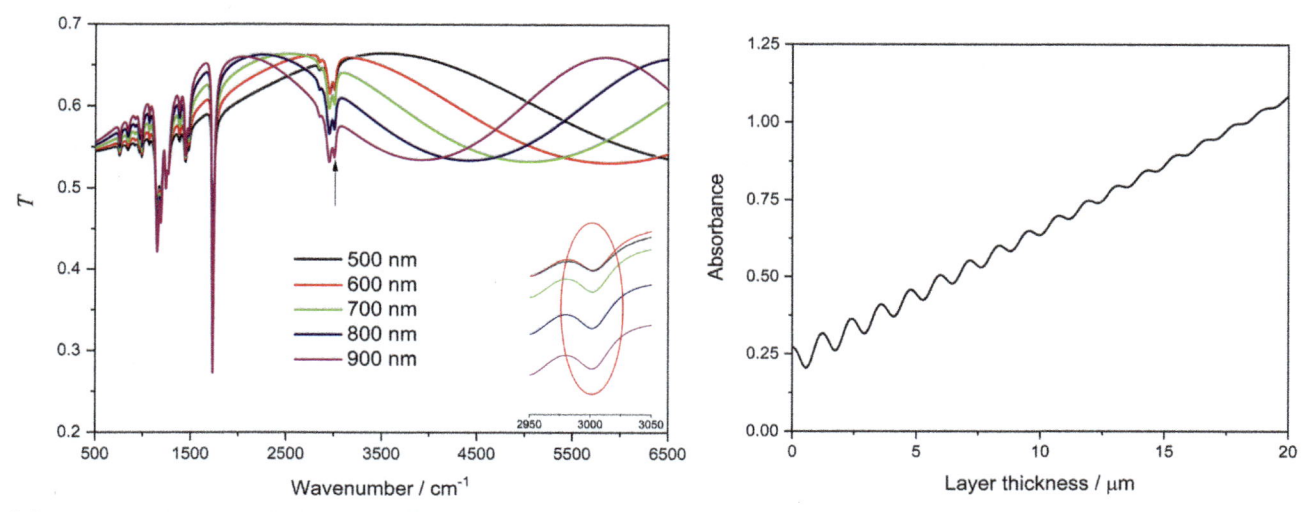

FIG. 6.13 Simulated transmittance spectra of PMMA layers on Si (left side) and change of the uncorrected absorbance of the indicated peak with thickness (right side).

removal. Looking on the spectra in Fig. 6.13, one is tempted to assume that the fringes in the absorbance on the right side are very comparable with the situation of an object swimming in the sea: Due to waves, this object will be continuously elevated and lowered. Now we remove the waves (i.e., the interference fringes) and see what remains. The effect of the naïve approach discussed previously is illustrated in Fig. 6.14.

Obviously, by ratioing the transmittance of the absorbing layer on the substrate to a corresponding nonabsorbing layer on the same substrate, the pronounced fringes can be completely removed from the spectrum. However, if we examine the absorbance change with the thickness in the corrected spectra, then we see that nevertheless the fringing remains (green curve in Fig. 6.14). As already mentioned, a second physics-based approach consists of adding the reflectance and the transmittance, the effect of which is shown in Fig. 6.15.

Adding the reflectance and transmittance has, in addition to the removal of the fringes, a further beneficial aspect, which is correcting the band shape for dispersion effects introduced by the reflectance. This can best be seen by comparing the shapes of the strongest band at $1727\,cm^{-1}$ (green curve of the left panel in Fig. 6.14 and blue curve of the left panel in Fig. 6.15) close to the baseline. The fringing of the peak value, however, due to the changing values of E^2 as a consequence of interference is again not corrected. To demonstrate this kind of fringing, we have used a film on a Si substrate, which

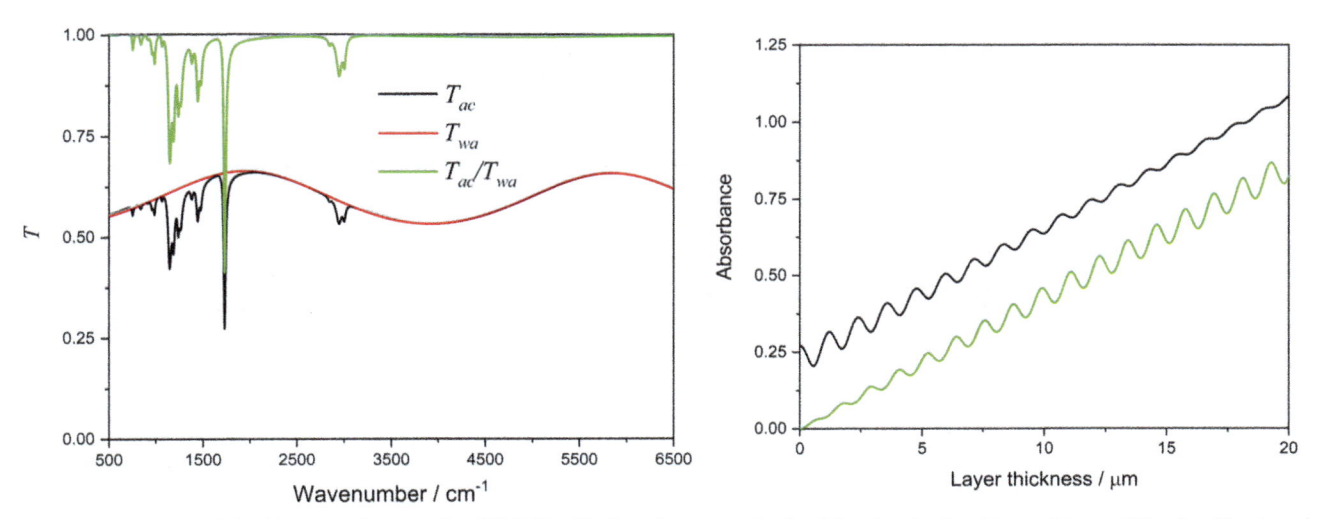

FIG. 6.14 Removal of the fringes on the example of PMMA with the naive approach of ratioing the simulated transmittance of the absorbing layer/substrate (black spectrum in the left panel) to the corresponding simulated transmittance of the nonabsorbing layer/substrate (red spectrum in the left panel). The resulting spectrum (green spectrum) still shows interference effects as the comparison in the right panel shows, where the variances of the absorbance of the peak with the highest wavenumber with the thickness are shown (*black*: without correction, *green*: correction via rationing).

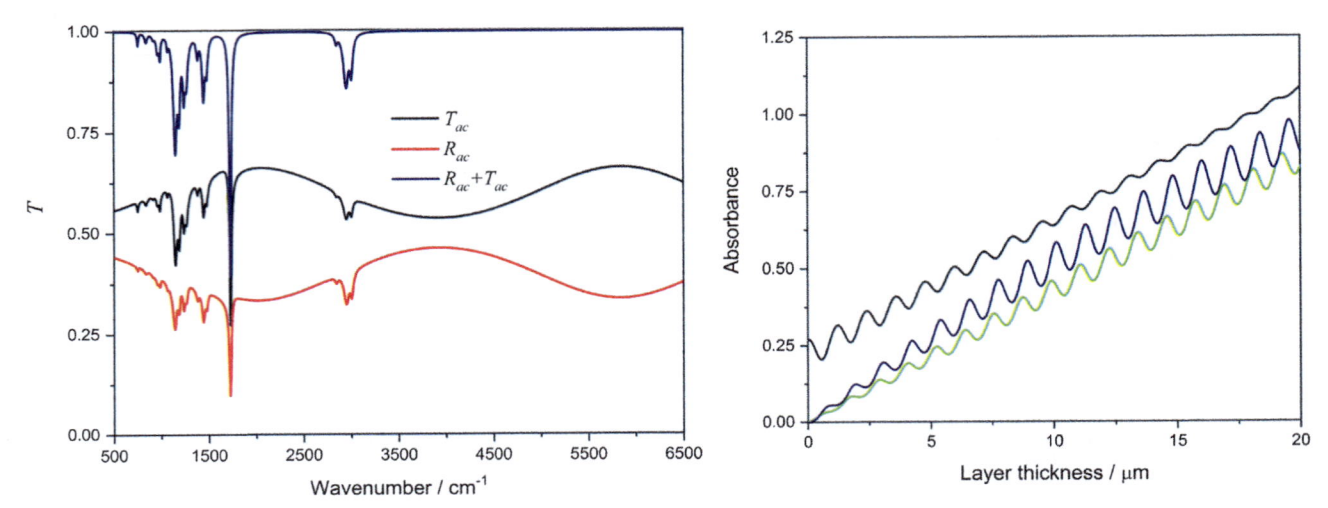

FIG. 6.15 Removal of the fringes on the example of PMMA with the naive approach of adding the simulated transmittance and the simulated reflectance of the absorbing layer/substrate *(black and red spectra in the left panel)*. The resulting spectrum *(blue spectrum)* still shows interference effects as the comparison in the right panel shows, where the variances of the absorbance of the peak with the highest wavenumber with the thickness are shown *(black: without correction, green: correction via rationing, cf. Fig. 6.14, blue: correction via adding of reflectance and transmittance)*.

obviously provides the largest changes of E^2; therefore, one would expect that CaF_2 is the ideal substrate for infrared spectroscopy [36] and, indeed, this assumption seems to have a lot of supporters. Since we discussed in Chapter 5 extensively the unity of absorption and dispersion, it is obvious where this assumption has its flaws. For example, while the index of refraction of the substrate matches that of PMMA well in nonabsorbing spectral regions, this match is destroyed in particular for the stronger absorption bands of the PMMA. Let us investigate this hypothesis in more details. Accordingly, Fig. 6.16 shows a number of spectra of PMMA layers on CaF_2 with thicknesses ranging from 0.5 to 2.5 µm. These spectra are actually absorption coefficient spectra, which were gained by ratioing the absorbance to the film thickness. They are compared to the true absorption coefficient spectrum, which was calculated from the absorption index spectrum.

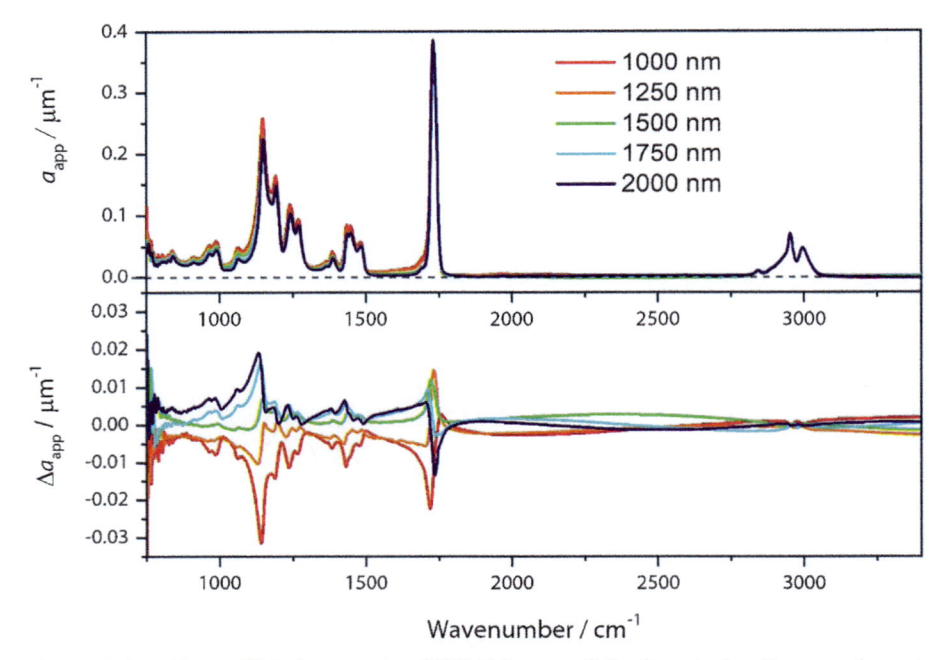

FIG. 6.16 Upper part: Apparent absorption coefficient a_{app} spectra of PMMA layers on CaF_2 after naive baseline correction and normalization of experimental spectra. Lower part: Variation of the absorption coefficient relative to the average value of all five layers. *(Reprinted with permission from T.G. Mayerhöfer, S. Pahlow, U. Hübner, J. Popp, CaF_2: an ideal substrate material for infrared spectroscopy?, Anal. Chem. 92 (2020) 9024–9031. Copyright 2020 American Chemical Society.)*

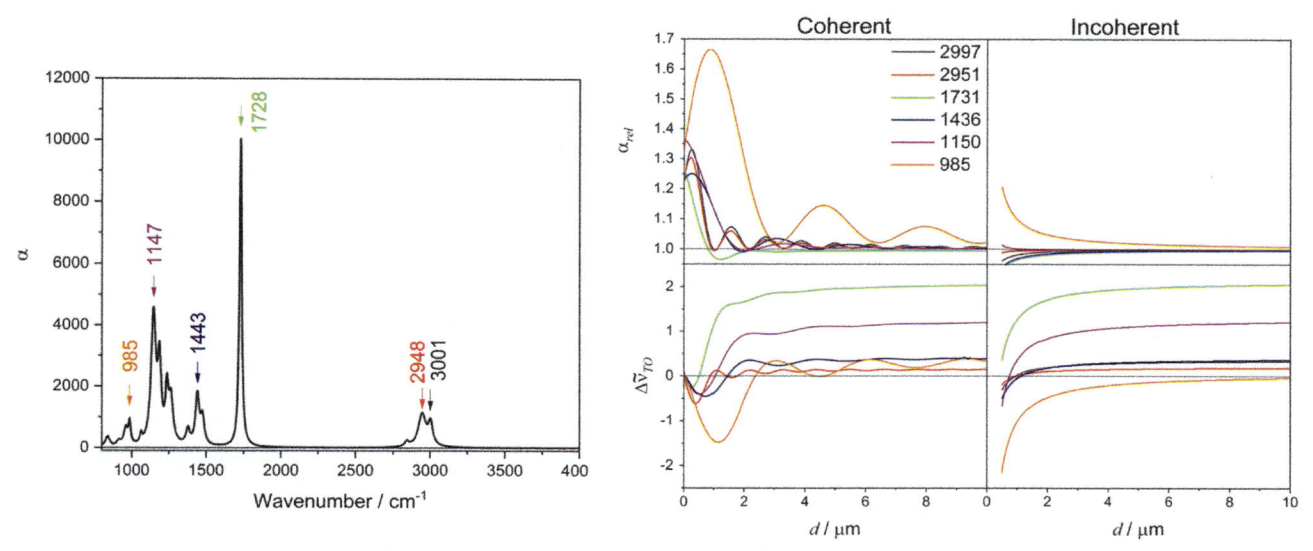

FIG. 6.17 Comparison of the calculated thickness dependence of some apparent α-peak values relative to the true α-values and the thickness dependence of the apparent wavenumber positions of these peaks relative to the true values for PMMA layers on CaF$_2$. *(Adapted with permission from T.G. Mayerhöfer, S. Pahlow, U. Hübner, J. Popp, CaF2: an ideal substrate material for infrared spectroscopy?, Anal. Chem. 92 (2020) 9024–9031. Copyright 2020 American Chemical Society.)*

The consequence of the index of refraction mismatch due to dispersion is best seen again by inspecting the 1727 cm^{-1} band. Apart from this, we also clearly see, that using a CaF$_2$ substrate does not prevent completely from introducing interference effects. In particular between 1000 and 1200 cm^{-1}, it is obvious that there are still deviations from the BBL approximation. It is worth to inspect those in some more detail and to compare the effects on different bands of PMMA. Accordingly, I have calculated the thickness dependence of the apparent absorption coefficient and the position of the band maximum for some of the bands. The results presented in Fig. 6.17 suggest that it might not be advisable to use those apparent values and that some correction is also necessary if CaF$_2$ is used as substrate. Before we jump into the discussion, let me first bring up some other things which I find worth mentioning. If I assume that coherence of the waves is absent in layers (I call those incoherent layers), then this assumption does not make sense before the layer reaches a certain thickness, not only because it is the thickness differences (the deviations of the interfaces from being plane parallel) that cause the incoherence. What I find actually more striking is that in the calculation scheme for incoherent layers, those have still a reflectance even in the limit of zero thickness, as already mentioned, which really does not make sense. In fact, the reflectance does not change at all with the thickness.

Changes of the peak positions for small thicknesses are therefore in the incoherent case due to changing the relative influences of reflectance in the transmittance spectra. How does that compare to the situation if a coherent superposition is assumed? In this case, the reflectance is actually zero for zero thickness and is very small for layers that are just a couple of nanometers thick. The reason is the same as for free-standing films: Every layer has two interfaces and if these interfaces are close to each other, then there is nearly a phase difference of π between the waves reflected from the first interface and those that are (back-)reflected from the second interface and transmitted through the first interface. Accordingly, those waves cancel each other and the overall reflectance is zero. When becoming thicker, then on the one hand the phase difference changes and, on the other, absorption decreases the intensity of the back-reflected waves. This also means that the deviations from the BBL approximation due to dispersion are strongest for CaF$_2$, but absorption decreases these deviations with increasing layer thickness. Exactly, this is what can be seen in Fig. 6.17. Certainly, due to dispersion of the index of refraction of the substrate and since the phases depend on the wavelengths/wavenumbers, the changes are different for every peak, which means that the peak intensity ratios can undergo strong changes. Strikingly, for very thin films, all peaks show a strong positive deviation. The reason for this is that the waves reflected from the layer surfaces and from the substrate surface are more or less in phase nearly independent of wavelength, so that absorption is generally stronger for very thin layers (e.g., $d < 20$ nm). For such layer thicknesses, an optical modeling is a must due to the strong deviations from the BBL approximation. Since the peak intensity ratios are very important in relation to indicating changes of the material itself, but also for quantitative analysis, it is very important to correct those for optical influences to avoid misinterpretation. Overall, whenever thin films are to be measured, deviations from the BBL approximation have to be considered and neither the band intensity, nor shape or position are reliable and need to be corrected, even if substrate and layer are index matched, since this match is lost in the surrounding of an absorption.

6.5 Attenuated total reflection

As already mentioned, infrared ATR spectroscopy is an exception from the rule that optical deviations from the BBL approximation are not mentioned in the textbooks of infrared spectroscopy. Nevertheless, like in case of transflection, many reflectance absorbance spectra, in particular of organic and biologic materials, look on the first view very much like transmittance absorbance spectra. Sometimes, it does not matter that there can be larger deviations as long as no comparison with spectra from spectral data bases is intended and just small relative changes are investigated. But once a spectrum is used to determine band positions and intensities, the peculiarities of ATR spectra need to be considered and a proper correction is necessary. Another exception is quantitative determinations as deviations usually increase with increasing oscillator strengths and, therefore, concentrations. As you will see in the course of this section, correcting ATR spectra is not a simple task, not because the theory would be hard to comprehend, but because modern ATR accessories lack the possibility to attach a polarizer. This seems to be a knockout criterion, because knowledge about the polarization state of the incident light is required to apply a proper correction (a yet not well-known way around may be a kind of calibration with a material of known optical constants, e.g., water [37] to obtain the polarization state of light for a certain ATR accessory). You might ask yourself, how is this possible, since your ATR accessory came with a software featuring an "Advanced ATR Correction Algorithm"—if this algorithm does not need the polarization state as input parameter (which is certainly not natural, i.e., it is not a 1:1 mixture of s- and p-polarized light due to the optical elements inside a spectrometer), then it cannot be that advanced. But let us first discuss the historical development of ATR spectroscopy before we get lost in details.

It seems that the first paper about the use of ATR infrared spectroscopy is that of Fahrenfort from 1961 [38]. It provides theoretical details, but it does only superficially compare the data obtained by ATR and different IR techniques, in particular transmission and external reflection spectroscopy. Above all, a direct and quantitative comparison is not provided. Fahrenfort concludes that "The new reflection spectra can be used directly in combination with collections of standard transmission spectra for identification purposes. Direct application for quantitative analysis is nearly as easy as that of transmission spectra. Furthermore, the special features of this ATR technique might be important in cases where the investigation of sample layers of arbitrary thickness is unavoidable."

Fahrenfort's claims were relativized 4 years later by Hansen in 1965 [1]. Hansen's goal was to compare reflectance absorbance with transmittance absorbance. To reach this goal, he recast Fresnel's equations in a form, which allows to easier calculate reflectance absorbance from the optical constants by using series expansions and determining the number of terms one needs to retain a certain accuracy. He then examined the functional dependence of reflectance absorbance and of the attenuation index $\kappa = k/n$ and found it to be linear only for small κ. One of his conclusions is therefore that ATR absorbance will resemble transmittance absorbance spectra only if $\kappa \ll 1$. He also pointed out already that for larger κ band, distortions can be expected. This means that ATR can only be used for weakly absorbing substances. But those can also be investigated by transmittance spectroscopy—what would then be the special advantage of ATR? This seeming advantage is, not only as pointed out by Fahrenfort, that thicker samples can be investigated. But is it also possible that the spectra of stronger absorbing materials can be quantitatively evaluated?

In fact, both views are correct. What has to be kept in mind is that $\kappa \ll 1$ is a necessary condition if ATR absorbance spectra are to resemble transmittance absorbance spectra. If stronger absorbing materials are to be investigated, their spectra cannot be interpreted in the same simple ways like absorbance spectra—they need more sophisticated evaluation. Where I disagree with Hansen is that he thought that transmittance absorbance generally follows the BBL approximation and that this approximation is actually exact. From this assumption, he developed the so-called effective thickness d_{eff}, which would be the sample thickness used in a transmittance experiment so that reflectance absorbance resembles transmittance absorbance.

The following derivations of the effective thicknesses are analogous to that of Milosevic [39]. We start with Eqs. (4.25), (4.36) and employ Eq. (4.40), $R_j = |r_j|^2$. Furthermore, we use $\varepsilon = n^2$ and assume that the incidence medium is nonabsorbing:

$$R_s = \left| \frac{n_1 \cos\alpha_i - \sqrt{\hat{n}_2^2 - n_1^2 \sin^2\alpha_i}}{n_1 \cos\alpha_i + \sqrt{\hat{n}_2^2 - n_1^2 \sin^2\alpha_i}} \right|^2$$

$$R_p = \left| \frac{\hat{n}_2^2 \cos\alpha_i - n_1 \sqrt{\hat{n}_2^2 - n_1^2 \sin^2\alpha_i}}{\hat{n}_2^2 \cos\alpha_i + n_1 \sqrt{\hat{n}_2^2 - n_1^2 \sin^2\alpha_i}} \right|^2$$

(6.22)

To arrive at the formula provided by Milosevic, we employ that $i^2 = -1$:

$$R_s = \left|\frac{n_1 \cos\alpha_i - i\sqrt{n_1^2 \sin^2\alpha_i - \hat{n}_2^2}}{n_1 \cos\alpha_i + i\sqrt{n_1^2 \sin^2\alpha_i - \hat{n}_2^2}}\right|^2$$

$$R_p = \left|\frac{\hat{n}_2^2 \cos\alpha_i - in_1\sqrt{n_1^2 \sin^2\alpha_i - \hat{n}_2^2}}{\hat{n}_2^2 \cos\alpha_i + in_1\sqrt{n_1^2 \sin^2\alpha_i - \hat{n}_2^2}}\right|^2$$

(6.23)

In the following, we perform series expansions for both, R_s and R_p, and remove the terms of higher order than linear in k. This means that if $k \ll 1$, we can approximate the root in the form $A + Bi$:

$$A = \sqrt{n_1^2 \sin^2\alpha_i - n_2^2}$$

$$B = -\frac{nk}{\sqrt{n_1^2 \sin^2\alpha_i - n_2^2}}$$

(6.24)

With these assumptions, Eq. (6.23) simplifies to:

$$R_s \approx 1 - \frac{4n_1 n_2 k \cos\alpha_i}{(n_1^2 - n_2^2)\sqrt{n_1^2 \sin^2\alpha_i - n_2^2}}$$

$$R_p \approx 1 - \frac{4n_1 n_2 k (2n_1^2 \sin^2\alpha_i - n_2^2) \cos\alpha_i}{(n_1^2 - n_2^2)[(n_1^2 + n_2^2)\sin^2\alpha_i - n_2^2]\sqrt{n_1^2 \sin^2\alpha_i - n_2^2}}$$

(6.25)

Next, we use the definition of reflectance absorbance and that $-\log_{10}(1-x)/\log_{10}e \approx x$ to arrive at:

$$A_s = -\log_{10}\left(1 - \frac{4n_1 n_2 k \cos\alpha_i}{(n_1^2 - n_2^2)\sqrt{n_1^2 \sin^2\alpha_i - n_2^2}}\right) \approx \log_{10}e \cdot \frac{4n_1 n_2 k \cos\alpha_i}{(n_1^2 - n_2^2)\sqrt{n_1^2 \sin^2\alpha_i - n_2^2}}$$

$$A_p = -\log_{10}\left(1 - \frac{4n_1 n_2 k (2n_1^2 \sin^2\alpha_i - n_2^2) \cos\alpha_i}{(n_1^2 - n_2^2)[(n_1^2 + n_2^2)\sin^2\alpha_i - n_2^2]\sqrt{n_1^2 \sin^2\alpha_i - n_2^2}}\right)$$

$$\approx \log_{10}e \cdot \frac{4n_1 n_2 k (2n_1^2 \sin^2\alpha_i - n_2^2) \cos\alpha_i}{(n_1^2 - n_2^2)[(n_1^2 + n_2^2)\sin^2\alpha_i - n_2^2]\sqrt{n_1^2 \sin^2\alpha_i - n_2^2}}$$

(6.26)

Would the BBL approximation hold perfectly for reflectance absorbance under ATR conditions, it could be written as:

$$A_s = 4\pi(\log_{10}e) \cdot \tilde{\nu} \cdot k \cdot d_{\text{eff},s}$$

$$A_p = 4\pi(\log_{10}e) \cdot \tilde{\nu} \cdot k \cdot d_{\text{eff},p}$$

(6.27)

If we compare Eq. (6.26) with Eq. (6.27), we see that the effective thicknesses are given by:

$$d_{\text{eff},s} = \frac{n_1 n_2 \cos\alpha_i}{\pi\tilde{\nu} \cdot (n_1^2 - n_2^2)\sqrt{n_1^2 \sin^2\alpha_i - n_2^2}}$$

$$d_{\text{eff},p} = \frac{n_1 n_2 (2n_1^2 \sin^2\alpha_i - n_2^2) \cos\alpha_i}{\pi\tilde{\nu} \cdot (n_1^2 - n_2^2)[(n_1^2 + n_2^2)\sin^2\alpha_i - n_2^2]\sqrt{n_1^2 \sin^2\alpha_i - n_2^2}}$$

(6.28)

Before we continue, please always keep in mind that these equations are approximations for vanishing absorption. Nevertheless, one could assume that from two ATR spectra recorded with s-polarized and two different angles of incidence it could be possible to determine n_2:

$$n_2 = n_1 \frac{\sqrt{\left(\frac{A_s(\alpha_{i,1})}{A_s(\alpha_{i,2})}\right)^2 \cos^2\alpha_{i,2} \tan^2\alpha_{i,1} - \sin^2\alpha_{i,2}}}{\sqrt{\left(\frac{A_s(\alpha_{i,1})}{A_s(\alpha_{i,2})}\right)^2 \cos^2\alpha_{i,2} \sec^2\alpha_{i,1} - 1}}.$$

(6.29)

Using artificial spectra, this works indeed very good in weekly absorbing spectral regions. Unfortunately, in practice, the errors are so large that the practical value seems to be close to zero, even when a polarizer and an ATR accessory with variable angle of incidence would be available.

The usual way of comparing transmittance absorbance and reflectance absorbance based on ATR would now be to relate the absorbances according to Eq. (6.27) to those resulting from taking the negative decadic logarithm of Eq. (6.21) in dependence of the absorption index. Since we know already that the absorption index is not linearly related to absorbance, I will offer a different kind of comparison in which I specify a molar oscillator strength S^{*2} and increase the concentration. Furthermore, I will not only calculate the ATR-related reflectance absorbance, but also the transmittance absorbance of a layer with the effective thickness on a CaF_2 substrate. Since absorbance according to Eq. (6.27) will also deviate from linearity in such a plot (cf. Section 6.4), the usual way of comparing transmittance absorbance and reflectance absorbance based on ATR would now be to relate the absorbances according to Eq. (6.27) to those resulting from taking the negative decadic logarithm of Eq. (6.23) in dependence of the absorption index. Since we know already that the absorption index is not linearly related to absorbance, I added a straight line with the slope being the same as that of the plot according to Eq. (6.27) for small concentrations. Since we know that the underlying physics can be described by an oscillator model, it also makes sense to investigate the changes of the absorbance not only at the oscillator position, but also at lower and higher wavenumber positions. For these two other positions, I chose to use $\tilde{\nu}_{TO} \pm \gamma/2$, which, for not too strong oscillators, should be the positions of the (local) maximum and the minimum of the refractive index of the band. The band itself shall resemble that of a C=O vibration, e.g., approximately that of PMMA $\left(c \cdot S^{*2} = S^2 = 200^2 \, cm^{-2}, \ c_{max} = 10 \, mol/L, \ \gamma = 17.5 \, cm^{-1}, \ \tilde{\nu}_{TO} = 1730 \, cm^{-1}, \ \varepsilon_\infty = 1.47^2\right)$. Furthermore, we use ZnSe as ATR crystal ($n_1(\tilde{\nu}_{TO}) \approx 2.426$) and an angle of incidence $\alpha_i = 45°$ like many commercially available ATR accessories (diamond has practically the same index of refraction at this wavenumber, so the results are also characteristic for this ATR crystal). For this particular angle of incidence, $A_p = 2A_s$, as can be seen in the upper panels of Fig. 6.18. Accordingly, in every

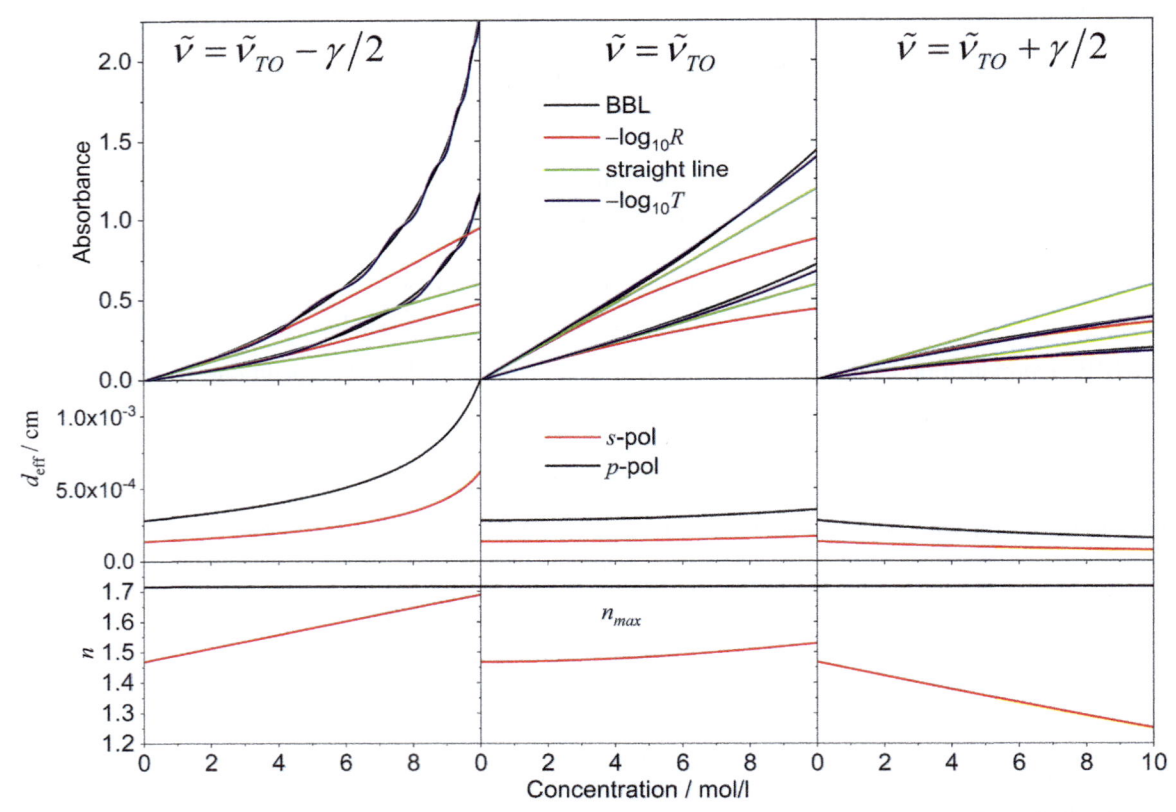

FIG. 6.18 Upper panels: Comparison of the reflectance absorbance *(red lines)* with the transmittance absorbance *(blue lines)*, the absorbance according to the BBL approximation *(green lines)*, and a line having a slope, which is the same as the initial slope of the curve following the BBL approximation *(black line)*. Medium panels: Effective thicknesses of the absorbing layers *(black—p-polarized, red—s-polarized)*. Lower panels, index of refraction *(red)*, and maximum index of refraction *(black)*. Left panels: Calculated at $\tilde{\nu} = \tilde{\nu}_{TO} - \gamma/2$. Center panels: Calculated at $\tilde{\nu} = \tilde{\nu}_{TO}$. Right panels: Calculated at $\tilde{\nu} = \tilde{\nu}_{TO} + \gamma/2$.

one of the upper panels, eight curves are displayed, the upper four curves for p-polarization and the lower four curves for s-polarization. Since for very low concentration, these four curves have very similar values, it is easy to differentiate which of them are based on which kind of polarization.

The green curve is a line which has the same slope as the initial slope of the curve belonging to the BBL approximation (black curve). That this latter curve is no straight line is based on two reasons, one of which is the principal deviation from linearity discussed in Section 5.3, the other reason is based on the change of the effective thickness (cf. medium panels). This can easily be seen if the upper left panel is compared with the medium left panel: The blue and black lines increase nearly in the same way as the effective thicknesses, with the peculiarity that the blue curve oscillates around the black curve. The reason for this oscillation has been discussed in the previous section (Section 6.4). Despite of the use of a CaF_2 substrate, this is an interference effect, which is only obvious, because for $\tilde{\nu} = \tilde{\nu}_{TO} - \gamma/2$, the index of refraction is high (lower panel), while the absorption index is comparably low (on this level of theory, the peak position is shifted to a higher wavenumber). Other interesting differences between the curves in the upper panels are that the reflectance transmittance is higher for $\tilde{\nu} = \tilde{\nu}_{TO} - \gamma/2$ than the green curve, while it is lower for the other two cases. This indicates a peak redshift, which is due to the fact that the reflectance absorbance becomes increasingly modulated by the refractive index for higher oscillator strength and thus the maximum shifts to the red. At least for low concentrations (or weak oscillators), the ATR reflectance absorbance is approximately linearly depending on oscillator strength/concentration. It may be necessary to emphasize that I chose the parameter in a way that the refractive index at all wavenumbers stayed below the maximum or critical index of refraction, which is characterized by the fact that if the sample reaches it, then the ATR condition breaks down. In other words, if the sample (medium 2) were not absorbing and its index higher than this critical index, then for this particular ATR crystal and angle of incidence, the reflectance would no longer be total (internal), but external (external reflection), something which is often not considered. If the index of the sample begins to reach the critical index, the effective thickness has a very steep increase before a discontinuity results (cf. medium left panel).

How strong is the peak shift? It seems that it is a function of the difference between the maximum possible subcritical index of refraction and the actual index of refraction. For an ATR crystal made out of diamond or ZnSe and an angle of incidence of 45°, which refers to most of the ATR accessories in the laboratories, this difference is not very large, and therefore, large shifts can be expected as shown in Fig. 6.19. Since these shifts are depending on the position of the local maximum of the index of refraction, they are also a function of oscillator strength. For the stronger oscillator in the lower part of Fig. 6.19, tremendous shifts result, which are still considerable even when a Ge ATR crystal is used. In addition, the band shapes can be changed, so that inorganic materials can have ATR spectra with not much resemblance to their transmittance counterparts.

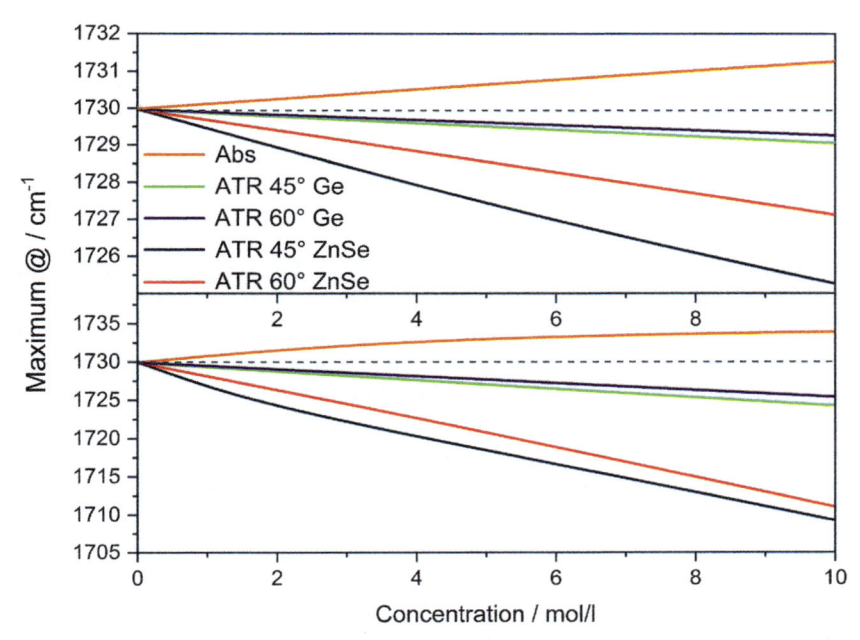

FIG. 6.19 Upper panel: Wavenumber position of the peak maximum for the same oscillator as in Fig. 6.18 in dependence of the angle of incidence and ATR crystal. For comparison, the peak shift due to dispersion in a true absorbance spectrum is shown. Lower panel: Same as in the upper panel for an oscillator with larger oscillator strength (all other parameters have been left unchanged; $c_{max} \cdot S^{*2} = S^2 = 500^2 \, cm^{-2}$).

I want to close this section with a statement by Hansen: "In general, polarization is found to be important. In some cases, absorbance is linear for both TM (*p*-polarized) and TE (*s*-polarized) but not for natural light or for partial polarization found in spectrometers without polarizers." What does this mean for (ATR) spectrometers without polarizers? They are not suitable for a simple evaluation of the resultant spectrum in terms of the BBL approximation, but it is also not possible to correct or evaluate the spectrum by wave optics, since the polarization state is not known (except if you calibrate them as suggested in Ref. [37]). So, whatever you use the corresponding spectra for, you should be able to check some of the results with reliable literature values and see if you really can put them to a sophisticated use.

6.6 Mixing rules

So far, we have investigated the optical properties of homogenous and scalar media. In practice, these assumptions limit the materials and corresponding spectra that can be investigated and quantitatively interpreted strongly. In the following, we will be a little less restrictive and consider media, which consist of two or more materials that are itself scalar, but are mixed in the area or volume of the samples. Typical example would be glass ceramics like Ceran, polymer blends, or biologic samples. To describe the optical properties of these materials adequately, it would be necessary in many cases to include scattering effects. We will not go thus far, since treating scattering effects alone would require a book of its own (see, e.g., Refs. [40,41]). Instead, I restrict myself to the peculiarities that I describe in the following.

Under many circumstances, apertures are used in infrared spectroscopy. If you have performed measurements yourself, then you know the not very surprising fact that apertures limit the irradiance and, therefore, the intensity the detector measures as long as the beam diameter is larger than the aperture. If we assume that the beam is homogenous, then halving the area of the aperture means halving the intensity that is measured at the detector. The same principle applies to samples, which are microheterogenous. By this term, I mean that the heterogeneity can be detected by a microscope using light of the same wavelength as the spectrometer with which the sample is measured. Let us presume that we have such a microheterogeneous sample, which consists of two different materials. In the extreme case, the sample would be completely unmixed and both materials would be concentrated in two different areas or volumes of the sample. If I would place the sample appropriately under an aperture, I would thus be able to measure a spectrum of the neat substance 1 or 2. When I take a larger aperture, I would measure a spectrum that contains a mixture of the spectra of both neat substances. But in this case, I have to be more precise which kind of spectrum I mean: Since it is the irradiance or intensity that is additive, and reflectance and transmittance are directly derived from intensities, the following relations hold [42]:

$$\begin{aligned} T &= \varphi_1 T(\hat{n}_1) + \varphi_2 T(\hat{n}_2) \\ R &= \varphi_1 R(\hat{n}_1) + \varphi_2 R(\hat{n}_2) \end{aligned}. \tag{6.30}$$

In Eq. (6.30), φ_i represents the area or volume fraction of material i (Eq. (6.30) is readily extended to more than 2 materials). If you are still waiting for the clou, here it is:

$$\begin{aligned} \log_{10}[\varphi_1 T(\hat{n}_1) + \varphi_2 T(\hat{n}_2)] &\neq \varphi_1 \log_{10} T(\hat{n}_1) + \varphi_2 \log_{10} T(\hat{n}_2) \\ \log_{10}[\varphi_1 R(\hat{n}_1) + \varphi_2 R(\hat{n}_2)] &\neq \varphi_1 \log_{10} R(\hat{n}_1) + \varphi_2 \log_{10} R(\hat{n}_2) \end{aligned}. \tag{6.31}$$

In words, if transmittances and reflectances are additive, absorbances ($-\log_{10}(T/T_0)$ or $-\log_{10}(R/R_0)$) cannot be additive. This obviously contradicts Beer's approximation. Before I resolve this paradox, let us first investigate the order of magnitude of the corresponding errors. To do this, we assume two different materials, each with one oscillator having the same oscillator strength and damping, both with values comparable to that of a C=O vibration. The oscillators shall have oscillator positions that differ by $50\,\text{cm}^{-1}$, 2.5 times the shared damping constant, so that there is small overlap. For n_∞, I assume a value of 1.47 and place the material in form of a layer onto CaF_2 to limit interference effects. The transmittance is calculated wave optics compatible (cf. Section 6.4) and then mixed in the ratio 0.25:0.75 following either Eq. (6.30) (black curve in Fig. 6.20) or by adding the transmittance absorbance directly in the same ratio following a mixing rule according to Beer's approximation (red curve). If we compare the black curve with the red curve, we observe a strong band flattening, so strong that the intensity is more than halved at the oscillator positions. In addition, one band looks like a shoulder without a local maximum and the other is redshifted by $2\,\text{cm}^{-1}$. Such band flattening is frequently observed in colloidal solutions, often attributed to shadowing, which is nonsense since shadows are something that belongs to ray optics and not to wave optics (compare Fig. 1.1B) with Fig. 1.1D; let me emphasize again that absorption in the wave optics limit does not know shadows!. Nevertheless, the dimension of the heterogeneity must somehow play a role. We must use these concepts to explain why the mixing rule according to Eq. (6.30) and the following mixing rules can both be correct [42]:

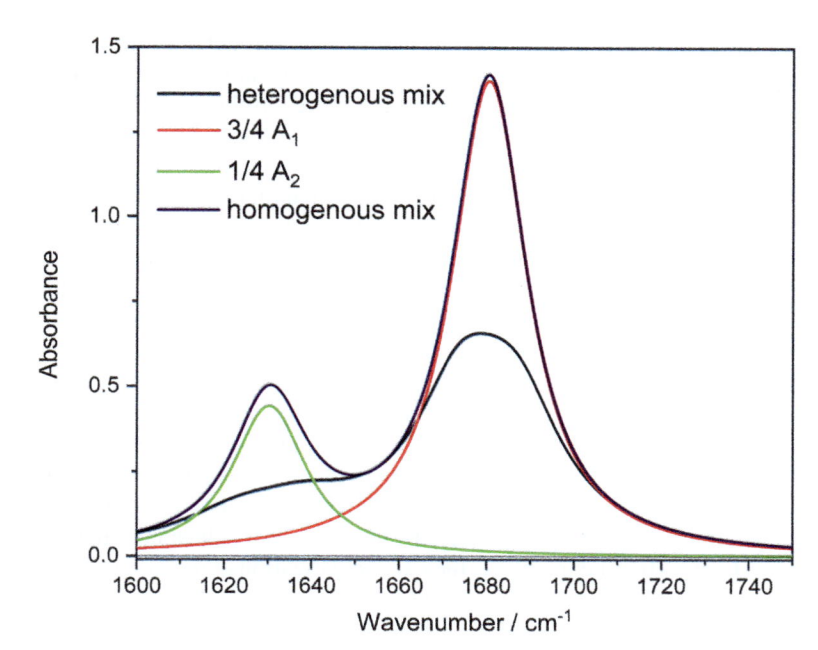

FIG. 6.20 Transmittance absorbance for a hypothetic 5-μm-thick layer on CaF_2. The layer is consisting of two different materials, each of which has one oscillator with $S = 200\,cm^{-1}$, $\gamma = 20\,cm^{-1}$, and $n_\infty = 1.47$, but one is located at $1730\,cm^{-1}$ and the other at $1780\,cm^{-1}$. The *green and the red curve* describe the transmittance absorbance of the neat materials. The *blue curve* assumes additivity of the individual absorbances, whereas the *black curve* is calculated according to Eq. (6.30) with $\varphi_1 = 0.75$ and $\varphi_2 = 0.25$.

$$\begin{aligned}
\hat{n} &= \hat{n}_1 \varphi_1 + \hat{n}_2 \varphi_2 & \text{(I)} \\
\hat{n}^2 &= \hat{n}_1^{\,2} \varphi_1 + \hat{n}_2^{\,2} \varphi_2 & \text{(II)} \\
\frac{\hat{n}^2 - 1}{\hat{n}^2 + 2} &= \varphi_1 \frac{\hat{n}_1^{\,2} - 1}{\hat{n}_1^{\,2} + 2} + \varphi_2 \frac{\hat{n}_2^{\,2} - 1}{\hat{n}_2^{\,2} + 2} & \text{(III)}
\end{aligned} \qquad (6.32)$$

Before doing this, I want to state that again the preceding mixing rules can readily be extended to more than two materials. Eq. (6.32) (I) is more or less on the theoretical level of Beer's approximation, whereas (III) is derived from the Lorentz-Lorenz relation (polarizability is additive, independent if the oscillator belongs to the same kind of molecule or a different one), as is (II), where we assume that the local field is equal to the applied electric field (cf. Section 5.3). One problem that has to be restated, though, is that the Lorentz-Lorenz relation is itself an approximation assuming a spherical molecule in a continuum, which is, at least for the three ideal systems investigated in Section 5.3, not a very good one.

So how is it possible that both, Eqs. (6.30), (6.32), make sense? Previously I have used the term microheterogenous, which is key to the explanation. A related term would be microhomogenous which I use in the sense that a mix of two or more materials appears to be homogenous under a microscope working at a certain wavelength (monochromatic light). Therefore, if the inhomogeneities are larger than the resolution limit, the sample is microheterogenous and Eq. (6.30) applies, whereas if the homogeneities are smaller than the resolution limit, Eq. (6.32) must be applied. Originally, I have investigated and proved experimentally this relationship in the context of randomly oriented polycrystalline samples (a topic to which we will revert to at the end of the second part of this book in Section 15.1) [43]. When I first applied the additivity of transmittance in the context of Beer's approximation, I noticed in the literature that for the special case that the second material is vacuum or air, deviations from Beer's approximation were already investigated in 1952 [44]. It is mere speculation, but it might be possible because the errors are small in pressed pellets [45], and that this was never discussed later on and, thus, cannot be found in textbooks.

How does nature know when Eq. (6.30) must be applied or when it has to pick Eq. (6.32)? First of all, nature certainly simply knows and does things at its convenience, while we do not know and only have the leisure to find a somehow fitting description of what is going on. It might seem that this is evasive, but actually there is no reason to avoid this point. Eq. (6.30) is the basic principle. If a sample is microhomogeneous, then at every location the spectrum is influenced by the same average index of refraction $\hat{n}$ in the sense of Eq. (6.32). Therefore, the spectrum of a macroscopic sample in the sense of Eq. (6.30) is an average of the microscopic spectra, which are all the same. Therefore, Eq. (6.30) always applies, and it just does not become obvious for microhomogeneous samples like liquid or solid solutions. For microheterogeneous samples, things, on the other hand, can become very complicated, not only because sometimes demixing is not complete, meaning that a surface or volume can consist of both microhomogeneous and microheterogeneous areas and volumes [42]. Furthermore, there is an additional mathematical complication. While the left part and the right part in Eq. (6.31) generally do not agree, the left and the right sides of the first line become the more similar, and the closer T is to unity. In other words, even for microheterogeneous samples, it is possible that the weaker bands (weak because of small oscillator strength and/or thickness and/or concentration) follow roughly the Beer-Lambert approximation, while the stronger bands with higher oscillator strengths do not do this any longer. This can render a quantitative spectrum evaluation very complicated. This is known for pellet spectra with powders the crystallites of which are not of submicron diameter or for spectra of films containing micron-sized holes (a concrete example for both featuring the same material can be found in Ref. [46]). The effect leads to the problem that bands of components with up to 20 vol% can vanish under the baseline as we showed recently [42]. This also happens in IR-imaging applications, where the samples of interest are necessarily microheterogeneous. It is known as optical dilution effect, but the vanishing of signals of minor components with decreasing resolution is wrongly explained by averaging of the absorbances (I refrain from giving any references, even if those at least report about detecting the effect and tried to explain it in contrast of the majority of users). This is exactly what does not happen! I also want to provide you with a simple possibility to check if you are allowed to apply Beer's approximation. Eq. (6.32) (I) leads to so-called isosbestic points in the spectra. These points are characterized by the two components having the same index of absorption and the same molar absorption coefficient (and very similar or even the same refractive indices). Therefore, if molecules A are exchanged by molecules B, the absorbance does not change. Such common points do no longer occur for Eq. (6.32) (II) and (III), and therefore, they are an indication of deviations from Beer's approximation. On the other hand, an isosbestic point can also seize to exist when a third chemical species comes into existence, e.g., via a reaction between the other two. Note that for small deviations from Beer's approximation, an isosbestic region is formed. Sometimes, it is possible to decide by optical modeling if the deviations are caused by the local field of Lorentz [47]. As a last remark, please keep in mind that Eq. (6.30) is also of the form to lead to isosbestic points, but in reflectance or transmittance spectra, and not in absorbance spectra (except in the limit of vanishing absorption).

6.7 How to correct the deviations and to obtain a wave-optics conform solution

6.7.1 Correction of the apparent absorbance

Now that you know what can go wrong if the BBL approximation is used to quantitatively interpret spectral data, you might ask yourself, how can I correct the deviations? In other words, how do I obtain wave-optics conform solutions? The preceding sections showed you how to solve the direct problem, which consists in computing reflectance and transmittance spectra for sample geometries, which are analytically solvable based on the dielectric function. Now, it is time think about how to solve the inverse problem, which consists in obtaining the dielectric function and the oscillator parameters from experimental spectra. If you like philosophy, you might have heard of Plato's cave: Plato has Socrates describe a group of people who have lived chained to the wall of a cave all of their lives, facing a blank wall. The people watch shadows projected on the wall from objects passing in front of a fire behind them and give names to these shadows (see Fig. 6.21). What we have to do to solve the inverse problem is somewhat similar. Except that names are not sufficient, since we want to go one step further and understand the inner essence of the things that cast the shadows. In other words, we eventually do not want to stop at the level of the dielectric function or the oscillator parameters, and we want to learn something about the materials. If we want to stay with Plato's cave, looking at the shadows with a limited view might easily obscure the true nature of the things which cast the shadows.

In the following, we will focus on two different methods to solve the inverse problem. The first method is less flexible and will work only for materials with low oscillator strengths like organic or biologic materials. But it has the advantage that algorithms are possible which allow to perform the corrections fast and without the need for an experienced user. One of their disadvantages is, however, that we do not directly obtain the oscillator parameters—this would require an additional step. The second method is called dispersion analysis, a term that was coined in 1961 by Spitzer and Kleinman [48], but the procedure was applied much earlier (the earliest application is to my knowledge that by Marianus Czerny in 1930 [49]). It is highly flexible, but requires experience, already because meaningful starting values for the oscillator parameters have to be estimated; otherwise, success is improbable.

The first method goes back to the year 1973 and was first suggested by Jones et al. [50] (the first author is the same who described the special case for microheterogeneous samples where the second component is vacuum mentioned in the preceding section). It is based on the approximative calculation of the index of absorption function from absorbance according to Eq. (5.65):

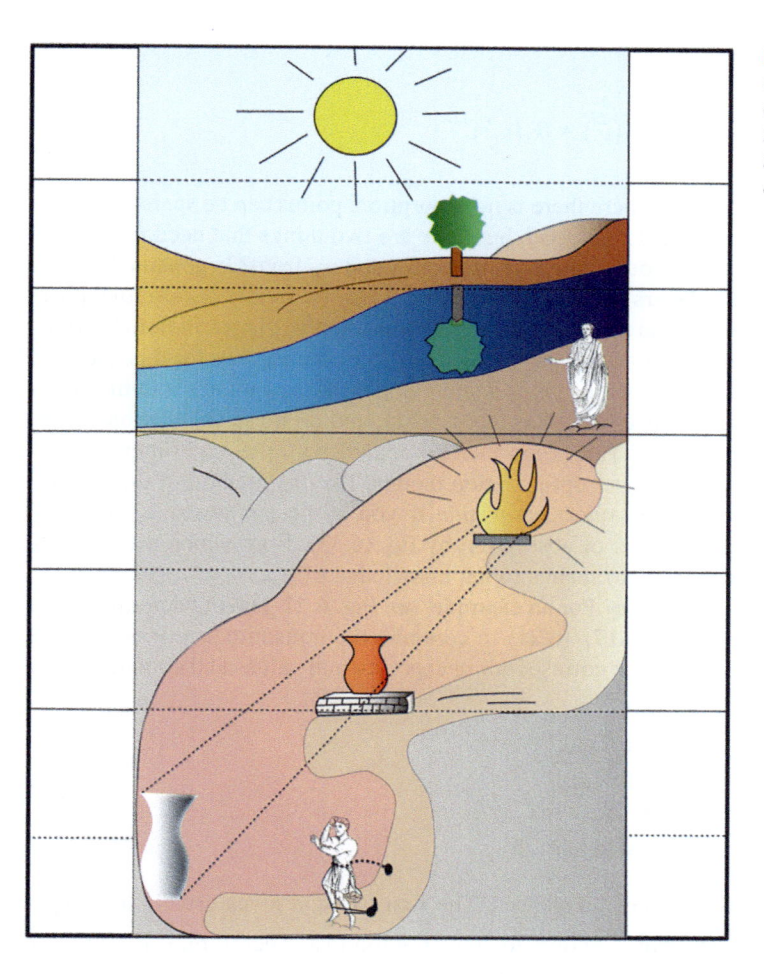

FIG. 6.21 Allegory of the cave. Left (from top to bottom): The sun; natural things; reflections of natural things; fire; artificial objects; shadows of artificial objects; Allegory level. *(From https://commons.wikimedia.org/wiki/File:Allegory_of_the_Cave_blank.png under the license Gothika /CC BY-SA (https://creativecommons.org/licenses/by-sa/4.0).)*

$$k(\tilde{\nu}) = \frac{A(\tilde{\nu})}{4\pi(\log_{10}e)d\tilde{\nu}}. \tag{6.33}$$

Note that for nonzero angles of incidence, d must be replaced by $2d/\sqrt{1 - \frac{1}{n_\infty^2}\sin^2\alpha}$.

Having the apparent index of absorption function, it is possible to calculate the corresponding index of refraction function by the suitable Kramers-Kronig relation. To that goal, we use Poor Man's Kramers-Kronig Analysis (cf. Section 5.8) and describe the index of absorption functions by triangular basis functions:

$$k_i(\tilde{\nu}_i) = \begin{cases} \frac{\tilde{\nu} - \tilde{\nu}_{i-1}}{\tilde{\nu}_i - \tilde{\nu}_{i-1}}, & \tilde{\nu}_{i-1} < \tilde{\nu} \leq \tilde{\nu}_i \\ \frac{\tilde{\nu}_{i+1} - \tilde{\nu}}{\tilde{\nu}_{i+1} - \tilde{\nu}_i}, & \tilde{\nu}_i < \tilde{\nu} \leq \tilde{\nu}_{i+1} \\ 0 & \text{otherwise} \end{cases}. \tag{6.34}$$

The corresponding index of refraction function can then be obtained by (cf. also Section 5.8) [35]:

$$\varepsilon_i(\tilde{\nu}_i)' = -\frac{1}{\pi}\left[\frac{g(\tilde{\nu} - \tilde{\nu}_{i-1})}{\tilde{\nu}_i - \tilde{\nu}_{i-1}} - \frac{(\tilde{\nu}_{i+1} - \tilde{\nu}_{i-1})g(\tilde{\nu} - \tilde{\nu}_i)}{(\tilde{\nu}_i - \tilde{\nu}_{i-1})(\tilde{\nu}_{i+1} - \tilde{\nu}_i)} + \frac{g(\tilde{\nu} - \tilde{\nu}_{i+1})}{\tilde{\nu}_{i+1} - \tilde{\nu}_i}\right]$$
$$+ \frac{1}{\pi}\left[\frac{g(\tilde{\nu} + \tilde{\nu}_{i-1})}{\tilde{\nu}_i - \tilde{\nu}_{i-1}} - \frac{(\tilde{\nu}_{i+1} - \tilde{\nu}_{i-1})g(\tilde{\nu} + \tilde{\nu}_i)}{(\tilde{\nu}_i - \tilde{\nu}_{i-1})(\tilde{\nu}_{i+1} - \tilde{\nu}_i)} + \frac{g(\tilde{\nu} + \tilde{\nu}_{i+1})}{\tilde{\nu}_{i+1} - \tilde{\nu}_i}\right]. \tag{6.35}$$
$$g(x) = x\ln|x|$$

The complex index of refraction function is then given by:

$$\hat{n}_j(\tilde{\nu}_j) = n_\infty + \sum_{j=2}^{N-1} A_j \left(n_j(\tilde{\nu}_j) + i k_j(\tilde{\nu}_j) \right). \tag{6.36}$$

The big advantage of using PMKKA is that in spectral regions where there is no absorption, points can be sparse, which is usually not possible for numerical KKR-based procedures. So far, so good, but there are two things that need further discussion. Assume that you have a layer on a substrate, where do you get the layer thickness and n_∞ from? In general, both can be obtained by independent measurements or methods. For layers on CaF_2 this is the only way (at least n_∞ is required [36]). For layers on gold or on Si, there is another way, since there are more pronounced interference fringes in nonabsorbing regions. These fringes cannot only be used to estimate the layer thickness, but also to determine n_∞, but for this, we need to invoke another procedure. For this procedure, we use rough estimates for d and n_∞, put these estimates into the corresponding equations that calculate us transmittance or reflectance (e.g., Eqs. 6.17–6.21), and do these calculations in the spectral region where the sample is transparent. The calculated values are then compared with the experimental ones, and then, the method of least squares is used together with a proper optimization method like the method of the steepest decent to find d and n_∞ (alternatively, you can use the *2T2D smart error sum* if you suspect systematic errors, cf. Section 8.3). Using d, one can now determine the apparent index of absorption by Eq. (6.33), from which the apparent complex index of refraction can be calculated based on Eq. (6.36) thanks to the knowledge of n_∞.

How good does the determination of d and n_∞ work in practice? For an example, see Fig. 6.22. Having determined the complex index of refraction function, we can put it into Eqs. (6.17)–(6.21) to compute the transmittance or reflectance predicted by wave optics. We can then compare the values with the corresponding experimental values and compute better estimations from one of the following two formulas:

$$\begin{aligned} k_{i+1}(\tilde{\nu}) &= k_i(\tilde{\nu}) \frac{\lg T_{\text{meas}}}{\lg T_{\text{calc},i}} \\ k_{i+1}(\tilde{\nu}) &= k_i(\tilde{\nu}) + \frac{(-\lg T_{\text{meas}} + \lg T_{\text{calc},i})}{4\pi(\log_{10} e) d \tilde{\nu}} \end{aligned}. \tag{6.37}$$

d must be replaced by $2d/\sqrt{1 - \frac{1}{n_\infty^2} \sin^2\alpha}$ for nonzero angles of incidences. The first formula gives usually a stronger improvement at the beginning of the iterative optimization, but is known to converge too fast. Therefore, it is advisable to use the second formula once the first does no longer decrease the residual sum of squares (RSS, or, alternatively, the *smart error sum*). Scheme 6.5 displays the sequence of the correction steps.

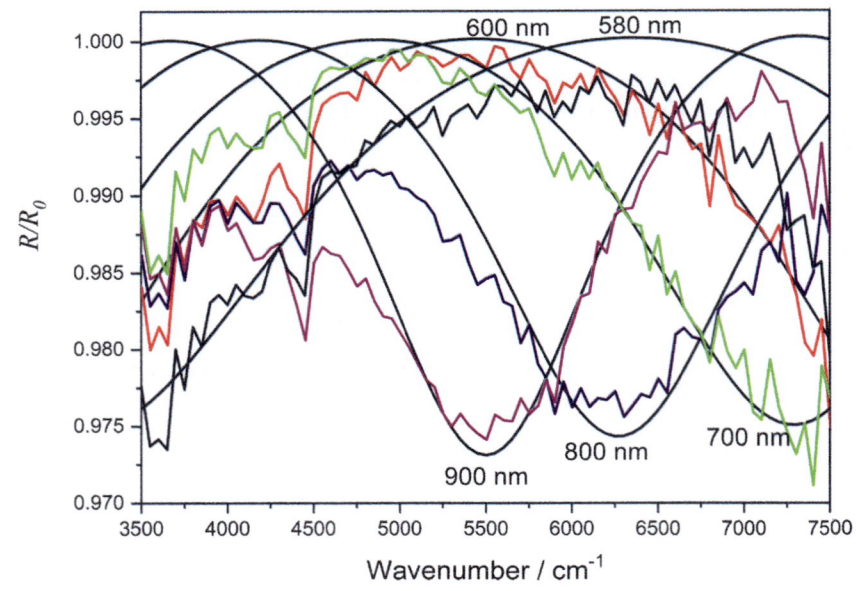

FIG. 6.22 Experimental interference fringes of layers of PMMA on gold and modeled spectra. The values of n_∞ determined by the corresponding procedure lie between 1.45 and 1.52 (according to literature, $n_\infty \approx 1.47$). The errors of the determined layer thicknesses (indicated in the figure are the nominal values) are of the same order of magnitude [51].

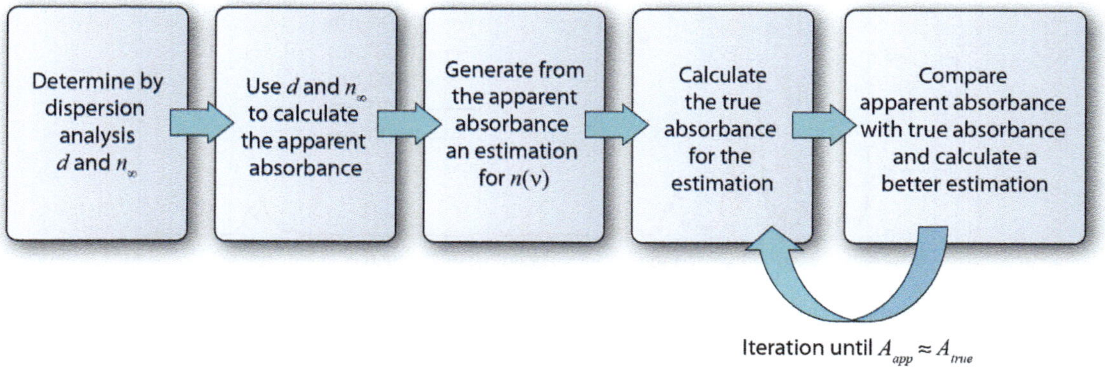

SCHEME 6.5 Correction scheme for the absorbance according to Ref. [51].

It is also possible to use a fixed n_∞ (preferably, when it is known. Since dispersion is usually small, also n_D^{20} can be used instead) and just calculate the thickness in the first step, or, for layers on CaF$_2$, to use both, an externally determined d and n_∞. Just to illustrate what kind of improvement is possible even when an index-matched substrate like CaF$_2$ is used, we compare the absorption coefficients of PMMA layers of different thicknesses before and after correction in Fig. 6.23.

Note that the same scheme can also be used to correct ATR measurements, in a more advanced way, than with commercial advanced ATR correction scripts. Since the incidence angle is not zero, however, for this scheme to be applicable, the polarization state of the incoming light needs be known. Since commercial ATR accessory no longer allow to apply a polarizer, it is necessary to determine the polarization state of the spectrometer accessory combination. As already mentioned in Section 6.5, a corresponding scheme was suggested in Ref. [37], which I want to introduce to you in the following. It is based on a known set of optical constants of a chemical substance omnipresent in labs, namely, water. As a liquid, it has the advantage that it automatically establishes intimate contact with the ATR crystal, which is a big advantage as it allows to record high-quality spectra. Once you have recorded a spectrum, since you did not use a polarizer, the reflectance values, according to Malus' law, are given by,

$$R = R_s \cos^2\varphi + R_p \sin^2\varphi, \tag{6.38}$$

and the goal is to determine the φ. Since you know the complex index of refraction function, you can take Eq. (6.22) to calculate R_s and R_p for water and then fit φ according to previous equation so that the right side optimally agrees with your measured ATR spectrum. To get a good fit, it also has to be taken into account that there is some spread concerning the angle of incidence. As suggested in Ref. [37], this spread can be modeled by assuming a Gaussian distribution:

$$s(\alpha_i) = c_N \exp\left[-\frac{\alpha_i - \alpha_\mu}{2\sigma^2}\right], c_N = \int_0^\infty \exp\left[-\frac{\alpha_i - \alpha_\mu}{2\sigma^2}\right] d\alpha_i. \tag{6.39}$$

If you miss a factor of $\sin\alpha_i$ in c_N compared to eq. (8) in Ref. [37], have a look at Ref. [52]. The angle spread should not be that large, since for $\alpha_\mu = 45°$ and a ZnSe or Diamond ATR crystal, the ATR condition is already easily violated for somewhat stronger oscillators and typical values of $n_\infty \approx 1.5$. For the accessory spectrometer combination, we use I tried, $\sigma \approx 1.0°$, which could in practice be disregarded completely for organic and biologic materials, which have oscillators with usually lower oscillator strengths than the H-O-stretching vibration in water. For the accessory used in Ref. [37], a value of $\sigma \approx 4.7°$ was determined, which would mean that a comparably large fraction of light beams violates the ATR condition.

How large would be the error of the polarization determination? We found that the water spectra can strongly alter from one day to the other for the same accessory spectrometer combination, maybe because of temperature differences. By strongly alter, I do not mean that bands shift or change their shapes considerably, but that on one day $-\log_{10}R$ could by a factor of up to 1.075 be larger than on other days. This higher absorbance leads to a higher value of the determined value of φ; since then, a higher fraction of p-polarization is assumed (remember, for $\alpha = 45°$, $A_p = 2 A_s$). In the previous case, φ increased from 12° to 30° if the conventional RSS is used for the fit. This means that the fraction of p-polarized light would increase from about 4.3% to 25%! Partially, the problem can be removed by the use of the 2T2D *smart error sum*, cf. Section 8.3. In this case, φ grew from 12° to 18°. The fraction of p-polarized light then increased from about 4.3% to 10%. The reason that some error remains is most probably that the dielectric function for water I used did not reproduce the band shape of the H-O-stretching vibration well (the calibration certainly depends on the quality of the dielectric function of the material used).

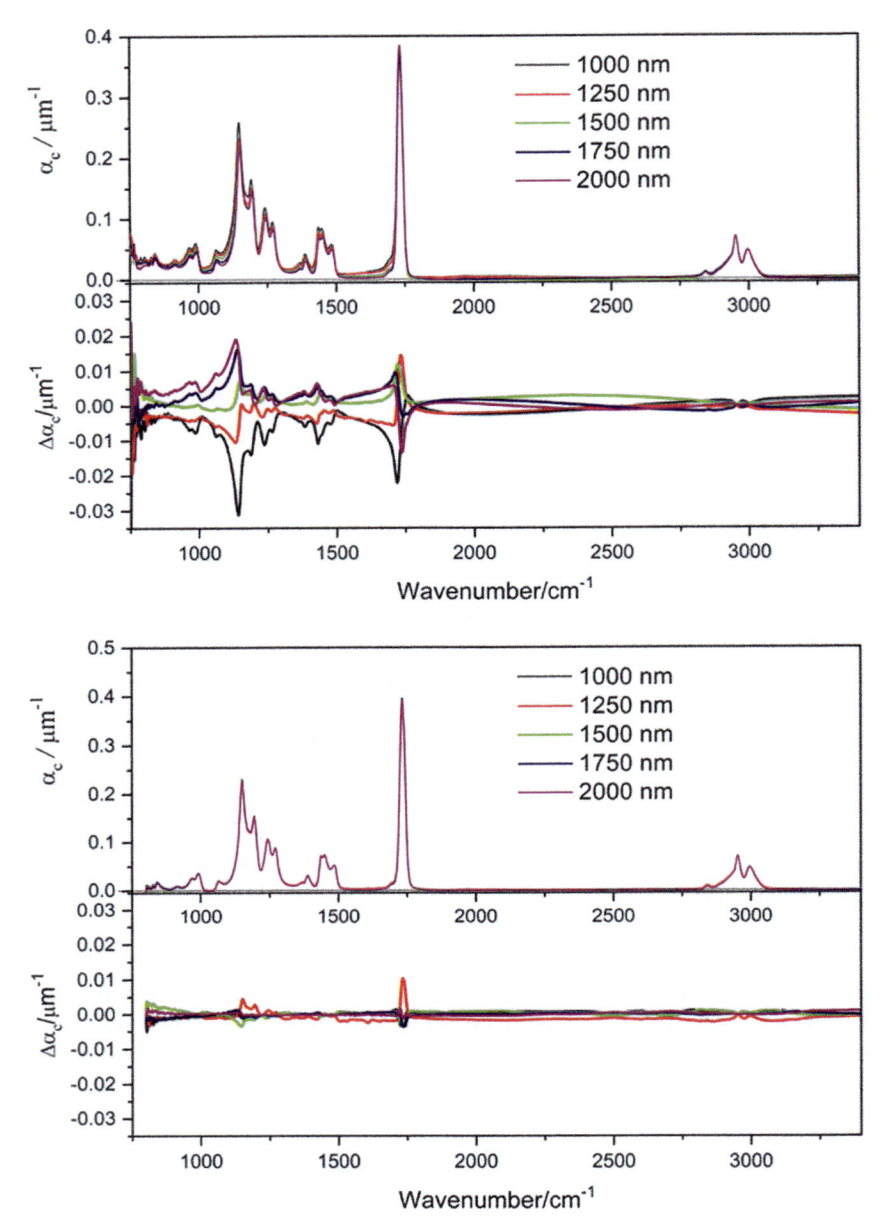

FIG. 6.23 Experimental absorption coefficient of PMMA layers with different thicknesses on CaF_2 before (upper panels) and after correction (lower panels) [36].

Once the polarization angle is determined, a first guess for the absorption index function can be calculated from the ATR absorbance by Eq. (6.33) if you determine d approximately by:

$$d \approx d_{\text{eff},s} \cos^2\varphi + d_{\text{eff},p} \sin^2\varphi = \frac{n_1 n_2 \cos\alpha_i}{\pi\tilde{\nu} \cdot (n_1^2 - n_2^2)\sqrt{n_1^2 \sin^2\alpha_i - n_2^2}} \left(\cos^2\varphi + \frac{2n_1^2 \sin^2\alpha_i - n_2^2}{(n_1^2 + n_2^2)\sin^2\alpha_i - n_2^2} \sin^2\varphi \right). \quad (6.40)$$

Certainly, this is just an approximation, since from Eq. (6.38) it does not follow that $A = A_s\cos^2\varphi + A_p\sin^2\varphi$. While the latter equation is a cornerstone of linear dichroism theory, it conflicts with Malus' law and is correct only in the limit of vanishing absorption/oscillator strength (cf. Chapter 11).

For the calculation of the refractive index function from the absorption index function, it is vital to have a good value for n_∞. It would be ideal if the angle of incidence could be switched to be subcritical, then n_∞ could be determined simply by the inversion of the external reflection formula (infrared refraction spectroscopy, cf. Section 7.1). Since ATR accessories usually do not allow this, one could measure the external reflection on a corresponding accessory if the sample is solid. In any way, a reliable n_∞ is vital, since in ATR spectra, n_∞ is not responsible for an offset as this is the case for external reflection, but influences directly the peak values. Therefore, any error in n_∞ will propagate and influence the absorption index function in the next iteration.

Overall, reflectance is usually above 0.1, which means that the scheme to determine the true absorbance, and, by that, the complex refractive index and/or the dielectric function will work fast and smoothly. In general, using the first alternative in Eq. (6.37) will be sufficient or do most of the work. If not, for the thickness d in the second alternative again, Eq. (6.40) can be employed.

All in all, since n_∞ must be known also for advanced ATR correction, and since the previous formalism can be made automatic and is based on wave optics, it would be an ideal replacement once the polarization of the spectrometer-accessory combination is known. On the downside, it is comparably slow, as it takes some minutes on a typical office PC and cannot correct stronger bands, where the condition $n_2 > n_1 \sin^2 \alpha_i$ is violated. There is, however, a much faster (up to two orders of magnitude) method, which is able to correct strong bands, as long as $n_1 > n_2 \sin^2 \alpha_i$ at the low and high wavenumber end of the experimental spectral range [53]. It only works if $\alpha_i = 45°$, but most modern ATR accessories use this angle of incidence. At this angle, $R_s^2 = R_p$ based on Fresnel's equations, therefore:

$$R(45°) = R_s(45°) \cos^2\varphi + R_s(45°)^2 \sin^2\varphi. \tag{6.41}$$

This leaves, as the only unknown, $R_s(45°)$, which we can determine after some algebraic manipulations according to Ref. [53]:

$$R_s(45°) = \frac{1}{2}\left(-\cot^2\varphi + \csc^2\varphi\sqrt{\cos^4\varphi + 4R(45°)\sin^2\varphi}\right). \tag{6.42}$$

Once $R_s(45°)$ is calculated, a direct Kramers-Kronig analysis following Eq. (5.162) in combination with Eq. (5.164) provides the complex refractive index function in a fast, efficient, and accurate way.

6.7.2 Dispersion analysis

For the second method, dispersion analysis, I actually already explained the principle to you in the last section, except that we used it for a case without noticeable dispersion, namely, for the spectral region without (perceptible) absorption. The principles for its use in regions with absorption are pretty much the same, except that we use dispersion relations to describe the dielectric function in absorbing regions. Accordingly, we use one oscillator per peak in a spectrum. This is already where experience comes into play, because some oscillators may not reveal their presence by a peak, some show up even as dips in a stronger band. A good strategy in these cases is to first only setup oscillators for strong bands, and, once the fit for these strong bands is no longer improving, to model the fine structure by adding additional oscillators (in certain cases, there is no obvious fine structure, like, e.g., in case of the C=O vibration of PMMA. Then, it can be helpful to scan the second derivative spectrum, which better reveals the number of peaks). Note that so far, all media in a layer stack must be not only isotropic, but also scalar to be modeled. This does not sound as a big limitation. For reasons I explain in the next part of the book, we actually have to go one step further.

A scalar medium is one that can be described by a (scalar) dielectric function. However, in case of randomly oriented polycrystalline media with anisotropic crystallites, even when the crystallites are much smaller than the wavelength, the bands cannot be described by Lorentz oscillators, because due to the averaging the resulting band shapes are strongly altered. In principle, the same is true for glasses, except that the regions of order are very small, which increases the damping constants. Since, in case of larger crystallites, the medium is no longer scalar, the usefulness seems to be strongly limited and only cubic materials seem to remain in single crystalline or polycrystalline form. Actually, when properly oriented, also optically uniaxial (tetragonal, hexagonal, and trigonal) and orthorhombic crystals can be analyzed, since in so-called principal orientations, the spectra depend only on one of the principal dielectric functions. For simplicity, I describe first the procedure in more detail employing it to a cubic crystal as an example. I select MgO, which according to symmetry analysis should only have one active IR vibration, which is threefold degenerated. In fact, we need two such oscillators to describe MgO's spectrum, because what symmetry analysis cannot achieve is to consider higher harmonics and we see not only the fundamental, but also the first overtone in the spectrum. Since the crystal is highly absorbing in the infrared, we can use the simplest optical model, which is available, namely, a semiinfinite incidence medium (vacuum) and a semiinfinite exit medium (the crystal). With only two oscillators, the first estimation does not need to be very sophisticated. This is displayed in Fig. 6.24 (upper left panel). After less than 15 s on a contemporary PC, the fit is achieved. It was carried

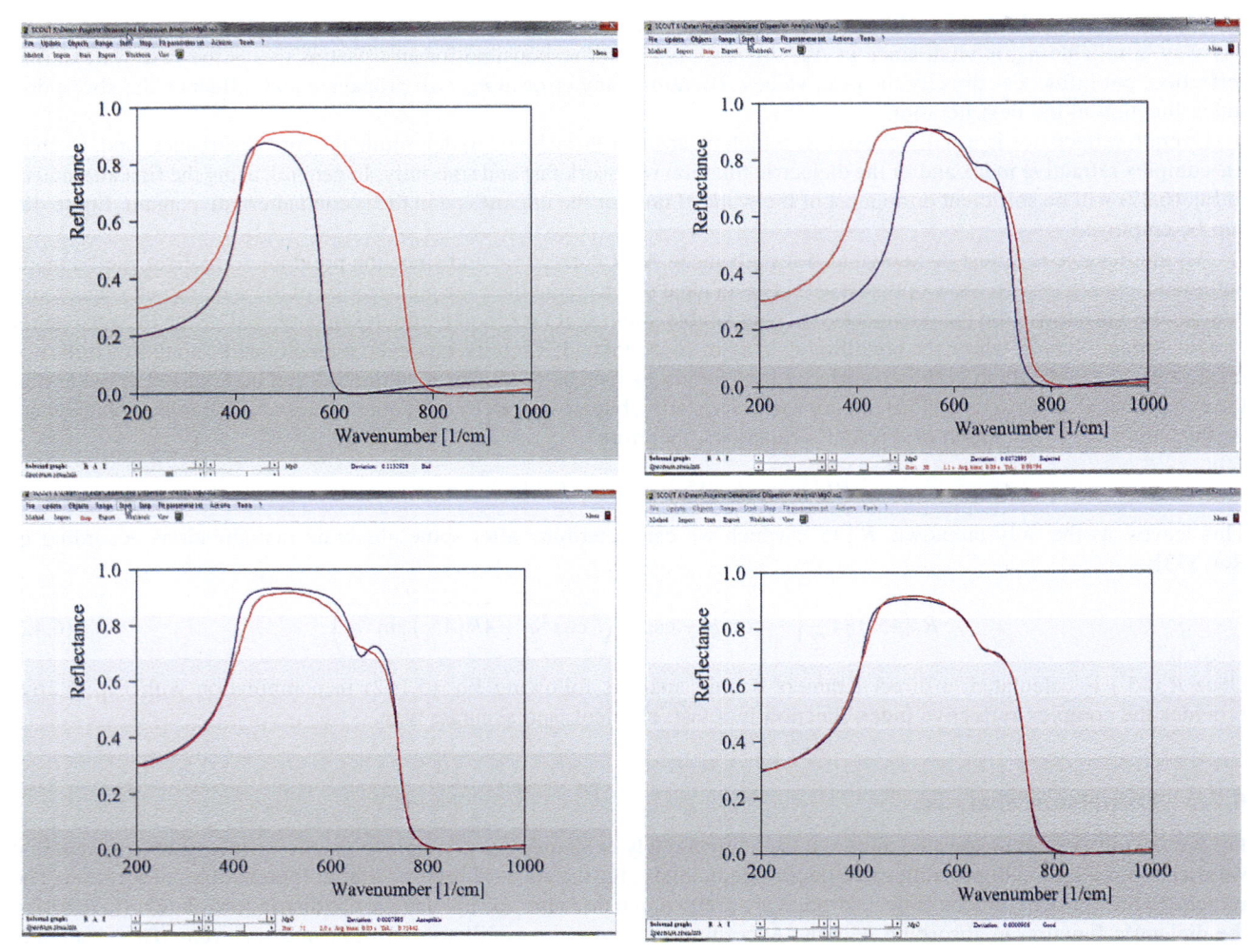

FIG. 6.24 Different stages of the dispersion analysis of MgO using SCOUT.

out on a commercial software, which I have received for free since I had been one of the beta testers of the original program Scout_98 back in the middle of the 90s [54]. A free alternative is the program RefFit [55]. The results of the fit are the dispersion parameters, i.e., two sets of three parameters for a Lorentz-oscillator and ε_∞, since the program uses for the fit Eq. (5.27). The results are displayed in Table 6.1.

You might think that the shoulder around $700\,cm^{-1}$ reveals the first overtone. But actually, it is the dip and not the shoulder, which is also obvious from the oscillator position in Table 6.1. If this is not obvious, then look at Fig. 6.25: The oscillator strength of the first mode is so large, that the real part of the dielectric function is negative between TO_1 and LO_1, which are the outer TO-LO pair. Accordingly, the inner TO-LO pair, TO_2 and LO_2, is inverted, therefore the dip (cf. Section 5.6). Keep in mind that all information you need to display the curves in Fig. 6.25, except the experimental data of course, can be reproduced by the dispersion parameters from Table 6.1.

Before I continue, a word of caution. Here is a passage from the book Light and Matter 1d edited by Ludwig Genzel from 1984 [56]: "It was discussed more than 60 years ago that the structure in the reflection bands of these crystals ... might be related to a frequency-dependent damping mechanism of the dispersion oscillator (...). Nevertheless, attempts are still being made to describe the structure of infrared absorption bands in alkali halides or other diatomic crystals by assuming two or three infrared active 'oscillators' with classical damping constants. Such a procedure always leads to an incorrect description of the absorption in the wings... Furthermore, it gives an unphysical picture of the lifetime of the dispersion oscillator, measured, say, by infrared absorption." In fact, I already pointed out in the previous chapter that damping is actually frequency-dependent (cf. Ref. [57]). Unfortunately, the models that were developed until the middle of the 1980s are so complex, that it looks like there was a kind of singularity afterward, and, very much like in case of the

TABLE 6.1 Oscillator parameters of MgO from dispersion analysis $\varepsilon_\infty = 3.01$.

i	$\tilde{\nu}_i$ (cm^{-1})	S_i (cm^{-1})	γ_i (cm^{-1})
1	413.7	1050	22.2
2	652.2	90	64.6

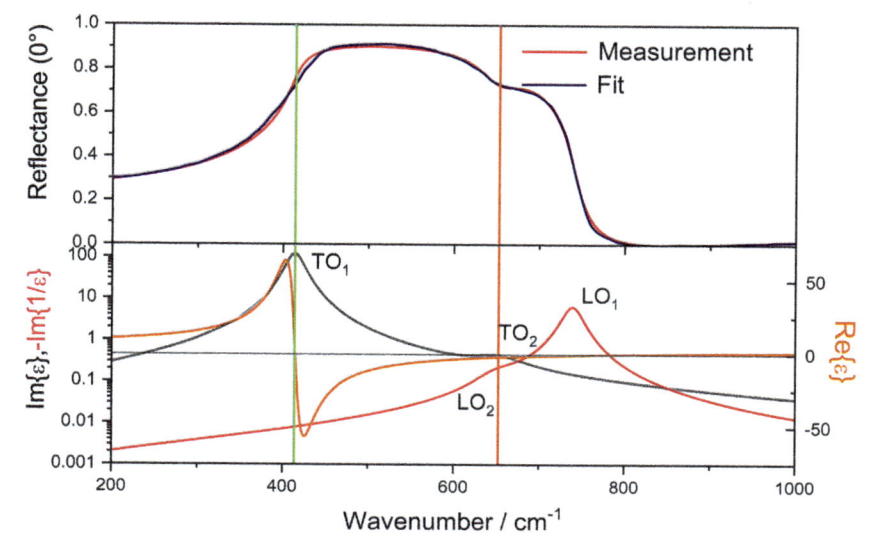

FIG. 6.25 Result of the dispersion analysis of MgO.

Lorentz-Lorenz model, further development and even the use of these models stopped completely. It seems that it was accepted, that we cannot even fully understand the spectra of the simplest compounds consisting of two different atoms in a unit cell of cubic symmetry, as this was accepted for the spectra of thermodynamically ideal binary mixtures, which we also cannot fully understand (more about this in Chapter 10). Therefore, if you use dispersion analysis, keep always in mind that the CDHO model helps to gain and parameterize the dielectric function and usually does a very good job, but the parameters themselves are of limited usefulness as are all our attempts to describe the world around us.

As a second example, I chose the same as the one in Section 6.7.1, namely, an isotropic layer on a thick isotropic substrate.

The first step of dispersion analysis is always to determine which optical model fits best to the sample at hand. In the general case, the incidence medium is either vacuum or air with $\varepsilon_0 = n_0 = 1$ (note that air has a dielectric function that is only about 0.5 per mille larger than unity, which we can safely disregard) or an ATR crystal, all of which are semiinfinite media. The simplest optical model would result if the sample can also be considered a semiinfinite medium like in the example of bulk MgO above. Such a semiinfinite medium is characterized by the absence of multiple reflections. Accordingly, the waves transmitted into such a medium do never reach another interface from which they would be reflected back. In practice, this can also be the case due to strong absorption as for MgO. A further good example in the infrared spectral region is a CaF$_2$ substrate around its phonon mode at 257 cm^{-1} [58], or a gold layer, which is thicker than 50 nm [59].

In contrast to the gold layer, at higher wavenumbers, above about 800 cm^{-1}, the CaF$_2$ substrate begins to be noticeably transparent. As a consequence, the substrate must then be seen as a thick, i.e., incoherent layer embedded in the incidence medium, since for thick substrates, the thickness deviations are usually greater than the wavelength of light. In principle, the same should be true for Si substrates, but those can be cut with high precision. Therefore, for a mm-thick substrate, interference fringes can be seen as long as the spectral resolution is better than 4 cm^{-1}. For lower resolutions, the fringes vanish, since due to the comparably large thickness of the substrates, the fringe period is small and the fringes are suppressed.

Prior to dispersion analysis of layers on such thick substrates, initially the optical constants of the substrate materials must be known. For gold, good optical constants are available in literature. I recommend the data provided by Johnson and Christy [60] and parameterized by Etchegoin et al. [61,62].

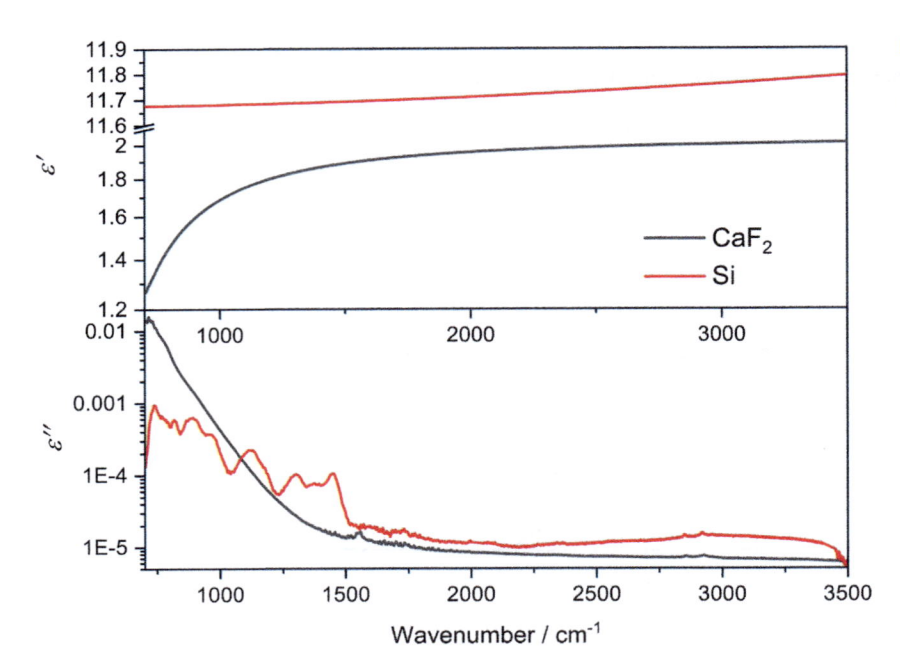

FIG. 6.26 Real (upper panel) and imaginary parts (lower panel) of the dielectric functions of Si and CaF_2 substrates [35,36].

For CaF_2, in principle, optical constants from the dispersion analysis of reflectance data from Ref. [58] could be taken. Since the surface of Si is usually oxidized, Si substrates show bands below about $1200\,cm^{-1}$. Because the values of the wavenumber-dependent absorption index change with the doping state of Si, it is advisable to determine these for the type of Si substrate, which is actually used. I would advise anyway to do the same for CaF_2 substrates, in particular if they are employed for transmission experiments (what they usually are). Dispersion parameters derived from reflectance measurements for very weak absorptions can easily be too inaccurate for the purpose. In particular, if the spectral region is extended down as far as possible toward the phonon(s) of CaF_2. As long as one is interested only in accurate dielectric functions, the method used in the previous section is more than adequate (the corresponding dielectric functions are depicted in Fig. 6.26) [35,36].

When the dielectric function of the substrate is known, the next step is the determination of ε_∞ and the thickness of the layer (in this case PMMA). Thin layers usually represent coherent layers and have typical thicknesses between 0.5 and $2\,\mu m$. Therefore, interference effects cannot be neglected. The corresponding optical model is correspondingly incidence medium/coherent layer/incoherent layer/exit medium. Note that for gold as substrate, the calculations can nevertheless be carried out with this more complex model, even when the gold layer is thicker as 50 nm and therefore represents a semi-infinite medium. In this case, the reflectance simply is not influenced by the substrate behind the gold layer.

In principle, it is possible in the next step to determine both, layer thickness and ε_∞, simultaneously by dispersion analysis, as I have done this for PMMA layers on gold in Ref. [51]. However, even when the fringes are intense like for PMMA on Si, variations of the determined ε_∞ occur, although comparably small ones. Even then, it can be advisable to use a literature value for ε_∞ (cf. also Section 5.8.4).

The fit to determine thickness and/or ε_∞ is expediently carried out in a broader spectral region without absorption, e.-g., for PMMA between 4000 and $7000\,cm^{-1}$. If ε_∞ is known, this value is kept constant and used as input for Eqs. (6.17)–(6.21) together with the dielectric function of the substrate and its thickness (for gold, as discussed previously, any thickness $d_2 > 50\,nm$ leads virtually to the same results, since the layer is then a semiinfinite medium) to calculate the reflectance or the transmittance in dependence of the wavenumber. Initially, d_1, the thickness of the PMMA layer, is set to the nominal thicknesses, the reflectance in the featureless spectral region was calculated and the RSS was determined.

This sum was then iteratively reduced until a minimum was reached with the thickness d_1 as the only free parameter. To give you an impression how good this works, the corresponding results for PMMA layers on gold are displayed in Table 6.2 together with those resulting from individual fits of the spectra from Ref. [51].

In addition to individual fits, it can also make sense to fit samples, which share the same optical model and the same materials together. To that end, the individual RSSs are summed up to provide a common error criterion. Also, I have added results of different thickness fits, where I fitted the spectral regions with absorptions based on a known dielectric function

TABLE 6.2 Layer thicknesses for PMMA layers on gold obtained by conventional dispersion analysis with collective or individual [51] fitting of the different spectra in the nonabsorbing region above 4000 cm^{-1}.

	$d_1(1)$/nm	$d_1(2)$/nm	$d_1(3)$/nm	$d_1(4)$/nm	$d_1(5)$/nm
Nominal d_1	580	600	700	800	900
Collective fit	554	603	683	809	939
Individual fit	553	573	686	781	938
k-Fit (CaF$_2$ data)	563	603	687	787	900

These thicknesses are compared to the ones obtained by fitting the absorbing spectral regions using the absorption index gained from layers on CaF$_2$.

(certainly based on Eqs. (6.17)–(6.21) and *not* based on the Beer-Lambert approximation!). Overall, it looks like the thickness values are accurate within an error margin of less than 2%.

Note that while I used a custom-made program based on Wolfram Mathematica employing a self-implemented version of the classical downhill simplex method [63] (with more options to customize and control the fits than offered by Mathematica's built-in function) to minimize the RSS, it is in principle possible to carry out dispersion analysis with a spreadsheet software, including Excel, LibreOffice, and GoogleSheets that features a nonlinear solver (this is how my former supervisor Helga Dunken performed her dispersion analyses).

In case of PMMA on CaF$_2$, it seems that the fringes are too weak to be useful for thickness fits. In this case, you either assume that the layers are equally thick to those you coat other substrates like Si with under the same conditions and coating parameters [35], or you need to fit the thickness via known dielectric functions of the coating material in the absorbing regions as described previously. In any way, as a word of caution, do always cross-check your results—a good resemblance between fitted and experimental spectra is not sufficient as you will see later on (unfortunately, most examples in literature do not follow this recommendation, so be warned!). In particular for layers on metallic substrates, this is very important, because sometimes if you do not wait long enough to equilibrate the conditions before the measurement, reflectance will be above unity and you need to correct spectra, which is not easy since the correction factor is usually wavenumber-dependent. It is very important that you realize that dispersion analysis, very much as any fitting method, relies on the absence of systematic errors (as already shortly mentioned, in Sections 8.2 and 8.3, I will introduce you to a method to avoid corresponding problems by a fitting method that allows to compensate systematic errors by the *smart error sum*).

Having determined the thickness it can be kept constant for the oscillator fits (for strong oscillators as they appear in inorganic materials, it may be necessary to refine iteratively the thicknesses together with the oscillator parameters. Often in these cases, absorption indices do not decrease rapidly enough away from the oscillator positions, but for organic materials like PMMA, the C—H vibrations are sufficiently weak and can safely be ignored for the thickness determination). For these fits, it is necessary to first estimate the oscillator parameters, which strongly decreases not only the time needed for dispersion analysis, but also the number of necessary interventions from the operator. Estimating oscillator parameters is not a difficult task per se, as far as the oscillator positions $\tilde{\nu}_{0,j}$ and, for separated peaks, the damping constants γ_j are concerned. At least, if the oscillators are comparably weak as in most organic and biologic materials, since then the oscillator positions deviate only weakly from the peak positions and the damping constants are similar to the width of the peaks at half height [64]. Unfortunately, in case of the oscillator strengths S_j, there is no rule of thumb. As a consequence, coming up with good estimates is less straightforward, but knowing that the oscillators of organic or biologic materials usually have strengths with $S_j \leq 200$ cm^{-1} and some practicing helps. Usually, one oscillator is assumed for each peak and shoulder. In complex cases, taking first or second derivatives of spectra can be helpful (cf. Section 7.1).

Once good estimates have been determined, the fitting procedure can be initiated. Having inserted the oscillator parameters into the dispersion relation (for PMMA into the classical damped harmonic oscillator model, Eq. (5.27)), a first estimation of the dielectric function of the layer is calculated. If more spectra are fitted together, the same dielectric function is used for each, since we will assume that the optical properties of the material, in this case PMMA, do not change with the layer thickness (therefore, we would use one common set of oscillator parameters for all samples/spectra). The estimated parameters allow us to calculate the complex refractive indices and the z-component of the wavevector. These are the input values for Eqs. (6.17)–(6.21) to calculate the reflectance $R(\tilde{\nu})$ and/or the transmittance $T(\tilde{\nu})$ with help of the layer thickness that has been determined earlier, which are compared to the experimental values via the RSS. Subsequently, better values

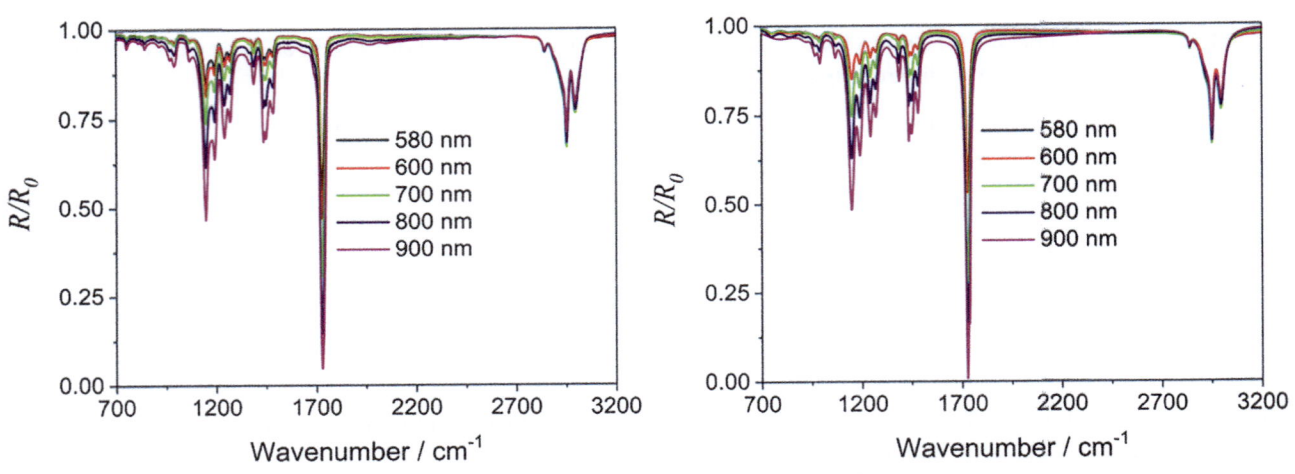

FIG. 6.27 Corrected experimental (left panel) and best fit (right panel) spectra of PMMA on gold.

for the oscillator parameters are calculated from the minimization conditions of the RSS. This is iteratively repeated, until the RSS is at its global minimum or below a certain preselected value.

Initially, it is advisable to fit adjacent oscillators that are separated from other groups of oscillators together, keeping their positions in the first run fixed. It is meaningful to start with the group with the highest wavenumber positions (this usually requires a more flexible program than a typical spreadsheet calculation program can offer, in particular since the RSS needs to be calculated only in the limited spectral region where the oscillators are located), since the dielectric background ε_∞ and, with it, n_∞ increase from high to low wavenumbers. Once all oscillators have been fitted, this can be repeated until the optima have been found.

Finally, all oscillators are fitted together. Results are presented in Fig. 6.27, where in large ranges, experimental and simulated spectra agree within line thickness. I know that I repeat myself, but please keep in mind that this is a necessary but not sufficient condition for a correct optical model and correct results, even when the conventional RSS is employed. To assure correctness, independent experiments are needed to compare the results, e.g., the dielectric functions for the same material on different substrates.

If you are willing to give away the advantage of having all data in your hand when using a few dispersion parameters, then you are also able to gain the dielectric function of a glassy material or of that of a randomly oriented polycrystalline material with small crystallites, which are also scalar materials, but do not show Lorentz-type bands in their dielectric function. The idea goes back to Eq. (5.141),

$$\varepsilon(\tilde{\nu}_j) = \varepsilon_\infty + \sum_{j=2}^{N-1} A_j \left[\varepsilon_j(\tilde{\nu}_j)' + i\varepsilon_j(\tilde{\nu}_j)'' \right], \tag{6.43}$$

where you can treat the A_j is free parameters that are varied during a dispersion analysis (cf. Section 5.8). Note that originally Kuzmenko suggested, not to use the triangular function bases alone, just to accompany Lorentz oscillators. The clou that it is now possible to use Eq. (6.43) for dispersion analysis is the correction of the error in Eq. (5.140) by Eq. (5.143) (this makes it in essence equal to a spline analysis if only splines of the first degree are used [65]). Alternatively, you can use a similar procedure based directly on the KKR invented by Nitsche und Fritz [66,67].

6.8 Further reading

I would love to provide you further books, which discuss the errors you make when using the BBL approximation in relation to wave optics and dispersion theory, but the fact that such are not available was my motivation to put together this first part of the book. One book I can warmly recommend to you is "Internal Reflection and ATR Spectroscopy" by Milan Milosevic [39], (and, certainly, along with it, "Internal Reflection Spectroscopy" by Nicolas J. Harrick! [68]), which covers, at least to a large part, the problems in ATR spectroscopy. I can also offer you Takeshi Hasegawa's book "Quantitative Infrared Spectroscopy for Understanding of a Condensed Matter" [69], but it follows a different philosophy by interweaving wave optics and the BBL law and disregarding dispersion theory. So, I am afraid you have to potluck with this book.

References

[1] W.N. Hansen, Expanded formulas for attenuated total reflection and the derivation of absorption rules for single and multiple ATR spectrometer cells, Spectrochim. Acta 21 (1965) 815–833.

[2] P. Lichvar, M. Liska, D. Galusek, What is the true Kramers-Kronig transform? Ceramics-Silikáty 46 (2002) 25–27.

[3] C. Pecharromán, J.E. Iglesias, Modeling particle size and clumping effects in the IR absorbance spectra of dilute powders, Appl. Spectrosc. 50 (1996) 1553–1562.

[4] T.G. Mayerhöfer, H. Mutschke, J. Popp, Employing theories far beyond their limits—the case of the (Boguer-) Beer–Lambert law, ChemPhysChem 17 (2016) 1948–1955.

[5] G.B. Airy, VI. On the phænomena of Newton's rings when formed between two transparent substances of different refractive powers, Lond. Edinb. Dublin Philos. Mag. J. Sci. 2 (1833) 20–30.

[6] S.H. Gordon, R.E. Harry-O'kuru, A.A. Mohamed, Elimination of interference from water in KBr disk FT-IR spectra of solid biomaterials by chemometrics solved with kinetic modeling, Talanta 174 (2017) 587–598.

[7] J.M. Chalmers, P.R. Griffiths, Handbook of Vibrational Spectroscopy, J. Wiley, Chichester, 2002.

[8] P.R. Griffiths, J.A. De Haseth, Fourier Transform Infrared Spectrometry, Wiley, 2007.

[9] J. Judek, K. Wilczyński, J.K. Piotrowski, Measurements of optical properties of liquids in a quartz cuvette: rigorous model, uncertainty analysis and comparison with popular approximations, Measurement 174 (2021) 109069.

[10] M. Planck, Zur elektromagnetischen Theorie der selectiven Absorption in isotropen Nichtleitern, in: Sitzungsberichte der Königlich Preussischen Akademie der Wissenschaften, vol. I, 1903, pp. 480–498.

[11] Y. Masasi, A note on the measurement of the absorption coefficient in infra-red region. Influence of the thickness of layer on the measurement of absorption coefficient, Bull. Chem. Soc. Jpn. 28 (1955) 489–492.

[12] W. Brügel, Einführung in die Ultrarotspektroskopie, D. Steinkopff, 1957.

[13] J.P. Hawranek, P. Neelakantan, R.P. Young, R.N. Jones, The control of errors in i.r. spectrophotometry—III. Transmission measurements using thin cells, Spectrochim. Acta A: Mol. Spectrosc. 32 (1976) 75–84.

[14] G.K. Ribbegård, R.N. Jones, The measurement of the optical constants of thin solid films in the infrared, Appl. Spectrosc. 34 (1980) 638–645.

[15] T. Tano, J. Umemura, Deviation from the Lambert law in ultrathin films such as black soap films, Appl. Spectrosc. 51 (1997) 944–948.

[16] M.K. Gunde, Z.C. Orel, Thickness-dependent optical effects in infrared reflection-absorption spectra of a fairly thick polymer layer, Appl. Spectrosc. 56 (2002) 24–30.

[17] P. Bassan, J. Lee, A. Sachdeva, J. Pissardini, K.M. Dorling, J.S. Fletcher, A. Henderson, P. Gardner, The inherent problem of transflection-mode infrared spectroscopic microscopy and the ramifications for biomedical single point and imaging applications, Analyst 138 (2013) 144–157.

[18] E. Hecht, Optics, fourth ed., 2002. Pearson Education.

[19] S.W. King, M. Milosevic, A method to extract absorption coefficient of thin films from transmission spectra of the films on thick substrates, J. Appl. Phys. 111 (2012) 073109.

[20] T.G. Mayerhöfer, H. Mutschke, J. Popp, The electric field standing wave effect in infrared transmission spectroscopy, ChemPhysChem 18 (2017) 2916–2923.

[21] M.K. Gunde, B. Aleksandrov, Thickness-dependent frequency shift in infrared spectral absorbance of silicon oxide film on silicon, Appl. Spectrosc. 44 (1990) 970–974.

[22] M.K. Gunde, Optical effects in IR spectroscopy: thickness-dependent positions of absorbance lines in spectra of thin films, Appl. Spectrosc. 46 (1992) 365–372.

[23] H. Dannenberg, J.W. Forbes, A.C. Jones, Infrared spectroscopy of surface coatings in reflected light, Anal. Chem. 32 (1960) 365–370.

[24] J. Pacansky, C. England, R.J. Waltman, Complex refractive indexes for polymers over the infrared spectral region: specular reflection IR spectra of polymers, J. Polym. Sci. B Polym. Phys. 25 (1987) 901–933.

[25] J. Filik, M.D. Frogley, J.K. Pijanka, K. Wehbe, G. Cinque, Electric field standing wave artefacts in FTIR micro-spectroscopy of biological materials, Analyst 137 (2012) 853–861.

[26] J. Lee, On the non-existence of the so-called "electric field standing wave effect" in transflection FTIR spectra, Vib. Spectrosc. 90 (2017) 104–111.

[27] T.G. Mayerhöfer, J. Popp, The electric field standing wave effect in infrared transflection spectroscopy, Spectrochim. Acta A Mol. Biomol. Spectrosc. 191 (2018) 283–289.

[28] K.C. Park, The extreme values of reflectivity and the conditions for zero reflection from thin dielectric films on metal, Appl. Opt. 3 (1964) 877–881.

[29] K. Wehbe, J. Filik, M.D. Frogley, G. Cinque, The effect of optical substrates on micro-FTIR analysis of single mammalian cells, Anal. Bioanal. Chem. 405 (2013) 1311–1324.

[30] D. Chandler-Horowitz, P.M. Amirtharaj, High-accuracy, midinfrared ($450 cm^{-1} \leq \omega \leq 4000 cm^{-1}$) refractive index values of silicon, J. Appl. Phys. 97 (2005) 123526.

[31] J. Connolly, B. diBenedetto, R. Donadio, Specifications of Raytran Material, SPIE, 1979.

[32] H.H. Li, Refractive index of alkaline earth halides and its wavelength and temperature derivatives, J. Phys. Chem. Ref. Data 9 (1980) 161–290.

[33] O. Stenzel, V. Hopfe, P. Klobes, Determination of optical parameters for amorphous thin film materials on semitransparent substrates from transmittance and reflectance measurements, J. Phys. D. Appl. Phys. 24 (1991) 2088.

[34] B. Harbecke, Coherent and incoherent reflection and transmission of multilayer structures, Appl. Phys. B Lasers Opt. 39 (1986) 165–170.

[35] T.G. Mayerhöfer, S. Pahlow, U. Hübner, J. Popp, Removing interference-based effects from infrared spectra—interference fringes re-revisited, Analyst 145 (2020) 3385–3394.

[36] T.G. Mayerhöfer, S. Pahlow, U. Hübner, J. Popp, CaF2: an ideal substrate material for infrared spectroscopy? Anal. Chem. 92 (2020) 9024–9031.
[37] M.S. Azam, M.D. Ranson, D.K. Hore, Temperature-dependent infrared refractive index of polymers from a calibrated attenuated total reflection infrared measurement, Appl. Spectrosc. 76 (2022) 1254–1262.
[38] J. Fahrenfort, Attenuated total reflection, Spectrochim. Acta 17 (1961) 698–709.
[39] M. Milosevic, Internal Reflection and ATR Spectroscopy, Wiley, 2012.
[40] C.F. Bohren, D.R. Huffman, Absorption and Scattering of Light by Small Particles, Wiley, 1983.
[41] M.I. Mishchenko, L.D. Travis, A.A. Lacis, Scattering, Absorption, and Emission of Light by Small Particles, Cambridge University Press, 2002.
[42] T.G. Mayerhöfer, J. Popp, Beyond Beer's law: spectral mixing rules, Appl. Spectrosc. 74 (2020) 1287–1294.
[43] T.G. Mayerhöfer, Modelling IR-spectra of single-phase polycrystalline materials with random orientation—a unified approach, Vib. Spectrosc. 35 (2004) 67–76.
[44] R.N. Jones, The absorption of radiation by inhomogeneously dispersed systems, J. Am. Chem. Soc. 74 (1952) 2681–2683.
[45] J. Jakeš, M. Slabina, B. Schneider, Influence of the mosaic effect on the intensity of infrared bands, Appl. Spectrosc. 26 (1972) 389–395.
[46] A. Garton, Infrared Spectroscopy of Polymer Blends, Composites and Surfaces, Hanser Publishers, 1992.
[47] I.A. Vinokurov, J. Kankare, Beer's law and the isosbestic points in the absorption spectra of conductive polymers, J. Phys. Chem. B 102 (1998) 1136–1140.
[48] W. Spitzer, D. Kleinman, Infrared lattice bands of quartz, Phys. Rev. 121 (1961) 1324–1335.
[49] M. Czerny, Messungen am Steinsalz im Ultraroten zur Prüfung der Dispersionstheorie, Z. Phys. 65 (1930) 600–631.
[50] R.N. Jones, D. Escolar, J.P. Hawranek, P. Neelakantan, R.P. Young, Some problems in infrared spectrophotometry, J. Mol. Struct. 19 (1973) 21–42.
[51] T.G. Mayerhöfer, S. Pahlow, U. Hübner, J. Popp, Removing interference-based effects from the infrared transflectance spectra of thin films on metallic substrates: a fast and wave optics conform solution, Analyst 143 (2018) 3164–3175.
[52] T.G. Mayerhöfer, Symmetric Euler orientation representations for orientational averaging, Spectrochim. Acta A Mol. Biomol. Spectrosc. 61 (2005) 2611–2621.
[53] T.G. Mayerhöfer, W.D. Costa, J. Popp, Sophisticated attenuated total reflection correction within seconds for unpolarized incident light at 45°, Appl. Spectrosc. 78 (3) (2023) 321–328, https://doi.org/10.1177/000370282312195.
[54] https://wtheiss.com/.
[55] https://reffit.ch/.
[56] L. Genzel, H. Bilz, Handbuch der Physik: Licht und Materie : 1d/Bd.-Hrsg. L. Genzel. Von H. Bilz …. Bd. 25. 2d, Springer, 1984.
[57] M. Born, K. Huang, Dynamical Theory of Crystal Lattices, Clarendon Press, 1954.
[58] W. Kaiser, W. Spitzer, R. Kaiser, L. Howarth, Infrared properties of CaF2, SrF2, and BaF2, Phys. Rev. 127 (1962) 1950–1954.
[59] T.G. Mayerhöfer, R. Knipper, U. Hübner, D. Cialla-May, K. Weber, H.-G. Meyer, J. Popp, Ultra sensing by combining extraordinary optical transmission with perfect absorption, ACS Photonics 2 (2015) 1567–1575.
[60] P.B. Johnson, R.W. Christy, Optical constants of the noble metals, Phys. Rev. B 6 (1972) 4370–4379.
[61] P.G. Etchegoin, E.C. Le Ru, M. Meyer, An analytic model for the optical properties of gold, J. Chem. Phys. 125 (2006) 164705.
[62] P.G. Etchegoin, E.C. Le Ru, M. Meyer, Erratum: "An analytic model for the optical properties of gold" (J. Chem. Phys. 125, 164705 (2006)), J. Chem. Phys. 127 (2007) 189901.
[63] J.A. Nelder, R. Mead, A simplex-method for function minimization, Comput. J. 7 (1965) 308–313.
[64] T.G. Mayerhöfer, J. Popp, Quantitative evaluation of infrared absorbance spectra—Lorentz profile versus Lorentz oscillator, ChemPhysChem 20 (2019) 31–36.
[65] M. Gilliot, Inversion of ellipsometry data using constrained spline analysis, Appl. Opt. 56 (2017) 1173–1182.
[66] R. Nitsche, T. Fritz, Determination of model-free Kramers-Kronig consistent optical constants of thin absorbing films from just one spectral measurement: application to organic semiconductors, Phys. Rev. B 70 (2004) 195432.
[67] R. Forker, M. Gruenewald, T. Fritz, Optical differential reflectance spectroscopy on thin molecular films, Ann. Rep. C (Phys. Chem.) 108 (2012) 34–68.
[68] N.J. Harrick, Internal Reflection Spectroscopy, Interscience Publishers, New York, 1967.
[69] T. Hasegawa, Quantitative Infrared Spectroscopy for Understanding of a Condensed Matter, Springer Japan, 2017.

Chapter 7

Additional insights gained by wave optics and dispersion theory

7.1 Infrared refraction spectroscopy

This section is all about dispersion theory. If you worked through this book, the following may seem completely trivial and it actually is. Therefore, infrared refraction spectroscopy [1] may develop itself to a versatile tool for the molecular spectroscopists and, maybe, even beyond this community. Apart from that, it can be considered as a valuable training exercise and an eye-opener. You might not have heard of the term infrared refraction spectroscopy, because it was me who coined it in 2021. The idea behind is that you quite often hear about absorption spectroscopy, even though nobody can really measure absorbance. What you actually measure is either transmittance or reflectance, or, if your method of detection is not light-based or indirect, absorptance. Nevertheless, everyone understands what is meant by absorption spectroscopy.

Refraction spectroscopy is a term generated based on a similar idea. The quantity you actually measure is reflectance, but if you use a high refractive index medium you have to stay well below the critical angle and you have to avoid using a metallic substrate (as in case of absorption spectroscopy, there are certain limitations, one of which is that the sample needs to be optically thick, i.e., semiinfinite). This means that for molecular substances, you end up with seemingly very poor spectra that look refractive index-like. While I heard this term a couple of times, I never gave it much thought until I wrote Section 5.8.2. In fact, when I saw that the logarithm of the reflectance results in a function that, apart from a constant and a factor, resembled the refractive index, I thought that this must be key to reveal the refractive index function, much like you can unveil the absorption index function by ATR spectroscopy. After some meddling with the formulas, I came to a much simpler solution. In absorbance spectroscopy, what you do is to ignore the refractive index—in refraction spectroscopy you are reciprocal and ignore the absorption index. How does this work? Let us restrict ourselves to normal incidence. Then, the reflectance is given by (cf. Chapter 2),

$$R = \left(\frac{1-\widehat{m}}{1+\widehat{m}}\right)^2, \tag{7.1}$$

where $\widehat{m}$ is the relative index of refraction given by,

$$\widehat{m} = \frac{\widehat{n}_{Sample}}{n_{inc}}. \tag{7.2}$$

As I already stated previously, let us ignore the absorption index of the sample in $\widehat{n}_{Sample}$ and set $\widehat{n}_{Sample} = n_{Sample}$. Since we know the refractive index of the incidence medium n_{inc}, we can then invert Eq. (7.1) and, thereby, obtain m:

$$m = \frac{-1 \pm 2\sqrt{R} - R}{R - 1}. \tag{7.3}$$

To obtain the correct solution, we employ the positive sign if $n_{inc} > n_{Sample}$ and, otherwise, the negative sign. If we can really ignore the absorption index, we should regain the refractive index of the sample. Before we check this, let us first discuss the usability of Eq. (7.1). As long as we use naturally polarized light and have a comparably small refractive index, Eq. (7.1) is actually valid far beyond normal incidence, easily up to 25° with errors less than about 1 %. This is obvious from Fig. 4.1 and means that one can even use a common infrared microscope objective with NA = 0.4 (This restricts the angle of incidence distribution to, e.g., between 9.8° and 23.6° for a Bruker Hyperion 2000.) In general, such reflectance measurements are pretty straightforward and can be carried out with standard equipment—no fancy spectroscopic ellipsometry [2], quantum cascade laser-based Mach-Zehnder interferometer [3,4], or complex reflectance microspectroscopy equipment [5] is needed. This is pretty much comparable concerning simplicity to absorbance spectroscopy, even though different incidence media are considered (but those would certainly require some new accessories).

You certainly might interpose, what is the refractive index function good for? Could you, e.g., also determine concentration? Here is what you can find about this in the Handbook of Infrared Spectroscopy, in the chapter about quantitative analysis in Volume 3, Classical Methods of quantitative analysis by John Coates on page 5: "Note that the spectral response of specular reflection is often dominated by complex refractive index terms, and is not readily amenable to quantitative measurements" [6]. On the other hand, in the same book, I could not find any explanation why a first derivative of an absorbance spectrum looks refractive index-like—the similarity is not even acknowledged. In the same way, the first derivative of a refractive index spectrum with respect to wavelength looks similar to the second derivative of an absorption index spectrum or an absorbance spectrum, probably because refractive index and absorption index are linked via the Kramers-Kronig relations [1,7].

If you remember, it is the oscillator strength S^2 that is proportional to the number of oscillators of a certain kind in a unit volume. Also, there is a twin of Beer's approximation, stating that changes of the refractive index around absorptions are approximately linearly depending on concentration (cf. Section 5.3). To see how good refraction spectroscopy works, we assume a single oscillator with variable oscillator strength, a damping constant of 30 cm^{-1} and an oscillator position at 1500 cm^{-1}. In addition, we assume that the refractive index at the high wavenumber end of the spectrum n_∞ equals 1.5. The results are depicted in Fig. 7.1.

Accordingly, for typical oscillator strengths in organic and biological matter, Eq. (7.3) provides results for the refractive index functions, which can barely be differentiated from the true refractive index functions. If the incidence medium is ZnSe and Ge, the curves agree within line thickness. The same is true for the differential refractive index. More importantly, the minimum values stay virtually the same, even if the curve is slightly shifted as for vacuum as incidence medium. This is important, because the minimum value is, just as the peak absorbance for absorbance spectroscopy, an approximative measure of concentration. This becomes obvious for the two larger oscillator strengths the refractive index curves of which are also depicted in Fig. 7.1: When the oscillator strengths doubles, so does approximately also the minimum value of the differential refractive index (note that for spectroscopist used to absorbance, we inverted the lower part of Fig. 7.1 in the corresponding paper to make it more appealing).

In fact, using the refractive index function, it would not even be necessary to use the differential refractive index. Much better would be to use the difference of the refractive index after and before the absorption $n(\tilde{\nu}) = n(\tilde{\nu}) - n_\infty$ (before here means at wavenumbers higher than the oscillator position). This difference is (nearly) equivalent to the integral absorbance

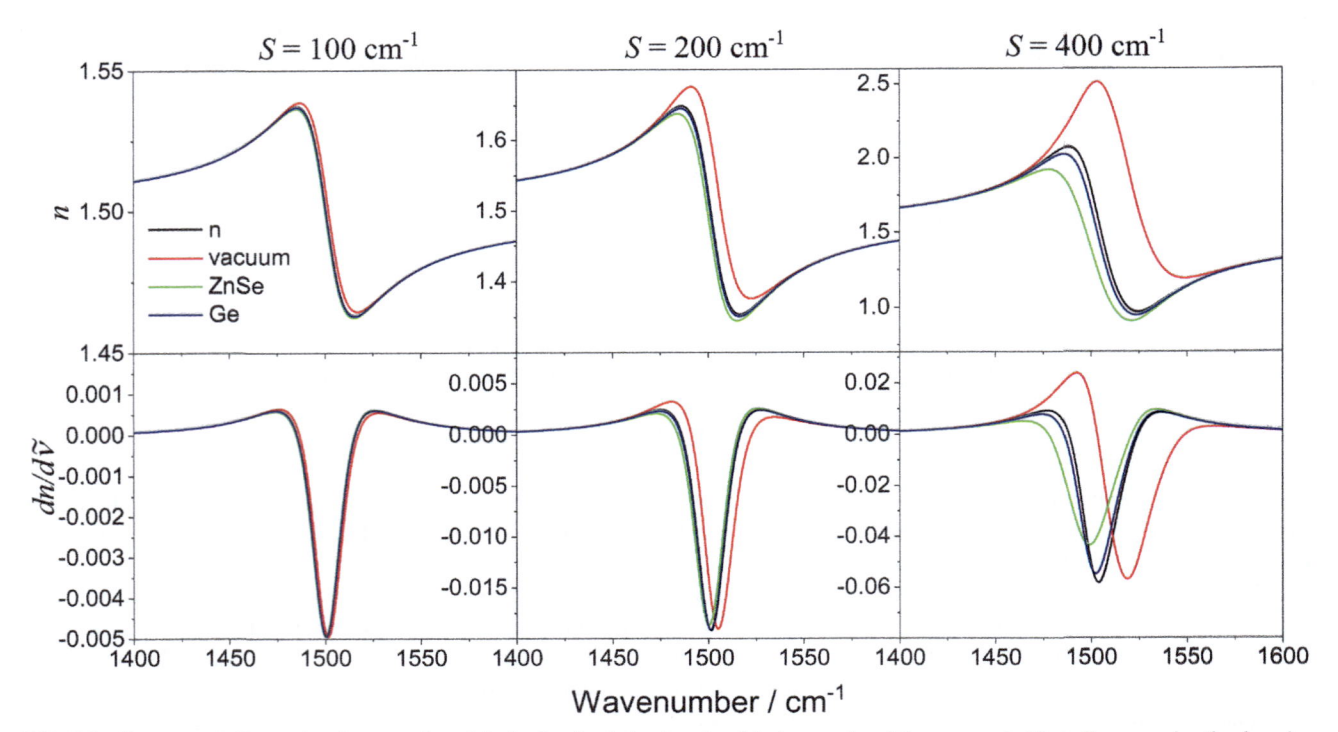

FIG. 7.1 Upper panel: Comparison between the original refractive index function (black curves) and those computed from the approximation based on Eq. (7.3) for different incidence media (vacuum, ZnSe, and Ge) and oscillator strengths. Lower panel: The equivalent curves for the derivatives with respect to wavenumber [1].

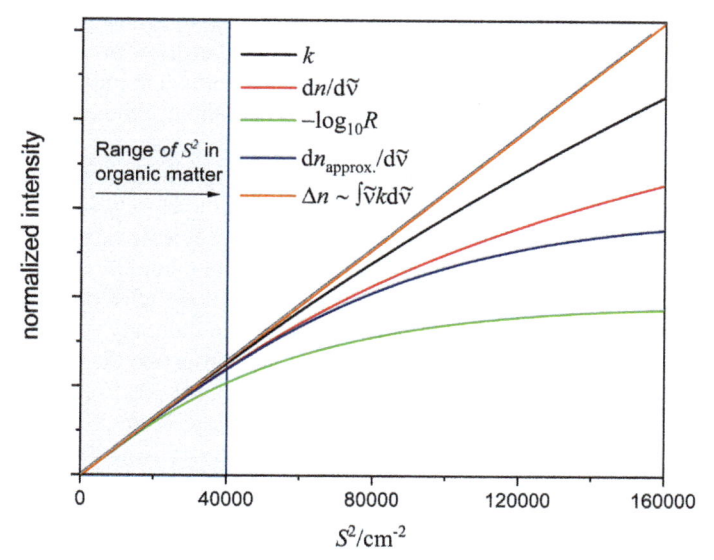

FIG. 7.2 Comparison of the adherence to Beer's approximation for the peak value of the absorption index (black curve), the differential refractive index (red curve), the differential refractive index regained from the reflectance via Eq. (7.3), the absorbance from ATR using a ZnSe crystal and 45° angle of incidence (green curve), and the integrated absorbance and Δn (orange curve) [1].

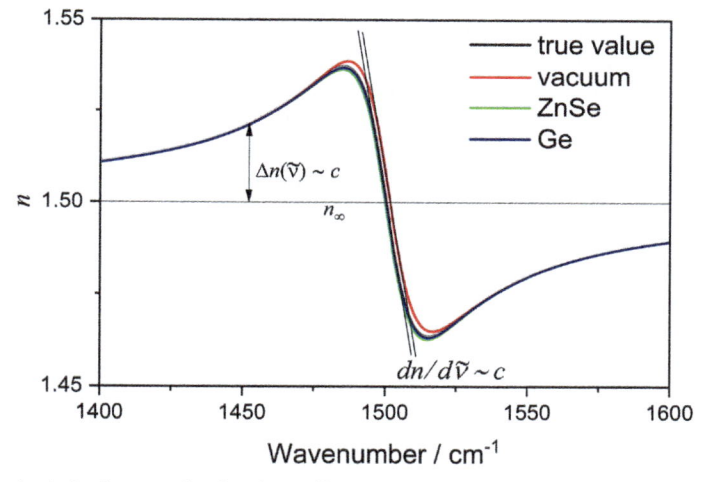

FIG. 7.3 The change of the refractive index is proportional to the oscillator strength and, therefore, to the concentration of oscillators. The same holds for the slope of the tangent at the refractive indices around the oscillator position, and this slope is independent of the incidence medium if the oscillator stays in the weak limit.

if the difference is taken for a refractive index at a wavenumber where dispersion is already low again. This integral absorption is strictly linear to the concentration as long as local field effects are neglected [8], which can be seen in Fig. 7.2 (an illustration of $n(\tilde{\nu})$ is provided in Fig. 7.3). It is also easy to prove this by assuming that we have a single oscillator (which may represent a number of oscillators in a transparency region below these oscillators as shown in Section 5.8.4):

$$n_0 = \lim_{\tilde{\nu}\to 0}\sqrt{1 + c\frac{S^{*2}}{\tilde{\nu}_0^2 - \tilde{\nu}^2 - i\tilde{\nu}\gamma}} = \sqrt{1 + c\frac{S^{*2}}{\tilde{\nu}_0^2}} \approx 1 + c\frac{S^{*2}}{2\tilde{\nu}_0^2} \to n_0 - n_\infty = \Delta n = c\frac{S^{*2}}{2\tilde{\nu}_0^2}. \tag{7.4}$$

For those of you who still cannot believe that n is proportional to the molar concentration, let me (seemingly) digress somewhat from the topic. Before the advent of spectroscopy, it was refractometry that was used to investigate structure. One example is the investigation of the structure of six-membered rings with double bonds as carried out by Brühl in 1887 (Brühl proved Kekulé's structure for Benzene!) [9]. In fact, refractometry in case of organic molecules was based on the concept of molar refraction, which led to deriving increments that could be assigned even to individual atoms in a molecule, although this does not seem to make any sense. Actually, if you know the density of an organic compound and its chemical formula, you can calculate its refractive index based on these increments with quite impressive accuracy. This remains true even for mixtures (so the influence of diverse functional groups and potential interactions between them cannot be that high!) [10].

How is this possible? Since refractometry was no longer particularly interesting after the triumph of spectroscopy, it looks like nobody took care to solve this conundrum. In fact, the solution is closely linked to Δn and, by that, to Beer's approximation [11]. Actually, Brühl's dissertation adviser was Landolt and Beer had suggested to Landolt much earlier to investigate the refractive indices of liquid organic compounds that constitute homologous series [12]. As a consequence, Landolt came up with the molar refraction in 1862. However, he and nobody after him thought this through to the end (August Beer might have been able to discover the connection, but he died already 1 year later).

If you consider Eq. (7.4), then the chromophore in homologous series must be the C—H bonding, which has a strong absorption due to $\sigma \to \sigma^*$ transitions in the vacuum UV around 150 nm [13,14]. According to Beer's approximation and its twin law for Δn, n_D^{20} should be approximately proportional to the concentration of C—H bonds in homologous series. In fact, as can be seen in Fig. 7.4, this works quite well, including for the carboxylic acids, which actually show a seemingly very odd decrease of n_D^{20} for increasing density.

How good does infrared refraction spectroscopy work in practice? Actually, very good! To prove this, we used a comparably old spectrometer, which had not seen a service for quite some time. The result is depicted in Fig. 7.5. I guess you agree with me that the correspondence is quite satisfying. With the advent of the new extremely bright light sources like quantum cascade lasers and supercontinuum "lasers," I guess any noise problems as displayed in Fig. 7.5 will be history anyway.

Maybe you think that introducing another method, which seems to be not superior to absorption spectroscopy, is actually senseless. In fact, a concept like refraction spectroscopy hopefully helps you to understand the main principles behind dispersion theory and its connection to spectroscopy. In addition, there actually might also be interesting applications on the horizon, since I think it is unlikely that we can perform transmission or attenuated total reflection measurements with a smartphone in the future. An alternative may be reflection measurements at near-normal incidence using a highly refracting incidence medium. Until dispersion analysis can be carried out by neuronal networks, it might also come in handy to be able to quantitatively evaluate reflection spectra by an approximate method. If you nevertheless dislike the idea of infrared refraction spectroscopy, you can alternatively apply the improved version of Kramers-Kronig analyses of reflectance spectra as introduced in Section 5.8.2.

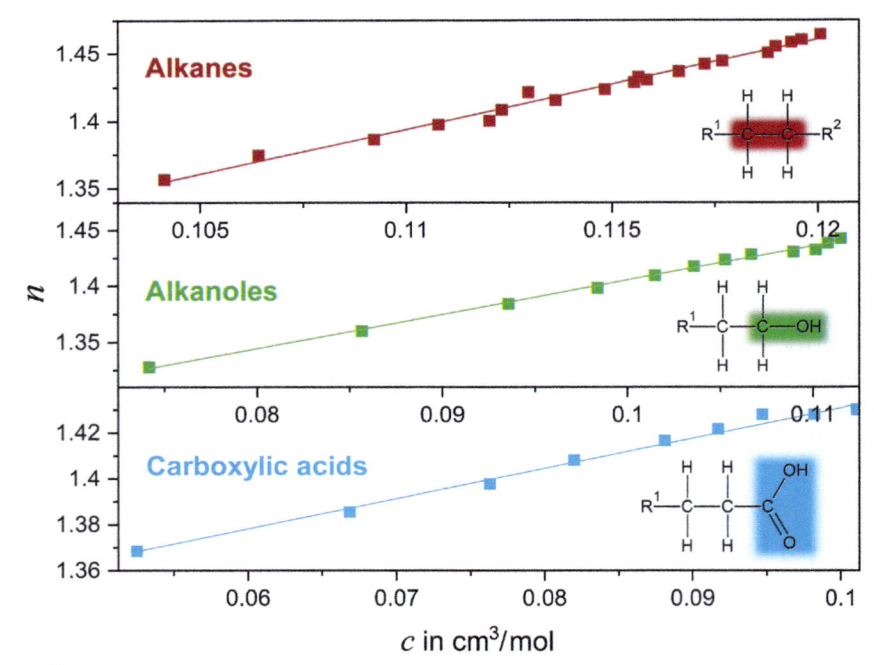

FIG. 7.4 Refractive index n_D^{20} as function of the concentration of C—H bonds for liquid n-alkanes, n-alkanoles (with terminal —OH), and carboxylic acids (with terminal —COOH) [11].

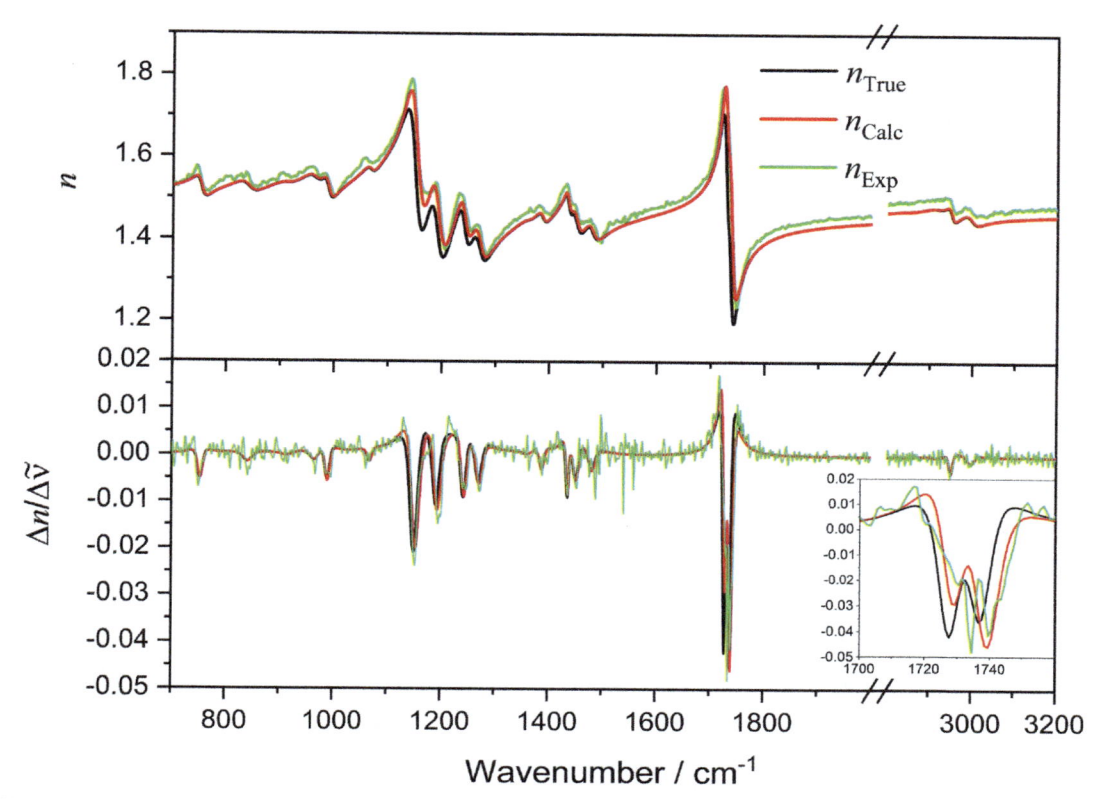

FIG. 7.5 Upper panel: The index of refraction of PMMA and the regained index of refraction from literature values compared to experimental values. Lower panel: The same for the derivates with respect to wavenumber [1].

7.2 Surface-enhanced infrared absorption (SEIRA)

Since I restrict myself in this book to methods, the resulting spectra of which can be evaluated analytically, I will leave out the predominant part of SEIRA that deals either with metal nanoparticles or with structured arrays (for more information about the latter, have a look at [15]), albeit it may also be possible to employ a proper oscillator model [16]. Quite often you read as motivation for the need for plasmonic structures that infrared spectroscopy is not sensitive enough to detect monolayers. In the following, I will show you that this is a fabrication. Even if the statement were correct, there are much simpler methods to enhance IR signals, which we will discuss in the following.

Before we start, I like to express my thanks to Dmytro Solonenko who made me aware of the possibility to use interference effects for enhanced spectroscopy during a get-together evening at the ICAVS in Vienna in 2015. Dmytro was a PhD student at that time and worked on interference-enhanced Raman spectroscopy. He used the transfer matrix method (Section 4.5) to calculate the electric field intensity enhancement of transparent layers on highly reflective substrates (in his case SiO_2 on Si) and compared it with experimental values [17]. Dmytro's work did not only inspire me to work on what I present to you in the following, but led ultimately somehow also to the views expressed in Sections 6.2–6.4.

In SEIRA, two materials are used mainly to prove that a given structure has its merits for signal enhancements. The first one is PMMA (poly(methyl methacrylate)) because every cleanroom facility has employees who are very experienced in fabricating layers of defined thicknesses. The second one is octadecanethiol (ODT), which binds well on gold surfaces and establishes monomolecular layers. For PMMA, we know the optical constants well, in contrast to ODT. Luckily, the optical constants of polyethylene are well-known and must be very similar to the one of ODT due to large spectral similarities. Therefore, we can perform the following calculations for a 2-nm-thick layer of PE instead and the results will be fairly similar to experimental spectra obtained for ODT.

Fig. 7.6 shows simulations of conventional IR spectra of a 2-nm layer of PE and a 30-nm layer of PMMA (for the latter, the chosen thickness is a kind of lower limit of what can be fabricated in a cleanroom facility). Looking at the spectra, my conclusion is that conventional infrared spectroscopic techniques should suffice to produce good spectra for the PMMA

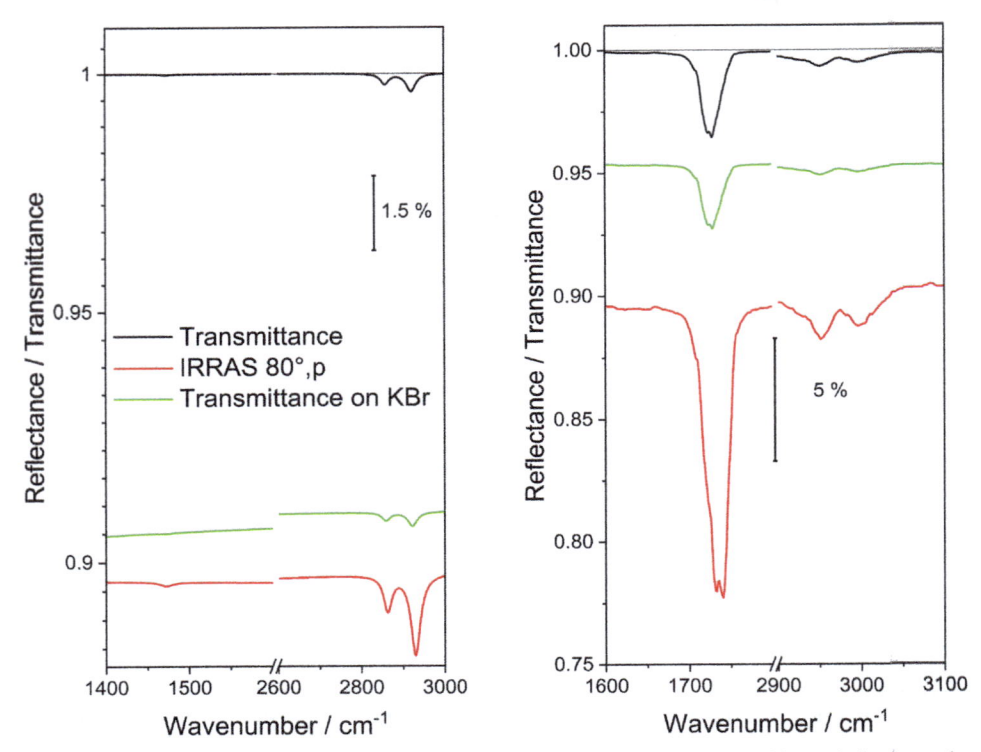

FIG. 7.6 Calculated transmittance of a free-standing film, a film on KBr, and reflectance of a film on gold (p-polarization and an angle of incidence of 80°). Left panel: 2 nm PE. Right panel: 30 nm PMMA [15].

film, even in transmission and if a transparent substrate is used. For the 2-nm PE film, the latter may be questionable, but if the film is placed on a gold substrate, with p-polarized light and a large angle of incidence (grazing incidence), the signal should be strong enough to be recorded without problems. Therefore, if you have fabricated a SEIRA substrate and you test it with the equivalent of a monolayer of ODT or a 30-nm layer of PMMA and you do not get an enhancement of the signal of an order of magnitude, I would classify your substrate as disappointing, all the more, because with much less effort a much higher signal can be achieved as you will see in the following.

The simplest possible structure to obtain signal enhancement is actually a transparent layer on a highly reflecting substrate. This technique has been recently rediscovered, but it is already known since 1988 and its use was then termed buried metal layer infrared reflection absorption spectroscopy (BML-IRRAS) [18]. Like for Raman spectroscopy, it consists of a transparent layer, e.g., Al_2O_3, CaF_2, or SiO_2 on a highly reflective substrate, like Al or gold. In contrast to the case where a monolayer is deposited directly on the metallic substrate, normal incidence is possible as is s-polarization for high angles of incidence (remember that a metal suppresses electric field components parallel to its surface). Instead of BML-IRRAS, it would therefore better in my opinion to call this method interference-enhanced IR spectroscopy (IEIRS) in analogy to interference-enhanced Raman spectroscopy.

Fig. 7.7 shows a field map of a 600-nm-thick film of SiO_2 on gold in part (A). Since the electric field can only penetrate a couple of 10 nm into the gold, the electric field intensity is practically zero in it. The SiO_2 layer is indicated by the two black lines. The electric field intensity on the surface of this layer is shown in part (B) together with the absorbance. In this case, the absorbance values are so small that to a good approximation absorbance ≈ absorptance, and since absorptance scales with the electric field intensity, absorbance in this particular case scales in the same way. Part (B) shows that the signal amplification is comparably broadband, but to cover the whole spectral range at least two different substrates would be necessary, one for the fingerprint region and the other one to cover the range of the hydrogen stretching vibrations. A very innovative solution to this problem is to sandwich a phase change material like VO_2 between two other layers of the same material, which is transparent, since VO_2 can be forced to change from the transparent phase into a highly reflective one [19]. Thus, two different thicknesses of the transparent layer result, which allows to cover both spectral regions.

The signal enhancement is about the same as that which is possible at grazing incidence with gold as substrate and p-polarized incident radiation. It is also quite instructive to see how the Si—O stretching vibrations change the refractive

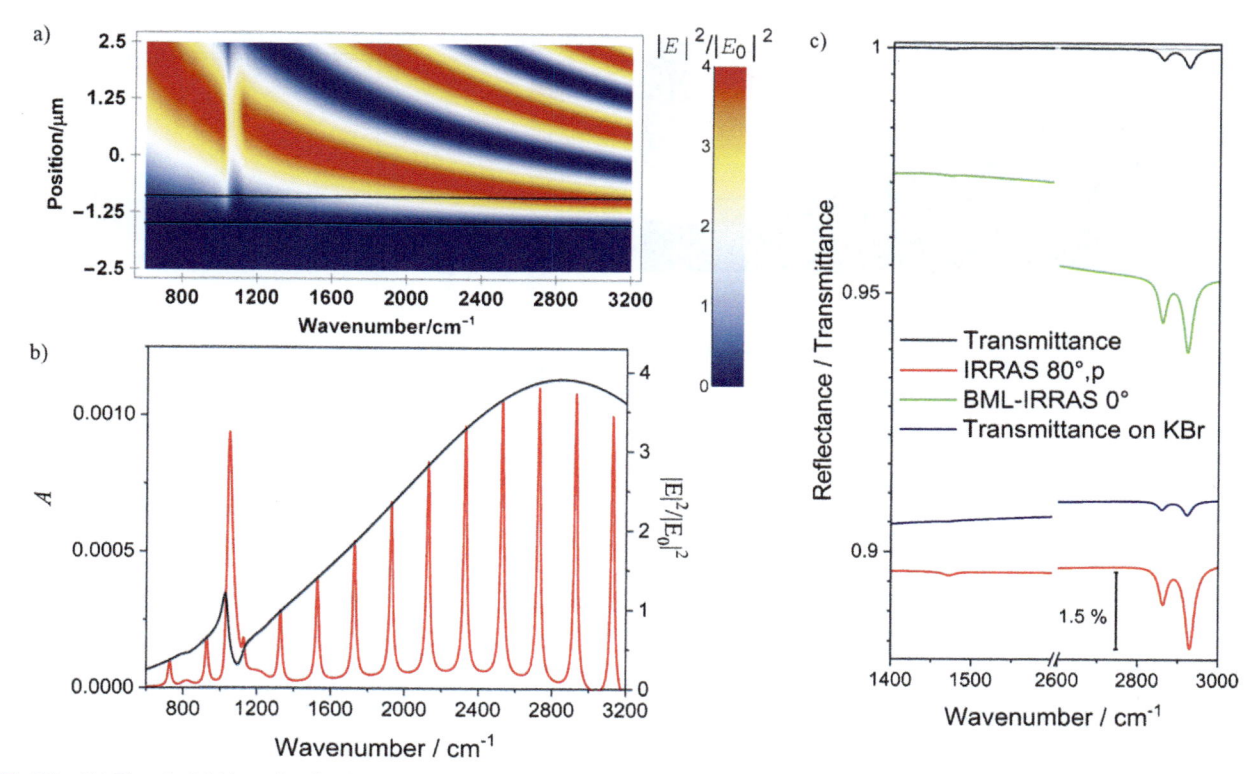

FIG. 7.7 (A) Electric field intensity distribution for a 600-nm-thick layer of amorphous SiO_2 on gold for normal incidence. (B) Electric field intensity on the surface of the SiO_2 layer in dependence of the wavenumber (in black) and absorbance difference spectrum of a 2-nm-thick layer of a fictive material with equally weak absorptions every 200 cm^{-1}. (C) Comparison of the simulated reflectance signal of a 2-nm layer of polyethylene on this substrate with conventional IR methods (cf. Fig. 7.6).

index so that it is higher at lower wavenumbers, which leads to an earlier onset of the electric field intensity curve. The enhancement can be further increased if additionally a larger angle of incidence is chosen as can be seen in Fig. 7.8.

Due to the high angle of incidence, the enhancement becomes much larger (though not the enhancement factor, which is in case of interference enhancement limited by 4 as maximum), but, as a disadvantage, its bandwidth is also much narrower. In addition, interference leads to a strong absorbance of the incident radiation by the gold leading to a large background (not visible in Fig. 7.8, cf. Ref. [20]). To remove this background is somewhat tricky, since even though the layer is just 2 nm thick, its elimination shifts the background band by about 5 cm^{-1}. In calculations, it is an old trick for obtaining the background spectrum to leave the layer in place, but to remove the real part of its dielectric function, since it is the real part that causes the shift. Experimentally, this is certainly not an option, so if you do not want to analyze your spectrum by dispersion analysis (remember, everything that can be forward calculated analytically, can certainly be analyzed by dispersion analysis, a very big advantage of employing interference enhancement!), a background-free method would come in handy. Such a method indeed exists, but it has not been used for a long time, even though this should be unproblematic.

What I am talking about is what I call interference-enhanced ATR spectroscopy [21,22]. As the name implies, one chooses a high refractive index material like Germanium as incidence medium and furnishes this crystal with a layer of a nonabsorbing material with a comparably low refractive index. If this layer would not be present, the critical angle would be about 14.5° for air or vacuum. If we are slightly above this angle and the layer is in place, what happens is that the light is still transmitted into the layer, but at the interface layer/air total reflection occurs. Accordingly, the layer becomes a cavity with one perfectly reflecting interface, but unlike in case of a highly reflecting metal there is no background, since behind the interface is nothing that could absorb light and generate this background. The result is depicted in Fig. 7.9.

Although the electric field intensity is much higher, the performance is not as good as the one of the previous substrates. However, this performance can still be improved by choosing a material for the layer with a lower refractive index-like, e.g., NaF. Close to the second critical angle, which is for above system at about 22° (for larger angles, light is no longer transmitted into the layer), an even stronger enhancement can be reached; since then, also the second interface is strongly reflecting. However, this means that less light reaches the layer/air interface and the mode is mostly concentrated within

204 PART | I Scalar theory

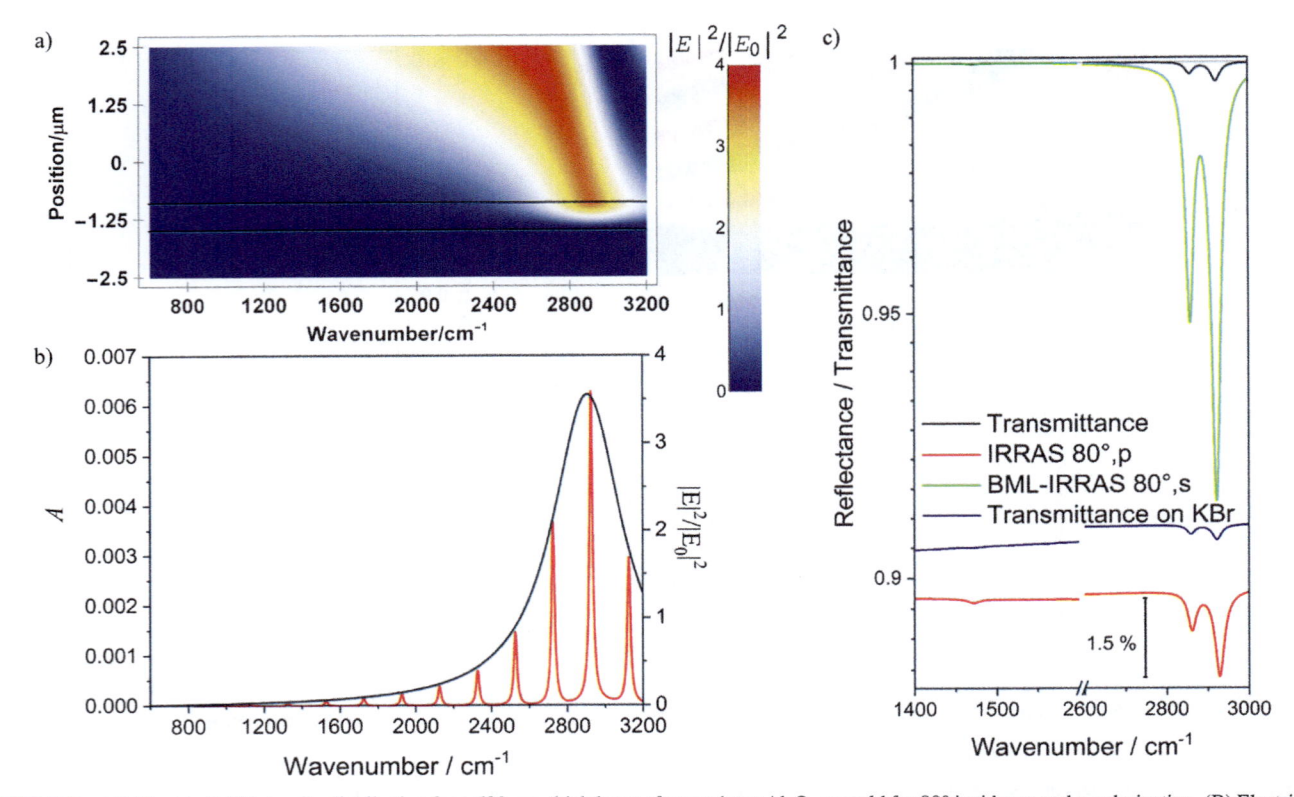

FIG. 7.8 (A) Electric field intensity distribution for a 600-nm-thick layer of amorphous Al_2O_3 on gold for 80° incidence and s-polarization. (B) Electric field intensity on the surface of the Al_2O_3 layer in dependence of the wavenumber (in black) and absorbance difference spectrum of a 2-nm-thick layer of a fictive material with equally weak absorptions every 200 cm^{-1}. (C) Comparison of the reflectance signal of a 2-nm layer of polyethylene on this substrate with conventional IR methods (cf. Fig. 7.6).

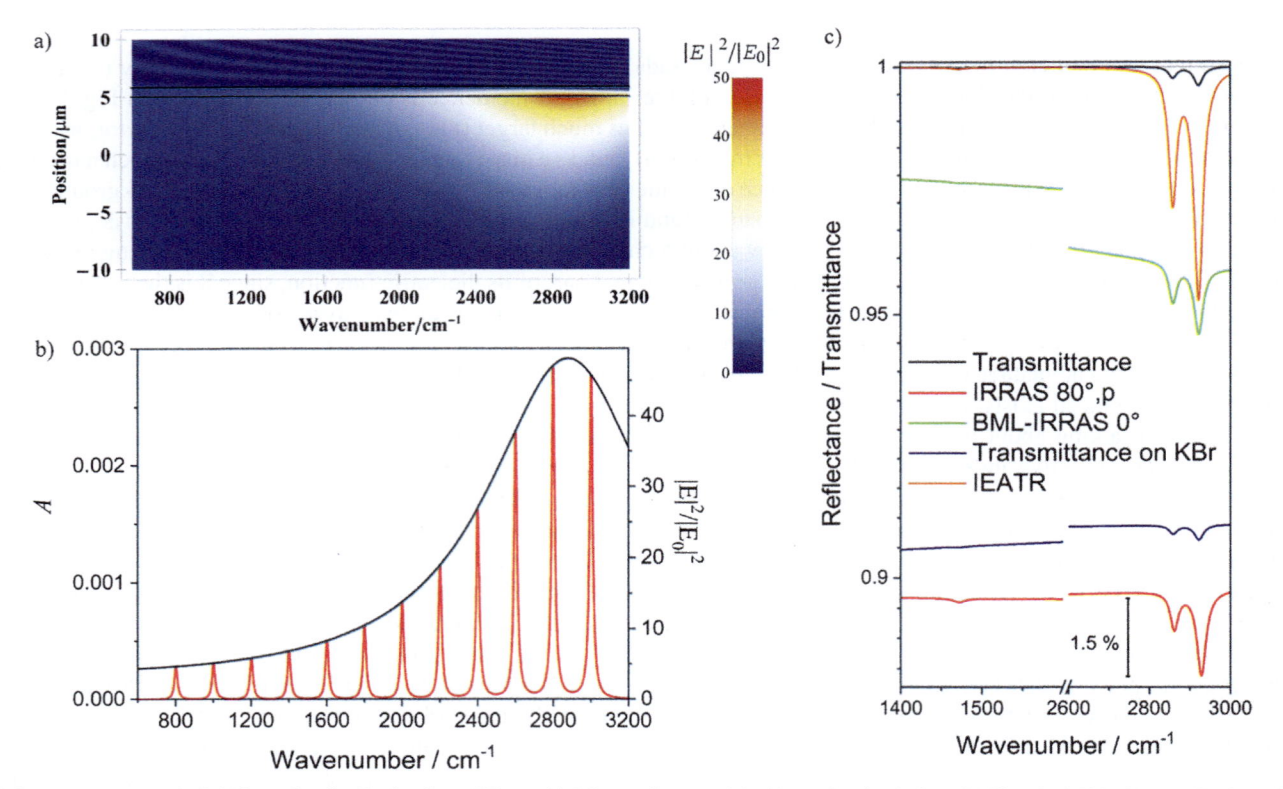

FIG. 7.9 (A) Electric field intensity distribution for a 600-nm-thick layer of a material with a refractive index of 1.5 and 14.5° incidence. (B) Electric field intensity on the surface of the layer in dependence of the wavenumber (in black) and absorbance spectrum of a 2-nm-thick layer of a fictive material with equally weak absorptions every 200 cm^{-1}. (C) Comparison of the reflectance signal of a 2-nm layer of polyethylene on this substrate with conventional IR methods (cf. Fig. 7.6) [21].

the layer. The enhancement can even get so large that a very strong band can occur in a virtually nonabsorbing layer material as long as $k > 0$ [21].

Note that above configuration is in principle the same as the Kretschmann configuration. If you are not familiar with the Kretschmann and the Otto configurations, these are two configurations, which allow to launch surface plasmons on unstructured metal surfaces. I have already introduced plasmons shortly in Section 5.6 in the context of inverse dielectric functions. The special thing about a plasmon is that it is a longitudinal vibration of the electrons on the surface, meaning that the direction of the movement and the polarization direction within the medium coincide. The excitation of such plasmons requires p-polarized radiation and large angles of incidences. In addition, the wavevector of the incident light must also be matched with that of the plasmon—this is not possible for light incident from air or vacuum. This match can, however, be achieved if the light is incident from a high refractive index medium, that is in case of internal reflection.

In case of the Kretschmann configuration, the layer stack is ATR crystal/gold/air, which means that in comparison with above configuration, the nonabsorbing low index material is replaced by a highly absorbing material with an also comparably high index of refraction that even exceeds that of Ge in parts of the mid-infrared spectral region. The light has to pass through the gold to generate an evanescent wave at the gold/air interface. Therefore, the layer thickness has to be comparably small, e.g., 20 nm; otherwise, not much light would be left to interact with a sample on the gold surface. On the other hand, the fact that the configuration is the same suggests that the plasmon is created close to an incidence angle of 14.5°, which is the critical angle for Ge/Air. Indeed, this is the case, but the problem is that the angle must be adjusted extremely precisely as can be seen in Fig. 7.10.

In contrast to interference-enhanced ATR spectroscopy, where the enhancement stays about the same for ±0.5° around the critical angle for a Ge/Air interface, for the Kretschmann configuration large changes already take place if the angle is varied by only ±0.01°—given the usual spread of the incidence angles this is practically impossible, whereas ±0.5° is achievable for current accessories [25]. Most probably, this is the reason why the Kretschmann configuration is practically not used in infrared spectroscopy, together with the fact that the shape of the bands varies through the spectral range in

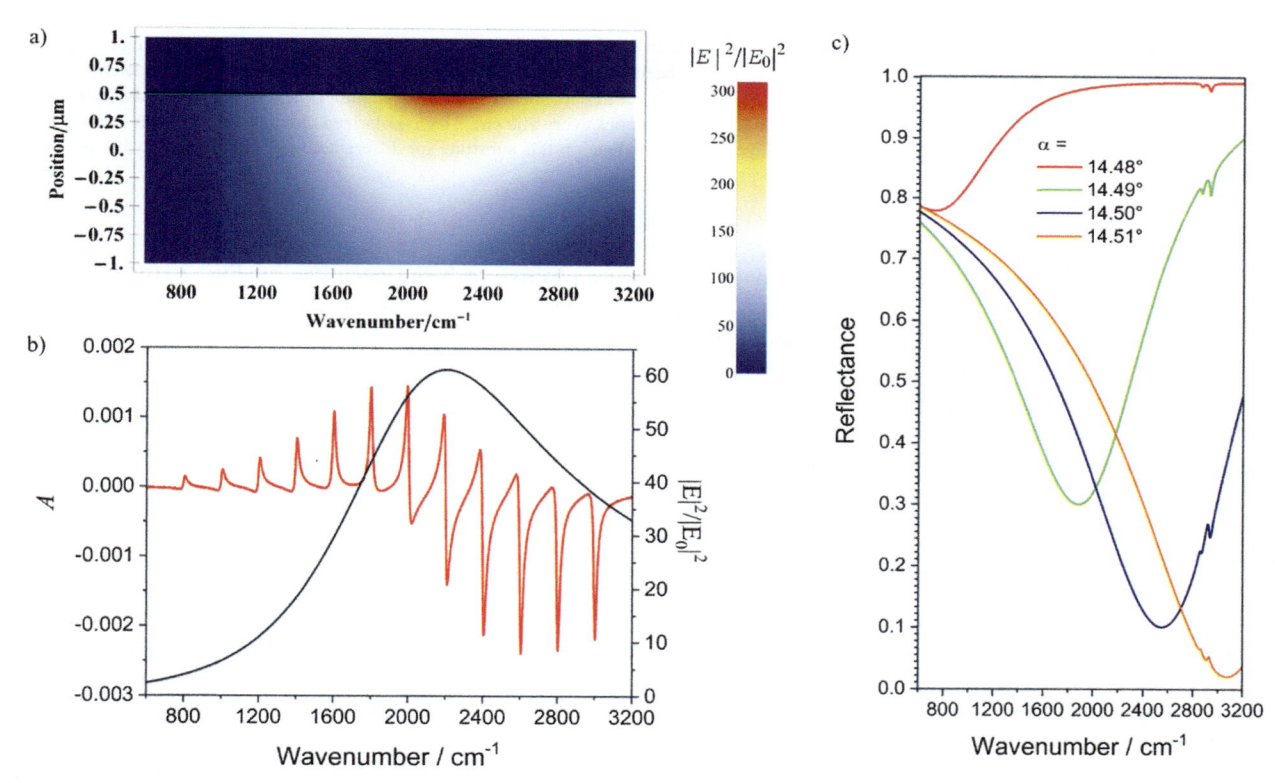

FIG. 7.10 Kretschmann configuration: (A) Electric field intensity distribution for a 20-nm-thick layer of gold on Ge (angle of incidence 14.49°, p-polarized incident radiation; the optical constants of gold have been extrapolated in the infrared spectral range using the parameterization from Etchegoin [23,24] since directly in this range determined optical constants seem less reliable). (B) Electric field intensity on the surface of the gold layer in dependence of the wavenumber (in black) and absorbance difference spectrum of a 2-nm-thick layer of a fictive material with equally weak absorptions every 200 cm^{-1} for p-polarized radiation and an angle of incidence of 14.49°. (C) Comparison of the reflectance signal of a 2-nm layer of polyethylene on this substrate for different angles of incidence using p-polarized radiation [21].

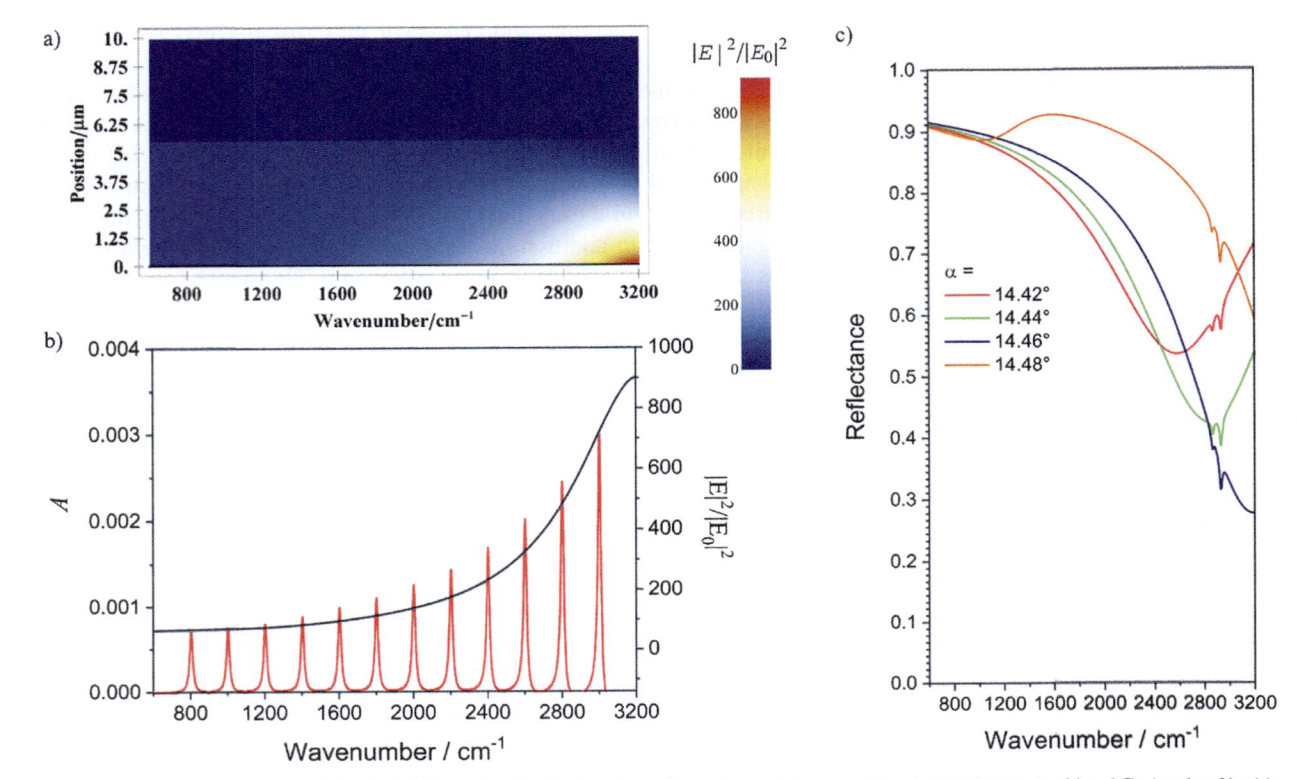

FIG. 7.11 Otto configuration: (A) Electric field intensity distribution for an 8-μm air gap between a 50-nm-thick layer of gold and Ge (angle of incidence 14.48°, p-polarized incident radiation; the optical constants of gold have been extrapolated in the infrared spectral range using the parameterization from Etchegoin [23,24]). (B) Electric field intensity on the surface of the gold layer in dependence of the wavenumber (in black) and absorbance difference spectrum of a 2-nm-thick layer of a fictive material with equally weak absorptions every 200 cm^{-1} for p-polarized radiation and an angle of incidence of 14.48°. (C) Comparison of the reflectance signal of a 2-nm layer of polyethylene on this substrate for different angles of incidence using p-polarized radiation [21].

which there is an enhancement and that they change from being positive to being negative, cf. Fig. 7.10B). While the enhancement factor of the electric field intensity is comparably large, the signal enhancement is clearly inferior to that possible in interference-enhanced ATR spectroscopy, because a large part of the light is lost by absorption in the gold layer.

This problem is avoided in the Otto configuration where the layer stack is ATR crystal/air/gold (the evanescent wave is also generated at the air/gold interface). The problem with the Otto configuration is that the air gap has to be properly tuned and that it shares with the Kretschmann configuration the same need for a very precise adjustment of the incidence angle and the absence of any spread. The performance is, however, better than that of the Kretschmann configuration and roughly comparable with that of interference-enhanced ATR spectroscopy (Fig. 7.11).

Overall, if your intention is just to enhance the signal of a monolayer, my opinion is that it is not worth the effort to use the Kretschmann or the Otto configuration.

As a last example in this section, I want to show something that could also be seen as the Otto configuration, although there is no air gap, which means that the monolayer is placed between Ge and gold. In addition, the angle of incidence is very high, say 80°, and p-polarized incident light is employed, like in the other example in this section where we had grazing incidence. Since the effect is continuously increasing with the angle of incidence, a plasmon resonance is not at play. Instead, it again must be an interference effect, the same which occurs if we use very high angles of incidence in transmission. As can be seen in Fig. 7.12, the effect is immense and it is comparable with the best available structured array plasmonic substrate where the same amount of material is incorporated into nanoslits [26]. Although I believed for some time I was the first to discover this sweet spot, its existence had also been realized earlier [27].

Overall, my conclusion is that for monolayers, there is no need to design and prepare special structured array substrates, because with less effort similar signal strengths are achievable. Again, I want to emphasize one big advantage for all substrates/methods discussed in this section, namely, that dispersion analysis can be applied to quantitatively evaluate the spectra. For periodic structured array substrates with structures small compared to the wavelength such an evaluation

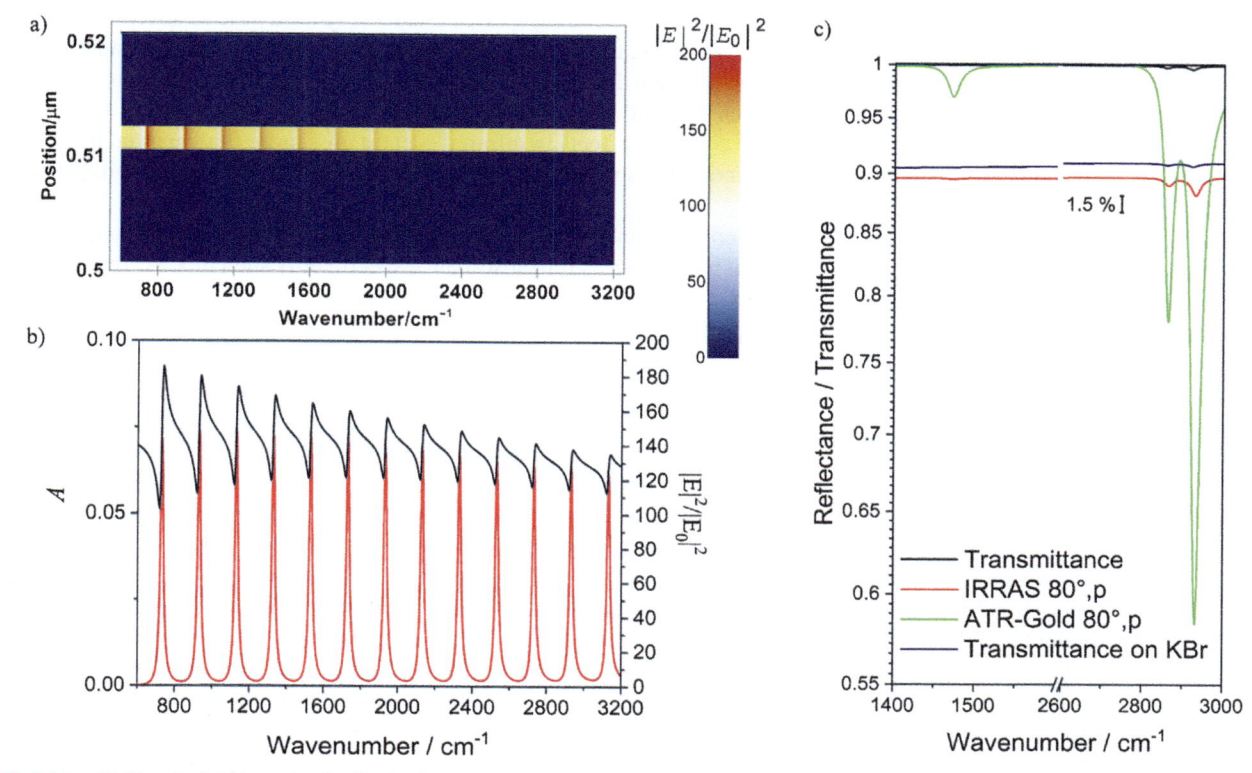

FIG. 7.12 (A) Electric field intensity distribution for a 2-nm-thick layer of a material with a refractive index of 1.5 between Ge as incidence medium and gold for light with 80° angle of incidence and p-polarization. (B) Electric field intensity on the surface of the layer in dependence of the wavenumber (in black) and absorbance spectrum of a 2-nm-thick layer of a fictive material with equally weak absorptions every 200 cm^{-1}. (C) Comparison of the reflectance signal of a 2-nm layer of polyethylene in this configuration with conventional IR methods.

may also be possible, since rigorous coupled-wave analysis (RCWA) can be seen as an analytical method when the number of spatial harmonics is truncated. However, to my best knowledge, until now nobody has combined RCWA with dispersion analysis and performed such an analysis.

7.3 Investigation of coupling effects

7.3.1 Indirect coupling

Strong coupling is a term that is related to quantum field theory (QFT). From a practical point of view, weak coupling can be treated on this level by perturbation theory, whereas this is no longer possible in case of strong coupling. Related to light-matter interaction, weak coupling means that we can treat light and matter as two different entities, whereas for strong coupling, the quantum states of light and matter hybridize and a polariton is generated. In other words, in case of strong coupling, the interaction between light and matter is so strong that the state of matter is strongly altered by the interaction. For inorganic materials, the oscillator strengths are often so large that strong coupling is not the exception but the rule. This is different for organic and biologic materials, which is why the BBL approximation has some merits in this case. Since strong coupling is supposed to be able to change the energy landscape of matter and, thereby, e.g., chemical reaction pathways and reactivity, there is much interest in the question if it is possible to force materials with comparably weak oscillators into the strong coupling regime by IR radiation. In fact, strong coupling is assumed to have an effect even when no light is incident, because of vacuum fluctuations. This quantum effect is sufficient to bring about said changes in the energy landscape (if you are interested in more details, have a look at these reviews [28–31]).

A potential way to change weak coupling into strong coupling is by interference effects or, as they are termed by the corresponding community, optical cavity effects. If you ask yourself, if this is all it is needed, why have I not discussed such effects in the preceding chapter where we investigated interference in great detail? The bottom line is that it seems to require stronger effects than the ones we discussed so far. Maybe, even yet stronger ones than those employed so far, although strong coupling and even a changed energy landscape for organic substances caused by IR radiation has already been reported. [32,33] I have to admit that when I first saw the corresponding spectra I had a very hard time to interpret them and their changes with thickness and angle of incidence. Luckily, and, maybe somewhat surprising, the spectra can be calculated with the conventional transfer matrix formalism we have discussed in Section 4.5. This possibility offers us a tool with which we can explore the spectral signatures, which are said to be connected with strong coupling. Before we do this, however, I think we should investigate and examine coupling effects in infrared spectra on a more fundamental level. This will allow us to be better prepared to disentangle the complex spectral changes caused by the special optical cavities used to cause strong coupling.

For the molecular spectroscopist, the concept of coupling in infrared spectra must be an extremely strange one. Used to the notion of normal and, thus, noncoupling modes, it must seem as if in absorbance spectra only such modes are present (as stated in the beginning of this book, I disregard any nonlinear effects, which are related to very strong light sources and anharmonic oscillators). In fact, the use of Lorentz profiles instead of classical damped harmonic oscillators is fully in line with and completely accommodates the normal mode concept. Lorentz profiles are additive with regard to the absorption index and, therefore, also concerning absorbance. However, it is important to keep in mind that this additivity is a forced one and is actually an approximation based on the assumption of weak oscillators (and, thus, weak coupling). In the classical damped harmonic oscillator model, polarizabilities of the individual molecules are assumed to be additive to result in the macroscopic polarization. Consequently, additivity occurs in this approximation on the level of the individual contributions of the oscillator terms to the dielectric function. To get to the Lorentz profile, we have to approximate the basic result of Maxwell's wave equation for scalar media $\varepsilon = 1 + \chi = n^2$ by $\sqrt{\varepsilon} = \sqrt{1+\chi} \approx 1 + \chi/2$, where χ is the electric susceptibility. As shown in Section 5.2, this is a very good approximation for a weak oscillator.

Actually, it is a very good approximation for a single oscillator, but how about two oscillators that are spectrally close to each other? The problem reminds a little bit of the probability of the presence of two electrons that are located at two bonded hydrogen atoms. In this case, this probability is enhanced between the two atoms compared to the sum of probabilities of individual atoms. Something similar is true for the spectrally neighbored oscillators, namely, for the absorption index. Nevertheless, in this case, it is the refractive index function we have to focus on. Remember the anomalous dispersion of the refractive index around the oscillator position and that the absorption index is the imaginary part of the dielectric function ε'' divided by 2 times the refractive index ($\varepsilon''/(2n) = k$). Due to this anomalous dispersion, the refractive index has a minimum on the high wavenumber side and has low values in the vicinity of this minimum. If you place a second, weaker oscillator in the corresponding wavenumber region, its absorption will be higher than expected, cf. Fig. 7.13.

In Fig. 7.13, the change of the absorption index is depicted if for one oscillator the oscillator position is varied while the second is fixed at 1700 cm^{-1}. Both oscillators have the same oscillator strength and the same damping constant. If Lorentz profiles are used to describe the complex index of refraction, the oscillators are decoupled and the effects of the oscillators on the absorption index are simply additive. This is no longer the case if we assume the classical damped harmonic oscillator model. As just derived, the oscillator positioned at higher wavenumber shows a higher peak intensity than that with the lower wavenumber (at least as long as the individual bands can be resolved). Accordingly, without a sophisticated analysis it is not possible to conclude that one oscillator is stronger than the other, at least if their strength is not very different [34].

This means that in absorbance spectra, oscillators will be coupled in general although this seems to be counterintuitive. In fact, on the level of wave optics and dispersion theory, all bands will be coupled. This is already a consequence of the fact that reflectance and transmittance are not linear functions of the dielectric function.

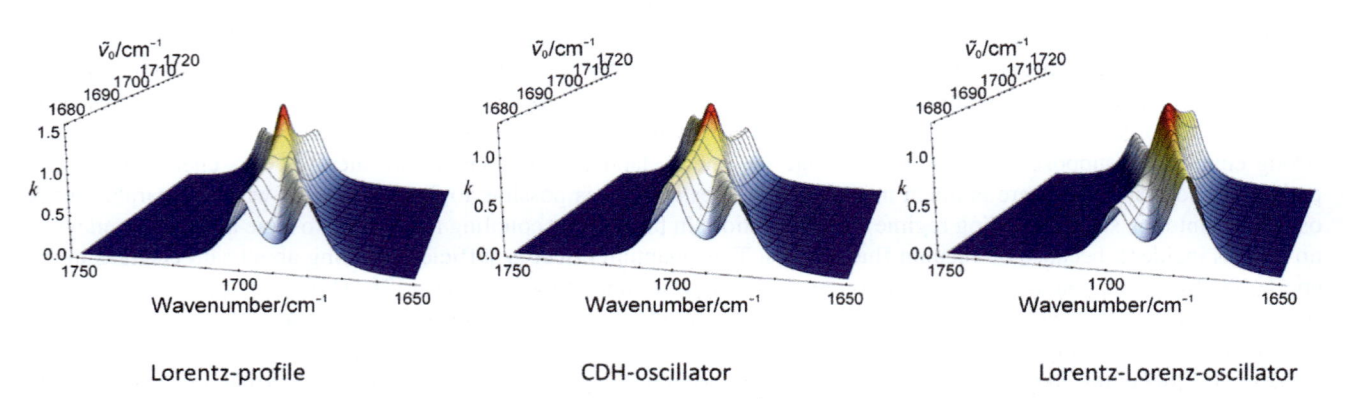

FIG. 7.13 Absorption index of two equally strong oscillators, one fixed at 1700 cm^{-1} the other with variable position using Lorentz profiles, the CDHO model, and the Lorentz-Lorenz model [34].

On top of this, we have to consider another effect which we already discussed in Sections 5.1 and 5.3, namely, the fact that the local electric field intensity is different from the one applied. In the simplest case (spherical molecule in a continuum with properties determined by the other molecules), the Lorentz-Lorenz approach can be used to model the oscillators. If we still use the CDHO model, the complex refractive index function $\widehat{n}(\tilde{\nu})$ would be given by (cf. Eq. 5.39):

$$\widehat{n}(\tilde{\nu}) = \left(\frac{\varepsilon_\infty + \left(1 - \frac{1}{3}\varepsilon_\infty\right) \sum_{j=1}^{N} \frac{S_j^2}{\tilde{\nu}_{0,j}^2 - \tilde{\nu}^2 - i\tilde{\nu}\gamma_j}}{1 - \frac{1}{3} \sum_{j=1}^{N} \frac{S_j^2}{\tilde{\nu}_{0,j}^2 - \tilde{\nu}^2 - i\tilde{\nu}\gamma_j}} \right)^{\frac{1}{2}}. \tag{7.5}$$

As we have discussed in Section 5.1, this leads to a redshift of the peak for a single oscillator. For two oscillators, the situation is more complex, since not only the band positions, but also the (apparent) oscillator strengths are altered. This is actually nothing new and can be found in the Born and Wolf [35] Section 2.3.4, but it looks like that the consequences were never investigated in detail. For two oscillators, it is obvious from Fig. 7.13 that the local field effects now lead to an apparent transfer of oscillator strength to the band with the lower wavenumber position. In practice, this means that we can expect everything from a transfer of oscillator strength to the band with the higher wavenumber position in the case of no local field effect to a transfer of oscillator strength to the band with the lower wavenumber position. The effects in the latter case can also be stronger than those in Fig. 7.13, if the local field strengths are beyond what is predicted by the Lorentz-Lorenz theory. This certainly introduces a kind of arbitrariness into quantitative spectrum interpretation since the general expression from which the Lorentz-Lorenz relation could be derived is not known (and probably also too complex to be meaningfully derived; just have a closer look at Bötcher's books [36,37], or read through Chapter 10). Note that it does not matter for these considerations if the two bands belong to the same or to two different materials, which means that it is probably not a good idea to select two neighbored bands to derive from them the content of mixtures. On the other hand, for pure materials, quantum mechanical methods may be helpful to derive real oscillator strength if interactions between light and matter are properly included into the model (but usually they are not; in particular, typically molecules are assumed to be in the gas phase).

How large would be the errors if we would naïvely analyze the bands in Fig. 7.13? Such a naïve analysis would, e.g., consist in using Lorentz profiles to analyze CDHO-based or Lorentz-Lorenz-based bands. The corresponding results are depicted in Fig. 7.14.

Obviously, if the spectra series generated with CDH oscillators is analyzed using Lorentz profiles, the transfer of oscillator strength from the peak with the lower wavenumber to the peak with the higher wavenumber is only weakly changing except when the two oscillators have less than 5 cm^{-1} distance. For small distances, the transfer is inverted, which can, however, not be noticed in Fig. 7.13, because then the individual peaks can no longer be resolved by the eye. For the two other cases depicted in Fig. 7.14, the situation is different. In both cases, the analysis yields that oscillator strength is

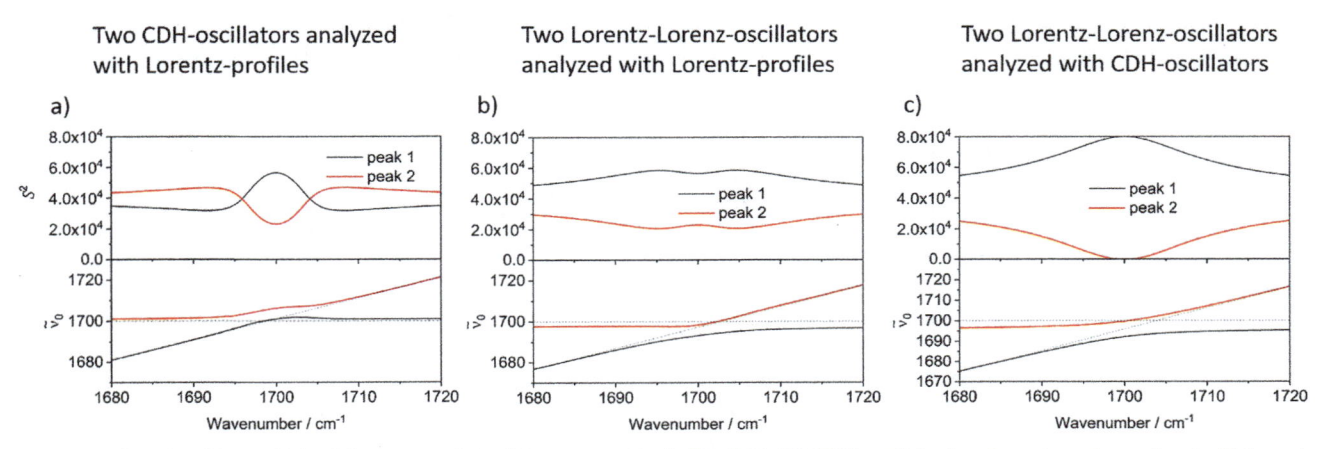

FIG. 7.14 Results of the analysis of the center and the right spectra series in Fig. 7.13. (A) CDH oscillator-based spectra series analyzed with Lorentz profiles. (B) Lorentz-Lorenz-based spectra series analyzed with Lorentz profiles. (C) Lorentz-Lorenz-based spectra series analyzed with CDHOs. Upper panels: Apparent oscillator strengths. Lower panels: Apparent oscillator positions [34].

transferred from the oscillator with the higher wavenumber position to the one with the lower wavenumber position, even when the spectral distance between the oscillators is low.

More interesting are the apparent oscillator positions. In this case, something occurs, which is inherently attributed to strong coupling, namely, the so-called anticrossing behavior. Accordingly, the apparent oscillator position can never be the same for both oscillators, even when the actual oscillator position is identical. One could state that the oscillators repel each other, which is actually an advantage since it automatically makes it easier to detect both.

Since what I basically did to obtain Fig. 7.14 was band fitting, you can imagine that this technique is one where you can expect larger consequences of this kind of coupling effect. Not so much concerning the band positions, although it should be clear by now why you will never find two Lorentz profiles with the same position. But in particular with regard to the individual band areas, which are assumed to be proportional to the oscillator strengths (remember, this is correct only in the absence of coupling; otherwise, the corresponding Kramers-Kronig sum rule is only valid strictly over the whole frequency region, from zero to infinite wavenumbers), thereby also with regard to the concentration of the absorbing species. A particular thought-provoking example for this, which could be interesting both for the molecular IR spectroscopist and for those mainly dealing with remote sensing and/or infrared spectroscopy of inorganic materials, is water. In the decade before I wrote this chapter (July 2022), many quantum mechanical calculations were carried out to better understand the structure of liquid water [38–45]. Attempts were made to resemble the band shapes as closely as possible by varying and amending quantum mechanical models. To support the colleagues in their quest (and for futures ones that are similar), I think the following considerations may be helpful.

A simple idea to see which oscillator model may have more physical relevance is to apply all three of them to fit bands. In particular, one of the bands of liquid H_2O is very demanding as it extends over a thousand wavenumbers from about 2800 to 3800 cm^{-1}. This band is attributed to the stretching vibrations of water of which there are two, both infrared-active. With two oscillators, only a comparably unsatisfactory fit can be achieved, which is why I used three. Remember that in water hydrogen bonding is important so that 3–4 water molecules form a kind of complex at ambient temperature. In addition, the second harmonic of the bending vibration is also located in this spectral region and may play a larger role due to Fermi resonance. In any way, using a third oscillator to model this spectral region should be justifiable.

Concerning the type of oscillator, I found that neither the usual Lorentz profiles nor the CDH oscillators would fit, except when you use Kim oscillators (cf. Section 5.5.2). To keep the number of varied parameters low, I applied a Gauss-Lorentz switch fixed at $\sigma_j = 0$, which means the corresponding bands will be purely Gaussian. The result is shown in Fig. 7.15.

As can be seen in Fig. 7.15, the absorption index function falls off strongly at the wings of the band. In addition, only when Gaussian bands are used, the band maximum can be modeled satisfyingly—note that its peak does not coincide with the peak of an individual band! Concerning the best model, it seems that the Kim-Lorentz-Lorentz oscillators can somewhat

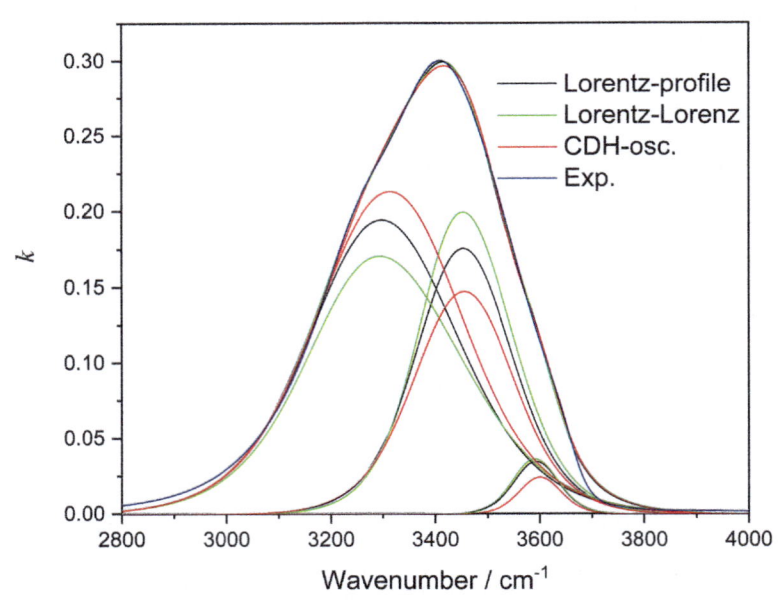

FIG. 7.15 Experimental water stretching vibration band (blue line) fitted with three Kim-Lorentz profiles (black lines), Kim-CDH oscillators (red lines), and Kim-Lorentz-Lorentz oscillators (green lines). The Gauss-Lorentz switch was fixed for all three bands at $\sigma_j = 0$ [34].

better model the experimental values. Keep in mind, however, that the eye can be delusive and just a good agreement between experiment and model does not mean that the model is correct. Also, because Kim oscillators have the problematic property that they are not completely Kramers-Kronig compatible (and for Kim-Lorentz profiles, this comes on top, as Lorentz profiles are also not Kramers-Kronig compatible) [46]. Anyway, the vital point is that obviously the band areas of the two important bands are strongly dependent on the chosen model. For the Kim-CDH oscillators, the area ratio is close to 2:1 for the low wavenumber band to the centered one, whereas it is close to 1:1 for the Kim-Lorentz-Lorenz model. When you inspect Fig. 7.15, please keep in mind that only for the Kim-Lorentz profiles, the sum curve is indeed the sum of the individual contributions—for the two other types coupling prevents this. In the latter cases, oscillators can add to the sum curve in wavenumber ranges where they would have individually close to zero contributions.

What does this mean for a quantum mechanical model? First of all, there is a small problem that quite often is disregarded anyway when quantum mechanical calculations of infrared spectra are compared to experimental ones. What is actually calculated is the following product of $n(\omega)\, a(\omega)$:

$$n(\omega)\, a(\omega) = \frac{4\pi\omega \tanh(\beta\hbar\omega/2)}{3\hbar cV} \times \int_{-\infty}^{+\infty} dt \exp[-i\omega t]\langle \mathbf{M}(t) \cdot \mathbf{M}(0)\rangle. \tag{7.6}$$

Here, V is the volume, β is given by $\beta = 1/(k_B T)$, and the $\mathbf{M}$s are the transition moments. The product of the refractive index and the decadic absorption coefficient are related to the imaginary part of the dielectric function ε'' by $n(\omega)\, a(\omega) = 2\pi\tilde{\nu}/\ln 10 \cdot \varepsilon''$. This means that what results from the quantum mechanical calculations resembles much more the imaginary part of the dielectric function than an absorbance spectrum. On the other hand, if I take the same oscillator parameters, feed them into the Lorentz-Lorenz relation and into the CDHO model and calculate the absorption constant functions and compare them with that of water (properly calculated from the complex refractive index function), I obtain what is shown in Fig. 7.16.

This means that it makes a big difference which model I use to represent the IR spectrum of the material. Correspondingly, it does not make much sense to try and optimize quantum mechanical calculations to resemble perfectly an experimental absorbance spectrum when, what is calculated, actually is the dielectric function. Before we go on, note that most of what I presented to you in this section can be found in Ref. [34].

Previously, we discussed that indirect coupling is caused even in the absence of local fields due the fact that Fresnel's equations predict that reflectance and transmittance are not linearly related to the dielectric function. In addition, we discussed in Section 5.6 that for p-polarized light transmittance and reflectance also depend on the inverse of the dielectric function. This is another cause for indirect coupling, even if we assume that Lorentz oscillators are uncoupled,

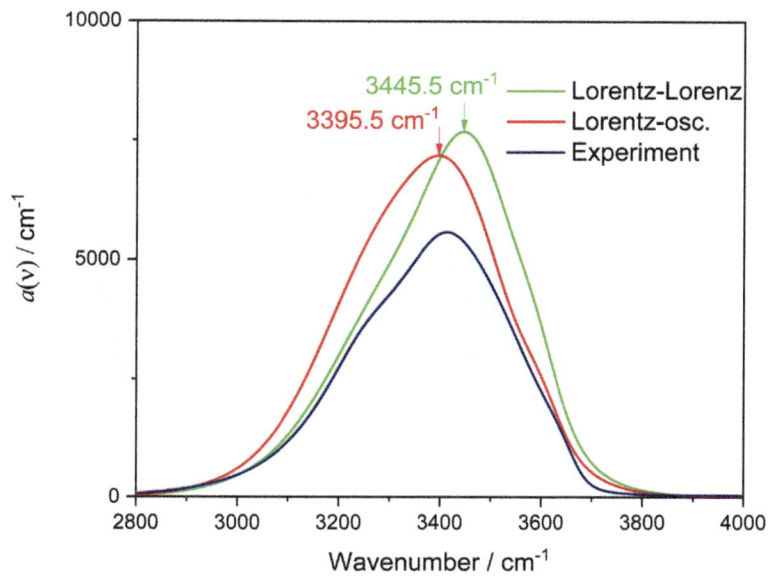

FIG. 7.16 Comparison of the absorption constant functions calculated from the oscillator parameters obtained by the Kim-Lorentz profile fit and fed into the Lorentz-Lorenz relation and the CDHO model in comparison with the experimental absorption constant function of water [34].

$$\varepsilon(\tilde{\nu}) = \varepsilon_\infty + \sum_{j=1}^{N} \frac{S_{\text{TO},j}^2}{\tilde{\nu}_{\text{TO},j}^2 - \tilde{\nu}^2 - i\tilde{\nu}\gamma_j}, \tag{7.7}$$

and that we can represent the inverse dielectric function also on the basis of Lorentz oscillators:

$$\left(\varepsilon_\infty + \sum_{j=1}^{N} \frac{S_{\text{TO},j}^2}{\tilde{\nu}_{\text{TO},j}^2 - \tilde{\nu}^2 - i\tilde{\nu}\gamma_j}\right)^{-1} = \varepsilon_\infty^{-1} - \sum_{j=1}^{N} \frac{S_{\text{LO},j}^2}{\tilde{\nu}_{\text{LO},j}^2 - \tilde{\nu}^2 - i\tilde{\nu}\gamma_j}. \tag{7.8}$$

In Section 5.6, I actually did only briefly discuss what happens if we have more than one oscillator. Remember that for one oscillator the transversal optical (TO) oscillator strength $S_{\text{TO},j}^2$ is the same as the longitudinal optical (LO) oscillator strength $S_{\text{LO},j}^2$ if $\varepsilon_\infty = 1$ [47]:

$$S_j^2 = S_{\text{TO},j}^2 = \varepsilon_\infty^2 S_{\text{LO},j}^2. \tag{7.9}$$

If there is, however, more than one oscillator, additional complications arise since the oscillators are always coupled in the inverse dielectric function. This means, while Eq. (7.8) still holds and allows a description with independent LO oscillators, an individual LO-oscillator strength actually depends on all TO-oscillator strengths, which is reminiscent of what happens if the oscillators are coupled in the dielectric function by the local field effect.

Fig. 7.17 illustrates the situation. $\tilde{\nu}_{\text{TO},1}$ and $\tilde{\nu}_{\text{TO},2}$ are fixed at 1000 and 1030 cm^{-1} (orange lines), while S_1 is varied from 0 to 500 cm^{-1} ($S_2 = 200$ cm^{-1}). Were the LO modes uncoupled, the position of the first mode $\tilde{\nu}_{\text{LO},1u}$ would vary according to $\tilde{\nu}_{\text{LO},1u} = \sqrt{\tilde{\nu}_{\text{TO},1}^2 + S_1^2}$, while $\tilde{\nu}_{\text{LO},2u}$ would be fixed, since S_2 is not altered (green curves). In particular, $\tilde{\nu}_{\text{LO},1u}$ would cross both $\tilde{\nu}_{\text{TO},2}$ and $\tilde{\nu}_{\text{LO},2u}$. In fact, this never happens because of the TO-LO rule (cf. Section 5.6), stating that $\tilde{\nu}_{\text{TO},1} < \tilde{\nu}_{\text{LO},1} < \tilde{\nu}_{\text{TO},2} < \tilde{\nu}_{\text{LO},2}\ldots$ This rule can be seen at work in Fig. 7.17, where $\tilde{\nu}_{\text{LO},1}$ approaches $\tilde{\nu}_{\text{TO},2}$, but never reaches it. Instead, $\tilde{\nu}_{\text{LO},2}$ begins to increase. In fact, with increasing oscillator strength, the LO modes begin to hybridize, and when the real part of the dielectric function starts to become negative, $\tilde{\nu}_{\text{LO},2}$ must be attributed (mostly) to $\tilde{\nu}_{\text{TO},1}$ and $\tilde{\nu}_{\text{LO},1}$ to $\tilde{\nu}_{\text{TO},2}$. Accordingly, $\tilde{\nu}_{\text{TO},1}$ and $\tilde{\nu}_{\text{LO},2}$ form an outer TO-LO pair, whereas $\tilde{\nu}_{\text{LO},1}$ and $\tilde{\nu}_{\text{TO},2}$ form an inner TO-LO pair. Let me emphasize again that such an attribution somehow neglects the fact that an LO oscillator couples to all other LO modes and is a kind of hybrid mode for stronger oscillators [47]. In any way, do not misinterpret the TO-LO rule as a rule that attributes $\tilde{\nu}_{\text{LO},i}$ automatically to $\tilde{\nu}_{\text{TO},i}$ as this is erroneously done in newer literature. Note that in case of an inner pair, the LO mode has its position at lower wavenumber than the TO mode. This inversion is caused by the real part of the dielectric

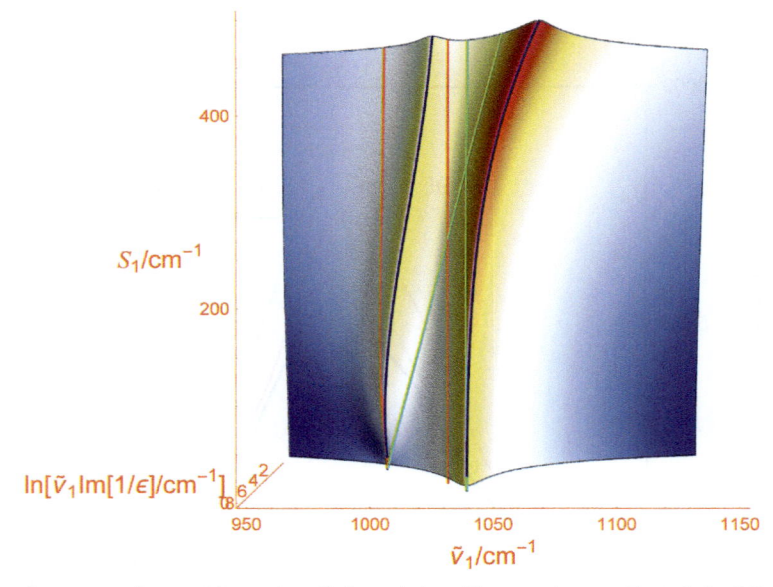

FIG. 7.17 Change of LO-mode wavenumber positions when S_1 is varied to illustrate the coupling of the LO modes. Oscillator parameters: $S_1 = 0 - 500$ cm^{-1}, $S_2 = 200$ cm^{-1}, $\gamma_1 = 10$ cm^{-1}, $\gamma_2 = 10$ cm^{-1}, $\tilde{\nu}_{\text{TO},1} = 1000$ cm^{-1}, $\tilde{\nu}_{\text{TO},2} = 1030$ cm^{-1}, and $\varepsilon_\infty = 2.25$. The 2D-plot results from the inversion of the dielectric function calculated by the CDHO model according to Eq. (7.7), whereas the blue lines are calculated by Eq. (7.10). The green lines would result if the LO modes were uncoupled with $\tilde{\nu}_{\text{LO},1u} = \sqrt{\tilde{\nu}_{\text{TO},1}^2 + S_1^2}$ and constant $\tilde{\nu}_{\text{LO},1u}$, since S_2 is not varied.

function and the polarization becoming negative around the second mode. This mode now causes a dip in the broad band of the first mode (cf. Fig. 5.35, SrTiO$_3$ has two such weak modes).

The positions of the coupled oscillator pair, $\tilde{\nu}_{LO,1}$ and $\tilde{\nu}_{LO,2}$, can be described as usual by the equations of motions and lead in this case assuming zero damping to, [48].

$$\tilde{\nu}^2_{LO,\pm} = \frac{1}{2}\left\{\tilde{\nu}^2_{LO,1u} + \tilde{\nu}^2_{LO,2u} \pm \left[\left(\tilde{\nu}^2_{LO,1u} - \tilde{\nu}^2_{LO,2u}\right)^2 + 4\frac{S_1^2 S_2^2}{\varepsilon_\infty^2}\right]^{\frac{1}{2}}\right\}, \tag{7.10}$$

where $\tilde{\nu}_{LO,-} = \tilde{\nu}_{LO,1}$ and $\tilde{\nu}_{LO,+} = \tilde{\nu}_{LO,2}$. You find the corresponding results in Fig. 7.17 represented by the blue lines. Given the fact that damping was neglected to derive the aforementioned formula, the correspondence of the LO positions according to Eq. (7.10) with the maxima of the negative imaginary part of the inverse dielectric function is close to perfect.

In addition to the coupling of two LO modes based on vibrations/phonons (or electronic excitations), these modes can also couple to plasmons, which are only longitudinal in nature, resulting in longitudinal plasmon-polaritons (LPP). A good example in this case is β-Ga$_2$O$_3$. In newer literature, this coupling is usually misunderstood, also because of the misinterpretation of the TO-LO rule. Note that β-Ga$_2$O$_3$ has a monoclinic structure. To simplify things, we are focusing in the following on the transition moments along the crystallographic b-axis, which are perpendicular to those of the a-c plane and can thus be separated from the latter [49].

From the upper part of the left panel of Fig. 7.18, it is obvious that Eq. (7.10) can well describe the actual LPP-mode positions (turquoise and blue dashed lines) as long as $\tilde{\nu}_p \leq 400\,\text{cm}^{-1}$ holds. For higher plasmon wavenumbers, the plasmon begins to stronger hybridize with the second phonon mode and the first phonon mode is largely described by the dashed black curve. The stronger hybridization is also reflected in the crossing between the green and the blue line, which should not take place according to Eq. (7.10), since the turquoise line stays always above the green line. This crossing is therefore an indication that for $\tilde{\nu}_p > 500\,\text{cm}^{-1}$, the blue line represents phonon mode 2 (LPP mode 3). Accordingly, in Fig. 7.18, the colors of the modes have been switched and a grey zone has been introduced. The corresponding grey zone begins at the curve defined by $\tilde{\nu}_{LO,1} = \sqrt{\tilde{\nu}_p^2/\varepsilon_\infty}$. It serves as a kind of approximate indicator for the start of the hybridization of the modes.

As soon as the mode positions begin to increase with increasing plasma frequency, hybridization sets in. The TO-LO rule still holds, which means that for crystal symmetries up to orthorhombic and for the scalar dielectric function of the monoclinic b-axis, no LPP mode crosses the line of a TO mode. Through hybridization, such crossings are avoided. Nevertheless, LPP modes finally end up having a lower wavenumber than corresponding TO modes. As a consequence, at high plasma frequencies, the corresponding LPP mode is the one with the highest wavenumber. This is also clearly seen in

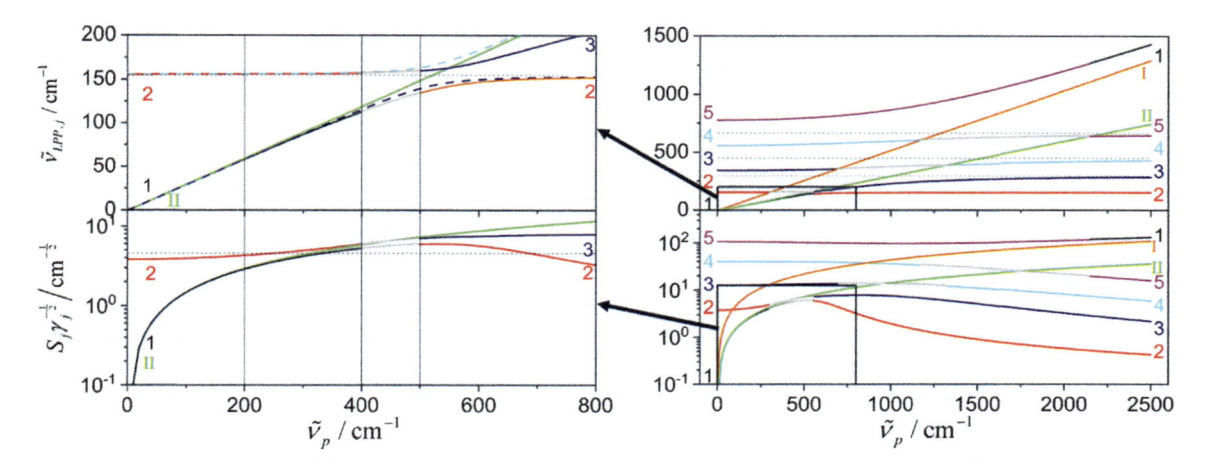

FIG. 7.18 Left panels: Modeled LO-oscillator positions and square root of the product of peak values times the wavenumber for the principal dielectric function along the b-axis of doped β-Ga$_2$O$_3$ up to a plasma frequency of 800 cm^{-1} (detail magnifications of the right panels). Right panels: Modeled LO-oscillator positions and square root of the product of peak values times the wavenumber for the principal dielectric function along the b-axis of doped β-Ga$_2$O$_3$ up to a plasma frequency of 2500 cm^{-1}. Upper panel: LO wavenumber positions, i.e., the wavenumber locations of the maxima of the negative imaginary part of the inverse of the model dielectric function. The thin dashed black lines indicate the TO oscillator positions of the four phonon modes. The orange and the green lines specify the plasmon-mode positions assuming uncoupled modes. Lower panel: square root of the negative imaginary part of the inverse of the model dielectric function and the model inverse dielectric function multiplied by the LP wavenumber position as a function of the plasma frequency ($S_{LO} = \sqrt{-\tilde{\nu}_{LO}\gamma\text{Im}\{1/\varepsilon\}}$). The green and the orange line represent the same quantity for uncoupled plasmons screened by $\varepsilon_{0,1}$ (green line) and ε_∞ (orange line) following Eq. (7.8) [49].

Fig. 7.18, right part, where for large plasma frequencies/high mode strengths, the highest wavenumber mode increases both in strength and with regard to the mode position, very much like the plasmon mode in the uncoupled model (orange line). The same behavior is shown by the coupled plasmon-phonon modes in the a-c plane of monoclinic materials like β-Ga_2O_3. For sufficiently high plasmon frequencies, the two LO phonon modes with the highest wavenumber positions will acquire increasingly plasmon character [49].

7.3.2 Direct coupling of oscillators

You may wonder why we need a section about direct coupling when this topic was actually already treated in Section 5.4. The simple answer is that we left out something. In Section 5.4, we treated oscillators that were coupled via a honeypot—in this section, we treat oscillators coupled by a spring. It makes a large difference if the coupling takes places via the former or the latter. This is already obvious from the corresponding equations of motions of the coupled system, cf. Eq. (5.79) with Eq. (7.11):

$$\frac{d^2 x_1}{dt^2} + \gamma_1 \frac{dx_1}{dt} + \omega_{0,1}^2 x_1 + v_{12} x_2 = q_1 E_x \exp(-i\omega t)$$
$$\frac{d^2 x_2}{dt^2} + \gamma_2 \frac{dx_2}{dt} + \omega_{0,2}^2 x_2 + v_{12} x_1 = q_2 E_x \exp(-i\omega t). \quad (7.11)$$

Here, v_{12} is the coupling constant. Actually, the term involving the coupling constant should read $v_{12}(x_2 - x_1)$. So, in Eq. (7.11), I have already considered that $\omega_{TO,1}^2 - v_{12} = \omega_{0,1}^2$ and $\omega_{TO,2}^2 - v_{12} = \omega_{0,2}^2$, where $\omega_{TO,j}$ would be the eigenfrequency of oscillator j if uncoupled and not driven by an external field.

When I carry out the derivations, introduce the wavenumber, and write the result more compactly, I arrive at:

$$\begin{pmatrix} A_1 & v_{12} \\ v_{12} & A_2 \end{pmatrix} \begin{pmatrix} x_1 \\ x_2 \end{pmatrix} = \begin{pmatrix} q_1 E_x \\ q_2 E_x \end{pmatrix}$$
$$A_1 = \tilde{\nu}_{0,1}^2 - \tilde{\nu}^2 - i\tilde{\nu}\gamma_1$$
$$A_2 = \tilde{\nu}_{0,2}^2 - \tilde{\nu}^2 - i\tilde{\nu}\gamma_2. \quad (7.12)$$

If I take the solutions for x_1 and x_2 and generate the dielectric function in the same way as in Section 5.4, the result is:

$$\varepsilon(\tilde{\nu}) = \varepsilon_1(\tilde{\nu}) + \varepsilon_2(\tilde{\nu}) = \frac{S_1^2(\tilde{\nu}_{0,2}^2 - \tilde{\nu}^2 - i\tilde{\nu}\gamma_2) - 2S_1 S_2 v_{12} + S_2^2(\tilde{\nu}_{0,1}^2 - \tilde{\nu}^2 - i\tilde{\nu}\gamma_1)}{(\tilde{\nu}_{0,1}^2 - \tilde{\nu}^2 - i\tilde{\nu}\gamma_1)(\tilde{\nu}_{0,2}^2 - \tilde{\nu}^2 - i\tilde{\nu}\gamma_2) - v_{12}^2}. \quad (7.13)$$

Eq. (7.13) is certainly similar to Eq. (5.84), which I reproduce here for an easier comparison:

$$\varepsilon_1(\tilde{\nu}) + \varepsilon_2(\tilde{\nu}) = \frac{S_1^2(\tilde{\nu}_{0,2}^2 - \tilde{\nu}^2 - i\tilde{\nu}(\gamma_2 + \gamma_{12})) + 2i S_1 S_2 \tilde{\nu}\gamma_{12} + S_2^2(\tilde{\nu}_{0,1}^2 - \tilde{\nu}^2 - i\tilde{\nu}(\gamma_1 + \gamma_{12}))}{(\tilde{\nu}_{0,1}^2 - \tilde{\nu}^2 - i\tilde{\nu}(\gamma_1 + \gamma_{12}))(\tilde{\nu}_{0,2}^2 - \tilde{\nu}^2 - i\tilde{\nu}(\gamma_2 + \gamma_{12})) + \tilde{\nu}^2 \gamma_{12}^2}. \quad (7.14)$$

There are three main differences:

1. In contrast to Eq. (7.13), the resonance wavenumbers in Eq. (7.14) are not redshifted; instead, the coupling constant introduces additional damping.
2. In Eq. (7.14), the term involving the product $S_1 S_2$ is imaginary. This leads to a phase shift. In the extreme case, the imaginary part looks like the real part and vice versa.
3. The negative sign before v_{12}^2 in the denominator and the positive sign in front of $\tilde{\nu}^2 \gamma_{12}^2$.

In particular, point 3 has a strong impact, because it is responsible for anticrossing behavior. In other words, there is no anticrossing behavior in case of honeypot coupling in contrast to spring coupling. Before we go on, you might ask yourself, why did I not report Eq. (7.13) in the chapter about dispersion theory (Chapter 5)? Actually, it is important to understand, e.g., why for CO_2 the symmetric stretching vibration is not IR-active, but for dispersion analysis, I see it as an advanced topic, as it describes anticrossing behavior and Fano-type bandshapes. This is of special importance for understanding the coupling of plasmons or interference effects with vibrational modes, or, to understand the fundament of VCD spectroscopy from a classical perspective (Chapter 16).

Let us first assume that we have a silent second mode,

$$\varepsilon(\tilde{\nu}) = \varepsilon_1(\tilde{\nu}) = \frac{S_1^2(\tilde{\nu}_{0,2}^2 - \tilde{\nu}^2 - i\tilde{\nu}\gamma_2)}{(\tilde{\nu}_{0,1}^2 - \tilde{\nu}^2 - i\tilde{\nu}\gamma_1)(\tilde{\nu}_{0,2}^2 - \tilde{\nu}^2 - i\tilde{\nu}\gamma_2) - v_{12}^2}, \tag{7.15}$$

which means that the oscillator strength of the second mode $S_2 = 0$. The difference compared to an uncoupled oscillator is the term $-v_{12}^2/(\tilde{\nu}_{0,2}^2 - \tilde{\nu}^2 - i\tilde{\nu}\gamma_2)$ in the denominator. Therefore, we can expect a corresponding feature due to the silent or dark mode (both terms are in use) in a spectrum. Let us first discuss how the imaginary part of the dielectric function changes if we have a strong IR-active mode at a fixed wavenumber and vary the position of the dark mode. We have to distinguish between weak coupling and strong coupling. We speak of strong coupling if $2v_{12}/\tilde{\nu}_{0,1} > (\gamma_1 + \gamma_2)/2$. In Fig. 7.19, you find examples for weak coupling and strong coupling. Note that for weak coupling, the band at $1000\,\mathrm{cm}^{-1}$ does not display a zero intensity, whereas this is the case for the strong coupling examples. The observation of zero intensity in-between the modes is equivalent with the abovementioned anticrossing behavior.

Usually in literature, only the case where the second mode is silent is investigated. To be comprehensive, however, the second case, where the second mode is also a driven one, should be investigated, too. For this second case, we assume that the damping constants of both modes and their strengths are equal. Under these assumptions, oscillator strength is increasingly transferred from the low wavenumber mode to the high wavenumber mode with higher coupling strength, cf. Fig. 7.20. This is surprising only at the first view. In fact, a typical example where two oscillators of equal strength and damping are strongly coupled is the CO_2 molecule. Thinking of normal modes, this molecule features two stretching vibrations, a symmetric one where both bond lengths change synchronously, and a second antisymmetric one where one bond length shrinks and the other increases. As molecular spectroscopists know well, only the antisymmetric mode is IR-active, whereas the symmetric one is inactive.

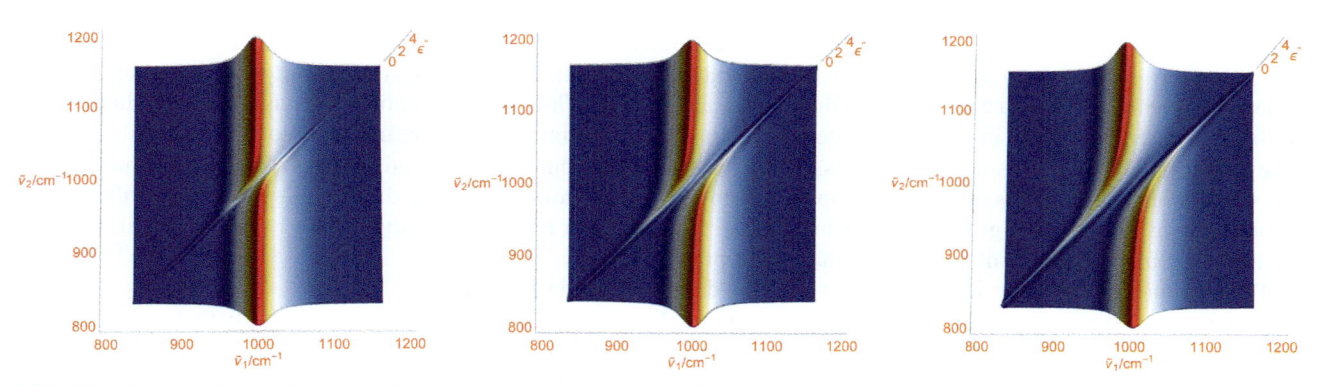

FIG. 7.19 Examples of weak (left side) and strong coupling (center and right side) based on the imaginary part of the dielectric function if the second mode is silent (based on Eq. 7.15). Oscillator parameters employed: $S_1 = 500\,\mathrm{cm}^{-1}$, $S_2 = 0\,\mathrm{cm}^{-1}$, $\gamma_1 = 50\,\mathrm{cm}^{-1}$, $\gamma_2 = 10\,\mathrm{cm}^{-1}$, $\tilde{\nu}_{0,1} = 1000\,\mathrm{cm}^{-1}$, $v_{12}/\tilde{\nu}_{0,1} = 20/40/60\,\mathrm{cm}^{-1}$.

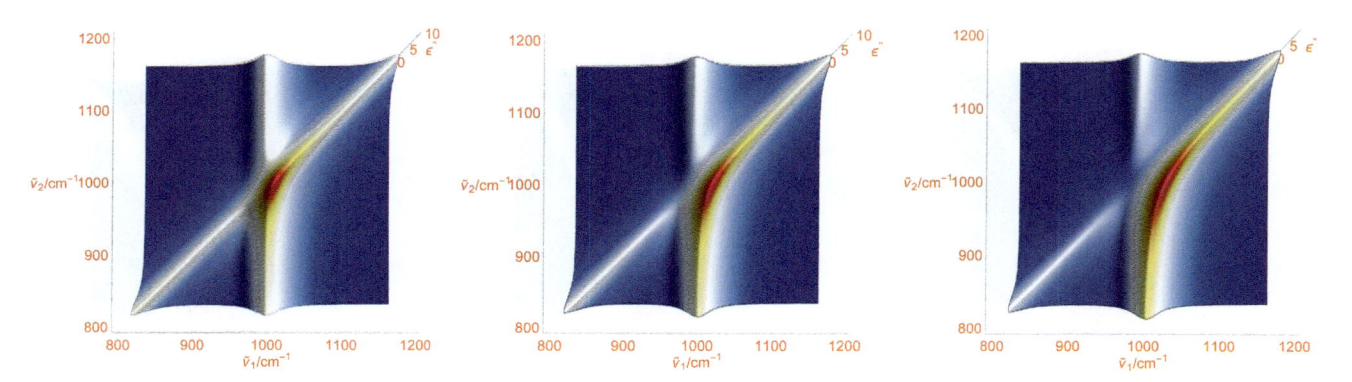

FIG. 7.20 Examples of weak coupling (left side) and strong coupling (center and right side) based on the imaginary part of the dielectric function for two modes with equal strength and damping constant. Oscillator parameters: $S_1 = 500\,\mathrm{cm}^{-1}$, $S_2 = 500\,\mathrm{cm}^{-1}$, $\gamma_1 = 50\,\mathrm{cm}^{-1}$, $\gamma_2 = 50\,\mathrm{cm}^{-1}$, $\tilde{\nu}_{0,1} = 1000\,\mathrm{cm}^{-1}$, $v_{12}/\tilde{\nu}_{0,1} = 20/40/60\,\mathrm{cm}^{-1}$.

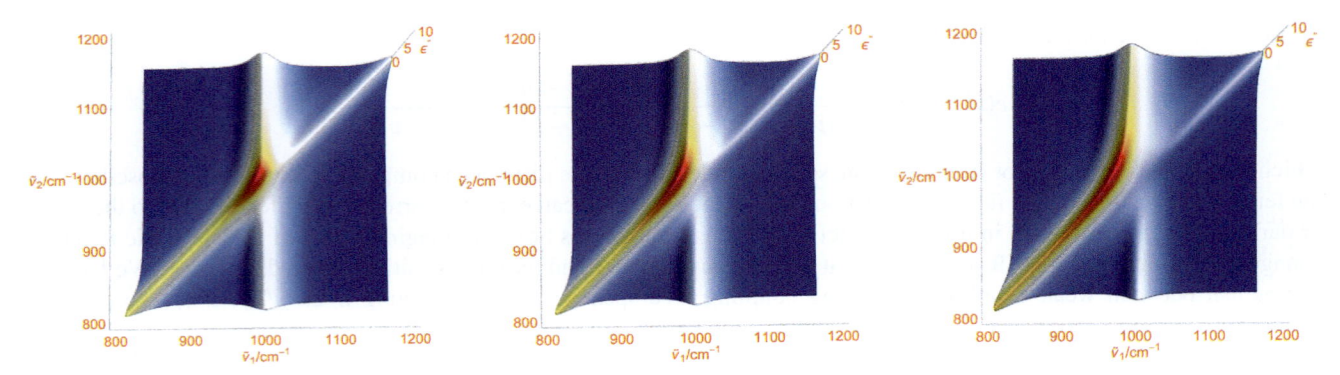

FIG. 7.21 Examples of weak coupling (left side) and strong coupling (center and right side) based on the imaginary part of the dielectric function. Oscillator parameters for two modes with equal absolute strength (but different signs) and damping constant: $S_1 = -500\,\mathrm{cm}^{-1}$, $S_2 = 500\,\mathrm{cm}^{-1}$, $\gamma_1 = 50\,\mathrm{cm}^{-1}$, $\gamma_2 = 50\,\mathrm{cm}^{-1}$, $\tilde{\nu}_{0,1} = 1000\,\mathrm{cm}^{-1}$, $v_{12}/\tilde{\nu}_{0,1} = 20/40/60\,\mathrm{cm}^{-1}$.

If both modes are driven, we have to remember that through indirect coupling (cf. Section 7.3.1), the mode with the higher wavenumber in the absorption index spectrum also gets the higher intensity. This means that both types of coupling are very hard to distinguish. On the other hand, we must not forget the mixed term $2S_1S_2v_{12}$ in Eq. (7.13). In case of honeypot coupling, the mixed term is sensitive to the sign of the effective charge. If one is negative, the sign of the term is changed and so is the coupling as you can see in Fig. 7.21. Certainly, if both effective charges are negative, you regain the case that is displayed in Fig. 7.20.

In practice, the situation can get even more complicated as already mentioned, since it is not possible to measure a pure absorbance spectrum. As emphasized a couple of times, it is either transmittance or reflectance that is measured, and neither is linearly related to the dielectric function nor to the complex index of refraction function. As a consequence, you can expect further types of coupling which we will investigate in the following. One last comment to spring coupling may be necessary before we take a closer look on the reflectance spectra. While spring coupling obviously has some importance for the explanation of modes in inorganic materials and in the fingerprint region of organic and biologic materials, it seems to be of less practical value, since it is not possible to observe the decoupled modes individually. On the other hand, as we will see in Chapter 16, it has an eminent importance to explain vibrational circular dichroism spectra. Accordingly, the introduction of a functional group in a prochiral molecule can much more change the infrared spectrum as if the same functional group enters a similar molecule that is not prochiral.

The influence of coupling in reflectance spectra is demonstrated in Fig. 7.22.

When you scrutinize Fig. 7.22, you should keep in mind that you usually do not have the possibility to move around the second oscillator as you see fit. Instead, an experimental spectrum will just represent a horizontal cut from which you would have to decide whether you see strong coupling or not. Keeping this in mind, it would be relatively easy to distinguish the case without coupling from the case involving coupling if both oscillators have the same position $\tilde{\nu}_{0,1} = \tilde{\nu}_{0,2} = 1000\,\mathrm{cm}^{-1}$. In this case, for strong coupling, the reflectance becomes close to zero at the oscillator position while it has larger values in

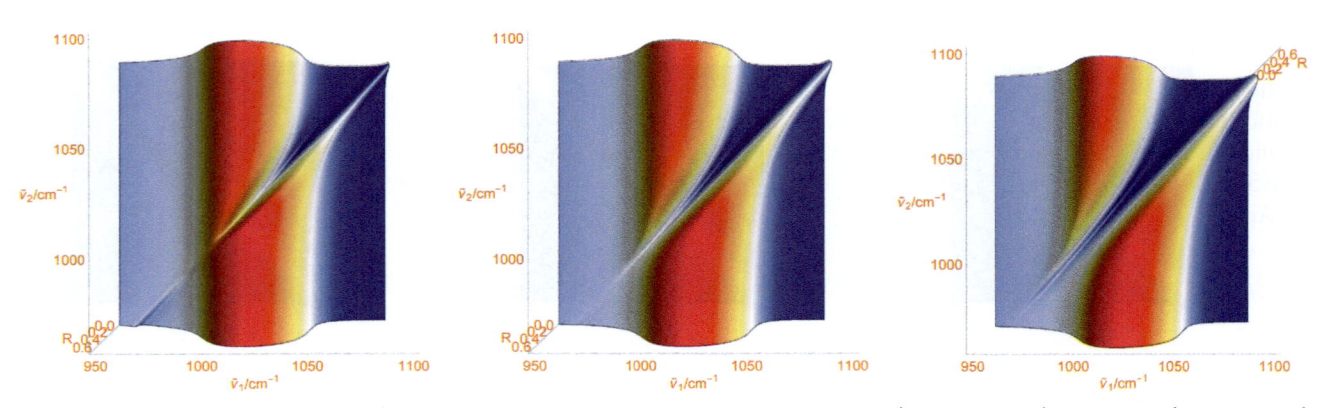

FIG. 7.22 Reflectance spectra displaying coupling effects. Oscillator parameters: $S_1 = 500\,\mathrm{cm}^{-1}$, $S_2 = 100\,\mathrm{cm}^{-1}$, $\gamma_1 = 10\,\mathrm{cm}^{-1}$, $\gamma_2 = 5\,\mathrm{cm}^{-1}$, $\tilde{\nu}_{0,1} = 1000\,\mathrm{cm}^{-1}$, $v_{12}/\tilde{\nu}_{0,1} = 0/10/20\,\mathrm{cm}^{-1}$.

the uncoupled case. However, if the position of the second oscillator is moved away from the first oscillator, the shapes of both bands become very similar. In contrast, if the second oscillator has a position located between the TO and LO position of the first oscillator, a dip comes up that does not easily allow to decide whether the oscillators are coupled or not. As we have discussed already a couple of times, this dip is in the first place a consequence of the fact that the second oscillator is situated in a region where the first oscillator causes the real part of the dielectric function to be very small or even to be negative. Coupling would amplify this effect, but it is not its primary cause. If you argue that coupling will anyway not be visible in reflectance spectra the peaks of which are caused by vibrations, you are certainly right, but the first band could also represent a plasmon due to a periodic metallic structure, which could couple to a molecular vibration (in this case, what is depicted in Fig. 7.22 would represent rather $1-R$ instead of R; a better approximation than coupled harmonic oscillators would require to take into account the self-interaction of the molecules within the cavity, which leads to higher order multipolar effects [16,50]).

7.3.3 Strong coupling between vibrations and the electric field—Polaritons

Strong coupling between excitations and the electric field is a comparably old concept pioneered by Kun, Fano, and Hopfield in the end of the 1950s that has been strongly hyped in the last years [51–54]. To be honest, while I had read about strong coupling quite often in older literature, at first, I did not realize the connection when I came across newer works on the topic. The reason is that strong coupling is ubiquitous for inorganic materials and has no immediate consequences, which means that all wave optics-based methods for the simulation and modeling of spectra remain applicable, in contrast to the switching from the BBL approximation to wave optics and dispersion theory. I have treated dispersion semiclassically in this book, meaning that I took into account that the oscillators in dispersion theory actually represent changes of energy from one vibrational quantum level to another. On the other hand, electromagnetic fields remain to be treated fully classical. This is no longer the case in the QFT. In this sense, strong coupling is a strong interaction between electromagnetic field and matter that requires QFT to be fully understood. To comprehend infrared spectra, we actually do not need QFT, but I want to make you familiar with the spectral signatures that are usually associated with it, so that you can identify them. The reason that makes strong coupling actually interesting for the infrared spectral range is the hope to alter the energy landscape of molecules so that chemical reactions may be influenceable. Accordingly, even if this concept may be strange, molecular substances could show a different reaction behavior when inside or outside a cavity. So far, corresponding experimental findings are somewhat inconsistent, but this is something I will not discuss in detail.

To understand the difference between strong and weak coupling, we can use relations we know from dispersion theory. To make things simpler, we assume a system with only one oscillator:

$$\varepsilon = \varepsilon_\infty + \frac{S_{TO}^2}{\tilde{\nu}_{TO}^2 - \tilde{\nu}^2 - i\gamma_{TO}\tilde{\nu}}. \tag{7.16}$$

If we assume $\tilde{\nu} \to 0$, then,

$$\varepsilon(\tilde{\nu}=0) = \varepsilon_0 = \varepsilon_\infty + \frac{S_{TO}^2}{\tilde{\nu}_{TO}^2} \to S_{TO}^2 = \tilde{\nu}_{TO}^2(\varepsilon_0 - \varepsilon_\infty). \tag{7.17}$$

This is a very interesting finding—I would consider it to be a different form of the Lyddane-Sachs-Teller (LDT) relation (cf. Section 5.6), which you obtain if you derive the same relation as the one above for S_{LO}^2 and then put both together. The LDT relation tells you that the contribution of a single excitation to the static dielectric constant is proportional to $S_{TO}^2/\tilde{\nu}_{TO}^2$. I tried to find out who derived it first, but could not track this down. Huang also does not state this, but did also not provide a derivation, so I concluded that it was already known before the publication of [52].

Anyway, accordingly, we can write Eq. (7.16) as:

$$\varepsilon_r = \varepsilon_\infty + \frac{\tilde{\nu}_{TO}^2(\varepsilon_0 - \varepsilon_\infty)}{\tilde{\nu}_{TO}^2 - \tilde{\nu}^2 - i\gamma_{TO}\tilde{\nu}}. \tag{7.18}$$

Now, we turn our attention to the light wave. In transparent regions of matter, the dispersion of light is given by:

$$n = \frac{c}{\omega}k = \frac{c}{2\pi\nu}k = \frac{k}{2\pi\tilde{\nu}} \to \tilde{\nu}^2 = \frac{k^2}{(2\pi)^2 n^2} = \frac{k^2}{(2\pi)^2 \varepsilon}. \tag{7.19}$$

Using Maxwell's wave equation, the result for the dependence of the mode position on the magnitude of the wavevector k is if zero damping is assumed:

$$\tilde{\nu}^2\left(\varepsilon_\infty + \frac{\tilde{\nu}_{TO}^2(\varepsilon_0 - \varepsilon_\infty)}{\tilde{\nu}_{TO}^2 - \tilde{\nu}^2}\right) = \frac{k^2}{(2\pi)^2}. \quad (7.20)$$

Solving this equation for $\tilde{\nu}$ returns four values the positive of which are plotted over k in Fig. 7.23 for a comparably strong and a weak oscillator. For the latter, the interaction between light and matter is weak and the dispersion of light is (nearly) the same as if the material would be transparent. In contrast, for strong interaction, light cannot move forward in the material between $\tilde{\nu}_{TO}$ and $\tilde{\nu}_{LO}$ because the real part of the dielectric function is negative. At the same time, the polarization of the material is so strong that in case of an ionic crystal, the lattice becomes distorted. Such a distortion is accompanied by an electromagnetic wave. In other words, the lattice vibration and the electromagnetic wave can no longer be distinguished and phonon and photon are hybridized into a polariton.

What does this mean for infrared spectra? Actually, nothing new. Remember that for (IR-)optical excitations, $k = 0$ (which means that for optically active vibrations all unit cells are in phase, cf. Section 5.1). This is different in Raman spectroscopy, with which the TO curve in Fig. 7.23 could be experimentally verified [55].

Fig. 7.23 also provides an illustrative way to understand the differences in the spectroscopy of inorganic and organic/biological matter. For inorganic materials, strong coupling is very common, whereas for organic/biological matter, the coupling between light and matter can usually be neglected. Correspondingly, it seems there is no way to influence reactions of organic/biological matter by light-matter coupling, whereas this seems to be the usual case for inorganic materials. In fact, it looks like there exists also a possibility for the former. The key in this case is to use light interference/cavity effects to couple light and matter strongly. As it seems, this is not possible with the usual interference effects, even if a gold substrate is used. Thankfully, we can enhance interference by improving the cavity. One way to do this is to apply a gold layer not only from one side, but to enclose the layer of interest by two gold layers. Fig. 7.24 shows a comparison between experimental and simulated spectra for this case.

First of all, it is possible to tune the interference enhancement that is supposed to lead to strong coupling by changing the thickness of the PMMA film. Since I have used PMMA several times throughout this book, in particular in Chapter 6, to illustrate certain changes in the spectra depending on the form of sample and type of substrate, you should be familiar with the spectra by now. Although the different substrate types introduced some changes, these were subtle compared to those that you see in Fig. 7.24. Actually, I have to admit that I would not have been able to identify PMMA based on the spectra in this figure. The reason is that the spectra are dominated by bands introduced by the interference effects and transmittance only takes on reasonable values when interference opens up an otherwise closed window. Surprisingly, strong coupling is almost always investigated using transmittance spectra, although reflectance spectra offer the same information, but with much higher signal intensities. Maybe the more symmetric bands in the transmittance spectra are more appealing, although one has to consider that they are actually upside down (in this case using absorbance certainly does not make sense). In any

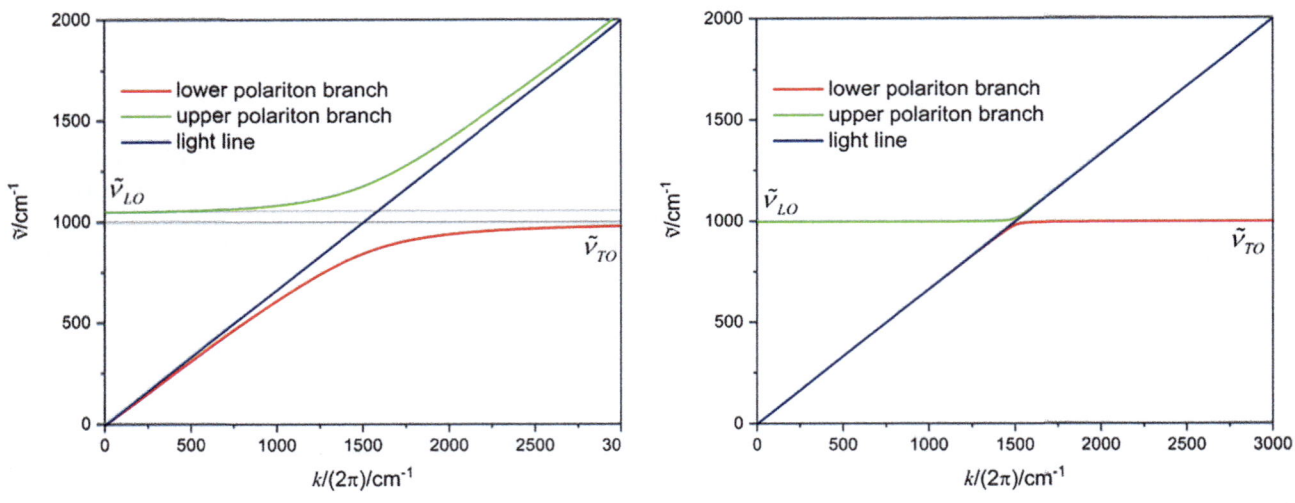

FIG. 7.23 Phonon-polariton dispersion. Oscillator parameters: $\gamma_1 = 10\,\text{cm}^{-1}$, $\tilde{\nu}_{0,1} = 1000\,\text{cm}^{-1}$, $\varepsilon_\infty = 2.25$. Left panel: Strong coupling, $S_1 = 500\,\text{cm}^{-1}$; right panel: weak coupling, $S_1 = 50\,\text{cm}^{-1}$.

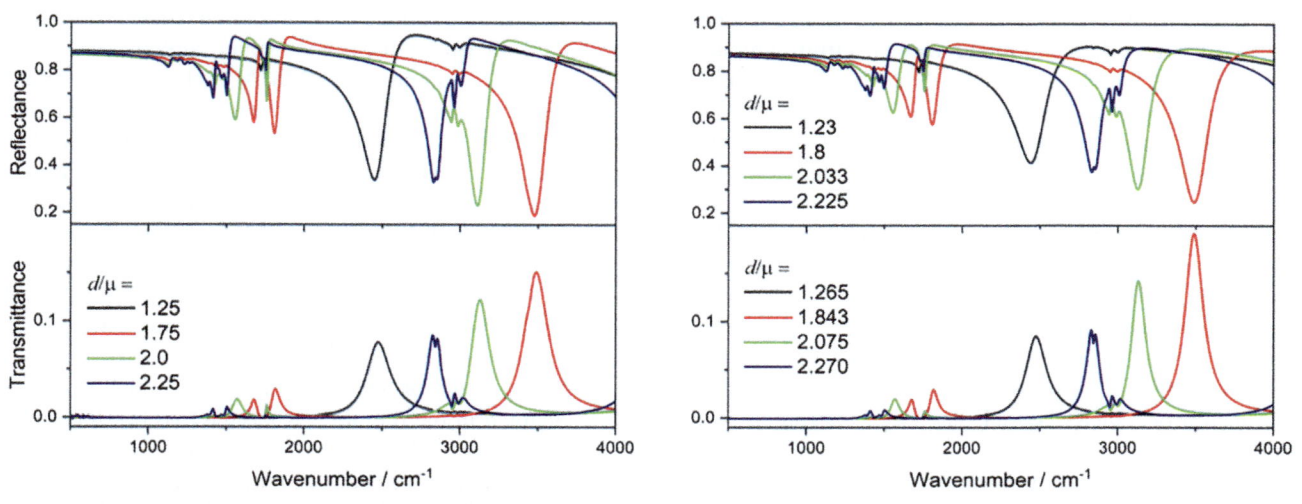

FIG. 7.24 Infrared spectra of PMMA films sandwiched between two gold layers of nominal 8 nm thickness. Left panel: Experimental spectra. Right panel: Calculated spectra. Upper panels: Reflectance spectra. Lower panels: Transmittance spectra.

way, interpreting these spectra seems to be very hard—we will come back to this problem. For the moment, the comparison between the experimental and the modeled spectra at least tells us that there is no new kind of optical problem that could not be understand with wave optics. The agreement is not perfect, but good enough that we can suspect smaller deviations are caused by the imperfect nature of the sample. Indeed, the gold films, e.g., are nominally 8 nm thin, which means that they are not homogenous, since the individual grains do not grow together before reaching larger thicknesses. Nevertheless, such inhomogeneous gold films are conducting as they represent an effective medium (cf. Chapter 10) consisting of metal and voids.

To understand what is happening, we have to go more into details and focus, e.g., on the band around 1730 cm^{-1}. A corresponding enlargement of the experimental spectra is provided in Fig. 7.25.

It is useful to first focus on the blue spectra and on the band at 1500 cm^{-1}. This band is a pure interference feature, where the otherwise zero transmittance takes on measurable values. The position of the 1730 cm^{-1} band is indicated by a vertical

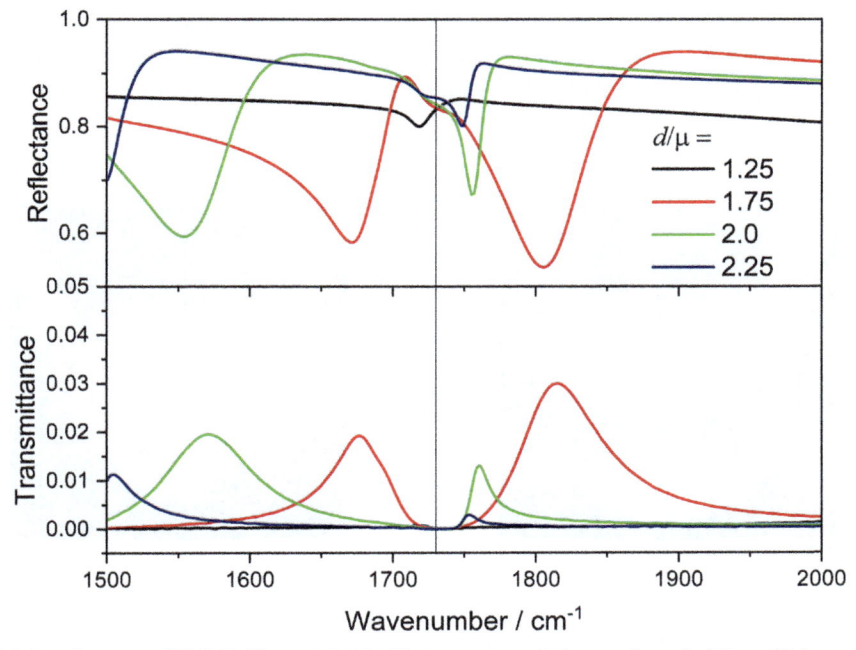

FIG. 7.25 Experimental infrared spectra of PMMA films sandwiched between two gold layers of nominal 8 nm thickness. Upper panel: Reflectance spectra. Lower panel: Transmittance spectra.

line. In the transmittance spectrum, there is a very small interference feature at about 1750 cm^{-1}, but there is no band visible. If you look at the reflectance spectrum, you actually see a kind of band, which is, however, already somewhat altered by the interaction with the interference feature at 1750 cm^{-1}. If we decrease the PMMA thickness to nominal 2 μm, the strong interference feature is shifted from 1500 cm^{-1} to about 1580 cm^{-1}. Correspondingly, the interference feature at originally 1750 cm^{-1} has increased, but is only slightly blue shifted. If we decrease the PMMA thickness further to nominal 1.75 μm (red spectra), the interference feature is at about the same position as the 1730 cm^{-1} band, and this is where it really gets tricky. The reason is that what you see can be interpreted in different ways. The prevalent interpretation is that there is a window due to interference, which is closed at the position of the C=O band of PMMA due to absorption. In this interpretation, the area of the interference feature is constant, which means that if you put an absorption in its middle, the window becomes bipartite, but keeps its area. In other words, the PMMA absorption divides the interference feature and redshifts one part and blueshifts the other. This is somewhat comparable to what we saw in Fig. 7.22, where we introduced a dip into the reflectance, which divides the reflectance band. Just because I introduce this dip does not mean that the two remaining reflectance bands indicate the generation of hybrid modes the position of which could be determined from the two local reflectance maxima. The latter interpretation is, however, adapted in the literature when the two transmittance minima around the 1730 cm^{-1} band are to be interpreted [32,56]. This is in particular strange, when you consider that the reflectance spectrum clearly shows the original structure of the 1730 cm^{-1} band as you can see it in absorbance or absorptance spectra. Just to finish the discussion of Fig. 7.25, if you decrease the PMMA thickness further to 1.25 μm, there is no longer an interference feature in above spectra, meaning the transmittance is simply zero. In the reflectance, however, you see the 1730 cm^{-1} band, although it is again inverted as in spectra of PMMA layers on a gold substrate.

Additional insights may be provided by Fig. 7.26. Since we have established that the experiment can be fully simulated by wave optics, we do no longer have to carry out each experiment. Instead, we can calculate the dependence of the transmittance on the angle of incidence. The results of this calculation are displayed in Fig. 7.26 for a PMMA thickness of 1.9 μm and s-polarized light. For the right panel, I have assumed that the two oscillators belonging to the 1730 cm^{-1} band have zero oscillator strengths. This means the right panel reflects the position dependence of the interference feature on the angle of incidence. Obviously, this is a parabola. Once I switch on the two oscillators, the left panel results. Obviously, for larger angles of incidence, the parabola is undisturbed, while it becomes flattened for lower angles of incidence, because it cannot cross the position of the 1730 cm^{-1} band. The position of the 1730 cm^{-1} band, on the other hand, is not shifted, but at lower angles of incidence, a second interference feature (which is said to be actually the 1730 cm^{-1} band, which becomes dispersive) occurs that mimics the behavior of the parabola on the right side. This means that we obviously have anticrossing behavior. But is this enough to indicate and prove strong coupling?

In Fig. 7.27, you see a simulation of the oscillator strength dependence of the feature at zero angle of incidence. For $f = 1$, the two oscillators have their original strength, while S_i^2 is doubled for $f = 2$ and zero for $f = 0$. The resulting feature reminds of the dip in the reflectance spectrum shown in Fig. 7.22. The dip occurs within the interference feature and pushes increasingly the borders outside the larger S_i^2 becomes. The question remains if the two features we see are indeed hybrid vibration-interference features since neither their intensity nor their area changes. In this regard, it is also instructive to do

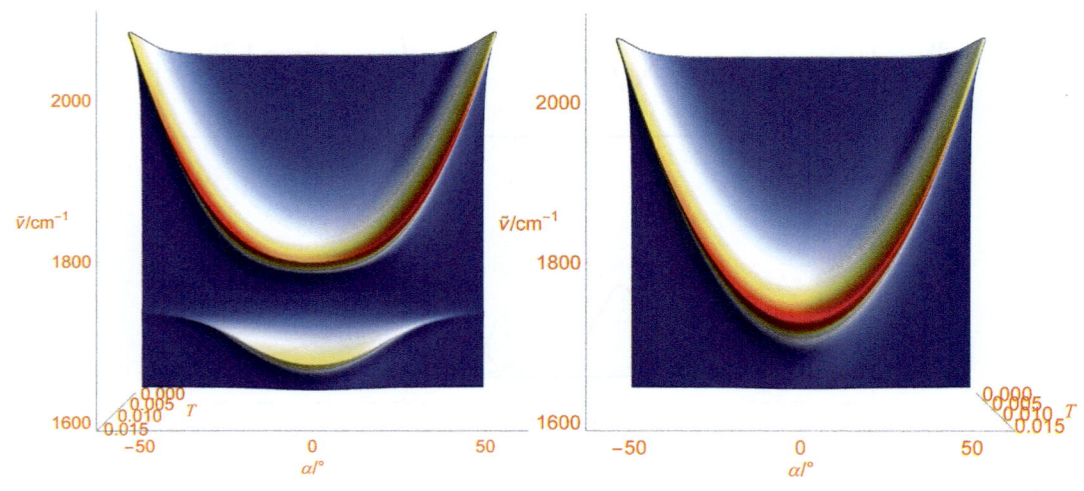

FIG. 7.26 Left panel: Dependence of the PMMA transmittance spectrum around the 1730 cm^{-1} band on the angle of incidence. Right panel: The same as in the left panel under the assumption that the band is not present.

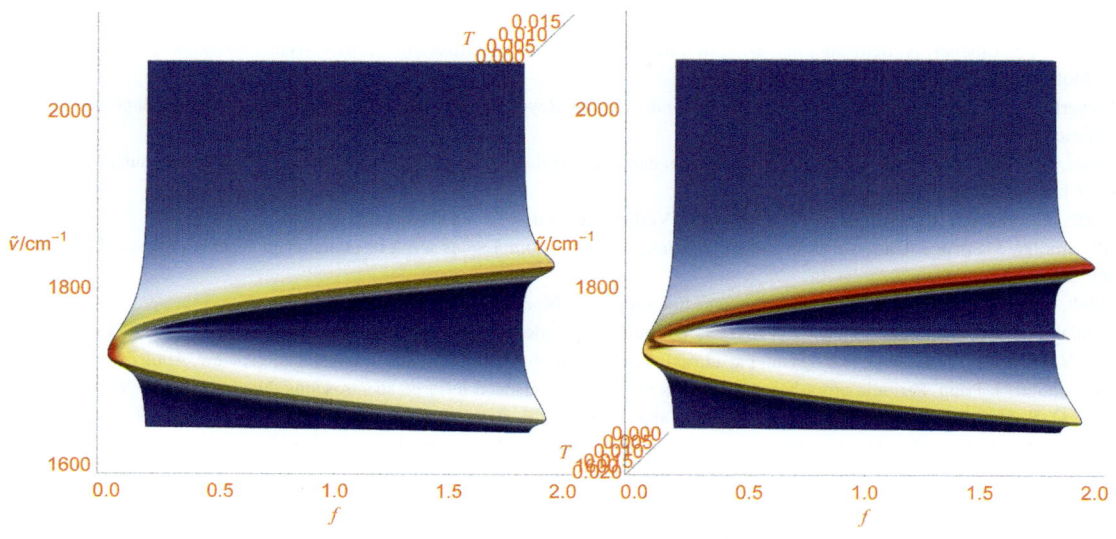

FIG. 7.27 Left panel: Dependence of the PMMA transmittance spectrum around the 1730 cm^{-1} band on the oscillator strength. Right panel: Same as left panel but with assuming a zero imaginary part of the dielectric function of the PMMA.

the same calculation with the imaginary part of the dielectric function of the PMMA set to zero. Apart from a new feature, the wavenumber position of which remains constant, the two branches of the interference feature seen in the left panel remain practically unchanged in the right one. Since we know that the position of the interference features depends on the refractive index function, it stands to reason that there is no real coupling between interference feature and the C=O band, but that the splitting is just caused by the refractive index function of the PMMA. The stronger the refractive index change, the larger is the split. I also checked how the features would depend on the damping constants of the oscillator at 1730 cm^{-1} and it turned out that even a fourfold increase did not change the overall feature. All in all, I am not convinced that what we see in case of films of organic materials in a cavity has anything to do with the polaritons the existence of which was verified for inorganic materials. It is my opinion that more investigations are needed to arrive at a final conclusion.

7.4 Further reading

With regard to infrared refraction spectroscopy, I cannot recommend any other literature since this is a method I suggested in 2021, and I do not expect it to appear in any book except this one on the short term.

Concerning SEIRA, I would recommend the Handbook of Infrared Spectroscopy of Ultrathin Films [18], also because it has in general a solid theoretical background. It is certainly limited to planar films. If you are interested in periodic arrays, no books that I could recommend exist so far. Instead, I would point you to two of my review articles [15,20]. In the first one, you also find a critical examination of enhancement factors, something I miss in other publications, which is why I hesitate to recommend any other literature.

For strong coupling in the IR, I also do not know of any books to recommend, but the following reviews may be helpful [28–31].

References

[1] T.G. Mayerhöfer, V. Ivanovski, J. Popp, Infrared refraction spectroscopy, Appl. Spectrosc. 75 (2021) 1526–1531.
[2] A. Ebner, R. Zimmerleiter, C. Cobet, K. Hingerl, M. Brandstetter, J. Kilgus, Sub-second quantum cascade laser based infrared spectroscopic ellipsometry, Opt. Lett. 44 (2019) 3426–3429.
[3] J. Hayden, S. Hugger, F. Fuchs, B. Lendl, A quantum cascade laser-based Mach-Zehnder interferometer for chemical sensing employing molecular absorption and dispersion, Appl. Phys. B Lasers Opt. 124 (2018).
[4] S. Lindner, J. Hayden, A. Schwaighofer, T. Wolflehner, C. Kristament, M. González-Cabrera, S. Zlabinger, B. Lendl, External cavity quantum cascade laser-based mid-infrared dispersion spectroscopy for qualitative and quantitative analysis of liquid-phase samples, Appl. Spectrosc. 74 (2020) 452–459.
[5] T. Huffman, R. Furstenberg, C. Kendziora, R. McGill, Infrared Complex Reflectance Micro-Spectroscopy, SPIE, 2021.

[6] J.M. Chalmers, P.R. Griffiths, Handbook of Vibrational Spectroscopy, J. Wiley, Chichester, 2002.
[7] A. Dabrowska, S. Lindner, A. Schwaighofer, B. Lendl, Mid-IR dispersion spectroscopy – a new avenue for liquid phase analysis, Spectrochim. Acta A Mol. Biomol. Spectrosc. 286 (2023) 122014.
[8] T.G. Mayerhöfer, A. Dabrowska, A. Schwaighofer, B. Lendl, J. Popp, Beyond Beer's law: why the index of refraction depends (almost) linearly on concentration, ChemPhysChem 21 (2020) 707–711.
[9] J.W. Brühl, Über den Einfluss der einfachen und der sogenannten mehrfachen Bindung der Atome auf das Lichtbrechungsvermögen der Körper, Z. Phys. Chem. 1U (1887) 307–361.
[10] W.A. Roth, F. Eisenlohr, Refraktometrisches Hilfsbuch, Verlag von Veit & Comp., Leipzig, 1911.
[11] T.G. Mayerhofer, S. Spange, Understanding refractive index changes in homologous series of unbranched organic compounds based on Beer's law, ChemPhysChem 24 (2023) e202300430.
[12] H. Landolt, Ueber die Brechungsexponenten flüssiger homologer Verbindungen, Ann. Phys. 193 (1862) 353–385.
[13] D.R. Salahub, C. Sandorfy, The far-ultraviolet spectra of some simple alcohols and fluoroalcohols, Chem. Phys. Lett. 8 (1971) 71–74.
[14] Y. Ozaki, Y. Morisawa, A. Ikehata, N. Higashi, Far-ultraviolet spectroscopy in the solid and liquid states: a review, Appl. Spectrosc. 66 (2012) 1–25.
[15] T.G. Mayerhöfer, J. Popp, Periodic array-based substrates for surface-enhanced infrared spectroscopy, Nano 7 (2018) 39–79.
[16] Y. Zhang, Q.-S. Meng, L. Zhang, Y. Luo, Y.-J. Yu, B. Yang, Y. Zhang, R. Esteban, J. Aizpurua, Y. Luo, J.-L. Yang, Z.-C. Dong, J.G. Hou, Sub-nanometre control of the coherent interaction between a single molecule and a plasmonic nanocavity, Nat. Commun. 8 (2017) 15225.
[17] D. Solonenko, O.D. Gordan, A. Milekhin, M. Panholzer, K. Hingerl, D.R.T. Zahn, Interference-enhanced Raman scattering of F16CuPc thin films, J. Phys. D Appl. Phys. 49 (2016) 115502.
[18] V.P. Tolstoy, I. Chernyshova, V.A. Skryshevsky, Handbook of Infrared Spectroscopy of Ultrathin Films, Wiley, 2003.
[19] G. Bakan, S. Ayas, A. Dana, Tunable enhanced infrared absorption spectroscopy surfaces based on thin VO_2 films, Optic. Mater. Express 8 (2018) 2190–2196.
[20] T.G. Mayerhöfer, S. Pahlow, J. Popp, Structures for surface-enhanced nonplasmonic or hybrid spectroscopy, Nanophotonics 9 (2020) 741.
[21] T.G. Mayerhöfer, J. Popp, Electric field standing wave effects in internal reflection and ATR spectroscopy, Spectrochim. Acta A Mol. Biomol. Spectrosc. 191 (2018) 165–171.
[22] N.J. Harrick, Internal Reflection Spectroscopy, Interscience Publishers, New York, 1967.
[23] P.G. Etchegoin, E.C. Le Ru, M. Meyer, An analytic model for the optical properties of gold, J. Chem. Phys. 125 (2006) 164705.
[24] P.G. Etchegoin, E.C. Le Ru, M. Meyer, Erratum: "An analytic model for the optical properties of gold" (J. Chem. Phys. 125, 164705 (2006)), J. Chem. Phys. 127 (2007) 189901.
[25] T.G. Mayerhöfer, W.D.P. Costa, J. Popp, Sophisticated attenuated total reflection correction within seconds for unpolarized incident light at 45°, Appl. Spectrosc. 78 (3) (2024) 321–328, https://doi.org/10.1177/00037028231219528.
[26] T.G. Mayerhöfer, R. Knipper, U. Hübner, D. Cialla-May, K. Weber, H.-G. Meyer, J. Popp, Ultra sensing by combining extraordinary optical transmission with perfect absorption, ACS Photonics 2 (2015) 1567–1575.
[27] P. Grosse, Conventional and unconventional infrared spectrometry and their quantitative interpretation, Vib. Spectrosc. 1 (1990) 187–198.
[28] T.W. Ebbesen, Hybrid light-matter states in a molecular and material science perspective, Acc. Chem. Res. 49 (2016) 2403–2412.
[29] D.S. Dovzhenko, S.V. Ryabchuk, Y.P. Rakovich, I.R. Nabiev, Light-matter interaction in the strong coupling regime: configurations, conditions, and applications, Nanoscale 10 (2018) 3589–3605.
[30] M. Hertzog, M. Wang, J. Mony, K. Börjesson, Strong light-matter interactions: a new direction within chemistry, Chem. Soc. Rev. 48 (2019) 937–961.
[31] X. Yu, Y. Yuan, J. Xu, K.-T. Yong, J. Qu, J. Song, Strong coupling in microcavity structures: principle, design, and practical application, Laser Photonics Rev. 13 (2019) 1800219.
[32] A. Shalabney, J. George, J. Hutchison, G. Pupillo, C. Genet, T.W. Ebbesen, Coherent coupling of molecular resonators with a microcavity mode, Nat. Commun. 6 (2015) 5981.
[33] A. Thomas, L. Lethuillier-Karl, K. Nagarajan, R.M.A. Vergauwe, J. George, T. Chervy, A. Shalabney, E. Devaux, C. Genet, J. Moran, T.W. Ebbesen, Tilting a ground-state reactivity landscape by vibrational strong coupling, Science 363 (2019) 615–619.
[34] T.G. Mayerhöfer, S. Pahlow, V. Ivanovski, J. Popp, Dispersion related coupling effects in IR spectra on the example of water and Amide I bands, Spectrochim. Acta A Mol. Biomol. Spectrosc. 288 (2023) 122115.
[35] M. Born, E. Wolf, A.B. Bhatia, Principles of Optics: Electromagnetic Theory of Propagation, Interference and Diffraction of Light, Cambridge University Press, 1999.
[36] C.J.F. Böttcher, Theory of Electric Polarization-Dielectrics in Static Fields, Elsevier Scientific Publishing Company, 1973.
[37] C.J.F. Böttcher, Theory of Electric Polarization-Dielectrics in Time-Dependent Fields, Elsevier, 1978.
[38] B.M. Auer, J.L. Skinner, IR and Raman spectra of liquid water: theory and interpretation, J. Chem. Phys. 128 (2008) 224511.
[39] S. Habershon, G.S. Fanourgakis, D.E. Manolopoulos, Comparison of path integral molecular dynamics methods for the infrared absorption spectrum of liquid water, J. Chem. Phys. 129 (2008) 074501.
[40] M. Yang, J.L. Skinner, Signatures of coherent vibrational energy transfer in IR and Raman line shapes for liquid water, Phys. Chem. Chem. Phys. 12 (2010) 982–991.
[41] G.R. Medders, F. Paesani, Infrared and Raman spectroscopy of liquid water through "first-principles" many-body molecular dynamics, J. Chem. Theory Comput. 11 (2015) 1145–1154.
[42] K.M. Hunter, F.A. Shakib, F. Paesani, Disentangling coupling effects in the infrared spectra of liquid water, J. Phys. Chem. B 122 (2018) 10754–10761.

[43] T.E. Li, J.E. Subotnik, A. Nitzan, Cavity molecular dynamics simulations of liquid water under vibrational ultrastrong coupling, Proc. Natl. Acad. Sci. 117 (2020) 18324–18331.
[44] F. Paesani, G.A. Voth, A quantitative assessment of the accuracy of centroid molecular dynamics for the calculation of the infrared spectrum of liquid water, J. Chem. Phys. 132 (2010) 014105.
[45] T. Hasegawa, Y. Tanimura, A polarizable water model for intramolecular and intermolecular vibrational spectroscopies, J. Phys. Chem. B 115 (2011) 5545–5553.
[46] C.D. Keefe, Curvefitting imaginary components of optical properties: restrictions on the lineshape due to causality, J. Mol. Spectrosc. 205 (2001) 261–268.
[47] T.G. Mayerhöfer, S. Höfer, V. Ivanovski, J. Popp, Understanding longitudinal optical oscillator strengths and mode order, Phys. B Condens. Matter 597 (2020) 412398.
[48] F. Gervais, Infrared dispersion in several polar-mode crystals, Opt. Commun. 22 (1977) 116–118.
[49] T.G. Mayerhöfer, J. Popp, Revisiting longitudinal optical modes in materials with plasmon and plasmon-like absorptions – $SrTiO_3$ and $\beta\text{-}Ga_2O_3$, Phys. B Condens. Matter 590 (2020) 412229.
[50] R. Arul, D.-B. Grys, R. Chikkaraddy, N.S. Mueller, A. Xomalis, E. Miele, T.G. Euser, J.J. Baumberg, Giant mid-IR resonant coupling to molecular vibrations in sub-nm gaps of plasmonic multilayer metafilms, Light Sci. Appl. 11 (2022) 281.
[51] K.U.N. Huang, Lattice vibrations and optical waves in ionic crystals, Nature 167 (1951) 779–780.
[52] K. Huang, On the interaction between the radiation field and ionic crystals, Proc. R. Soc. Lond. A Math. Phys. Sci. 208 (1951) 352–365.
[53] U. Fano, Atomic theory of electromagnetic interactions in dense materials, Phys. Rev. 103 (1956) 1202–1218.
[54] J.J. Hopfield, Theory of the contribution of excitons to the complex dielectric constant of crystals, Phys. Rev. 112 (1958) 1555–1567.
[55] C.H. Henry, J.J. Hopfield, Raman scattering by polaritons, Phys. Rev. Lett. 15 (1965) 964–966.
[56] J.P. Long, B.S. Simpkins, Coherent coupling between a molecular vibration and Fabry-Perot optical cavity to give hybridized states in the strong coupling limit, ACS Photonics 2 (2015) 130–136.

Chapter 8

2D correlation analysis

8.1 Basics

If you are not familiar with 2D correlation spectroscopy (2D-COS), you might be surprised that you find something about this technique in this book. For some of you, it might not seem to be more than a toy with which pleasant two- and three-dimensional illustrations can be created that seem to deliver hardly more than semiquantitative information. In fact, in what follows, I hope I will be able to convince you that 2D-COS can be extremely valuable. But first, after a short introduction of the theoretical framework, we will leave what is safely known and investigate the 2D correlation spectra of the Lorentz oscillator and the concept of local fields and see how this infusion changes the world of 2D-COS. This is very important, much more so than for conventional spectra, because as you will see, 2D-COS is extremely sensitive to nonlinear effects. Therefore, I guarantee you that you will never see an asynchronous map generated from real-world absorbance data (this is certainly an oxymoron, as absorbance is a mere artificial quantity) that has only zero values as would be the case if Beer's approximation would hold strictly (the asynchronous map, as you will see soon, compiles the nonlinear, or, better, the non-proportional effects in a spectral series).

The starting point is generally a series of spectra, but, as you will learn later, such a series can, in extreme cases, be composed of only two spectra, one of which will not even have to be an experimental one. If you want to draw conclusions about the correlation of bands in the spectra of a compound or a mixture, it is necessary that such a series is generated by varying one parameter a spectrum is sensitive to systematically and portion-wise, with every difference between two steps being the same. In the language of 2D-COS, you need a perturbation, a term, which has a historical background. Originally, Isao Noda, who is the inventor of 2D-COS and who was also involved in the development of polymers, had the idea to stretch and compress these polymers periodically. This is how he generated the first series of infrared spectra used for 2D-COS. Later on, he generalized the idea so that much more and different perturbations can be used [1]. As I have done this earlier, I generalize this to the point where a perturbation is every parameter that can be systematically varied and has an influence on a spectrum (any kind of and not even limited to!). In addition, if you are not interested in investigating the correlation in detail, the necessity of varying the parameter in equal steps can also be relieved—in principle, to the point where you do not even have to care about ordering the spectra in a way that changes the parameter systematically [2]. But let us first introduce the basics before we get lost in details. Say, we have a series of dynamic spectra $\tilde{y}(\tilde{\nu}_i, t_j)$ generated by varying a parameter t in a systematic way and put them into a matrix $\mathbf{Y}$:

$$\mathbf{Y} = \begin{bmatrix} \tilde{y}(\tilde{\nu}_1, t_1) & \tilde{y}(\tilde{\nu}_2, t_1) & \ldots & \tilde{y}(\tilde{\nu}_n, t_1) \\ \tilde{y}(\tilde{\nu}_1, t_2) & \tilde{y}(\tilde{\nu}_2, t_2) & \ldots & \tilde{y}(\tilde{\nu}_n, t_2) \\ \ldots & \ldots & \ldots & \ldots \\ \tilde{y}(\tilde{\nu}_1, t_m) & \tilde{y}(\tilde{\nu}_2, t_m) & \ldots & \tilde{y}(\tilde{\nu}_n, t_m) \end{bmatrix}, \tag{8.1}$$

where n is the number of discrete wavenumber points, m is the number of spectra in the series, $\Delta t = t_{j+1} - t_j$ and $\tilde{y}(\tilde{\nu}_i t_j) = y(\tilde{\nu}_i t_j) - \bar{y}(\tilde{\nu}_i t_j)$. $\bar{y}(\tilde{\nu}_i t_j)$ can, but not necessarily is, the average of all values in the series. In fact, sometimes it is necessary to set $\bar{y}(\tilde{\nu}_i, t_j) = 0$, and in other cases, it may be necessary to choose a particular spectrum of the series to be the reference, e.g., as we will see in the case of binary mixtures, the spectrum of one of the neat components, i.e., an end-member spectrum. In this example, t would be the volume fraction. If t would be the layer thickness, it would be of advantage or sometimes even necessary to set $\bar{y}(\tilde{\nu}_i, t_j) = 0$. In any way, for this particular example, it is required for $t = 0$ that $\tilde{y}(\tilde{\nu}_i, t = 0) = 0$, as we will see later.

So far, we have not specified the type of infrared spectrum $y(\tilde{\nu}_i, t_j)$, reflectance, transmittance, or absorbance. If you asked someone working in the field of molecular spectroscopy, absorbance would nowadays be the logical choice. However, since 2D-COS is based on matrices and matrices have the property that their symmetry does not change if they are multiplied by a factor, it is sometimes advantageous to use reflectance or transmittance because you do not have to get rid of the baseline. The magnitude of the baseline is given by a ratio (value for the sample divided by the empty channel value) for these two (at least to a first approximation), which transforms into a difference in case of absorbance due to the logarithmic dependence. In the latter case, the symmetry of the matrices becomes distorted (if you ask yourself, why should I care about the symmetry of this matrix, please be patient, I will come to that).

In the next step, we will generate the variance-covariance matrix $\boldsymbol{\Phi}_{\tilde{\nu}\tilde{\nu}}$:

$$\boldsymbol{\Phi}_{\tilde{\nu}\tilde{\nu}} = \frac{1}{m-1}\mathbf{Y}^T\mathbf{Y}, \tag{8.2}$$

which is nothing else but the so-called synchronous spectrum. The synchronous spectrum comprises the changes linear in or, better, proportional to t. Those changes that are not proportional are captured by the asynchronous spectrum $\boldsymbol{\Psi}_{\tilde{\nu}\tilde{\nu}}$. To generate it, the Hilbert-Noda transformation matrix $\mathbf{N}$ must be computed:

$$N_{xy} = \begin{cases} 0 & \text{if } x = y \\ \dfrac{1}{\pi(y-x)} & \text{otherwise} \end{cases}. \tag{8.3}$$

The asynchronous spectrum is then given by

$$\boldsymbol{\Psi}_{\tilde{\nu}\tilde{\nu}} = \frac{1}{m-1}\mathbf{Y}^T\mathbf{N}\mathbf{Y}. \tag{8.4}$$

How do these spectra look like? Let's first start with something simple which is well-known, namely the synchronous and the asynchronous spectrum of absorbance bands assuming Lorentz profiles (cf. Section 5.2). Spectra with two peaks are shown in Fig. 8.1.

The peaks on the diagonal from $(1500\,\text{cm}^{-1}, 1500\,\text{cm}^{-1})$ to $(1600\,\text{cm}^{-1}, 1600\,\text{cm}^{-1})$ in the synchronous spectrum are called the auto(correlation)peaks. These peaks are contained in every synchronous spectrum since every peak is trivially fully correlated to itself. Correspondingly, these peaks are always positive and their intensity is related to the spectral change with changing t—trivially, they would vanish if the peak intensity in the normal spectrum were not a function of t. The two peaks off this diagonal are called the cross-peaks and can either be negative or positive. The intensities of the cross-peaks indicate either a simultaneous or coincidental change, and their sign depends on whether the intensities of the corresponding autopeaks change either in the same way (positive sign) or in the opposite direction (negative sign). The asynchronous spectrum in our particular example is null since the use of Lorentz profiles implicitly assumes the validity of Beer's approximation (cf. Section 5.2). As a consequence, molecular spectroscopists have a hard time interpreting experimental asynchronous spectra since 2D-COS is so sensitive that every deviation from Beer's approximation (and, certainly, also from Lambert's approximation) will be detected, even if it cannot be spotted in conventional spectra. The clue is that we know there will always be a deviation since oscillators should better be described by classical damped harmonic (Lorentz) oscillators (CDHO, cf. Section 5.1). Therefore, it is not surprising that the asynchronous spectra of such oscillators are generally rich, while the synchronous spectra are nearly the same as for Lorentz profiles, see Fig. 8.2.

Before we scrutinize the asynchronous spectrum, I want to point out that synchronous spectra are always symmetric to the diagonal from low to high wavenumbers, whereas asynchronous spectra are always antisymmetric. This is a very important feature that can be exploited in a very surprising way as we will see later. A typical pair of cross-peaks is very beautifully expressed for the peak located at $1530\,\text{cm}^{-1}$ (note that the average spectra are usually depicted above and on one

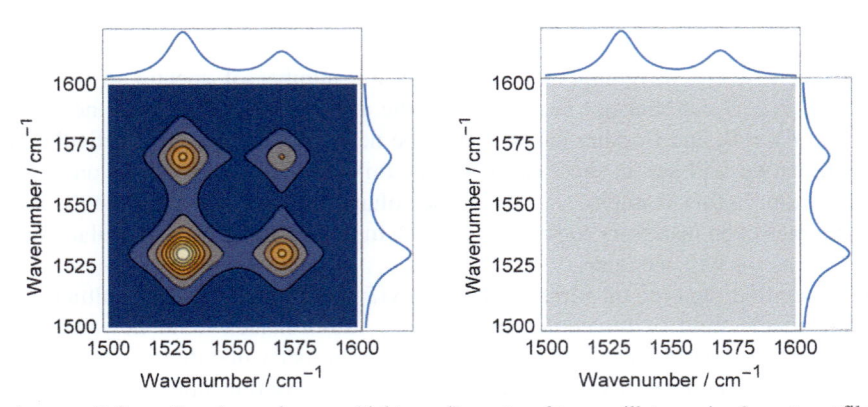

FIG. 8.1 Simulated synchronous (left panel) and asynchronous (right panel) spectra of two oscillators using Lorentz profiles.

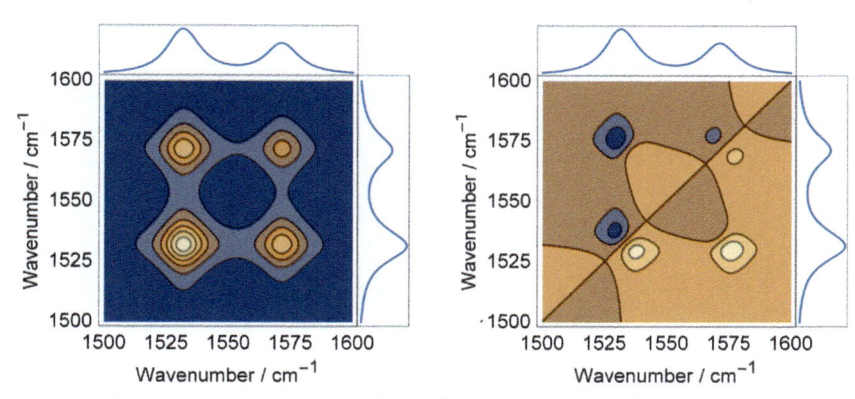

FIG. 8.2 Simulated synchronous (left panel) and asynchronous (right panel) of two oscillators using classical damped harmonic (Lorentz) oscillators.

side of the 2D correlation map) and indicates a peak shift of $1.2\,\text{cm}^{-1}$. You will see later that much smaller peak shifts, far below the spectral resolution, make themselves noticeable in the asynchronous spectrum with the same pattern. The reason that the second oscillator leaves a much smaller footprint in this spectrum, even though its peak shift is with $0.9\,\text{cm}^{-1}$ not much smaller, is due to the lower relative intensity. Correspondingly, if the first oscillator was removed, the same pattern as for the first oscillator would become visible also for the second oscillator. Concerning the cross-peaks, it is very obvious that they are also influenced by the peak shift. As we know from Section 5.2, a larger oscillator strength leads to a blue shift of the peaks. On the other hand, if we used the Lorentz-Lorenz equation, the dispersion relation would be:

$$\frac{\hat{n}^2-1}{\hat{n}^2+2} = \frac{1}{3}\sum_{j=1}^{N}\frac{S_j^2}{\tilde{\nu}_j^2-\tilde{\nu}^2-i\tilde{\nu}\gamma_j} \rightarrow \hat{n}^2 = \frac{\frac{2}{3}\sum_{j=1}^{N}\frac{S_j^2}{\tilde{\nu}_j^2-\tilde{\nu}^2-i\tilde{\nu}\gamma_j}+1}{1-\frac{1}{3}\sum_{j=1}^{N}\frac{S_j^2}{\tilde{\nu}_j^2-\tilde{\nu}^2-i\tilde{\nu}\gamma_j}}. \tag{8.5}$$

The local field of Lorentz leads to a redshift of the peaks (Fig. 8.3). Correspondingly, our pair of cross-peaks has been mirrored with regard to the diagonal compared to Fig. 8.2. Obviously, an asynchronous pair of peaks belonging to an autopeak is positive below the diagonal for blue shifts, whereas it is negative below the diagonal for redshifts. For cross-peaks, the general finding is that a cross-peak is positive if the intensity change at $\tilde{\nu}_1$ occurs "before" that at $\tilde{\nu}_2$ and negative if the opposite is true. For all this to work as described earlier the spectra of a series must be in consecutive order and the perturbations must be changed in equal steps.

As already mentioned, an asynchronous null spectrum derived from an experiment is impossible (maybe, if you disregard experimental errors, with the exception of vibrational circular dichroism spectra in dependence of the thickness [3], cf. Chapter 16). Actually, my interest in 2D-COS was caused by a visit of Prof. Noda in January 2020. I had the opportunity to talk to him about nonlinearities in absorbance spectra during lunch. I do not remember exactly what he said, but he confirmed indirectly that many more nonlinearities are present in 2D-COS as expected based on the BBL approximation. My biased expectation, which I did not express, was actually, that it will never be possible to record a series of spectra, e.g.,

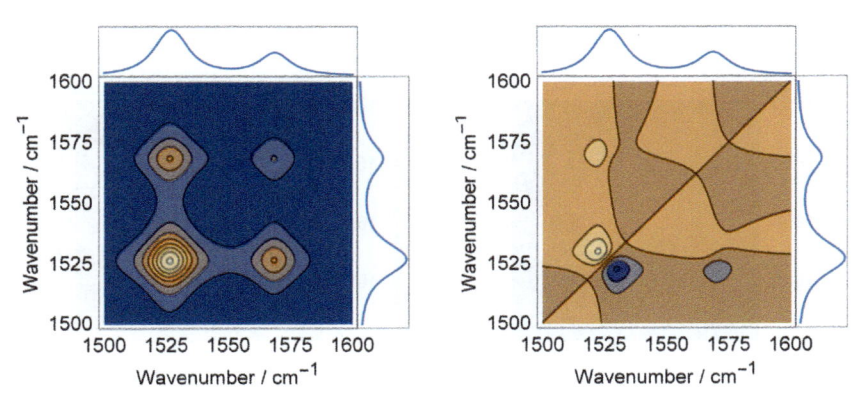

FIG. 8.3 Simulated synchronous (left panel) and asynchronous (right panel) of two oscillators using Lorentz oscillators and assuming a local field of Lorentz following Eq. (8.5).

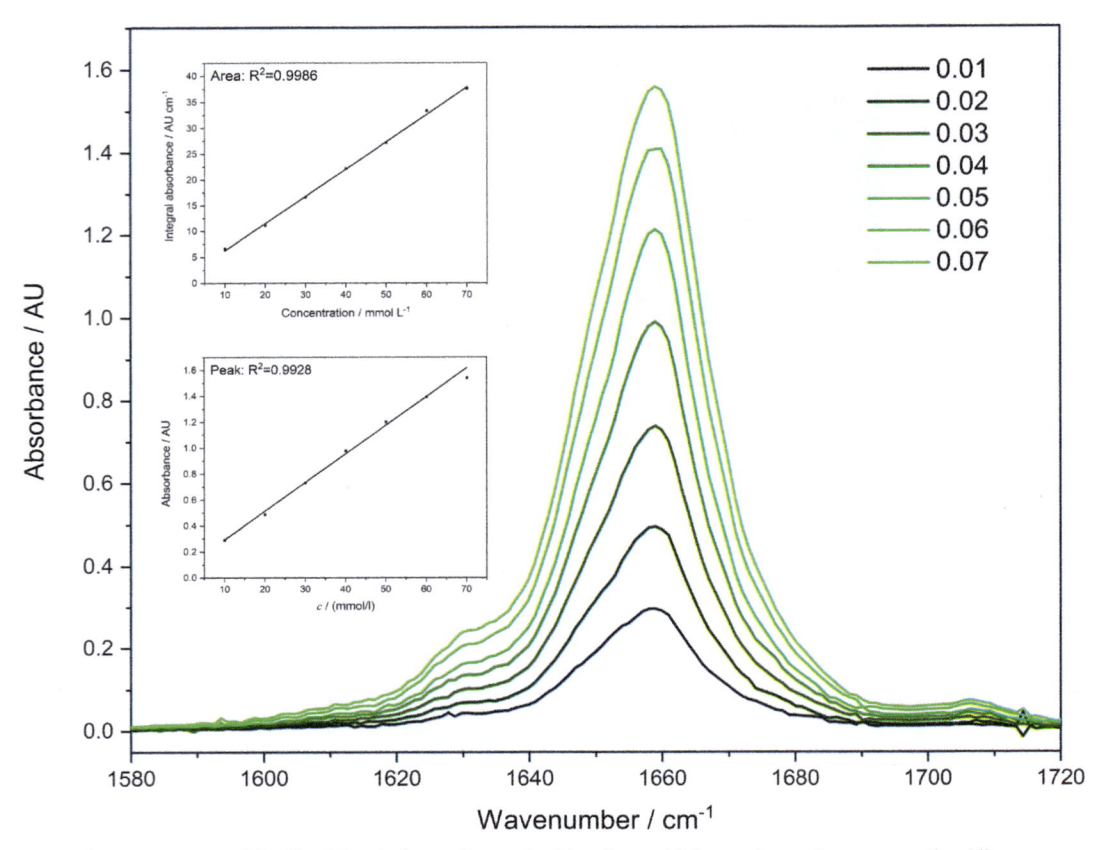

FIG. 8.4 The absorbance spectrum of the C=O band of caprolactam in chloroform with increasing molar concentration [4].

for a concentration calibration, that would lead to an asynchronous null spectrum. A couple of hours before, I had already investigated such a series by 2D-COS, namely caprolactam in chloroform, and I had confirmed the existence of nonlinearities. We had used this series originally to prove that there is a twin of Beer's approximation for the index of refraction, which changes approximately linearly with molar concentration (cf. Section 5.3) [4]. To verify the results, in addition to the refractive index changes, also the absorbance changes with molar concentration were determined. The results are provided in Fig. 8.4. Obviously, as can be seen from the insets, both the peak maxima as well as the area of the bands increase linearly with molar concentration. The R^2 values, i.e., the percentage of variance explained, are with values >99% very high, and the small part of nonexplained variance by the linear models can easily be explicated by changing chemical interactions. Let us have a look at the 2D correlation spectra (cf. Fig. 8.5). The synchronous spectrum does not have any surprises in store. Maybe the single autopeak is somewhat less symmetric than in the earlier examples because of the somewhat surprising

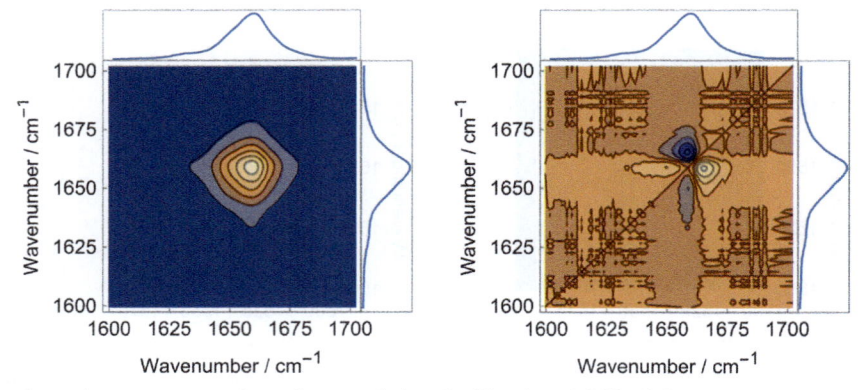

FIG. 8.5 Synchronous and asynchronous spectra of caprolactam solutions in chloroform (cf. Fig. 8.4).

shoulder in the spectrum, but Alicja Dabrowska and Andreas Schwaighofer, who were in charge of the experiments [4], verified that this shoulder is not an effect of insufficient purity.

The asynchronous spectrum is much more interesting because there is obviously a kind of cross-peak structure at the peak position which reveals a blueshift. If one just looks at the conventional spectra this shift is hard to notice. Actually, it seems that the peaks do not shift. When I normalized the curve, I noticed that the bands somewhat broadened, but only to higher wavenumbers. This can be an effect due to an increase of oscillator strength while the local electric field due to the molecules of the solvent stays constant, but the cause can certainly be also a change of the interaction between the solute molecule and the molecules of the solvent.

Such a change of interaction can be safely excluded only for a thermodynamically ideal system. As we have discussed already in Section 5.3, a real system, which comes close to such an ideal system, is the benzene-toluene system. A very interesting spectral range for this system is the one between 1425 and 1525 cm^{-1}, which is depicted in Fig. 8.6 [5].

It is really hard to make out any differences between the experimental and the forward calculated spectra. It seems that this is the point where Beer's approximation is unrestrictedly valid, also because the results obtained by the Lorentz-Lorenz relation seem to have converged to the ones obtained assuming Beer's approximation. If we take a look at the corresponding 2D correlation spectra in Fig. 8.7, it becomes immediately obvious that Beer's approximation still does not hold and that local field effects are at play, which can approximately be described by the Lorentz-Lorenz relation (the reasons why the Lorentz-Lorenz relation remains an approximation, although a better one compared to Beer's approximation, are discussed in detail in Chapter 9).

Before continuing this discussion, I want to point out again one important detail that has to be considered. That the synchronous map illustrates proportionate changes, while the asynchronous reveals disproportionate changes is only true if there is no constant term in the relation that describes the perturbation. If the perturbation variable is the volume fraction, then there is a corresponding problem. Say, the absorbance of the mixture i is given by the absorbance of the neat substances A_1 and A_2:

$$A_i = A_1 \varphi_1 + A_2 \varphi_2, \tag{8.6}$$

and, since we have a binary system, the relation $\varphi_2 = \varphi_1 - 1$ between the volume fractions holds.

Therefore,

$$A_i = A_1 \varphi_1 + A_2(1 - \varphi_1) = (A_1 - A_2)\varphi_1 + A_2. \tag{8.7}$$

This means that we would have to set $\bar{y}(\tilde{\nu}_i, t_j) = A_2$, otherwise the asynchronous spectrum would always be nonzero, even when Eq. (8.6) holds. But, as we have already seen, it does not remain valid.

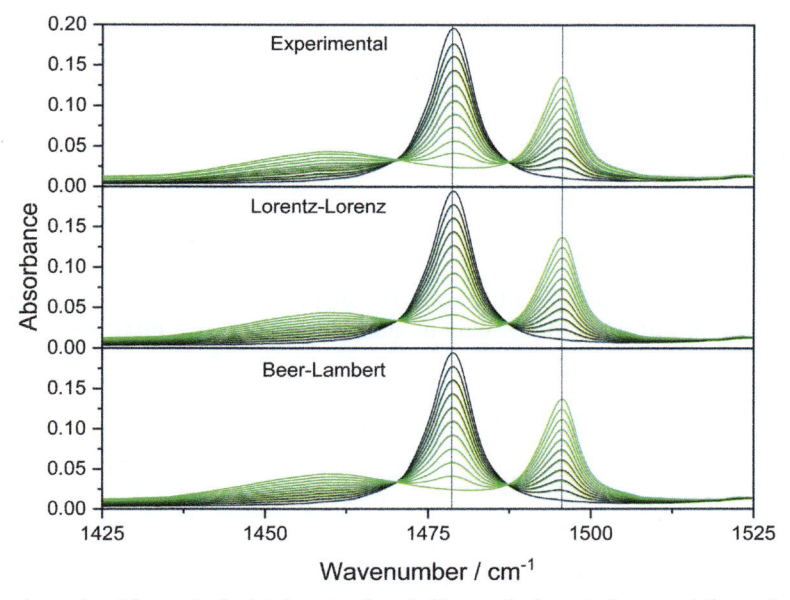

FIG. 8.6 Comparison of experimental and forward calculated spectra (based either on the Lorentz-Lorenz relation or the BBL approximation) of mixtures in the quasi-ideal benzene-toluene system [5].

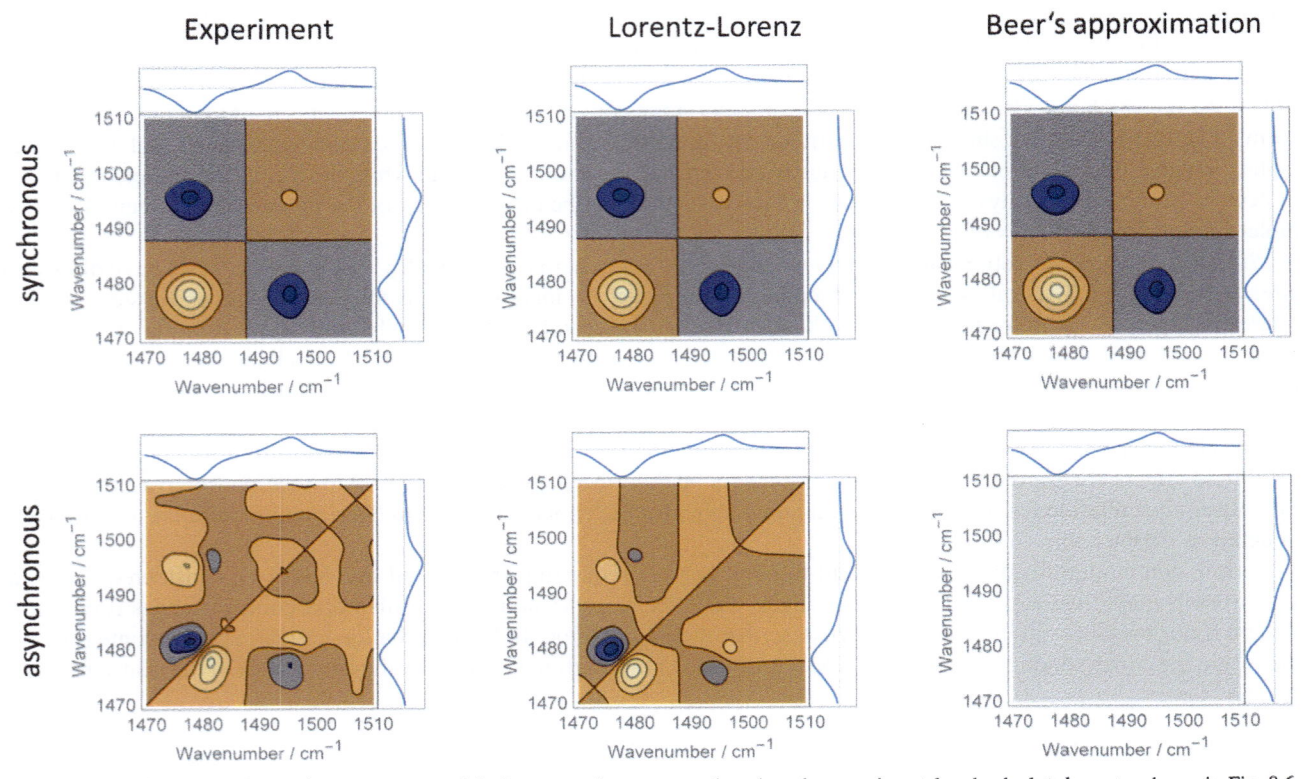

FIG. 8.7 Synchronous and asynchronous spectra of the benzene-toluene system based on the experimental and calculated spectra shown in Fig. 8.6.

In fact, if we go back to Fig. 8.6, and inspect the figure more carefully looking for differences, then a very small blueshift of the band at 1478.5 cm^{-1} of about 0.5 cm^{-1} becomes visible. Nevertheless, I can assure you that even when you go over to increasingly less intense bands, 2D correlation spectra will always reveal that there is a disproportionality which can approximately be described by the Lorentz-Lorenz relation. This should raise an alarm since conventional chemometrics is built on the validity of Beer's approximation. Those who use chemometrics will tell you that these disproportionalities/nonlinearities are not a problem for chemometrics because there are procedures which are able to tackle them. At least conventional chemometrics definitely has a larger problem as you will learn in the next chapter. While it seems that this chapter is close to its end, I promise you that it actually begins to get interesting because in the next section, we will turn around 2D correlation analysis and change it from a semiquantitative method into a fully quantitative tool with a more than surprising application for dispersion analysis.

8.2 Smart error sum

To develop such a fully quantitative tool, I will introduce you first to the concept of hetero/hybrid 2D correlation analysis (this section is based on Ref. [6]). In this kind of analysis, we fuse two different spectral series 1 and 2 in our 2D correlation maps:

$$\begin{aligned} \boldsymbol{\Phi}_{\tilde{\nu}\tilde{\nu}} &= \frac{1}{m-1}\mathbf{Y}_1^T\mathbf{Y}_2 \\ \boldsymbol{\Psi}_{\tilde{\nu}\tilde{\nu}} &= \frac{1}{m-1}\mathbf{Y}_1^T\mathbf{N}\mathbf{Y}_2 \end{aligned}. \tag{8.8}$$

If we had two different spectroscopic series and the same perturbation, the technique would be called hetero 2D correlation analysis. Another possibility is that the same spectroscopic technique is employed, but that two different perturbations are used. This is then called hybrid 2D correlation analysis. What if we splice two series of, say, infrared spectra, using the same perturbation, but one series contains experimental, whereas the other one modeled spectra? Have a look at Fig. 8.8.

Unoptimized hybrid map

Optimized hybrid map based on smart error sum

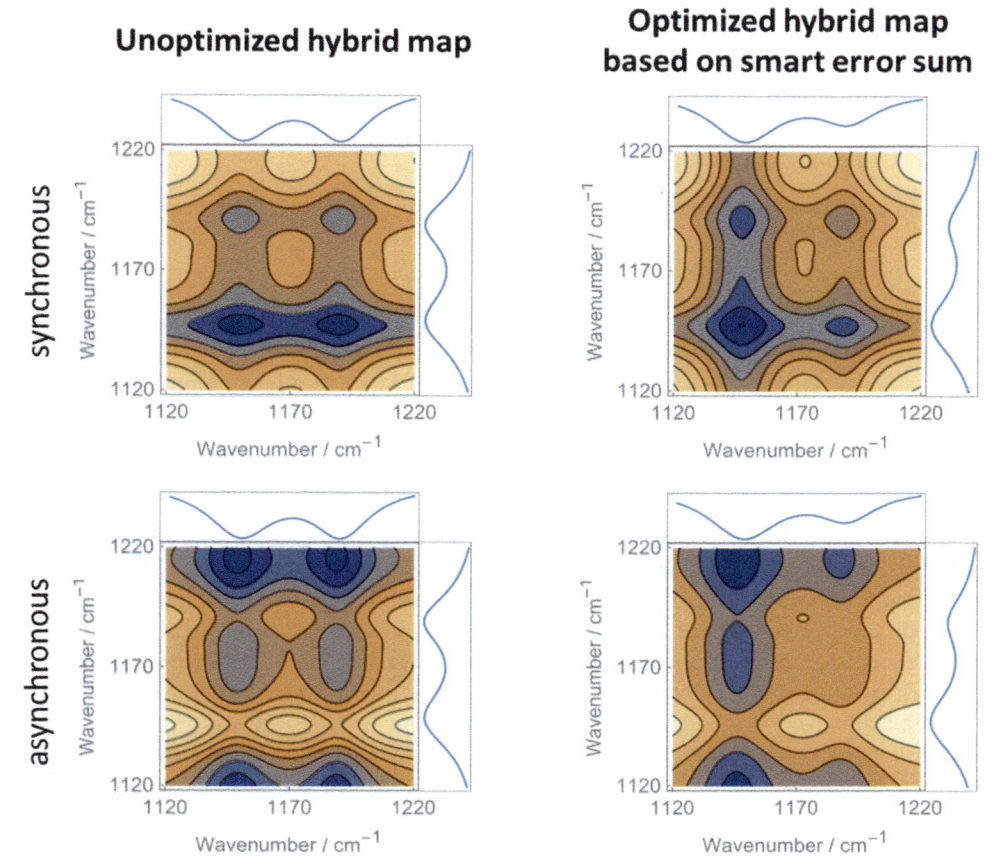

FIG. 8.8 Synchronous and asynchronous hybrid maps where one series contains experimental and the other modeled spectra [6].

If the modeled data deviate from the experimental ones, as in the left part of Fig. 8.8, the symmetry of the synchronous and the antisymmetry of the asynchronous maps are strongly disturbed. After the fit, symmetry and antisymmetry, respectively, are not perfect, but clearly improved.

Again, it seems this is semiquantitative at best, but mathematically we can express the symmetry in a synchronous map by

$$\Phi(\tilde{\nu}_j, \tilde{\nu}_k) = \Phi(\tilde{\nu}_k, \tilde{\nu}_j), \tag{8.9}$$

from which it follows that

$$\sum_{k=1}^{l} \sum_{j=k}^{l} \left[\Phi(\tilde{\nu}_j, \tilde{\nu}_k) - \Phi(\tilde{\nu}_k, \tilde{\nu}_j)\right] = 0. \tag{8.10}$$

In the case of hybrid 2D correlation, synchronous spectra do not necessarily obey the above condition. Accordingly, the difference between the elements $\Phi(\tilde{\nu}_j, \tilde{\nu}_k)$ and $\Phi(\tilde{\nu}_k, \tilde{\nu}_j)$ is a measure of spectral dissimilarity, which can be generally formulated as

$$D_S^p = \sum_{k=1}^{l} \sum_{j=k}^{l} \left[\Phi(\tilde{\nu}_j, \tilde{\nu}_k) - \Phi(\tilde{\nu}_k, \tilde{\nu}_j)\right]^p, \tag{8.11}$$

where D_S is the so-called Minkowski distance and, for $p=2$, the Euclidian distance. The Minkowski distance has been used previously to compare quantitatively two different synchronous spectra [6]—now we are able to do this directly using only one matrix. This is certainly nice, but we still have the yet unsolved problem what to do with the asynchronous maps. With the above idea, it is, however, not complicated to derive an analog expression for asynchronous maps. We know that conventional asynchronous maps are antisymmetric:

$$\Psi(\tilde{\nu}_j, \tilde{\nu}_k) = -\Psi(\tilde{\nu}_k, \tilde{\nu}_j). \tag{8.12}$$

Therefore, if we add up all elements linked by antisymmetry, the result is

$$\sum_{k=1}^{l} \sum_{j=k}^{l} \left[\Psi(\tilde{\nu}_j, \tilde{\nu}_k) + \Psi(\tilde{\nu}_k, \tilde{\nu}_j) \right] = 0. \tag{8.13}$$

For hybrid maps of two series that are nonidentical, the above double sum will be different from zero:

$$D_A^p = \sum_{k=1}^{l} \sum_{j=k}^{l} \left[\Psi(\tilde{\nu}_j, \tilde{\nu}_k) + \Psi(\tilde{\nu}_k, \tilde{\nu}_j) \right]^p. \tag{8.14}$$

If we set $p = 2$, we can see D_S^2 and D_A^2 as special residual sums of squares, which we call in the following the synchronous and the asynchronous sums of squares, SRSS and ARSS. The latter allows not only to compare asynchronous maps quantitatively—it also helps perform inverse modeling and, in particular, carrying out dispersion analysis. We can combine both SRSS and ARSS in one residual sum of squares by [7]:

$$\ln(\text{SRSS}) + \ln(\text{ARSS}) = \text{C2DCRSS}. \tag{8.15}$$

C2DCRSS stands for combined 2D correlation residual sum of squares and the logarithm is needed to acknowledge the different magnitudes of SRSS and ARSS. Since C2DCRSS is somehow an awkward term, I call it a *smart error sum*. Why should this error sum be smart? Actually, it is as smart as an artificial neural network is intelligent.

When we perform dispersion analysis (or band fitting if you use Lorentz profiles to fit absorbance spectra), we assume that the data we use, i.e., the spectra, are free of any systematic error. If the spectra contained systematic errors, it would be smart not to force the fit spectrum to agree with the experimental spectrum at all costs, but this is exactly what the use of a conventional error sum leads to.

What the smart error sum causes is to force the fit to increase the symmetry of the synchronous hybrid 2D correlation map and the antisymmetry of the asynchronous map. In other words, it tries to establish the same correlations in the set of fitted spectra as those that can be found in the experimental spectra. How does this help? Imagine the Beer-Lambert approximation was correct and you have a set of 5 spectra of thin films on a substrate with the thicknesses $(500 + (i - 1) \cdot 100)$ nm and the absorbance of a band would be given by $A_{\text{real}} = 1 + (i - 1) \cdot 0.2$. Now think of a systematic error that increases absorbance by 10%: $A_{\text{measured}} = (1 + (i - 1) \cdot 0.2) \cdot 1.1$. In the latter example, the correlation between the absorbances is different from the true values. If you now generate modeled spectra with the BBL approximation using the smart error sum and force the thicknesses to follow the relation $(a + (i - 1) \cdot b)$, the fitted spectra could always agree with A_{measured} [7]. But if there is a disproportionate relationship, which in reality exists, then the correlations between the fitted spectra could never agree with the experimental ones. If I have confused you with this example, let us forget the smartness for a moment and focus instead on the symmetry properties of the hybrid synchronous and asynchronous spectra. First, assume that we take the experimental reflectance spectra of PMMA layers on gold and generate a second series by multiplying the original series by a factor of two. The result is depicted in Fig. 8.9.

On a first view, it seems that the multiplication by a factor of 2 has not altered the 2D correlation maps. However, this is true only for the synchronous spectrum, while the asynchronous spectrum shows that the disproportionate part has changed somewhat. What is still strictly preserved, however, is the antisymmetry of the asynchronous spectrum. To generate the second pair of hybrid maps on the right side of Fig. 8.9, the original series was not multiplied by a common factor, but divided by individual factors. For conventional dispersion analysis, it would be necessary to do this because all spectra have spectral ranges where the reflectance was higher than unity and, thus, I corrected them by dividing them by their highest value after denoising (the spectra were only denoised to find the highest value. Then the original spectra were divided by this value and used for dispersion analysis). The uncorrected and the corrected spectra are depicted in Fig. 8.10. Note that this mathematical procedure does certainly not correct the problem. The reason for the values above unity are changes in the reference spectrum, which are exemplary shown in Fig. 8.11. These changes are wavenumber dependent, which is why a mere division of a spectrum by its highest value cannot remove the systematic error. This error is obviously even changing from measurement to measurement. It may be necessary to add that we did not try to improve the experimental situation, since we actually wanted to show that even with low quality spectra the interference effects in these spectra can be corrected (cf. Section 6.3 and Ref. [8]).

The division of the spectra by individual numbers still leads to symmetric synchronous maps. The antisymmetry of the asynchronous map, however, is no longer perfect, but it seems to be at least close to antisymmetry so that we can also in this case assume that the use of the *smart error sum* should be able to make a difference. For the following procedure, a proper

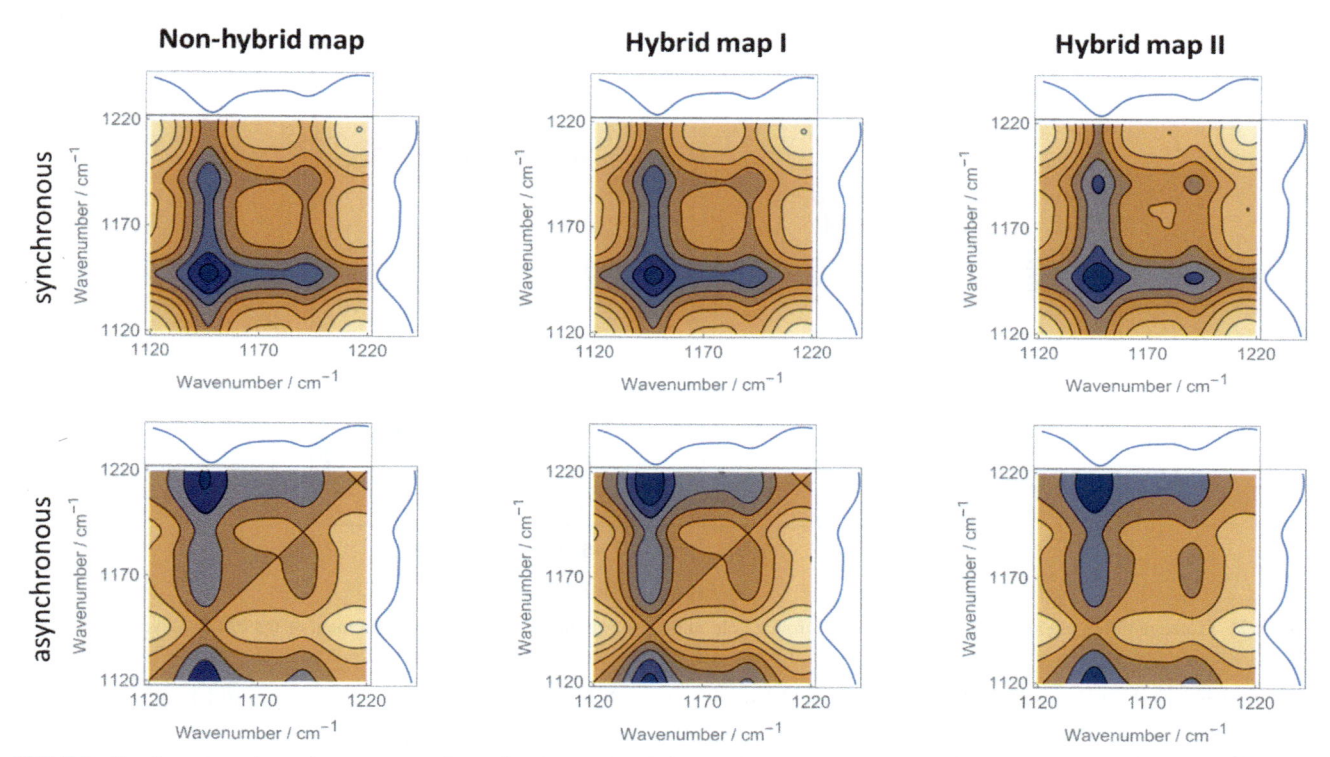

FIG. 8.9 Synchronous and asynchronous maps of experimental spectra of PMMA layers on gold. For the hybrid maps I, a second series was generated by multiplying the original one by a factor of 2. For the hybrid maps II, the experimental spectra were divided by their highest values after denoising.

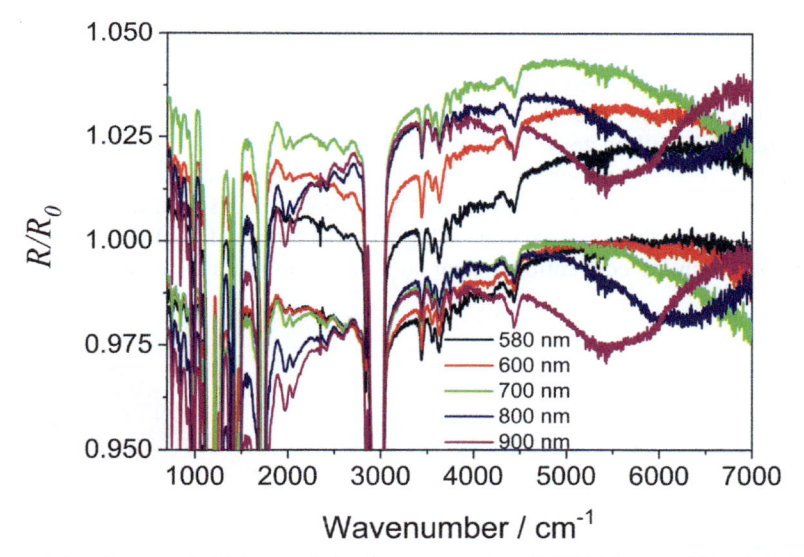

FIG. 8.10 Corrected (below unity) and uncorrected (above unity) reflectance spectra of PMMA layers with various thicknesses as indicated in the figure on gold.

underlying physical model is of high importance, i.e., in this particular case the use of wave optics on the one hand and the employment of dispersion theory on the other hand. This ensures that reflectance cannot be larger than unity and that non-linear relationships, e.g., between thickness or concentration and absorption index function, are properly established. Accordingly, the physical model together with the use of the smart error sum assures that the correlation found in the spectral series is also established in the series of the simulated spectra during the fitting process. To check this assumption,

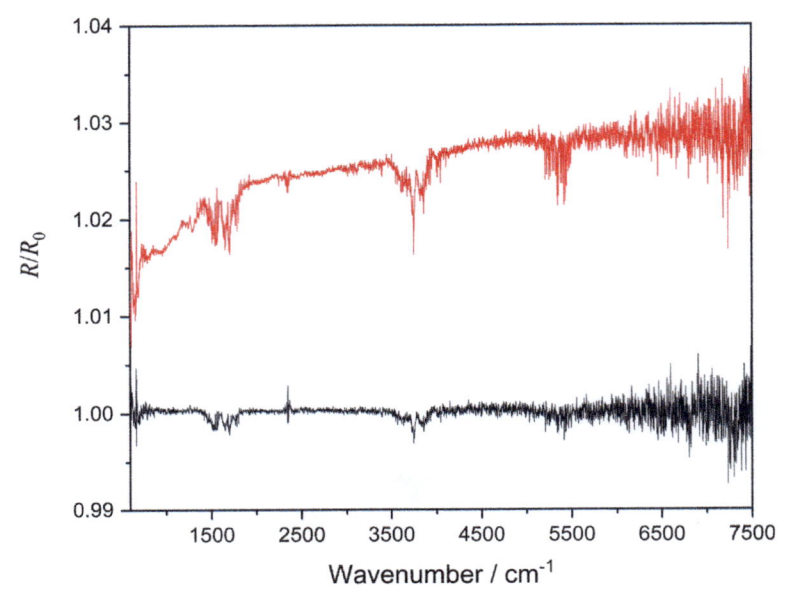

FIG. 8.11 Change of the reference spectrum (reflection from a gold mirror) during the course of the measurement of the PMMA layers on gold [8].

the experimental spectra of PMMA layers on gold were very helpful, since on the one hand we could be sure that they contain systematic errors, on the other hand, we could crosscheck the results of dispersion analysis with those obtained from PMMA layers on Si and CaF_2. Also, we could check how dispersion analysis of spectra corrected with an individual but wavelength-independent factor would perform.

But before we look into the fits, let us first check if we can further illuminate how and why the smart error sum works. Imagine a single peak in an artificial spectrum of a layer having an oscillator with the oscillator parameters $S^2 = 200\,cm^{-1}$, $\gamma = 30\,cm^{-1}$, and $\tilde{\nu}_0 = 1730\,cm^{-1}$ on gold. Based on these oscillator parameters and the suitable formula for the reflectance (cf. Section 6.3) we generate a series of spectra by varying the layer thickness. Then we calculate the squared differences between this series and one that is generated while we vary S between 180 and $220\,cm^{-1}$. Certainly, with an increasing difference between S and the value $S = 200\,cm^{-1}$ with which we computed the series, the conventional error sum will increase, while at $S = 200\,cm^{-1}$ it will be zero. Now we generate a new series of spectra by multiplying each spectrum with the same factor F, say $F = 1.05$ and varying S again between 180 and $220\,cm^{-1}$. This time we do not obtain the value zero, but a minimum instead. This minimum, however, will not be located at $S = 200\,cm^{-1}$, but at a lower value since the corresponding reflectance minima in the spectra will be at higher reflectance values. Indeed, introducing an error of 5% will decrease the value for S for which the conventional error sum has its minimum by nearly 10%, cf. Fig. 8.12.

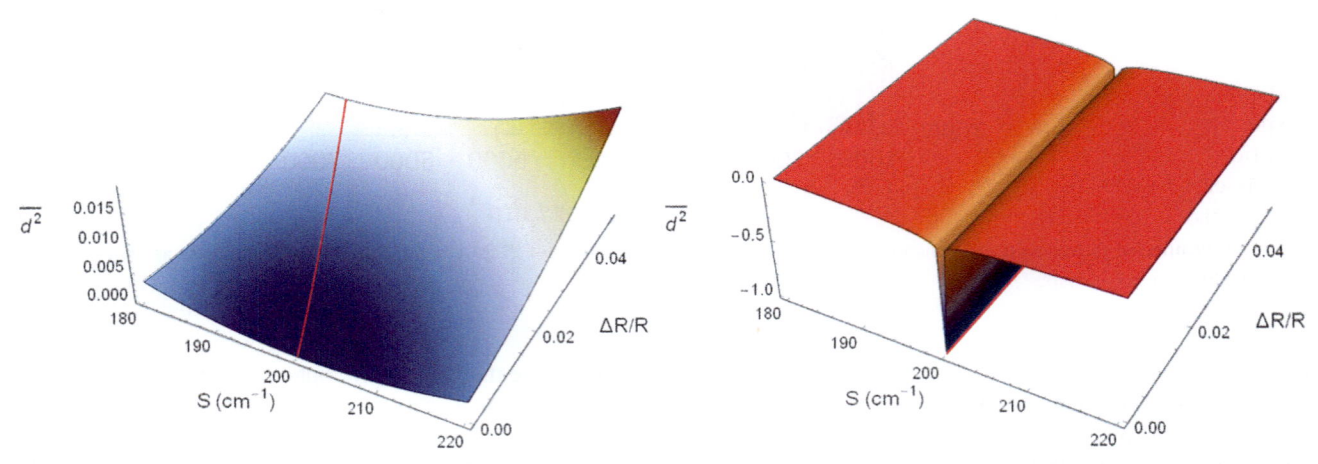

FIG. 8.12 Variance of the minimum of the error sums with S and the relative error in the reflectance. Left: Conventional error sum. Right: Smart error sum.

In contrast, the minimum of the smart error sum stays at $S=200\,cm^{-1}$, unperturbed from the error introduced into the reflectance spectra. In reverse, when the erroneous series of spectra is fitted with the smart error sum, the result will provide the true value irrespective of the error as long as it is multiplicative. Since experimental reflectance and transmittance are calculated as ratios of intensity of the sample to intensity of the reference (either a gold mirror for reflectance or an empty beam for transmittance), the errors usually are multiplicative. Note that converting reflectance or transmittance to absorbance would render such multiplicative errors into additive ones as already mentioned. Therefore, employing fits based on the smart error sum to absorbance spectra would not work. If you are still confused about how the smart error sum works— no worries, we will examine this problem again later on, but from a somewhat different angle. But before, I will show you some proof that it does work.

It should be clear that we run into serious problems when we try to fit a spectrum above unity by dispersion analysis and the conventional sum of squared errors because, as I already mentioned, the simulated spectrum cannot exceed unity. Accordingly, it is generally problematic to fit weak bands because their minima might not even be below unity. Nevertheless, to illustrate the problem, I tried to do this. But before we examine the results, let me remind you that even when we use the conventional error sum, this is not an ordinary dispersion analysis as we discussed in Section 6.7. The reason is that we talk about a series of spectra, in this particular case of 5 PMMA layers on gold, the strong reflective properties of which lead to strong interference effects inside the layers (cf. Section 6.3). The usual way would be to perform dispersion analysis individually on each of the spectra (and we will do this later on also in this chapter). Instead, I wrote a program, which allows to fit a series of spectra simultaneously, with a common set of oscillator parameters. This is something that is very useful in this particular case. Why? Because I noticed that the individual fits I did beforehand had problems in the regions of the C—H stretching vibrations (above $3000\,cm^{-1}$). The best-fit spectra and the experimental spectra deviated strongly and the band shapes in the best-fit spectra looked odd. As soon as I used the new program, suddenly the fit quality around the bands increased clearly. Obviously, the fit algorithm (I use a custom-made version of the simplex algorithm by Nelder and Mead [9], because it is much more flexible than the built-in functions of Mathematica, combined with thermal annealing [10], which I, however, usually deactivate), could not find the correct minimum, when only one spectrum was fitted, but was able to make it out, when the fit was based on a series. Certainly, just because a series is used, random errors are also reduced, but systematic errors remain, which brings us back to the *smart error sum*. When I replaced the conventional error sum by the smart one, (which only involved a few lines of code), and started the fit, it certainly took longer, since double sums have to be calculated. This means that the computational effort increases with the square of the wavenumber points. The result, however, was very recompensing as can be seen in Fig. 8.13.

If the conventional error sum is used, a number of weak bands cannot be fitted as already discussed, simply because these bands are located above the baseline. In addition, the stronger bands show intensities that are systematically too low. In contrast, on the first view, the comparison between experimental and best-fit average spectrum seems to be odd for the *smart error sum* because the spectra do not agree. The simulated average spectrum is downward shifted relative to the experimental spectrum, but with the relative intensities and band shapes preserved. As a consequence, when we compare the index of absorption spectra obtained by the fit with that obtained from transmittance spectra of PMMA layers on CaF_2 without systematic errors, we understand that the absorption index spectra agree much better than if we compare the result with that obtained by the conventional error sum. How do we know that the transmittance spectra of PMMA on CaF_2 are more or less error-free? First of all, because the corresponding spectra had transmittances slightly below that of the pure substrate, the conventional fit spectra and those obtained by the *smart error sum* largely agree. In fact, in Fig. 8.13 there is a small difference visible because CaF_2 has a strong phonon mode, which renders it nearly nontransparent below about $800\,cm^{-1}$. Therefore, the corresponding band at about $750\,cm^{-1}$ shows large differences between the left and the right side in this figure. In addition, we also had a set of transmittance spectra of PMMA layers on Si (which I do not show here, since also in this case dispersion analysis with the conventional and the *smart error sum* led to essentially the same result), which also agreed with those from the spectra of PMMA layers on CaF_2.

How would this comparison look like if the conventional and the *smart error sum* were applied for dispersion analysis of the naively corrected spectra? We know from Fig. 8.11, that the error changes nonlinearly with the wavenumber. It is therefore interesting to see how the *smart error sum* deals with the naive correction. The results for the conventional and the *smart error sum* are compared in Fig. 8.14. On the first view, the fit based on the conventional error sum leads to meaningful results, but a comparison of the absorption index spectrum with the reference shows that the predicted values are systematically overestimated, in particular in the lower wavenumber range. The average best-fit spectrum based on the *smart error sum* on the other hand again deviates from the average experimental spectrum, but this time the average best-fit spectrum is situated above the (corrected) average experimental spectrum. Above about $1050\,cm^{-1}$, the result agrees extremely well with the reference spectrum. In the range below there are some deviations, but these are much smaller than

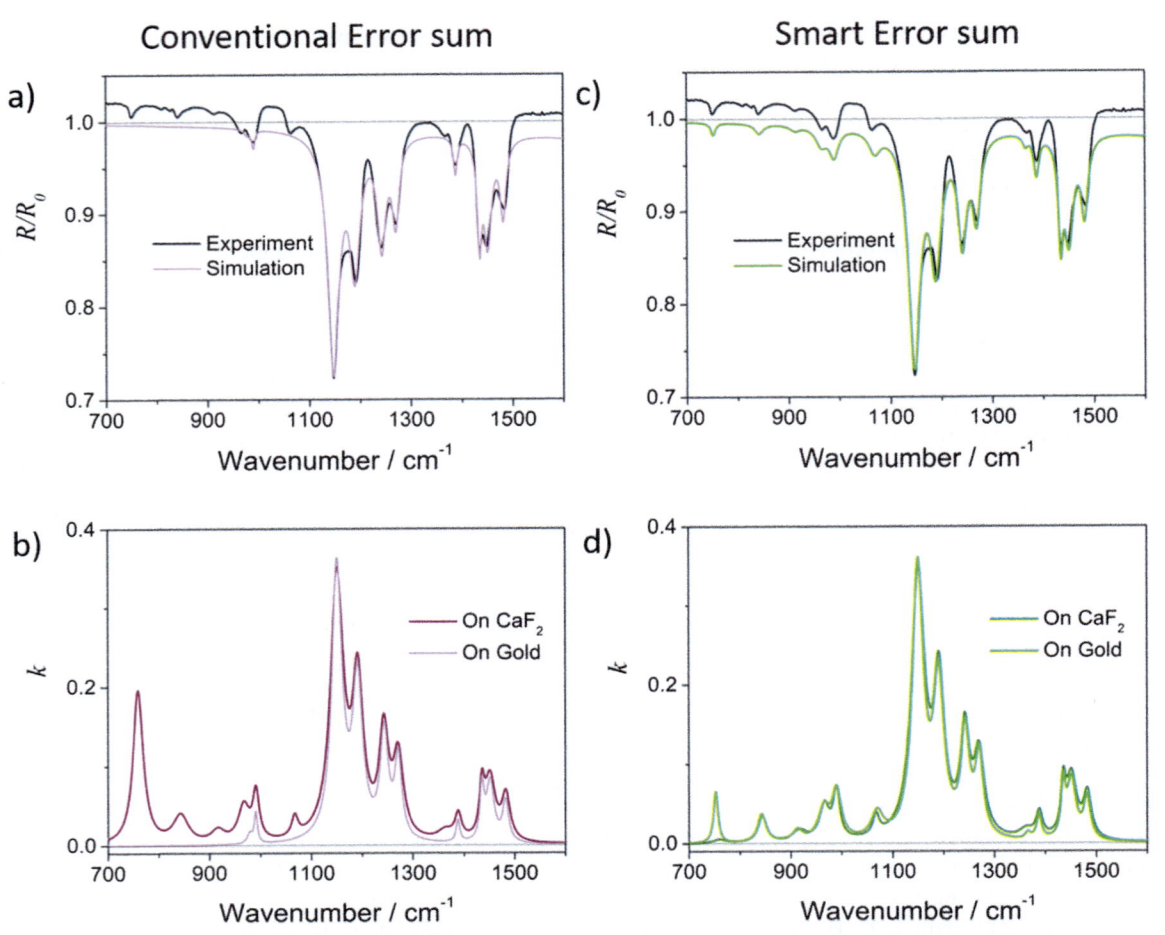

FIG. 8.13 Upper row: Average experimental and best-fit reflectance spectra of PMMA on Gold. Lower Row: Resulting index of absorption spectra compared to absorption index spectra obtained from series fits of PMMA layers on CaF$_2$. Left: Conventional error sum. Right: *Smart error sum* [6].

those attained with the conventional error sum. All in all, also in this case the *smart error sum* does not lead to a perfect fit, but to one that is clearly superior compared to the one obtained with the conventional error sum.

8.3 2T2D smart error sum

The biggest downside of the *smart error sum* is that a series of spectra is required. Since it does not work with a single spectrum like the conventional error sum, you cannot use it in general for dispersion analysis.

When I discussed this issue with Marie Richard-Lacroix, she suggested extending the concept of the *smart error sum* to a two-trace two-dimensional (2T2D) correlation analysis. 2T2D-COS is a relatively new idea of Isao Noda, which dates back to 2018 [11], and decreases 2D-COS to the minimal amount of two experimental spectra. In this case, synchronous and asynchronous spectra are calculated according to

$$\begin{aligned} \Phi(\tilde{\nu}_j \tilde{\nu}_k) &= \tfrac{1}{2}[s(\tilde{\nu}_j) \cdot s(\tilde{\nu}_k) + m(\tilde{\nu}_j) \cdot m(\tilde{\nu}_k)] \\ \Psi(\tilde{\nu}_j \tilde{\nu}_k) &= \tfrac{1}{2}[s(\tilde{\nu}_j) \cdot m(\tilde{\nu}_k) - s(\tilde{\nu}_k) \cdot m(\tilde{\nu}_j)] \end{aligned}, \quad (8.16)$$

where s and m are the two dynamical spectra. Noda showed that even when 2T2D COS uses only two spectra, intrinsically a third spectrum is always onboard, which is the null spectrum [12]—this is why one has to take care that for $t=0$ the resulting spectrum is the null spectrum as in the case of mixtures (cf. Eq. 8.7). Therefore, to make proper use of 2D correlation maps, the choice of reference is not free. This is also the reason why the *smart error sum* only works for multiplicative errors.

Using Eq. (8.16) for the *smart error sum* means that one spectrum would be the experimental one and the other the simulated spectrum. There is one problem, though. While the classical hybrid correlation maps can deviate from being

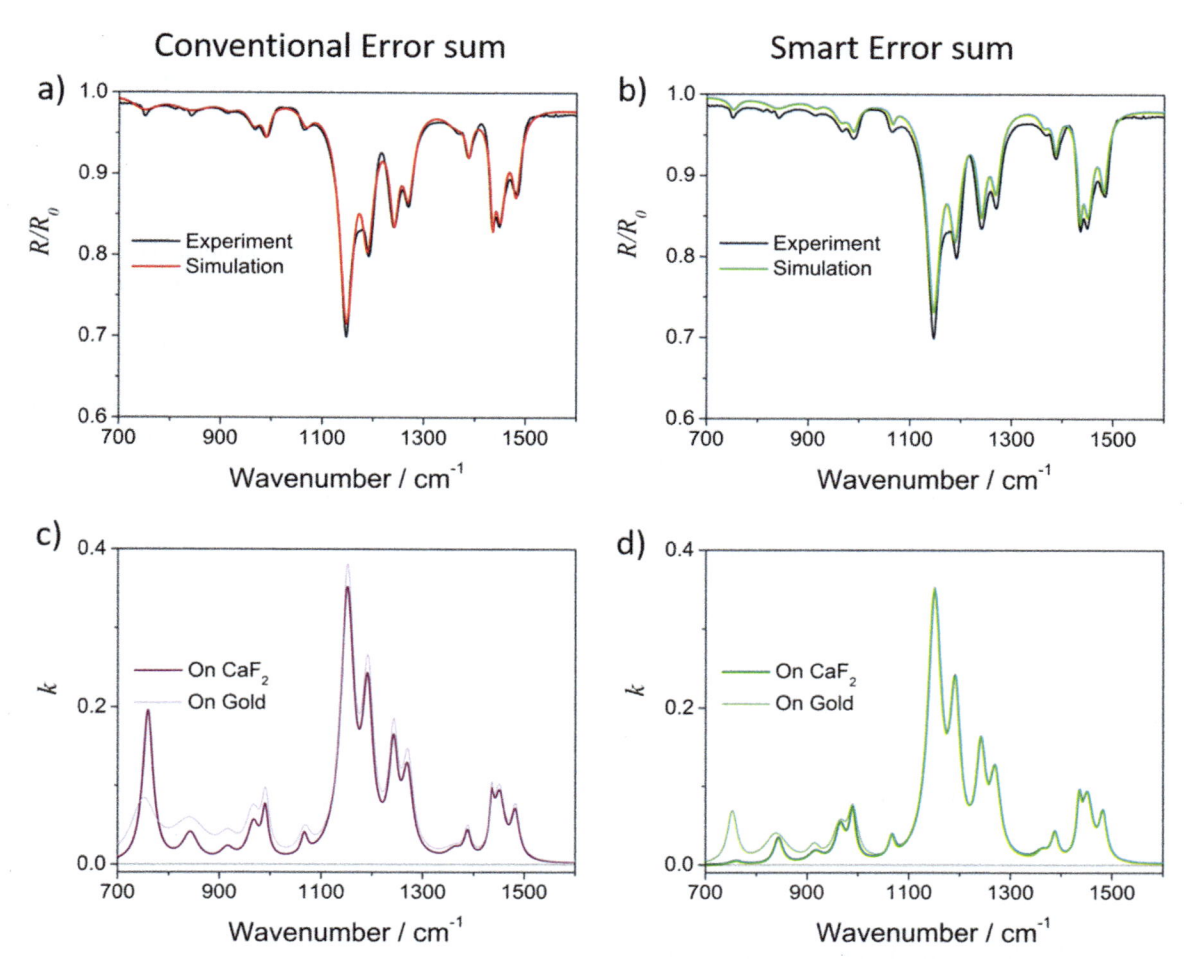

FIG. 8.14 Upper row: Average corrected experimental and corresponding best-fit reflectance spectra of PMMA on Gold. Lower Row: Resulting index of absorption spectra compared to absorption index spectra obtained from series fits of PMMA layers on CaF$_2$. Left: Conventional error sum. Right: *Smart error sum* [6].

symmetric and antisymmetric, respectively, this is not possible for the hybrid 2T2D system. So how could we use increases in symmetry and antisymmetry, when there could be none? Richard-Lacroix suggested a different approach which is to focus on the amplitudes in the asynchronous spectrum, which would become zero if both spectra were identical. This is certainly correct. However, the important point is that the asynchronous spectrum not only vanishes, if both spectra, $s(\tilde{\nu})$ and $m(\tilde{\nu})$, are identical, but also if one spectrum is linearly dependent on the other, i.e., $s(\tilde{\nu}) = F\, m(\tilde{\nu})$ with the factor F being independent of the wavenumber. This indeed leads to the same idea that is underlying the series-based *smart error sum*. To illustrate this, one can also make use of the so-called phase angle, which is defined as

$$\Theta(\tilde{\nu}_1, \tilde{\nu}_2) = \arctan\left\{\frac{\Psi(\tilde{\nu}_1, \tilde{\nu}_2)}{\Phi(\tilde{\nu}_1, \tilde{\nu}_2)}\right\}. \tag{8.17}$$

$\Theta(\tilde{\nu}_1, \tilde{\nu}_2)$ specifies the angle between the real part $\Phi(\tilde{\nu}_1, \tilde{\nu}_2)$ and the imaginary part $\Psi(\tilde{\nu}_1, \tilde{\nu}_2)$. If we force by fitting $\Psi(\tilde{\nu}_1, \tilde{\nu}_2)$ to be zero, then experimental and simulated spectrum can still differ by a factor. Accordingly, we define the 2T2D *smart error sum* as [13]

$$D^p_{A2T} = \sum_{k=1}^{l}\sum_{j=k}^{l} \left[\Psi(\tilde{\nu}_j, \tilde{\nu}_k)\right]^p, \tag{8.18}$$

with $p = 2$. Note that it is sufficient to focus on the points above the diagonal due to the antisymmetry of the asynchronous map. It is also possible to understand the properties of the conventional *smart error sum* fully with the help of Eq. (8.17). Accordingly, the phase angle of the series of measured and experimental spectra is forced by a fit to agree, which means that the ratio of the series of simulated spectra $\Psi_2(\tilde{\nu}_1, \tilde{\nu}_2)/\Phi_2(\tilde{\nu}_1, \tilde{\nu}_2)$ needs to be the same as the one of the experimental set

$\Psi_1(\tilde{\nu}_1,\tilde{\nu}_2)/\Phi_1(\tilde{\nu}_1,\tilde{\nu}_2)$. But numerator and divisor can deviate by a common factor, which is the additional degree of freedom that allows experimental and best-fit spectra to disagree if this leads to more similarity with regard to the correlations caused by the perturbation t. An alternative form of the original *smart error sum* is therefore

$$D_{\text{SES}}^p = \sum_{k=1}^{l} \sum_{j=k}^{l} \left[\Theta_{\text{ex}}(\tilde{\nu}_k,\tilde{\nu}_j) - \Theta_{\text{sim}}(\tilde{\nu}_j,\tilde{\nu}_k) \right]^p, \tag{8.19}$$

where $\Theta_{\text{ex}}(\tilde{\nu}_k,\tilde{\nu}_j)$ are the phase angles of the original data and $\Theta_{\text{sim}}(\tilde{\nu}_k,\tilde{\nu}_j)$ are those of the simulated curves. This form, assuming $p=2$, without consideration of the symmetry properties of the maps and therefore slower by a factor of 2, has originally been introduced by Shinzawa et al. [14,15].

When I derived the original *smart error sum*, I had no knowledge of Shinzawa et al.'s work, probably because it had been used exclusively for the method of alternating least squares. This is somewhat speculative, but I think its use impeded realizing its full potential. Anyway, in this form, a theoretical problem of Eq. (8.15) is avoided, which could potentially occur if either SRSS or ARSS becomes zero (in practice, this is impossible, due to the mere fact that numerical errors related to the conversion of numbers to the binary system will occur).

For the 2T2D *smart error sum*, the phase angles become zero when experiment and simulation agree within a factor, since then the components of the asynchronous map are supposed to become zero. Using this idea, an alternative form for the 2T2D-based *smart error sum* can be formulated as

$$D_{\text{A2T}}^p = \sum_{k=1}^{l} \sum_{j=k}^{l} \left[\Theta(\tilde{\nu}_k,\tilde{\nu}_j) \right]^p. \tag{8.20}$$

From all different *smart error sums* it seems that this formulation has the fastest convergence [7]. However, as Eq. (8.18) can be reformulated (for $p=2$ and after giving up the advantage of having to calculate only one-half of the additions due to the antisymmetry) to scale with the number of wavenumber points instead of scaling with its square which leads to a dramatic speed increase, after which the conventional error sum is no longer much faster:

$$\begin{aligned}
D_{\text{A2T}}^2 &= \sum_{k=1}^{l} \sum_{j=1}^{l} \left[\Psi(\tilde{\nu}_j,\tilde{\nu}_k) \right]^2 \\
&= \sum_{k=1}^{l} \sum_{j=1}^{l} \left[\frac{1}{2} \left[s(\tilde{\nu}_j) \cdot m(\tilde{\nu}_k) - s(\tilde{\nu}_k) \cdot m(\tilde{\nu}_j) \right] \right]^2 \\
&= \frac{1}{4} \sum_{k=1}^{l} \sum_{j=1}^{l} \left[s(\tilde{\nu}_j)s(\tilde{\nu}_j) \cdot m(\tilde{\nu}_k)m(\tilde{\nu}_k) + s(\tilde{\nu}_k)s(\tilde{\nu}_k) \cdot m(\tilde{\nu}_j)m(\tilde{\nu}_j) - 2 \cdot s(\tilde{\nu}_j) \cdot m(\tilde{\nu}_j)s(\tilde{\nu}_k) \cdot m(\tilde{\nu}_k) \right] \\
&= \frac{1}{4} \left(\sum_{j=1}^{l} s(\tilde{\nu}_j)s(\tilde{\nu}_j) \sum_{k=1}^{l} m(\tilde{\nu}_k)m(\tilde{\nu}_k) + \sum_{k=1}^{l} s(\tilde{\nu}_k)s(\tilde{\nu}_k) \sum_{j=1}^{l} m(\tilde{\nu}_j)m(\tilde{\nu}_j) - 2 \sum_{j=1}^{l} m(\tilde{\nu}_j)s(\tilde{\nu}_j) \sum_{k=1}^{l} m(\tilde{\nu}_k)s(\tilde{\nu}_k) \right) \\
&= \frac{1}{4} \left(\sum_{j=1}^{l} s(\tilde{\nu}_j)^2 \sum_{k=1}^{l} m(\tilde{\nu}_k)^2 + \sum_{k=1}^{l} s(\tilde{\nu}_k)^2 \sum_{j=1}^{l} m(\tilde{\nu}_j)^2 - 2 \left(\sum_{j=1}^{l} m(\tilde{\nu}_j)s(\tilde{\nu}_j) \right)^2 \right) \\
&= \frac{1}{2} \left(\sum_{j=1}^{l} s(\tilde{\nu}_j)^2 \sum_{k=1}^{l} m(\tilde{\nu}_k)^2 - 2 \left(\sum_{j=1}^{l} m(\tilde{\nu}_j)s(\tilde{\nu}_j) \right)^2 \right)
\end{aligned} \tag{8.21}$$

This important simplification was Rainer Heintzmann's idea when we discussed the 2T2D *smart error sum*.

How good does the 2T2D *smart error sum* perform in comparison with the conventional *smart error sum*? This comparison is depicted in Fig. 8.15. Obviously, the results are not identical, which means that we still do not know the real extent of the systematic error accurately—the *smart error sums* can certainly only help to reduce it, but not remove it completely.

One problem of the *smart error sum*, which is resolved by the 2T2D *smart error sum*, has not been discussed so far. This problem is related to the fit of the dielectric background in bulk samples and, additionally, the layer thickness in the case of films. The absence of features in the first case and their relative weakness in the latter case as well as the presence of noise, render the original *smart error sum* quasi useless for these cases. This means that ε_∞ and, in the case of a layer, its thickness

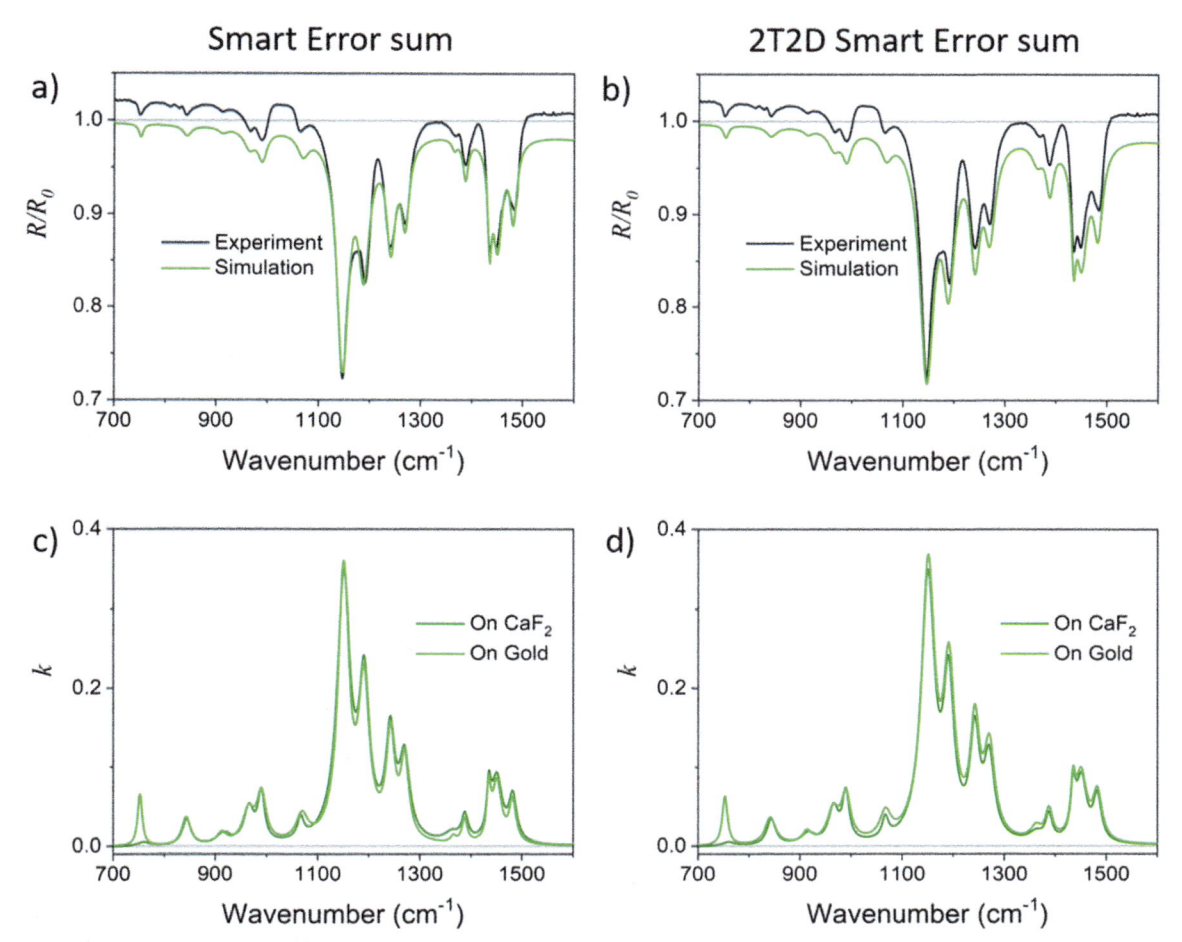

FIG. 8.15 Upper row: Average experimental and best-fit reflectance spectra of PMMA on Gold. Lower Row: Resulting index of absorption spectra compared to absorption index spectra obtained from series fits of PMMA layers on CaF$_2$. Left: Conventional *smart error sum*. Right: 2T2D *smart error sum* [13].

would have to be known beforehand, otherwise dispersion analysis with the *smart error sum* would not be possible. One can circumvent this problem by dividing the smoothed spectra by their highest values and apply conventional dispersion analysis (on the unsmoothed spectra as usual) to obtain ε_∞ and, if applicable, the layer thickness and then go back to the uncorrected spectra to continue the analysis using the *smart error sum*. Alternatively, one uses the 2T2D *smart error sum* together with the uncorrected spectra. Both methods lead to nearly the same results as is demonstrated in Fig. 8.16.

Fig. 8.17 demonstrates the ability of dispersion analysis based on the 2T2D *smart error sum* to fit individual spectra. Again, the results in the nonabsorbing spectral range are very similar to those obtained from conventional dispersion analysis on corrected spectra. That the 2T2D *smart error sum* allows individual dispersion analysis of uncorrected spectra with improved results compared to conventional dispersion analysis is obvious from the comparison of the absorption index spectra.

Certainly, the results are still somewhat scattered and only the absorption index spectra for the two thicker layers compare well with the reference. However, the scattering is much more obvious in the absorption index spectra obtained with conventional dispersion analysis. More importantly, the obvious incorrect increase toward lower wavenumbers, which is a relict from the improper correction, is virtually absent thanks to correction with the 2T2D *smart error sum*.

In some instances, I found that the simulated spectra become unrealistically small. In such cases, the introduction of a counterbalance could be of advantage. Such a counterbalance could be the conventional residual sum of squares (RSS). It is, in fact, simple to combine the RSS with the 2T2D *smart error sum* by calculating the product:

$$\text{PDRSS} = D^2_{A2T} \cdot \text{RSS}. \tag{8.22}$$

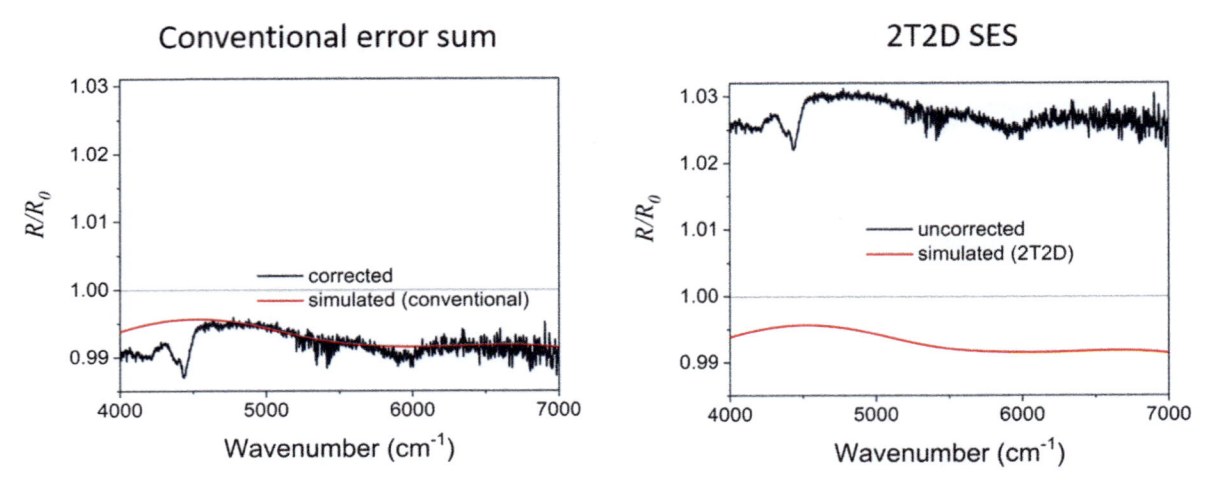

FIG. 8.16 Average experimental and best-fit reflectance spectra of PMMA on Gold in the spectral range of the interference fringes.

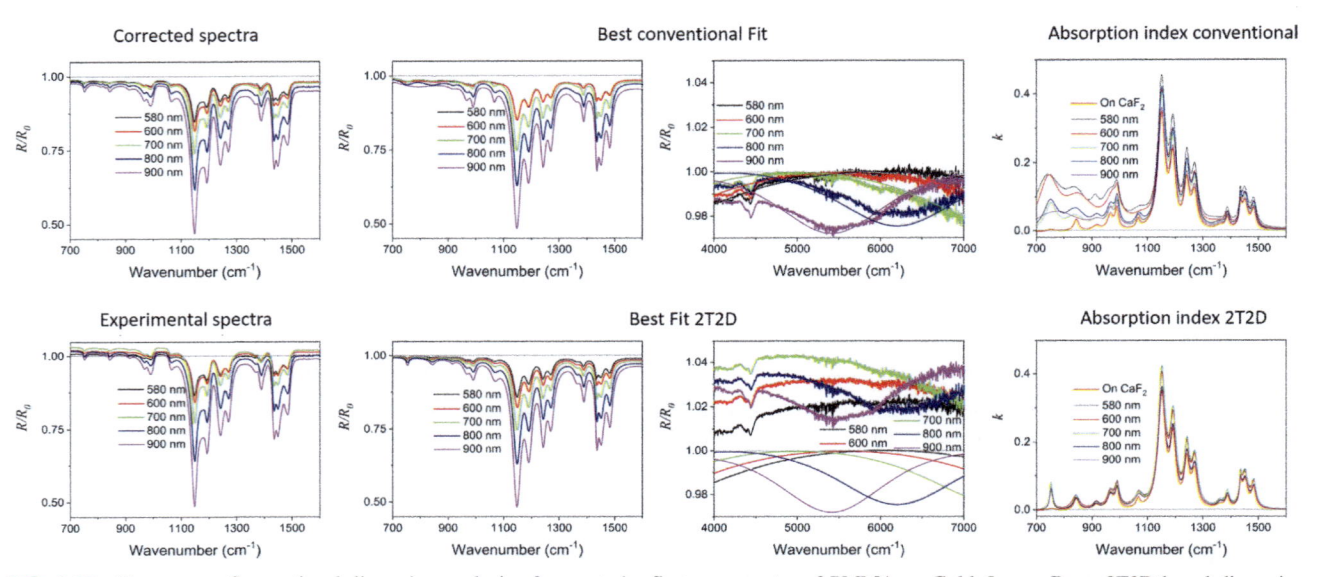

FIG. 8.17 Upper row: Conventional dispersion analysis of corrected reflectance spectra of PMMA on Gold. Lower Row: 2T2D-based dispersion analysis of uncorrected reflectance spectra of PMMA on Gold. In the right column, resulting index of absorption spectra are compared to absorption index spectra obtained from series fits of PMMA layers on CaF$_2$.

With this product, I have, e.g., fitted the perpendicular principal component of the dielectric function tensor. In contrast to the 2T2D *smart error sum*, the PDRSS allowed me to gain the parameters of one oscillator which did not leave a visible footprint in the corresponding spectrum of a *z*-cut single crystal, cf. Section 14.4.

8.4 Further reading

Concerning 2D correlation spectroscopy and its theoretical foundations the clear recommendation is the book of Isao Noda and Yukihiro Ozaki "Two-dimensional Correlation Spectroscopy: Applications in Vibrational and Optical Spectroscopy" [1]. It is an excellent read, but take into account that the book is rooted in the world of molecular infrared spectroscopy. As a consequence, the parts that are concerned with IR spectroscopy are all focused on absorbance. If an asynchronous map resulting from conventional absorbance spectra does not contain much features, a viable alternative is to use transmittance or reflectance spectra instead.

With regard to the *smart error sums* there is no alternative to this book yet. In principle, this could be a very interesting topic not only for infrared spectroscopy. Probably not so much with regard to classical inverse modeling of nonlinear problems, but I see potential as a quality parameter for deep learning applications. Why? Because, on the one hand, improper linear relations will automatically be discarded even when the linear term dominates because of the then missing asynchronous part. On the other hand, correlations in the data will be emphasized which might allow to reduce training effort. In addition, it could be a tool to prevent overfitting, i.e., memorization. Correspondingly, it can already be seen nowadays that for the training of neural networks 2D correlation spectra are employed.

References

[1] I. Noda, Y. Ozaki, Two-Dimensional Correlation Spectroscopy: Applications in Vibrational and Optical Spectroscopy, Wiley, 2005.
[2] N. Isao, Two-dimensional correlation analysis of unevenly spaced spectral data, Appl. Spectrosc. 57 (2003) 1049–1051.
[3] T.G. Mayerhöfer, A.K. Singh, J.-S. Huang, C. Krafft, J. Popp, Unveiling chiral optical constants of α-pinene and propylene oxide through ATR and VCD spectroscopy in the mid-infrared range, Spectrochim. Acta A Mol. Biomol. Spectrosc. 302 (2023) 123136.
[4] T.G. Mayerhöfer, A. Dabrowska, A. Schwaighofer, B. Lendl, J. Popp, Beyond Beer's law: why the index of refraction depends (almost) linearly on concentration, ChemPhysChem 21 (2020) 707–711.
[5] T.G. Mayerhöfer, O. Ilchenko, A. Kutsyk, J. Popp, Beyond Beer's law: quasi-ideal binary liquid mixtures, Appl. Spectrosc. 76 (2022) 92–104.
[6] T.G. Mayerhöfer, M. Richard-Lacroix, S. Pahlow, U. Hübner, R. Heintzmann, J. Popp, Hybrid 2D correlation-based loss function for the correction of systematic errors, Anal. Chem. 94 (2022) 695–703.
[7] T.G. Mayerhöfer, I. Noda, S. Pahlow, R. Heintzmann, J. Popp, Correcting systematic errors by hybrid 2D correlation loss functions in nonlinear inverse modelling, PLoS One 18 (2023) e0284723.
[8] T.G. Mayerhöfer, S. Pahlow, U. Hübner, J. Popp, Removing interference-based effects from the infrared transflectance spectra of thin films on metallic substrates: a fast and wave optics conform solution, Analyst 143 (2018) 3164–3175.
[9] J.A. Nelder, R. Mead, A simplex-method for function minimization, Comput. J. 7 (1965) 308–313.
[10] S. Kirkpatrick, C.D. Gelatt, M.P. Vecchi, Optimization by simulated annealing, Science 220 (1983) 671–680.
[11] I. Noda, Two-trace two-dimensional (2T2D) correlation spectroscopy—a method for extracting useful information from a pair of spectra, J. Mol. Struct. 1160 (2018) 471–478.
[12] I. Noda, Closer examination of two-trace two-dimensional (2T2D) correlation spectroscopy, J. Mol. Struct. 1213 (2020) 128194.
[13] T.G. Mayerhöfer, M. Richard-Lacroix, S. Pahlow, U. Hübner, J. Popp, *Smart error sum* based on hybrid two-trace two-dimensional (2T2D) correlation analysis, Appl. Spectrosc. 77 (2023) 583–592.
[14] H. Shinzawa, J.-H. Jiang, M. Iwahashi, I. Noda, Y. Ozaki, Self-modeling curve resolution (SMCR) by particle swarm optimization (PSO), Anal. Chim. Acta 595 (2007) 275–281.
[15] H. Shinzawa, M. Iwahashi, I. Noda, Y. Ozaki, A convergence criterion in alternating least squares (ALS) by global phase angle, J. Mol. Struct. 883-884 (2008) 73–78.

Chapter 9

Chemometrics

9.1 Introduction

In this chapter, we will discuss the simplest formalisms of classical, i.e., conventional chemometrics so that you can form an opinion about its merits and drawbacks. Mainly, chemometrics is applied in the field of molecular infrared spectroscopy, for good reasons, as you will see later on. In this part of infrared spectroscopy, it seems to me that chemometrics has become so popular that all knowledge about alternatives has somehow faded and that using it is a must. In fact, nowadays, whole chairs are concerned with this topic. To be fair, they are usually also dealing with nonlinear methods like neuronal networks, which are approaches that I think are not only quite interesting and promising but also a must for practical applications. In fact, reading through books about conventional chemometrics, you sometimes get the impression that linear problems or problems that can be linearized make up 99% of all problems in physics and chemistry, while the application of 2D correlation analysis (see Chapter 8) proves that it is quite the opposite.

In the following, I will restrict myself mainly to the part of chemometrics that deals with calibration methods, i.e., methods that are assumed to allow you to determine the concentration of a component in a mixture based on a set of measurements of mixtures with known concentrations. The basis of all calibration methods is Beer's approximation (cf. Section 5.3):

$$A(\tilde{\nu}) = \varepsilon^*(\tilde{\nu})cd, \qquad (9.1)$$

where $A(\tilde{\nu})$ is the absorbance, $\varepsilon^*(\tilde{\nu})$ the molar absorption coefficient, c the molar concentration, and d is the length of the light path or the thickness of the sample. The simplest way of calibration is to determine $\varepsilon^*(\tilde{\nu})$ for a single wavenumber from a number of measurements of samples with different concentrations to reduce random errors. In practice, usually peak values of absorbance bands are used. If Beer's approximation is valid, the peak values for different concentrations are situated on a straight line and the task is to determine the slope of this line in a way that the errors are minimized. This problem, which is called linear regression, can be solved analytically. To do this, we divide the measured peak absorbance A_i by the thickness so that we obtain

$$A_i/d = \varepsilon^* c_i. \qquad (9.2)$$

ε^* is then given by

$$\varepsilon^* = \frac{1}{d} \frac{n \sum_{i=1}^{n} A_i c_i - \sum_{i=1}^{n} c_i \sum_{i=1}^{n} A_i}{n \sum_{i=1}^{n} c_i^2 - \left(\sum_{i=1}^{n} c_i\right)^2}. \qquad (9.3)$$

Certainly, in this case, we assume that there is no systematic error. In fact, for strongly diluted solutions, the method usually works satisfactorily. To demonstrate this, I reproduce Fig. 8.4 from the previous chapter.

The quantity R^2 provided in the inset of Fig. 9.1 is the coefficient of determination given by

$$R^2 = \frac{\sum_{i=1}^{n} (A_i - \overline{A})^2}{\sum_{i=1}^{n} (\widehat{A}_i - \overline{A})^2}, \quad \overline{A} = \frac{1}{n} \sum_{i=1}^{n} A_i, \qquad (9.4)$$

where $\widehat{A}_i$ are the predicted values. R^2 is a measure for the goodness of the fit, to be more precise, a measure of how well data are replicated by a linear model. Since the values of R^2 can be between 0 and 1, 0.9928, as in the earlier example, seems to imply that the fit is actually very good, but it does not mean that a linear relationship between concentration and absorbance

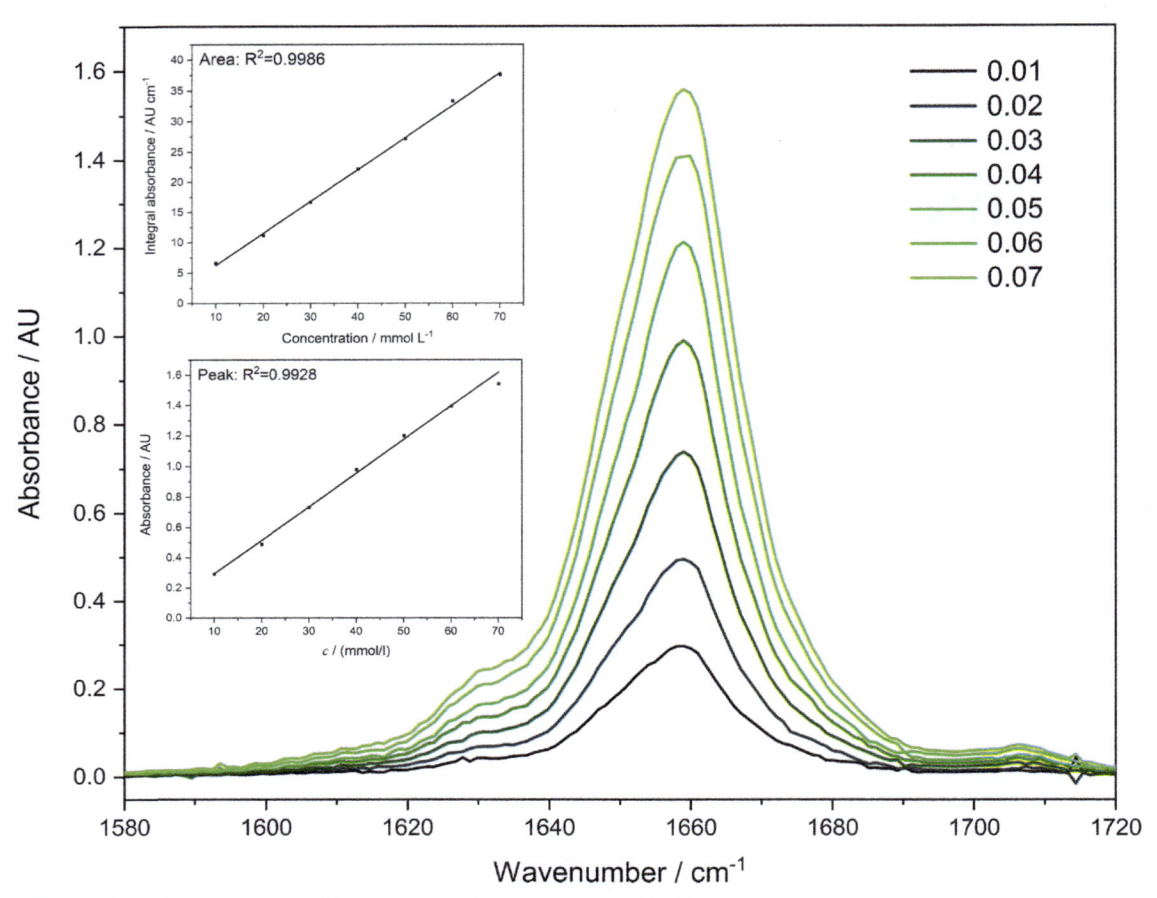

FIG. 9.1 The C=O band of caprolactam with increasing molar concentration with chloroform as solvent [1].

is established beyond doubt. On a closer look, there is a sigmoidal deviation between the points and the line, which does not seem to be a random error, and we already know that actually a nonlinear relation is at play. Beer's approximation is certainly of high value for analytical applications—the question is how large the systematic errors are in relation to the random errors. And how does Beer's approximation perform for mixtures?

Assuming that mixtures obey Beer's approximation is certainly a considerable extension of the original experimental findings of August Beer since we no longer deal with diluted solutions. There is a famous quote from John William Strutt, the later Lord Rayleigh: *"In many departments of science a tendency may be observed to extend the field of familiar laws beyond their proper limits"* [2]. Let us put the assumption of linearity to the test. I will use for this test the systems, benzene-toluene, benzene-tetrachloromethane, and benzene-cyclohexane, and spectra which I got from Oleksii Ilchenko and Andrii Kutsyk, colleagues who I met at ICAVS-10 in Auckland. These systems are ideal from a thermodynamical point of view, which means that the interaction between any two molecules in the mixture is the same, which can be checked, e.g., by the adherence to Raoult's law, but also by linear behavior with regard to properties like density, mutual- and self-diffusion coefficients as well as shear viscosity. Since it is in general believed that any deviation from Beer's approximation results from changes of chemical interactions, these systems are ideal test candidates as those changes are virtually absent. For the first test, I picked bands that seem to follow Beer's approximation (but the 2D correlation spectra nevertheless revealed nonlinearities, cf. Section 8.1). Since there are 11 spectra per series overall, I made a linear fit to the absorbance values for 10 mixtures and tried to predict the composition of the 11th from this calibration (Leave One Out Cross-Validation). The result is depicted in Fig. 9.2.

The benzene-toluene system shows comparably small relative errors, but particularly in the benzene-tetrachloromethane system, the deviations are tremendous. Still, it is possible that the peaks I have chosen may simply not be well suited for the analysis or not representative and show particularly strong deviations. One could assume that it would be better if the calibration did not only take into account the absorbance at a single wavenumber but instead the whole absorbance spectrum. This implies that Beer's approximation should hold at every wavenumber (remember

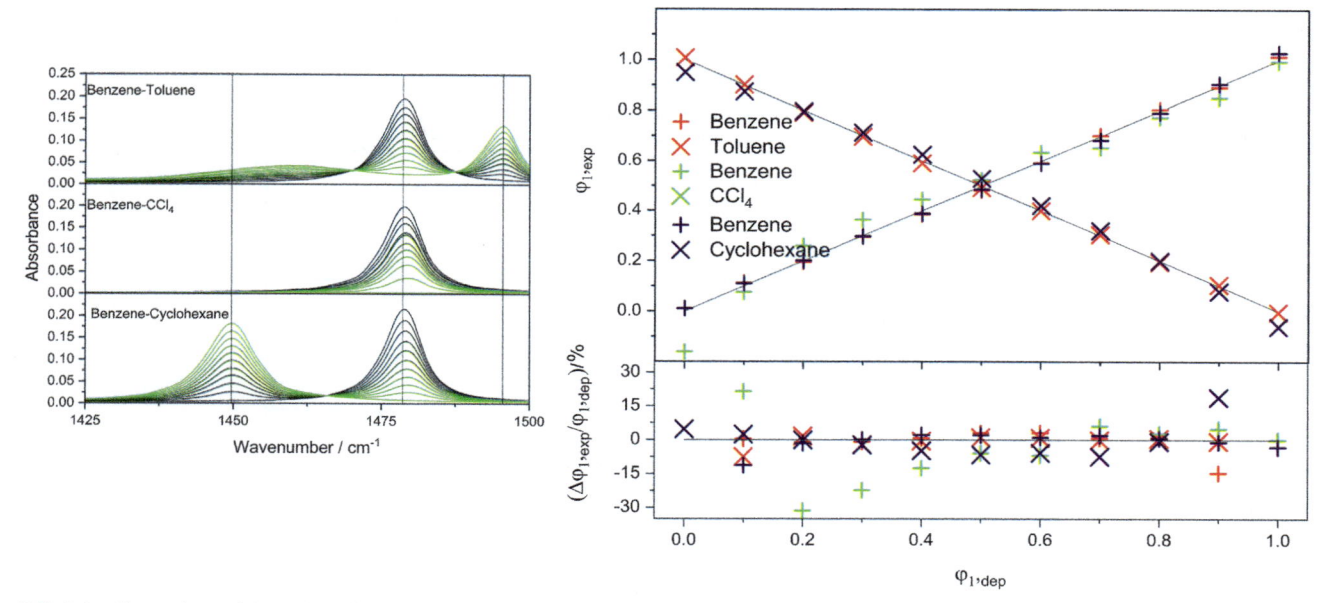

FIG. 9.2 Comparison of the volume fractions of the mixtures, benzene-toluene, benzene-tetrachloromethane, and benzene-cyclohexane (right panel), as prepared and as determined from peak values indicated in the spectra of the mixtures (left panel) [3].

Section 5.3, wherein I showed that the deviations from Beer's approximation increase with the oscillator strength of the corresponding vibrational transition!). This is the argumentation that leads from a univariate method, as in Fig. 9.2, to multivariate calibration methods that we will focus on in the following.

9.2 Classical least squares (CLS) regression

For univariate regression, the absorbance values (or the absorbance values divided by the thickness) are a vector, and this is also the case for the concentrations and the volume fractions. Now that we employ not one absorbance value per concentration but a whole spectrum, the vectors of absorbance values constitute the rows of a matrix $\mathbf{A}$. If you familiarized yourself with Section 8.1, this matrix $\mathbf{A}$ is in principle the same as the matrix $\mathbf{Y}$, except that the perturbation is in this case the concentration or the volume fraction $\boldsymbol{\varphi}$ and that this perturbation does not have to change linearly (but for the three systems we investigate in the following, this is actually the case):

$$\underbrace{\begin{pmatrix} A_1(\tilde{\nu}_1) & A_1(\tilde{\nu}_2) & . & . & A_1(\tilde{\nu}_l) \\ A_2(\tilde{\nu}_1) & . & & & . \\ . & & & & . \\ . & & & & . \\ A_m(\tilde{\nu}_1) & A_m(\tilde{\nu}_2) & . & . & A_m(\tilde{\nu}_l) \end{pmatrix}}_{\mathbf{A}} = \underbrace{\begin{pmatrix} \varphi_{11} & \varphi_{12} & . & . & \varphi_{1j} \\ \varphi_{21} & . & & & . \\ . & & & & . \\ . & & & & . \\ \varphi_{m1} & \varphi_{m2} & . & . & \varphi_{mj} \end{pmatrix}}_{\boldsymbol{\varphi}} \cdot \underbrace{\begin{pmatrix} K_1(\tilde{\nu}_1) & K_1(\tilde{\nu}_2) & . & . & K_1(\tilde{\nu}_l) \\ K_2(\tilde{\nu}_1) & . & & & . \\ . & & & & . \\ . & & & & . \\ K_j(\tilde{\nu}_1) & K_j(\tilde{\nu}_2) & . & . & K_j(\tilde{\nu}_l) \end{pmatrix}}_{\mathbf{K}}. \quad (9.5)$$

The matrix $\mathbf{A}$ has as many columns as each of the spectra contains absorbance values, in this case l values depending on the spectral resolution. For example, if the resolution is $1\,\mathrm{cm}^{-1}$ and the spectral range extends from 400 to $4000\,\mathrm{cm}^{-1}$, $l = 3601$. Furthermore, the number of lines corresponds to the number of different mixtures m, in our case 11 (this includes the spectra of the neat components). This also means that we have m different sets of volume fractions φ_{mj}, each for j different components K. Correspondingly, the matrix $\boldsymbol{\varphi}$ has m lines and j columns. Finally, the matrix $\mathbf{K}$ is supposed to contain the spectra of the pure components and has therefore j lines and l columns.

After having prepared the m mixtures, their spectra are recorded so that the matrices $\mathbf{A}$ and $\boldsymbol{\varphi}$ are known. The a priori unknown matrix $\mathbf{K}$ can simply be obtained from the known matrices by

$$\mathbf{K} = \left(\boldsymbol{\varphi}^{\mathrm{T}}\boldsymbol{\varphi}\right)^{-1}\boldsymbol{\varphi}^{\mathrm{T}}\mathbf{A}. \quad (9.6)$$

For those of you not familiar with matrix operations, Eq. (9.6), which is called the generalized inverse, may look odd. Why not just multiply both sides of $\mathbf{A} = \boldsymbol{\varphi}\mathbf{K}$ from the left with $\boldsymbol{\varphi}^{-1}$? This only works for square matrices, and since the number

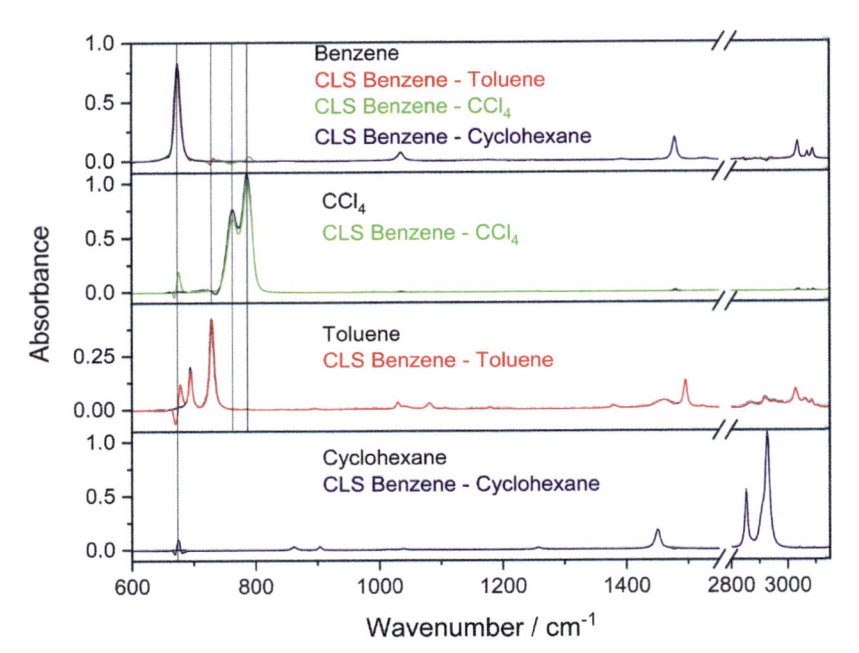

FIG. 9.3 Comparison of the experimental pure component spectra with those obtained from the spectra of the mixtures, benzene-toluene, benzene-CCl$_4$, and benzene-cyclohexane, based on Eq. (9.6).

of wavenumber points, mixtures, and components will never be the same, we first have to multiply both sides from the left with the transpose of φ where lines and columns are interchanged.

Before we continue, it is instructive to take a look at the component spectra of our model systems, cf. Fig. 9.3. On the first view, experimental and calculated component spectra show some differences in the low wavenumber region, where the strongest oscillators are located (with the exception of cyclohexane, where the C—H vibrations are also strong, simply by sheer numbers). In particular, the γ C—H vibration of benzene at 674 cm^{-1}, which has A$_{2u}$ symmetry, leads to dispersion-like artifacts in the calculated spectra of the other components with negative absorbance values. These artifacts and the lower intensities of the C—Cl vibrations in tetrachloromethane are the consequence of band shifts due to local field effects (cf. Section 5.1 and Chapter 10) and indicate the limits of the linear approach CLS is based upon. In literature, these are usually interpreted as being due to changing chemical interactions between different molecules, i.e., nonideality. As long as nonideal systems are being investigated, this is certainly understandable, but why did it take so long before someone investigated ideal systems? Or did someone and did not publish the results because they were not explainable by Beer's approximation? Anyway, these irregularities are often seen as not problematic because CLS is viewed as being able to cope with nonlinearities, which is something we will investigate in the next step.

This next step is to use the results of the calibration, which are the computed component spectra, to predict the concentrations/volume fractions. Before we investigate the accuracy of the prediction, I first want to illustrate that the matrix **K** is equivalent to the molar absorption coefficient in Eq. (9.3) if molar concentrations are used instead of volume fractions. In the ordinary least square problem, we want to minimize the mean squared error s:

$$s = \frac{1}{m} \sum_{j=1}^{m} \left(\frac{A_j}{d} - \varepsilon^* c_j - R \right)^2. \tag{9.7}$$

Here, R represents a residual error and n is the number of different molar concentrations and measurements. s is at minimum if the derivatives with respect to ε^* and R are zero:

$$\frac{\partial s}{\partial \varepsilon^*} = -\frac{2}{m} \sum_{j=1}^{m} \left(\frac{A_j}{d} - \varepsilon^* c_j - R \right) c_j = 0,$$

$$\frac{\partial s}{\partial R} = -\frac{2}{m} \sum_{j=1}^{m} \left(\frac{A_j}{d} - \varepsilon^* c_j - R \right) = 0. \tag{9.8}$$

These equations can be written as

$$\left(\sum_{j=1}^{m} c_j^2\right)\varepsilon^* + \left(\sum_{j=1}^{m} c_j\right)R = \sum_{j=1}^{m} \frac{A_j}{d} c_j,$$
$$\left(\sum_{j=1}^{m} c_j\right)\varepsilon^* + mR = \sum_{j=1}^{m} \frac{A_j}{d}. \tag{9.9}$$

After elimination of R this leads to Eq. (9.3). On the other hand, the right-hand side of Eq. (9.9) can be written as

$$\begin{pmatrix} \sum_{j=1}^{m} \frac{A_j}{d} c_j \\ \sum_{j=1}^{m} \frac{A_j}{d} \end{pmatrix} = \frac{1}{d} \begin{pmatrix} c_1 & c_2 & \cdots & c_m \\ 1 & 1 & \cdots & 1 \end{pmatrix} \begin{pmatrix} A_1 \\ A_2 \\ \vdots \\ A_m \end{pmatrix} = \frac{1}{d} \mathbf{c}^T \mathbf{A}. \tag{9.10}$$

After reorganization, this yields a number of equations that represent m lines:

$$\frac{1}{d}\mathbf{A} = \mathbf{c}\begin{pmatrix} \varepsilon^* \\ R \end{pmatrix} \rightarrow \begin{pmatrix} A_1 \\ A_2 \\ \vdots \\ A_m \end{pmatrix} = \begin{pmatrix} c_1 & 1 \\ c_2 & 1 \\ \vdots & \vdots \\ c_m & 1 \end{pmatrix}\begin{pmatrix} \varepsilon^* \\ R \end{pmatrix} \rightarrow \begin{array}{l} A_1/d = c_1\varepsilon^* + R \\ A_2/d = c_2\varepsilon^* + R \\ \vdots \\ A_m/d = c_m\varepsilon^* + R \end{array} \tag{9.11}$$

On the other hand,

$$\mathbf{c}^T\mathbf{c} = \begin{pmatrix} c_1 & c_2 & \cdots & c_m \\ 1 & 1 & \cdots & 1 \end{pmatrix}\begin{pmatrix} c_1 & 1 \\ c_2 & 1 \\ \vdots & \vdots \\ c_m & 1 \end{pmatrix} = \begin{pmatrix} \sum_{j=1}^{m} c_j^2 & \sum_{j=1}^{m} c_j \\ \sum_{j=1}^{m} c_j & m \end{pmatrix}, \tag{9.12}$$

therefore,

$$\mathbf{c}^T\mathbf{c}\begin{pmatrix} \varepsilon^* \\ R \end{pmatrix} = \frac{1}{d}\mathbf{c}^T\mathbf{A} \rightarrow \begin{pmatrix} \varepsilon^* \\ R \end{pmatrix} = \frac{1}{d}(\mathbf{c}^T\mathbf{c})^{-1}\mathbf{c}^T\mathbf{A}. \tag{9.13}$$

If we apply this to volume fractions and several components instead of only one, we arrive at Eq. (9.6). Some care must be taken in interpreting the component spectra. Certainly, the component spectra are not identical to the molar absorption coefficients because, first of all, absorbance would have to be divided by the thickness, and the use of volume fractions does not lead to molar quantities. It is actually quite simple: The component spectra would be identical to the spectra of the pure components if Beer's approximation was exact. This brings us finally back to the original question, namely, how large are the errors if we use the calibration to predict the volume fraction based on a single spectrum $\mathbf{A}_u$:

$$\boldsymbol{\varphi}_u = \mathbf{A}_u \mathbf{K}^T (\mathbf{K}^T\mathbf{K})^{-1}. \tag{9.14}$$

Fig. 9.4 shows a comparison of the results of the least square method for single peaks with no discernible deviation from Beer's approximation and the multivariate CLS method, where I have again performed the calibration with 10 mixtures to predict the composition of the 11th. The goal of this comparison is to dispel the widespread assumption in chemometrics that a multivariate calibration using the whole spectrum is automatically superior to one that uses just a peak value. Obviously, if this peak belongs to a comparably weak oscillator, the errors can be even smaller than the ones produced by CLS [3]. In this respect, I want to emphasize that this was a well-known fact in the beginnings of chemometrics, where bands that obviously did not follow Beer's approximation were not used for the analysis [4]. This is just another typical example where it is clearly of advantage to understand the underlying physics. In addition, this also clearly dispels a second assumption, namely, that CLS can cope with nonlinearities. Certainly, there might also be some random errors that could play a role, but

248 PART | I Scalar theory

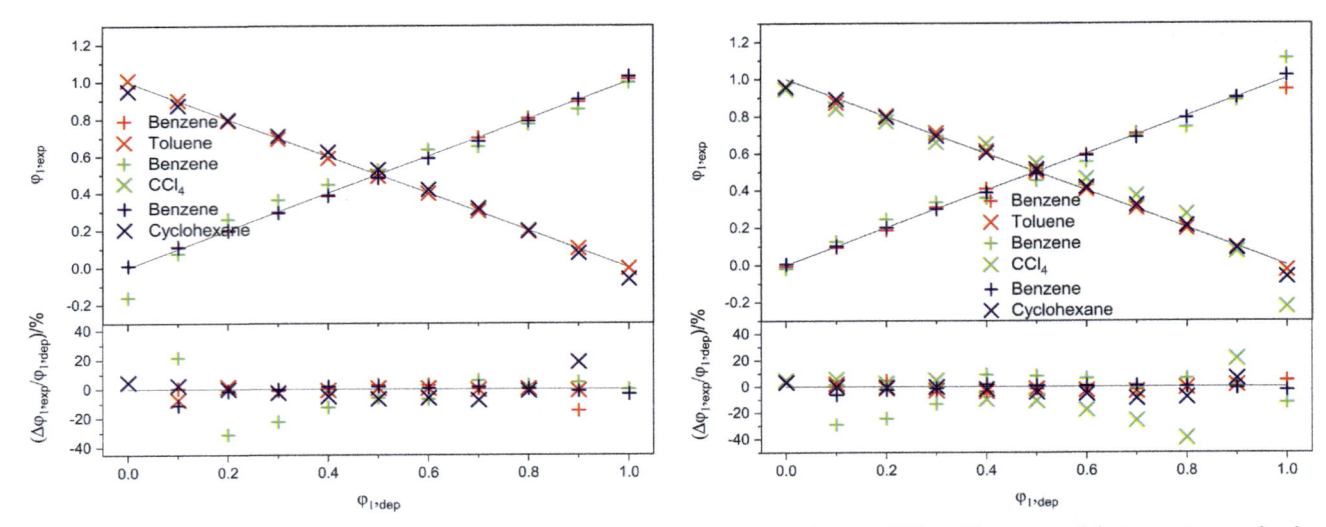

FIG. 9.4 Left panel: Comparison of the volume fractions of the mixtures, benzene-toluene, benzene-CCl$_4$, and benzene-cyclohexane, as prepared and as determined from peak values indicated in the spectra (taken over from Fig. 9.2). Right panel: The same comparison for CLS.

the preparation of mixtures of ideal binary liquids should not lead to such big random errors as displayed in Fig. 9.4 when we just deal with measuring up the necessary volumes of the two components and mixing them together. In particular, as stated earlier, for ideal mixtures the volumes are additive so that a systematic error due to volume compression can be excluded.

Maybe the problem could have been avoided or at least reduced if we had not first calculated component spectra but tried to solve the problem directly without the need for two matrix inversions. This assumption can be easily tested by the use of inverse least squares (ILS) regression. But before we introduce this method, let us first have a look at how large the errors would be if the related deviations from linearity were not described by the Lorentz-Lorenz relation. To that goal, I have calculated the spectra of the series from the endmember data using Eq. (6.32). The result is depicted in Fig. 9.5. Since the Lorentz-Lorenz relation cannot quantitatively describe the effects (the blueshifts of the bands and the intensities are for some bands larger and for others smaller than calculated, cf. also Chapter 10), the component spectra show, e.g., stronger deviations for the C—Cl bands as a result of bands shifts that are less severe in reality. Some behavior is, however, also

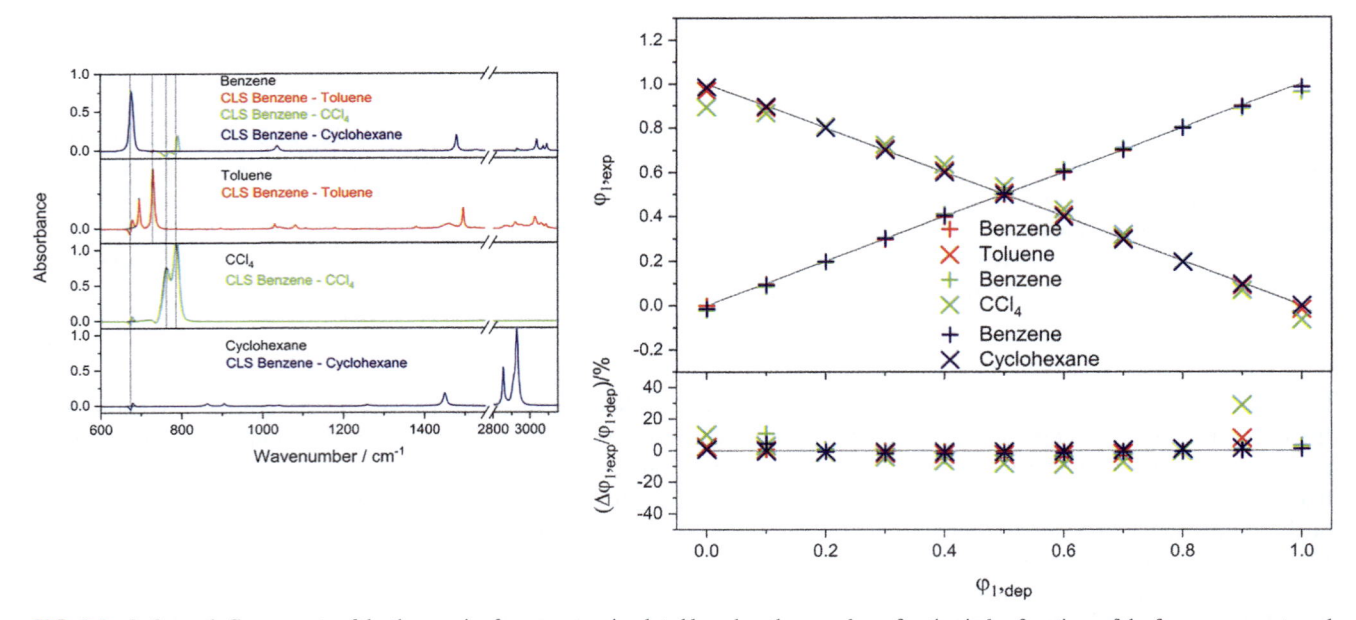

FIG. 9.5 Left panel: Components of the three series from spectra simulated based on the complex refractive index functions of the four components and the Lorentz-Lorenz relation as determined from CLS. Right panel: Results of the corresponding volume fraction determination.

correctly reproduced like the sigmoid curve shape which would result if the volume fractions for CCl_4 were fitted (green crosses in Figs. 9.4 and 9.5), albeit the deviations are stronger. While the Lorentz-Lorenz relation cannot fully explain the observed deviations from Beer's approximation, the results show that there are indeed local field effects at play which are just too complex to explain them by this comparably simple theory. In any way, the observed deviations are not random, so errors concerning the preparation of the mixtures can be excluded as the main cause.

9.3 Inverse least squares (ILS) regression

This method is usually motivated by a problem that exists only for nonideal mixtures with potential complex formation or mixtures, the components of which are a priori unknown. In this case, it is not possible to employ CLS. Instead, we assume that the volume fractions can be calculated according to

$$\boldsymbol{\varphi} = \mathbf{A}\mathbf{P}_{ILS}. \tag{9.15}$$

$\mathbf{P}_{ILS}$ is a matrix without any obvious physical meaning and can be calculated by

$$\mathbf{P}_{ILS} = \left(\mathbf{A}^T\mathbf{A}\right)^{-1}\mathbf{A}^T\boldsymbol{\varphi}, \tag{9.16}$$

which is just the generalized inverse of Eq. (9.15). There is, however, one problem that inevitably results from above relation, namely, the problem of matrix inversion of a (nearly) singular matrix. The matrix product $\mathbf{A}^T\mathbf{A}$ generates a matrix of the size $l \times l$ (remember, l was the number of wavenumber points), but in the limiting case where Beer's approximation is valid, the number of linearly independent lines or columns would be very small. Put into mathematical terms, the matrix generated by $\mathbf{A}^T\mathbf{A}$ still has a rank of m, which is the number of mixtures/spectra. Therefore, we would either need to carry out a large number of measurements to change the matrix $\mathbf{A}$ from having a landscape shape toward possessing a square shape, or we reduce the number of wavenumber points to achieve the same. Indeed, it is the latter that is usually preferred, albeit this seems to be counterintuitive for the avowed chemometrician since this means willingly putting away information and moving from multivariate analysis toward the initial univariate analysis. Luckily, you know already from the previous section that this may indeed be meaningful because the first multivariate analysis method I introduced you to did actually not perform better than univariate analysis as long as the peak is chosen with some care (had I instead chosen the strong γ C—H vibration of benzene at $674 \, cm^{-1}$, which shows appreciable shifts and deviations from Beer's approximation, the univariate analysis would have had a poorer result [3]). This leads to the fact that we can be optimistic concerning the performance of ILS, in contrast to the typical chemometrician—in the following we will check if this optimism is indeed called for.

Once the matrix $\mathbf{P}_{ILS}$ has been determined, and a spectrum of the mixture with unknown composition $\boldsymbol{\varphi}_u$ has been recorded, this composition can be determined by

$$\boldsymbol{\varphi}_u = \mathbf{A}_u \mathbf{P}_{ILS}, \tag{9.17}$$

keeping in mind, that $\mathbf{A}_u$ must not contain more entries than the number of spectra used for calibration. The error certainly depends on the number of chosen wavenumber points, but it is not simply the more the better as this is sometimes suggested in the contemporary chemometric literature. As in the case of the univariate calibration, comparably weak bands that better adhere to Beer's approximation are better suited than stronger peaks, at least as long as spectral noise does not play a role. Therefore, ALS based on two weaker bands may perform better than if it is based on three stronger peaks. For the following figure, I did not choose the peaks manually but used a peak-finding algorithm, which I adjusted to employ only peaks with an absorbance $A > 0.05$.

Overall, as can be seen in Fig. 9.6, this choice of peak absorbance values leads to relative errors that are about halved compared to CLS, which means that this is indeed an improvement compared to univariate calibration in contrast to CLS.

9.4 Principal component analysis (PCA)/principal component regression (PCR)

What is a principal component? Before I saw the mathematical definition, this was a very abstract term for me. The usual description of the method is that first the direction of the largest change of the data is determined and the vector in this direction is the first principal component. Subsequently, the influence of the first component is removed and, again, the direction of the largest change is determined, which provides the vector of the second component. Successively, this is continued until all components are identified or up to the point where the components are thought to be no longer important.

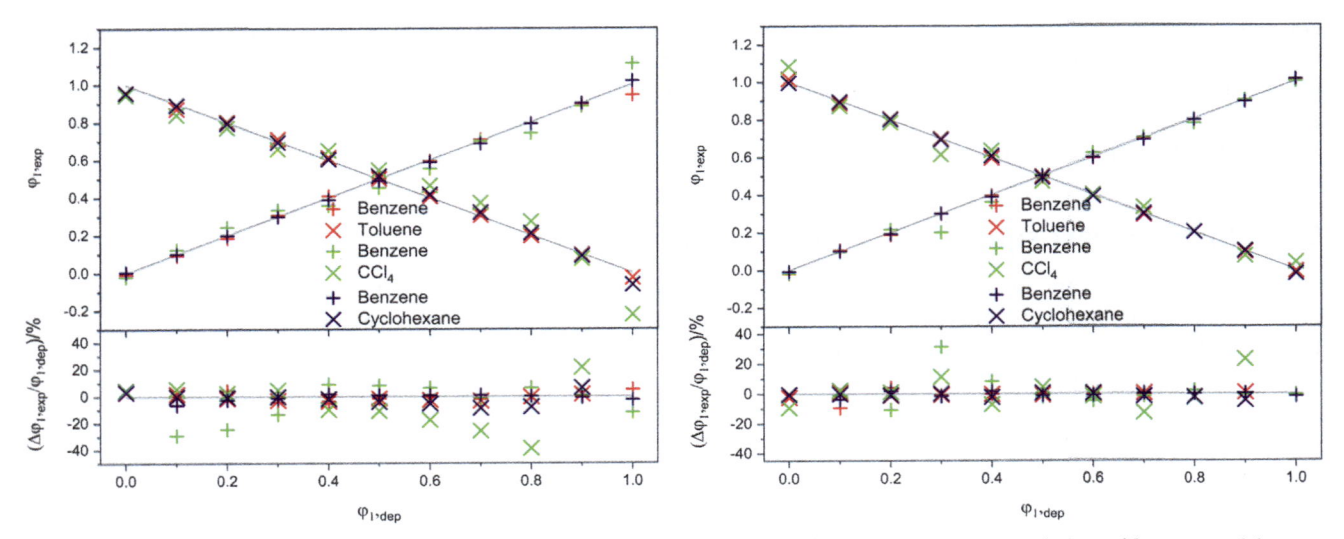

FIG. 9.6 Left panel: Comparison of the volume fractions of the mixtures, benzene-toluene, benzene-tetrachloromethane, and benzene-cyclohexane, as determined by CLS (taken over from Fig. 9.4). Right panel: The same for ILS.

Do you get any idea from this description: how PCA is actually performed? My first idea was that the method of determining the vectors follows some kind of geometrical algorithm. But actually, what you need is also important for computing the transmittance and reflectance of anisotropic materials, to be more specific, the polarization of waves inside the anisotropic medium and their speed and damping (absorption), namely, to determine eigenvalues and eigenvectors/eigenwaves. As a chemist, the first time I stumbled over eigenvalues was in Quantum Chemistry, but the concept seemed strange and abstract. On the other hand, there is a very intuitive way to understand eigenvalues and eigenvectors. Assume we have a very simple matrix $\mathbf{B}$:

$$\mathbf{B} = \begin{bmatrix} 3 & 0 \\ 0 & -2 \end{bmatrix}. \tag{9.18}$$

If it is applied to an arbitrary vector $\mathbf{u} = [x, y]$, the result is $\mathbf{Bu} = [3x, -2y]$, which means that the original vector is stretched to thrice its lengths in the x- and twice in the y-direction, where it is flipped by the negative sign. This means that its direction is no longer parallel (or antiparallel) to the original vector. This does, however, not happen to every vector. For some, only their length changes; some become additionally inverted, but they are still parallel to the former vector before $\mathbf{B}$ is applied. For above matrix, these are vectors parallel either to the x- or the y-direction, thanks to the off-diagonal zeros. Such vectors are *eigenvectors* and the factors by which the length changes are the *eigenvalues*.

In the general case, the corresponding equation is

$$\mathbf{Bu} = \gamma \mathbf{u}, \tag{9.19}$$

with the eigenvalues γ and the eigenvectors $\mathbf{u}$. To determine the eigenvalues, one has to solve the following equation:

$$(\mathbf{B} - \gamma \mathbf{I})\mathbf{u} = 0. \tag{9.20}$$

To solve this equation, the length in one principal direction must become zero, which means it cannot be inverted and, therefore, the determinant is zero:

$$\det(\mathbf{B} - \gamma \mathbf{I}) = 0. \tag{9.21}$$

Once the eigenvalues are known, the eigenvectors can be determined, e.g., by Cramer's rule. The largest eigenvalue corresponds to the eigenvector in whose direction the largest change happens. If we put all these vectors as columns into a matrix $\mathbf{P}$ and the eigenvalues on the diagonals of a matrix $\mathbf{\Lambda}$ and fill up the rest with zeros, then it follows that

$$\mathbf{BP} = \mathbf{P\Lambda}. \tag{9.22}$$

Therefore, it is possible to decompose $\mathbf{B}$ into

$$\mathbf{B} = \mathbf{P\Lambda P}^{-1}, \tag{9.23}$$

as long as all eigenvectors are linearly independent; otherwise, $\mathbf{P}$ would not be invertible.

Where is the connection to PCA? In CLS, we have to know beforehand how many components there are, whereas in PCA we can find out their actual number, namely, the number of eigenvectors and corresponding eigenvalues. At least, this is what is usually assumed by chemometricians. In fact, as we will see, if we assume Beer's approximation to be valid, generate corresponding spectra of different mixtures and apply PCA, we indeed see that there are just two principal components as the other eigenvalues are virtually zero.

As we already know from the previous chapter, the matrix $\mathbf{A}^T\mathbf{A}$ is not invertible. To make nevertheless full use of the spectral information from all wavenumber points, we determine the eigenvalues and the eigenvectors of this matrix, which is apart from a factor and the fact that we have not mean-centered the absorbance values, the variance-covariance matrix you know already from Section 8.1 about 2D correlation analysis (cf. Eq. 8.2):

$$\mathbf{A}^T\mathbf{A}\mathbf{P} = \mathbf{\Lambda}\mathbf{P}, \tag{9.24}$$

where $\mathbf{P}$ is a matrix that contains the eigenvectors as columns, called loading vectors, and $\mathbf{\Lambda}$ a diagonal matrix with the eigenvalues as elements. When the matrix $\mathbf{P}$ is multiplied by the matrix $\mathbf{T}$, which contains the score vectors, the matrix $\mathbf{A}$ is regained:

$$\mathbf{A} = \mathbf{T} \cdot \mathbf{P}. \tag{9.25}$$

Note that the big difference between the matrix $\mathbf{Y}$ known from 2D Correlation analysis and the matrix $\mathbf{A}^T\mathbf{A}$ is that for the latter you are in principle free to choose concentrations or volume fractions as you see fit, whereas for the former usually a linear change of the perturbation is chosen.

In principle, the number of eigenvalues is equal to the number of wavenumber points. However, only the first values up to the number of different mixtures are relevant (up to the rank of the matrix), later values are usually more than 10 orders of magnitude smaller. Let's have a look at the eigenvalues of our three ideal mixtures, cf. Fig. 9.7 [5].

As already stated, the number of principal components is not two, despite the fact that these are ideal binary mixtures without any tendency to form compounds. As we know from the 2D correlation spectra, cf. Section 8.1, the nonlinearities are responsible for deluding into thinking that there are more chemical components than two. In fact, earlier eigenvalues show a pattern very similar to nonideal two-component mixtures, including the two linear ranges (in this log plot) with different slopes. The usual interpretation in the literature is that there are actually three components, the two starting components and a new compound that is formed by complexation. Earlier results show that such conclusions are highly questionable, even if they also rely on a further method which we will discuss in the next section, called multivariate curve resolution (MCR), the idea of which is to determine the spectra of the seemingly formed compound. The need for this method derives from the fact that the eigenvectors are not useful for this goal. I also want to demonstrate how a PCA of modeled spectra will look like in comparison with the experimental spectra.

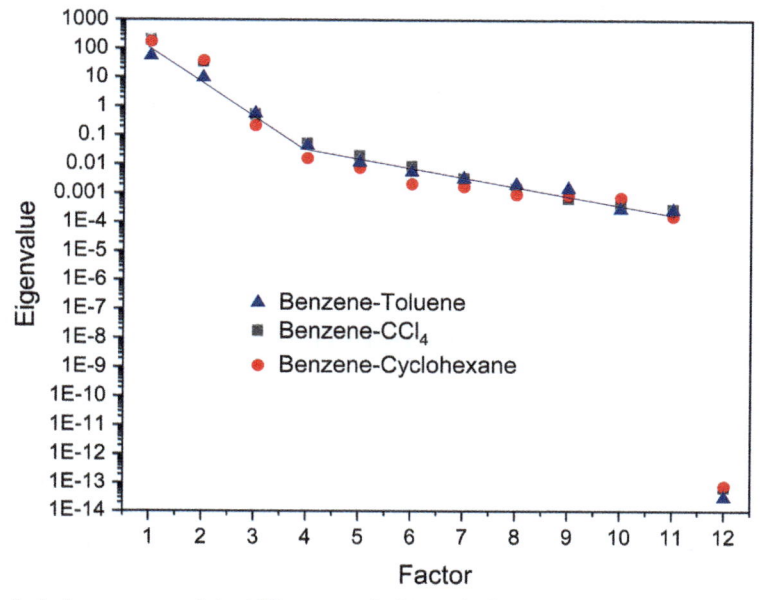

FIG. 9.7 Eigenvalues of the principal components of the ATR spectra of mixtures in the systems: benzene-toluene, benzene-tetrachloromethane, and benzene-cyclohexane. The 12th eigenvalue of the benzene-tetrachloromethane system is negative, therefore its magnitude is shown.

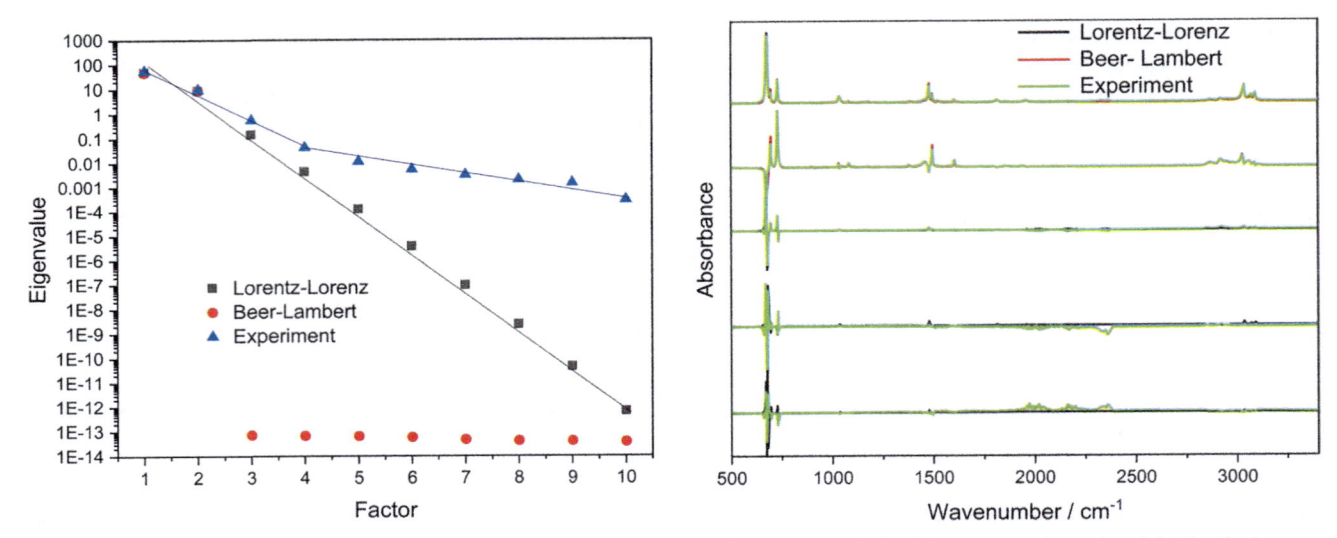

FIG. 9.8 Left panel: Eigenvalues of the principal components in the benzene-toluene system, derived from experiment and modeled by the Lorentz-Lorenz relation and Beer's approximation. Right panel: The first five principal component spectra which are the first five eigenvectors [5].

Fig. 9.8 demonstrates how the eigenvalues will look like if Beer's approximation sufficed to describe ideal systems. Only two eigenvalues would be meaningful and the number would indeed be equal to the number of chemical components. The assumption of local fields already destroys this way of interpretation, as the example of the eigenvalues of the Lorentz-Lorenz relation (cf. Section 5.1) shows. Furthermore, the eigenvalues of the experimental spectra demonstrate that there are even stronger deviations in reality from linearity than those predicted by the Lorentz-Lorenz relation. Nevertheless, it is interesting to see that the first two eigenvalues are more or less the same for the different models and the experimental spectra and that the third eigenvalue predicted by Lorentz-Lorenz is of roughly the right order to mimic experimental spectra.

The same is true for the principal component spectra. If the data are not mean-centered, the first component spectrum agrees with the average spectrum of the series. Already the second component spectrum cannot be seen as an absorbance spectrum since it contains negative values around the bands, even more pronounced than the component spectra in the case of CLS.

The fact that the eigenvalues decrease much less than expected is a staggering blow for PCA since one of the main arguments for using this method is that few components can describe a system as nearly as good as the full number if one selects the most important ones (i.e., the largest ones). This dogma is obviously wrong and this also has consequences for PCA regression. In principle, this regression relies on the fact that the eigenvectors are linearly independent. The concentration or volume fraction of a mixture can be calculated according to

$$\boldsymbol{\varphi}_u = \left(\mathbf{A}_u \mathbf{P}^T\right) \mathbf{P}_{ILS}. \tag{9.26}$$

If we make full use of 10 mixture spectra for calibration and then determine the 11th, PCR is not much better than ILS, as can be seen in Fig. 9.9.

9.5 Multivariate curve resolution (MCR)-alternating least squares (ALS)

As already stated, many molecular spectroscopists believe that PCA allows to determine the number of chemical components. With CLS it is believed to be possible to determine the spectra of a neat component (which would be correct if there were no nonlinearities, but as you know this is not the case), but you still would need to know the concentration in addition to the spectra, which is not possible if new compounds are formed in a mixture. It would certainly come in handy if it were possible to determine the spectra of such newly formed compounds, and it is believed that this is possible by a method called alternating least squares. For this method, it is also necessary to know the number of compounds, but this seems to be no problem because this number is believed to be computed by PCA. Based on this number, a tentative concentration or volume fraction matrix $\boldsymbol{\varphi}$ is assumed. Some think that this matrix should have random entries (cf., e.g. [6]), but in my own experiences, it is better if these entries are close to the actual values with which the starting components are mixed

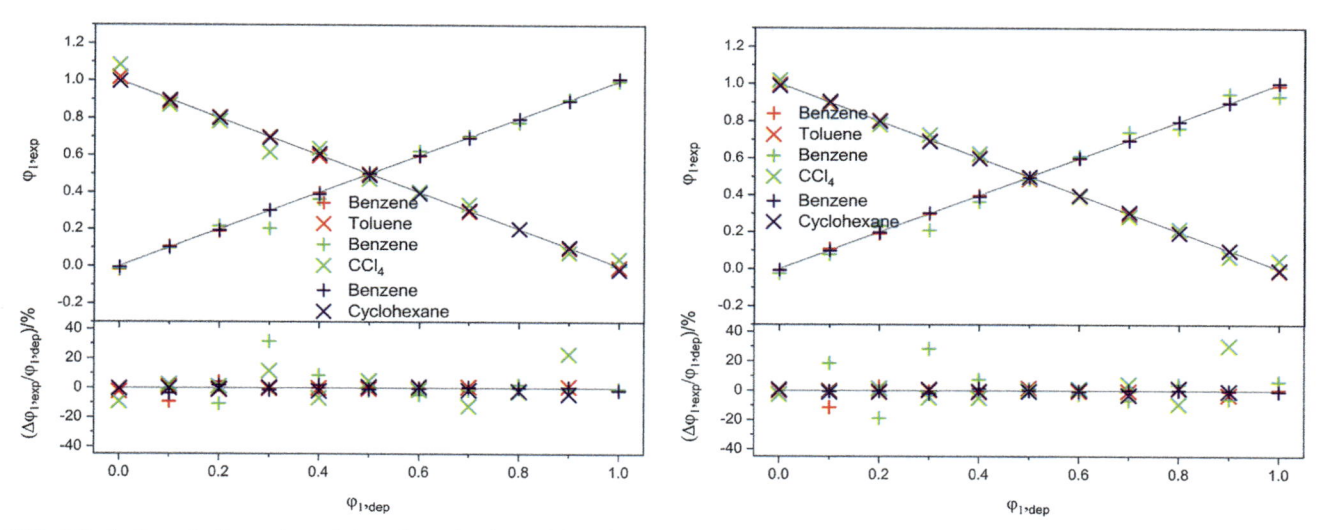

FIG. 9.9 Left panel: Comparison of the volume fractions of the mixtures, benzene-toluene, benzene-tetrachloromethane, and benzene-cyclohexane, as determined by ILS (taken over from Fig. 9.6). Right panel: The same for PCR.

minus a small amount which is assumed for the newly formed component. With these assumptions, a first matrix $\mathbf{K}'$ can be calculated according to

$$\mathbf{K}' = \left(\boldsymbol{\varphi}^T \boldsymbol{\varphi}\right)^{-1} \boldsymbol{\varphi}^T \mathbf{A}. \tag{9.27}$$

Did Beer's approximation hold, the true $\mathbf{K}$ would consist of spectra that have only positive absorbance values. We have seen in the section about CLS that this is not true (cf. Section 9.2), but we play along for a moment and assume that this would be correct. Since $\mathbf{K}'$ is based on incorrect entries for $\boldsymbol{\varphi}$, it is likely that some parts of the spectra contained in $\mathbf{K}'$ are negative. The corresponding values are increased for the next step to very small negative numbers. The correspondingly modified $\mathbf{K}'$ will then be used as $\mathbf{K}''$ to calculate a refined $\boldsymbol{\varphi}'$:

$$\boldsymbol{\varphi}' = \mathbf{A}\mathbf{K}''^T \left(\mathbf{K}''\mathbf{K}''^T\right)^{-1}. \tag{9.28}$$

Very much like $\mathbf{K}'$, $\boldsymbol{\varphi}'$ will have negative entries, which again is not meaningful. Therefore, in the next step, these negative values will be increased to very small negative numbers as this was done previously for $\mathbf{K}'$, and, following Eq. (9.27), improved component spectra will be computed, from which, again, an improved concentration/volume fraction matrix will be obtained. These steps are iteratively repeated until negative values will no longer appear in both $\mathbf{K}$ and $\boldsymbol{\varphi}$. In practice, it can happen that this point will never be reached, due to the fact that in reality $\mathbf{K}$ can become negative due to said nonlinearities. There is, however, an easy way to avoid this problem. Since strong bands are sometimes in the lower wavenumber regions (e.g., for benzene and toluene), these regions are tacitly cut off and only the rest of the spectrum is used for analysis. Anyway, the performance of MCR-ALS is really amazing. I performed it for the three model systems that we have discussed in this chapter and the results are depicted in Fig. 9.10 [5].

The assumption was for each system that it has three components. Indeed, if you go back to Fig. 9.7, I have interpreted the result of the PCAs just as this was done in literature, e.g., in Ref. [7], where the eigenvalues could also be seen as belonging to two different linear curves and the slope changed after component 3. While in the system investigated in Ref. [7] complex formation is not unlikely, it is completely unlikely in the quasiideal systems the component spectra of which are shown in Fig. 9.10. Nevertheless, MCR-ALS returns what you put in, i.e., if you assume a third component, it will produce three spectra. If you inspect and scrutinize the component spectra, then you see that the spectra of the third component are more or less linear combinations of the spectra of the other two components. Unfortunately, nowadays symmetry analysis is no longer fashionable in vibrational spectroscopy, even though it can be very useful at times. This is such a time since symmetry analysis tells us that any complex built of two different molecules would have to have a lower symmetry than each of its components, except if both components already had the lowest possible symmetry. The latter is not the case for the three ideal systems that we have discussed in this chapter. Lower symmetry, however, means that overall more vibrations become IR-active while others may become inactive (except for the lowest symmetry for which all vibrations become IR-active). Therefore, it is impossible that a component spectrum in these three systems would show the same

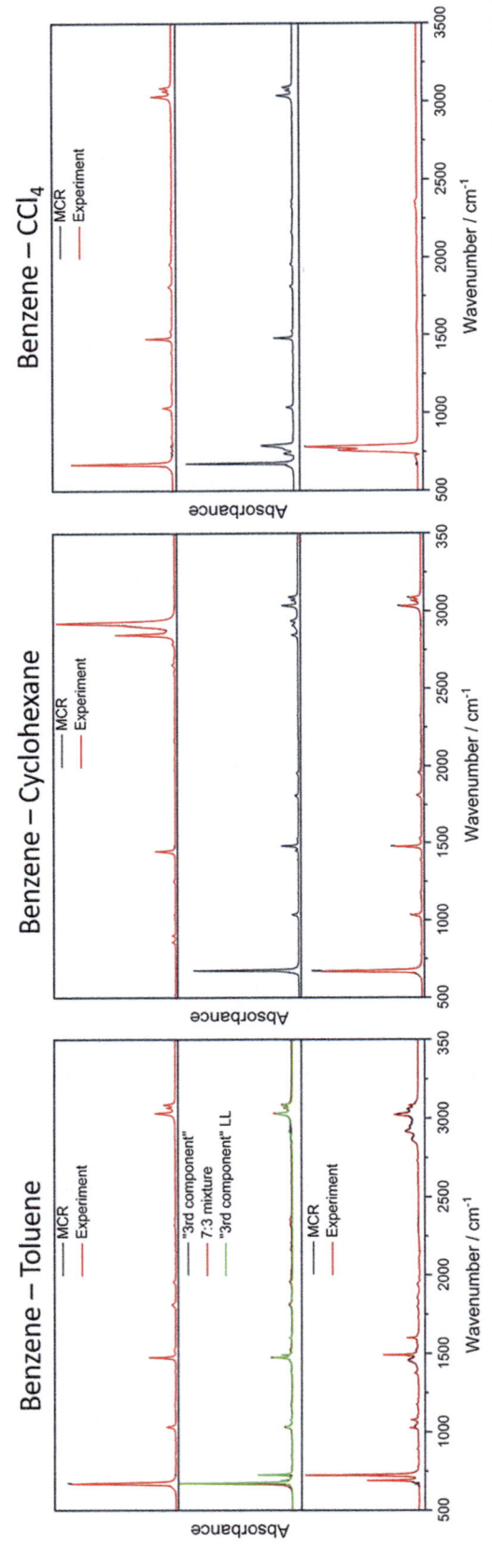

FIG. 9.10 Results of MCR-ALS in the systems, benzene-toluene, benzene-tetrachloromethane, and benzene-cyclohexane, assuming three components for each system. The spectra in the first and third row are indeed very similar to those of the neat components (those are also shown in the left panel and virtually congruent with the calculated spectra), whereas the spectra in the second row are similar to linear combinations of the first and third spectrum (in case of the left panel a 7:3 mixture of the neat spectra produces the spectrum of the nonexisting complex).

bands as both components. As a consequence, all 3rd component spectra shown in Fig. 9.10 are bogus. In fact, if you calculate the spectra of the mixtures by Lorentz-Lorenz theory and you apply MCR-ALS you find more or less the same third component spectra (remember, the third principal components of the experimental spectra are very similar to those derived from Lorentz-Lorenz theory, cf. Fig. 9.8). Therefore, if you apply MCR-ALS, be very critical concerning the results, since just because you get a spectrum of a potential third component does not mean that this component indeed exists.

9.6 Further reading

There are numerous books about chemometrics, but I have not found a single one that would explain what happens if the assumed linearity did not hold right from the start. Maybe I have not read enough, and there are—the increasing use of artificial intelligence and, in particular, deep learning methods show that spectroscopists know that there are nonlinearities (deep learning is in particular known for the ability to handle nonlinearities). Specifically, the problems introduced by polarization and local fields are too complex in my opinion to be handled analytically (as you will see in Chapter 10, you would need a separate theory for each quasiideal binary system, and those are the simplest systems that can be imagined, apart from gas mixtures). Therefore, the importance of deep learning will steadily increase, and I invite you to study further in this direction instead of wasting too much time with conventional chemometric methods.

References

[1] T.G. Mayerhöfer, A. Dabrowska, A. Schwaighofer, B. Lendl, J. Popp, Beyond Beer's law: why the index of refraction depends (almost) linearly on concentration, ChemPhysChem 21 (2020) 707–711.
[2] J.W. Strutt, XXXVI. On the light from the sky, its polarization and colour, Philos. Mag. Ser. 4 (41) (1871) 274–279.
[3] T.G. Mayerhöfer, O. Ilchenko, A. Kutsyk, J. Popp, Infrared spectroscopy of quasi-ideal binary liquid mixtures: the challenges of conventional chemometric regression, Spectrochim. Acta A Mol. Biomol. Spectrosc. 280 (2022) 121518.
[4] D.M. Haaland, E.V. Thomas, Partial least-squares methods for spectral analyses. 1. Relation to other quantitative calibration methods and the extraction of qualitative information, Anal. Chem. 60 (1988) 1193–1202.
[5] T.G. Mayerhöfer, O. Ilchenko, A. Kutsyk, J. Popp, Beyond Beer's law: quasi-ideal binary liquid mixtures, Appl. Spectrosc. 76 (2022) 92–104.
[6] T. Hasegawa, Quantitative Infrared Spectroscopy for Understanding of a Condensed Matter, Springer Japan, 2017.
[7] T. Shimoaka, T. Hasegawa, Molecular structural analysis of hydrated ethylene glycol accounting for the antifreeze effect by using infrared attenuated total reflection spectroscopy, J. Mol. Liq. 223 (2016) 621–627.

Chapter 10

Spectral mixing rules

10.1 Introduction

A book about infrared spectroscopy would not be complete without mixing rules. Nevertheless, as far as I know, this is the first such book including a chapter about such rules. How is this possible? The simple reason is that if a book is based on the BBL approximation, absorbance is assumed to be additive, so that if one mixes different gases, liquids, or solids, the resulting spectra should be simple weighted sums of the spectra of the neat substances irrespective of the state of matter and structure. In case of gases under low pressure or very diluted solutions, this works to a good approximation also from the perspective of wave optics. For mixtures in the condensed state, however, this is generally not the case. As I have already touched upon a couple of times in this book, light changes matter as both are coupled, and it is this changed state that we see in the spectra (more precisely, for condensed matter, the molecular dipole moment of an isolated molecule times the number density of molecules does not equal the polarization). It might be that the coupling is weak, but even then, we would be able to see such local field effects, e.g., if we look at spectra with tools introduced in the preceding chapters like 2D correlation spectroscopy or principal component analysis (cf. Chapters 8 and 9).

Nevertheless, I also was originally very hesitant to believe the approach the theories introduced in this chapter are based on are valid. The reason is that I expected science to be exact. So, when I first ran into certain theories that predict how spectra of mixtures should look like, the results were not very encouraging. Quite the opposite, actually. Not only, that the predicted spectra did not resemble the experimental ones, there is a multitude of different theories. Certainly, there are certain aspects that allow you to select the most suitable one, like for particles in a matrix, or for a pellet of two kinds of particles, etc. But even when you know the structure of your sample and you pick the theory that should be able to describe the spectrum of the mixture, you often end up disappointed. Also, I usually try in this book to show you as much as possible comparisons between theory and experiments, which is unfortunately not always possible for the examples in this chapter. The reason is that there is no possibility at the moment to prepare samples with desired structures. As an example, we tried to fabricate for several years samples that would have a simple chessboard structure as the one simulated at the end of this chapter, but until the print of this book we failed. For example, small clefts between white and black squares (which stand for two different materials), originating from imperfect fabrication, led to strong scattering effects, which were much larger than the expected effects caused by the mixing.

I have to admit that my skepticism against mixing rules, because of the often-poor agreement between experiment and simulation, prevailed over most of my scientific career and just began to crumble somewhat over the last years. Basically, what one needs is a different point of view. If you are lucky, you may find a theory that allows you to model your experimental spectrum to a good degree, but there is no way to formulate one theory that will be able to cope with most of the complications possible. This is a limit of infrared spectroscopy (and optical spectroscopy in general) that has to be accepted for the time being.

For most of this chapter, we will be dealing with samples where the mixing occurs on a nanoscopic level. In other words, if you take an infrared microscope, you should not be able to see any structure, i.e., the samples are assumed to appear homogenous relative to the resolving power of light or, in other words, inhomogeneities must be much smaller than the wavelengths (so-called microhomogeneous materials). Roughly, structures should be smaller than about 1/10 of the wavelength, i.e., between 2.5 and 0.25 μm in the MIR. Actually, one has to keep in mind that the theories for nanoscopic structures I introduce in the following are derived for the electrostatic limit. From this perspective, they can be seen as to be relatively useful even when we are not close to this limit. Later on in this chapter, we will also deal with microheterogenous structures, i.e., structures that are revealed using an IR microscope. For such samples, different mixing rules apply, and, for some probably somewhat surprisingly, reflectances or transmittances become additive, so that the latter have to be averaged to understand their spectra.

10.2 Lorentz-Lorenz theory

The derivation in this and the following sections will be somewhat simplified since I want to focus on the main results and the similarities and dissimilarities of the different approaches. Also, I see no reason to derive approximations rigorously, but if you are interested, I will provide you with some corresponding literature that does at the end of this chapter.

First of all, it is important to emphasize that the externally applied or the mean or observed electric field **E** is not the local or effective field $\mathbf{E}_{loc}$ that is acting on a molecule or a particle. The reason for the difference is that matter is not homogeneous, i.e., in the gaps between the molecules and particles, the polarizability is different. To estimate this difference, I assume that the molecule or particle of interest is surrounded by a sphere large compared to the molecule but small compared to the overall sample. I will separately consider the effect of the molecules inside the sphere on the molecule of interest and those outside the sphere. Both the molecules inside and outside the sphere are assumed to be distributed homogenously, be it randomly or in an ordered manner. For the molecules inside the sphere, it was shown for special configurations like a cubic crystal lattice that the symmetry leads to the situation that the action of each molecule on the molecule of interest is canceled by another so that it makes no difference if we assume their presence or absence. In fact, I will assume that they are absent and that the molecule of interest is surrounded by vacuum inside the sphere. This is equivalent to assuming that there is a single dipole inside the sphere. Concerning the molecules outside the sphere, I will assume that they are far away enough from the molecule of interest to view those as a continuum, which can be described with a constant dielectric function. Correspondingly, the polarization **P** outside the sphere is constant and we are left with a well-known problem of electrostatics of determining the potential of a dipole in a sphere of vacuum surrounded by a dielectric material. Accordingly, it follows that this gives an extra field $\mathbf{E}_L$ that has to be added to the applied field $\mathbf{E}_0$:

$$\mathbf{E}_{loc} = \mathbf{E}_0 + \mathbf{E}_L = \mathbf{E} + \frac{1}{3\varepsilon_0}\mathbf{P}. \tag{10.1}$$

To determine **P**, we assume that each molecule has an electric dipole moment **p** caused by the local electric field according to,

$$\mathbf{p} = \alpha \mathbf{E}_{loc}, \tag{10.2}$$

where α is the polarizability which I assume to be scalar. This means that the molecule must be isotropic, i.e., of tetrahedral or octahedral symmetry. Sometimes, it is stated in the literature that an overall random orientation of the molecules or crystallites is sufficient, but experimental results point to the opposite, so that the condition of isotropy of the individual molecule may be a serious constraint concerning the applicability of the Lorentz-Lorenz (LL) relation.

Overall, it is assumed that the microscopic dipole moments simply add up, so that the actual macroscopic polarization is given by:

$$\mathbf{P} = N \cdot \mathbf{p} = N \cdot \alpha \cdot \mathbf{E}_{loc}. \tag{10.3}$$

If we use Eq. (10.1) to eliminate $\mathbf{E}_{loc}$, the result is:

$$\mathbf{P} = N \cdot \alpha \cdot \left(\mathbf{E}_0 + \frac{1}{3\varepsilon_0}\mathbf{P}\right) = \frac{N \cdot \alpha}{1 - \dfrac{N \cdot \alpha}{3\varepsilon_0}} \cdot \mathbf{E}_0. \tag{10.4}$$

Therefore, we obtain for the effective dielectric function ε_{eff} as shown in the following result:

$$\varepsilon_{\text{eff}} = 1 + \frac{N \cdot \alpha}{\varepsilon_0 - \dfrac{1}{3}N \cdot \alpha} = \frac{1 + \dfrac{2}{3}\dfrac{N \cdot \alpha}{\varepsilon_0}}{1 - \dfrac{1}{3}\dfrac{N \cdot \alpha}{\varepsilon_0}}. \tag{10.5}$$

Correspondingly, after some algebraic manipulations, we find that,

$$\frac{\varepsilon_{\text{eff}} - 1}{\varepsilon_{\text{eff}} + 2} = \frac{1}{3}\frac{N \cdot \alpha}{\varepsilon_0}. \tag{10.6}$$

If there is more than one kind of molecule, we can rewrite Eq. (10.6) as:

$$\frac{\varepsilon_{\text{eff}} - 1}{\varepsilon_{\text{eff}} + 2}V = \frac{1}{3}\sum_i \frac{f_i \cdot \alpha_i}{\varepsilon_0} = \sum_i \frac{\varepsilon_i - 1}{\varepsilon_i + 2}V_i. \tag{10.7}$$

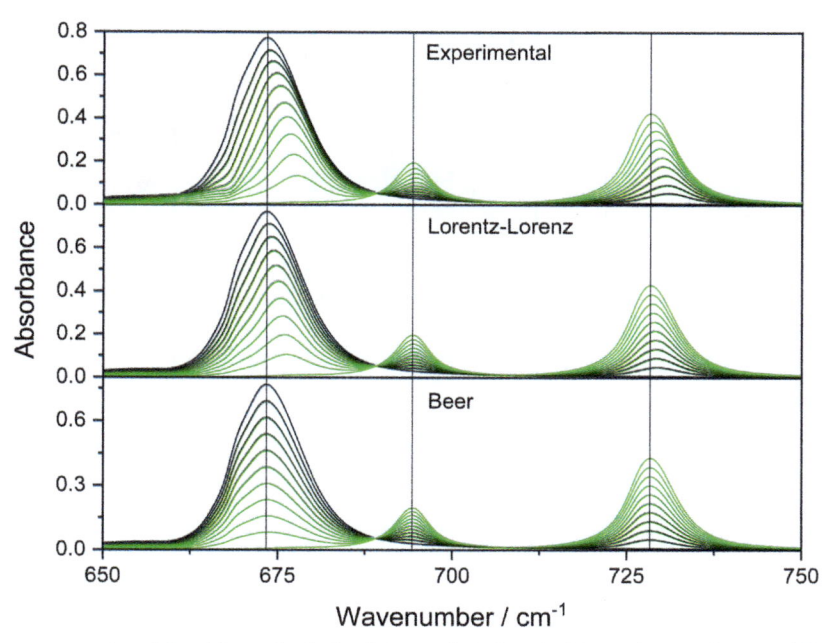

FIG. 10.1 Comparison between experimental and forward calculated spectra (based on the Lorentz-Lorenz theory and Beer's approximation) of mixtures in the ideal system benzene-toluene [1].

Here, I have made use of the additivity of the polarizabilities of the different molecules in the volume V and that the volumes V_i are additive (assumption of an ideal mixture; note that $N_i = f_i \cdot V_i$, with f_i the number of dipole moments belonging to molecule i and V_i the volume all molecules of this kind occupy). If we take into account that the volume fraction $\varphi_i = V_i/V$, we obtain the final form:

$$\frac{\varepsilon_{\text{eff}} - 1}{\varepsilon_{\text{eff}} + 2} = \sum_i \frac{\varepsilon_i - 1}{\varepsilon_i + 2} \varphi_i. \tag{10.8}$$

The ε_i are the dielectric functions of the pure components, which can easily be determined from reflection or transmission measurements followed by dispersion analysis or other methods as detailed in Chapter 6. ε_{eff} can be used to calculate reflectance, transmittance, or absorbance and compare it with results obtained by experiments. One example is a mixture between benzene and toluene the experimental spectra of which are shown in Fig. 10.1 together with those simulated according to Eq. (10.8) and Beer's approximation. Obviously, the Lorentz-Lorenz approximation predicts correctly the presence of a blue shift for lower volume fractions, but is unable to calculate the full extent of this blueshift, presumably because neither the benzene nor the toluene molecule is isotropic and spherical. The same is true for the extent of the peak height, which is predicted to be higher if Lorentz-Lorenz theory is used compared to if Beer's approximation is applied. However, the experimental peaks have an even higher absorbance than predicted by Lorentz-Lorenz theory. So, overall, the Lorentz-Lorenz approximation is a step in the right direction, but obviously important aspects are missed. Are these aspects properly addressed by other approaches?

10.3 Maxwell-Garnett approximation

The Maxwell-Garnett (MG) approximation was not introduced by Garnett as you may sometimes read, but by Maxwell-Garnett, which is a double surname [2,3]. In principle, the MG approximation seems to be simple to derive based on the Lorentz-Lorenz relation by assuming that the molecules or particles are not suspended in vacuum, but in a host medium with a dielectric function ε_h different from unity, cf. Scheme 10.1.

Accordingly,

$$\mathbf{E}_{loc} = \mathbf{E}_0 + \mathbf{E}_L = \mathbf{E}_0 + \frac{1}{3\varepsilon_h}\mathbf{P}. \tag{10.9}$$

Therefore, following the same path as in the previous section, we obtain,

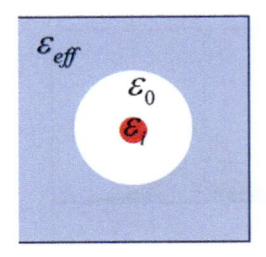

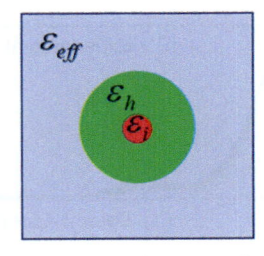

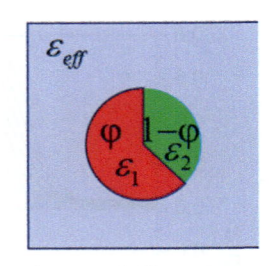

SCHEME 10.1 *Left panel*: Lorentz-Lorenz relation—a molecule or particle characterized by ε_i is suspended in a sphere containing vacuum ($\varepsilon_0 = 1$). *Center panel*: Maxwell-Garnett approximation—a molecule or particle characterized by ε_i is suspended in a sphere with ε_h. *Right panel*: Bruggeman approximation—the sphere consists of different molecules or particles.

$$\frac{\varepsilon_{\mathit{eff}} - \varepsilon_h}{\varepsilon_{\mathit{eff}} + 2\varepsilon_h} = \frac{1}{3}\frac{N \cdot \alpha}{\varepsilon_h}, \tag{10.10}$$

and, analogously,

$$\frac{\varepsilon_{\mathit{eff}} - \varepsilon_h}{\varepsilon_{\mathit{eff}} + 2\varepsilon_h} = \sum_i \frac{\varepsilon_i - \varepsilon_h}{\varepsilon_i + 2\varepsilon_h}\varphi_i. \tag{10.11}$$

For the case that we only have one kind of inclusion, Eq. (10.11) can easily be solved for $\varepsilon_{\mathit{eff}}$ [4]:

$$\varepsilon_{\mathit{eff}} = \varepsilon_h + 3\varphi\, \varepsilon_h \frac{\varepsilon_i - \varepsilon_h}{\varepsilon_i + 2\varepsilon_h - \varphi(\varepsilon_i - \varepsilon_h)}. \tag{10.12}$$

If ε_h, ε_i, and the volume fraction are known, the effective dielectric function can be computed. In principle, for LL, MG, and the Bruggeman approximation (which is introduced in the next section), if the materials are fixed, the volume fraction is the only parameter that determines the resulting spectra, regardless of the actual shape of the inclusion or the surrounding matrix, which is a strongly simplifying assumption. Concerning real samples, one could also characterize them with regard to if they have separated particle structure as this is implied in Scheme 10.1, or if the particles can also touch each other. Certainly, for increased volume fractions, the probability that particle touch each other increases. In fact, if you consider the matrix as a dielectric and the particles as metallic, above a certain threshold, a path will be created on which the free electrons can travel through the effective medium, rendering it a conductor. For our problem at hand, i.e., calculating spectra, I would say volume fractions of the inclusions below this threshold, the so-called percolation limit, are a requirement for the MG approximation to hold. Otherwise, it is not assured that regardless where a microscope is pointed at, the same spectra ensue, because areas with dimensions close or even above the resolution limit result (maybe it would not be possible with a microscope to resolve the structures, but to discern that there are structures). The same argumentation also limits the volume fractions in the other way, because if the matrix is dominant (which one would assume that it usually is), then you easily might hit with a focused light beam a region where only matrix material is present. Even if you assume that the particles are aligned regularly, at a certain volume fraction, the distances between particles will be too large.

Why should this be a problem for macroscopic measurements? Because as already discussed in Section 6.6, reflectance and transmittance are additive. If the effective dielectric function is everywhere the same, this does not matter, but if this is not the case then the resulting (average) spectra might be altered. If the signal only comes from the inclusions, this might not be a problem. If you consider the original use case of the MG approximation, i.e., plasmonic spheres in a dielectric, the plasmonic effect caused by the sphere modes might be dominant and the approximation nevertheless acceptable. But if the matrix also has a spectrum, the results gained by MG may not be acceptable. A concrete example is displayed in Fig. 10.2. Again, we compare the experimental spectra of the mixture benzene-toluene with spectra simulated based on the MG approximation, assuming first toluene as matrix and then benzene. Note that for volume fractions $\varphi \leq 0.5$, this makes strictly speaking no sense. The reason for this is that the MG approximation is asymmetric. While Eq. (10.12) correctly reduces to yield the endmembers for $\varphi = 0$ and $\varphi = 1$, an exchange of host and guest results in different spectra for the same volume fractions as can easily be seen in Fig. 10.2 (just focus on the benzene band at $673\,\mathrm{cm}^{-1}$). If toluene is seen as the matrix, the band shape is comparable to those determined experimentally, except that the band does not blueshift as strong as for the Lorentz-Lorenz relation (cf. Fig. 10.1). This is not surprising since the sphere in the middle of which the benzene molecule is situated usually has an increased dielectric function in the MG approximation compared to vacuum in the LL approximation. This results in a reduced blueshift, as well as a decreased peak maximum. If benzene is assumed to be the

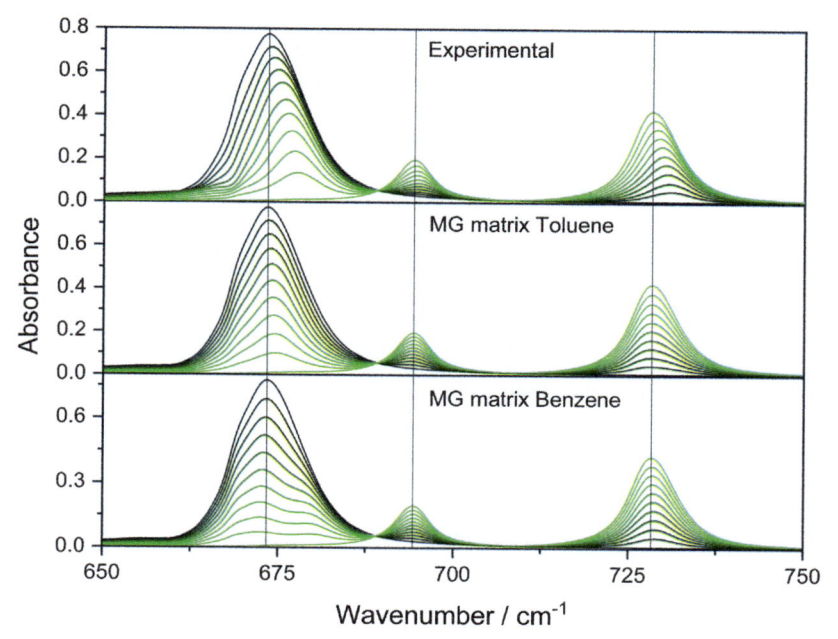

FIG. 10.2 Comparison between experimental *(upper panel)* and forward calculated spectra of mixtures in the ideal system benzene-toluene. *Center panel*: Maxwell-Garnett approximation with toluene as matrix and benzene as inclusion. *Lower panel*: Maxwell-Garnett approximation with benzene as matrix and toluene as inclusion.

matrix, the peak maximum redshifts instead, but there is also a shoulder appearing on the higher wavenumber side of the band. This is obviously not reflected in the experimental spectra. Also, the weaker bands do not change much. This was to be expected, since the effects should increase with the oscillator strength, just as for the LL relation. For inorganic materials, it can thus be expected that the deviations from the BBL approximation become very strong and the differences between the different effective medium approximations become very large.

Consider the case that we have a diluted solution of different compounds in a transparent solvent. From spectrophotometric experiments, we know that in highly diluted solutions, the absorbances of the different compounds are indeed to a good approximation additive, which is reasonable since in the limit of vanishing oscillator strengths, Beer's approximation seems to hold. Nevertheless, as we also know, the dielectric function of the solvent and its rate of change influences the molar absorption coefficient (the LL approximation, as well as MG, and the Bruggeman approximation predict solvatochromism [5]!). If we assume that the latter could be neglected, it looks like MG should be favorable over LL. This seems, however, not to be the case.

10.4 Bruggeman approximation

As discussed in the preceding section, the MG approximation is not symmetric, because there is a host material and one guest material (or more than one), which are treated differently. In the case of mixtures, however, all the different constituents may be similar and of similar volume fractions, so that a host cannot be identified. Therefore, a theory is needed in which the different components can be exchanged arbitrarily without changing the result. Such a theory has been provided by Dirk Anton Georg Bruggeman in 1935 [6]. The principle idea is that the molecule or particle is neither suspended in vacuum as the LL approximation suggests, nor in the matrix material as assumed by the MG approximation, but in the effective medium that results as consequence of the different materials suspended in it (cf. Scheme 10.1). To derive the corresponding mixing rule, we start from Eq. (10.11) and assume that the host material is equal to that of the effective dielectric function, $\varepsilon_{eff} = \varepsilon_h$,

$$\frac{\varepsilon_{eff} - \varepsilon_h}{\varepsilon_{eff} + 2\varepsilon_h} = \sum_i \frac{\varepsilon_i - \varepsilon_h}{\varepsilon_i + 2\varepsilon_h} \varphi_i \xrightarrow{\varepsilon_{eff}=\varepsilon_h} 0 = \sum_i \frac{\varepsilon_i - \varepsilon_{eff}}{\varepsilon_i + 2\varepsilon_{eff}} \varphi_i. \tag{10.13}$$

In other words, the dipole moments of all the different constituents cancel each other. This is an equation of degree *i*. For two components, it consequently leads to a quadratic equation that is particular easy to solve:

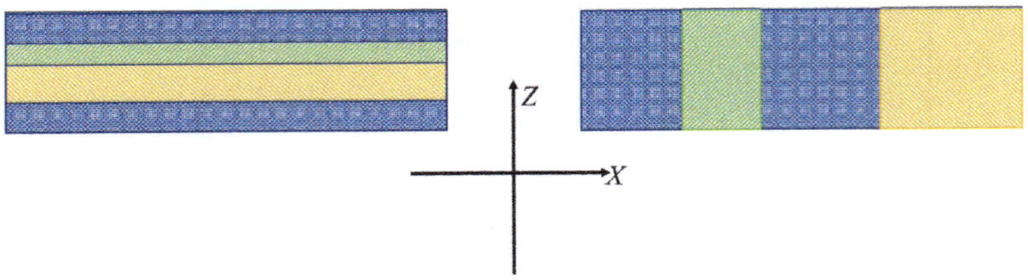

SCHEME 10.2 Illustration of the Wigner bounds. *Left panel*: Series connection with alternating layers in Z-direction. *Right panel*: Parallel connection of the same layers.

$$\varepsilon_{eff} = \frac{1}{4}\left(\beta + \sqrt{\beta^2 + 8\varepsilon_1\varepsilon_2}\right), \quad \beta = (3\varphi_1 - 1)\varepsilon_1 + (3\varphi_2 - 1)\varepsilon_2 = 3\varphi_1(\varepsilon_1 - \varepsilon_2) - \varepsilon_1 + 2\varepsilon_2. \tag{10.14}$$

Note that I have already picked the correct solution. Help to select the correct solution is provided by the so-called Wigner bounds, which are also reported in [6]. These bounds are given by:

$$\varepsilon_{eff} = \varphi_1\varepsilon_1 + \varphi_2\varepsilon_2 \quad \text{(I)}$$
$$\varepsilon_{eff} = \frac{1}{\frac{\varphi_1}{\varepsilon_1} + \frac{\varphi_2}{\varepsilon_2}} \quad \text{(II)}. \tag{10.15}$$

The idea behind these bounds is easy to grasp, in particular if you know the formulas for parallel and series connection of resistors in electric circuits. The first formula describes the series connection and the second the parallel connection. Accordingly, one assumes that to derive the first formula material one and two (or, in general, the different materials) are arranged in alternating layers, whereas in the second case each of the different materials penetrate from the top to the bottom in Z-direction as this is illustrated in Scheme 10.2.

The Wigner bounds reflect the extreme values that are possible for the effective property given the properties of the individual materials that constitute the effective medium. If, e.g., the effective property would be the conductivity, and the blue material in in Scheme 10.2 would be conductive, whereas the two other materials would be insulating, the series connection leads to an insulating effective material, whereas the parallel connection leads to a conductive effective material, but its conductivity could not be higher as the individual conductivities. For IR spectroscopy, the important conclusion is that for an effective medium, under the assumption that electrostatics holds, peaks can shift from the TO position (Eq. 10.15 (I)) to the LO position (Eq. 10.15(II)).

I again use the system benzene-toluene to investigate whether the Bruggeman effective medium model leads to a better description of the experimental spectra. Have a look at Fig. 10.3 where the results are depicted. First of all, it can be seen that the Bruggeman model leads correctly to blueshifts like the LL relation does. This means that obviously the shift is a function of the difference between the dielectric function of the molecule and its surrounding medium as already mentioned. If the surrounding medium is nonabsorbing, it cannot take on smaller values of the dielectric function than that of vacuum, which leaves us with the open question, how the experimental blueshift can be actually larger than predicted by the LL relation, but I (and literature as far as I know it) cannot answer this question. Compared to the spectra simulated by MG with toluene as matrix, the blueshifts are about the same for the benzene band, whereas the maximum intensities are always predicted to be smaller than they are experimentally. In contrast to MG, however, the Bruggeman model also predicts blueshifts for the toluene bands since benzene and toluene are treated symmetrically.

I want to emphasize that the fact that the Bruggeman model does not predict the spectra of the benzene–toluene mixture well does not mean that you should abandon this model. For different systems, it might do in fact a better job. In any way, such shortcomings have led to the fact that there are a multitude of different extensions of the effective medium models, like, e.g., considering different shapes for the molecules or inclusions, in particular an ellipsoidal shape. Such a shape could lead to an anisotropic behavior if the ellipsoids are aligned, even though the constituent materials can be characterized by a scalar dielectric function. On the other hand, the assumption that molecules or particles are ellipsoidal introduces additional degrees of freedom, which generally help to fit experimental spectra better without the certainty that it is indeed this aspect that improves the simulations. This is why I do not introduce you into this extension in detail, but follow a different path, which was also pursued by Wolfgang Theiß in his doctoral thesis (note that the following section is strongly influenced by this thesis and is dedicated to Wolfgang—R.I.P.) [7]. Accordingly, I will provide you with an introduction into the so-called Bergman representation [8] in the following.

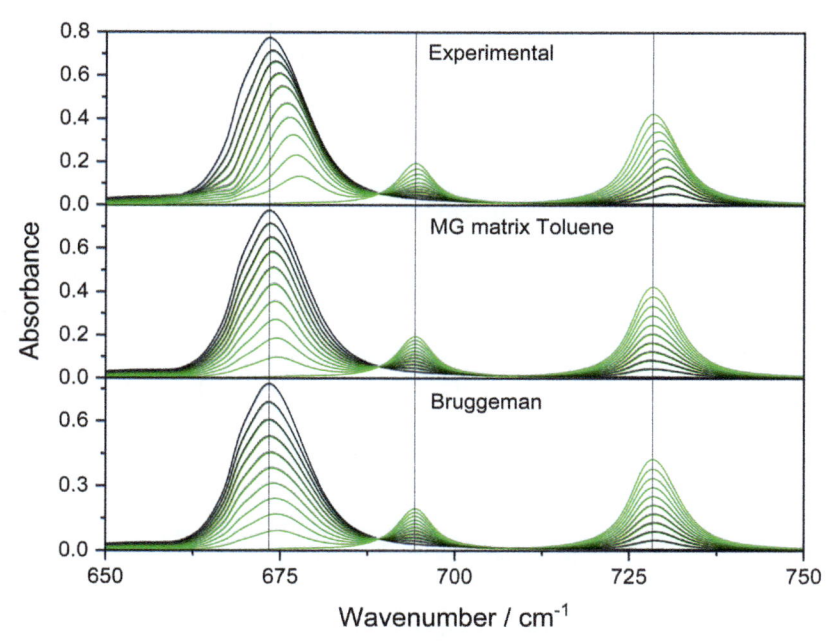

FIG. 10.3 Comparison between experimental *(upper panel)* and forward calculated spectra of mixtures in the ideal system benzene-toluene. *Center panel*: Maxwell-Garnett approximation with toluene as matrix and benzene as inclusion. *Lower panel*: Bruggeman approximation.

10.5 The Bergman representation

According to the Bergman theorem [8],

$$\varepsilon_{\mathit{eff}} = \varepsilon_h \left(1 - \varphi \int_0^1 \frac{g(n)}{t-n} dn\right) \quad \text{with} \quad t = \frac{\varepsilon_h}{\varepsilon_h - \varepsilon_i}, \quad (10.16)$$

it is possible to model all resonances that result from the geometry of an inhomogeneous medium consisting of two different compounds or molecules for real values of the variable t in the interval $[0,1]$. In particular, this includes the geometry of particles and molecules. To finally understand this theorem, we will start with some preliminary considerations about the polarization in many-particle systems.

10.5.1 Dipole interactions and resulting polarization in many-particle systems

In the following, we assume a system consisting of many spherical particles or molecules that interact exclusively via their dipole moments. Note that we will allow particles to touch each other, but nevertheless restrict ourselves to dipole interactions, although multipole interactions should result and are supposed to be important in this case (if you are interested in results that include multipole interactions, have a look at [9]). However, I will not try to derive an exact theory, but instead I want to illustrate what is the principal idea behind the Bergman representation. Actually, an exact theory would anyway require that we accurately know position and shape of each particle, which is in practice impossible. Instead, we will investigate the most important properties using systems with some 20 particles. Simply the fact that we reduce the distance between these particles will lead to resonances that are not present if the dipoles are separated by large distances. If the particles were far away from each other, an externally applied field would polarize each particle independent of each other like molecules in a gas under low pressure.

If we decrease the interparticle distance, the polarization of each particle will influence the polarization of each other, and this mutual interaction will lead to an effective field that can be very different from the external one. To keep things comparably simple, we will assume that all particles are spheres (again, we are not looking for an exact theory in the first place), but can have different volumes V_i. The positions of the particles will by described by their position vector $\mathbf{r}_i$ and their dipole moments by $\mathbf{p}_i$. It is certainly not possible to ascribe to a single molecule a dielectric function, but if many molecules build up a condensed phase, their dielectric function would be ε_i (in this case, the subscript i stands for inclusion as in Eq. (10.16)). For particles, it is easier to imagine that a dielectric function could be assigned to them, and, as I will discuss later in Section 15.1, this dielectric function would not be much different from that of a condensed phase of the same material, even if the particles are smaller than 100 nm (the main difference would be, that the damping constants are somewhat larger by a factor not much higher than unity). To simplify things, we will assume that for these preliminary considerations, the molecules or particles are embedded in vacuum, i.e., $\varepsilon_h = 1$.

The polarizability of the ith sphere is given by:

$$\alpha_i = 3\frac{\varepsilon_i - 1}{\varepsilon_i + 2} V_i. \tag{10.17}$$

Similar to Eq. (10.1), the electric field at the location of a particle or molecule is the applied electric field $\mathbf{E}_0$ plus the fields that originate from the other dipoles. The field originating from dipole j at the location i $\mathbf{E}_j(\mathbf{r}_i)$ is given by (under negligence of higher multipoles):

$$\mathbf{E}_j(\mathbf{r}_i) = \frac{1}{4\pi\varepsilon_0} \frac{1}{(r_i - r_j)^3} \left(3\mathbf{p}_j \cdot (\mathbf{r}_i - \mathbf{r}_j) \frac{(\mathbf{r}_i - \mathbf{r}_j)}{(r_i - r_j)^2} - \mathbf{p}_j \right). \tag{10.18}$$

What needs to be solved are the equations,

$$\mathbf{p}_i = \varepsilon_0 \alpha_i \left(\mathbf{E}_0 + \sum_{j \neq i} \mathbf{E}_j(\mathbf{r}_i) \right). \tag{10.19}$$

To do so, it is helpful to reformulate Eq. (10.19) by dividing the dipole moments by the square root of the sphere volume, $\boldsymbol{\pi}_i = \mathbf{p}_i / \sqrt{V_i}$:

$$\boldsymbol{\pi}_i = 3\varepsilon_0 \frac{\varepsilon_i - 1}{\varepsilon_i + 2} \sqrt{V_i} \mathbf{E}_0 + \frac{\varepsilon_i - 1}{\varepsilon_i + 2} \sum_{j \neq i} \mathbf{W}_{ij} \boldsymbol{\pi}_j. \tag{10.20}$$

Herein, the matrix $\mathbf{W}_{ij}$ is given by:

$$\mathbf{W}_{ij} = \frac{9}{4\pi r_{ij}^5} \sqrt{V_i V_j} \begin{pmatrix} x_{ij}^2 - r_{ij}^2/3 & x_{ij}y_{ij} & x_{ij}z_{ij} \\ x_{ij}y_{ij} & y_{ij}^2 - r_{ij}^2/3 & y_{ij}z_{ij} \\ x_{ij}z_{ij} & y_{ij}z_{ij} & z_{ij}^2 - r_{ij}^2/3 \end{pmatrix}. \tag{10.21}$$

Here, I have introduced the short notation $r_i - r_j = r_{ij}$, etc. Additionally, I will use in the following the $3N$-dimensional vectors (where N is the number of molecules/particles),

$$\mathbf{p}^{3N} = \begin{pmatrix} \boldsymbol{\pi}_1 \\ \vdots \\ \boldsymbol{\pi}_N \end{pmatrix} \quad \text{and} \quad \mathbf{E}^{3N} = 3\varepsilon_0 \begin{pmatrix} \sqrt{V_1} \mathbf{E}_0 \\ \vdots \\ \sqrt{V_N} \mathbf{E}_0 \end{pmatrix}, \tag{10.22}$$

and the $3N \times 3N$ matrix (note that $\mathbf{W}_{ij} = \mathbf{W}_{ji}$),

$$\mathbf{M} = \begin{pmatrix} 0 & \mathbf{W}_{12} & \cdots & \mathbf{W}_{1N} \\ \mathbf{W}_{21} & 0 & \cdots & \mathbf{W}_{2N} \\ \vdots & \vdots & \ddots & \vdots \\ \mathbf{W}_{N1} & \mathbf{W}_{N2} & \cdots & 0 \end{pmatrix}, \tag{10.23}$$

to be able to write the final equation in very compact form:

$$\mathbf{p}^{3N} = \frac{\varepsilon_i - 1}{\varepsilon_i + 2} \mathbf{E}^{3N} + \frac{\varepsilon_i - 1}{\varepsilon_i + 2} \mathbf{M} \mathbf{p}^{3N}. \tag{10.24}$$

To investigate the properties of Eq. (10.24), let us first assume that there is no externally applied electric field (the sample is in the dark), which means that the first term vanishes. Accordingly, one finds that:

$$\frac{\varepsilon_i + 2}{\varepsilon_i - 1} \mathbf{p}^{3N} = \mathbf{M} \mathbf{p}^{3N}. \tag{10.25}$$

This is an eigenvalue equation for $\mathbf{M}$, which has solutions if the term $(\varepsilon_i+2)/(\varepsilon_i-1)$ is equal to an eigenvalue of the matrix $\mathbf{M}$. This means that the nature of the dielectric function itself does not influence the matrix $\mathbf{M}$, and it just depends on the arrangement of the particles or molecules relative to each other. This is the reason why the corresponding resonances are called geometric resonances. Accordingly, if you leave the geometric arrangement the same, you can exchange the materials without causing a change in these resonances. In other words, if you have a certain configuration of your particles and you have determined the eigenvalues of $\mathbf{M}$, you do not need to determine the eigenvalues again if you can construct the same configuration/geometry employing other particles with a different dielectric function. A second interesting aspect is that the symmetry of $\mathbf{M}$ calls for real eigenvalues λ_k:

$$\frac{\varepsilon_i + 2}{\varepsilon_i - 1} = \lambda_k \rightarrow \varepsilon_i = -\frac{2 + \lambda_k}{1 - \lambda_k}. \tag{10.26}$$

What range of values can the eigenvalues take on? The elements of $\mathbf{M}$ become small very fast if the distances between the spheres increase (approximately $\propto r^{-3}$, cf. Eq. (10.21)). Therefore, the result for Eq. (10.26) is,

$$\lim_{\lambda_k \to 0} \varepsilon_i = -\lim_{\lambda_k \to 0} \frac{2 + \lambda_k}{1 - \lambda_k} = -2. \tag{10.27}$$

This is simply the resonance of isolated spheres well-known from plasmonics. If the spheres are close to each other, the eigenvalues are of the order of unity (you have to believe this as to prove it is out of the scope of this short introduction, since in this case, you have to take into account also multipoles). If we assume the latter statement to be correct, the eigenvalues are in the interval $[1,-2]$, which means that based on Eq. (10.26), resonances can only occur if the dielectric function is real and negative. The dispersion relations, however, prove that this is not possible (cf. Chapter 5), because dielectric functions can only become negative if there is absorption, which means that there always will be an imaginary part. For weak oscillators, the real part of the dielectric function stays positive. It becomes negative for strong oscillators between the transverse optical (TO) and the longitudinal optical (LO) wavenumber. For metals, this is the case for wavenumbers smaller than the plasma wavenumber. Note that in the Bergman representation, the need for negative real parts of the dielectric function is transformed by using $t = \varepsilon_h/(\varepsilon_h - \varepsilon_i)$, where $t \in [0,1]$.

Once the eigenvalues λ_k and the corresponding eigenvectors $\mathbf{\Pi}_i$ have been determined, Eq. (10.24) can be solved. To that goal, one can employ the series expansions,

$$\mathbf{p}^{3N} = \sum_{i=1}^{3N} a_i \mathbf{\Pi}_i, \quad \mathbf{E}^{3N} = \sum_{i=1}^{3N} b_i \mathbf{\Pi}_i E_0. \tag{10.28}$$

Thereby, Eq. (10.24) is transformed into

$$\sum_{i=1}^{3N} a_i \mathbf{\Pi}_i = \frac{\varepsilon_i - 1}{\varepsilon_i + 2} \left(\sum_{i=1}^{3N} b_i \mathbf{\Pi}_i E_0 + \sum_{i=1}^{3N} a_i \lambda_i \mathbf{\Pi}_i \right), \tag{10.29}$$

which, after multiplication by $\mathbf{\Pi}_j$ and employing that $\mathbf{\Pi}_i \cdot \mathbf{\Pi}_j = \delta_{ij}$, leads to the solutions,

$$a_j = \frac{b_j}{\frac{\varepsilon_i + 2}{\varepsilon_i - 1} - \lambda_j} E_0. \tag{10.30}$$

From Eq. (10.30), it is possible to compute $\mathbf{p}^{3N}$:

$$\mathbf{p}^{3N} = \sum_{i=1}^{3N} \frac{b_i}{\frac{\varepsilon_i + 2}{\varepsilon_i - 1} - \lambda_i} E_0 \mathbf{\Pi}_i. \tag{10.31}$$

Based on Eq. (10.31), the dipole moments of all spheres can be calculated and, by that, the polarization of the system. What was unclear to me, when I scrutinized Wolfgang's PhD thesis and came across Eq. (10.31), was how to calculate the b_i when I tried to program the problem. This might be clear to you, but to be on the safe side, you need to compare Eq. (10.28) with the right part of Eq. (10.22) to obtain the coefficients:

$$\mathbf{E}^{3N} = \sum_{i=1}^{3N} b_i \mathbf{\Pi}_i E_0 = 3\varepsilon_0 \begin{pmatrix} \sqrt{V_1} E_0 \\ \vdots \\ \sqrt{V_N} E_0 \end{pmatrix}. \tag{10.32}$$

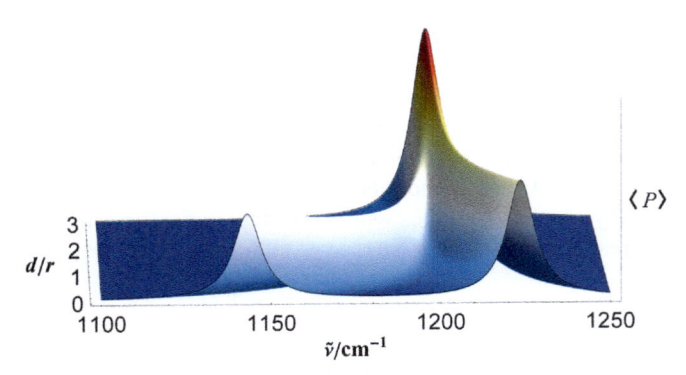

FIG. 10.4 Dependence of the averaged polarization of two spheres with radius r in dependence of the distance d between the spheres. The oscillator position is at $800\,\text{cm}^{-1}$, the damping constant equals $10\,\text{cm}^{-1}$, and the oscillator strength is very large ($S = 2000\,\text{cm}^{-1}$).

Once you have programmed everything, you are able to investigate how the dipole coupling between the spheres change the resonances of the system. In particular, you can use the spheres to form different geometric shapes.

First, let us investigate how the modes of two spheres of equal radius change when they approach each other. Have a look at Fig. 10.4.

To generate Fig. 10.4, I have assumed a very strong oscillator with a (TO) position at $800\,\text{cm}^{-1}$ and $S = 2000\,\text{cm}^{-1}$ (for reststrahlen bands S is usually around $1000\,\text{cm}^{-1}$, e.g., in case of MgO). Accordingly, the LO position is at about $1400\,\text{cm}^{-1}$. What you can see are two modes for touching spheres ($d = 0$) that degenerate to the isolated sphere mode, sometimes also called Fröhlich mode [10], when the distance d becomes larger than about $2r$. Note that we have already discussed two touching spheres, in the introduction, cf. Fig. 1.1. In fact, part (A) in this figure represented the isolated sphere and in (B) and (D) the polarization was perpendicular to the line that connected the centers of the spheres. Since, for the example, the oscillator strength was comparably low (amorphous SiO_2 with $S \approx 500\,\text{cm}^{-1}$), the shift of this mode could not be seen, but the redshift of the mode parallel to the connecting line was discernible (cf. [11]).

As already mentioned, touching spheres are quite interesting in general, because you can use them to generate other particles with special shapes like needles (prolate spheroids), disks (oblate spheroids), cubes, etc. Corresponding spectra (actually, what is shown is the imaginary part of the averaged polarization and the polarization along a high symmetry axis for two spheres, a 5×5 layer of spheres, and a needle-shaped configuration of spheres and perpendicular to it) are provided in Fig. 10.5. The less symmetric the geometric form is, the more different modes occur. The first two spectra for a single sphere and two touching spheres have already been discussed. The next shape is approximately sphere-like and consists of an arrangement of spheres, which could be seen as taken from a face-centered cubic close packing of spheres. Accordingly, the spectrum is dominated by one mode that is only slightly shifted compared to that of the single sphere. For the cube consisting of $3 \times 3 \times 3$ spheres, five modes result, whereas a more accurate model predicts six modes for a cube [12,13]. The 5×5 layer and the needle shape are interesting, because like in the case of two touching spheres, the resulting shape is anisotropic as a consequence of the anisometric shape, although the dielectric function of the spheres is assumed to be scalar. What is important is that all modes are always located between the TO and the LO positions, which means that peak shifts in real mixtures beyond these limits cannot be explained by geometric resonances. Also, it is important to note that for organic or biologic materials, such shape effects should not play an important role, because the oscillator strengths are generally too low. One exception may be the C=O mode, which is often broadened, but to resolve different geometric modes the difference between the TO and the LO positions is too small. On the other hand, for metallic particles, the LO position is equal to the plasma wavenumber. This means, e.g., for gold and silver particles, such shape effects are easy to observe. Also, keep in mind that the effects are due to geometric resonances, which means that exchanging dielectric functions does not alter them. Therefore, the peaks and their relative intensities look the same as in Fig. 10.5, only the wavenumber region in which they appear is different.

How does the Bergman representation relate to these results? The overall dipole moment can be, after the introduction of t, averaging over all orientations and some smaller modifications, expressed as:

$$\langle P \rangle = \sum_{i=1}^{3N} \frac{g_i}{t - (1-\lambda_i)/3} E_0. \tag{10.33}$$

This has already some similarity with Eq. (10.16) except that for the Bergman representation, usually an integral formulation is employed.

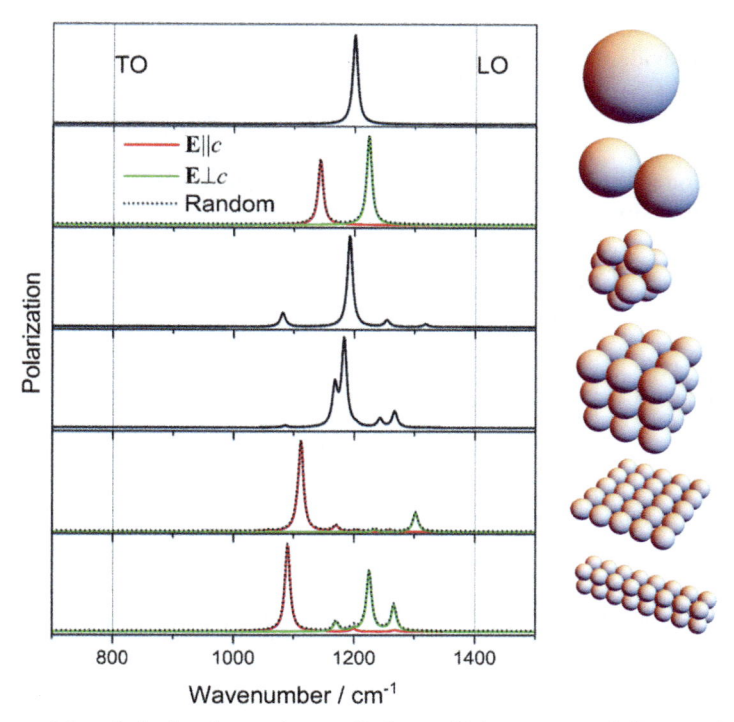

FIG. 10.5 Averaged polarization and the polarization along and perpendicular to a high symmetry axis for two spheres, a 5×5 layer of spheres, and needle-shaped configuration for touching spheres. The assumed oscillator position is at $800\,\text{cm}^{-1}$, the damping constant equals $10\,\text{cm}^{-1}$, and the oscillator strengths is very large ($S = 2000\,\text{cm}^{-1}$).

In any way, the comparison shows that geometric resonances are very important in effective medium models. In this respect, the Bergman representation is seen as a kind of generalization that incorporates all previous models, which have only one free parameter (the volume fraction), because those are already based on the assumption of a certain kind of geometry (I discuss this later in more detail). In this respect, the Bergman representation is much more flexible. Nevertheless, I must emphasize again that all geometric resonances are located between the TO and the LO positions, which somehow limits the flexibility of the Bergman representation. To better understand this, one can imagine that it consists of contributions of the form,

$$\frac{g_i}{t - n_i}, \qquad (10.34)$$

where g_i is an amplitude and a kind of strength of a resonance and n_i describes its location relative to t. Just to remind you, $t = \varepsilon_h/(\varepsilon_h - \varepsilon_i)$, which means that the influence of a geometric resonance on a spectrum depends on the frequency dependence of the in general complex dielectric functions of host and inclusion and how close the corresponding complex t-value gets to real values that can cause the geometric resonance.

10.5.2 Basic properties of the spectral density

For a real system, it is certainly impossible to know in detail how the spectral density $g(n)$ looks like. Instead, the idea is to fit this function, which means in practice to describe it with an analytic function with some free parameters, e.g., a polynomial function or Gaussian functions. Before we take a closer look, I will first introduce you to some basic properties of the spectral density. To that end, we use the fact that the geometric information encoded in the Bergman representation, i.e., in the spectral density, is decoupled from the properties of the dielectric functions of host and inclusion. When we assume that these dielectric functions are comparably similar, in other words, that the difference $\varepsilon_h - \varepsilon_i$ is small, a series expansion in this difference is possible. In the simplest case, the zeroth-order approximation, it follows that $\varepsilon_h = \varepsilon_i$, which is not helpful. The first-order approximation is given by:

$$\varepsilon_{\text{eff}} = \varepsilon_h - \varphi(\varepsilon_h - \varepsilon_i) + \mathcal{O}\!\left((\varepsilon_h - \varepsilon_i)^2\right) = (1 - \varphi)\varepsilon_h + \varphi\varepsilon_i + \mathcal{O}\!\left((\varepsilon_h - \varepsilon_i)^2\right). \qquad (10.35)$$

This is nothing but a simple additive mixture of the two dielectric functions weighted by their volume factors. For isotropic systems, it is also possible to provide the second-order approximation [14]:

$$\varepsilon_{eff} = \varepsilon_h - \varphi(\varepsilon_h - \varepsilon_i) - \varphi\frac{1-\varphi}{3}\frac{1}{\varepsilon_h}(\varepsilon_h - \varepsilon_i)^2 + \mathcal{O}\Big((\varepsilon_h - \varepsilon_i)^3\Big). \tag{10.36}$$

We compare this series expansion with the Bergman representation, for which t becomes very large if $\varepsilon_h - \varepsilon_i$. Therefore, it is possible to perform a series expansion for $1/t$:

$$\begin{aligned}
\varepsilon_{eff} &= \varepsilon_h \left(1 - \varphi \int_0^1 \frac{g(n)}{t-n}dn\right) = \\
&= \varepsilon_h \left(1 - \varphi\frac{1}{t} \int_0^1 \frac{g(n)}{1-n/t}dn\right) = \\
&= \varepsilon_h \left(1 - \varphi\frac{1}{t} \int_0^1 g(n)\Big[1 + n/t + (n/t)^2 + \mathcal{O}(n/t)^3\Big]dn\right) = \\
&= \varepsilon_h - \varphi(\varepsilon_h - \varepsilon_i) \int_0^1 g(n)dn - \varphi\frac{1}{\varepsilon_h}(\varepsilon_h - \varepsilon_i)^2 \int_0^1 ng(n)dn + \mathcal{O}\Big((\varepsilon_h - \varepsilon_i)^3\Big).
\end{aligned} \tag{10.37}$$

When Eq. (10.37) is compared with Eq. (10.36), it follows that,

$$\int_0^1 g(n)dn = 1, \tag{10.38}$$

and that,

$$\int_0^1 ng(n)dn = \frac{1}{3}(1-\varphi). \tag{10.39}$$

From Eq. (10.38), it can be concluded that the spectral density is a normalized distribution function the weighting factor of which decreases with growing volume fraction from 1/3 for $\varphi=0$ to zero for $\varphi=1$. While this is a conclusion based on the assumption of small differences between the dielectric functions of host and inclusions, the results can be extended to arbitrary dielectric functions of host and inclusions, since they only refer to the geometry of the system. Accordingly, every fit function of the spectral density must obey Eqs. (10.38), (10.39). In addition, every effective medium theory must also obey the corresponding conditions. Their differences obviously are caused by differences in the higher orders of which we could not (and cannot) gain information from considerations as the ones in this section.

10.5.3 Percolation

In Section 10.4, I have already mentioned that effective medium theories can also be used to describe the conductivity (and related properties like thermal conductivity, etc.) and that percolation is important in this respect. Percolation means, e.g., that there is a path of touching inclusions within the nonconducting host, so that an overall conductivity of the medium results if the inclusions are conductive. Since the spectral density represents the geometry of the system, it also must include and explain percolation.

Accordingly, we focus in the following on metallic particles with a DC conductivity σ_i in an insulating dielectric host ($\sigma_h \ll \sigma_i$). As a consequence, $t=\sigma_h/(\sigma_h-\sigma_i)$ will be negative and small. The corresponding Bergman representation is:

$$\sigma_{eff} = \sigma_h - \sigma_h\varphi \int_0^1 \frac{g(n)}{t-n}dn. \tag{10.40}$$

How can σ_{eff} become dominated by σ_i once percolation sets in at a certain volume fraction? Obviously, with percolation, σ_{eff} must not change if one would change σ_h, since the conducting path of the inclusion would not be affected by this change. If we would assume that $\sigma_h \to 0$ the value of the integral on the right side of Eq. (10.40) would have to increase as $1/\sigma_h$; otherwise, σ_{eff} could not stay constant. As a consequence, the spectral density must contain a term $g_0 \delta(n)$ ($\delta(0) = 1$). Accordingly, it must be possible to split the spectral density into two parts:

$$g(n) = g_0 \delta(n) + g_{Res}(n). \tag{10.41}$$

Consequently,

$$\sigma_{eff} = \sigma_h \left(1 - \varphi g_0 - \varphi \int_0^1 \frac{g_{Res}(n)}{t-n} dn\right) + \varphi g_0 \sigma_i. \tag{10.42}$$

Therefore, if $\sigma_h \to 0$, $\sigma_{eff} = \varphi g_0 \sigma_i$. This offers a possibility to determine g_0 by measuring the conductivity of the metal-dielectric mixture. Remember, since $g(n)$ only depends on the geometry of the system, Eq. (10.42) can be used for the effective dielectric function as well.

10.5.4 Dependence of the effective dielectric function on concrete spectral densities

In Section 10.5.1, I showed you that any feature caused by an oscillator related to the geometry of the system must be contained within its TO and LO position. For metals, this is still a vast wavenumber range, from zero to the plasma frequency, but for dielectrics, this means that for weak oscillators with small TO-LO shifts every effective medium theory is practically as good as any other. But when is an oscillator weak in this sense? A representation of a dielectric function in the t-plane (i.e., of its real and imaginary part) might be useful to better understand the problem, which I want to demonstrate in the following, where I again follow the paths of ref. [7].

To generate Fig. 10.6, I assumed that $g(n)$ is a normal distribution with the center at 0.3 and a standard deviation of 0.1. The corresponding spectral density is depicted in Fig. 10.6C. Note that it is very convenient that the normal distribution always has a unit area; otherwise, you would have to assure that this is indeed the case. In Fig. 10.6A and B, you find the corresponding real and imaginary parts of the effective dielectric functions in dependence of the real and imaginary parts of $t(=t_1 + i \cdot t_2)$. As can be seen, the effective dielectric function is mainly influenced around $t_2 = 0$ and for $t_1 > 1$. Obviously, the real part of the effective dielectric function is symmetrical to the plane $t_2 = 0$, t_1, whereas the imaginary part is

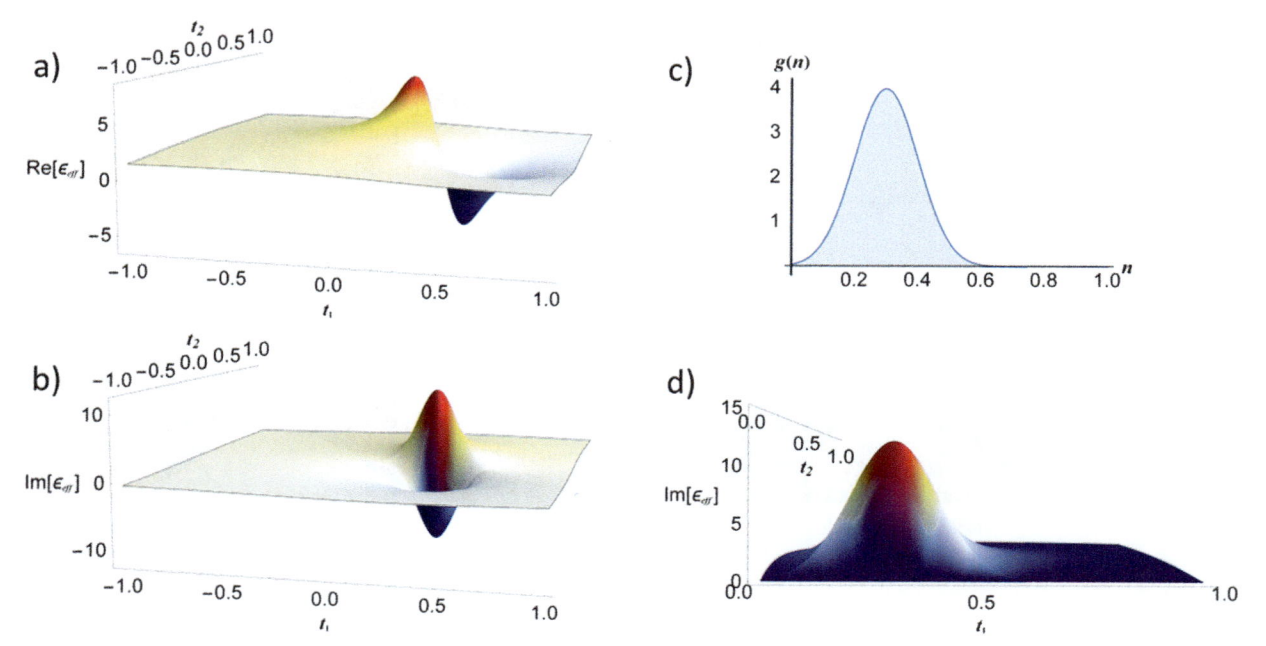

FIG. 10.6 (A) Real part of the effective dielectric function in dependence of t ($t = t_1 + i \cdot t_2$). (B) Imaginary part of the same effective dielectric function as in (A). (C) Spectral density used to compute the effective dielectric function. (D) Enlarged part of (B).

antisymmetric (this derives from the Bergman representation if it is split into its real and imaginary parts). What is striking is that the imaginary part of the effective dielectric function resembles the spectral density in the range $t_2=0$, $t_1 \in [0,1]$, as can be seen in Fig. 10.6D. If you would compute the effective dielectric function for other spectral densities, you would find that this seems to be generally true and that in some distance from $t_1, t_2=0$ all spectral densities would lead to similar values for the effective dielectric function. This is equivalent with stating that all effective medium approaches lead to similar values for the effective dielectric function away from the resonances. Actually, this means you can choose whatever effective medium approach you think is appropriate in spectral regions far from resonance. Those include linear combinations of the individual dielectric functions or absorption index functions. In other words, away from resonances, which also includes weak modes, because the real part of the dielectric function then does not become negative, absorbance becomes approximately additive.

I have to admit that even with the introductory section (Section 10.5.1), it was not easy for me to understand what exactly the spectral density stands for. Therefore, in the following, I depart from literature and try a somewhat different approach, which I hope is more illustrative. This approach is based on assuming that the guest material features a single oscillator, which is accommodated in a dielectric and featureless host medium. I will, in particular, investigate how the effective dielectric function changes under the assumptions of different spectral densities and oscillator strengths.

Fig. 10.7A shows the three Gaussian distributions used for the spectral density to calculate Fig. 10.7B. If the Gaussian distribution is centered at $n=0$ (all three distributions have the same standard deviation $\sigma=0.01$. The oscillator parameters are $\gamma=10\,\text{cm}^{-1}$ and $\tilde{\nu}_0=800\,\text{cm}^{-1}$ and, for Fig. 10.7C and D $S=500\,\text{cm}^{-1}$), the corresponding peak is at $\tilde{\nu}_0$, irrespective of oscillator strength. For $n=0.5$ and $n=1$, the peak position depends on the oscillator strength. In case of $n=0.5$, the peak position is between the TO and the LO position, whereas for $n=1$, the peak position is equal to the LO position. This agrees actually with the limits that I would expect based on the Wigner bounds.

Not only do the peak positions vary from the TO to the LO position, but the intensities also change accordingly. For $n=0$, the values of the effective dielectric function are much higher than for $n=1$, which reminds one of the facts that the values of the dielectric function are usually much higher than those of its inverse. This is also obvious from Fig. 10.7C, where the position of the Gaussian distribution is varied from 0 to 1. First, the values increase, which is due to the fact that for $n=0$, half of the distribution is out of the integration range. Accordingly, increasing n first leads to an increase of the area of the Gaussian distribution within the integration range and, thus, to an increase of the values of the imaginary part of the effective dielectric function. Once this effect wears off, these values continuously decrease while the peak shifts to the

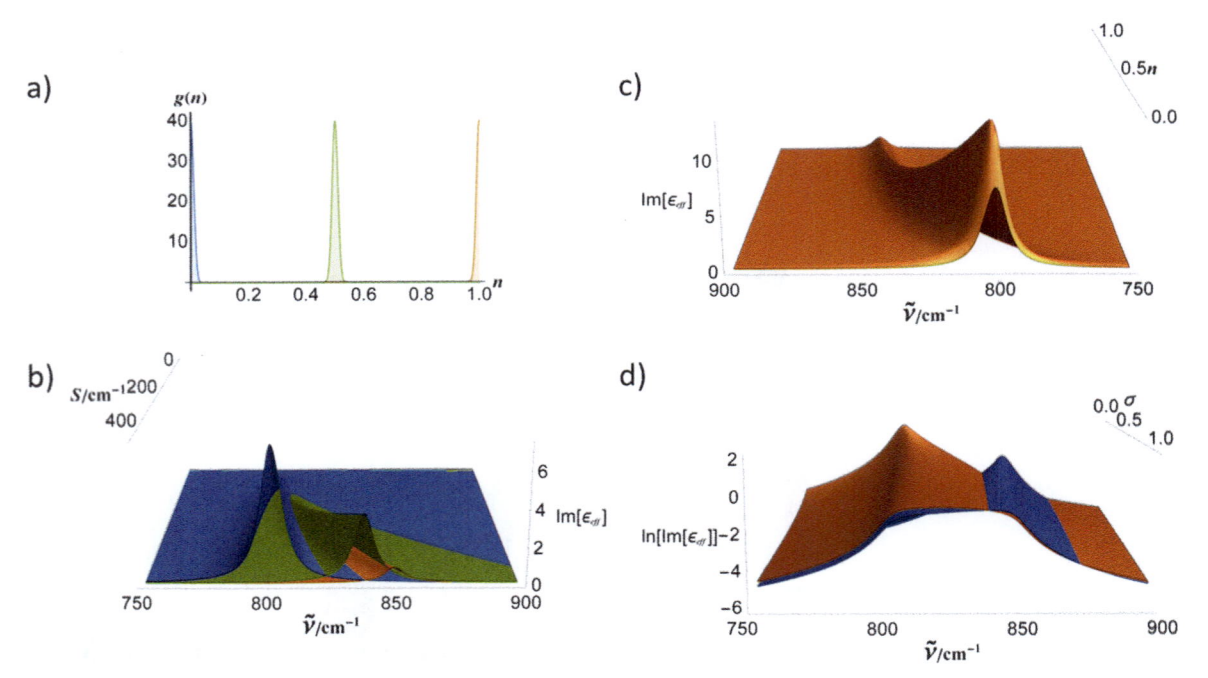

FIG. 10.7 (A) The three spectral densities that were used to calculate the effective dielectric function the imaginary parts of which are displayed in (B). (C) Imaginary part of the effective dielectric function calculated by varying the position of the center of the distributions shown in (A) from $n=0$ (*blue curve* in A) to 1 (*orange curve* in A). (D) Based on Gaussian distributions centered in $n=0$ and $n=1$ as displayed in (A), the standard deviation is varied from 0.005 to 1.

LO position. Fig. 10.7D shows what happens if for a fixed position of the Gaussian distribution, the standard deviation σ increases. Accordingly, the band begins to broaden until it covers the whole spectral range between the TO and the LO position with a nearly constant value of the effective dielectric function in this range.

This should give you a good picture of what to expect from the application of effective medium theories and if these theories can be useful to model a particular problem that you face. By employing the Bergman representation, you are most flexible to model experimental bands with shapes that do not need to resemble Lorentzians, but their onsets must be between the TO and the LO position. Before we proceed, one word of caution. We already discussed that the Maxwell-Garnett theory is not symmetric, which means that host and inclusion have different properties. This will also be in general the case for the Bergman representation. In fact, the Bergman representation contains all other effective medium theories, including the Maxwell-Garnett theory. This means that the symmetry properties of the Bergman representation depend critically on what kind of spectral density is chosen. I will not go too much into details, but it can be shown that every effective medium theory can be represented by the following form:

$$\varepsilon_{eff} = \varepsilon_h (1 - G(t)). \tag{10.43}$$

By comparison with Eq. (10.16), one finds that

$$G(t) = \varphi \int_0^1 \frac{g(n)}{t - n} dn. \tag{10.44}$$

Obviously, $G(t)$ is a function of φ, which means that effective medium approximations in general change their spectral density with φ. This certainly makes sense. In practice, it would mean that a band's shape, as well as its position, changes with the volume fraction, which is exactly what we observe, e.g., in case of the mixture of benzene and toluene (cf. Fig. 10.1).

For the Maxwell-Garnett theory, one finds by bringing it in the form of Eq. (10.43) and by comparing it with Eq. (10.44) that:

$$g(n) = \delta(n - (1 - \varphi)/3). \tag{10.45}$$

This is a very simple function, which for $\varphi=0$ results in $g(0)=1$ if $n=1/3$ (and the spectral density is zero for all other values). For increasing volume fractions, since $n=1/3 - \varphi/3$, n becomes increasingly smaller and reaches zero for $\varphi=1$. This means that if hosted in a nonabsorbing medium, a peak of the imaginary part of the effective dielectric function somewhat shifts to the LO position, but keeps its shape. The same spectral density also results from the Lorentz-Lorenz theory. For the Bruggeman approximation, the dependence of the spectral density on n and φ is more complex. It is given by [14]:

$$g(n) = \frac{3\varphi - 1}{2\pi\varphi} \delta(n)\Theta(3\varphi - 1) + \frac{3}{4\pi\varphi} \sqrt{(n - n_-)(n_+ - n)}, \quad n_\pm = \frac{1}{3}\left(1 + \varphi \pm 2\sqrt{2\varphi - 2\varphi^2}\right) \tag{10.46}$$

Here, $\delta(x)$ is the Dirac delta function and $\Theta(x)$ the Heaviside step function (Fig. 10.8).

This implies that for very low volume fractions, the sphere modes will dominate, which means that the peaks will be slightly shifted toward the LO position, whereas for higher volume fractions, the spectral density is dominated by the

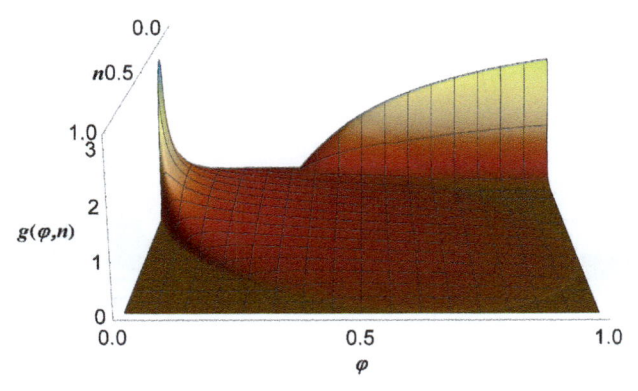

FIG. 10.8 Spectral densities of the Bruggeman approximation.

features at $n=0$. Accordingly, the peak practically stays at the TO position and a blueshift is nearly not recognizable until it becomes obvious below $\varphi=1/3$. Also, some broadening takes place. This is exactly what you can see in the lower part of Fig. 10.3. Note that discussing the Bruggeman approximation in relation to the Bergman representation implies that a formula that is symmetric with regard to the treatment of matrix and inclusions is somehow included in a formula, which does not seem to have the same symmetry. In fact, it has not, and this is problematic. We will come back to this problem later.

While it seems that the Bergman representation was derived mainly with microhomogeneous solids in mind, nothing should prevent its use for liquid mixtures. In contrast, the fact that the mixture is close to perfect is reflected in their radial distribution functions, which suggests an ordering with regard to the first coordination sphere. Looking at these functions, it is tempting to assume that the spectral density for liquid mixtures should be describable in a simple way, e.g., by assuming that $g(n)$ is characterized by a single normal distribution:

$$g(n) = \frac{1}{\sigma\sqrt{2\pi}} \exp\left[-\frac{1}{2}\left(\frac{n-\mu}{\sigma}\right)^2\right]. \tag{10.47}$$

Accordingly, $0 \leq \mu \leq 1$, and the value of the mean would be determined by the peak value, where $\mu=0$ would mean that the peak would be found at the TO position and, correspondingly, $\mu=1$ that the peak appears at the LO position. The standard deviation σ would influence the band shape (cf. Fig. 10.7). This means that apart from the volume fraction, I have introduced two additional parameters, where a small value of the second parameter, the standard deviation, would signal that the band shape remains practically unchanged and the normal distribution could be replaced by a Dirac delta function. Indeed, based on spectral densities according to Eq. (10.47), the modeling of the spectra of the mixtures in the system benzene-toluene is superior to the simpler effective medium approximations, cf. Fig. 10.9. The peak shifts of the 674 cm^{-1} band are, in contrast to these simpler theories, suitably addressed. The peak shapes agree nearly perfectly, only with regard to the intensities the agreement is less satisfactory for lower volume fractions. For the weaker toluene band, the agreement is nearly perfect, which means that Beer's approximation is valid. For the stronger toluene band, the agreement is less good in comparison with the benzene band, but again much better than for the simpler theories. Quite interesting is a view of the resulting spectral densities, which is provided in Fig. 10.10. The peak shift in the spectra translates into a shift of the mean as discussed previously, which was expected. The standard deviation is comparable small, and, indeed, it would be possible to replace the normal distribution with the Dirac delta function without degrading the quality of the fit considerably. Overall, it seems that the peak shift is to a good approximation linearly related to the volume fraction. Taking this together, it seems like it should be possible to describe the system toluene-benzene with a much simpler approximation than the Bergman

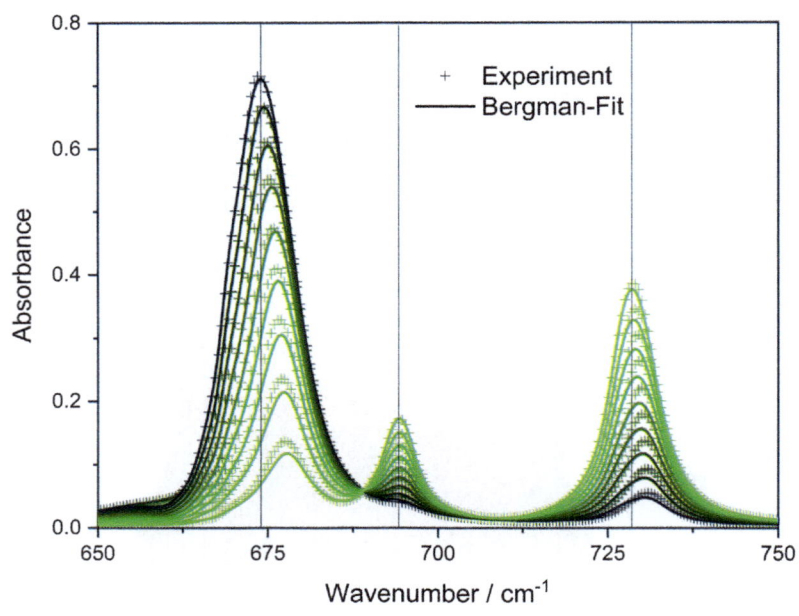

FIG. 10.9 Experimental and modeled spectra in the system benzene-toluene ($\varphi=0.1–0.9$). The modeling was performed using the Bergman representation assuming spectral densities based on Eq. (10.47) assigning toluene as matrix material.

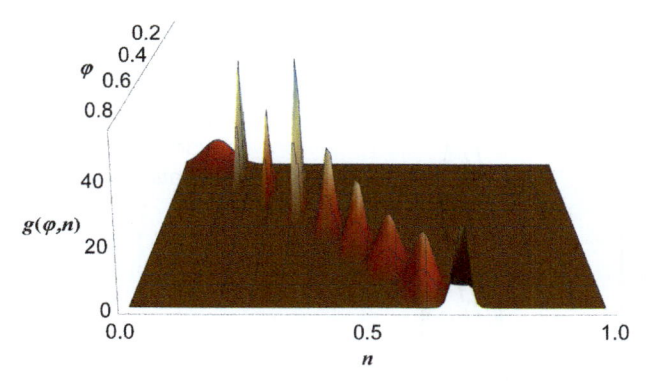

FIG. 10.10 Spectral densities resulting from the fit of the experimental spectra in the system benzene-toluene ($\varphi=0.1$–0.9) with the Bergman representation based on Eq. (10.47).

representation. This simpler approximation just has to take care of the peak shifts in an appropriate way, in contrast to the other approximations.

Indeed, this is possible. What I did not discuss so far in detail, is that there are extensions for the simpler approximations, e.g., if instead of spherical inclusions or particles ellipsoidal ones are considered. In this case, the so-called depolarization factor is not $1/3$ as for spheres, but variable. If you are not familiar with the definition of the polarization factor, it is the ratio of the internal electric field of an inclusion induced by surface charges when an external field is applied to the polarization of the dielectric [15]. For example, it stands to reason that the polarization of a spheroidal inclusion is different along the two different main axes:

$$E_{x,\text{int}} = E_{x,\text{ext}} - \overline{L}_x P_{x,\text{int}}. \tag{10.48}$$

Here, $E_{x,\text{ext}}$ is the x-component of the applied electric field, $E_{x,\text{int}}$ the x-component of the internal electric field, $P_{x,\text{int}}$ the polarization, and $\overline{L}_x$ the depolarization factor along x. If the depolarization factor of a sphere, $1/3$, is replaced by $\overline{L}_i$ with $i=x, y, z$ in the Bruggeman approximation, it reads:

$$\varphi_1 \frac{\varepsilon_1 - \varepsilon_{\text{eff},i}}{\varepsilon_1 + \left(\overline{L}_i^{-1} - 1\right)\varepsilon_{\text{eff},i}} + \varphi_2 \frac{\varepsilon_2 - \varepsilon_{\text{eff},i}}{\varepsilon_2 + \left(\overline{L}_i^{-1} - 1\right)\varepsilon_{\text{eff},i}} = 0. \tag{10.49}$$

Eq. (10.49) leads to a dielectric tensor, which would need to be appropriately averaged if the effective medium can be described by a scalar dielectric function. This seems to be trivial, but in Chapter 15, I show that this is everything else but simple. But we can apply empirically the following form of Eq. (10.49),

$$\varphi_1 \frac{\varepsilon_1 - \varepsilon_{\text{eff}}}{\varepsilon_1 + \left(\overline{L}^{-1} - 1\right)\varepsilon_{\text{eff}}} + \varphi_2 \frac{\varepsilon_2 - \varepsilon_{\text{eff}}}{\varepsilon_2 + \left(\overline{L}_i^{-1} - 1\right)\varepsilon_{\text{eff}}} = 0, \tag{10.50}$$

which has the solutions:

$$\varepsilon_{\text{eff}} = \frac{1}{4}\left(\beta \pm \sqrt{\beta^2 + 4\left(\overline{L}^{-1} - 1\right)\varepsilon_1 \varepsilon_2}\right), \quad \beta = \left(\overline{L}^{-1}\varphi_1 - 1\right)\varepsilon_1 + \left(\overline{L}^{-1}\varphi_2 - 1\right)\varepsilon_2. \tag{10.51}$$

The additional parameter $\overline{L}^{-1}$ allows to adapt the peak position. If this form of the Bruggeman approximation is employed for the fit, a value for $\overline{L}^{-1}$ results that is very similar to the one that can be obtained for n by fitting a line through the peak positions in Fig. 10.10 and extending it to $\varphi=1$. The corresponding simulation is shown in Fig. 10.11. Overall, it is of nearly similar quality as the one by the Bergman representation, except that the overall intensity of the benzene band is for all volume fractions somewhat smaller than the experimental ones. From this point of view, it seems that the Bergman representation still has its merits. However, for the fit shown in Fig. 10.9, toluene was assigned as matrix material and the benzene molecules had the role of inclusions. What happens if the roles are swapped? A closer view on Fig. 10.12 shows that this swapping of the roles is not a good idea. While the fit of the weak toluene band is also immaculate, and the fit of the stronger toluene band is slightly improved, a clear degradation can be observed for the benzene band, in particular for low volume fractions. This trend is increased for the also ideal mixtures benzene-CCl_4 and benzene-cyclohexane: The band shapes for the matrix material become distorted, the more the higher the oscillator strength is. This is also true for the general example depicted in Fig. 10.7. This means that the Bergman representation should only be used for cases where

274 PART | I Scalar theory

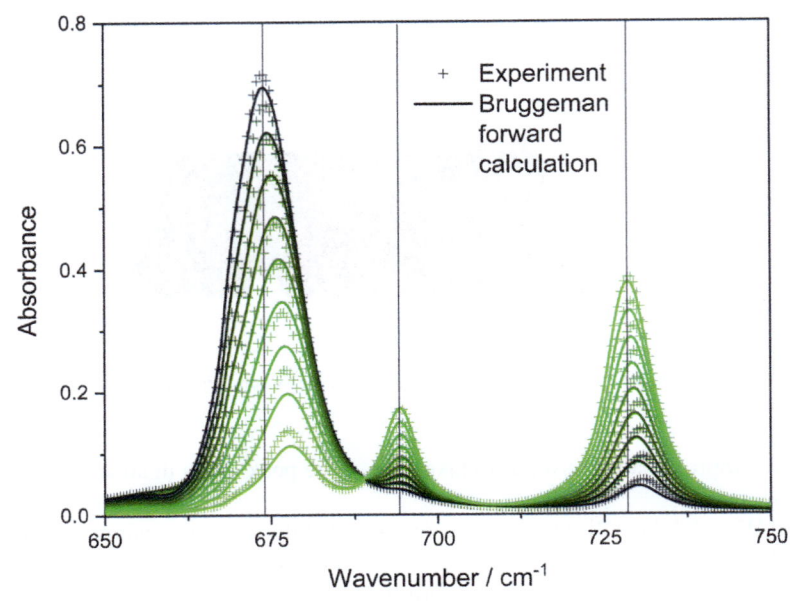

FIG. 10.11 Experimental spectra in the system benzene–toluene ($\varphi=0.1$–0.9) and simulated spectra using the Bruggeman approximation according to Eq. (10.50).

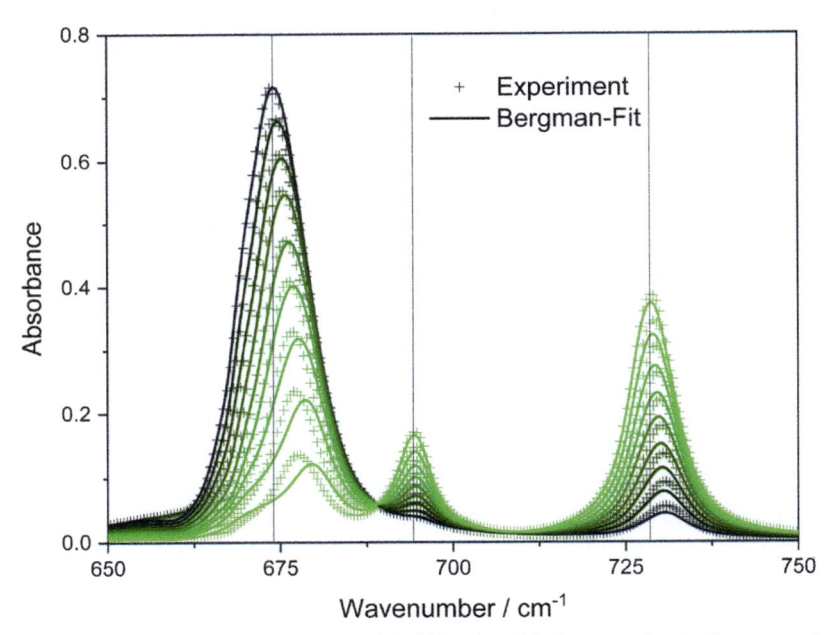

FIG. 10.12 Experimental spectra in the system benzene-toluene ($\varphi=0.1$–0.9) and modeled spectra using the Bergman representation assuming spectral densities based on Eq. (10.47) and assigning benzene as matrix material.

the roles of matrix and inclusions are clear with the inclusions having a lower volume fraction than the matrix. Another disadvantage is that to my best knowledge, the Bergman representation is at present limited to binary systems.

In general, one has to keep in mind that in this chapter, we assume the media to be scalar media. CCl_4 is a molecule that is intrinsically isotropic, but this is not the case for toluene and benzene. Certainly, we can assume the liquids to be scalar media as well, but it might be that we have to take into account that the molecules are both, anisometric and anisotropic, and devise an effective medium theory that takes this into account properly to improve simulation and modeling of corresponding spectra (the same is certainly true for solids and crystallites). In any case, the existing theories are definitely not good enough to allow quantitative predictions in general.

10.6 Microheterogeneity and size dependence of spectral features

As a reminder, the preconditions for effective medium approximations to hold are that the dimensions of heterogeneous structures must be small compared to wavelength. This means that for solids particles or crystallites, as well as the distances between them must be small. The reason was that the electric field should be approximately constant over space, since the relations that we used had their origins in electrostatics where these conditions are fulfilled. Vehement proponents of effective medium theories sometimes think they could easily extend these theories by allowing the electric field to have phase differences. A simple counterargument would be a situation where an experiment shows that heterogeneity does not always lead to an averaged dielectric function. In fact, it is very easy to devise or come across this situation—you just have to use a microscope operating in the wavelength range you are interested in. If the sample you are looking at can be described by an effective dielectric function, it will be featureless if inspected under the microscope. For samples for which the microscope reveals structures, this possibility is obviously not a given. Accordingly, it is the resolving power of light that is important. If you still cannot imagine what is going on, have a look at Fig. 10.13, where a microheterogeneous and a microhomogeneous mixture employing ¼ of green and ¾ of red are displayed. Now, if you imagine you would you measure the (average) reflectance spectrum of both samples, you would see a difference. The reason is that in general intensities are additive, i.e., reflectances, absorptances, or transmittances (note that absorbance is not an intensity, since it is not proportional to the electric field intensity, which seems not only counterintuitive, but also produce a kind of paradox). Therefore, if you record a reflectance spectrum, this spectrum is the average of the reflectance of green and red according to their respective area (or volume) fractions:

$$T = \varphi_1 T(\varepsilon_1) + \varphi_2 T(\varepsilon_2)$$
$$R = \varphi_1 R(\varepsilon_1) + \varphi_2 R(\varepsilon_2). \tag{10.52}$$

This seems to make sense for the microheterogeneous sample, but what happens in case of the microhomogeneous sample? For this sample, there is only one kind of material and the averaging according to Eq. (10.52) is without consequences. This is because the colors have been mixed already for every pixel producing an effective dielectric function as discussed in the previous sections.

In the general case, one could imagine that even when only two substances are present with overall volume fractions, these could be different at different locations leading to varying locally microhomogeneous mixtures with different effective dielectric functions that constitute an overall microheterogeneous mixture. A simplified illustration of this situation is provided in Scheme 10.3.

The edge length of the squares would be equal to the resolution limit of light. Accordingly, Eq. (10.52) could be generalized to:

$$T = \sum_j \varphi_j T(\varepsilon_j)$$
$$R = \sum_j \varphi_j R(\varepsilon_j). \tag{10.53}$$

While this is a very fundamental finding, it is generally not discussed in textbooks of infrared spectroscopy. In fact, the division in microhomogeneous and microheterogeneous samples is absent in general in literature, although hidden hints of

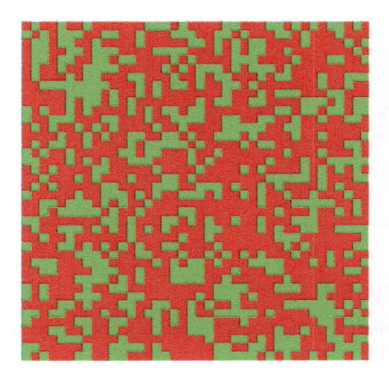

FIG. 10.13 A microheterogeneous *(left side)* and a microhomogeneous *(right side)* mixture of ¼ green and ¾ of red.

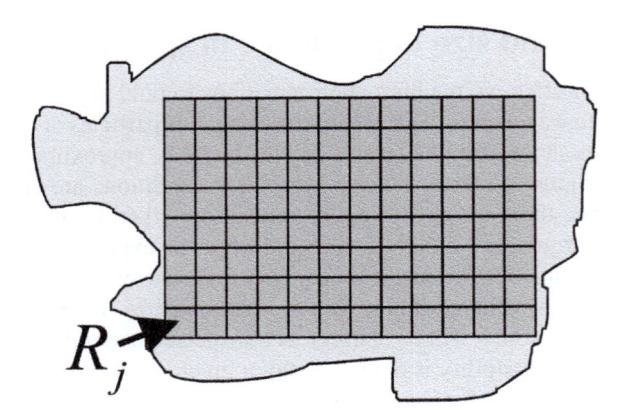

SCHEME 10.3 Grid on the surface of a sample.

the existence of this problem can be found. One of those have been provided by R. Norman Jones in 1952 [16]. He investigated pellet spectra and the question what happens if the sample particles are not microhomogeneously distributed. In this case, the light beam would traverse through regions where, in the extreme case, would be no particles and, therefore, where the light would not be attenuated by absorption. In his model, he averaged transmittances of sample volumes with and without particles and found that this leads to deviations from the BBL approximation. In fact, the result is equivalent to Eq. (10.52), except that Jones assumed only one colored material instead of two.

Another hint is provided when dealing with polycrystalline materials where you could look upon the different orientations of the crystallites as different colors. One model to calculate the spectra of such materials is correspondingly to assume a mosaiced surface similar to that in Scheme 15.1 and average the reflectance for many different orientations [17].

To give you an impression of the effects, Fig. 10.14 shows a simulation assuming two different materials each with, say, a C=O vibration with an oscillator position 50 cm^{-1} apart placed as 5-micron-thick layers on top of a CaF$_2$ substrate. If you like, you can visualize the samples as having a chessboard structure with white and black representing the two different materials. In case of the microheterogeneous sample, you would be in the geometric optics limit with edge lengths of the squares of, e.g., 100 μm. On the other hand, for effective medium approximations to apply, the edge length would have to be smaller than about one tenth of the wavelength, i.e., roughly smaller than 500 nm.

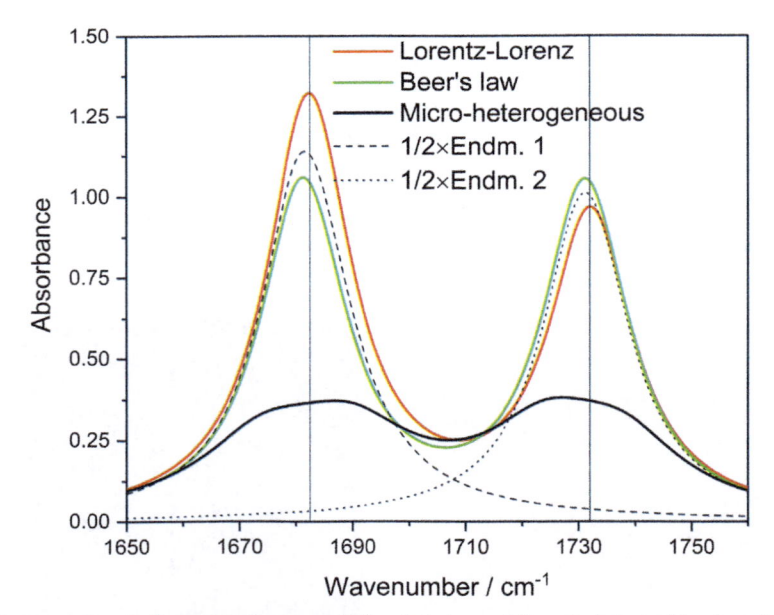

FIG. 10.14 Simulated absorbance spectra of a 5-μm-thick layer on a CaF$_2$ substrate consisting of mixtures of two hypothetic materials featuring a C=O vibration with oscillator positions indicated by the *vertical lines*. The *dashed* and *dotted lines* are the spectra of the pure (homogeneous) materials multiplied with the factor ½. The *orange* spectrum assumes a microhomogeneous mixture following the Lorentz-Lorenz relation, while the *green* spectrum assumes that Beer's approximation holds. The *black* spectrum assumes microheterogeneity.

Unfortunately, such samples are hard to prepare with the accurateness necessary to avoid unwanted effects, e.g., due to scattering, and we tried for several years without success.

For more complex anisotropic samples, we actually succeeded to produce samples with crystallite sizes larger and smaller than the resolution limit. In this case, however, the simulation is much more complex, but the results fully prove that eq. (10.53) is a valid approach (we come back to this problem in Chapter 15). When thinking about this model, I wondered for a long time what would happen if the size of the edges of the squares would be close to the resolution limit. Would there be an abrupt change in the spectra or would spectra begin to change already well before this limit, so that there would be a continuous transition from microhomogeneity to microheterogeneity? Again, producing corresponding samples seems to be extremely hard and we did not succeed until the printing of this book. Because of the lack of a better alternative, I decided to apply finite-difference time-domain (FDTD) simulations to investigate the problem at hand. The result is depicted in Fig. 10.15.

For those of you who have not heard of this method, searching for its application on the web of sciences returns 10 times the number of results compared to mid-infrared spectroscopy; accordingly, it is a well-established method. In principle, it solves Maxwell's equations locally applying a grid and using finite differences. Thus, it becomes exact in the limit of vanishing differences. In other words, choosing the differences small enough, errors would be too small to notice. More importantly, this method is not based on Fresnel's equations, the matrix formalism, the assumption that dielectric functions should average, the Lorentz-Lorenz relation, or the averaging of intensities. On the other hand, if those formulas can be derived from Maxwell's equations assuming certain geometries, they should automatically be included in the FDTD results when these geometries can be described by cubic grids. We come back and check this soon, but before doing this, what can be seen in the FDTD simulations?

Obviously, the spectra change continuously with decreasing edge length. The peak intensities strongly increase. They do so gradually for the lower wavelength band while the peak blueshifts. For the higher wavelength band, the peak also blueshifts, but the peak intensity decreases if the edge lengths become smaller than about 3 μm, which is approximately the wavelength of the light at the peak position. This means that IR spectra do indeed depend on the structure size and change continuously with it! While this is extremely surprising from the point of view of literature, it is not surprising at all given the fact that light has wave properties. Since we are, however, more used to allowing electrons and atoms to have wave properties (we accept that their wavefunctions change in different potential wells) than to think, light could change by interaction with structures with different sizes and dielectric properties. Once the latter assumption is taken into consideration, it is easy to find traces of the structural dependence of the light-matter interaction in the literature. A typical

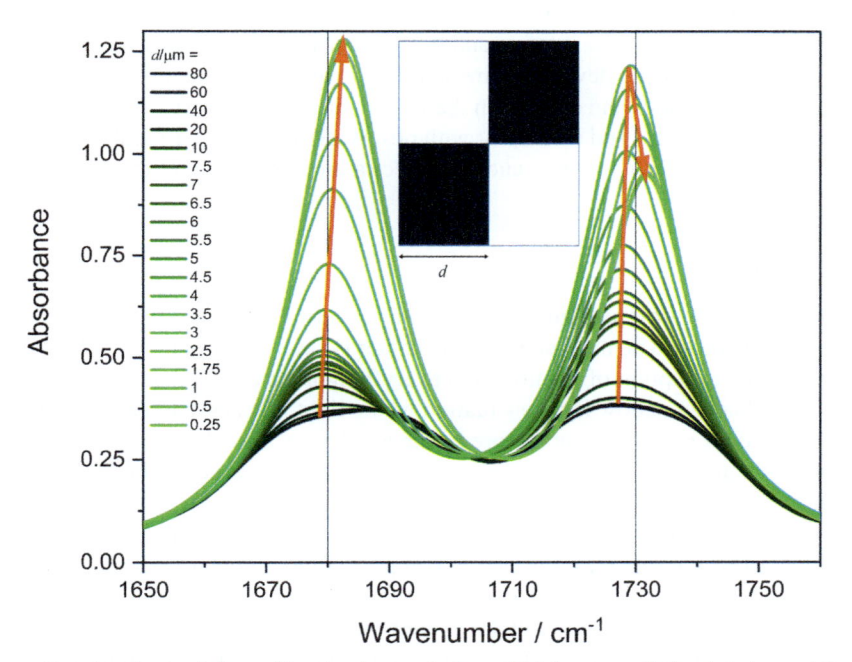

FIG. 10.15 Finite-difference time-domain simulations of the absorbance of a 5-μm-thick layer on a CaF_2 substrate consisting of mixtures of two hypothetic materials, each featuring a C=O vibration with similar damping and oscillator strength but different oscillator positions, in form of a chessboard structure with edge lengths d varying from 0.25 to 80 μm. The checkerboard structure is assumed to be infinite in the directions along the edges. Accordingly, a plane wave source is employed.

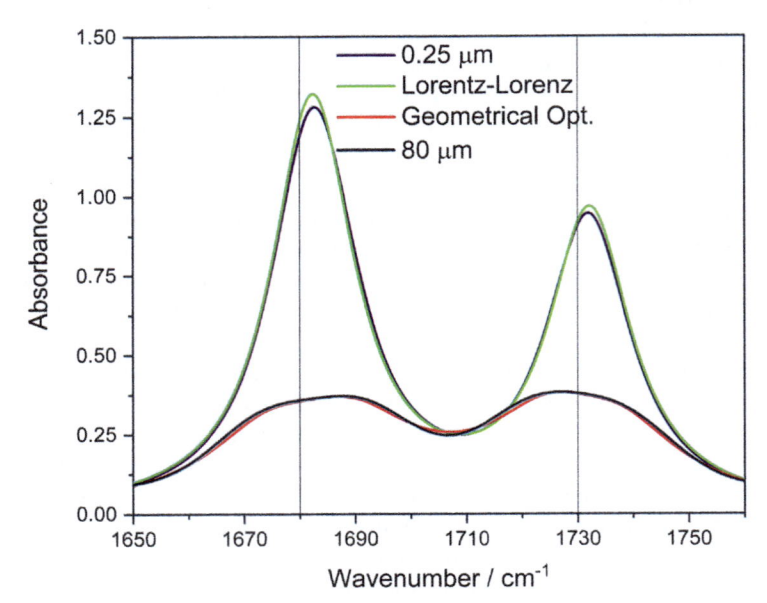

FIG. 10.16 Finite-difference time-domain simulations of the absorbance of a 5-μm-thick layer on a CaF_2 substrate consisting of mixtures of two hypothetic materials, each featuring a C=O vibration with similar damping and oscillator strength but different oscillator positions, in form of a chessboard structure with edge lengths d of 0.25 and 80 μm. These simulations are compared with those based on the Lorentz-Lorenz relation and the geometrical optics limit.

example are biological samples featuring protein structures. Often the absorbance differences between the strong amid bands and the other bands are leveled out, indicating microheterogeneity detected by the probing light.

Until corresponding samples could be prepared that prove exactly the results shown in Fig. 10.15, you certainly may doubt these results (actually you should always doubt anything and check it by yourself), but you should keep in mind that these results perfectly fit in with all results that follow from analytical expressions mentioned previously, like Fresnel's formulas, etc. Have a look at Fig. 10.16.

In this figure, you find a comparison between the results of the FDTD simulations for the extreme edge lengths 0.25 and 80 μm and those based on the Lorentz-Lorenz relation and the geometrical optics limit. While the agreement between corresponding pairs is not perfect, I think the deviations are small enough to be explained by numerical errors. Correspondingly, this at least proves that wave optics and dispersion theory are internally consistent. In other words, if you allow wave optics and dispersion theory to be helpful and consistent with observations and experiments, you have to allow the thought that light also probes the dimensions of structures and that spectra are depending on those dimensions.

10.7 Further reading

You may have asked yourself where is the section that deals with scattering effects that are quite often seen in the spectra of biological samples. The reason you also do not find it in this chapter is not that it would not belong to it—quite the contrary. However, consider that with the exception of Section 10.6 everything in this chapter is based on electrostatics. To understand, scattering effects would need to allow for the situation where light that is scattered from different scattering centers has phase differences. This would require a much more advanced theory, e.g., the T-matrix formalism (Mie-theory only describes scattering from a single object and its application would require that scattering centers are far away from each other—often quite the opposite of the situations in real samples the spectra of which show scattering effects) [18]. Introducing the T-matrix formalism requires a book of its own (see, e.g. [19]) and even with analytical relations in hand, I do not see a possibility to solve the inverse problem for samples as complex as biological tissue (but maybe it is just me lacking the necessary imaginative power—please do not feel discouraged!). The same is certainly true concerning understanding diffuse reflectance spectra. Personally, I think multiple scattering is something methods based on neural networks can excel. Or, you can step down to the theoretical level of the BBL approximation, assume the additivity of absorbance, and use something like extended multiplicative scatter correction (EMSC). To me, this is a little bit like using the heliocentric world view in order to repair the strange movements of the planets in the geocentric world view, but for a concrete problem this solution might be just good enough.

If you want more background with regard to effective medium theory, I suggest the books of Sihvola [4] and Choy [18] or the review article of Theiss [14]. Concerning the structure size dependence of infrared spectra, there is as far as I know no alternative to this chapter.

References

[1] T.G. Mayerhöfer, O. Ilchenko, A. Kutsyk, J. Popp, Beyond Beer's law: quasi-ideal binary liquid mixtures, Appl. Spectrosc. 76 (2022) 92–104.
[2] J.C.M. Garnett, J. Larmor XII, Colours in metal glasses and in metallic films, Philos. Trans. R. Soc. Lond. Ser. A 203 (1904) 385–420.
[3] J.C.M. Garnett, J. Larmor VII, Colours in metal glasses, in metallic films, and in metallic solutions.—II, Philos. Trans. R. Soc. Lond. Ser. A 205 (1906) 237–288.
[4] A.H. Sihvola, Electromagnetic Mixing Formulas and Applications, Institution of Electrical Engineers, 1999.
[5] S. Spange, T.G. Mayerhöfer, The negative solvatochromism of Reichardt's dye B30 – a complementary study, ChemPhysChem 23 (2022) e202200100.
[6] D.A.G. Bruggeman, Berechnung verschiedener physikalischer Konstanten von heterogenen Substanzen. I. Dielektrizitätskonstanten und Leitfähigkeiten der Mischkörper aus isotropen Substanzen, Ann. Phys. 416 (1935) 636–664.
[7] W. Theiss, Optische Eigenschaften inhomogener Materialien, na, 1989.
[8] D.J. Bergman, The dielectric constant of a composite material—a problem in classical physics, Phys. Rep. 43 (1978) 377–407.
[9] V.A. Markel, V.N. Pustovit, S.V. Karpov, A.V. Obuschenko, V.S. Gerasimov, I.L. Isaev, Electromagnetic density of states and absorption of radiation by aggregates of nanospheres with multipole interactions, Phys. Rev. B 70 (2004) 054202.
[10] H. Fröhlich, Theory of Dielectrics: Dielectric Constant and Dielectric Loss, Clarendon Press, 1949.
[11] T.G. Mayerhöfer, S. Höfer, J. Popp, Deviations from Beer's law on the microscale – nonadditivity of absorption cross sections, Phys. Chem. Chem. Phys. 21 (2019) 9793–9801.
[12] R. Fuchs, Theory of the optical properties of ionic crystal cubes, Phys. Rev. B 11 (1975) 1732–1740.
[13] R. Fuchs, Infrared absorption in MgO microcrystals, Phys. Rev. B 18 (1978) 7160–7162.
[14] W. Theiß, The use of effective medium theories in optical spectroscopy, in: R. Helbig (Ed.), Advances in Solid State Physics, Vol. 33, Springer Berlin Heidelberg, Berlin, Heidelberg, 1993, pp. 149–176.
[15] C.F. Bohren, D.R. Huffman, Absorption and Scattering of Light by Small Particles, Wiley, 1983.
[16] R.N. Jones, The absorption of radiation by inhomogeneously dispersed systems, J. Am. Chem. Soc. 74 (1952) 2681–2683.
[17] G. Doll, J. Steinbeck, G. Dresselhaus, M. Dresselhaus, A. Strauss, H. Zeiger, Infrared anisotropy of La(1.85)Sr(0.15)CuO(4-y), Phys. Rev. B 36 (1987) 8884–8887.
[18] T.C. Choy, Effective Medium Theory: Principles and Applications, OUP Oxford, 2015.
[19] M.I. Mishchenko, L.D. Travis, A.A. Lacis, Scattering, Absorption, and Emission of Light by Small Particles, Cambridge University Press, 2002.

Part II

Tensorial theory

Chapter 11

What is wrong with linear dichroism theory

If you think that the last chapters were hefty and math-heavy, the next chapters will not be a positive surprise. I fully understand how you will feel because when a colleague first made me aware of the 4 × 4 matrix formalism introduced by Yeh, which was, I think, in 1997, I looked through the corresponding paper [1] and understood absolutely nothing. My PhD work was centered on oriented glass ceramics. My job was to determine the orientation of these glass ceramics by infrared spectroscopy. From a theoretical point of view, my supervisor told me to apply linear dichroism theory (LDT), but this theory led to obvious inconsistencies. For example, it could not predict certain band shifts that I observed in my spectra. On the other hand, there seemed to be no simple alternatives in the literature. So, I was trapped between a theory that did not work and a lot of formulas I could not comprehend. Even worse, the latter formulas just described reflection and transmission of completely ordered systems, like single crystals or perfectly oriented films, and could not be used for disordered or completely randomly oriented polycrystalline materials. My glass ceramics, however, were not even homogeneous. They consisted of a glass matrix which was actually co-oriented to the crystalline phase consisting of fresnoite ($Ba_2TiSi_2O_8$). My solution was to develop my own theory [2], which was actually a good approximation to the wave optics-based formalisms as I discovered later.

The basic problem with my theory was that it predicted that infrared spectra of oriented samples needed two angles to describe orientation effects (like the wave optics-based formalisms), but LDT features only one, the angle between the polarization direction of the incoming light and the transition moment.

According to LDT, an electronic or vibrational transition between the states i and j that is light-induced requires a change in the dipole moment. For a vibration this transition moment $\mathbf{M}_{ij}$ is defined as the change of the dipole moment $\mathbf{\mu}$ with the distance r:

$$\left(\frac{\partial \mathbf{\mu}}{\partial r}\right)_{r_0} = \mathbf{M}_{ij}. \tag{11.1}$$

The probability of a transition and, thereby, absorbance, is assumed to be proportional to the square of the transition moment:

$$A \sim \mathbf{M}_{ij}^2. \tag{11.2}$$

If we use polarized light it is assumed by LDT that:

$$A \sim (\mathbf{M}_{ij} \cdot \mathbf{E})^2 = M_{ij}^2 E^2 \cos^2 \vartheta, \tag{11.3}$$

and the angle ϑ between polarization direction and transition moment seems to be all that counts. From the point of view of the first section of this book, Eq. (11.3) is already wrong, simply because it is not absorbance that is proportional to the electric field intensity, but absorptance, but this is another issue, let us focus on LDT and its shortcomings. Usually, there are some simple assumptions as follows [3]:

1. The sample, which is assumed to be of uniaxial symmetry, like, e.g., a stretched polymer, is oriented either parallel or perpendicular with its main axis to the polarization direction X of linear polarized light to avoid a change of the polarization.
2. Absorbance is proportional to the square of $\mathbf{M}_X$ according to Eq. (11.2).
3. For partly oriented samples orientational averaging leads to $\langle \mathbf{M}_X^2 \rangle = \mathbf{M}^2 \langle \cos^2 \vartheta \rangle$.
4. The orientation distribution function is expanded into a sum of Legendre polynomials. Absorbance is only sensitive to $P_2(\cos \vartheta)$.

Important consequences are that orientation distributions with the same $P_2(\cos\vartheta)$ cannot be distinguished and that only the magnitude of ϑ can be determined, but not its sign. This means that random orientation cannot be distinguished from the situation where one-third of the transition moments are perfectly co-oriented to the polarization direction and the two other thirds are perpendicularly oriented. Furthermore, if all transition moments are at the magic angle:

$$\langle \cos^2\vartheta \rangle = 1/3 \rightarrow P_2(\cos\vartheta) = 1/2(3\cos^2\vartheta - 1) = 0, \tag{11.4}$$

and $\vartheta = \arccos\sqrt{1/3} \approx 54.7°$ (this is the orientation of a space diagonal of a cube, which is in polar coordinates $\varphi = 45°$, $\theta = 54.7°$), then the spectra should also be the same as for the two aforementioned situations.

This is actually already highly questionable, just because of the initial assumption of a uniaxial sample. You cannot draw a conclusion for orientation distributions (randomly oriented sample, perfectly oriented sample, magic angle), which are different from the one you assume! Indeed, it is easy to prove that if I investigate a cube-shaped optically uniaxial sample with the optical axis oriented along one of the space diagonals, then the linear polarized light will not stay linearly polarized (simply put the sample between crossed polarizers and look through the arrangement) and the first statement earlier is not fulfilled. Also, even if a polymer is uniaxially stretched, above assumption about the polarization is not correct. I checked this by putting a stretched PET (polyethylene terephthalate) foil between crossed polaroids and measured the UV-Vis spectrum. (I used this spectral range because of the polaroids and their very good polarizing properties with a degree of polarization exceeding 99%.) The result is depicted in Fig. 11.1.

The high quality of the polaroids is obvious from the very low transmittance if there is no sample between the polars ("empty Pol."), which is for most of the spectrum about 0.01%. In contrast, if the stretched polymer is between the crossed polarizers, no orientation exists where the transmittance values come even close to this value (the fringes stem from interference effects). On the other hand, for certain wavelengths, the transmittance values can be very high, which will lead to more leakage and errors of orientation determination using Raman spectroscopy.

Apart from these practical problems and much more serious ones, when Zbinden provided his theoretical considerations concerning LDT in the form of a book (actually, it was just a chapter in a book about the infrared spectroscopy of polymers, albeit, with 67 pages, the longest one) [4], they were inferior to those known already for a long time. Do not get me wrong—there is no principal problem with using approximate theories, but as a scientist, you should always point out the existence of more sophisticated ones and discuss merits and limits.

Paul Drude published already in 1887 a theory for calculating reflection and refraction for absorbing crystals of arbitrary symmetries, which originated from his PhD thesis from the same year [5]. The equations were not based on Maxwell's equations, very much like Fresnel's equation originally. And, also very much like the latter, they were nevertheless correct, which was proved by Frank Matossi and Florenz Dane in 1927 [6]. The latter colleagues compared measurements and calculations based on Drude's theory for calcite and found excellent agreement, which is actually not what I find remarkable. What is remarkable though are the orientations of the crystals the authors investigated. Calcite is optically uniaxial and, as I will discuss in the following sections, for such materials the orientation of the optical axis, not only

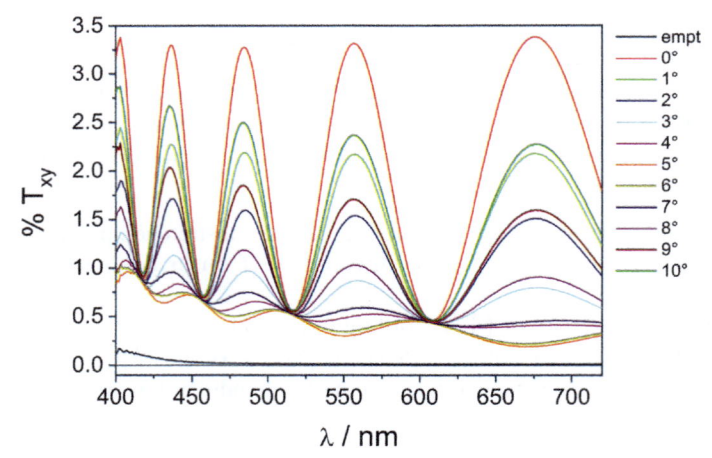

FIG. 11.1 Transmittance of a stretched PET foil between crossed polaroids. 0° indicates that stretching direction and polarization direction agree. The effect is much larger than what passes through crossed polarizers without the sample (black spectrum).

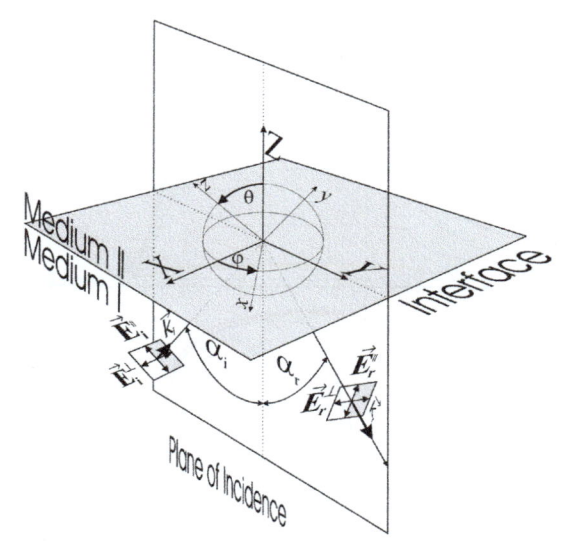

FIG. 11.2 Illustration of the optical model and the meaning of the angles φ and θ.

relative to the polarization direction, but also relative to the surface of the sample is pivotal. As we will see later, for perpendicular incidence, the formula for reflection is comparably simple:

$$R_X = R_1(n_1)\cos^2\varphi + R_2(n_2)\sin^2\varphi$$
$$R_i(n_i) = |r_i(n_i)|^2 = \left|\frac{n_i-1}{n_i+1}\right|^2$$
$$r_i = \frac{n_i-1}{n_i+1}, \quad n_1 = n_{ord} = \sqrt{\varepsilon_a}, \quad n_2 = \sqrt{\frac{\varepsilon_a \varepsilon_c}{\varepsilon_a \sin^2\theta + \varepsilon_c \cos^2\theta}}. \tag{11.5}$$

The important point to realize is that the reflectance for light polarized along X depends on two angles φ and θ, which obviously produces some contradiction to LDT where only one angle is considered. These two angles are the same as for the spherical coordinate system (cf. Fig. 11.2). For $\varphi, \theta = 0°$, the optical axis is oriented along Z and the spectrum is only influenced by ε_a. Eq. (11.5) then simply becomes

$$R_X = \left|\frac{\sqrt{\varepsilon_a}-1}{\sqrt{\varepsilon_a}+1}\right|^2. \tag{11.6}$$

We now have two principal possibilities to render R_X to depend on ε_c. Either we vary φ from $0°$ to $90°$ for $\theta = 90°$ (certainly we can also vary both at the same time, but let us keep the situation simple) or we set $\varphi = 90°$ and vary θ from $0°$ to $90°$. If we choose the first alternative, then Eq. (11.5) becomes

$$R_X = \left|\frac{\sqrt{\varepsilon_a}-1}{\sqrt{\varepsilon_a}+1}\right|^2 \cos^2\varphi + \left|\frac{\sqrt{\varepsilon_c}-1}{\sqrt{\varepsilon_c}+1}\right|^2 \sin^2\varphi. \tag{11.7}$$

On the other hand, for the second possibility, we find

$$R_X = \left|\frac{n_2-1}{n_2+1}\right|^2, \quad n_2 = \sqrt{\frac{\varepsilon_a \varepsilon_c}{\varepsilon_a \sin^2\theta + \varepsilon_c \cos^2\theta}}. \tag{11.8}$$

The endpoint is in both cases $\varphi, \theta = 90°$. This means that in the end the optical axis, which is in the internal coordinate system parallel to the z-axis, is oriented along X. As a consequence, Eq. (11.5) becomes

$$R_X = \left|\frac{\sqrt{\varepsilon_c}-1}{\sqrt{\varepsilon_c}+1}\right|^2, \tag{11.9}$$

and we obtain a pure spectrum which is only dependent on ε_c.

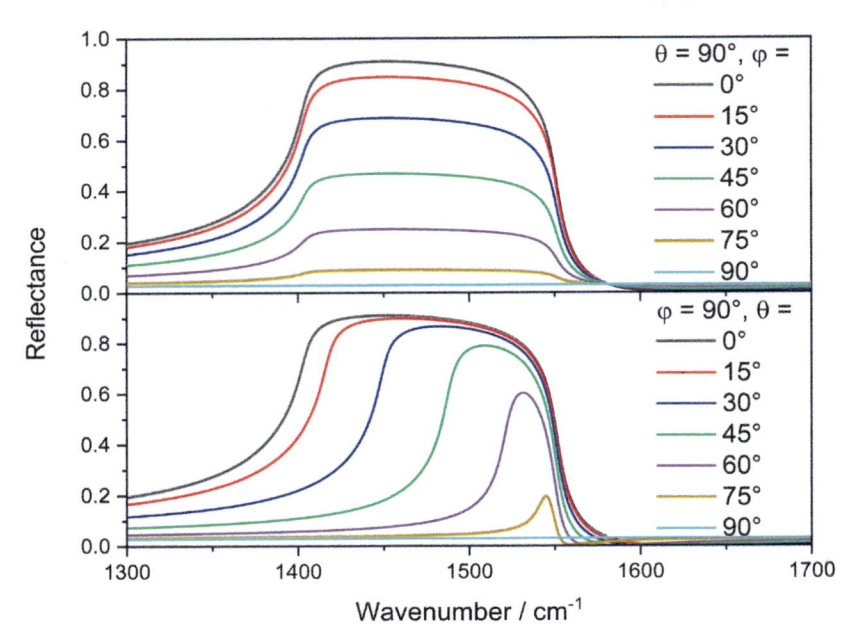

FIG. 11.3 Band of the ν_3 vibrational mode in the simulated reflectance spectra of calcite for different orientations of the optical axis.

Matossi and Dane focused on the spectral region of calcite where the ν_3 vibration of the oxyanion is located, i.e., the region around a wavelength of 7 μm or 1430 cm^{-1}. The corresponding band has a very high oscillator strength, which makes the related effects very obvious. The ν_3 vibration is doubly degenerated and has its transition moment perpendicular to the optical axis. This means that it will only vanish if both φ and θ equal 90°. What then remains is only an offset due to $\varepsilon_{c,\infty}$ as can be seen in Fig. 11.3 (turquoise spectrum). As detailed earlier, start and endpoint of the two spectral series shown in this figure are the same. The starting point is a spectrum influenced only by ε_a according to Eq. (11.6) (black spectrum), and the endpoint is a spectrum which can be calculated by employing Eq. (11.9) setting $\varepsilon_c = \varepsilon_{c,\infty}$. The first spectral series in the upper panel of Fig. 11.3 has been calculated from Eq. (11.7), which is nothing else but a linear combination of the two end-member spectra, i.e., the black and the turquoise spectrum. The second series in the lower panel of Fig. 11.3, which is essentially the one Matossi and Dane investigated, is based on Eq. (11.8). The difference between the first series is immediately obvious. In this series, the band maximum shifts by nearly 150 cm^{-1} to higher wavenumbers before it vanishes (this shift is called a TO-LO shift since the band shifts toward the LO position, cf. Section 5.6). Another difference between the two spectral series is that the full width at half height decreases only for the second series. From these observations alone, it is already clear that LDT must fail for strong oscillators. Obviously, the angles φ in the first series and the angles θ in the second series are equal to the angle ϑ, i.e., they equal the angle between the polarization direction and the transition moment. According to LDT this should render both spectra series indistinguishable, which they obviously are not. Had Zbinden known Matossi and Dane's work, it should have been clear to him that his theoretical considerations are incomplete and approximate, but there is no hint in his book that he knew of it. So, how come he (and all that came after him and propagated LDT in their books) was not aware of the obvious shortcomings of LDT?

The simple answer is that the weaker an oscillator is, the less pronounced is above effect, and in organic and biological matter most of the oscillators are much weaker. Apart from the hydroxyl groups, which are often broad, the strongest oscillator belongs to the C=O group. So how strong would be the effect in this particular case (actually, what we assume in the following is an oscillator that has the same strength as a C=O group when the latter is randomly oriented; when it is oriented, much higher oscillator strengths are possible [7])? To investigate it further and to better compare it with literature we switch over to transmittance absorbance, also because reflectance is not very high for most of the organic and biological samples and looks odd if you are used to absorbance (lnR looks like the refractive index function). Accordingly, we would find based on the formalisms provided in the next chapter that [8]

$$T_X = T_1(n_1 d)\cos^2\varphi + T_2(n_2 d)\sin^2\varphi;$$

$$T_i(n_i d) = |t_i(n_i d)|^2 = \left|\frac{(1 - r_{1,i})(1 - r_{2,i})\exp(i2\pi\tilde{\nu} n_i d)}{n_i + 1}\right|^2$$

$$r_{1,i} = \frac{n_i - 1}{n_i + 1}, \quad r_{2,i} = -r_{1,i}, i = 1, 2; \quad n_1 = n_{ord} = \sqrt{\varepsilon_a}, \quad n_2 = \sqrt{\frac{\varepsilon_a \varepsilon_c}{\varepsilon_a \sin^2\theta + \varepsilon_c \cos^2\theta}},$$

(11.10)

and absorbance is given by $A = -\log_{10}T$. To avoid most of the problems with absorbance that we discussed in the preceding part of the book, we will make a number of simplifying assumptions. First of all, we will assume that we have only one oscillator with a transition moment oriented parallel to the crystallographic *c-axis* and none perpendicular to it. In addition, we assume that $\varepsilon_{\infty, i} = 1$. By doing so, we virtually remove the interface and nearly all interference effects, except around the oscillator, which cannot be avoided because of dispersion [8,9]. In addition, our hypothetic material does not alter the polarization of incoming light at wavenumbers above the resonance frequency. As a consequence, Eq. (11.10) is simplified to

$$T_X = \cos^2\varphi + T_2(n_2 d)\sin^2\varphi;$$

$$T_2(n_i d) = |t_2(n_i d)|^2 = \left|\frac{(1 - r_{1,2})(1 - r_{2,2})\exp(i 2\pi \tilde{\nu} n_2 d)}{n_2 + 1}\right|^2 \tag{11.11}$$

$$r_{1,2} = \frac{n_2 - 1}{n_2 + 1}, \quad r_{2,2} = -r_{1,2}, \quad n_1 = n_{ord} = 1, \quad n_2 = \sqrt{\frac{\varepsilon_c}{\sin^2\theta + \varepsilon_c \cos^2\theta}}.$$

Because of the different choice that I had to make, namely to have an oscillator with a transition moment along the optical axis as the vibration of the C=O group is always nondegenerate, $\vartheta = 90° - \varphi$ in Fig. 11.4. Anyway, in the corresponding configuration, we have

$$T_X = \cos^2\varphi + T_2(n_2 d)\sin^2\varphi;$$

$$T_2(n_i d) = |t_2(n_i d)|^2 = \left|\frac{(1 - r_{1,2})(1 - r_{2,2})\exp(i 2\pi \tilde{\nu}\sqrt{\varepsilon_c} d)}{\sqrt{\varepsilon_c} + 1}\right|^2 \tag{11.12}$$

$$r_{1,2} = \frac{\sqrt{\varepsilon_c} - 1}{\sqrt{\varepsilon_c} + 1}, \quad r_{2,2} = -r_{1,2}.$$

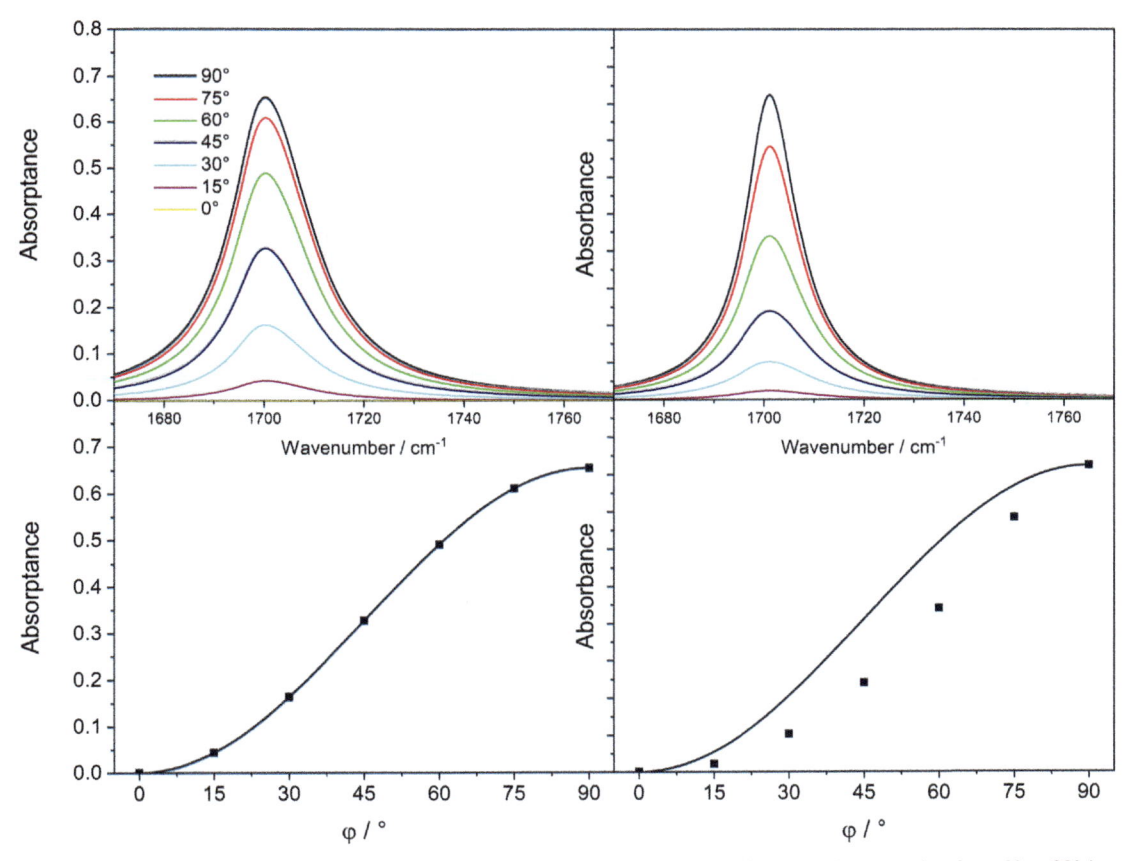

FIG. 11.4 Upper panels: Variance of the absorptance (left) and the absorbance (right) for different angles φ ranging from 0° to 90° in steps of 15° ($\theta = 90°$). Lower panels: Comparison of the peak values from the spectra of the upper panels (filled squares) with $A_{max}\cos^2\vartheta$, $\vartheta = 90° - \varphi$ of the peak values for $\varphi, \theta = 90°$ (continuous lines) [8].

Since the absorptance is $A = 1 - R - T \approx 1 - T$ it is clear that it is not absorbance, but absorptance in this case that is proportional to $\cos^2 \vartheta$, which is well known from Malus' law for about 200 years, and can also be seen in Fig. 11.4, a further reason why LDT is inappropriate in general.

Based on what has been presented so far, one could come to the conclusion that it is generally the absorptance that needs to be considered for orientation studies as this has indeed been done in the literature [10]. But in the same way as just focusing on absorbance, this would miss the point. Because if we decide to fix the angle φ at $\varphi = 90°$ and vary the angle θ, we obtain the following relation:

$$T_X = |t_2(n_i d)|^2 = \left| \frac{(1 - r_{1,2})(1 - r_{2,2}) \exp(i 2 \pi \tilde{\nu} n_2 d)}{n_2 + 1} \right|^2$$

$$r_{1,2} = \frac{n_2 - 1}{n_2 + 1}, \quad r_{2,2} = -r_{1,2}, \quad n_2 = \sqrt{\frac{\varepsilon_c}{\sin^2 \theta + \varepsilon_c \cos^2 \theta}}.$$

(11.13)

While reflection will still affect the transmittance spectrum and, thereby, absorbance, for this configuration the term that influences the exponential function through n_2 is depending on the angle between transition moment and polarization direction. This is a decisive difference the consequences of which we can observe in Fig. 11.5. Accordingly, for this conformation, it is approximately absorbance that follows $\cos^2 \vartheta$, but only if we take the peak values. The reason we have to take the peak values is that obviously there is a peak shift which increases with θ and can be as large as about $10 \, \text{cm}^{-1}$ in our model system. In the case of a real material, though, the peak shift will be less since the index of refraction (to be more specific, n_∞) will be different from unity. If we assume $\varepsilon_{a,\infty} \approx \varepsilon_{c,\infty} \approx 2.25$ the peak shift will be on the order of $5 \, \text{cm}^{-1}$, which is nevertheless clearly detectable.

It also seems to be known in the community that uses LDT, that it is not preferable to use the stronger bands in oriented organic materials to determine the orientation. Why should LDT work better if you use a weak band? For such

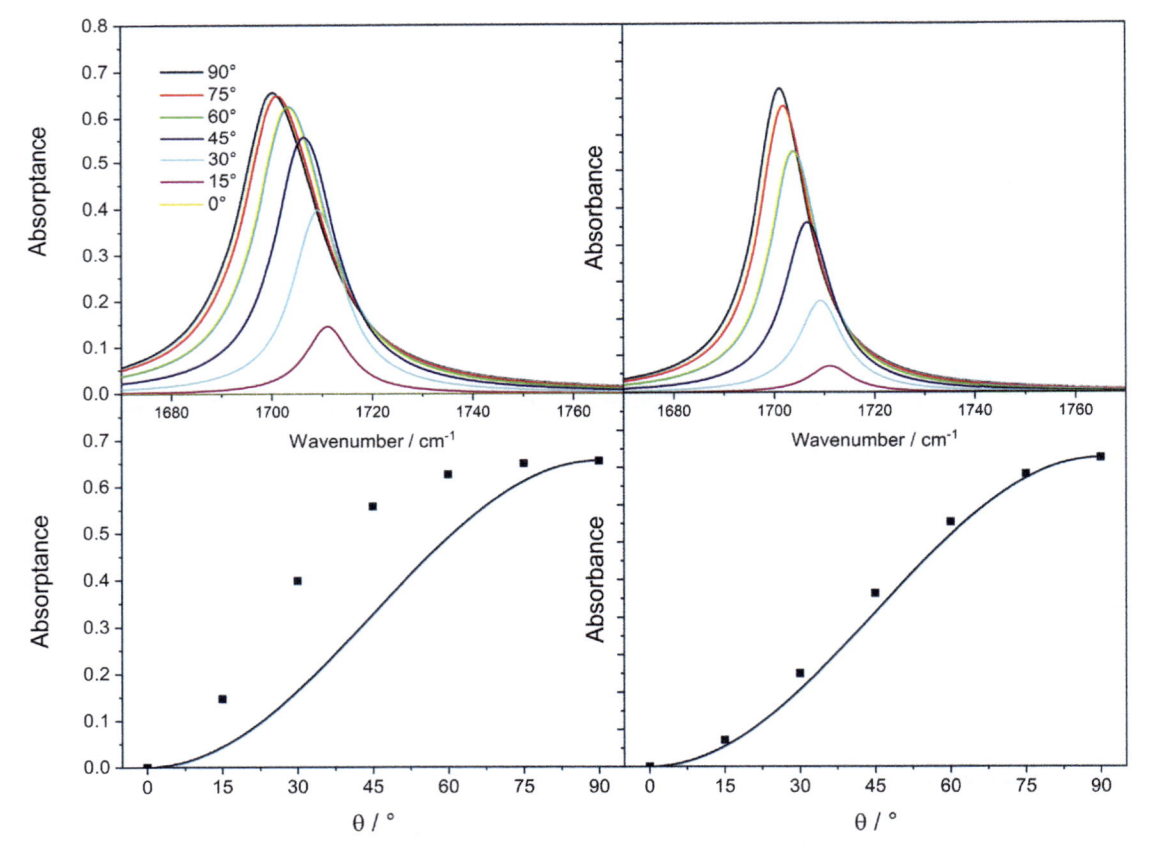

FIG. 11.5 Upper panels: Variance of the absorptance (left) and the absorbance (right) for different angles θ ranging from 0° to 90° in steps of 15° ($\varphi = 90°$). Lower panels: Comparison of the peak values from the spectra of the upper panels with $A_{max} \cos^2 \vartheta$, $\vartheta = 90° - \theta$ of the peak values for $\varphi, \theta = 90°$ [8].

bands, also absorbance and absorptance become small. If you assume that $T=\exp(-A)$, you can perform a series expansion:

$$T = \exp(-A) \rightarrow$$
$$1 - R - A = 1 - A + \frac{A^2}{2} - \ldots \xrightarrow{A \ll 1}$$
$$1 - A = 1 - A \rightarrow$$
$$A = A.$$
(11.14)

Therefore, if absorbance becomes small and reflectance can be neglected, absorptance ≈ absorbance. This can actually be achieved in two different ways. If you have the chance to change the thickness of the sample, reduce this thickness (but always keep in mind that interference effects will occur in real samples; note that for the simulations above and below I assumed a thickness of 1 μm)! The alternative would be to choose bands the oscillator strength of which is clearly below that of C=O bands.

If we, for example, choose an oscillator with $S = 75\,\text{cm}^{-1}$, which might resemble the vibration in PMMA (poly(methyl methacrylate)) at $1476\,\text{cm}^{-1}$, we should expect a clear degradation of the effect. Let's put this example to the test and have a look at Fig. 11.6.

Indeed, this is what is to be expected. For such a weak oscillator, LDT is a theory that will describe the observations to a satisfying degree, also because the TO-LO shift is small and the shift of the band with the variation of the angle θ is hardly noticeable in practice when there is a difference between $\varepsilon_{i,\infty}$ and vacuum. Therefore, at least for the IR spectroscopy of organic materials the employers of LDT are on the safe side as long as you do not employ 2D correlation spectroscopy, which is sensitive enough to detect even small peak shifts [9].

On the other hand, we should not forget that for LDT there are completely different orientation distributions that lead to more or less the same spectra. Therefore you should not be able to distinguish these distributions when you focus on weak

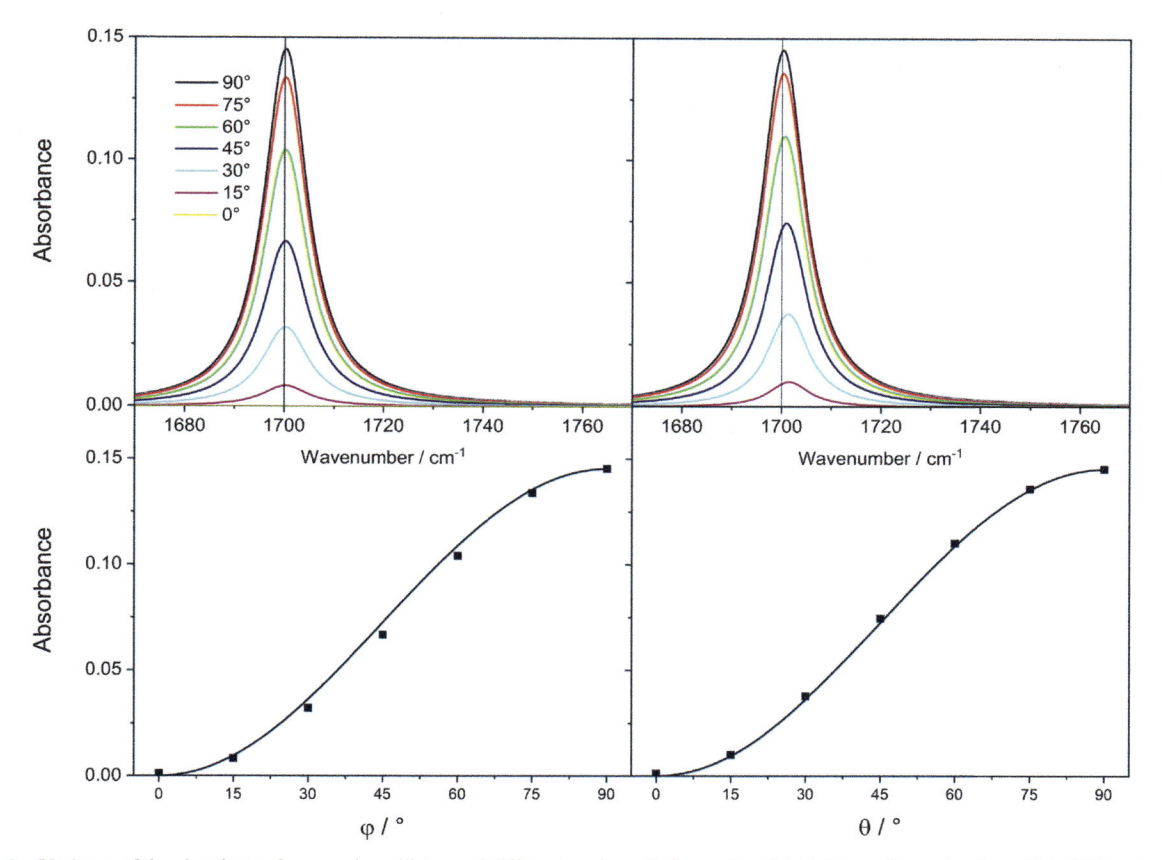

FIG. 11.6 Variance of the absorbance for a weak oscillator and different angles φ (left panel) and θ (right panel) ranging from 0° to 90° in steps of 15°. Lower panels: Comparison of the peak values from the spectra of the upper panels with $A_{max}\cos^2\vartheta$ of the peak values for $\varphi = 90°$, $\vartheta = 90° - \theta$ for the left panel and $\theta = 90°$, $\vartheta = 90° - \varphi$ for the right panel [8].

bands (again, if you do not want to prove LDT wrong, do not use 2D correlation spectroscopy! [9]). Would this be possible when a medium-strong band, like a C=O band, is used? To investigate this, we go back to the original example and see if we can distinguish between random orientation, the so-called magic angle configuration and the configuration where one-third of the transition moments are oriented perfectly with the polarization direction whereas the other two-thirds have their transition moments oriented perpendicular to this direction. There is, however, still a pitfall the existence of which we have to realize. What is actually averaged if you have a random orientation? In the literature you read that for random orientation the dielectric tensor is averaged and, thereby, reduced to a scalar ("the medium becomes isotropic") [11]. This seems to be clear enough, so the resulting dielectric function is simply [12]

$$\varepsilon = \frac{1}{3}\varepsilon_a + \frac{1}{3}\varepsilon_b + \frac{1}{3}\varepsilon_c. \tag{11.15}$$

For LDT, however, the following would be random orientation:

$$k = \frac{1}{3}k_a + \frac{1}{3}k_b + \frac{1}{3}k_c. \tag{11.16}$$

This would mean that we actually assume that [12]

$$\hat{n} = \frac{1}{3}\hat{n}_a + \frac{1}{3}\hat{n}_b + \frac{1}{3}\hat{n}_c. \tag{11.17}$$

Since $\varepsilon = \hat{n}^2$, we have the problem that both, Eqs. (11.15), (11.17) cannot be correct at the same time unless $\varepsilon = n = 1$, i.e., in the limit of zero absorption. But actually, this problem does not matter, because one can easily prove by experiment that neither formula is correct in general, because they are valid only if additionally vanishing anisotropy is assumed. To imitate nature requires more complicated average procedures [13]. But even if these are used, there remains a further problem in practice, as these procedures can only be employed if the crystallites are very small compared to wavelength, i.e., if the samples are micro-homogeneous (crystallites small compared to the wavelength of light and it is not possible to detect the crystallites with a microscope operating at this wavelength). A rule of thumb says that this means that the crystallites have to be smaller than one-tenth of the wavelength, which means that for the mid-IR the crystallites need to be smaller than 0.25 μm on the upper end (4000 cm^{-1}) and 2.5 μm on the lower end (400 cm^{-1}). For the community interested in remote sensing, this means that most of their samples, which are usually polycrystalline, have crystallite sizes that are considered as large.

How drastic are the resulting effects? Again, they depend in the end on the oscillator strengths. For inorganic materials the effects can in no way be disregarded. An example to which we will come back in later chapters is fresnoite (Ba$_2$TiSi$_2$O$_8$). Experimental spectra of samples with (mostly) small or large crystallites can be found in Fig. 11.7.

Obviously, the differences are very strong, not only concerning band shapes but also with regard to band positions. While the reflectance peaks of the sample with small crystallites are untypically located close to the oscillator frequency

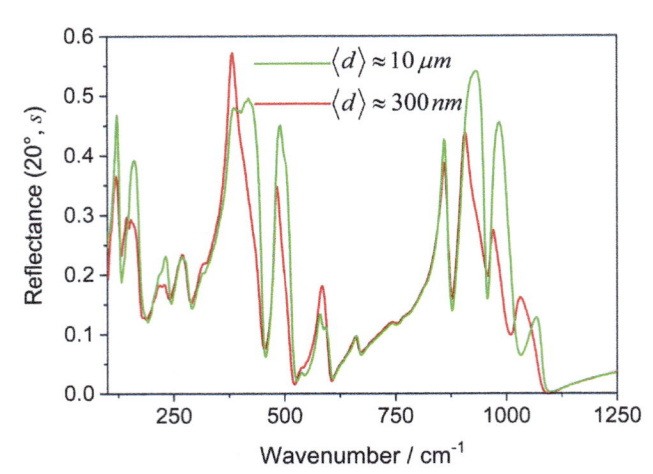

FIG. 11.7 Comparison of the reflectance spectra of polycrystalline fresnoite with optically large (green spectrum) and small (red spectrum) crystallites [13–16].

and their bands seem not to have the typical Lorentz-oscillator shape, the peaks for the large crystallite sample are shifted toward the LO positions (but do not reach them) and the band shapes seem not to be extraordinary.

Quite instructive is also to investigate the spectral region between about 600 and 900 cm^{-1}. In this region, the spectra are nearly identical. The reason is that in this spectral region in a first approximation $\varepsilon_a \approx \varepsilon_c$ holds [17], which means that fresnoite behaves like a crystal of cubic symmetry and crystallite size effects cannot occur.

The reason for the spectral changes is that for large crystallites the dielectric tensor does not reduce to a scalar. In contrast, there is a kind of mosaic effect and one has to calculate the spectra for a large number of orientations (the more orientations are necessary the stronger the anisotropy is) and take the average [18]. This mosaic effect is essentially the same that I introduced to you with regard to micro-heterogeneous samples in the first part of the book (Section 6.6), the existence of which I actually concluded from the spectral behavior of polycrystalline samples with random orientations and optically large crystallites. My contribution to solve the latter problem was to apply for the first time the 4×4 matrix formalism I will introduce in the next section. This formalism allows you to calculate not only spectra recorded with polarized light but also such spectra where you additionally use an analyzer. Correspondingly, I predicted that randomly oriented polycrystalline samples with optically large crystallites should show a nonvanishing cross-polarization despite being optically isotropic, something which I also proved experimentally [14,15,19]. Actually, if you have ever seen a recording of an image of a corresponding sample between crossed polarizers, you surely remember the different colors of the crystallites caused by the different orientations. Then you intuitively know that if you decrease the magnification and more and more crystallites come into the field of view, the image will never get dark, right?!

As I already stated, inorganic materials show large effects in reflectance. For organic or biological matter, we could expect that these effects were much smaller. Indeed, I would expect reflectance spectra to be less strongly influenced. However, is the same valid for transmittance? We can go back to our hypothetic material with a typical C=O vibration. Assuming again a thickness of 1 μm, I have calculated different scenarios using an arithmetic average of the principal indices of refraction according to Eq. (11.17) ($\frac{1}{3}n_a + \frac{1}{3}n_b + \frac{1}{3}n_c$), the situation that one-third of the transition moments would be oriented parallel to the polarization direction and the other two-thirds perpendicular to it (1/3 $R(\varphi = 90°, \theta = 90°)$), the magic angle configuration ($\varphi = 45°, \theta = (90-54.74)°$) and the situation that all crystallites are large compared to the resolution limit ($\langle R(\varphi,\theta) \rangle$). The result is depicted in Fig. 11.8.

Remember, all these orientation distributions should result in the same spectrum according to LDT, but they are obviously easy to distinguish if one uses wave optics! I hope this can motivate you enough to work through the following chapters even if the going will get tough.

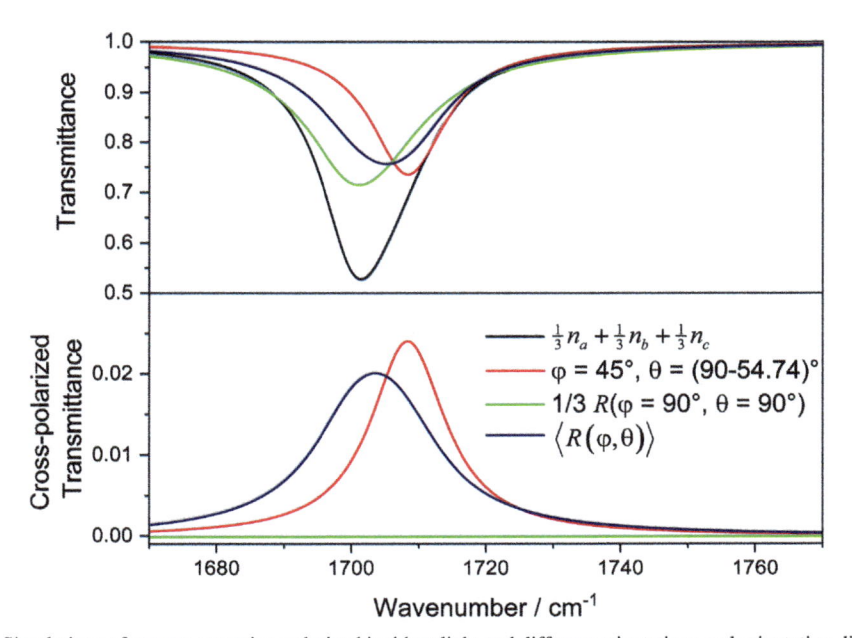

FIG. 11.8 Upper panel: Simulations of spectra assuming polarized incident light and different orientations and orientation distributions that should not be distinguishable according to LDT. Lower panel: Simulations of the cross-polarized spectra of the same orientations and orientation distributions (the black curve is hidden below the green one).

References

[1] P. Yeh, Electromagnetic propagation in birefringent layered media, J. Opt. Soc. Am. 69 (1979) 742–756.

[2] T.G. Mayerhöfer, H.H. Dunken, R. Keding, C. Rüssel, Determination of the crystallographic orientation of oriented fresnoite glass ceramics by infrared reflectance spectroscopy, Phys. Chem. Glasses 42 (2001) 353–357.

[3] J. Michl, E.W. Thulstrup, Spectroscopy with Polarized Light: Solute Alignment by Photoselection, Liquid Crystal, Polymers, and Membranes Corrected Software Edition, Wiley, Deerfield Beach, FL, 1995.

[4] R. Zbinden, Infrared Spectroscopy of High Polymers, Academic Press, New York-London, 1964.

[5] P. Drude, Ueber die Gesetze der Reflexion und Brechung des Lichtes an der Grenze absorbirender Krystalle, Ann. Phys. 268 (1887) 584–625.

[6] F. Matossi, F. Dane, Reflexion, dispersion und absorption von Kalkspat im Absorptionsgebiet bei 7 μ, Z. Phys. 45 (1927) 501–507.

[7] V. Ivanovski, T.G. Mayerhöfer, J. Stare, M.K. Gunde, J. Grdadolnik, Analysis of the polarized IR reflectance spectra of the monoclinic α-oxalic acid dihydrate, Spectrochim. Acta A: Mol. Biomol. Spectrosc. 218 (2019) 1–8.

[8] T.G. Mayerhöfer, Employing theories far beyond their limits—linear dichroism theory, ChemPhysChem 19 (2018) 2123–2130.

[9] T.G. Mayerhöfer, I. Noda, J. Popp, The footprint of linear dichroism in infrared 2D-correlation spectra, Spectrochim. Acta A: Mol. Biomol. Spectrosc. 304 (2024) 123311.

[10] V.G. Gregoriou, S. Tzavalas, S.T. Bollas, Angular dependence in infrared linear dichroism: a reevaluation of the theory, Appl. Spectrosc. 58 (2004) 655–661.

[11] M. Born, E. Wolf, A.B. Bhatia, Principles of Optics: Electromagnetic Theory of Propagation, Interference and Diffraction of Light, Cambridge University Press, 1999.

[12] T.G. Mayerhöfer, J. Popp, Effective optical constants: a fundamental discrepancy, Vib. Spectrosc 42 (2006) 118–123.

[13] T.G. Mayerhöfer, New method of modeling infrared spectra of non-cubic single-phase polycrystalline materials with random orientation, Appl. Spectrosc. 56 (2002) 1194–1205.

[14] T.G. Mayerhöfer, Modelling IR spectra of single-phase polycrystalline materials with random orientation in the large crystallites limit—extension to arbitrary crystal symmetry, J. Opt. A: Pure Appl. Opt. 4 (2002) 540.

[15] T.G. Mayerhöfer, Modelling IR-spectra of single-phase polycrystalline materials with random orientation—a unified approach, Vib. Spectrosc. 35 (2004) 67–76.

[16] T. Mayerhöfer, Z. Shen, R. Keding, J. Musfeldt, Optical isotropy in polycrystalline $Ba_2TiSi_2O_8$: testing the limits of a well established concept, Phys. Rev. B 71 (2005).

[17] T.G. Mayerhöfer, H.H. Dunken, Single-crystal IR spectroscopic investigation on fresnoite, Sr-fresnoite and Ge-fresnoite, Vib. Spectrosc. 25 (2001) 185–195.

[18] G. Doll, J. Steinbeck, G. Dresselhaus, M. Dresselhaus, A. Strauss, H. Zeiger, Infrared anisotropy of $La_{\{1.85\}}Sr_{\{0.15\}}CuO_{\{4-y\}}$, Phys. Rev. B 36 (1987) 8884–8887.

[19] T.G. Mayerhöfer, Z. Shen, R. Keding, T. Höche, Modelling IR-spectra of single-phase polycrystalline materials with random orientation—supplementations and refinements for optically uniaxial crystallites, Optik 114 (2003) 351–359.

Chapter 12

Reflection and transmission of plane waves from and through anisotropic media—Generalized 4×4 matrix formalism

To understand the optical properties of anisotropic media, we directly start with the corresponding matrix formalism in its most general form, from which special cases can easily be regained. We have already discussed in Chapter 4 how the 2×2 matrices for scalar media can be incorporated into a 4×4 matrix so that the layer stacks of cubic and anisotropic materials can be computed readily at the same time. In fact, we generalize the matrix formalism in a way that such a distinction is no longer necessary. We begin with introducing the 4×4 matrix formalisms by Berreman and Yeh, which we combine in one common formalism because it is instructive and the result is easier to apply and more stable than the original formalisms. Nevertheless, in the combined formalism it is still possible that certain singularities occur, which we will discuss and treat in detail. The formalism is able, in principle, to treat magnetic and dielectric anisotropy up to the lowest, i.e., triclinic symmetry. Since most materials are nonmagnetic, we will use this assumption and derive at the end of this chapter simplifications for special cases such as for normal incidence, the *a-c* plane of a monoclinic crystal, and a generally oriented uniaxial material, both under normal and nonnormal incidence.

4×4 matrix formalisms allow a systematic approach to calculate the reflectance and transmittance of layered materials and crystals [1–3]. A couple of different approaches have been introduced so far (actually, much more than I will quote because, in this part of the literature, the wheel is frequently reinvented). Teitler and Henvis employed an integral formulation of the problem [3]. Berreman used a differential formulation that gave him the possibility to describe and compute the reflectance and transmittance of continuously varying media such as cholesteric liquid crystals [1]. Both approaches are based on first-order Maxwell equations. In contrast, Yeh introduced a 4×4 matrix formalism by employing second-order wave equations [4].

Yeh's original approach is limited to media with dielectric anisotropy only. On the other hand, it can be applied to stratified structures on an anisotropic substrate (absorbing and therefore semiinfinite), whereas an isotropic exit medium is needed to employ Berreman's original formalism. Both limitations can be removed using a combined approach which is mostly based on Berreman's formalism [4]. In the following, we will introduce a combined formalism which is closer to Yeh's formalism.

A problem with all these formalisms is the occurrence of singularities, in particular, if orientational averages of the reflectance or of the transmittance are to be calculated. One reason for this occurrence is the high symmetry of certain media (especially in case of cubic and uniaxial magnetic point groups). In such media, the polarization directions of the waves are generally (cubic media), or for movement along high-symmetry axes, not constrained. Because of the occurrence of singularities, the calculation of eigenpolarizations in the same way as for lower symmetry orientations fails. This type of singularity has already been treated exhaustively in the literature (for a survey, cf. Ref. [5] and references therein). An additional cause of singularities, which is actually more frequent than the first type, is linked to the choice and the arrangement of the eigenpolarizations in the so-called Dynamical Matrix.

In the following, I present a combined 4×4 matrix formalism, which preserves the forms of the original formalisms of Berreman and Yeh as far as possible. This formalism is applicable to general bianisotropic-layered media or crystals/domains with linear response. As a consequence, an arbitrary incidence medium can be handled and the form of the formalism is optimized with regard to applicability. Accordingly, it allows easy switching between Berreman's and Yeh's formalism, as long as the media or domains are of only dielectric anisotropy. In addition, the solutions for the eigenvalues of the wavevector components in the direction of stratification and for the eigenpolarizations of the waves can derived analytically. To circumvent singularities, explicit solutions for all cases of degenerate eigenvalues will be provided. These are based on the tables of the point-magnetic groups of symmetry derived by Dmitriev [6].

In addition, I derive a general and robust solution for those singularities that are caused by an improper choice of eigenpolarizations in the Dynamical Matrix. This solution will be based on a suitable linear combination of the eigenpolarizations. Throughout this chapter, I will use the time dependence $\exp(-i\omega t)$. The original derivation can be found in Ref. [7].

12.1 Berreman's formalism: Maxwell equations and constitutive relations

Following Berreman's chapter, Maxwell's equations can be provided in compact form as

$$\mathbf{RG} = \frac{1}{c}\frac{\partial}{\partial t}\mathbf{C}. \tag{12.1}$$

R is a matrix operator which has the form

$$\mathbf{R} = \begin{bmatrix} \mathbf{0} & -\mathbf{curl} \\ \mathbf{curl} & \mathbf{0} \end{bmatrix}, \tag{12.2}$$

with **0**, which is the 3×3 null matrix and **curl** stands for the curl operator, which is in this special case given by

$$\mathbf{curl} = \begin{pmatrix} 0 & -\frac{\partial}{\partial Z} & \frac{\partial}{\partial Y} \\ \frac{\partial}{\partial Z} & 0 & -\frac{\partial}{\partial X} \\ -\frac{\partial}{\partial Y} & \frac{\partial}{\partial X} & 0 \end{pmatrix}, \tag{12.3}$$

with the laboratory coordinates X, Y, and Z.

The vectors **G** and **C** contain the components electromagnetic-field vectors **E**, **H**, **D**, and **B**:

$$\mathbf{G} = (E_X,\ E_Y,\ E_Z,\ H_X,\ H_Y,\ H_Z)^T, \qquad \mathbf{C} = (D_X,\ D_Y,\ D_Z,\ B_X,\ B_Y,\ B_Z)^T. \tag{12.4}$$

If nonlinear optical effects and spatial dispersion do not play a role, the material equations or constitutive relations between **G** and **C** can be written as

$$\mathbf{C} = \mathbf{M}_{X,Y,Z}\mathbf{G}. \tag{12.5}$$

The 6×6 matrix $\mathbf{M}_{X,Y,Z}$ consists of four 3×3 submatrices, namely ε, μ, ρ, and ρ':

$$\mathbf{M}_{X,Y,Z} = \begin{pmatrix} \varepsilon & \rho \\ \rho' & \mu \end{pmatrix}. \tag{12.6}$$

ε and μ are the permittivity and the permeability tensor, respectively, and ρ and ρ' are the two so-called magnetic-electro tensor dyadics. The latter are identical to the 3×3 null matrix in the absence of optical activity.

All tensors are usually specified with regard to an intrinsic-coordinate system x,y,z fixed inside a single crystal or a crystallite (a natural choice would be to employ the crystal's crystallographic axes a,b,c if the crystal's symmetry is orthorhombic or higher; for a monoclinic crystal the b-axis would be selected as y-axis and, e.g., a as the x-axis, while in this case the z-axis would not coincide with the c-axis. For a triclinic crystal there is no natural choice). To determine the values of $\mathbf{M}_{x,y,z}$ in the laboratory coordinates, which represent the reference frame X,Y,Z, an orthogonal transformation is necessary according to [8]

$$\mathbf{M}_{X,Y,Z} = \mathbf{P} \cdot \mathbf{M}_{x,y,z} \cdot \mathbf{P}^{-1}. \tag{12.7}$$

In Eq. (12.7), **P** is the mapping operator which is generally given by

$$\mathbf{P} = \begin{pmatrix} \mathbf{A} & \mathbf{0} \\ \mathbf{0} & \det|\mathbf{A}| \times \mathbf{A} \end{pmatrix}, \tag{12.8}$$

where **A** will in the following represent only rotation matrices. Thus, since $\det|\mathbf{A}| = 1$ for rotation matrices, $\mathbf{P} = \mathbf{A}$ in Eq. (12.7) in what follows. **A** is usually expressed using one of the 24 different Euler angle orientation representations [8].

Eqs. (12.1), (12.5) can be combined and thus yield the spatial wave equation in which $\boldsymbol{\Gamma}$ represents the time-independent part of **G** ($\exp[-i\omega t]\boldsymbol{\Gamma} = \mathbf{G}$) and $k_0 = \omega/c$ the free-space wavenumber:

$$\mathbf{R}\boldsymbol{\Gamma} = -ik_0 \mathbf{M}_{X,Y,Z}\boldsymbol{\Gamma}. \tag{12.9}$$

As in the scalar part of this book, we assume that the samples consist of layers where the properties do not change in the X- and Y-directions, i.e., the layer stack is stratified in the Z-direction. As a consequence, $\mathbf{M}_{X,Y,Z}$ stays constant along X and Y and changes only along Z. Therefore, the conservation of momentum requires that all waves inside the sample which are excited by incident waves have the same spatial dependence in the X- and Y-directions. Therefore, the components k_X and k_Y of the wavevector $\mathbf{k}_i = k_0(k_X, k_Y, k_Z)^T$ must be constant throughout the sample. As a consequence, $\partial/\partial X = ik_0 k_X$ and $\partial/\partial Y = ik_0 k_Y$. Correspondingly, the curl operator simplifies to

$$\mathbf{curl} = \begin{pmatrix} 0 & -\dfrac{\partial}{\partial Z} & ik_0 k_Y \\ \dfrac{\partial}{\partial Z} & 0 & -ik_0 k_X \\ -ik_0 k_Y & ik_0 k_X & 0 \end{pmatrix}. \tag{12.10}$$

If Eq. (12.10) substitutes Eq. (12.3) in Eq. (12.9), then two linear algebraic equations in the third and sixth row of $\boldsymbol{\Gamma}$ result. Thus, it is possible to eliminate the field components $E_Z = \Gamma_3$ and $H_Z = \Gamma_6$. The results can then be put into the remaining four differential equations to obtain a system of four linear homogeneous first-order differential equations for the four residual field variables E_X, E_Y, H_X, H_Y:

$$\frac{\partial}{\partial Z} \begin{pmatrix} E_X \\ H_Y \\ E_Y \\ -H_X \end{pmatrix} = ik_0 \begin{pmatrix} U_{51} & U_{55} & U_{52} & -U_{54} \\ U_{11} & U_{15} & U_{12} & -U_{14} \\ -U_{41} & -U_{45} & -U_{42} & U_{44} \\ U_{21} & U_{25} & U_{22} & -U_{24} \end{pmatrix} \begin{pmatrix} E_X \\ H_Y \\ E_Y \\ -H_X \end{pmatrix}. \tag{12.11}$$

Eq. (12.11) is usually abbreviated in the form

$$\frac{\partial}{\partial Z} \boldsymbol{\Psi} = ik_0 \boldsymbol{\Delta} \boldsymbol{\Psi}. \tag{12.12}$$

In Eq. (12.12), $\boldsymbol{\Delta}$ represents the 4×4 matrix from Eq. (12.11) and $\boldsymbol{\Psi} = (E_X, H_Y, E_Y, -H_X)^T$. The elements of the 4×4 matrix U_{ij} are given in terms of the elements of $\mathbf{M}$ and k_X, k_Y as follows, where we have employed the abbreviations $f' = M_{33} M_{66} - M_{36} M_{63}$, $g = (0, 0, 0, -k_Y, k_X)$ and $h = (k_Y, -k_X, 0, 0, 0)$:

$$U_{ij} = M_{ij} + (M_{i3} + g_i) a_{3j} + (M_{i6} + h_i) a_{6j}$$

$$a_{3j} = \frac{(M_{6j} + h_j) M_{36} - \left(M_{3j} + g_j\right) M_{66}}{f'} \tag{12.13}$$

$$a_{6j} = \frac{\left(M_{3j} + g_j\right) M_{63} - (M_{6j} + h_j) M_{33}}{f'}.$$

This formulation of the $\boldsymbol{\Delta}$-matrix has the disadvantage that it tends to obscure the underlying physics. However, it is of a much more convenient form than that provided in Refs. [4, 9].

12.2 Berreman's formalism: Calculation of the refractive indices and the polarization directions

For normal incidence, the eigenvalues γ_i of the $\boldsymbol{\Delta}$-matrix according to Eq. (12.14) agree with the indices of refraction of the two forward and the two backward-traveling waves (the latter two eigenvalues are characterized by a negative sign for nonabsorbing media; for absorbing media they are identified by imaginary parts with negative signs):

$$\text{Det}(\boldsymbol{\Delta} - \gamma \mathbf{I}) = 0. \tag{12.14}$$

In Eq. (12.14), $\mathbf{I}$ is the identity matrix. The equation leads to a quartic equation in γ, which is called the Booker quartic:

$$\gamma^4 + \alpha_1 \gamma^3 + \alpha_2 \gamma^2 + \alpha_3 \gamma + \alpha_4 = 0. \tag{12.15}$$

The four coefficients $\alpha_1, \alpha_2, \alpha_3, \alpha_4$ are given by

$$\alpha_1 = -(\Delta_{11} + \Delta_{22} + \Delta_{33} + \Delta_{44})$$
$$\alpha_2 = \Delta_{11}\Delta_{22} + \Delta_{11}\Delta_{33} + \Delta_{11}\Delta_{44} - \Delta_{12}\Delta_{21} - \Delta_{13}\Delta_{31} - \Delta_{14}\Delta_{41} + \Delta_{22}\Delta_{33} + \Delta_{22}\Delta_{44} - \Delta_{23}\Delta_{32} - \Delta_{24}\Delta_{42} + \Delta_{33}\Delta_{44} - \Delta_{34}\Delta_{43}$$
$$\alpha_3 = \Delta_{31}(\Delta_{13}(\Delta_{22} + \Delta_{44}) - \Delta_{12}\Delta_{23} - \Delta_{14}\Delta_{43}) + \Delta_{32}(\Delta_{23}(\Delta_{11} + \Delta_{44}) - \Delta_{13}\Delta_{21} - \Delta_{24}\Delta_{43})$$
$$+ \Delta_{33}(\Delta_{12}\Delta_{21} + \Delta_{14}\Delta_{41} + \Delta_{24}\Delta_{42} - \Delta_{11}\Delta_{22} - \Delta_{44}(\Delta_{11} + \Delta_{22})) + \Delta_{34}(\Delta_{43}(\Delta_{11} + \Delta_{22}) - \Delta_{13}\Delta_{41} - \Delta_{23}\Delta_{42})$$
$$+ \Delta_{41}(\Delta_{14}\Delta_{22} - \Delta_{12}\Delta_{24}) + \Delta_{42}(\Delta_{11}\Delta_{24} - \Delta_{14}\Delta_{21}) + \Delta_{44}(\Delta_{12}\Delta_{21} - \Delta_{11}\Delta_{22})$$
$$\alpha_4 = \Delta_{41}(\Delta_{12}(\Delta_{24}\Delta_{33} - \Delta_{23}\Delta_{34}) + \Delta_{13}(\Delta_{22}\Delta_{34} - \Delta_{24}\Delta_{32}) + \Delta_{14}(\Delta_{23}\Delta_{32} - \Delta_{22}\Delta_{33}))$$
$$+ \Delta_{42}(\Delta_{11}(\Delta_{23}\Delta_{34} - \Delta_{24}\Delta_{33}) + \Delta_{13}(\Delta_{24}\Delta_{31} - \Delta_{21}\Delta_{34}) + \Delta_{14}(\Delta_{21}\Delta_{33} - \Delta_{23}\Delta_{31}))$$
$$+ \Delta_{43}(\Delta_{11}(\Delta_{24}\Delta_{32} - \Delta_{22}\Delta_{34}) + \Delta_{12}(\Delta_{21}\Delta_{34} - \Delta_{24}\Delta_{31}) + \Delta_{14}(\Delta_{22}\Delta_{31} - \Delta_{21}\Delta_{32}))$$
$$+ \Delta_{44}(\Delta_{11}(\Delta_{22}\Delta_{33} - \Delta_{23}\Delta_{32}) + \Delta_{12}(\Delta_{23}\Delta_{31} - \Delta_{21}\Delta_{33}) + \Delta_{13}(\Delta_{21}\Delta_{32} - \Delta_{22}\Delta_{31})).$$

(12.16)

It is possible to derive an analytical solution and, correspondingly, the four solutions of γ are found to be

$$\gamma_1 = -\frac{1}{12}\left(3\alpha_1 + \sqrt{3K_4} + \sqrt{6(K_5 + K_6)}\right)$$
$$\gamma_2 = -\frac{1}{12}\left(3\alpha_1 + \sqrt{3K_4} - \sqrt{6(K_5 + K_6)}\right)$$
$$\gamma_3 = -\frac{1}{12}\left(3\alpha_1 - \sqrt{3K_4} + \sqrt{6(K_5 - K_6)}\right)$$
$$\gamma_4 = -\frac{1}{12}\left(3\alpha_1 - \sqrt{3K_4} - \sqrt{6(K_5 - K_6)}\right),$$

(12.17)

where the abbreviations $K_1 - K_6$ are given by

$$K_1 = 2\alpha_2^3 - 9\alpha_1\alpha_2\alpha_3 + 27\alpha_3^2 + 27\alpha_1^2\alpha_4 - 72\alpha_2\alpha_4$$
$$K_2 = \alpha_2^2 - 3\alpha_1\alpha_3 + 12\alpha_4$$
$$K_3 = \left(K_1 + \sqrt{K_1^2 - 4K_2^3}\right)^{1/3}$$
$$K_4 = 3\alpha_1^2 - 8\alpha_2 + \frac{4 \times 2^{1/3}K_2}{K_3} + 2 \times 2^{2/3}K_3$$
$$K_5 = 3\alpha_1^2 - 8\alpha_2 - \frac{2 \times 2^{1/3}K_2}{K_3} - 2^{2/3}K_3$$
$$K_6 = \frac{3\sqrt{3}(\alpha_1^3 - 4\alpha_1\alpha_2 + 8\alpha_3)}{\sqrt{K_4}}.$$

(12.18)

In practice, it is usually not advisable to use explicit solutions since on the one hand using them is slower than solving Eq. (12.14) numerically. On the other hand, often numerical errors will add up to the point where the analytical solutions are not accurate enough for practical problems, in particular, if one deals with strongly absorbing materials. I nevertheless decided to present the analytical solutions here because how they simplify for special symmetries and special configurations of material and incoming light wave is still instructive. It is also instructive to realize that it is not possible to predict the sign of the solutions in Eq. (12.17) by inspecting them only visually. Therefore, the identification of the forward-traveling waves with the eigenvalues γ_{+I} and γ_{+II} the corresponding solutions for backward-traveling waves by γ_{-I} and γ_{-II} must be repeated for every interesting wavelength/frequency.

The eigenvectors, which give the polarization directions of the waves, can be found by putting each γ_i into the homogeneous system of linear equations:

$$\begin{pmatrix} \Delta_{11} - \gamma & \Delta_{12} & \Delta_{13} & \Delta_{14} \\ \Delta_{21} & \Delta_{22} - \gamma & \Delta_{23} & \Delta_{24} \\ \Delta_{31} & \Delta_{32} & \Delta_{33} - \gamma & \Delta_{34} \\ \Delta_{41} & \Delta_{42} & \Delta_{43} & \Delta_{44} - \gamma \end{pmatrix} \cdot \begin{pmatrix} \Psi_1 \\ \Psi_2 \\ \Psi_3 \\ \Psi_4 \end{pmatrix} = \mathbf{0},$$

(12.19)

and by solving simultaneously three of the four equations in Eq. (12.19). This yields solutions of three components of Ψ in terms of a fourth component, which is arbitrary. With help of Cramer's rule, the following four solutions for the eigenpolarizations can been obtained (these solutions are unique only up to an arbitrary factor; as a consequence, any component of an eigenvector can be specified in a denominator-free form. Therefore, it is not necessary to treat cases where the denominator becomes zero as this was done in Ref. [10]):

$$\Psi_{1,2,3,i} = \begin{pmatrix} \det \begin{vmatrix} \Delta_{14} & \Delta_{12} & \Delta_{13} \\ \Delta_{24} & \Delta_{22}-\gamma_i & \Delta_{23} \\ \Delta_{34} & \Delta_{32} & \Delta_{33}-\gamma_i \end{vmatrix} \\ \det \begin{vmatrix} \Delta_{11}-\gamma_i & \Delta_{14} & \Delta_{13} \\ \Delta_{21} & \Delta_{24} & \Delta_{23} \\ \Delta_{31} & \Delta_{34} & \Delta_{33}-\gamma_i \end{vmatrix} \\ \det \begin{vmatrix} \Delta_{11}-\gamma_i & \Delta_{12} & \Delta_{14} \\ \Delta_{21} & \Delta_{22}-\gamma_i & \Delta_{24} \\ \Delta_{31} & \Delta_{32} & \Delta_{34} \end{vmatrix} \\ \det \begin{vmatrix} \Delta_{11}-\gamma_i & \Delta_{12} & \Delta_{13} \\ \Delta_{21} & \Delta_{22}-\gamma_i & \Delta_{23} \\ \Delta_{31} & \Delta_{32} & \Delta_{33}-\gamma_i \end{vmatrix} \end{pmatrix}, \quad \Psi_{1,2,4,i} = \begin{pmatrix} \det \begin{vmatrix} -\Delta_{14} & \Delta_{12} & \Delta_{13} \\ -\Delta_{24} & \Delta_{22}-\gamma_i & \Delta_{23} \\ -\Delta_{44}+\gamma_i & \Delta_{42} & \Delta_{43} \end{vmatrix} \\ \det \begin{vmatrix} \Delta_{11}-\gamma_i & -\Delta_{14} & \Delta_{13} \\ \Delta_{21} & -\Delta_{24} & \Delta_{23} \\ \Delta_{41} & -\Delta_{44}+\gamma_i & \Delta_{43} \end{vmatrix} \\ \det \begin{vmatrix} \Delta_{11}-\gamma_i & \Delta_{12} & -\Delta_{14} \\ \Delta_{21} & \Delta_{22}-\gamma_i & -\Delta_{24} \\ \Delta_{41} & \Delta_{42} & -\Delta_{44}+\gamma_i \end{vmatrix} \\ \det \begin{vmatrix} \Delta_{11}-\gamma_i & \Delta_{12} & \Delta_{13} \\ \Delta_{21} & \Delta_{22}-\gamma_i & \Delta_{23} \\ \Delta_{41} & \Delta_{42} & \Delta_{43} \end{vmatrix} \end{pmatrix},$$

$$\Psi_{1,3,4,i} = \begin{pmatrix} \det \begin{vmatrix} \Delta_{14} & \Delta_{12} & \Delta_{13} \\ \Delta_{34} & \Delta_{32} & \Delta_{33}-\gamma_i \\ \Delta_{44}-\gamma_i & \Delta_{42} & \Delta_{43} \end{vmatrix} \\ \det \begin{vmatrix} \Delta_{11}-\gamma_i & \Delta_{14} & \Delta_{13} \\ \Delta_{31} & \Delta_{34} & \Delta_{33}-\gamma_i \\ \Delta_{41} & \Delta_{44}-\gamma_i & \Delta_{43} \end{vmatrix} \\ \det \begin{vmatrix} \Delta_{11}-\gamma_i & \Delta_{12} & \Delta_{14} \\ \Delta_{31} & \Delta_{32} & \Delta_{34} \\ \Delta_{41} & \Delta_{42} & \Delta_{44}-\gamma_i \end{vmatrix} \\ \det \begin{vmatrix} \Delta_{11}-\gamma_i & \Delta_{12} & \Delta_{13} \\ \Delta_{31} & \Delta_{32} & \Delta_{33}-\gamma_i \\ \Delta_{41} & \Delta_{42} & \Delta_{43} \end{vmatrix} \end{pmatrix}, \quad \Psi_{2,3,4,i} = \begin{pmatrix} \det \begin{vmatrix} \Delta_{24} & \Delta_{22}-\gamma_i & \Delta_{23} \\ \Delta_{34} & \Delta_{32} & \Delta_{33}-\gamma_i \\ \Delta_{44}-\gamma_i & \Delta_{42} & \Delta_{43} \end{vmatrix} \\ \det \begin{vmatrix} \Delta_{21} & \Delta_{24} & \Delta_{23} \\ \Delta_{31} & \Delta_{34} & \Delta_{33}-\gamma_i \\ \Delta_{41} & \Delta_{44}-\gamma_i & \Delta_{43} \end{vmatrix} \\ \det \begin{vmatrix} \Delta_{21} & \Delta_{22}-\gamma_i & \Delta_{24} \\ \Delta_{31} & \Delta_{32} & \Delta_{34} \\ \Delta_{41} & \Delta_{42} & \Delta_{44}-\gamma_i \end{vmatrix} \\ \det \begin{vmatrix} \Delta_{21} & \Delta_{22}-\gamma_i & \Delta_{23} \\ \Delta_{31} & \Delta_{32} & \Delta_{33}-\gamma_i \\ \Delta_{41} & \Delta_{42} & \Delta_{43} \end{vmatrix} \end{pmatrix}. \quad (12.20)$$

In Eq. (12.20), the subscripts of the Ψ indicate the choice of the equations from Eq. (12.19). It may seem wasteful to state all possible solutions where one seems to be enough. However, since certain of the solutions become singular at special orientations, it is advisable to combine at least two of the solutions linearly. These linear combinations are also solutions, pretty much in the same way as this is done, e.g., in quantum mechanics, which I found inspiring to avoid this kind of singularities. One of the possible combinations is given by

$$\Psi_i = \Psi_{1,2,3,i} + \Psi_{2,3,4,i}, \quad i = +I, -I, +II, -II. \quad (12.21)$$

From the eigenvectors we finally construct the matrix $\mathbf{D}_\Psi$, which is called the Dynamical Matrix, according to

$$\mathbf{D}_\Psi = \begin{pmatrix} \Psi_{+I,1} & \Psi_{-I,1} & \Psi_{+II,1} & \Psi_{-II,1} \\ \Psi_{+I,2} & \Psi_{-I,2} & \Psi_{+II,2} & \Psi_{-II,2} \\ \Psi_{+I,3} & \Psi_{-I,3} & \Psi_{+II,3} & \Psi_{-II,3} \\ \Psi_{+I,4} & \Psi_{-I,4} & \Psi_{+II,4} & \Psi_{-II,4} \end{pmatrix}. \quad (12.22)$$

12.3 Yeh's formalism: Maxwell equations and constitutive relations

Yeh's formalism, as I introduce it here, is restricted to media with a scalar permeability μ and zero cross-coupling tensors $\boldsymbol{\rho}$ and $\boldsymbol{\rho}'$ (an extension, which we will not discuss in the following, is nevertheless easily possible [11]).

From Maxwell's equations the existence of electromagnetic waves can be deduced as discussed in Section 3.4. Plane waves are represented by either

$$\mathbf{E}(\mathbf{r},\omega) = \mathbf{E}_0 \exp(i(\mathbf{k}\cdot\mathbf{r} - \omega t)), \tag{12.23}$$

or

$$\mathbf{H}(\mathbf{r},\omega) = \mathbf{H}_0 \exp(i(\mathbf{k}\cdot\mathbf{r} - \omega t)). \tag{12.24}$$

If we substitute Eqs. (12.23), (12.24) into Eq. (3.2), the following set of equations result (cf. Eq. 3.30):

$$\begin{aligned}\mathbf{k}\times\mathbf{E} &= \omega\mu\mathbf{H}\\ \mathbf{k}\times\mathbf{H} &= -\omega\varepsilon\mathbf{E} = -\omega\mathbf{D}.\end{aligned} \tag{12.25}$$

If we solve the first equation in Eq. (12.25) for $\mathbf{H}$ and put the result into the second, we obtain

$$\mathbf{k}\times(\mathbf{k}\times\mathbf{E}) + \omega^2\mu\varepsilon\mathbf{E} = 0. \tag{12.26}$$

When we substitute k_Z by γ, and use the relative dielectric tensor, Eq. (12.26) reads in explicit form:

$$\begin{pmatrix} \mu\varepsilon_{XX} - k_Y^2 - \gamma^2 & \mu\varepsilon_{XY} + k_X k_Y & \mu\varepsilon_{XZ} + k_X\gamma \\ \mu\varepsilon_{YX} + k_X k_Y & \mu\varepsilon_{YY} - k_X^2 - \gamma^2 & \mu\varepsilon_{YZ} + k_Y\gamma \\ \mu\varepsilon_{ZX} + k_X\gamma & \mu\varepsilon_{ZY} + k_Y\gamma & \mu\varepsilon_{ZZ} - k_X^2 - k_Y^2 \end{pmatrix} \cdot \begin{pmatrix} E_X \\ E_Y \\ E_Z \end{pmatrix} = 0. \tag{12.27}$$

With this equation, we can determine the eigenvalues γ_i and the corresponding eigenvectors $\mathbf{p}_i$.

12.4 Yeh's formalism: Calculation of the refractive indices and the polarization directions

In order to determine the eigenvalues, the determinant of the 3×3 matrix in Eq. (12.27) must be zero:

$$\det\begin{vmatrix} \mu\varepsilon_{XX} - k_Y^2 - \gamma^2 & \mu\varepsilon_{XY} + k_X k_Y & \mu\varepsilon_{XZ} + k_X\gamma \\ \mu\varepsilon_{YX} + k_X k_Y & \mu\varepsilon_{YY} - k_X^2 - \gamma^2 & \mu\varepsilon_{YZ} + k_Y\gamma \\ \mu\varepsilon_{ZX} + k_X\gamma & \mu\varepsilon_{ZY} + k_Y\gamma & \mu\varepsilon_{ZZ} - k_X^2 - k_Y^2 \end{vmatrix} = 0. \tag{12.28}$$

This again leads to the Booker quartic

$$\gamma^4 + \alpha_1\gamma^3 + \alpha_2\gamma^2 + \alpha_3\gamma + \alpha_4 = 0, \tag{12.29}$$

but the four coefficients $\alpha_1, \alpha_2, \alpha_3, \alpha_4$ are now given by

$$\alpha_1 = \frac{1}{\varepsilon_{ZZ}}(k_X(\varepsilon_{XZ} + \varepsilon_{ZX}) + k_Y(\varepsilon_{YZ} + \varepsilon_{ZY}))$$

$$\alpha_2 = \frac{1}{\varepsilon_{ZZ}}\left(k_X k_Y(\varepsilon_{XY} + \varepsilon_{YX}) + k_X^2(\varepsilon_{XX} + \varepsilon_{ZZ}) + k_Y^2(\varepsilon_{YY} + \varepsilon_{ZZ}) + \mu(\varepsilon_{XZ}\varepsilon_{ZX} + \varepsilon_{YZ}\varepsilon_{ZY} - (\varepsilon_{XX} + \varepsilon_{YY})\varepsilon_{ZZ})\right)$$

$$\alpha_3 = \frac{1}{\varepsilon_{ZZ}}\begin{pmatrix} k_X^3(\varepsilon_{XZ} + \varepsilon_{ZX}) + k_X^2 k_Y(\varepsilon_{YZ} + \varepsilon_{ZY}) + k_X\left(k_Y^2(\varepsilon_{XZ} + \varepsilon_{ZX}) + \mu(\varepsilon_{XY}\varepsilon_{YZ} - \varepsilon_{YY}(\varepsilon_{XZ} + \varepsilon_{ZX}) + \varepsilon_{YX}\varepsilon_{ZY})\right) + \\ k_Y\left(k_Y^2(\varepsilon_{YZ} + \varepsilon_{ZY}) + \mu(\varepsilon_{XZ}\varepsilon_{YX} + \varepsilon_{XY}\varepsilon_{ZX} - \varepsilon_{XX}(\varepsilon_{YZ} + \varepsilon_{ZY}))\right) \end{pmatrix}$$

$$\alpha_4 = \frac{1}{\varepsilon_{ZZ}}\begin{pmatrix} k_X^4\varepsilon_{XX} + k_Y^4\varepsilon_{YY} + k_X^3 k_Y(\varepsilon_{XY} + \varepsilon_{YX}) + \mu k_Y^2(\varepsilon_{XY}\varepsilon_{YX} + \varepsilon_{YZ}\varepsilon_{ZY} - \varepsilon_{YY}(\varepsilon_{XX} + \varepsilon_{ZZ})) + \\ k_X k_Y\left(\mu(\varepsilon_{YZ}\varepsilon_{ZX} + \varepsilon_{XZ}\varepsilon_{ZY}) + (\varepsilon_{XY} + \varepsilon_{YX})(k_Y^2 - \mu\varepsilon_{ZZ})\right) + \\ \mu^2(\varepsilon_{XZ}(\varepsilon_{YX}\varepsilon_{ZY} - \varepsilon_{YY}\varepsilon_{ZX}) + \varepsilon_{XY}(\varepsilon_{YZ}\varepsilon_{ZX} - \varepsilon_{YX}\varepsilon_{ZZ}) + \varepsilon_{XX}(\varepsilon_{YY}\varepsilon_{ZZ} - \varepsilon_{YZ}\varepsilon_{ZY})) + \\ k_X^2\left(k_Y^2(\varepsilon_{XX} + \varepsilon_{YY}) + \mu(\varepsilon_{XY}\varepsilon_{YX} + \varepsilon_{XZ}\varepsilon_{ZX} - \varepsilon_{XX}(\varepsilon_{YY} + \varepsilon_{ZZ}))\right) \end{pmatrix}. \tag{12.30}$$

Again, the eigenvectors can be determined by Cramer's rule:

$$\mathbf{p}_{1,2,i} = N_{1,2,i} \begin{pmatrix} \det \begin{vmatrix} -\mu\varepsilon_{XZ} - k_X\gamma_i & \mu\varepsilon_{XY} + k_Xk_Y \\ -\mu\varepsilon_{YZ} - k_Y\gamma_i & \mu\varepsilon_{YY} - k_X^2 - \gamma_i^2 \end{vmatrix} \\ \det \begin{vmatrix} \mu\varepsilon_{XX} - k_Y^2 - \gamma_i^2 & -\mu\varepsilon_{XZ} - k_X\gamma_i \\ \mu\varepsilon_{YX} + k_Xk_Y & -\mu\varepsilon_{YZ} - k_Y\gamma_i \end{vmatrix} \\ \det \begin{vmatrix} \mu\varepsilon_{XX} - k_Y^2 - \gamma_i^2 & \mu\varepsilon_{XY} + k_Xk_Y \\ \mu\varepsilon_{YX} + k_Xk_Y & \mu\varepsilon_{YY} - k_X^2 - \gamma_i^2 \end{vmatrix} \end{pmatrix}, \quad \mathbf{p}_{1,3,i} = N_{1,3,i} \begin{pmatrix} \det \begin{vmatrix} -\mu\varepsilon_{XZ} - k_X\gamma_i & \mu\varepsilon_{XY} + k_Xk_Y \\ -\mu\varepsilon_{ZZ} + k_X^2 + k_Y^2 & \mu\varepsilon_{ZY} + k_Y\gamma_i \end{vmatrix} \\ \det \begin{vmatrix} \mu\varepsilon_{XX} - k_Y^2 - \gamma_i^2 & -\mu\varepsilon_{XZ} - k_X\gamma_i \\ \mu\varepsilon_{ZX} + k_X\gamma_i & -\mu\varepsilon_{ZZ} + k_X^2 + k_Y^2 \end{vmatrix} \\ \det \begin{vmatrix} \mu\varepsilon_{XX} - k_Y^2 - \gamma_i^2 & \mu\varepsilon_{XY} + k_Xk_Y \\ \mu\varepsilon_{ZX} + k_X\gamma_i & \mu\varepsilon_{ZY} + k_Y\gamma_i \end{vmatrix} \end{pmatrix},$$

$$\mathbf{p}_{2,3,i} = N_{2,3,i} \begin{pmatrix} \det \begin{vmatrix} -\mu\varepsilon_{YZ} - k_Y\gamma_i & \mu\varepsilon_{YY} - k_X^2 - \gamma_i^2 \\ -\mu\varepsilon_{ZZ} + k_X^2 + k_Y^2 & \mu\varepsilon_{ZY} + k_Y\gamma_i \end{vmatrix} \\ \det \begin{vmatrix} \mu\varepsilon_{YX} + k_Xk_Y & -\mu\varepsilon_{YZ} - k_Y\gamma_i \\ \mu\varepsilon_{ZX} + k_X\gamma_i & -\mu\varepsilon_{ZZ} + k_X^2 + k_Y^2 \end{vmatrix} \\ \det \begin{vmatrix} \mu\varepsilon_{YX} + k_Xk_Y & \mu\varepsilon_{YY} - k_X^2 - \gamma_i^2 \\ \mu\varepsilon_{ZX} + k_X\gamma_i & \mu\varepsilon_{ZY} + k_Y\gamma_i \end{vmatrix} \end{pmatrix}.$$

(12.31)

The normalization factors N_i, which are determined by the condition $\mathbf{p}_i \cdot \mathbf{p}_i = 1$, have been adopted from Yeh's chapter (usually, it is not necessary to normalize the $\mathbf{p}_i$). The eigenvectors originally provided in this chapter are the $\mathbf{p}_{2,3,i}$. One important difference to the solution derived in this chapter is that we have not introduced the constraint of $\boldsymbol{\varepsilon}$ being a dielectric tensor which can be brought into diagonal form. This is important since we also want to calculate the reflectance and transmittance spectra of monoclinic and triclinic crystals for which the dielectric tensor cannot be diagonalized (even if $\boldsymbol{\varepsilon}$ would be real, the rotation matrices to diagonalize the tensor would change with wavenumber).

The corresponding polarization directions $\mathbf{q}_i$ of the $\mathbf{H}$-fields can be computed by the following relation, where we use that these directions must be both normal to the direction of propagation and to the $\mathbf{p}_i$s

$$\mathbf{q}_i = \frac{1}{\mu} \begin{pmatrix} k_X \\ k_Y \\ \gamma_i \end{pmatrix} \times \mathbf{p}_i. \qquad (12.32)$$

To provide the Dynamical Matrix, we only need the X- and Y-components of the $\mathbf{p}_i$ and $\mathbf{q}_i$. If we again use superpositions of two eigenpolarizations, the vectors $\boldsymbol{\Psi}_i$ take on the following form:

$$\boldsymbol{\Psi}_i = \begin{pmatrix} \mathbf{p}_{X,(1,3),i} + \mathbf{p}_{X,(2,3),i} \\ \frac{1}{\mu}\left(\left(\mathbf{p}_{X,(1,3),i} + \mathbf{p}_{X,(2,3),i}\right)\gamma_i - \left(\mathbf{p}_{Z,(1,3),i} + \mathbf{p}_{Z,(2,3),i}\right)k_X\right) \\ \mathbf{p}_{Y,(1,3),i} + \mathbf{p}_{Y,(2,3),i} \\ -\frac{1}{\mu}\left(\left(\mathbf{p}_{Z,(1,3),i} + \mathbf{p}_{Z,(2,3),i}\right)k_Y - \left(\mathbf{p}_{Y,(1,3),i} + \mathbf{p}_{Y,(2,3),i}\right)\gamma_i\right) \end{pmatrix}, \quad i = +I, -I, +II, -II. \qquad (12.33)$$

In Eq. (12.33), I have additionally deviated slightly from the original form provided by Yeh by using $-\mathbf{q}_{i,\,X}$ instead of $\mathbf{q}_{i,\,X}$ in the fourth row of $\boldsymbol{\Psi}_i$. I did this to assure positive energy flow in the direction of propagation and compatibility with Berreman's formalism. The $\boldsymbol{\Psi}_i$ as defined in Eq. (12.33) can be put in Eq. (12.22) to yield again the Dynamical Matrix.

12.5 The transfer matrix

Like in Section 4.5, we assume a layered structure bound by two semiinfinite media, incidence, and exit medium. The four amplitudes $A_i(j)$ of the two forward and the two backward-traveling waves at the left side of the interface between the layers

j and $j+1$ are linked with the amplitudes $A_i(j-1)$ of the corresponding waves in medium $j-1$ at the boundary between medium $j-1$ and j by

$$\begin{pmatrix} A_1(j-1) \\ A_2(j-1) \\ A_3(j-1) \\ A_4(j-1) \end{pmatrix} = \mathbf{D_\Psi}^{-1}(j-1)\mathbf{D_\Psi}(j)\mathbf{P}(j) \begin{pmatrix} A_1(j) \\ A_2(j) \\ A_3(j) \\ A_4(j) \end{pmatrix}. \qquad (12.34)$$

The matrices $\mathbf{P}(j)$ are again diagonal matrices called Propagation Matrices and are given by

$$\mathbf{P}(j) = \begin{pmatrix} \exp(ik_0 d_j \gamma_{+I}) & 0 & 0 & 0 \\ 0 & \exp(ik_0 d_j \gamma_{-I}) & 0 & 0 \\ 0 & 0 & \exp(ik_0 d_j \gamma_{+II}) & 0 \\ 0 & 0 & 0 & \exp(ik_0 d_j \gamma_{-II}) \end{pmatrix}. \qquad (12.35)$$

The transfer matrix $\mathbf{T_p}$ links the amplitudes of the incident waves and the reflected waves at the interface medium 0/medium 1 with the amplitudes of the transmitted waves in the exit medium $(J+1)$ according to

$$\begin{pmatrix} A_s \\ B_s \\ A_p \\ B_p \end{pmatrix} = \underbrace{\mathbf{D_\Psi}^{-1}(0) \prod_{j=1}^{J} \mathbf{T_p}(j)\, \mathbf{D_\Psi}(J+1)}_{\tilde{\mathbf{M}}} \begin{pmatrix} C_s \\ 0 \\ C_p \\ 0 \end{pmatrix}. \qquad (12.36)$$

In Eq. (12.36) the transfer matrix $\mathbf{T_p}$ of the jth layer is defined by $\mathbf{T_p}(j) = \mathbf{D_\Psi}(j)\mathbf{P}(j)\mathbf{D_\Psi}^{-1}(j)$. The transfer matrices in the product are ordered according to their Z-coordinates. A_s and A_p represent the amplitudes of the incident waves with s- and p-polarization. The B_i and C_i stand for the amplitudes of the reflected and the transmitted waves, respectively. $\tilde{\mathbf{M}}$ is the overall transfer matrix from which the reflectances and transmittances can eventually be calculated. As there are no backward-traveling waves in the semiinfinite medium $J+1$, $\mathbf{D_\Psi}(J+1)$ simplifies to

$$\mathbf{D_\Psi}(J+1) = \begin{pmatrix} \Psi_{+I,1} & 0 & \Psi_{+II,1} & 0 \\ \Psi_{+I,2} & 0 & \Psi_{+II,2} & 0 \\ \Psi_{+I,3} & 0 & \Psi_{+II,3} & 0 \\ \Psi_{+I,4} & 0 & \Psi_{+II,4} & 0 \end{pmatrix}. \qquad (12.37)$$

If, e.g., the reflectances of an absorbing crystal are of interest, $\tilde{\mathbf{M}}$ is in this case represented by

$$\tilde{\mathbf{M}} = \mathbf{D_\Psi}^{-1}(0)\,\mathbf{D_\Psi}(1), \qquad (12.38)$$

as the numbers of layers J then equals zero. Therefore, Eq. (12.38) links the amplitudes of the incoming and reflected waves with those of the transmitted waves at a single interface. In the following, we will use only reflectance spectra of crystals to determine their dielectric tensor function for which Eq. (12.38) can be used even in seemingly nonabsorbing regions as long as they are millimeter-thick or thicker. In fact, at higher wavenumbers/frequencies, damped harmonic oscillators lead to a long tail in the spectrum where absorption is still strong enough so that light singly reflected from the backside of the crystal will not be able to reach the frontside. At the wavenumber/frequency where this starts to happen, there is a clear indication in the spectrum which consists of a step where the reflectance is clearly increasing at the high wavenumber/frequency side.

12.6 The treatment of singularities
12.6.1 Degenerate eigenvalues

There are two different reasons for the existence of singularities in the 4×4 matrix formalism leading to problems with the execution of corresponding computer code. The first problem that arises is that degenerate eigenvalues exist for certain forms of the $\tilde{\mathbf{M}}$-matrix. As a result, the number of independent equations for determining the eigenpolarizations in the Berreman and Yeh formalisms are reduced by one and the eigenpolarizations in Eqs. (12.20), (12.31) are either linearly dependent or identical in pairs for both the forward- and the backward-propagating waves. The physical reason for this failure of the formalisms is that the polarization directions are not restricted by the constitutive relations (Eq. 12.5). The majority of materials where this can happen are of high symmetry, such as those with scalar tensors (in this case the eigenvalues are degenerate for arbitrary orientations) or uniaxial materials, when the waves move along the axis with high symmetry. An overview of the affected magnetic groups can be found in Ref. [7].

Unfortunately, the solution to this problem is not as simple as in the case of scalar permeability, for which all exceptions can be treated with a common solution [10]. For magnetic materials, three different groups can be formed with regard to the treatment of singularities resulting from degenerate eigenvalues. The first group is formed by all magnetic groups that have nonvanishing elements in the first and fourth quadrants of the $\mathbf{\Delta}$-matrix when degenerate eigenvalues occur. As a consequence, s- and p-waves are decoupled for all group members and a common solution can exist. Thus, the general form of the $\mathbf{\Delta}$-matrix is as follows:

$$\mathbf{\Delta} = \begin{pmatrix} \Delta_{11} & \Delta_{12} & 0 & 0 \\ \Delta_{21} & \Delta_{22} & 0 & 0 \\ 0 & 0 & \Delta_{33} & \Delta_{34} \\ 0 & 0 & \Delta_{43} & \Delta_{44} \end{pmatrix}. \tag{12.39}$$

The occurrence of such a form of the $\mathbf{\Delta}$-matrix is not necessarily connected with the existence of singularities. The associated matrix $\mathbf{D_\Psi}$ is given by

$$\mathbf{D_\Psi} = \begin{pmatrix} \dfrac{\Delta_{11} - \Delta_{22} + \Lambda_1}{2\Delta_{21}} & \dfrac{\Delta_{11} - \Delta_{22} - \Lambda_1}{2\Delta_{21}} & 0 & 0 \\ 1 & 1 & 0 & 0 \\ 0 & 0 & \dfrac{\Delta_{33} - \Delta_{44} + \Lambda_2}{2\Delta_{43}} & \dfrac{\Delta_{33} - \Delta_{44} - \Lambda_2}{2\Delta_{43}} \\ 0 & 0 & 1 & 1 \end{pmatrix} \tag{12.40}$$

$$\Lambda_1 = \sqrt{4\Delta_{12}\Delta_{21} + (\Delta_{11} - \Delta_{22})^2}, \quad \Lambda_2 = \sqrt{4\Delta_{34}\Delta_{43} + (\Delta_{33} - \Delta_{44})^2}.$$

For the second and the third group, the corresponding general solution is the particular solution that belongs to one of the group members. For group 2 this is a biisotropic medium with $\rho = \rho'$ and the general solution:

$$\mathbf{D_\Psi} = \begin{pmatrix} -\dfrac{\Lambda_5 + \Lambda_1}{2\Lambda_4} & -\dfrac{\Lambda_5 + \Lambda_1}{2\Lambda_4} & -\dfrac{\Lambda_5 - \Lambda_1}{2\Lambda_4} & -\dfrac{\Lambda_5 - \Lambda_1}{2\Lambda_4} \\ -\dfrac{\sqrt{2}\sqrt{\Lambda_2 - \Lambda_1}(\Lambda_5 + \Lambda_1)}{-2\Lambda_3 + 2\Delta_{13}(-\Delta_{34}\Delta_{43} + \Lambda_1)} & \dfrac{\sqrt{2}\sqrt{\Lambda_2 - \Lambda_1}(\Lambda_5 + \Lambda_1)}{-2\Lambda_3 + 2\Delta_{13}(-\Delta_{34}\Delta_{43} + \Lambda_1)} & \dfrac{-\sqrt{\Lambda_2 + \Lambda_1}(\Lambda_5 - \Lambda_1)}{\sqrt{2}(\Lambda_3 + \Delta_{13}(\Delta_{34}\Delta_{43} + \Lambda_1))} & \dfrac{-\sqrt{\Lambda_2 + \Lambda_1}(\Lambda_5 - \Lambda_1)}{\sqrt{2}(\Lambda_3 + \Delta_{13}(\Delta_{34}\Delta_{43} + \Lambda_1))} \\ \dfrac{\sqrt{2}\Lambda_6\sqrt{\Lambda_2 - \Lambda_1}}{\Lambda_3 + \Delta_{13}(-\Delta_{34}\Delta_{43} + \Lambda_1)} & -\dfrac{\sqrt{2}\Lambda_6\sqrt{\Lambda_2 - \Lambda_1}}{\Lambda_3 + \Delta_{13}(-\Delta_{34}\Delta_{43} + \Lambda_1)} & \dfrac{\sqrt{2}\Lambda_6\sqrt{\Lambda_2 + \Lambda_1}}{\Lambda_3 + \Delta_{13}(\Delta_{34}\Delta_{43} + \Lambda_1)} & \dfrac{\sqrt{2}\Lambda_6\sqrt{\Lambda_2 + \Lambda_1}}{\Lambda_3 + \Delta_{13}(\Delta_{34}\Delta_{43} + \Lambda_1)} \\ 1 & 1 & 1 & 1 \end{pmatrix}$$

$$\Lambda_1 = \sqrt{\Delta_{12}^2 \Delta_{21}^2 + \Delta_{12}\left(4\Delta_{13}\Delta_{21}\Delta_{24} + 2\left(2\Delta_{24}^2 - \Delta_{21}\Delta_{34}\right)\Delta_{43}\right) + \Delta_{34}\left(4\Delta_{13}^2\Delta_{21} + 4\Delta_{13}\Delta_{24}\Delta_{43} + \Delta_{34}\Delta_{43}^2\right)}$$

$$\Lambda_2 = \Delta_{12}\Delta_{21} + 2\Delta_{13}\Delta_{24} + \Delta_{34}\Delta_{43}$$

$$\Lambda_3 = \Delta_{12}(\Delta_{13}\Delta_{21} + 2\Delta_{24}\Delta_{43})$$

$$\Lambda_4 = \Delta_{13}\Delta_{21} + \Delta_{24}\Delta_{43}$$

$$\Lambda_5 = -\Delta_{12}\Delta_{21} + \Delta_{34}\Delta_{43}$$

$$\Lambda_6 = \Delta_{12}\Delta_{24} + \Delta_{13}\Delta_{34}.$$

$$\tag{12.41}$$

The general solution for group 3 is the same as that obtained for the magnetic group $S_4(C_2)$:

$$\mathbf{D}_\Psi = \begin{pmatrix} \dfrac{-\Delta_{11} + \Delta_{22} - \Lambda}{2\Delta_{42}} & \dfrac{-\Delta_{11} + \Delta_{22} + \Lambda}{2\Delta_{42}} & \dfrac{\Delta_{12}}{\Delta_{42}} & \dfrac{\Delta_{12}}{\Delta_{42}} \\ -\dfrac{\Delta_{21}}{\Delta_{42}} & -\dfrac{\Delta_{21}}{\Delta_{42}} & \dfrac{-\Delta_{11} + \Delta_{22} + \Lambda}{2\Delta_{42}} & \dfrac{-\Delta_{11} + \Delta_{22} - \Lambda}{2\Delta_{42}} \\ 1 & 1 & 0 & 0 \\ 0 & 0 & 1 & 1 \end{pmatrix} \quad (12.42)$$

$$\Lambda = \sqrt{4\Delta_{12}\Delta_{21} + (\Delta_{11} - \Delta_{22})^2 - 4\Delta_{13}\Delta_{42}}.$$

In addition to symmetry reasons, degenerate eigenvalues can also occur if the relation $\mu_{ij}/\varepsilon_{ij} = \text{const.} = \mu/\varepsilon$ and $\rho_{ij} = \rho'_{ij} = 0$ is fulfilled for all possible combinations of i and j. As a consequence, the eigenvalues are always degenerate, independent of the orientation of the medium. Unfortunately, there are a plethora of different ways to meet the previous conditions (since ρ_{ij} and ρ'_{ij} need not necessarily be zero) and, because of mode coupling, a common solution cannot exist. However, the occurrence of this type of singularity in practice seems comparatively unlikely. Nevertheless, it is always possible to find an analytical, albeit mostly comparatively complex, solution. E.g., for the particular problem with $\boldsymbol{\mu}_{x,y,z} = \mu \times \text{diag}(1, 2, 3)$ and $\boldsymbol{\varepsilon}_{x,y,z} = \varepsilon \times \text{diag}(1, 2, 3)$, we obtain the $\boldsymbol{\Delta}$-matrix

$$\boldsymbol{\Delta} = \begin{pmatrix} \Delta_{11} & \Delta_{12} & 0 & \Delta_{14} \\ \Delta_{21} & \Delta_{11} & \Delta_{23} & 0 \\ 0 & \Delta_{14} & \Delta_{11} & \Delta_{34} \\ \Delta_{23} & 0 & \Delta_{43} & \Delta_{11} \end{pmatrix}, \quad (12.43)$$

and solutions which have the form

$$\mathbf{D}_\Psi = \begin{pmatrix} \dfrac{\Lambda_4 - \Delta_{12}(-2\Delta_{14}\Delta_{23} + \Delta_{34}\Delta_{43} + \Lambda_1)}{\sqrt{2}\Lambda_2\sqrt{\Lambda_3 - \Lambda_1}} & -\dfrac{\Lambda_4 - \Delta_{12}(-2\Delta_{14}\Delta_{23} + \Delta_{34}\Delta_{43} + \Lambda_1)}{\sqrt{2}\Lambda_2\sqrt{\Lambda_3 - \Lambda_1}} \\ -\dfrac{-\Delta_{12}\Delta_{21} + \Delta_{34}\Delta_{43} + \Lambda_1}{2\Lambda_2} & -\dfrac{-\Delta_{12}\Delta_{21} + \Delta_{34}\Delta_{43} + \Lambda_1}{2\Lambda_2} \\ -\dfrac{\Lambda_5 - \Delta_{14}(-\Delta_{34}\Delta_{43} + \Lambda_1)}{\sqrt{2}\Lambda_2\sqrt{\Lambda_3 - \Lambda_1}} & \dfrac{\Lambda_5 + \Delta_{14}(\Delta_{34}\Delta_{43} - \Lambda_1)}{\sqrt{2}\Lambda_2\sqrt{\Lambda_3 - \Lambda_1}} \\ 1 & 1 \\ \dfrac{\Lambda_4 + \Delta_{12}(2\Delta_{14}\Delta_{23} - \Delta_{34}\Delta_{43} + \Lambda_1)}{\sqrt{2}\Lambda_2\sqrt{\Lambda_3 + \Lambda_1}} & -\dfrac{\Lambda_4 + \Delta_{12}(2\Delta_{14}\Delta_{23} - \Delta_{34}\Delta_{43} + \Lambda_1)}{\sqrt{2}\Lambda_2\sqrt{\Lambda_3 + \Lambda_1}} \\ \dfrac{\Delta_{12}\Delta_{21} - \Delta_{34}\Delta_{43} + \Lambda_1}{2\Lambda_2} & \dfrac{\Delta_{12}\Delta_{21} - \Delta_{34}\Delta_{43} + \Lambda_1}{2\Lambda_2} \\ -\dfrac{\Lambda_5 + \Delta_{14}(\Delta_{34}\Delta_{43} + \Lambda_1)}{\sqrt{2}\Lambda_2\sqrt{\Lambda_3 + \Lambda_1}} & \dfrac{\Lambda_5 + \Delta_{14}(\Delta_{34}\Delta_{43} + \Lambda_1)}{\sqrt{2}\Lambda_2\sqrt{\Lambda_3 + \Lambda_1}} \\ 1 & 1 \end{pmatrix} \quad (12.44)$$

$$\Lambda_1 = \sqrt{\Delta_{12}^2\Delta_{21}^2 + \Delta_{12}(4\Delta_{23}(\Delta_{14}\Delta_{21} + \Delta_{23}\Delta_{34}) - 2\Delta_{21}\Delta_{34}\Delta_{43}) + \Delta_{43}(4\Delta_{14}(\Delta_{14}\Delta_{21} + \Delta_{23}\Delta_{34}) + \Delta_{34}^2\Delta_{43})}$$

$$\Lambda_2 = \Delta_{12}\Delta_{23} + \Delta_{14}\Delta_{43}$$

$$\Lambda_3 = \Delta_{12}\Delta_{21} + 2\Delta_{14}\Delta_{23} + \Delta_{34}\Delta_{43}$$

$$\Lambda_4 = \Delta_{12}^2\Delta_{21} + 2\Delta_{14}^2\Delta_{43}$$

$$\Lambda_5 = \Delta_{12}(\Delta_{14}\Delta_{21} + 2\Delta_{23}\Delta_{34}).$$

12.6.2 Singular form of the Dynamical Matrix

In addition to the occurrence of degenerate eigenvalues, singularities also exist if the matrix D_Ψ takes on one of the following two forms:

$$D_\Psi = \begin{pmatrix} D_{\Psi 11} & D_{\Psi 12} & D_{\Psi 13} & D_{\Psi 14} \\ D_{\Psi 21} & D_{\Psi 22} & D_{\Psi 23} & D_{\Psi 24} \\ 0 & 0 & 0 & 0 \\ 0 & 0 & 0 & 0 \end{pmatrix}, \quad D_\Psi = \begin{pmatrix} 0 & 0 & 0 & 0 \\ 0 & 0 & 0 & 0 \\ D_{\Psi 31} & D_{\Psi 32} & D_{\Psi 33} & D_{\Psi 34} \\ D_{\Psi 41} & D_{\Psi 42} & D_{\Psi 42} & D_{\Psi 42} \end{pmatrix}, \quad (12.45)$$

e.g., if $\rho_{ij} = \rho'_{ij} = 0$ and $\mu_{X, Y, Z} = \text{diag}(\mu_{11}, \mu_{22}, \mu_{33})$, $\varepsilon_{X, Y, Z} = \text{diag}(\varepsilon_{11}, \varepsilon_{22}, \varepsilon_{33})$ and only one of the eigenvectors from Eq. (12.20) or (12.31) is employed instead of two to form the Dynamical Matrix (the use of $\Psi_{1, 2, 3, i}$, $\Psi_{1, 2, 4, i}$, $p_{1, 2, i}$, and $p_{1, 3, i}$ usually leads to the first form, whereas the use of $\Psi_{1, 3, 4, i}$, $\Psi_{2, 3, 4, i}$, and $p_{2, 3, i}$ provides the second form of the Dynamical Matrix in Eq. (12.45)). As a consequence, either D_Ψ^{-1} is not defined or the denominator d' in Eq. (12.47) becomes zero. There are two possible solutions to this problem, namely either to linearly combine the eigenvectors or to choose an eigenvector of the first kind for Ψ_{+I} and Ψ_{-I} and one of the second kind for Ψ_{+II} and Ψ_{-II} as this was suggested in Ref. [10], of which I find the first solution to be more convenient. The reason is that the second solution requires to sort the eigenvalues since otherwise the Ψ_i can become either zero (in a denominator-free form) or singular due to a division by zero.

12.7 The calculation of reflectance and transmittance coefficients

The matrix $\tilde{M}$ obtained either from Eq. (12.36) for layer stacks or from Eq. (12.38) for crystals, links the amplitudes of the incident (A_s, A_p) and reflected (B_s, B_p) s- or p-polarized waves at the interface between the incidence medium and medium 1 to those (C_s, C_p) of the s- or p-polarized waves transmitted into the exit medium $J+1$ according to

$$\begin{pmatrix} A_s \\ B_s \\ A_p \\ B_p \end{pmatrix} = \begin{pmatrix} \tilde{M}_{11} & \tilde{M}_{12} & \tilde{M}_{13} & \tilde{M}_{14} \\ \tilde{M}_{21} & \tilde{M}_{22} & \tilde{M}_{23} & \tilde{M}_{24} \\ \tilde{M}_{31} & \tilde{M}_{32} & \tilde{M}_{33} & \tilde{M}_{34} \\ \tilde{M}_{41} & \tilde{M}_{42} & \tilde{M}_{43} & \tilde{M}_{44} \end{pmatrix} \begin{pmatrix} C_s \\ 0 \\ C_p \\ 0 \end{pmatrix}. \quad (12.46)$$

The reflection and transmission coefficients can then be computed from the following relations (with the abbreviation $d' = \tilde{M}_{11}\tilde{M}_{33} - \tilde{M}_{13}\tilde{M}_{31}$, their first subscript provides the orientation of the polarizer and the second that of the analyzer, cf. Fig. 12.1):

$$\begin{aligned} r_{ss} &= \left(\frac{B_s}{A_s}\right)_{A_p=0} = \frac{\tilde{M}_{21}\tilde{M}_{33} - \tilde{M}_{23}\tilde{M}_{31}}{d'}, & t_{ss} &= \left(\frac{C_s}{A_s}\right)_{A_p=0} = \frac{\tilde{M}_{33}}{d'}, \\ r_{sp} &= \left(\frac{B_p}{A_s}\right)_{A_p=0} = \frac{\tilde{M}_{41}\tilde{M}_{33} - \tilde{M}_{43}\tilde{M}_{31}}{d'}, & t_{sp} &= \left(\frac{C_p}{A_s}\right)_{A_p=0} = \frac{-\tilde{M}_{31}}{d'}, \\ r_{ps} &= \left(\frac{B_s}{A_p}\right)_{A_s=0} = \frac{\tilde{M}_{11}\tilde{M}_{23} - \tilde{M}_{21}\tilde{M}_{13}}{d'}, & t_{ps} &= \left(\frac{C_s}{A_p}\right)_{A_s=0} = \frac{-\tilde{M}_{13}}{d'}, \\ r_{pp} &= \left(\frac{B_p}{A_p}\right)_{A_s=0} = \frac{\tilde{M}_{11}\tilde{M}_{43} - \tilde{M}_{41}\tilde{M}_{13}}{d'}, & t_{pp} &= \left(\frac{C_p}{A_p}\right)_{A_s=0} = \frac{\tilde{M}_{11}}{d'}. \end{aligned} \quad (12.47)$$

If, e.g., only a polarizer and no analyzer is used, the reflection and transmission coefficients r_i and t_i ($i = s, p$) are given by

$$r_i = r_{ii} + r_{ij}, \quad t_i = t_{ii} + t_{ij}, \quad j = s, p; j \neq i. \quad (12.48)$$

Since the polarization of the waves for r_{ij} and t_{ij} is perpendicular to that which belongs to the reflection and transmission coefficients r_{ii} and t_{ii}, the following relations for the reflectance R and transmittance T are obtained:

$$R_i = R_{ii} + R_{ij} = |r_{ii}|^2 + |r_{ij}|^2 = |r_i|^2 \quad T_i = T_{ii} + T_{ij} = |t_{ii}|^2 + |t_{ij}|^2 = |t_i|^2. \quad (12.49)$$

Eq. (12.49) completes the formulation of the 4×4-matrix formalism.

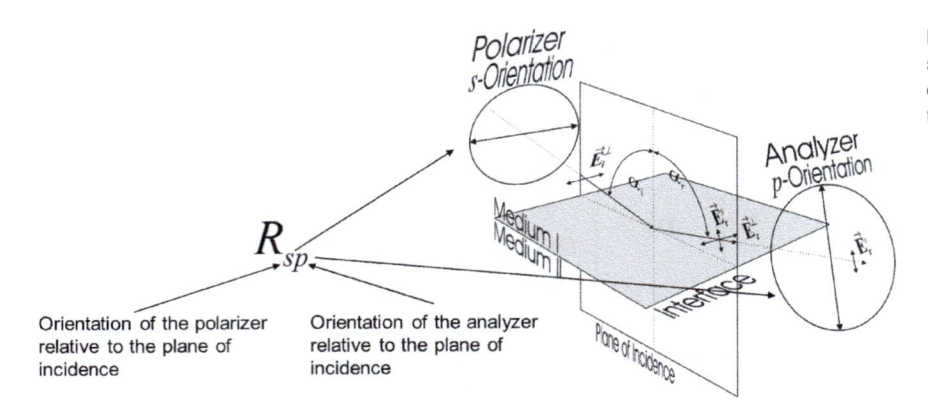

FIG. 12.1 Illustration of the meaning of the subscripts of the reflection and transmission coefficients as well as those of reflectance and transmittance.

12.8 Simplifications for special cases

Quantitative evaluation of single-crystal spectra, i.e., dispersion analysis, will require that the comparably complex computations based on the formulas introduced in the preceding section have to be repeated many times. Certainly, for every experimental point in a spectrum, but, in addition, the complete modeled spectrum will have to be calculated many times until the fit converges. Furthermore, for low-symmetry crystals a number of experimental spectra have to be fitted in parallel. So, what we talk about is to repeat previous calculations for one spectrum about a thousand times (depending on the spectral resolution), repeat this calculation for as much spectra as have to be calculated in parallel and reiterate this until modeled and experimental spectra coincide satisfactorily, which again means often a couple of thousand times. This means that a dispersion analysis of a crystal of the lowest possible symmetry can easily take months. On the other hand, for crystals of higher symmetry, special considerations allow us to speed-up dispersion analysis considerably, e.g., dispersion analysis of crystals of monoclinic symmetry can be carried out within a couple of hours if the operator has some experience. Further speed increases are possible for crystals of orthorhombic and uniaxial symmetry. For those of you, who need to be reminded of the different cases of crystal symmetry, have a look at Fig. 12.2, which gives you an overview over the principal crystal symmetries. The figure starts with cubic crystals which we have already discussed in Section 6.7.

Cubic crystals are isotropic in the infrared spectral range and, therefore, all orientations of a cubic crystal are equivalent. This is different for all other crystal symmetries. The next lower symmetry category is formed by the optical uniaxial symmetry which means that there is one direction with high symmetry. If a wave travels along this axis, its speed is independent of polarization as it only experiences the ordinary index of refraction n_o which is given by the square root of the element of

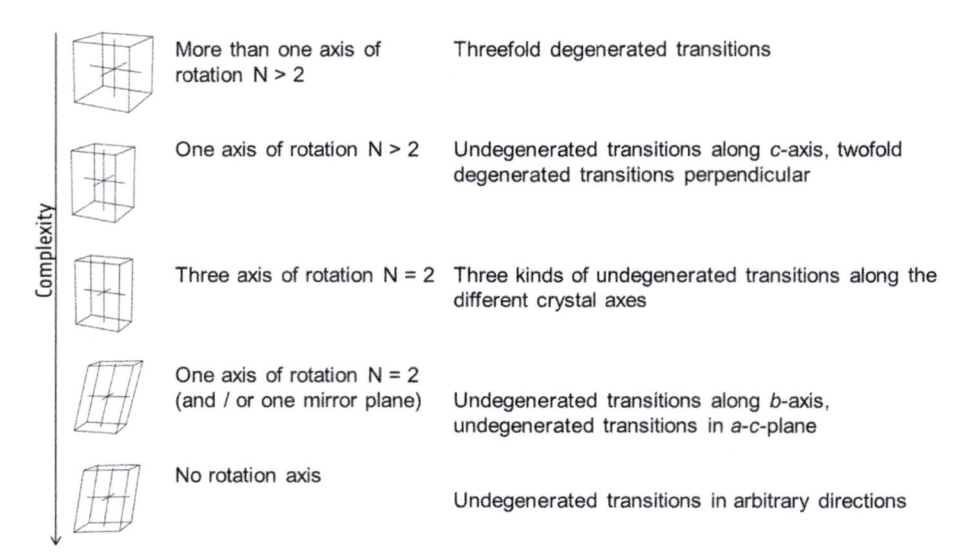

FIG. 12.2 Principal crystal symmetries and their properties.

the dielectric tensor that appears twice in its diagonal form $\varepsilon_a : n_o = \sqrt{\varepsilon_a}$. As a consequence, spectra that are taken with s-polarized light so that the polarization direction is perpendicular to the optical-(c-)axis are depending only on ε_a and can be quantitatively evaluated in the same way as spectra from cubic crystals. If then a second spectrum is taken so that the s-polarized incident light is parallel to the c-axis, the spectrum is only depending on ε_c and can again be evaluated like a spectrum from a cubic crystal. Such orientations are called principal orientations of a single crystal and the corresponding spectra are principal spectra of the crystal. For uniaxial crystals, all what is needed is a so-called a-cut, which has a surface which is parallel to the crystallographic c-axis. From such a cut both principal spectra can be gained if the orientation of the c-axis within the surface plane is known. Due to the symmetry in uniaxial crystals it is even possible to simplify the dielectric tensor somewhat for general orientations of the optical axis. This tensor is provided in Fig. 12.3, but we will also derive it in the following.

Such principal orientations are also possible in case of orthorhombic crystals. In case of such crystals, the recipe is the same as for obtaining ε_c in case of uniaxial crystals, i.e., a surface is needed, which is parallel to one of the crystallographic axes. If its orientation is known, again the corresponding principal spectrum can be obtained and evaluated (even if the orientation is not known, it can be determined by knowing that for proper orientation the cross terms r_{sp} and r_{ps} become zero).

In case of monoclinic crystals there is only one principal orientation along the b-axis. Therefore, only ε_b can be conventionally determined. The remaining 2×2 tensor can only be diagonalized in nonabsorbing spectral regions. Therefore, the determination of the full dielectric tensor function is more complex.

For triclinic crystals, unfortunately, there is no simplification possible by orienting the crystal in a special way. This is the reason why it took until 2014 until the first dispersion analysis of a triclinic crystal was successful and up to now (March 2024) only four of these have been carried out successfully.

12.8.1 Nonmagnetic ($\mu = 1$), dielectric anisotropic ($\varepsilon_{ij} = \varepsilon_{ji}$) material and normal incidence

The kind of simplification I discuss in the following is of interest for orthorhombic and monoclinic materials in general orientations as well as for triclinic crystals in general. In practice it seems that it is an example without value since measuring reflectance at normal incidence seems to be impossible at present—at least there is no accessory commercially available or custom-made of which I would know of. In the past, however, there has been at least one accessory the design of which and its employment was discussed in literature [12]. The authors of aforesaid reference were the first who have tried to determine the dielectric tensor function of a triclinic crystal and who have also published a paper about how to calculate the reflectance under the conditions mentioned previously [13]. As far as I know they were the first to treat this particular situation. Here, we will not follow their derivation, but simply use the $\mathbf{\Delta}$-matrix and simplify it accordingly.

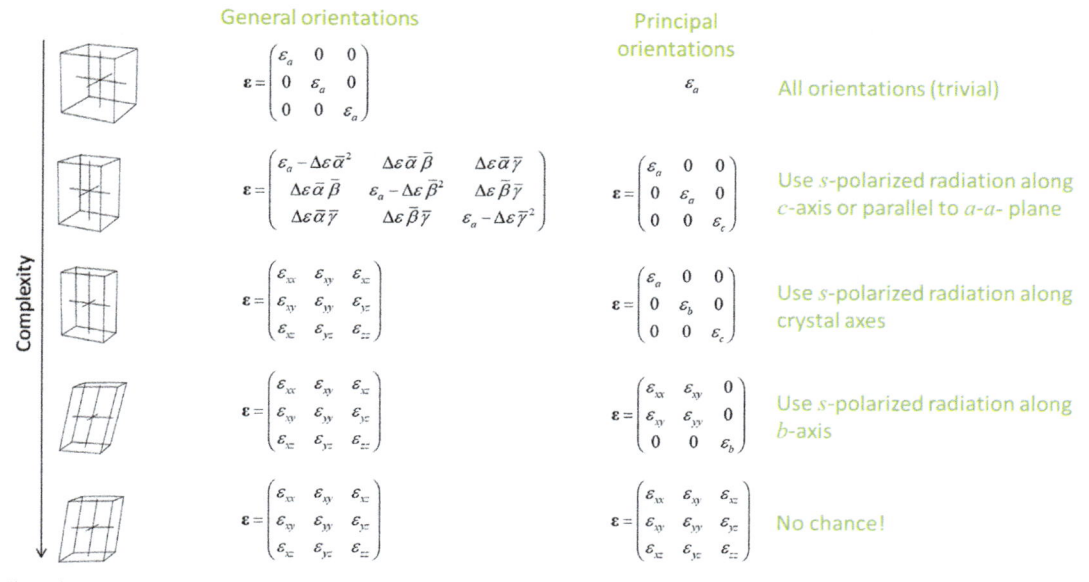

FIG. 12.3 Crystal symmetries, their dielectric tensors in general and principal orientations, and correspondingly optimized measurement conditions.

For normal incidence, the Dynamical Matrices are much easier and thus faster to compute. Since for normal incidence, the wavevector is $\mathbf{k}_i = k_0(0,0,1)^T$, the $\boldsymbol{\Delta}$-matrix (Eqs. 12.11–12.13) simplifies under the conditions ($\mu = 1$) and ($\varepsilon_{ij} = \varepsilon_{ji}$) to

$$\boldsymbol{\Delta} = \begin{pmatrix} 0 & 1 & 0 & 0 \\ \varepsilon_{XX} - \dfrac{\varepsilon_{XZ}^2}{\varepsilon_{ZZ}} & 0 & \varepsilon_{XY} - \dfrac{\varepsilon_{XZ}\varepsilon_{YZ}}{\varepsilon_{ZZ}} & 0 \\ 0 & 0 & 0 & 1 \\ \varepsilon_{XY} - \dfrac{\varepsilon_{XZ}\varepsilon_{YZ}}{\varepsilon_{ZZ}} & 0 & \varepsilon_{YY} - \dfrac{\varepsilon_{YZ}^2}{\varepsilon_{ZZ}} & 0 \end{pmatrix}. \qquad (12.50)$$

As a consequence, two of the coefficients of the Booker quartic become zero:

$$\begin{aligned} \alpha_1 &= 0 \\ \alpha_2 &= -\Delta_{21} - \Delta_{43} \\ \alpha_3 &= 0 \\ \alpha_4 &= \Delta_{21}\Delta_{43} - \Delta_{23}^2, \end{aligned} \qquad (12.51)$$

and the two others are drastically simplified. Correspondingly, the same is true for the eigenvalues γ_i, which are given by

$$\gamma_{\pm I} = \pm \frac{\sqrt{\Delta_{21} - \sqrt{4\Delta_{23}^2 + (\Delta_{21} - \Delta_{43})^2} + \Delta_{43}}}{\sqrt{2}}$$

$$\gamma_{\pm II} = \pm \frac{\sqrt{\Delta_{21} + \sqrt{4\Delta_{23}^2 + (\Delta_{21} - \Delta_{43})^2} + \Delta_{43}}}{\sqrt{2}}. \qquad (12.52)$$

In contrast to the general case (Eq. 12.17), the eigenvalues reveal their sign and keep it for every wavenumber/frequency. As a consequence, they do not need to be sorted. Therefore, it is not even necessary to compute them explicitly. The same is true for the polarization direction, two of which are

$$\boldsymbol{\Psi}_{1,2,3,i} = \begin{pmatrix} \Delta_{23} \\ \gamma_i \Delta_{23} \\ \gamma_i^2 - \Delta_{21} \\ \gamma_i^3 - \gamma_i \Delta_{21} \end{pmatrix}, \quad \boldsymbol{\Psi}_{2,3,4,i} = \begin{pmatrix} \gamma_i^3 - \gamma_i \Delta_{43} \\ \Delta_{23}^2 + \Delta_{21}(\gamma_i^2 - \Delta_{43}) \\ \gamma_i \Delta_{23} \\ \gamma_i^2 \Delta_{23} \end{pmatrix}. \qquad (12.53)$$

From these eigenpolarizations, the Dynamical Matrices can easily be calculated employing Eqs. (12.21), (12.22). If we assume that the second medium, i.e., the single crystal under investigation, is semiinfinite, its $\mathbf{D}$-matrix $\mathbf{D}_{cry}$ is given by

$$\mathbf{D}_{cry} = \begin{pmatrix} \gamma_{+I}^3 - \gamma_{+I}\Delta_{43} + \Delta_{23} & 0 & \gamma_{+II}^3 - \gamma_{+II}\Delta_{43} + \Delta_{23} & 0 \\ \Delta_{21}\gamma_{+I}^2 + \gamma_{+I}\Delta_{23} + \Delta_{23}^2 - \Delta_{21}\Delta_{43} & 0 & \Delta_{21}\gamma_{+II}^2 + \gamma_{+II}\Delta_{23} + \Delta_{23}^2 - \Delta_{21}\Delta_{43} & 0 \\ \gamma_{+I}^2 + \gamma_{+I}\Delta_{23} - \Delta_{21} & 0 & \gamma_{+II}^2 + \gamma_{+II}\Delta_{23} - \Delta_{21} & 0 \\ \gamma_{+I}^3 + \gamma_{+I}^2\Delta_{23} - \gamma_{+I}\Delta_{21} & 0 & \gamma_{+II}^3 + \gamma_{+II}^2\Delta_{23} - \gamma_{+II}\Delta_{21} & 0 \end{pmatrix}. \qquad (12.54)$$

This $\mathbf{D}$-matrix becomes only in rare cases singular, when all nondiagonal elements of the dielectric tensor are zero ($\Delta_{ij}=0$, $i \neq j$), which can only happen for uniaxial or orthorhombic crystals, or if $\varepsilon_{ZZ} = \frac{\varepsilon_{XZ}\varepsilon_{YZ}}{\varepsilon_{XY}}$. In both cases, $\mathbf{D}_{cry}$ has only four nonzero elements:

$$\mathbf{D}_{cry} = \begin{pmatrix} \dfrac{1}{\sqrt{\Delta_{21}}} & 0 & 0 & 0 \\ 1 & 0 & 0 & 0 \\ 0 & 0 & \dfrac{1}{\sqrt{\Delta_{43}}} & 0 \\ 0 & 0 & 1 & 0 \end{pmatrix}. \qquad (12.55)$$

If we assume further that the incidence medium is vacuum, then its **D**-matrix $\mathbf{D}_{inc}$ is given by

$$\mathbf{D}_{inc} = \begin{pmatrix} 1 & -1 & 0 & 0 \\ 1 & 1 & 0 & 0 \\ 0 & 0 & 1 & -1 \\ 0 & 0 & 1 & 1 \end{pmatrix}. \tag{12.56}$$

If we apply Eq. (12.38), $\tilde{\mathbf{M}} = \mathbf{D}_{inc}^{-1} \cdot \mathbf{D}_{cry}$ and use for $\mathbf{D}_{cry}$ Eq. (12.54), the explicit expression for $\tilde{\mathbf{M}}$ is

$$\tilde{\mathbf{M}} = \frac{1}{2} \begin{pmatrix} \gamma_{+I}^3 + \gamma_{+I}^2 \Delta_{21} + \Delta_{23} + \Delta_{23}^2 + & & \gamma_{+II}^3 + \gamma_{+II}^2 \Delta_{21} + \Delta_{23} + \Delta_{23}^2 + & \\ \gamma_{+I}(\Delta_{23} - \Delta_{43}) - \Delta_{21}\Delta_{43} & 0 & \gamma_{+II}(\Delta_{23} - \Delta_{43}) - \Delta_{21}\Delta_{43} & 0 \\ \gamma_{+I}^3 - \gamma_{+I}^2 \Delta_{21} + \Delta_{23} - \Delta_{23}^2 - & & \gamma_{+II}^3 - \gamma_{+II}^2 \Delta_{21} + \Delta_{23} - \Delta_{23}^2 - & \\ \gamma_{+I}(\Delta_{23} + \Delta_{43}) + \Delta_{21}\Delta_{43} & 0 & \gamma_{+II}(\Delta_{23} + \Delta_{43}) + \Delta_{21}\Delta_{43} & 0 \\ (\gamma_{+I} + 1)(\gamma_{+I}(\gamma_{+I} + \Delta_{23}) - \Delta_{21}) & 0 & (\gamma_{+II} + 1)(\gamma_{+II}(\gamma_{+II} + \Delta_{23}) - \Delta_{21}) & 0 \\ (\gamma_{+I} - 1)(\gamma_{+I}(\gamma_{+I} + \Delta_{23}) - \Delta_{21}) & 0 & (\gamma_{+II} - 1)(\gamma_{+II}(\gamma_{+II} + \Delta_{23}) - \Delta_{21}) & 0 \end{pmatrix}. \tag{12.57}$$

All that is required additionally for the calculation of the reflection coefficients is employing Eq. (12.47).

12.8.2 Nonmagnetic ($\mu = 1$), dielectric ($\varepsilon_{ij} = \varepsilon_{ji}$) monoclinic material—a-c-plane

While for triclinic crystals only four dispersion analyses have been carried out successfully in the infrared spectral range [14–17], there have been accomplished many more for monoclinic crystals. Whereas nearly all spectra on which the analyses are based, have been recorded at nonnormal incidence, no dispersion analysis before 2011 used the proper formulas for this case—all analyses were based on the formulas provided by Belousov and Pavinich [18], albeit there is an earlier work by Koch, Otto, and Kliewer where this problem had already been solved [19]. Although the errors of assuming normal incidence may be tolerable for angles of incidence below some limit (looking at Fig. 3.1, I would place this limit at about 8°), it seems that the comparably low increase of complexity of the formulas for nonnormal incidence allows to use these formulas once $\alpha_i \neq 0$ (for the computing speed increase I determined a factor of 4.5 compared to the general formulas, which increases only to 4.6 if additionally normal incidence is assumed). It is strange that it seems that nobody had investigated these formulas before I did this in 2011 [20].

If we assume that the plane of incidence is parallel to the Y-axis (for s-polarized light, the polarization direction is parallel to the X-axis), then the $\boldsymbol{\Delta}$-matrix for a general orientation is given by

$$\boldsymbol{\Delta} = \begin{pmatrix} 0 & 1 & 0 & 0 \\ -k_Y^2 + \varepsilon_{XX} - \dfrac{\varepsilon_{XZ}^2}{\varepsilon_{ZZ}} & 0 & \varepsilon_{XY} - \dfrac{\varepsilon_{XZ}\varepsilon_{YZ}}{\varepsilon_{ZZ}} & -\dfrac{k_Y \varepsilon_{XZ}}{\varepsilon_{ZZ}} \\ -\dfrac{k_Y \varepsilon_{XZ}}{\varepsilon_{ZZ}} & 0 & -\dfrac{k_Y \varepsilon_{YZ}}{\varepsilon_{ZZ}} & 1 - \dfrac{k_Y^2}{\varepsilon_{ZZ}} \\ \varepsilon_{YX} - \dfrac{\varepsilon_{YZ}\varepsilon_{XZ}}{\varepsilon_{ZZ}} & 0 & \varepsilon_{YY} - \dfrac{\varepsilon_{YZ}^2}{\varepsilon_{ZZ}} & -\dfrac{k_Y \varepsilon_{YZ}}{\varepsilon_{ZZ}} \end{pmatrix}. \tag{12.58}$$

For the following, we assume a block-diagonalized dielectric tensor of a monoclinic crystal

$$\boldsymbol{\varepsilon} = \begin{pmatrix} \varepsilon_{XX} & \varepsilon_{XY} & 0 \\ \varepsilon_{XY} & \varepsilon_{YY} & 0 \\ 0 & 0 & \varepsilon_{ZZ} \end{pmatrix}, \tag{12.59}$$

where $\varepsilon_{ZZ} = \varepsilon_b$ (the interface is the a-c crystal face).

Setting the angle of incidence zero leads to $k_Y = 0$ and Eq. (12.50). On the other hand, using that $\varepsilon_{XZ} = \varepsilon_{ZX} = \varepsilon_{YZ} = \varepsilon_{ZY} = 0$ Eq. (12.59), simplifies the $\boldsymbol{\Delta}$-matrix to

$$\Delta = \begin{pmatrix} 0 & 1 & 0 & 0 \\ -k_Y^2 + \varepsilon_{XX} & 0 & \varepsilon_{XY} & 0 \\ 0 & 0 & 0 & 1 - \dfrac{k_Y^2}{\varepsilon_{ZZ}} \\ \varepsilon_{XY} & 0 & \varepsilon_{YY} & 0 \end{pmatrix}. \tag{12.60}$$

The coefficients of the Booker quartic are correspondingly found to be

$$\begin{aligned}
\alpha_1 &= 0 \\
\alpha_2 &= \frac{1}{\varepsilon_{ZZ}}\left(k_Y^2(\varepsilon_{YY} + \varepsilon_{ZZ}) - \varepsilon_{ZZ}(\varepsilon_{XX} + \varepsilon_{YY})\right) \\
\alpha_3 &= 0 \\
\alpha_4 &= \frac{1}{\varepsilon_{ZZ}}\left(k_Y^4 \varepsilon_{YY} + k_Y^2(\varepsilon_{XY}^2 - \varepsilon_{YY}(\varepsilon_{XX} + \varepsilon_{ZZ}))\right) - \varepsilon_{XY}^2 \varepsilon_{ZZ} + \varepsilon_{XX}\varepsilon_{YY}\varepsilon_{ZZ}.
\end{aligned} \tag{12.61}$$

Eqs. (12.17), (12.18) are simplified correspondingly to

$$\begin{aligned}
K_1 &= 2\alpha_2^3 - 72\alpha_2\alpha_4 \\
K_2 &= \alpha_2^2 + 12\alpha_4 \\
K_3 &= \left(K_1 + \sqrt{K_1^2 - 4K_2^3}\right)^{1/3} \\
K_4 &= -8\alpha_2 + \frac{4 \times 2^{1/3} K_2}{K_3} + 2 \times 2^{2/3} K_3 \\
K_5 &= -8\alpha_2 - \frac{2 \times 2^{1/3} K_2}{K_3} - 2^{2/3} K_3 \\
K_6 &= 0,
\end{aligned} \tag{12.62}$$

and

$$\begin{aligned}
\gamma_1 &= -\frac{1}{12}\left(\sqrt{3K_4} + \sqrt{6K_5}\right) \\
\gamma_2 &= -\frac{1}{12}\left(\sqrt{3K_4} - \sqrt{6K_5}\right) \\
\gamma_3 &= -\frac{1}{12}\left(-\sqrt{3K_4} + \sqrt{6K_5}\right) \\
\gamma_4 &= -\frac{1}{12}\left(-\sqrt{3K_4} - \sqrt{6K_5}\right).
\end{aligned} \tag{12.63}$$

At first view, it looks like that despite the simplification it is still not possible to determine the signs of the eigenvalues. However, after some algebraic manipulation the eigenvalues can be recast into the following form:

$$\begin{aligned}
\gamma_1 &= \sqrt{-\frac{1}{2\varepsilon_{ZZ}}\left(K_1 + \sqrt{K_1^2 + K_2}\right)} \\
\gamma_2 &= -\sqrt{-\frac{1}{2\varepsilon_{ZZ}}\left(K_1 + \sqrt{K_1^2 + K_2}\right)} \\
\gamma_3 &= \sqrt{\frac{1}{2\varepsilon_{ZZ}}\left(-K_1 + \sqrt{K_1^2 + K_2}\right)} \\
\gamma_4 &= -\sqrt{\frac{1}{2\varepsilon_{ZZ}}\left(-K_1 + \sqrt{K_1^2 + K_2}\right)},
\end{aligned} \tag{12.64}$$

wherein K_1 and K_2 are altered and now given by

$$\begin{aligned}
K_1 &= -\varepsilon_{ZZ}(\varepsilon_{XX} + \varepsilon_{YY}) + k_Y^2(\varepsilon_{YY} + \varepsilon_{ZZ}) \\
K_2 &= -4\left(\varepsilon_{XY}^2 + \varepsilon_{YY}(k_Y^2 - \varepsilon_{XX})\right)\left(k_Y^2 - \varepsilon_{ZZ}\right)\varepsilon_{ZZ}.
\end{aligned} \tag{12.65}$$

Due to the form of Eq. (12.65) it is obvious that γ_1 and γ_3 represent the solutions for the forward-traveling waves whereas γ_2 and γ_4 belong to the backward-traveling waves. In order to determine the polarizations, we have to compute the eigenvectors, which we find as usual by Cramer's rule:

$$\Psi_{1,2,3,i} = \begin{pmatrix} \varepsilon_{XY}\left(1 - \dfrac{k_Y^2}{\varepsilon_{ZZ}}\right) \\ \varepsilon_{XY}\left(1 - \dfrac{k_Y^2}{\varepsilon_{ZZ}}\right)\gamma_i \\ \left(1 - \dfrac{k_Y^2}{\varepsilon_{ZZ}}\right)(\gamma_i^2 - (\varepsilon_{XX} - k_Y^2)) \\ \gamma_i^3 - \gamma_i(\varepsilon_{XX} - k_Y^2) \end{pmatrix}. \quad (12.66)$$

In this case, we just determine one of the four possible eigenvectors since due to the form of $\Psi_{1,2,3,i}$, a linear combination with another eigenvector is not necessary as all four are always different from zero as ε_{XY} is nonzero. From there it is easy to construct the Dynamical Matrix and the $\tilde{M}$-matrix as well as to calculate reflectance.

It seems that it should be possible to derive the normal incidence case directly from Eq. (12.64) in combination with Eq. (12.65), but it is not that simple. We have to start with the simplified Δ-matrix which results from setting $k_Y = 0$ in Eq. (12.60):

$$\Delta = \begin{pmatrix} 0 & 1 & 0 & 0 \\ \varepsilon_{XX} & 0 & \varepsilon_{XY} & 0 \\ 0 & 0 & 0 & 1 \\ \varepsilon_{XY} & 0 & \varepsilon_{YY} & 0 \end{pmatrix}. \quad (12.67)$$

The corresponding eigenvalues are then

$$\gamma_1 = \frac{\sqrt{\varepsilon_{XX} + \varepsilon_{YY} - \sqrt{4\varepsilon_{XY}^2 + (\varepsilon_{XX} - \varepsilon_{YY})^2}}}{\sqrt{2}}$$

$$\gamma_3 = \frac{\sqrt{\varepsilon_{XX} + \varepsilon_{YY} + \sqrt{4\varepsilon_{XY}^2 + (\varepsilon_{XX} - \varepsilon_{YY})^2}}}{\sqrt{2}}, \quad (12.68)$$

and do no longer depend on the perpendicular modes that constitute ε_{ZZ} (to be precise, $\varepsilon_{ZZ} - 1$ since ε_{ZZ} would be unity independent of frequency without modes that have their transition moment along ε_{ZZ}). Eq. (12.68) only provides the eigenvalues of the forward-traveling waves. Those of the backward-traveling waves are simply the negative values of γ_1 and γ_3. The corresponding eigenvectors are found by setting $k_Y = 0$ in Eq. (12.66):

$$\Psi_{1,2,3,i} = \begin{pmatrix} \varepsilon_{XY} \\ \varepsilon_{XY}\gamma_i \\ \gamma_i^2 - \varepsilon_{XX} \\ \gamma_i^3 - \gamma_i\varepsilon_{XX} \end{pmatrix}. \quad (12.69)$$

Again, one solution is sufficient since $\varepsilon_{XY} \neq 0$ always holds for a monoclinic crystal in contrast to crystals of higher symmetry where $\varepsilon_{XY} = 0$ for principal orientations.

12.8.3 Nonmagnetic ($\mu = 1$), dielectric uniaxial ($\varepsilon_{ij} = \varepsilon_{ji}$, $\varepsilon_a = \varepsilon_b$) material

For this case, the Δ-matrix is the same as in Eq. (12.58) for the monoclinic material as long as we do not discuss special orientations. Ok, what would be the point of discussing this special case if there is no simplification? Actually, because of the condition $\varepsilon_a = \varepsilon_b$ (ε_a and ε_c are the principal dielectric functions based on optical transitions which have their transition moments perpendicular and parallel to the optical axis), the formulas for the eigenvalues are strongly simplified, and the same is true also for the eigenvectors, so strong that relatively simple equations can be derived for the reflectance. As a consequence of $\varepsilon_a = \varepsilon_b$, of the general six components of a symmetric dielectric tensor, only four are independent. This can be shown if these six components are written in the following form:

$$\begin{aligned}
\varepsilon_{XX} &= \varepsilon_a - \Delta\varepsilon \sin^2\varphi \sin^2\theta = \varepsilon_a - \Delta\varepsilon\bar{\alpha}^2 \\
\varepsilon_{YY} &= \varepsilon_a - \Delta\varepsilon \cos^2\varphi \sin^2\theta = \varepsilon_a - \Delta\varepsilon\bar{\beta}^2 \\
\varepsilon_{ZZ} &= \varepsilon_a - \Delta\varepsilon \cos^2\theta = \varepsilon_a - \Delta\varepsilon\bar{\gamma}^2 \\
\varepsilon_{XY} &= \Delta\varepsilon \sin\varphi \cos\varphi \sin^2\theta = \Delta\varepsilon\bar{\alpha}\bar{\beta} \\
\varepsilon_{XZ} &= -\Delta\varepsilon \sin\varphi \sin\theta \cos\theta = \Delta\varepsilon\bar{\alpha}\bar{\gamma} \\
\varepsilon_{YZ} &= \Delta\varepsilon \cos\varphi \sin\theta \cos\theta = \Delta\varepsilon\bar{\beta}\bar{\gamma},
\end{aligned} \qquad (12.70)$$

where $\Delta\varepsilon = \varepsilon_a - \varepsilon_c$. $\bar{\alpha}, \bar{\beta}$, and $\bar{\gamma}$ are the direction cosines between the Z- and the optical axis (note that $\bar{\alpha}^2 + \bar{\beta}^2 + \bar{\gamma}^2 = 1$). Equivalently, the orientation of the optical axis can be specified by the two angles φ and θ of the spherical coordinate system (cf. Fig. 12.4). For the moment we stick with the direction cosines, but later we will change over to the latter description of orientation.

The equation to determine the eigenvalues is now given by [21]:

$$\det \begin{vmatrix} \varepsilon_a - \Delta\varepsilon\bar{\alpha}^2 - k_Y^2 - \gamma^2 & \Delta\varepsilon\bar{\alpha}\bar{\beta} & \Delta\varepsilon\bar{\alpha}\bar{\gamma} \\ \Delta\varepsilon\bar{\alpha}\bar{\beta} & \varepsilon_a - \Delta\varepsilon\bar{\beta}^2 - \gamma^2 & \Delta\varepsilon\bar{\beta}\bar{\gamma} + k_Y\gamma \\ \Delta\varepsilon\bar{\alpha}\bar{\gamma} & \Delta\varepsilon\bar{\beta}\bar{\gamma} + k_Y\gamma & \varepsilon_a - \Delta\varepsilon\bar{\gamma}^2 - k_Y^2 \end{vmatrix} = 0. \qquad (12.71)$$

The Booker quartic seems to be on the first view still comparably complex, but with help of the condition $\bar{\alpha}^2 + \bar{\beta}^2 + \bar{\gamma}^2 = 1$, this complexity is drastically reduced:

$$\begin{aligned}
\left(\gamma^2 - (\varepsilon_a - k_Y^2)\right)(\beta_0 \gamma^2 + \beta_1 \gamma + \beta_2) &= 0 \\
\beta_0 &= \varepsilon_{ZZ} = \varepsilon_a \sin^2\theta + \varepsilon_c \cos^2\theta \\
\beta_1 &= 2k_Y \varepsilon_{YZ} \\
\beta_2 &= k_Y^2 \varepsilon_{YY} - \varepsilon_a \varepsilon_c.
\end{aligned} \qquad (12.72)$$

From Eq. (12.72), to be more precise, from the first factor, it is immediately obvious that one set of eigenvalues is given by $\gamma_{\pm I} = \pm\sqrt{\varepsilon_a - k_Y^2}$. Accordingly, this set is independent of the orientation of the optical axis. Thus, this solution belongs to the so-called ordinary wave. This ordinary wave is correspondingly influenced only by ε_a, like any wave in a cubic crystal. The remaining quadratic equation adds two further solutions, which are $\gamma_{\pm II} = \left(\beta_1 \pm \sqrt{\beta_1^2 - 4\beta_0\beta_2}\right)/2\beta_0$. These eigenvalues belong to the extraordinary wave. The solutions in this form have first been derived by Birecki and Kahn [22].

The eigenvectors seem to be comparably complex as well, but they already contain the explicit eigenvalues:

$$\Psi_{1,2,3,i} = \begin{pmatrix} -\gamma_i \varepsilon_{XZ} k_Y + \varepsilon_{XY}(\varepsilon_a - k_Y^2) \\ \gamma_i\left(-\gamma_i \varepsilon_{XZ} k_Y + \varepsilon_{XY}(\varepsilon_a - k_Y^2)\right) \\ \gamma_i^2(\varepsilon_{ZZ} - k_Y^2) + (\varepsilon_a - k_Y^2)(\varepsilon_a + k_Y^2 - \varepsilon_{XX} - \varepsilon_{ZZ}) \\ \gamma_i(\varepsilon_a^2 - (\varepsilon_a - k_Y^2 - \gamma_i^2)\varepsilon_{ZZ} - \varepsilon_a\varepsilon_{XX}) - \varepsilon_{YZ}k_Y(\varepsilon_a - k_Y^2 - \gamma_i^2) \end{pmatrix}$$

$$\Psi_{2,3,4,i} = \begin{pmatrix} -\gamma_i(\varepsilon_{YY}(\varepsilon_a - k_Y^2) - 2\varepsilon_{YZ}k_Y\gamma_i + (\varepsilon_a - \gamma_i^2)\varepsilon_{ZZ} - \varepsilon_a^2) \\ \gamma_i^2(\varepsilon_a(\varepsilon_{XX} - \varepsilon_a) + \varepsilon_{ZZ}(\varepsilon_a - k_Y^2)) + 2\gamma_i k_Y \varepsilon_{YZ}(\varepsilon_a - k_Y^2) + (k_Y^2 \varepsilon_{YY} - \varepsilon_a\varepsilon_c)(\varepsilon_a - k_Y^2) \\ \gamma_i(-\varepsilon_{XZ}k_Y\gamma_i + \varepsilon_{XY}(\varepsilon_a - k_Y^2)) \\ \gamma_i \varepsilon_a(\varepsilon_{XY}\gamma_i - \varepsilon_{XZ}k_Y) \end{pmatrix}. \qquad (12.73)$$

Under normal incidence, a further drastic reduction of the complexity results. If $k_Y = 0$, then the eigenvalues are given by $\gamma_{\pm I} = \pm\sqrt{\varepsilon_a}$ and $\gamma_{\pm II} = \pm\sqrt{\frac{\varepsilon_a \varepsilon_c}{\varepsilon_{ZZ}}}$. It is then preferable to use the polarization vectors $\mathbf{p}_{2,3,i}$. These are found to be

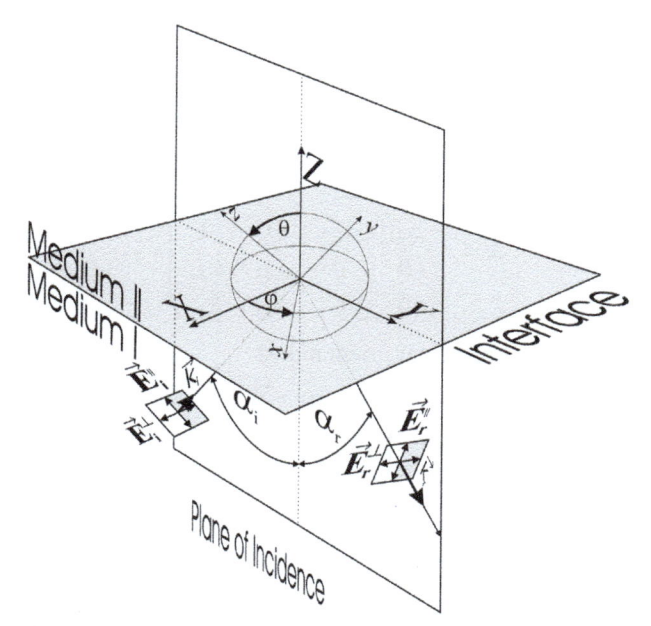

FIG. 12.4 Optical model and coordinate system which are used throughout this section. The two angles φ and θ denote the orientation of the optical axis which is assumed to be oriented parallel to the z-axis of the internal coordinate system of the uniaxial crystal.

$$\mathbf{p}_{2,3,i} = \begin{pmatrix} \bar{\gamma}(((1-\bar{\alpha}^2)\Delta\varepsilon - \varepsilon_a)\varepsilon_a + \gamma_i^2(\varepsilon_a - \bar{\gamma}^2\Delta\varepsilon)) \\ \bar{\alpha}\bar{\beta}\bar{\gamma}\Delta\varepsilon\varepsilon_a \\ \bar{\alpha}\bar{\gamma}^2\Delta\varepsilon(\gamma_i^2 - \varepsilon_a) \end{pmatrix}, \tag{12.74}$$

where I have switched back to the direction cosines. These allow a very compact and simple description of the polarization if we put in the eigenvalues and simplify:

$$\mathbf{p}_{2,3,I} = \begin{pmatrix} \bar{\beta} \\ \bar{\alpha} \\ 0 \end{pmatrix}, \quad \mathbf{p}_{2,3,II} = \begin{pmatrix} -\bar{\alpha} \\ \bar{\beta} \\ \bar{\gamma}(\varepsilon_c\varepsilon_{ZZ}^{-1} - 1) \end{pmatrix}. \tag{12.75}$$

Eq. (12.75) demonstrates that the X- and Y-components are oriented perpendicular to the wavevector and to each other (note that the coordinates of the optical axis are $(\bar{\alpha}, \bar{\beta}, \bar{\gamma})^T$). What is remarkable, though, is that the Z-component of the polarization of the extraordinary wave does not vanish in general, even for normal incidence, except if the optical axis is oriented perpendicularly to the interface because then $\varepsilon_c\varepsilon_{ZZ}^{-1} = 1$. This means that the wave has generally a mixed transversal-longitudinal character, something which is important and we will investigate in further detail later.

Using the 4 × 4 matrix formalism and setting $\sin\varphi = -\bar{\alpha}(\bar{\alpha}^2 + \bar{\beta}^2)^{-1}$ and $\cos\varphi = \bar{\beta}(\bar{\alpha}^2 + \bar{\beta}^2)^{-1}$ one finally arrives at the following formulas for the reflection coefficients [23]:

$$\begin{aligned} r_{XX} &= r_1 \cos^2\varphi + r_2 \sin^2\varphi \\ r_{XY} &= r_{YX} = (r_1 - r_2)\cos\varphi \sin\varphi \\ r_{YY} &= r_1 \sin^2\varphi + r_2 \cos^2\varphi \\ r_i &= \frac{n_i - 1}{n_i + 1}, \quad n_1 = n_{ord} = \sqrt{\varepsilon_a}, \quad n_2 = \sqrt{\frac{\varepsilon_a\varepsilon_c}{\varepsilon_a \sin^2\theta + \varepsilon_c \cos^2\theta}}. \end{aligned} \tag{12.76}$$

If no analyzer is used, a very simple formula for the reflectance is obtained (polarizer before the uniaxial medium):

$$R_X = R_1(n_1)\cos^2\varphi + R_2(n_2)\sin^2\varphi$$
$$R_i(n_i) = |r_i(n_i)|^2 = \left|\frac{n_i - 1}{n_i + 1}\right|^2. \tag{12.77}$$

12.8.4 Nonmagnetic ($\mu = 1$), dielectric ($\varepsilon_{ij} = \varepsilon_{ji}$) uniaxial or orthorhombic material with principal orientations

This case has actually already been treated, at least partly, in Section 12.6.1. As a consequence of the principal orientations, the s- and p-waves are decoupled. The general form of the Δ-matrix is as follows:

$$\Delta = \begin{pmatrix} \Delta_{11} & \Delta_{12} & 0 & 0 \\ \Delta_{21} & \Delta_{22} & 0 & 0 \\ 0 & 0 & \Delta_{33} & \Delta_{34} \\ 0 & 0 & \Delta_{43} & \Delta_{44} \end{pmatrix}. \tag{12.78}$$

The dielectric tensor is in diagonal form for principal orientations:

$$\boldsymbol{\varepsilon} = \begin{pmatrix} \varepsilon_X & 0 & 0 \\ 0 & \varepsilon_Y & 0 \\ 0 & 0 & \varepsilon_Z \end{pmatrix}. \tag{12.79}$$

If we use Eq. (12.58),

$$\Delta = \begin{pmatrix} 0 & 1 & 0 & 0 \\ -k_Y^2 + \varepsilon_{XX} - \dfrac{\varepsilon_{XZ}^2}{\varepsilon_{ZZ}} & 0 & \varepsilon_{XY} - \dfrac{\varepsilon_{XZ}\varepsilon_{YZ}}{\varepsilon_{ZZ}} & -\dfrac{k_Y \varepsilon_{XZ}}{\varepsilon_{ZZ}} \\ -\dfrac{k_Y \varepsilon_{XZ}}{\varepsilon_{ZZ}} & 0 & -\dfrac{k_Y \varepsilon_{YZ}}{\varepsilon_{ZZ}} & 1 - \dfrac{k_Y^2}{\varepsilon_{ZZ}} \\ \varepsilon_{YX} - \dfrac{\varepsilon_{YZ}\varepsilon_{XZ}}{\varepsilon_{ZZ}} & 0 & \varepsilon_{YY} - \dfrac{\varepsilon_{YZ}^2}{\varepsilon_{ZZ}} & -\dfrac{k_Y \varepsilon_{YZ}}{\varepsilon_{ZZ}} \end{pmatrix}, \tag{12.80}$$

and we employ that $\varepsilon_{XY} = \varepsilon_{YX} = \varepsilon_{XZ} = \varepsilon_{ZX} = \varepsilon_{YZ} = \varepsilon_{ZY} = 0$ we obtain

$$\Delta = \begin{pmatrix} 0 & 1 & 0 & 0 \\ -k_Y^2 + \varepsilon_X & 0 & 0 & 0 \\ 0 & 0 & 0 & 1 - \dfrac{k_Y^2}{\varepsilon_Z} \\ 0 & 0 & \varepsilon_Y & 0 \end{pmatrix}, \tag{12.81}$$

which is not only block-diagonalized, but all diagonal elements are zero. The eigenvalues therefore are simply

$$\gamma_{\pm I} = \pm\sqrt{\varepsilon_X - k_Y^2}$$
$$\gamma_{\pm II} = \pm\sqrt{(\varepsilon_Y \varepsilon_Z^{-1})(\varepsilon_Z - k_Y^2)}, \tag{12.82}$$

and the eigenvectors

$$\boldsymbol{\Psi}_{1,2,3} = \begin{pmatrix} (\varepsilon_X - k_Y^2)^{-\frac{1}{2}} \\ 1 \\ 0 \\ 0 \end{pmatrix}, \quad \boldsymbol{\Psi}_{2,3,4} = \begin{pmatrix} 0 \\ 0 \\ \left(1 - \dfrac{k_Y^2}{\varepsilon_Z}\right)^{\frac{1}{2}} \\ 1 \end{pmatrix}. \tag{12.83}$$

In this case, we find for the reflectance of crystals [24,25]:

$$R_s = \left| \frac{\cos\alpha - (\varepsilon_X - \sin^2\alpha)^{\frac{1}{2}}}{\cos\alpha + (\varepsilon_X - \sin^2\alpha)^{\frac{1}{2}}} \right|^2$$

$$R_p = \left| \frac{(\varepsilon_Y \varepsilon_Z)^{\frac{1}{2}} \cos\alpha - (\varepsilon_Z - \sin^2\alpha)^{\frac{1}{2}}}{(\widehat{\varepsilon}_Y \widehat{\varepsilon}_Z)^{\frac{1}{2}} \cos\alpha + (\varepsilon_Z - \sin^2\alpha)^{\frac{1}{2}}} \right|^2. \tag{12.84}$$

12.8.5 Nonmagnetic ($\mu = 1$), biisotropic medium

Even though a biisotropic medium is a scalar medium with $\varepsilon_{jj}=\varepsilon$ and $\varepsilon_{ji}=0$, I find it worth to discuss in this section since it is chiral with $\rho \neq \rho' \neq 0$ (ρ and ρ' are also scalars) and, thus, a 4×4 matrix formalism is needed since the polarization of the incoming light is changed upon transmission. Berreman originally treated this case, but assumed that $\rho'=0$. This agreed with Drude's assumption of what the constitutive relations for chiral media should look like [26]. In fact, the form of the constitutive relations is still under debate, but the two forms that are the ones most favorable, show only differences that are too small to experimentally decide the issue (see, e.g. [27]). In the following we use the Drude-Condon relation (and not the Drude-Born-Fedorov relations), as this has been done in Ref. [28]. But, in contrast to the latter, we keep the plane of incidence the Y-Z-plane and the time dependence as $\exp(-i\omega t)$. Therefore, $\rho = -ig$ and $\rho' = ig$, where g is the complex chirality admittance function.

Under these assumptions, the general form of the Δ-matrix is given by

$$\Delta = \begin{pmatrix} 0 & 1 & ig & 0 \\ \varepsilon - \dfrac{\varepsilon k_Y^2}{\varepsilon - g^2} & 0 & 0 & ig\left(1 + \dfrac{\varepsilon k_Y^2}{\varepsilon - g^2}\right) \\ -ig\left(1 + \dfrac{\varepsilon k_Y^2}{\varepsilon - g^2}\right) & 0 & 0 & 1 - \dfrac{\varepsilon k_Y^2}{\varepsilon - g^2} \\ 0 & -ig & \varepsilon & 0 \end{pmatrix}. \tag{12.85}$$

The eigenvalues are

$$\gamma_{\pm I} = \pm\sqrt{g^2(1+x) + \varepsilon - \frac{1}{2}x(1+\varepsilon) - \frac{1}{2}\sqrt{x^2(\varepsilon-1)^2 + 8g^2((2+x)\varepsilon - x)}}$$
$$\gamma_{\pm II} = \pm\sqrt{g^2(1+x) + \varepsilon - \frac{1}{2}x(1+\varepsilon) + \frac{1}{2}\sqrt{x^2(\varepsilon-1)^2 + 8g^2((2+x)\varepsilon - x)}}, \tag{12.86}$$

with $x = \varepsilon k_Y^2/(\varepsilon - g^2)$ and the corresponding eigenvectors

$$\Psi_{\gamma_{\pm I}} = \begin{pmatrix} -\dfrac{i(4g^2 - p + x(\varepsilon-1))}{g(p + x - (4+x)\varepsilon)} \\ \dfrac{i(p + x - x\varepsilon)\gamma_{\pm I}}{g(p + x - (4+x)\varepsilon)} \\ -\dfrac{4\gamma_{\pm I}}{p + x - (4+x)\varepsilon} \\ 1 \end{pmatrix}, \quad \Psi_{\gamma_{\pm II}} = \begin{pmatrix} \dfrac{i(4g^2 + p + x(\varepsilon-1))}{g(p - x + (4+x)\varepsilon)} \\ \dfrac{i(p - x + x\varepsilon)\gamma_{\pm II}}{g(p - x + (4+x)\varepsilon)} \\ \dfrac{4\gamma_{\pm II}}{p - x + (4+x)\varepsilon} \\ 1 \end{pmatrix}, \tag{12.87}$$

using $p = \sqrt{x^2(\varepsilon-1)^2 + 8g^2(x(\varepsilon-1) + 2\varepsilon)}$. Usually, experiments are carried out in transmittance under zero angle of incidence which considerably simplifies the situation further. Accordingly, the Δ-matrix is given by

$$\Delta = \begin{pmatrix} 0 & 1 & ig & 0 \\ \varepsilon & 0 & 0 & ig \\ -ig & 0 & 0 & 1 \\ 0 & -ig & \varepsilon & 0 \end{pmatrix}. \tag{12.88}$$

The eigenvalues are

$$\gamma_{\pm I} = \pm\sqrt{\varepsilon} + g$$
$$\gamma_{\pm II} = \pm\sqrt{\varepsilon} - g, \tag{12.89}$$

and the corresponding eigenvectors

$$\Psi_1 = \begin{pmatrix} \dfrac{i}{\sqrt{\varepsilon}} \\ -i \\ 1 \\ -\dfrac{1}{\sqrt{\varepsilon}} \\ 1 \end{pmatrix}, \quad \Psi_2 = \begin{pmatrix} -\dfrac{i}{\sqrt{\varepsilon}} \\ i \\ 1 \\ -\dfrac{1}{\sqrt{\varepsilon}} \\ 1 \end{pmatrix}, \quad \Psi_3 = \begin{pmatrix} -\dfrac{i}{\sqrt{\varepsilon}} \\ -i \\ 1 \\ \dfrac{1}{\sqrt{\varepsilon}} \\ 1 \end{pmatrix}, \quad \Psi_4 = \begin{pmatrix} \dfrac{i}{\sqrt{\varepsilon}} \\ i \\ 1 \\ \dfrac{1}{\sqrt{\varepsilon}} \\ 1 \end{pmatrix}. \tag{12.90}$$

The difference between γ_{+I} and γ_{+II} is thus $2g$. On the other hand, the eigenvectors are independent of the chirality admittance function. This means that for normal incidence, reflectance is not influenced by chirality since the $\mathbf{D_\Psi}$ matrices do not contain g, in contrast to transmittance which is depending on the propagation matrix. This would be different if I would have chosen to base the formalism on the Drude-Born-Fedorov relation. As mentioned previously, however, the difference would have been too small to measure because it is quadratic in g.

Since one is usually more interested in calculating the reflectance or transmittance for circularly polarized light than for linearly polarized light, we have to convert the reflection coefficients (Eq. (12.47)) correspondingly. For this conversion, I provide the one given by Lekner [27]:

$$\begin{aligned}
r_{++} &= \frac{1}{2}(r_{ss} - r_{pp}) - \frac{1}{2}i(r_{sp} + r_{ps}), & t_{++} &= \frac{1}{2}(t_{pp} + t_{ss}) + \frac{1}{2}i(t_{sp} - t_{ps}), \\
r_{+-} &= -\frac{1}{2}(r_{ss} + r_{pp}) - \frac{1}{2}i(r_{sp} - r_{ps}), & t_{+-} &= \frac{1}{2}(t_{pp} - t_{ss}) + \frac{1}{2}i(t_{sp} + t_{ps}), \\
r_{-+} &= -\frac{1}{2}(r_{ss} + r_{pp}) + \frac{1}{2}i(r_{sp} - r_{ps}), & t_{-+} &= \frac{1}{2}(t_{pp} - t_{ss}) - \frac{1}{2}i(t_{sp} + t_{ps}), \\
r_{--} &= \frac{1}{2}(r_{ss} - r_{pp}) + \frac{1}{2}i(r_{sp} + r_{ps}), & t_{--} &= \frac{1}{2}(t_{pp} + t_{ss}) - \frac{1}{2}i(t_{sp} - t_{ps}).
\end{aligned} \qquad (12.91)$$

Note that I adopt Lekner's way of speaking of positive and negative helicity to avoid the terminology of left and right circular polarized light for which two opposite conventions are in use, one focusing on the light source and the other on the observer. We will fix this issue once we are discussing vibrational circular dichroism in Chapter 16. Analogously, if only a circular polarizer and no analyzer is used, the reflection and transmission coefficients r_i and t_i ($i = +, -$) are given by

$$r_i = r_{ii} + r_{ij}, \quad t_i = t_{ii} + t_{ij} \quad j = +, -; j \neq i. \qquad (12.92)$$

Accordingly, we obtain the following relations for the reflectance R and transmittance T:

$$R_i = R_{ii} + R_{ij} = |r_{ii}|^2 + |r_{ij}|^2 = |r_i|^2 \quad T_i = T_{ii} + T_{ij} = |t_{ii}|^2 + |t_{ij}|^2 = |t_i|^2. \qquad (12.93)$$

12.9 Further reading

It is hard to suggest anything which could complement or extend this chapter. Maybe the book of Azzam and Bashara could provide a few aspects which may not have been tackled [9]. In particular, in IR spectroscopy books, anisotropy is only treated in the framework of Linear Dichroism Theory, which I consider to be inadequate for most real-world problems.

References

[1] D.W. Berreman, Optics in stratified and anisotropic media: 4 × 4-matrix formulation, J. Opt. Soc. Am. 62 (1972) 502–510.
[2] P. Yeh, Electromagnetic propagation in birefringent layered media, J. Opt. Soc. Am. 69 (1979) 742–756.
[3] S. Teitler, B.W. Henvis, Refraction in stratified, anisotropic media, J. Opt. Soc. Am. 60 (1970) 830–834.
[4] P.J. Lin-Chung, S. Teitler, 4 × 4 matrix formalisms for optics in stratified anisotropic media, J. Opt. Soc. Am. A 1 (1984) 703–705.
[5] M.V. Berry, The optical singularities of bianisotropic crystals, Proc. R. Soc. A Math. Phys. Eng. Sci. 461 (2005) 2071–2098.
[6] V. Dmitriev, Tables of the second rank constitutive tensors for linear homogeneous media described by the point magnetic groups of symmetry—abstract, J. Electromagn. Waves Appl. 14 (2000) 525–526.
[7] T.G. Mayerhöfer, J. Popp, Modelling IR spectra of polycrystalline materials in the large crystallites limit—quantitative determination of orientation, J. Opt. A Pure Appl. Opt. 8 (2006) 657–671.
[8] T.G. Mayerhöfer, Symmetric Euler orientation representations for orientational averaging, Spectrochim. Acta A Mol. Biomol. Spectrosc. 61 (2005) 2611–2621.
[9] R.M.A. Azzam, N.M. Bashara, Ellipsometry and Polarized Light, North-Holland, 1987.
[10] W. Xu, L.T. Wood, T.D. Golding, Optical degeneracies in anisotropic layered media: treatment of singularities in a 4 × 4 matrix formalism, Phys. Rev. B 61 (2000) 1740–1743.
[11] R. Ossikovski, O. Arteaga, Extended Yeh's method for optically active anisotropic layered media, Opt. Lett. 42 (2017) 3690–3693.
[12] J.R. Aronson, A.G. Emslie, P.F. Strong, Optical constants of triclinic anisotropic crystals: blue vitriol, Appl. Opt. 24 (1985) 1200–1203.
[13] A.G. Emslie, J.R. Aronson, Determination of the complex dielectric tensor of triclinic crystals: theory, J. Opt. Soc. Am. 73 (1983) 916–919.
[14] S. Höfer, J. Popp, T.G. Mayerhöfer, Determination of the dielectric tensor function of triclinic $CuSO_4 \cdot 5H_2O$, Vib. Spectrosc. 67 (2013) 44–54.

[15] S. Höfer, J. Popp, T.G. Mayerhöfer, Dispersion analysis of triclinic $K_2Cr_2O_7$, Vib. Spectrosc. 72 (2014) 111–118.
[16] R.M. Almeida, S. Höfer, T.G. Mayerhöfer, J. Popp, K. Krambrock, R.P.S.M. Lobo, A. Dias, R.L. Moreira, Optical phonon features of triclinic montebrasite: dispersion analysis and non-polar Raman modes, Vib. Spectrosc. 77 (2015) 10.
[17] C. Ye, M.J. Rucks, J.A. Arnold, T.D. Glotch, Mid-infrared optical constants of labradorite, a triclinic plagioclase mineral, earth and space, Science 6 (2019) 2410–2422.
[18] M.V. Belousov, V.F. Pavinich, Infrared reflection spectra of monoclinic crystals, Opt. Spektrosk. 45 (1978) 920–926.
[19] E.E. Koch, A. Otto, K.L. Kliewwer, Reflection spectroscopy on monoclinic crystals, Chem. Phys. 3 (1974) 362–369.
[20] T.G. Mayerhöfer, S. Weber, J. Popp, Simplified formulas for non-normal reflection from monoclinic crystals, Opt. Commun. 284 (2011) 719–723.
[21] J. Lekner, Reflection and refraction by uniaxial crystals, J. Phys. Condens. Matter 3 (1991) 6121.
[22] G.J. Sprokel, The Physics and Chemistry of Liquid Crystal Devices, Springer, US, 2013.
[23] J. Lekner, Normal-incidence reflection and transmission by uniaxial crystals and crystal plates, J. Phys. Condens. Matter 4 (1992) 1387.
[24] J.L.P. Mosteller, F. Wooten, Optical properties and reflectance of uniaxial absorbing crystals, J. Opt. Soc. Am. 58 (1968) 511–518.
[25] O.E. Piro, Optical properties, reflectance, and transmittance of anisotropic absorbing crystal plates, Phys. Rev. B 36 (1987) 3427–3435.
[26] P. Drude, Lehrbuch der Optik, Verlag von S, Hirzel, 1906.
[27] J. Lekner, Optical properties of isotropic chiral media, Pure Appl Optic. J. Eur. Opt. Soc. A 5 (1996) 417–443.
[28] E. Georgieva, Reflection and refraction at the surface of an isotropic chiral medium: eigenvalue–eigenvector solution using a 4×4 matrix method, J. Opt. Soc. Am. A 12 (1995) 2203–2211.

Chapter 13

Dispersion relations—Anisotropic oscillator models

Some of the considerations that are of importance in this chapter have already been made in the previous one—nevertheless, I will repeat them at the beginning of this chapter to bring them back to your attention. The main difference compared to Chapter 4 is that in this part the direction of **E** and **D** usually does not coincide. As a consequence, the dielectric tensor has the following general form which we have already assumed in the last chapter:

$$\mathbf{E} = \varepsilon_0 \begin{pmatrix} \varepsilon_{xx} & \varepsilon_{xy} & \varepsilon_{xz} \\ \varepsilon_{yx} & \varepsilon_{yy} & \varepsilon_{yz} \\ \varepsilon_{zx} & \varepsilon_{zy} & \varepsilon_{zz} \end{pmatrix} . \mathbf{D}. \tag{13.1}$$

We will presume for the following that the dielectric tensor is symmetric, i.e., $\varepsilon_{ij} = \varepsilon_{ji}$. As a consequence, there can be up to 6 independent components whose dispersion we need to describe somehow. Nevertheless, depending on the crystal symmetry, complexity can be dramatically decreased. If the crystal's elementary cell has axes of rotations and/or mirror planes, then their orientation is independent of frequency or wavenumber, which means that the dielectric tensor can either be fully diagonalized or, at least, block-diagonalized, and it seems that we can directly take over the classical damped harmonics oscillator model discussed in Chapter 4 (including the suggested modifications or other forms), at least if we have a principal cut for which a certain crystallographic axis is perpendicular to the surface. Fig. 13.1 provides an overview over the 9 different crystal systems.

The characteristic of the cubic crystal system is that all three crystallographic axes are perpendicular to each other and of the same length, which means that the dielectric tensor is reduced to a scalar. Such crystals are isotropic except for very short wavelengths for which intrinsic double refraction occurs. If you have never heard of intrinsic double refraction, it is not something that is well known, although it was already investigated by Lorentz [1]. In fact, in practice, it did not generate much interest until the semi-conductor roadmap detailed a switch from 193 nm to 157 nm in the beginning of the new millennium. SiO_2 glass is unfortunately no longer transparent at 157 nm so that another isotropic material had to be found to produce lenses for lithography in the wafer stepper. A promising candidate was quickly found, namely CaF_2, which is of cubic crystal symmetry and, additionally, has a high-laser resistance.

However, when the wavelength becomes shorter and goes toward the X-ray spectral region, a single crystal becomes increasingly an inhomogeneous medium. Accordingly, the dielectric tensor also shows spatial dispersion and the light begins to be scattered from the atoms—the principle of X-ray diffraction measurements is based on this inhomogeneity. Correspondingly, the directions through a cubic crystal begin already noticeably at 157 nm to be no longer equivalent and to influence the light transmitted through it in different ways. This phenomenon is called intrinsic double refraction [2]. Intrinsic double refraction was extensively researched and the result is well known: The corresponding problem could not be solved and the chip industry was stuck for a long time with 193 nm until the advent of EUV sources and optics (which took much longer than expected in the original roadmap—on the other hand, impressive progress was made to use lasers at 193 nm to an extent unforeseen in the beginning). The lesson to be learned is that inhomogeneity of matter is always something that can strongly alter optical properties and, thereby, spectra (cf. Section 6.6 and Chapter 10). We will come back to this problem when we discuss an optical crystallite size effect in polycrystalline spectra in Chapter 15.

For the remainder of this chapter, we will focus on crystals that are anisotropic because of the anisometric shapes of their elementary cells and/or nonorthogonality of their crystal axes, but we will nevertheless start with a short rehearsal of the cubic crystal system so that the differences to the other crystal systems can be better appreciated as we increasingly decrease symmetry.

Crystal system		Dielectric tensor (main axes system)
Cubic		$\begin{pmatrix} \varepsilon & 0 & 0 \\ 0 & \varepsilon & 0 \\ 0 & 0 & \varepsilon \end{pmatrix} = \varepsilon \begin{pmatrix} 1 & 0 & 0 \\ 0 & 1 & 0 \\ 0 & 0 & 1 \end{pmatrix}$
Tetragonal, Hexagonal, Trigonal (opt. uniaxial)		$\begin{pmatrix} \varepsilon_a & 0 & 0 \\ 0 & \varepsilon_a & 0 \\ 0 & 0 & \varepsilon_c \end{pmatrix}$
Orthorhombic		$\begin{pmatrix} \varepsilon_a & 0 & 0 \\ 0 & \varepsilon_b & 0 \\ 0 & 0 & \varepsilon_c \end{pmatrix}$
Monoclinic		$\begin{pmatrix} \varepsilon_{xx} & \varepsilon_{xy} & 0 \\ \varepsilon_{xy} & \varepsilon_{yy} & 0 \\ 0 & 0 & \varepsilon_b \end{pmatrix}$
Triclinic		$\begin{pmatrix} \varepsilon_{xx} & \varepsilon_{xy} & \varepsilon_{xz} \\ \varepsilon_{xy} & \varepsilon_{yy} & \varepsilon_{yz} \\ \varepsilon_{xz} & \varepsilon_{yz} & \varepsilon_{zz} \end{pmatrix}$

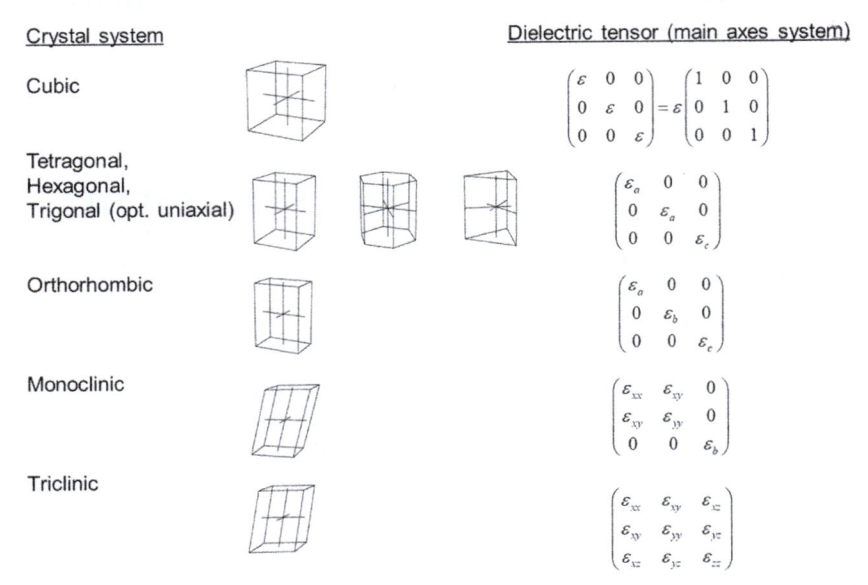

FIG. 13.1 Crystal symmetries and corresponding symmetric-dielectric tensors in principal orientations (main axes system).

13.1 Cubic crystal system

The crystal classes that belong to the cubic crystal system are either tetrahedral or octahedral classes, namely T, T_h, T_d, O, and O_h, which means that all infrared active vibrations are triply degenerated. Correspondingly, as already introduced, the dielectric tensor function reduces to a scalar according to

$$\begin{pmatrix} \varepsilon & 0 & 0 \\ 0 & \varepsilon & 0 \\ 0 & 0 & \varepsilon \end{pmatrix} = \varepsilon \mathbf{1}. \tag{13.2}$$

In most instances, the dispersion in the infrared-spectral region can be captured well enough by the classical damped harmonic oscillator model which leads to the following dispersion formula:

$$\varepsilon(\tilde{\nu}) = \varepsilon_\infty + \sum_{j=1}^{N} \frac{S_j^2}{\tilde{\nu}_{0,j}^2 - \tilde{\nu}^2 - i\tilde{\nu}\gamma_j}, \tag{13.3}$$

where S_j is the oscillator strength of the jth mode, γ_j is its damping constant, and $\tilde{\nu}_{0,j}$ is its resonance wavenumber. ε_∞ is a constant offset introduced by the modes at higher wavenumber beyond the transparency region that delineates the IR from the UV/vis-spectral region.

This oscillator model will be extended and generalized in the following to describe the dielectric tensor functions of the less-symmetric crystals.

13.2 Optically uniaxial: Tetragonal, hexagonal, and trigonal crystal systems

While the tetragonal, hexagonal, and trigonal crystal systems are certainly differentiable via X-ray diffraction analysis, their optical properties or, more precisely, their dielectric tensors all show the same structure. The important feature that renders them equal is the possession of a rotation axis with an N-fold symmetry with $N > 2$, cf. Fig. 13.2. The crystal classes they belong to are C_3, C_{3i}, C_{3v}, D_3, D_{3d}; C_6, C_{3h}, C_{6h}, C_{6v}, D_{3h}, D_6, D_{6h}; C_4, S_4, C_{4h}, C_{4v}, D_{2d}, D_4, D_{4h}.

This common characteristic leads to the fact that the dielectric tensor can be represented by a rotational ellipsoid around the symmetry axis with N-fold symmetry. As a consequence, a plane wave traveling along this high-symmetry axis will have the same speed and has the same refractive index irrespective of its polarization, which means that no birefringence and dichroism occur in this case: For such waves the crystals seem to be isotropic, which is why the high-symmetry axis is also called the optical axis. Vibrations that have their transition moment perpendicular to this axis are correspondingly

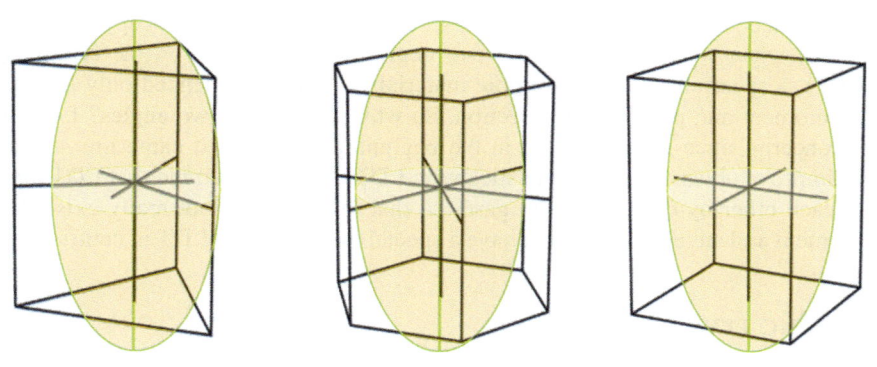

FIG. 13.2 Elementary cells of triclinic, hexagonal, and tetragonal crystals (from *left* to the *right*).

doubly degenerated, while the modes having their transition moments along the *c*-axis are undegenerated. It is important to realize that other orientations of transition moments are not possible, i.e., all transition moments must be either parallel or perpendicular to the optical axis. This is not only important for the structural interpretation of the spectra of a uniaxial crystal, but also to understand how we can model their dispersion in the infrared-spectral region. Correspondingly, as already introduced, the dielectric tensor has two principal components ε_a and ε_c, with ε_a appearing twice:

$$\boldsymbol{\varepsilon} = \begin{pmatrix} \varepsilon_a & 0 & 0 \\ 0 & \varepsilon_a & 0 \\ 0 & 0 & \varepsilon_c \end{pmatrix}. \tag{13.4}$$

The situation can be illustrated possibly best by a spring model (cf. Fig. 13.3), where the springs along *z* have a spring constant different from those along *x* and *y*. Accordingly, it is possible to separate the problem and describe every principal component of the dielectric tensor in Eq. (13.4) separately by the already known classical damped harmonic oscillator model:

$$\varepsilon_k(\tilde{\nu}) = \varepsilon_{\infty,k} + \sum_{j=1}^{N_k} \frac{S_{jk}^2}{\tilde{\nu}_{0,jk}^2 - \tilde{\nu}^2 - i\tilde{\nu}\gamma_{jk}} \quad k = a, c. \tag{13.5}$$

Correspondingly, N_k is the number of oscillators having their transition moment along either *a* or *c*.

General orientations of uniaxial crystals can be described by using either direction cosines or the two angles known from the spherical coordinate system. What I did not provide so far explicitly is how the dielectric tensor in Eq. (13.4), i.e., in its principal form, changes with the angles φ and θ:

$$\boldsymbol{\varepsilon} = \mathbf{A} \begin{pmatrix} \varepsilon_a & 0 & 0 \\ 0 & \varepsilon_a & 0 \\ 0 & 0 & \varepsilon_c \end{pmatrix} \mathbf{A}^{-1}, \mathbf{A} = \begin{pmatrix} \cos\varphi & -\cos\theta\sin\varphi & \sin\theta\sin\varphi \\ \sin\varphi & \cos\theta\cos\varphi & -\cos\varphi\sin\theta \\ 0 & \sin\theta & \cos\theta \end{pmatrix}. \tag{13.6}$$

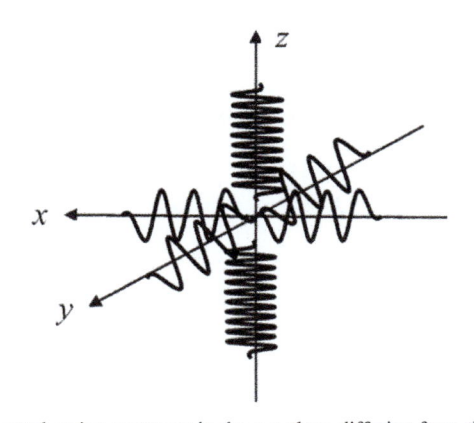

FIG. 13.3 Uniaxial spring model with degenerated spring constants in the *x*-*y*-plane differing from those along the *z*-axis.

Some of you who are more acquainted with the standard textbooks of infrared spectroscopy might have had a bad feeling already in Section 12.8.3, when we introduced these two angles, because based on the repeatedly mentioned Linear Dichroism Theory (LDT), a spectrum of an anisotropic material should be influenced only by one angle, which is the one between transition moment and polarization direction. So why do we need two angles? I can fully understand your bad feeling and your concerns since I shared them in the beginning when I first came into contact with wave optics. Do not worry, this will improve unless you decide to stick with LDT. Just keep in mind that LDT inherently assumes molecules separated from each other by distances as in gases so that interfaces do not really exist. Accordingly, the angle between a transition moment and an interface cannot have a special importance in LDT in contrast to higher level theories.

13.3 Orthorhombic crystals

One could say that the extension of the uniaxial to the orthorhombic case is not very demanding and I would fully agree. If we look at the spring model (Fig. 13.3), then the modification would be that in the orthorhombic case we now have three different sets of springs, but those are still oriented perpendicular to each other along the axes of the Cartesian-coordinate system. If we look at the elementary cells of orthorhombic crystals, they all coincide in having three twofold rotation axis which are perpendicular to each other. Accordingly, the corresponding crystal classes are C_{2v}, D_2, D_{2h}, which differ only concerning the mirror planes. As a consequence, the principal elements of the tensor in diagonal form are all different, but it is still possible to model each of them based on the classical damped harmonic oscillator model:

$$\varepsilon_k(\tilde{\nu}) = \varepsilon_{\infty,k} + \sum_{j=1}^{N_k} \frac{S_{jk}^2}{\tilde{\nu}_{0,jk}^2 - \tilde{\nu}^2 - i\tilde{\nu}\gamma_{jk}} \quad k = a,b,c, \tag{13.7}$$

with the only difference to Eq. (13.5) that now three different components need to be calculated, each with its own number of oscillators N_k. This corresponds to the fact that for orthorhombic crystals we have three sets of undegenerated vibrations having their transition moments along one of the crystallographic axes a, b, c. To address general orientations, the rotation matrix is more complex than in the uniaxial case, and is a function of three angles φ, θ, ψ, which are called the Euler angles. The first two angles can still be seen as the two angles of the spherical coordinate system. In this case the third rotation would be carried out around the rotated axis z ($= z''$), as this is indicated in Fig. 13.4.

The corresponding rotation matrix is given by [3].

$$\mathbf{A} = \begin{pmatrix} \cos\psi\cos\varphi - \cos\theta\sin\varphi\sin\psi & -\sin\psi\cos\varphi - \cos\theta\sin\varphi\cos\psi & \sin\theta\sin\varphi \\ \cos\psi\sin\varphi + \cos\theta\cos\varphi\sin\psi & -\sin\psi\sin\varphi + \cos\theta\cos\varphi\cos\psi & -\sin\theta\cos\varphi \\ \sin\theta\sin\psi & \sin\theta\cos\psi & \cos\theta \end{pmatrix}, \tag{13.8}$$

with the general form of the dielectric tensor

$$\boldsymbol{\varepsilon} = \mathbf{A} \begin{pmatrix} \varepsilon_a & 0 & 0 \\ 0 & \varepsilon_b & 0 \\ 0 & 0 & \varepsilon_c \end{pmatrix} \mathbf{A}^{-1}. \tag{13.9}$$

A word of caution: The definition of the Euler angles as provided in Fig. 13.4 is not unique—there are many more possible definitions. The first rotation can be carried out either around $X = x$, $Y = y$ or $Z = z$, which gives already three different possibilities. We can then either rotate around the axis of the laboratory coordinates or around the internal axes, and we can decide either to use 3 different axes or to use the first axis again for the third rotation [3]. Overall, this gives us 24 different possibilities, which must be always kept in mind if literature values are to be compared.

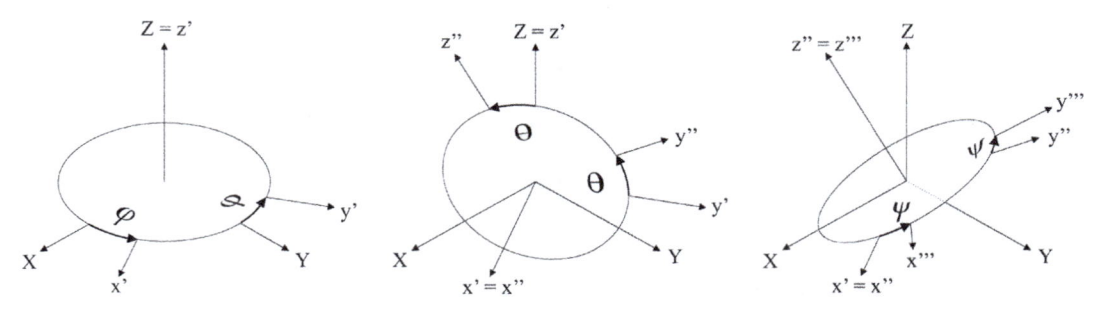

FIG. 13.4 One possible definition of the Euler angles using mobile rotation axes and one axis, the z-axis, twice [3].

13.4 Monoclinic crystals

For monoclinic crystals (crystal classes C_2, C_s, C_{2h}), the angle between the a- and the c-axis is different from 90°. This has far-reaching consequences for their spectroscopic properties because it means that the directions of the transition moments are no longer fixed to one of these axes. Instead, one kind of transition moment is always parallel to the a-c plane and the second kind is perpendicular to that plane and therefore still fixed to one of the crystallographic axes, namely the b-axis. As a consequence, the dielectric tensor can only be block-diagonalized

$$\varepsilon = \begin{pmatrix} \varepsilon_{xx} & \varepsilon_{xy} & 0 \\ \varepsilon_{xy} & \varepsilon_{yy} & 0 \\ 0 & 0 & \varepsilon_b \end{pmatrix}, \quad (13.10)$$

since when the real part is fully diagonalized, the imaginary part is not and vice versa. The principal orientations are therefore oriented either parallel or perpendicular to the b-axis. Compared to orthorhombic crystals, the same number of different faces is needed to obtain the full dielectric tensor function, which is one cut the surface of which is perpendicular to the b-axis and one cut where the surface and one edge is parallel to this axis. The latter cut serves to obtain a spectrum only influenced by ε_b in the same way as for optically uniaxial and orthorhombic crystals, i.e., with s-polarized light and the polarization direction parallel to the b-axis. ε_b can thus be described again by the conventional classical damped harmonic oscillator model. This model has to be modified for the oscillators having their transition moment in the a-c-plane. Obviously, we need a new parameter which describes the orientation of the transition moment, which is the angle Φ_j, cf. Fig. 13.5.

Certainly, the definition of the angle Φ_j is not unique, which means that one always needs to be careful when literature data are compared. In any way, the introduction of the new parameter allows us to provide a formula for the dispersion of the dielectric tensor based on the classical damped harmonic oscillator model. The 2×2 matrix can be formulated as follows:

$$\varepsilon_{x,y}(\tilde{\nu}) = \begin{pmatrix} \varepsilon_{\infty,xx} & \varepsilon_{\infty,xy} \\ \varepsilon_{\infty,xy} & \varepsilon_{\infty,yy} \end{pmatrix} + \sum_{j=1}^{N} \frac{S_j^2}{\tilde{\nu}_j^2 - \tilde{\nu}^2 - i\tilde{\nu}\gamma_j} \begin{pmatrix} \cos^2\Phi_j & \sin\Phi_j \cos\Phi_j \\ \sin\Phi_j \cos\Phi_j & \sin^2\Phi_j \end{pmatrix}. \quad (13.11)$$

Accordingly, not only does the wavenumber dependent part need to be modified but also ε_∞ becomes a tensor since the transition moments in the UV-vis are no longer necessarily parallel to the crystal axes (certainly, in wavenumber regions higher than the first transition, the dielectric background would vanish and be replaced by the unity tensor).

13.5 Triclinic crystals

For triclinic crystals, crystal classes C_1, C_i, the orientation of the transition moments is arbitrary, i.e., they are not bound to any directions inside the crystal. Correspondingly, the dielectric tensor always has the general form

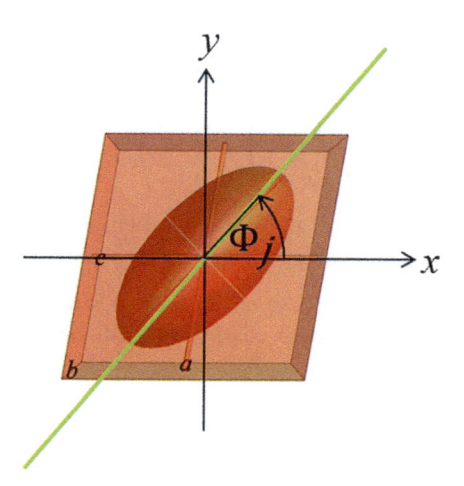

FIG. 13.5 Illustration of the a-c plane in a monoclinic crystal and the definition of the angle between transition moment and the c-axis.

$$\boldsymbol{\varepsilon} = \mathbf{A} \begin{pmatrix} \varepsilon_{xx} & \varepsilon_{xy} & \varepsilon_{xz} \\ \varepsilon_{yx} & \varepsilon_{yy} & \varepsilon_{yz} \\ \varepsilon_{zx} & \varepsilon_{zy} & \varepsilon_{zz} \end{pmatrix} \mathbf{A}^{-1}. \tag{13.12}$$

As a consequence, there is no natural way to define the internal coordinate system x, y, z. We usually started from a natural face and defined the rotation matrix relative to this face (or to its face normal). The direction of the transition moment is then given again by the two angles of the spherical coordinate system relative to the internal coordinate system and the dispersion of the dielectric tensor can be modeled by

$$\boldsymbol{\varepsilon}\tilde{\nu} = \begin{pmatrix} \varepsilon_{\infty,xx} & \varepsilon_{\infty,xy} & \varepsilon_{\infty,xz} \\ \varepsilon_{\infty,yx} & \varepsilon_{\infty,yy} & \varepsilon_{\infty,yz} \\ \varepsilon_{\infty,zx} & \varepsilon_{\infty,zy} & \varepsilon_{\infty,zz} \end{pmatrix} + \sum_{j=1}^{N} \frac{S_j^2}{\tilde{\nu}_j^2 - \tilde{\nu}^2 - i\tilde{\nu}\gamma_j} \begin{pmatrix} \sin^2\Theta_j \cos^2\Phi_j & \sin^2\Theta_j \sin\Phi_j \cos\Phi_j & \sin\Theta_j \cos\Theta_j \cos\Phi_j \\ \sin^2\Theta_j \sin\Phi_j \cos\Phi_j & \sin^2\Theta_j \sin^2\Phi_j & \sin\Theta_j \cos\Theta_j \sin\Phi_j \\ \sin\Theta_j \cos\Theta_j \cos\Phi_j & \sin\Theta_j \cos\Theta_j \sin\Phi_j & \cos^2\Theta_j \end{pmatrix}. \tag{13.13}$$

It may be worth mentioning that eq. (13.13) is in general not sufficient to describe the dielectric tensor in the UV-vis region because there are transitions in crystals that have a distribution of orientations of their moments [4]. Also, I again want to point out the need for two angles due to the importance of interfaces in wave optics—there is no way around. Note that the classical damped harmonic oscillator model may be replaced by some of the four parameter models we discussed in Chapter 5, but what cannot be used is the Berreman-Unterwal model.

13.6 Generalized oscillator models

It looks like that we have been treated all possible elementary cells and the corresponding dielectric tensors and we are ready to start. Indeed, for conventional dispersion analysis, this statement is correct. However, conventional dispersion analysis requires the orientation of the crystallographic axes to be known. It is certainly not difficult (or not that difficult) to determine these orientation by X-ray diffraction, i.e., by the Laue method, if you know how to perform X-ray diffraction analysis. Unfortunately, with increasing tendency of specialization, there are chemistry and physics faculties nowadays without the necessary equipment and know-how. In any way, it might come in handy to determine oscillator parameters of crystals the orientation (and, maybe, not even the crystal system) of which is not known a priori.

This is where generalized dispersion analysis comes into play. For this kind of dispersion analysis, which I will also introduce in the next chapter (Section 14.6), we need properly adapted oscillator models.

We do not need to discuss the triclinic system again because this is already a kind of generalized dispersion analysis since we do not know the orientation of the transition moment of the oscillators a priori. Even if we would not know that we investigate a triclinic crystal, the oscillator model given previously would work. If we however would apply it, say, to a crystal of the uniaxial system, we would realize that we would obtain two classes of oscillators:

$$\boldsymbol{\varepsilon}(\tilde{\nu}) = \mathbf{A}_\infty \cdot \begin{pmatrix} \varepsilon_{\infty,a} & 0 & 0 \\ 0 & \varepsilon_{\infty,a} & 0 \\ 0 & 0 & \varepsilon_{\infty,c} \end{pmatrix} \cdot \mathbf{A}_\infty^{-1} + \sum_{j=1}^{N_a} \frac{S_{j,a}^2}{\tilde{\nu}_{j,a}^2 - \tilde{\nu}^2 - i\tilde{\nu}\gamma_{j,a}} \left(\mathbf{A}_{j,a} \cdot \begin{pmatrix} 1 & 0 & 0 \\ 0 & 1 & 0 \\ 0 & 0 & 0 \end{pmatrix} \cdot \mathbf{A}_{j,a}^{-1} \right)$$
$$+ \sum_{j=1}^{N_c} \frac{S_{j,c}^2}{\tilde{\nu}_{j,c}^2 - \tilde{\nu}^2 - i\tilde{\nu}\gamma_{j,c}} \left(\mathbf{A}_{j,c} \cdot \begin{pmatrix} 0 & 0 & 0 \\ 0 & 0 & 0 \\ 0 & 0 & 1 \end{pmatrix} \cdot \mathbf{A}_{j,c}^{-1} \right). \tag{13.14}$$

If we write the dielectric tensor function in this way, all individual rotation matrices $\mathbf{A}_\infty$, $\mathbf{A}_{j,a}$, and $\mathbf{A}_{j,c}$ should be approximately equal, i.e., give approximately the same values for the angles Φ_j and Θ_j:

$$\mathbf{A}_j = \begin{pmatrix} \cos\Phi_j & -\cos\Theta_j \sin\Phi_j & \sin\Theta_j \sin\Phi_j \\ \sin\Phi_j & \cos\Theta_j \cos\Phi_j & -\cos\Phi_j \sin\Theta_j \\ 0 & \sin\Theta_j & \cos\Theta_j \end{pmatrix} = \mathbf{A}. \tag{13.15}$$

In other words, the angles Φ_j and Θ_j determine the orientation of the optical axis for each oscillator, be it parallel to it and the oscillator nondegenerated or be it perpendicular to it and the oscillator degenerated. The average value would then give the best estimation for this orientation. Note that the orientation of the a-axes cannot be determined by optical means, just the plane they span. (With Raman spectroscopy, however, it would be possible if the elementary cell has no center of inversion. Then, all IR-active vibrations are also Raman-active, a situation which leads to TO-LO effects in Raman spectra.

If the incident beam is oriented in the a-a-plane, the modes would take on a mixed TO-LO character which would vanish if the beam would be along one of the a-axis.)

For the orthorhombic system, the situation is more complicated since the rotation matrix is then a function of all three Euler angles. Accordingly, if the following form of the dielectric function is used,

$$\boldsymbol{\varepsilon}(\tilde{\nu}) = \mathbf{A}_\infty \cdot \begin{pmatrix} \varepsilon_{\infty,a} & 0 & 0 \\ 0 & \varepsilon_{\infty,b} & 0 \\ 0 & 0 & \varepsilon_{\infty,c} \end{pmatrix} \cdot \mathbf{A}_\infty^{-1} + \sum_{j=1}^{N_a} \frac{S_{j,a}^2}{\tilde{\nu}_{j,a}^2 - \tilde{\nu}^2 - i\tilde{\nu}\gamma_{j,a}} \left(\mathbf{A}_{j,a} \cdot \begin{pmatrix} 1 & 0 & 0 \\ 0 & 0 & 0 \\ 0 & 0 & 0 \end{pmatrix} \cdot \mathbf{A}_{j,a}^{-1} \right)$$

$$+ \sum_{j=1}^{N_b} \frac{S_{j,b}^2}{\tilde{\nu}_{j,b}^2 - \tilde{\nu}^2 - i\tilde{\nu}\gamma_{j,b}} \left(\mathbf{A}_{j,b} \cdot \begin{pmatrix} 0 & 0 & 0 \\ 0 & 1 & 0 \\ 0 & 0 & 0 \end{pmatrix} \cdot \mathbf{A}_{j,b}^{-1} \right) \quad (13.16)$$

$$+ \sum_{j=1}^{N_c} \frac{S_{j,c}^2}{\tilde{\nu}_{j,c}^2 - \tilde{\nu}^2 - i\tilde{\nu}\gamma_{j,c}} \left(\mathbf{A}_{j,c} \cdot \begin{pmatrix} 0 & 0 & 0 \\ 0 & 0 & 0 \\ 0 & 0 & 1 \end{pmatrix} \cdot \mathbf{A}_{j,c}^{-1} \right)$$

$$\mathbf{A}_j = \mathbf{A}_j(\Phi_j, \Theta_j, \Psi_j) = \mathbf{A},$$

three different classes of oscillators would be found. The rotation matrix with the average angles would be the one which allows to determine the diagonal form of the tensor. However, it would not be possible by optical means to tell which principal direction belongs to which crystallographic axis.

The monoclinic case is actually a mixture of the uniaxial and the orthorhombic case:

$$\boldsymbol{\varepsilon}(\tilde{\nu}) = \mathbf{A}_\infty \cdot \begin{pmatrix} \varepsilon_{\infty,xx} & \varepsilon_{\infty,xy} & 0 \\ \varepsilon_{\infty,xy} & \varepsilon_{\infty,yy} & 0 \\ 0 & 0 & \varepsilon_{\infty,b} \end{pmatrix} \cdot \mathbf{A}_\infty^{-1}$$

$$+ \sum_{j=1}^{N_{ac}} \frac{S_{j,ac}^2}{(\tilde{\nu}_{j,ac}^2 - \tilde{\nu}^2) - i\tilde{\nu}\gamma_{j,ac}} \left(\mathbf{A}_{j,ac} \cdot \begin{pmatrix} \cos^2\varphi_j & \cos\varphi_j \sin\varphi_j & 0 \\ \cos\varphi_j \sin\varphi_j & \sin^2\varphi_j & 0 \\ 0 & 0 & 0 \end{pmatrix} \cdot \mathbf{A}_{j,ac}^{-1} \right) \quad (13.17)$$

$$+ \sum_{j=1}^{N_b} \frac{S_{j,b}^2}{\tilde{\nu}_{j,b}^2 - \tilde{\nu}^2 - i\tilde{\nu}\gamma_{j,b}} \left(\mathbf{A}_{j,b} \cdot \begin{pmatrix} 0 & 0 & 0 \\ 0 & 1 & 0 \\ 0 & 0 & 0 \end{pmatrix} \cdot \mathbf{A}_{j,b}^{-1} \right)$$

$$\mathbf{A}_j = \mathbf{A}_j(\Phi_j, \Theta_j),$$

the angles Φ_j and Θ_j determine the orientation of the crystallographic b-axis for each oscillator and the angles φ_j determine additionally the orientation of the transition moments perpendicular to this axis.

13.7 Further reading

I am really sorry—in this chapter, I have to break with a tradition since, even after long considerations, no book came to my mind I could suggest to you. Maybe one could be helpful to some extent, but only for those of you who are really mathematically inclined: Born's and Huang's Dynamical Theory of Crystal Lattices [5].

References

[1] H.A. Lorentz, Double refraction by regular crystals, Proc. Koninklijke Akademie van Wetenschappen 24 (1922) 333–339.
[2] J.H. Burnett, Z.H. Levine, E.L. Shirley, Intrinsic birefringence in calcium fluoride and barium fluoride, Phys. Rev. B 64 (2001) 241102.
[3] T.G. Mayerhöfer, Symmetric Euler orientation representations for orientational averaging, Spectrochim. Acta A Mol. Biomol. Spectrosc. 61 (2005) 2611–2621.
[4] C. Sturm, S. Höfer, K. Hingerl, T.G. Mayerhöfer, M. Grundmann, Dielectric function decomposition by dipole interaction distribution: application to triclinic $K_2Cr_2O_7$, New J. Phys. 22 (2020) 073041.
[5] M. Born, K. Huang, Dynamical Theory of Crystal Lattices, Clarendon Press, 1954.

Chapter 14

Dispersion analysis of anisotropic crystals—Examples

14.1 Optically uniaxial crystals

The first dispersion analysis on a uniaxial crystal dates back to 1959 and was carried out by Spitzer, Kleinman, and Walsh, who also coined the term *dispersion analysis* [1].

As an example, I selected Fresnoite because it shows strong anisotropy over wide spectral ranges. As already discussed, as an optically uniaxial crystal, Fresnoite has two principal spectra which can both be obtained from a c-cut of an oriented single crystal, i.e., a single crystal plate, the surface of which is parallel to the c-axis. To obtain the principal spectra, it is advantageous if one edge of the plate is parallel to the c-axis and the other parallel to the a-a-plane. If we first orient this c-cut in a way that the latter edge is parallel to the polarization direction for s-polarized light, then we obtain a spectrum that is only depending on ε_a (cf. Fig. 14.1). From this principal spectrum, we can determine the principal dielectric function by conventional dispersion analysis. For Fresnoite, the experimental $\mathbf{E} \parallel \varepsilon_a$ spectrum (black) together with the correspondingly modeled spectrum (green) is depicted in Fig. 14.2. I guess we agree that the correspondence between both the experimental and the modeled spectrum is excellent. Given that I used only 1 CDH-oscillator per peak, the determined oscillator parameters should be accurate (certainly apart from the fact that all oscillators, from the UV/vis to the FIR, are coupled by local fields that cannot be considered properly in the dispersion relation, cf. Section 5.1). The same is true for the $\mathbf{E} \parallel \varepsilon_c$ spectrum (red) and the modeled spectrum (blue), which is also depicted in Fig. 14.2.

Conventional dispersion analysis usually ends at this point. Generally, it is argued that if the correspondence between experimental and modeled spectra is as good as that displayed in Fig. 14.2, then a further independent test of the accuracy of the obtained dielectric function tensor is not necessary. Hence, such tests, which can, e.g., consist of using the determined dielectric function tensor to predict the result of an independent experiment, are usually missing. Such an independent experiment could consist of simulating spectra of the single crystal for nonprincipal orientations.

Another possibility is to use the dielectric tensor function to predict the spectra of polycrystalline materials (note that this gives two different spectra, one assuming a randomly oriented microhomogeneous sample and one assuming a randomly oriented microheterogeneous sample; we will discuss how this works in detail in Section 15.1). I carried out this test, and the result is provided in Fig. 14.3. It goes without saying; nevertheless, I want to emphasize it here: These simulations are mere forward calculations, which means there is no free parameter one could use to improve the correspondence between experiment and calculation.

Overall, we find that the agreement between experimental- and forward-calculated spectra is excellent with regard to peak positions, relative intensities, and band shapes. For the randomly oriented polycrystalline sample with optically large crystallites, we find some discrepancies concerning the overall intensities, but those are to be expected due to the comparably high surface roughness of such samples, which leads to partly diffuse reflection that does not reach the detector. All in all, based on this test, we can conclude that the dielectric function tensor is good enough to do optical modeling with it, which is one of its main uses. For the following examples, I will not show results of such tests, but keep in mind that they are important and are an essential part of dispersion analysis—if they are missing, you cannot trust the result.

Before we start with the next section, however, I want to show you the result of another independent test of the dielectric function tensor. Since the elementary cell of the Fresnoite crystal has no inversion center, the rule of mutual exclusion does not apply and all IR-active vibrations are Raman active as well. This gives us the chance to compare IR and Raman spectra, and, by that, to check the TO- and the LO-positions. In particular, the A_1-modes, which are the ones having their transition moment along the c-axis, are very interesting. While it is not possible to excite a purely LO-mode via IR spectroscopy (just because you see some peak at or close to the LO-position, like as in the case of the Berreman effect, does not mean that you really excite a pure LO mode that is traveling in the same direction as the change of the transition moment is, remember and cf. Sections 5.5.1 and 5.6), you can do exactly this by Raman spectroscopy. What you require is a microscope because you

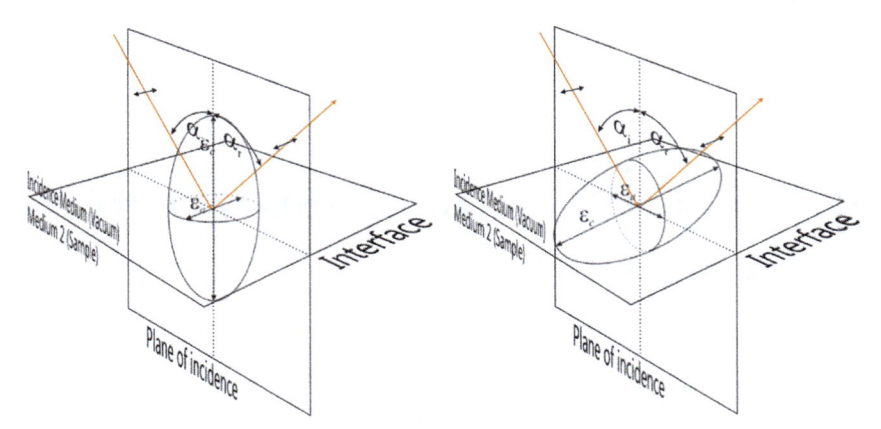

FIG. 14.1 Illustration of the measurement principle to determine the two principal dielectric functions using the two principal orientations of an optically uniaxial crystal.

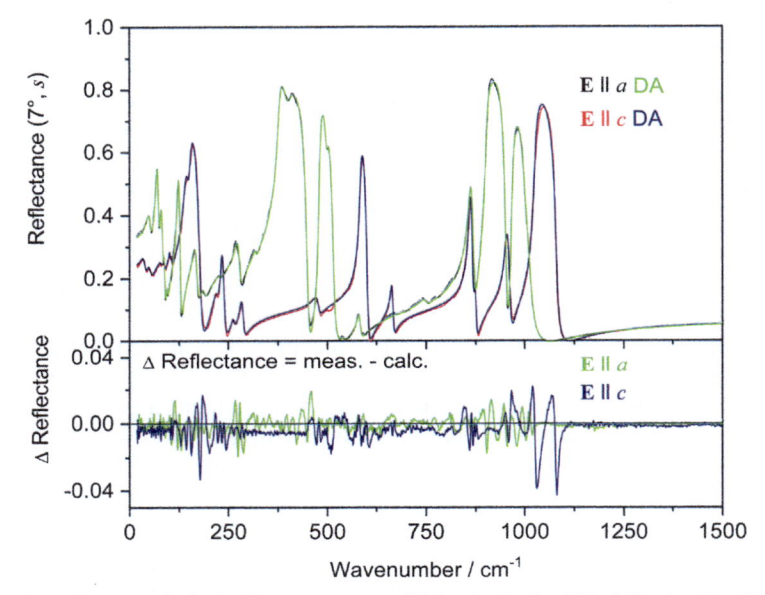

FIG. 14.2 Comparison between experimental principal reflectance spectra of Fresnoite *(red and black lines)* and modeled spectra *(green and blue line)* based on Eq. (13.5).

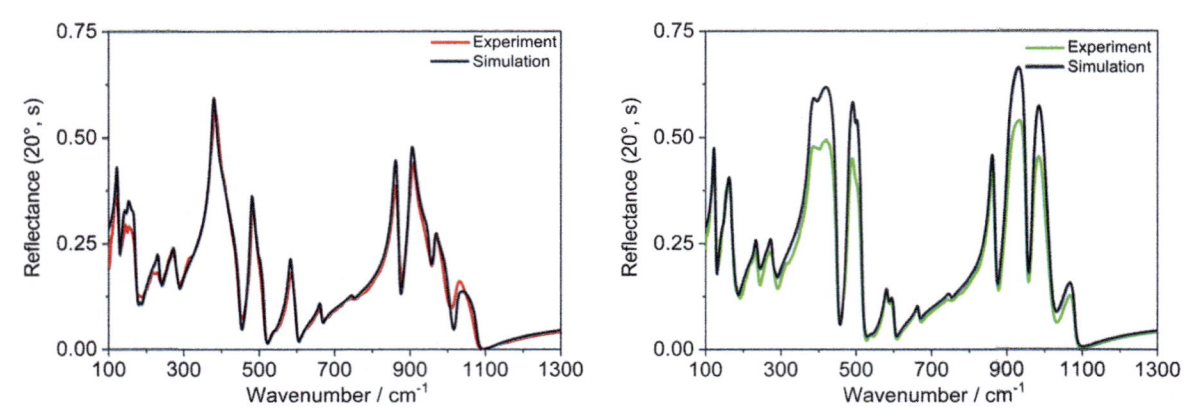

FIG. 14.3 Forward calculations based on the dielectric tensor function of Fresnoite for randomly oriented polycrystalline microhomogeneous *(left panel)* and microheterogeneous samples *(right panel)* as well as comparison with experimental data.

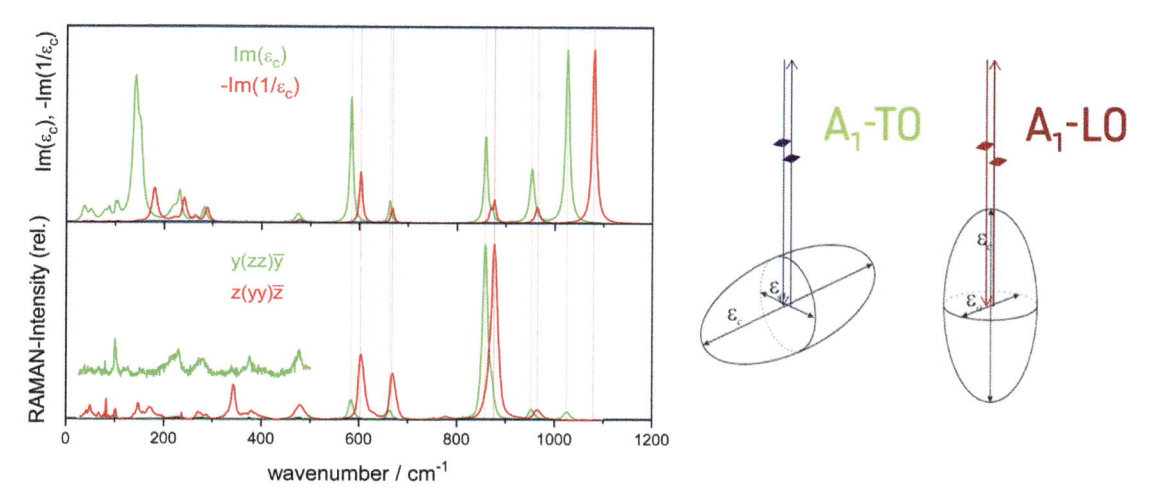

FIG. 14.4 *Left panel*: Comparison of Raman spectra of the Fresnoite single crystal belonging to the A_1-modes with the imaginary part of the corresponding principal dielectric function and the negative imaginary part of the corresponding inverse principal dielectric function. *Right panel*: Illustration of the orientation of the crystal relative to the incoming light and the orientation of polarizer and analyzer.

need a 180° configuration (i.e., the light that you detect is backscattered). The crystal must be oriented in a way that the optical axis is parallel to the direction of the incident light. If the analyzer is then oriented parallel to the polarizer (in Porto's notation [2], $z(yy)\bar{z}$, the bar indicates the backscattering geometry), then the A_1-modes are excited but as LO-modes, i.e., their positions in the Raman spectrum should be the same as the maxima of the dielectric loss function ($-\text{Im}(1/\varepsilon_c)$). The same can be repeated for the other principal orientation, where the optical axis is oriented perpendicular to the direction of the incoming light. If the polarization direction agrees with the orientation of the analyzer, still the A_1-modes are excited with the difference that in this case transition moment and light direction are perpendicular, like for TO-modes (in Porto's notation: $y(zz)y$). Therefore, these modes should be at the same positions as in IR spectra and the bands should have their peaks at the maxima of ε_c. Please have a look at Fig. 14.4. I guess we agree that the agreement is excellent.

14.2 Orthorhombic crystals

As already discussed in Section 12.8.4 and Section 13.3, there are no principal novelties for orthorhombic crystals compared to optically uniaxial crystals with regard to theory, except that there are now three principal orientations of a crystal that needs to be measured (cf. Fig. 14.5). So, the only complication is that you require a single crystal in form of a cuboid or a cube where all edges are parallel to one of the crystallographic axes. Or, if you want to save material, two cuts where the two edges are parallel to one crystallographic axis each. Again, using s-polarized light and orienting one of the edges parallel to the polarization direction of the incoming light, you obtain three different principal spectra, each of which is

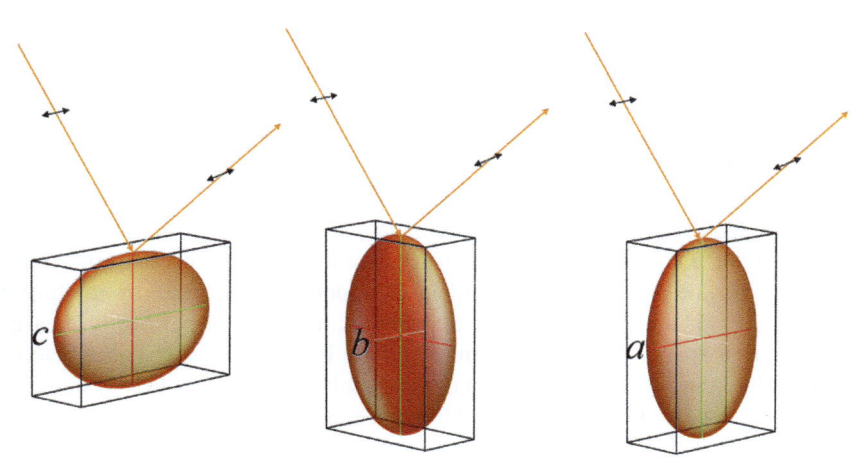

FIG. 14.5 Principal orientations and corresponding measurement geometries of an orthorhombic crystal.

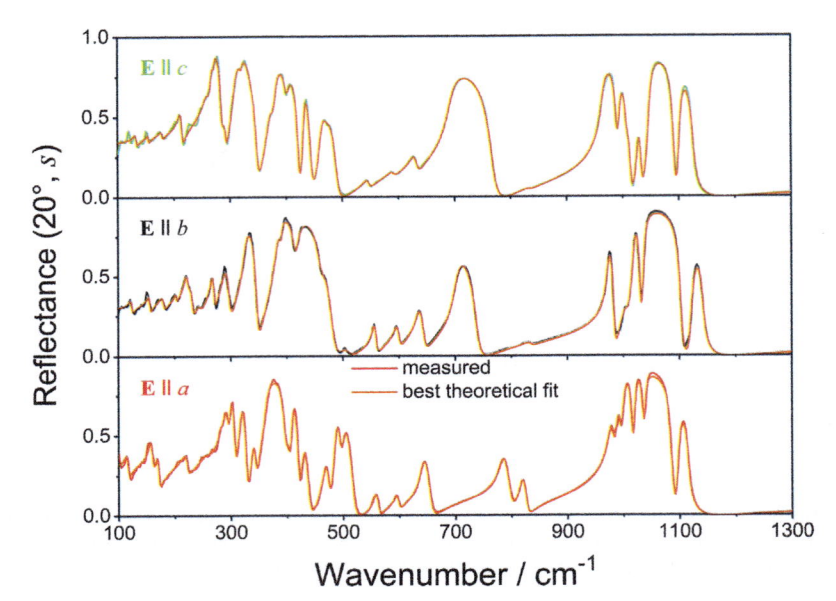

FIG. 14.6 Comparison between experimental principal reflectance spectra of single crystalline KTiOPO$_4$ *(red, black, and green lines)* and modeled spectra *(orange lines)* based on Eq. (13.7).

dependent on only one of the principal dielectric functions. Once you have performed conventional dispersion analysis on all three of them you have all you need. As an example, I want to show you results for KTiOPO$_4$, which are provided in Fig. 14.6 [3]. As for Fresnoite, only one oscillator has been used for each one of the peaks, which proves once more that CDH-oscillators are usually extremely useful for most of the bands in the infrared spectral region. One exception is probably NdGaO$_3$ because some of its oscillators are extremely strong [4]—I have discussed this example in detail already in Chapter 5.

14.3 Monoclinic crystals

The first dispersion analysis on a monoclinic crystal was performed by Belousov and Pavinich [5]. To date, more than 40 dispersion analyses have been carried out to which we also contributed [6–15], not the least with the first dispersion analysis taking into account a nonnormal angle of incidence [10]. I have not counted the ones that have been carried out for optically uniaxial or orthorhombic crystals, but I think, in comparison, it is a rather small number, which is due to the much higher complexity of the analysis for monoclinic crystals. Unfortunately, quite often in literature, the monoclinic crystal structure and the corresponding peculiar optical properties are neglected, e.g., in the case of the famous charge-transfer complexes consisting of tetrathiafulvalene (TTF) and tetracyanochinodimethan (TCNQ) and their derivates or relatives. These are often two-dimensional conductors or even superconductors and crystallize in the monoclinic (or triclinic) crystal system. It can possibly be understood based on the fact that the first salts were synthesized and measured before 1978. But as far as I know, there has still been to date (in the year 2024) no single dispersion analysis of the spectra of such a compound. How much the interpretation of the spectra of these compounds is worth without using the proper form of dispersion analysis is an open question.

Just to recap, the problem with monoclinic crystals is that the dielectric function tensor can only be block-diagonalized because you can only either diagonalize the real or the imaginary part but not both at the same time, cf. Fig. 14.7.

To determine the corresponding 2 × 2 block of the dielectric tensor function, at least three different measurements of the a-c crystal face are necessary, in line with the 3 independent components it has, e.g., by rotating the crystal around its surface normal by the angle $\varphi = 45°$ and 90°. From our experience, it is best to perform 4 measurements with $\varphi = 0°$, 45°, 90°, and 135° and fit all four spectra at the same time based on Eq. (13.11) as it is illustrated in Fig. 14.8.

Corresponding results for the example of one of the many Tuton-salts, K$_2$Ni(SO$_4$)$_2$·6H$_2$O, are shown in Fig. 14.9. These spectra were measured at an angle of incidence of 16° and the corresponding dispersion analysis was the first to consider this nonnormal angle of incidence. Overall, as has been discussed earlier, the minimal speed increase of neglecting nonnormal angles of incidence does not justify this neglection, while the accuracy of the fit is clearly increased [10].

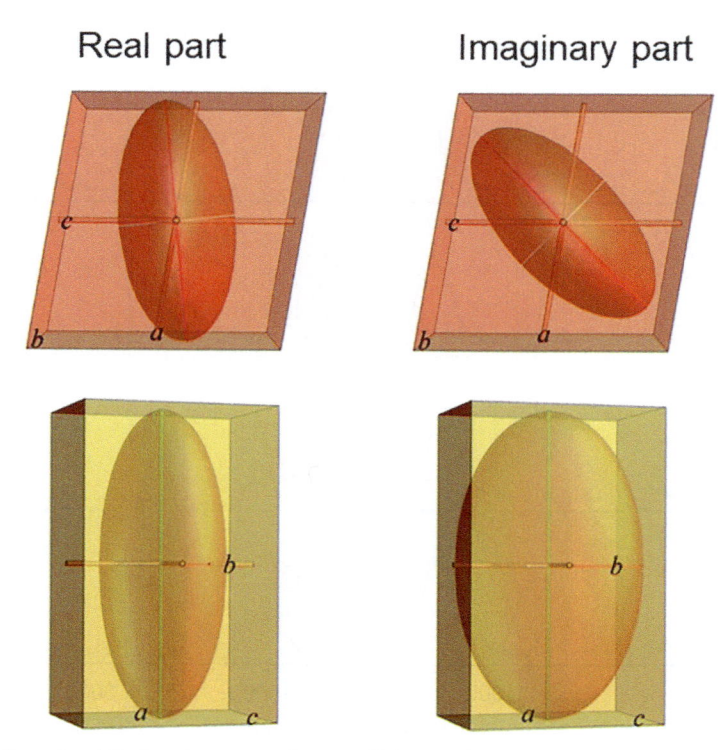

FIG. 14.7 Tensor ellipsoid for the real and the imaginary part of a monoclinic crystal.

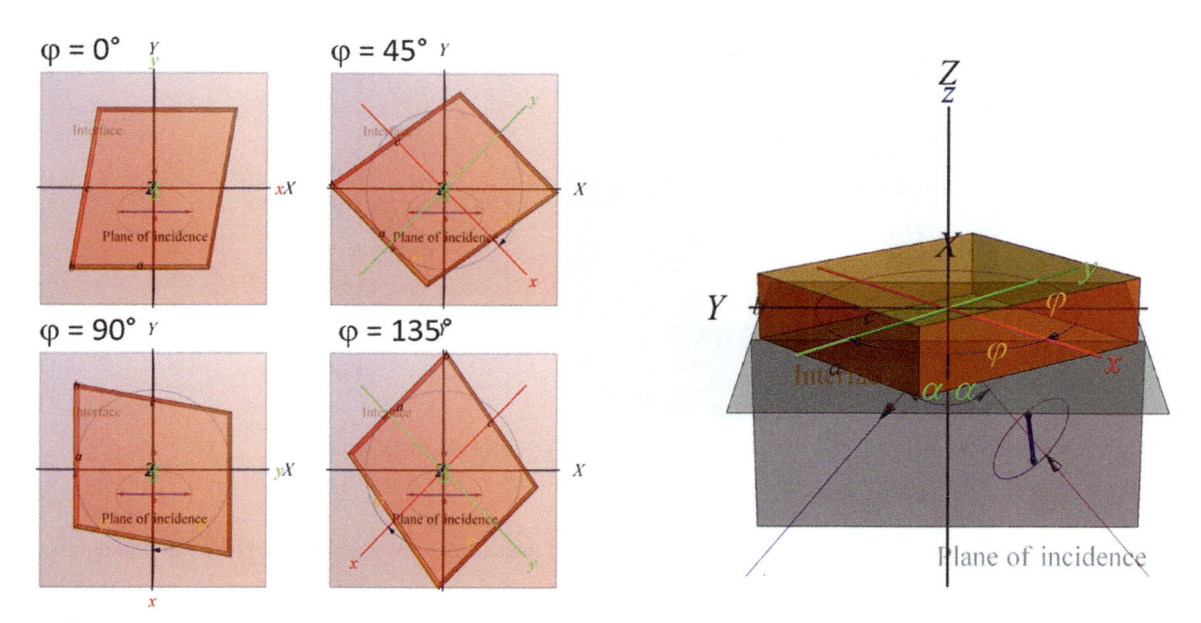

FIG. 14.8 The four different measurements that have to be performed to obtain the 2 × 2 block of the dielectric tensor relating to the *a-c*-plane.

14.4 Excursus: Perpendicular modes

You might have expected that triclinic crystals are next, but before I discuss their dispersion analysis, I want to introduce to you how to perform dispersion analysis in the case of so-called perpendicular modes. Such modes have their transition moment perpendicular to the surface of a sample. Quite often crystals grow anisometric which means that certain crystal faces are hard to obtain, of small dimension and therefore, hard to measure. Imagine you have an optically uniaxial crystal,

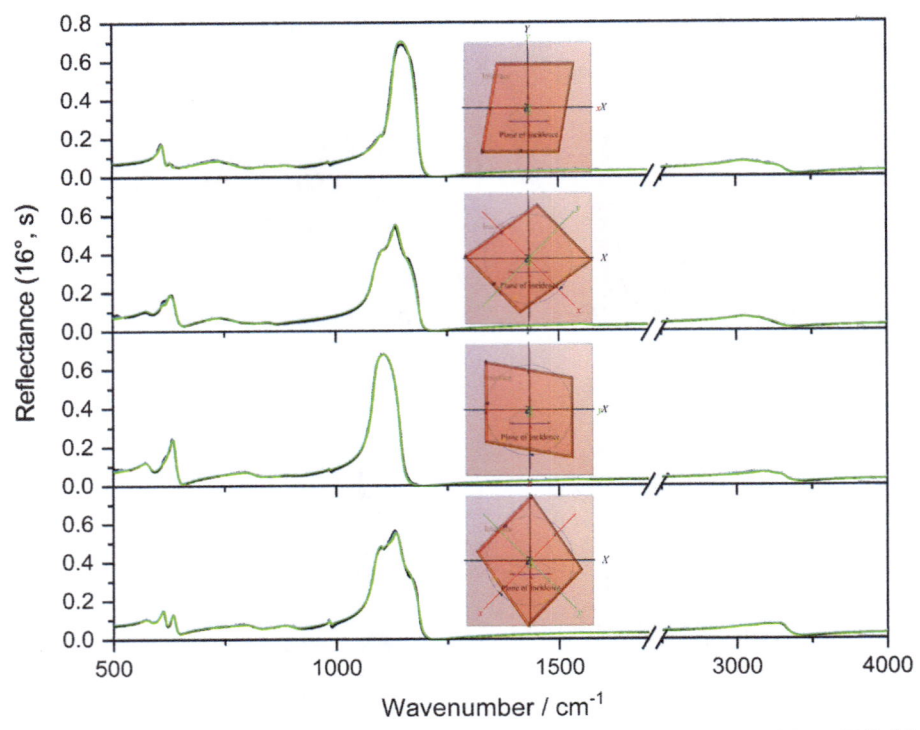

FIG. 14.9 Comparison between experimental reflectance spectra of the a-c-plane of single crystalline $K_2Ni(SO_4)_2 \cdot 6H_2O$ *(black lines)* and modeled spectra *(green lines)* to obtain the 2×2 block of the dielectric tensor for $\varphi = 0°$, $45°$, $90°$, and $135°$ (cf. Fig. 14.8) based on Eq. (13.11).

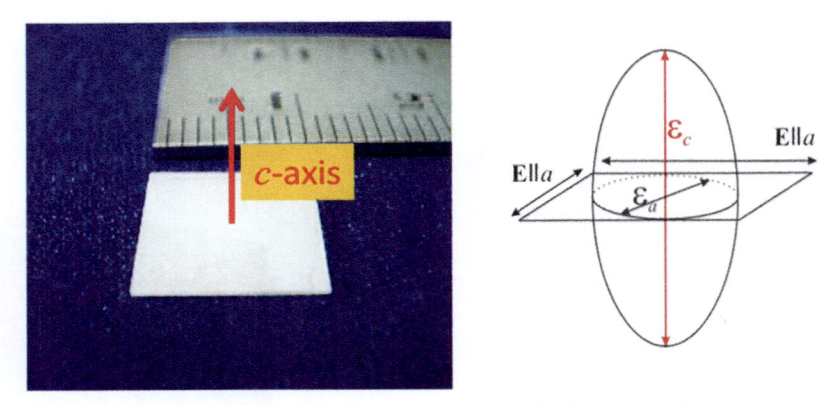

FIG. 14.10 An optically uniaxial sample (a-cut) with the optical axis oriented perpendicular to the surface.

a thin plate, where the optical axis is perpendicular to the surface of the sample (cf. Fig. 14.10). How can you determine ε_c in this case?

A natural solution to this problem seems to be to use p-polarized light in connection with large angles of incidence because this should enhance the intensities of the perpendicular modes in the spectrum (cf. Eq. 12.84):

$$R_s = \left| \frac{\cos \alpha - \sqrt{\varepsilon_a - \sin^2 \alpha}}{\cos \alpha + \sqrt{\varepsilon_a - \sin^2 \alpha}} \right|^2$$

$$R_p = \left| \frac{\sqrt{\varepsilon_a} \cos \alpha - \sqrt{1 - \frac{\sin^2 \alpha}{\varepsilon_c}}}{\sqrt{\varepsilon_a} \cos \alpha + \sqrt{1 - \frac{\sin^2 \alpha}{\varepsilon_c}}} \right|^2 .$$

(14.1)

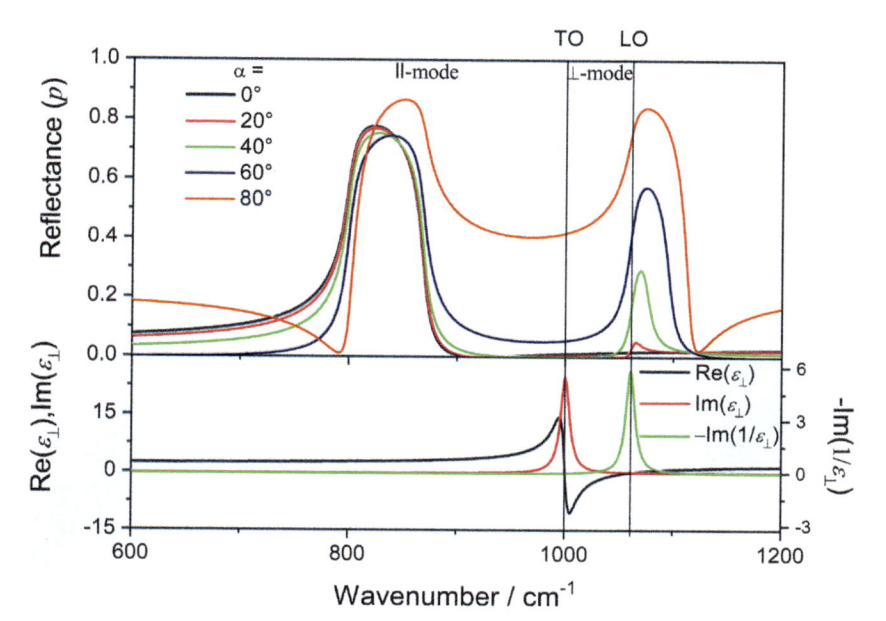

FIG. 14.11 *Upper panel*: Reflectance spectrum of a hypothetical uniaxial material where the c-mode is a perpendicular ($\perp$) mode. *Lower panel*: Real and imaginary part of the perpendicular principal dielectric function as well as negative imaginary part of the inverse perpendicular principal dielectric function [16].

Scrutinizing Eq. (14.1) it indeed looks like it makes sense to first employ a measurement with s-polarized light to determine ε_a by conventional dispersion analysis. Once this has been done, it seems that one can employ an experiment with p-polarized light and determine ε_c. The only problem is that it is actually the inverse of ε_c that appears in Eq. (14.1). You know already what this means from Section 5.6: Peaks belonging to ε_c will appear close to their LO-positions as this is illustrated in Fig. 14.11.

In fact, as Fig. 14.11 shows, the peaks are even shifted beyond the LO position for larger angles of incidence. In any way, it seems to be a good idea to carry out measurements at several different angles of incidence and fit those spectra together to obtain ε_c as suggested by Mosteller and Wooten [17]. To be more precise, Mosteller and Wooten wanted to determine the real and the imaginary part of the perpendicular optical constants function by varying it independently (two unknowns, therefore two different measurements at two different angles), i.e., without making use of dispersion analysis or the fact that real and imaginary part are related by the Kramers-Kronig relations. It was demonstrated shortly afterwards that knowing n_a exactly leads to errors of only 5% in n_c [18]. Can this accuracy be preserved for dispersion analysis, i.e., can the oscillator parameter be determined as accurately as the optical constants? Unfortunately this is not possible because the position of the band of a perpendicular mode is a function of not only the oscillator position but also the oscillator strength due to the TO-LO shift and, accordingly, also of the dielectric background. This means that there exists an infinite number of triples of these two variables that lead to comparably similar bands in the spectrum. This is a big problem not only because there is no indication in the spectrum where the TO position is located, but also because the perpendicular dielectric background cannot be determined accurately as is shown in Fig. 14.12. Even if $\varepsilon_{\infty,\perp}$ would be known accurately, the minimum in Fig. 14.12, left panel, is not well pronounced. As a consequence, even starting values comparably close to the true values do usually not lead to results close to these true values. In addition, a comparably small error of 5% shifts the minimum away from $S_\perp = 500$ cm^{-1}, $\tilde{\nu}_\perp = 1000$ cm^{-1} to about $S_\perp = 470$ cm^{-1}, $\tilde{\nu}_\perp = 1005$ cm^{-1} if the reflectances for two angles of incidence, 60° and 70° are combined. It seems that under these circumstances, dispersion analysis cannot be applied successfully. In fact, using specular reflectance spectra this problem seems not to be solvable. Therefore, I thought to circumvent the difficulties by employing ATR spectroscopy [19]. The usefulness of ATR spectroscopy becomes immediately obvious if we compare the spectra obtained by specular reflectance and attenuated total reflection as in Fig. 14.13.

Obviously, the minima in ATR spectra are located in the vicinity of the oscillator positions, including the perpendicular mode, which means that its otherwise unknown position is, at least approximately, revealed. Therefore, good starting values for the oscillator positions can be chosen. But this is not the only advantage. In fact, using one conventional reflectance and one ATR spectrum even fixes the problems that occur if $\varepsilon_{\infty,\perp}$ is not accurately known as this is obvious from the right panel of Fig. 14.14.

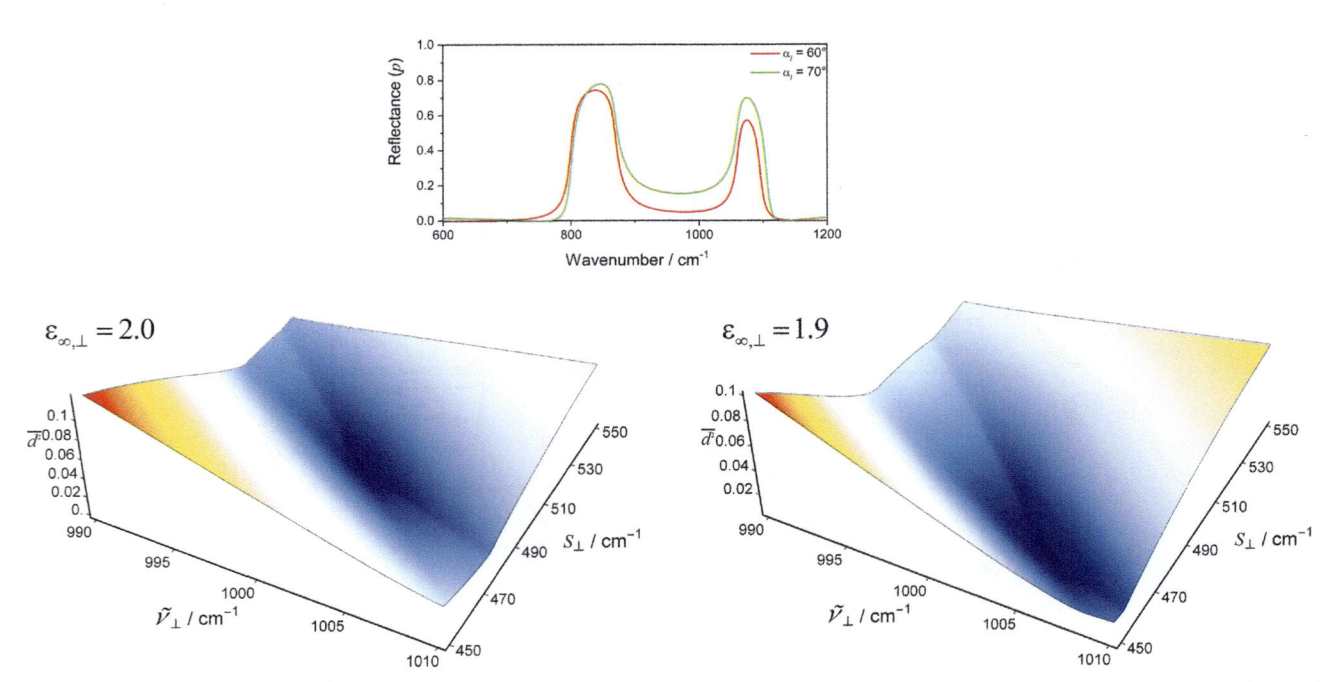

FIG. 14.12 Mean square error dependence on the oscillator position and the oscillator strength for the reflectance spectra with p-polarized light and incidence angles 60° and 70° for the case that the correct dielectric background $\varepsilon_{\infty,\parallel}=2.0$ is known *(left panel)* and the case that erroneously $\varepsilon_{\infty,\parallel}=1.9$ is assumed [19].

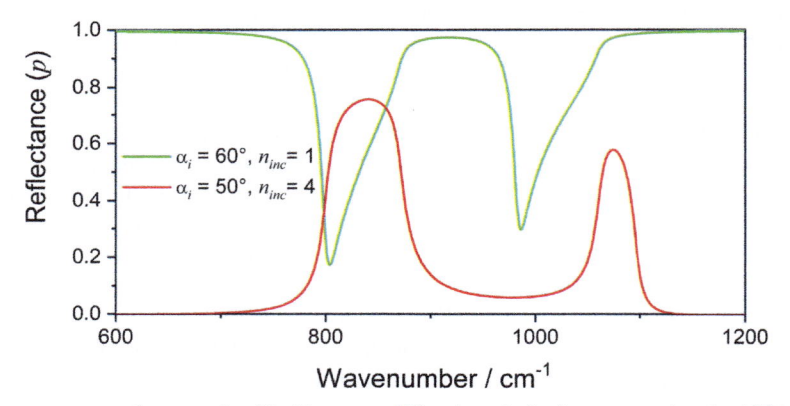

FIG. 14.13 Specular reflectance spectrum for an angle of incidence $\alpha_i=60°$ and p-polarization compared to the ATR spectrum for $\alpha_i=50°$.

Even when $\varepsilon_{\infty,\perp}$ is accurately known, the use of one ATR spectrum is of obvious advantage as it improves strongly the markedness of the minimum as the comparison between Fig. 14.14 and Fig. 14.12 illustrates. From this it can be concluded that dispersion analysis is now enabled in principle, which means that artificial spectra can be analyzed by it and the original oscillator parameters regained, at least for the previous case where the peaks belonging to the two different principal dielectric functions do not overlap. In practice, however, previous method is seriously flawed. The reason for this flaw is that intimate contact needs to be established between the ATR-crystal (the incidence medium) and the sample, which would be a single crystal or a layer. To assure this intimate contact, there is so far only one way, which is to let the crystal or the layer grow with the ATR-crystal as substrate like this has been done, e.g., for certain nitrates [20]. Even where this is possible and the correct growth of the crystal with one crystal axis perpendicular to the surface of the ATR-crystal can be assured and this crystal can be sacrificed, the mismatch between the structure of the ATR-crystal and the sample on the level of the elementary cells might change the spectra by the induced strain. Another approach is to prepare a liquid or a soft solid thin film onto the ATR-crystal which would improve the contact indirectly. We tried both CS_2 as well as polyethylene as contacting agents, but both solutions did not work well enough. As comparisons between the experimental and modeled

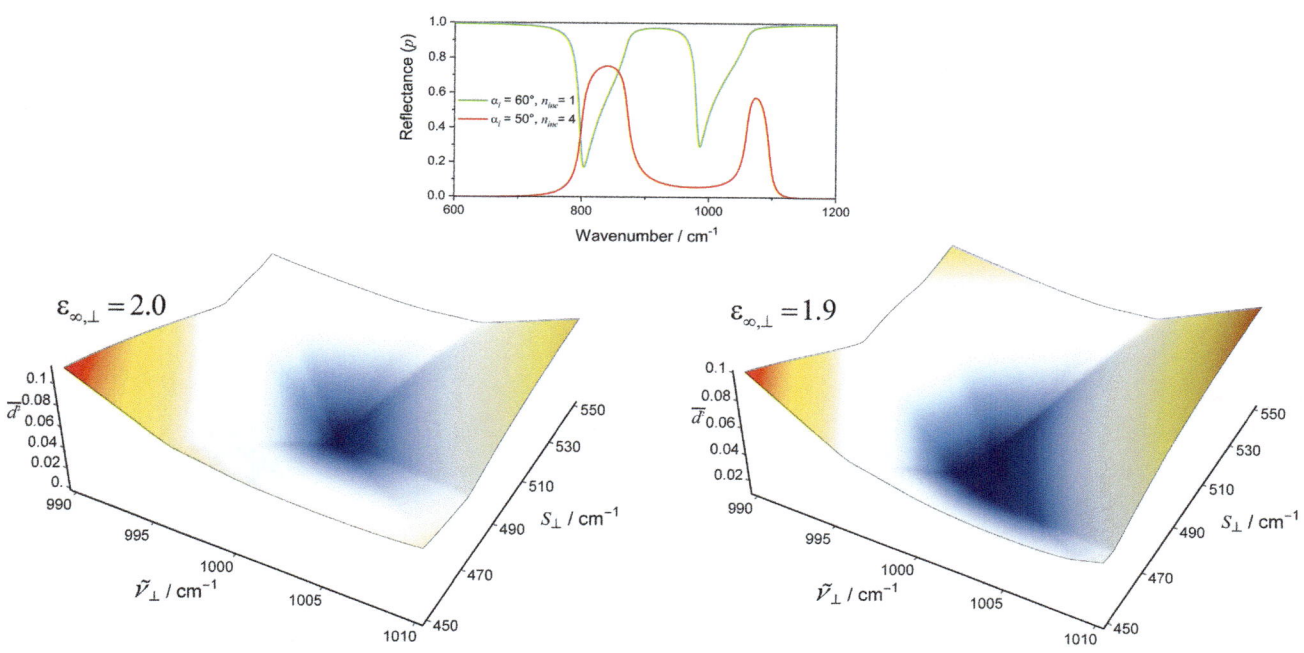

FIG. 14.14 Mean-square error dependence on the oscillator position and the oscillator strength for the case that the correct dielectric background $\varepsilon_{\infty,\parallel} = 2.0$ is known *(left panel)* and the case that erroneously $\varepsilon_{\infty,\parallel} = 1.9$ is assumed when one conventional reflectance and one ATR spectrum is employed (*p*-polarized light and incidence angles 50° and 60°) [19].

spectra showed that the former follow the predicted trends only in a semiquantitative way. Therefore, applying dispersion analysis is not useful (the only way the experimental spectra could indeed help is to provide a good estimate for the TO oscillator position!).

While I was working on how the oscillator strengths of direct and inverse dielectric function are connected (Section 5.6), I found a better solution, which does not require recording an ATR spectrum. As discussed earlier, we cannot determine from a conventional reflectance spectrum the TO-position of the perpendicular oscillator, but what we get from spectra is a good estimate of the LO-position [16]. It is therefore tempting to assume that modeling the principal component of the dielectric tensor along the perpendicular axis with an inverse dielectric function could help since in this case we only need the LO-, but not the TO-position. Indeed, as Fig. 14.15 shows, the squared error has a well-developed minimum, even if only one conventional reflectance spectrum is employed. When the LO-oscillator parameters are known, then $\varepsilon_{\infty,\perp}$ is obtained simply by inverting the inverse dielectric function. $\varepsilon_{\infty,\perp}$ can then be fitted in the conventional way and the TO-oscillator parameters are obtained.

It looks like as this would be a simple solution to the problem, but what works in theory certainly does not have to do this in practice. To see if it can work, I have used a spectrum of a Fresnoite single crystal recorded from a *c*-cut at 50° angle of incidence and with *p*-polarization. This spectrum is compared to the negative imaginary principal component of the inverse dielectric function tensor along the perpendicular direction in Fig. 14.16. As is obvious and as we have already discussed in Section 5.6, mode coupling of the LO-modes leads to the effect that most of the oscillator strength is concentrated in the mode with the highest wavenumber. Since the position of the corresponding peak is well beyond the peak position of the highest ε_a peak, it should be possible to determine its parameters by dispersion analysis. The other peaks, however, should be comparably weak and their footprints in the spectrum either hard to detect or not detectable at all.

This seems to be not the only problem. The lower part of Fig. 14.16 shows a simulation based on the dielectric function tensor derived from reflectance spectra of an *a*-cut [21]. The *a*-cut allows for the separation of the principal components of the dielectric tensor function when using *s*-polarized light since if the polarization direction is oriented parallel to the *c*-axis, the spectrum only depends on the component in *c*-direction and the same holds true for the *a*-axis if the crystal is rotated by 90° around the surface normal (cf. Section 14.1). The comparison between the experimental spectrum (black spectrum, upper part) and the simulation (green spectrum) shows that relative intensities, band positions, and band shapes are generally well predicted by the simulation. However, the overall intensities are about 15% higher in the experimental spectrum, at least in the MIR spectral region. If the experimental spectrum is divided by a factor of 1.15, the resemblance between the scaled experimental spectrum and the simulation is very good (lower part of Fig. 14.16), except at around 1075 cm^{-1} and

334 PART | II Tensorial theory

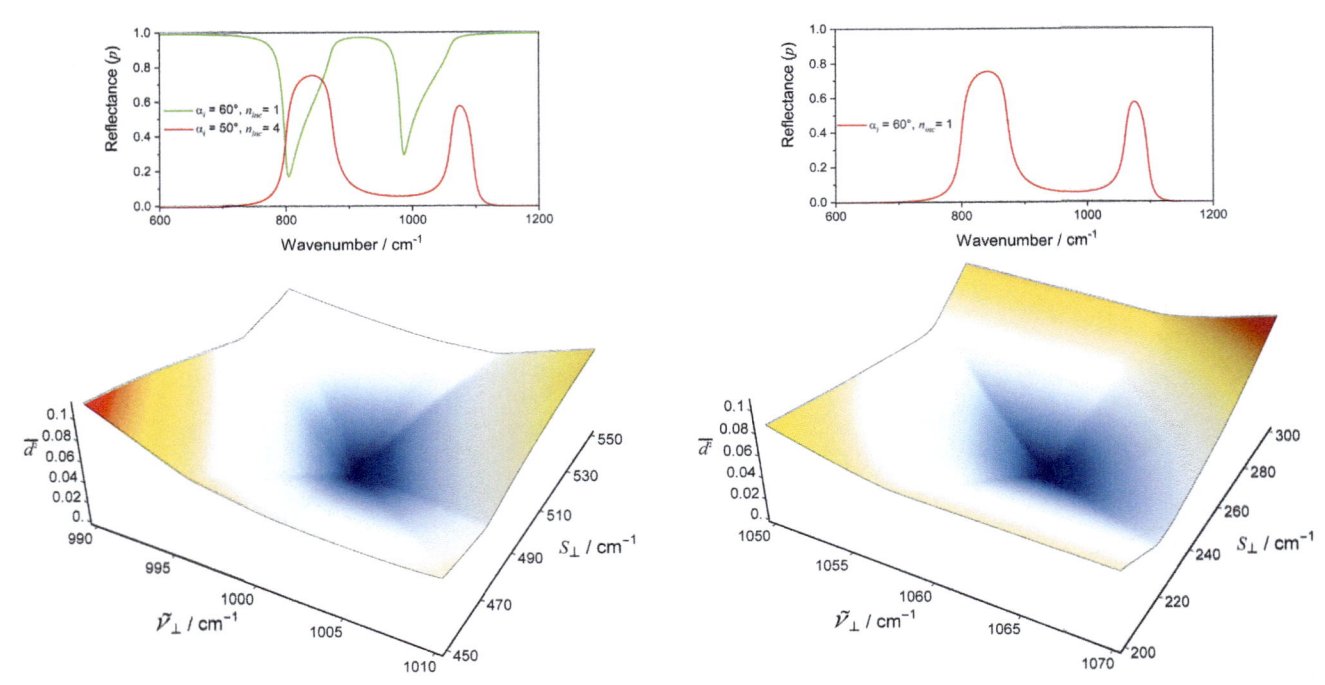

FIG. 14.15 Mean-square error dependence on the oscillator position and the oscillator strength for the case that one specular and one ATR spectrum is employed *(left panel)* and when only a single specular reflectance spectrum is available, but an inverse dielectric function model for the perpendicular principal component of the dielectric tensor is employed [16].

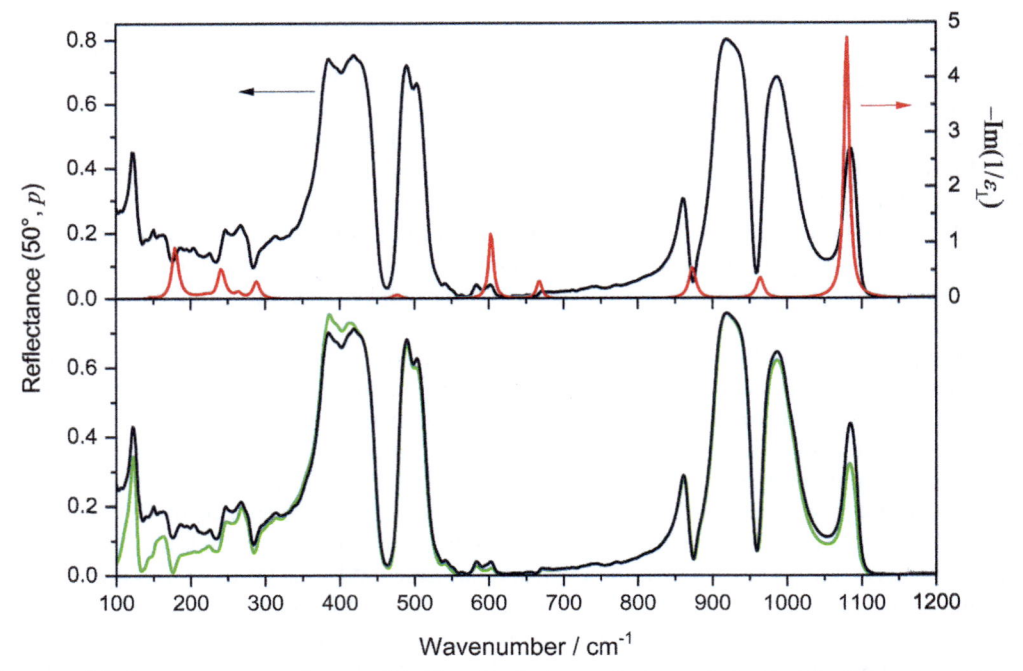

FIG. 14.16 *Upper panel*: Reflectance spectrum of a *c*-cut of the Fresnoite single crystal with *p*-polarized incident light and an angle of incidence of 50° *(black curve)* compared to the negative imaginary part of the inverse principal component of the dielectric tensor function **perpendicular** to the surface of the crystal. *Lower panel*: Comparison of the scaled experimental spectrum (factor: 0.87) and a spectrum simulated based on single crystal data [21].

below 400 cm^{-1}. The deviations below 400 cm^{-1} are likely due to larger experimental errors in the far-infrared region of the spectrum. The reason for the underestimation of the peak intensity of the band with the highest wavenumber is unclear. The dielectric tensor function may be somewhat less accurate for the component of the dielectric tensor function along the optical axis in this region, but the extent of the deviation around 1075 cm^{-1} cannot be explained by an error in the determination of the dielectric tensor function.

Overall, because of the LO-mode coupling, it is very difficult to identify any bands in the experimental spectrum that have their transition moments parallel to the surface normal, i.e., to the c-axis, apart from the band at around 1075 cm^{-1}. One could assume that at higher angles of incidence this would be easier, but since the inverse dielectric function is less intense than the dielectric function itself (roughly by a factor given by the dielectric background of the c-axis $\varepsilon_{\perp,\infty}$), due to the mode coupling and the overlap with the parallel modes, the perpendicular modes are hard to identify even at 80° angle of incidence.

In Fig. 14.17A, the principal reflectance spectrum belonging to ε_a can be seen for an angle of incidence of zero and how it is influenced by the c-axis modes for increasing angle of incidence. Even for larger angles of incidence, the spectrum is still heavily influenced by the parallel modes, as seen at the cut-off at 80° angle of incidence. The influence of the perpendicular modes is weak, except for the band with the highest wavenumber, which is beyond the parallel mode with the highest wavenumber around 1000 cm^{-1}. This is obvious from Fig. 14.17B where the parallel modes have been switched off for the simulation. In addition, the oscillator strengths of the inverse dielectric functions are, as already mentioned, strongly coupled, and a large part of the oscillator strengths is transferred to the mode located at the highest wavenumber, which is demonstrated in Fig. 14.17C.

This explains why the highest wavenumber mode is dominant in the experimental spectra. To create Fig. 14.17C, the TO-oscillator strengths of the two perpendicular modes adjacent to the highest wavenumber perpendicular mode were increased from 0% to 200% (while the parallel modes remained off). Even though the TO-oscillator strength of these two modes was higher than that of the highest wavenumber perpendicular mode at 200%, their spectral footprints stayed much weaker. Mode coupling is also evident, as the peak values of the highest wavenumber mode not only increases, but also its wavenumber position is increasingly blueshifted due to the transferred oscillator strength.

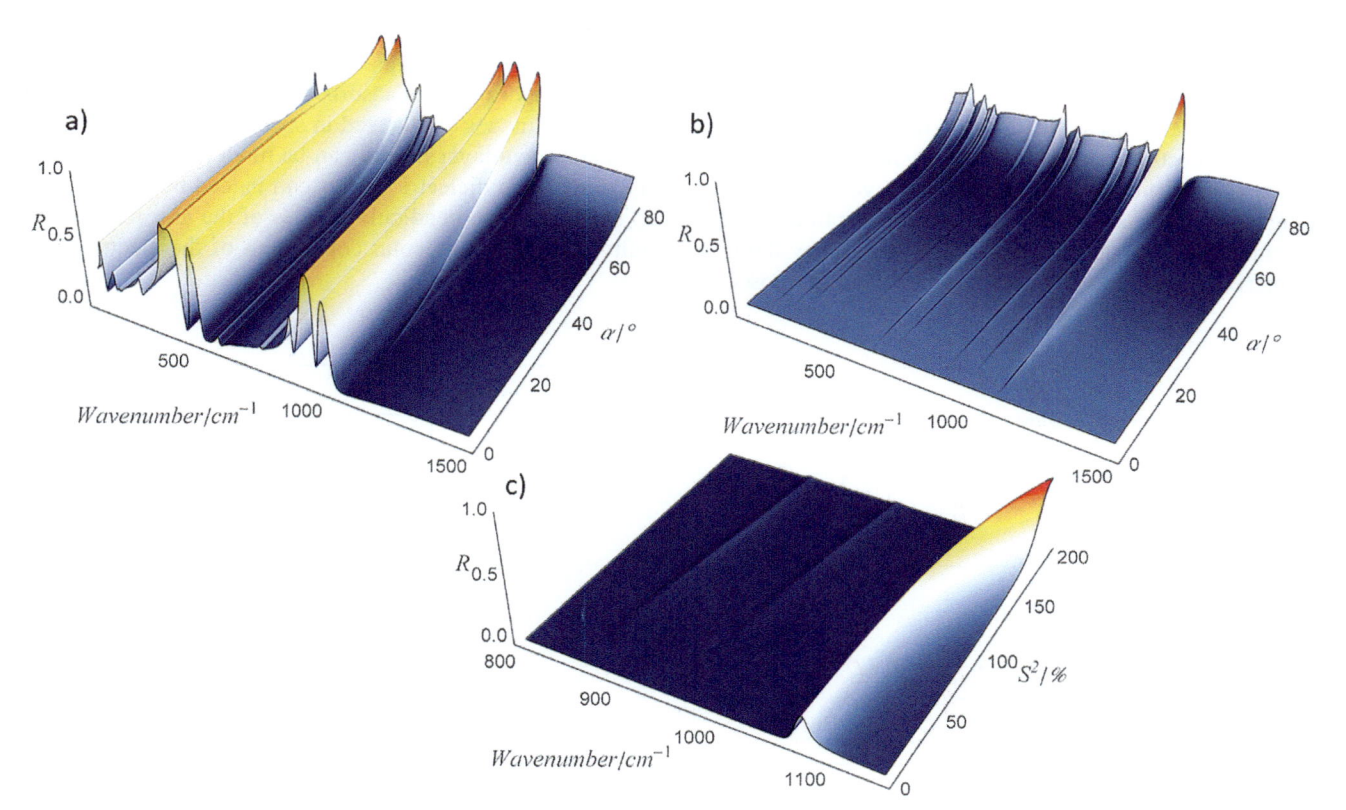

FIG. 14.17 (A) Dependence of the c-cut spectrum of Fresnoite on the angle of incidence. (B) Same as (A), but with the parallel modes switched off. (C) Mode coupling at 50° angle of incidence. The oscillator strengths of the two modes below the highest wavenumber perpendicular mode is varied from 0% to 200% (the parallel modes are still switched off for better visibility) [21].

This is why it is difficult to determine the oscillator parameters from c-cuts, even when $\varepsilon_a = \varepsilon_\parallel$ has been accurately determined beforehand. The perpendicular mode around 1075 cm^{-1} is easy to identify, but the following modes do not reveal their presence in the spectrum. The perpendicular mode around 965 cm^{-1} is close to a minimum in the spectrum and the reflectance begins to steeply increase without any indication of its existence. The perpendicular mode at around 875 cm^{-1} is difficult to pinpoint due to its similarity in position and strength to a doubly degenerated mode. The peak value is higher than it should be if only the parallel mode is present, suggesting that there is actually an accidentally triply degenerated mode at this position (although strength and position of perpendicular and parallel mode do not agree perfectly). The next perpendicular mode can be detected since there is a small feature at 667 cm^{-1}, and the same is true for the one after it, indicating its presence by a small peak at about 600 cm^{-1} (cf. Fig. 14.16). Unfortunately, this completes the list, while there are actually 7 more modes. One of them is close to the minimum at around 460 cm^{-1} and, like the second perpendicular mode, in the range of the steeply increasing band around 500 cm^{-1}. However, it is much weaker so that simply putting an oscillator in this range and seeing if it generates an improvement of the fit does not work. Even using prior knowledge and employing the parameters found out from dispersion analysis of the a-cut did not work for the second mode, as the oscillator would either lose strength until it reaches zero or the position would change until it is moved out of range. The same happened in general for any oscillator in the FIR spectral region. While there should be 6 oscillators in this region, none of added oscillators would lead to any improvement of the model. Obviously, the differences between expected and experimental spectrum are too strong in this spectral region (cf. Fig. 14.16). Therefore, the dispersion analysis was restricted to the five highest wavenumber modes in the MIR, but even in the case that an advanced *smart error sum* based on a hybrid 2T2D *smart error sum* (cf. Section 8.3) and the conventional residual sum of squares is applied, the result only semiquantitatively resembles the one from the analysis of an a-cut, see Fig. 14.18 [21].

The comparison of experimental and modeled spectrum in Fig. 14.18 shows that the agreement is quite good. The resulting dielectric function $\varepsilon_\perp$ however, only semiquantitatively resembles that determined from the a-cut. Peak heights and positions (which equal the oscillator positions) are definitely not exact matches to the reference values. I attempted to include $\varepsilon_{\perp,\infty}$ as a fit parameter, but it quickly increased to unrealistic values. I therefore have to conclude that the dielectric functions from this kind of dispersion analysis are only semiquantitative at best. What may be possible, however, is to use an ATR measurement to determine the TO-position and use the Berreman-Unterwal model (cf. Section 5.5.1) to fit the perpendicular component of the dielectric tensor function.

What also works (so far only simulations are published, but we have already established that modeling of experimental spectra is possible), is to model the dielectric tensor functions of monolayer of gases adsorbed onto dielectric substrates [22,23]. In these cases, the gases do not have overlapping modes and the modes are located in spectral regions where the dielectric substrates are nonabsorbing. It may be very surprising that a monolayer indeed can be modeled with the 4×4 matrix formalism, but despite the monolayers small thickness a macroscopic dielectric tensor can indeed be successfully assigned to it. This includes a correct description of the TO-LO shift of the perpendicular modes of the absorbed gases, proving again that the TO-LO shift is caused by the discontinuity of the perpendicular component of the electric field at interfaces and does not have some obscure microscopic origin.

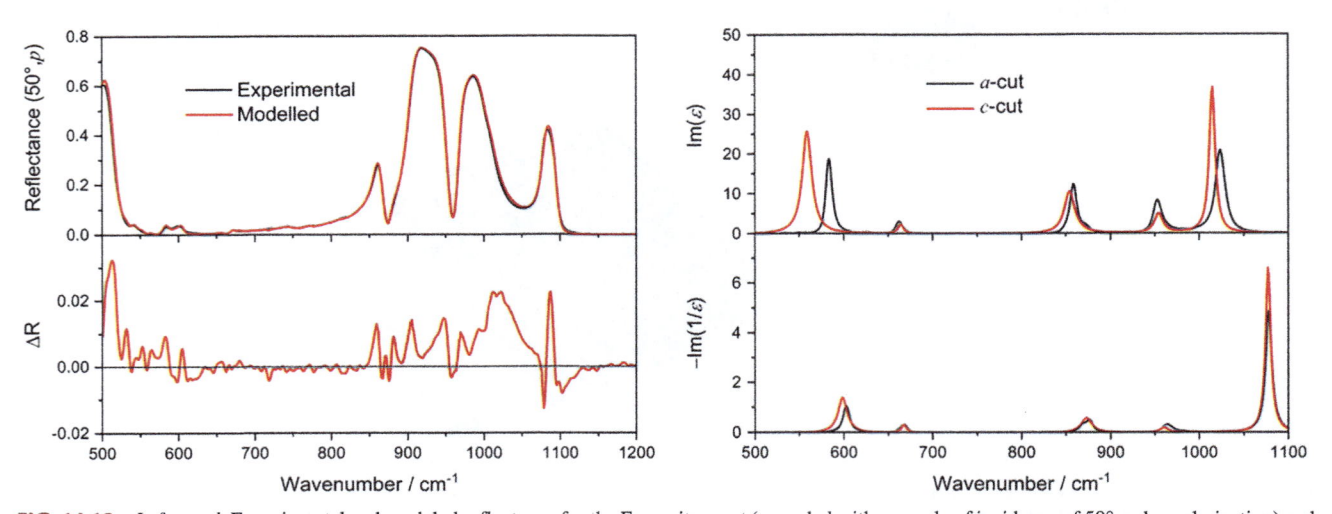

FIG. 14.18 *Left panel*: Experimental and modeled reflectance for the Fresnoite c-cut (recorded with an angle of incidence of 50° and p-polarization) and corresponding difference reflectance ΔR. *Right panel*: Resulting loss functions *(lower part)* and imaginary part of the dielectric function for the perpendicular component of the dielectric function tensor and comparison with reference values from an a-cut [21].

14.5 Triclinic crystals

For a long time, dispersion analysis of triclinic crystals was the supreme discipline in the sense that actually no one was able to perform it. There had been one attempt in 1985 where Aronson, Emslie, and Strong tried to determine the dielectric function tensor of $CuSO_4 \cdot 5\,H_2O$ (blue vitriol) [24], after they laid the ground for it by providing the necessary theory in 1983 [25]. I have to call it an attempt for two reasons: First of all, according to theory, Aronson and Emslie concluded that 6 different measurements are necessary to determine the 6 unknowns in the dielectric tensor, i.e., 3 diagonal elements and 3 off-diagonal elements (the other 3 off-diagonal elements are linked by symmetry as long as the crystal does not show optical activity). But when they performed their analysis, they only used 3 spectra from 2 crystal faces that were orthogonal to each other. Their crystal had the form of a cube, but they did not use the third orthogonal face because it "was a somewhat pitted face." It is therefore obvious that they could not have succeeded.

When we made our own attempt to analyze a similar crystal in 2010, we began to understand that it anyway would not have been easy to succeed in 1985, in particular because of the missing CPU power. However, we also soon learnt that six measurements are not enough, even when you make sure, that the spectra are as different as possible. To do this, Aronson's, Emslie's, and Strong's idea to use a cube is in my opinion the correct way. You then take two spectra from each of the three orthogonal faces (spectra from opposing faces with the same polarization direction are identical), with s-polarized light and the polarization directions along the edges as this is depicted in Fig. 14.19. Note that for an orthorhombic crystal recording two spectra with the polarization direction along the same edge would result in identical spectra, while these are supposed to be different for triclinic crystals.

We tried for some time to fit the 6 so-gained spectra of blue vitriol, but there were strong problems with convergence. I found the solution of the riddle by investigating the error sums in a way analogous to the one used to understand the results in the preceding section.

As Fig. 14.20 shows, 6 measurements are not enough. If the transition moment is oriented along the space diagonal, the global minimum cannot be discerned from the other three local minima, therefore the four possible orientations ($\Phi = 45°$, $\Theta = 45°$), ($\Phi = 45°$, $\Theta = 135°$), ($\Phi = 135°$, $\Theta = 45°$), and ($\Phi = 135°$, $\Theta = 135°$) cannot be distinguished. Even when the orientation is not one of these extreme cases, the problem of local minima still remains as long as the transition moment is not oriented along one of the edges. Since fitting 6 spectra together using the 4×4 matrix formalism without any simplification is already computational expensive and takes a lot of time, the existence of such convergence problems seemed to render dispersion analysis virtually impossible. On the other hand, the solution to the problem is quite simple, which is to use twelve measurements instead of six, the original six one and, additionally, six along the diagonals of the three cube surfaces as illustrated in Fig. 14.21. Obviously, the global minimum can then unambiguously be determined.

The results of the twelve measurements for blue vitriol are shown in Fig. 14.22. By inspecting this figure, it becomes obvious that blue vitriol is something I would call pseudo-orthorhombic since the spectra from different faces that share the same polarization direction (parallel to one of the edges) are remarkably similar to the point where some of the pairs cannot be distinguished by the naked eye like the first two red spectra from the left. While the results of dispersion analysis convinced us that we made the first successful analysis of a triclinic crystal [26], we thought it might be necessary to search for a

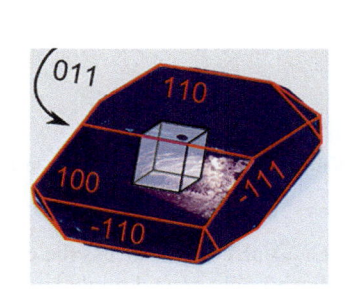

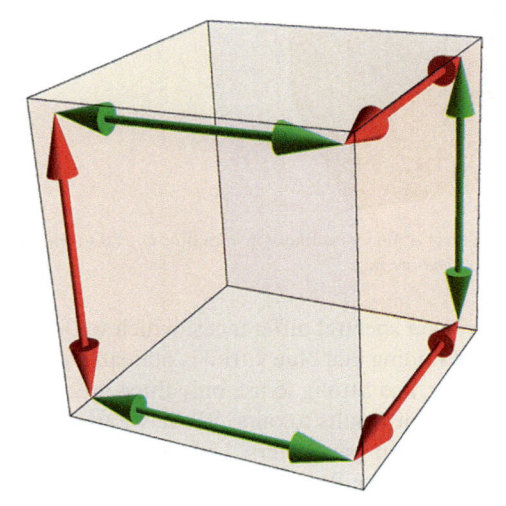

FIG. 14.19 Blue vitriol single crystal with indicated orientation of the faces and how we cut a cube out of it *(left panel)*. Two measurements from the three orthogonal faces of the cube were recorded with polarization directions indicated by the *red* and *green arrows (right panel)*.

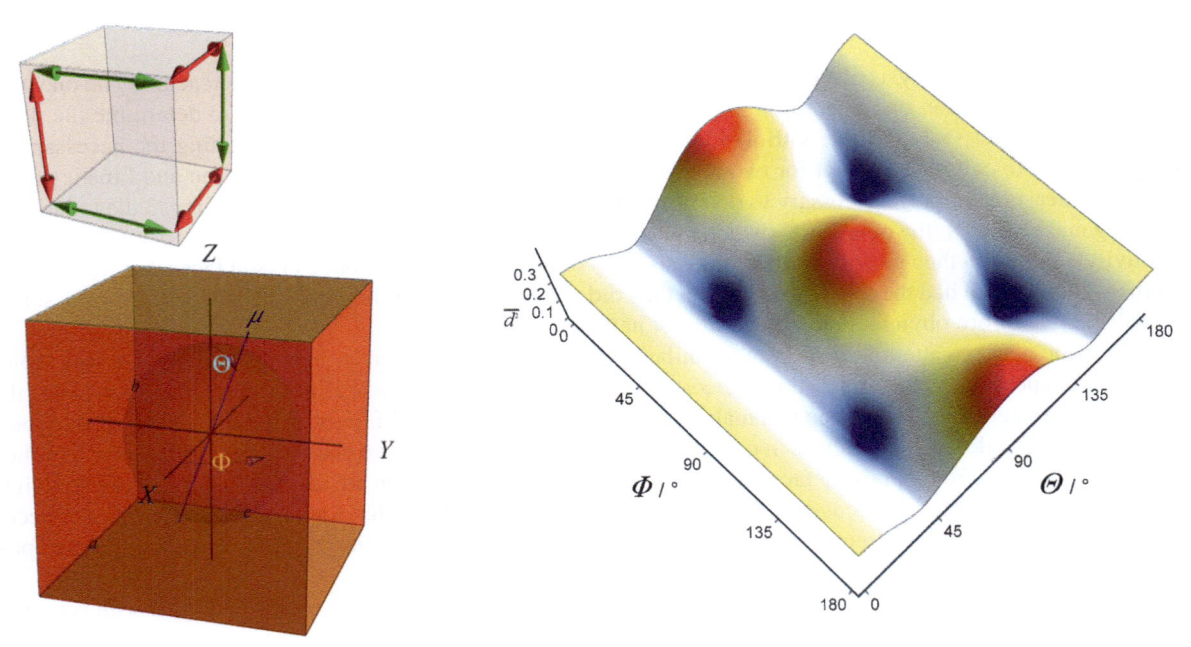

FIG. 14.20 Mean-squared error $\overline{d^2}$ for an oscillator in a triclinic crystal in dependence of the orientation of its transition moment using 6 measurements as indicated by the *green* and *red* arrows.

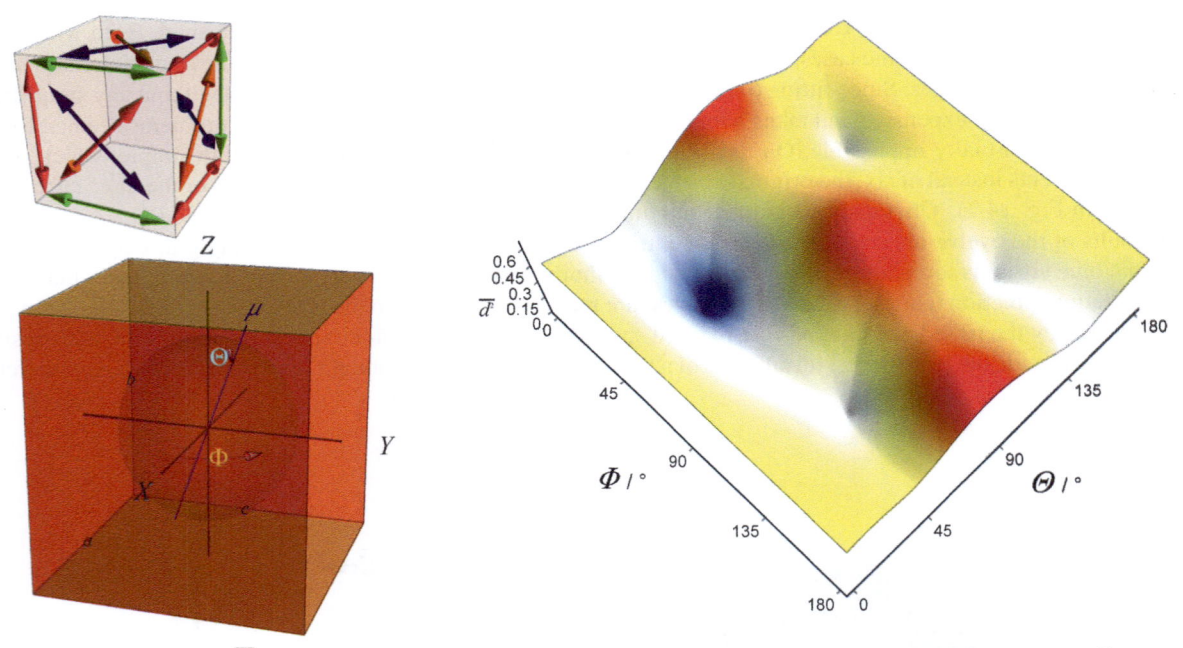

FIG. 14.21 Mean-squared error $\overline{d^2}$ for an oscillator in a triclinic crystal in dependence of the orientation of its transition moment using 12 measurements as indicated by the *green* and *red* arrows.

triclinic crystal with stronger spectral differences, which we found in $K_2Cr_2O_7$. Before I discuss the corresponding spectra, I want to express that the finding that blue vitriol is optically a pseudo-orthorhombic crystal might have also played a role to convince Aronson, Emslie, and Strong to use only three spectra for their dispersion analysis. I also want to reveal why we never directly compared their results to ours: While we tried hard and repeatedly, we never could get meaningful optical constants out of their oscillator parameters. Unfortunately, they used an oscillator position normalized form of the dispersion equation so that their oscillator strengths and damping constants cannot be directly compared. Using conversed parameters, it looks like their damping constants are much too large.

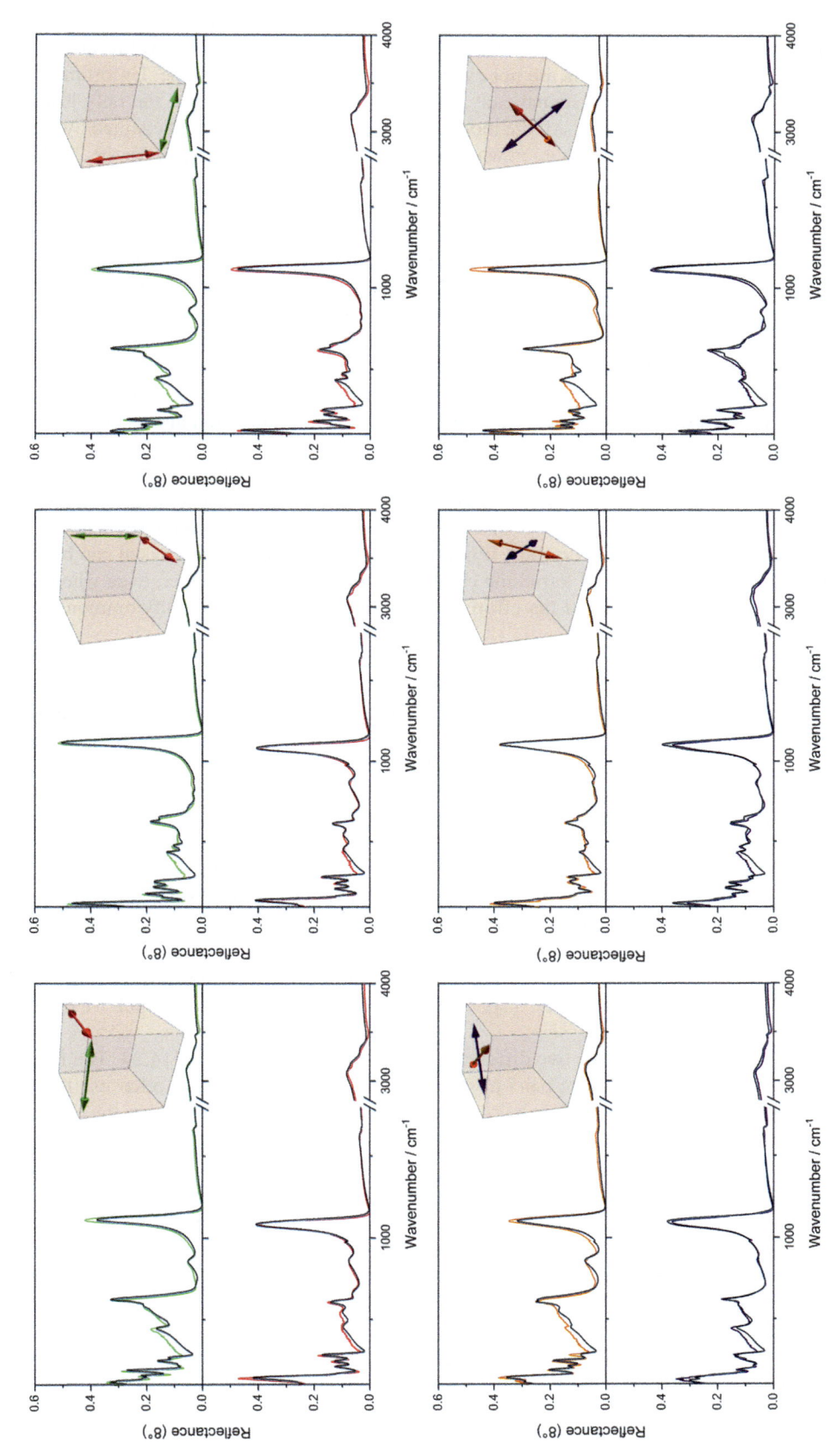

FIG. 14.22 The twelve experimental reflectance spectra and corresponding modeled spectra (*in black*) for *blue* vitriol.

One additional remark to pitted faces. It looks like the surface quality of the different faces is generally for all crystals not the same. Therefore, we found it necessary to adapt the reflectance by a factor. Remarkably, this factor is also depending on the polarization direction so that the factors are not even the same for a particular face. Unfortunately, I never came across a work where this would have been systematically investigated. As a consequence, the only justification of introducing these correction factors is that it works. All the more it is important to check the data gained by dispersion analysis by an independent experiment. Note that this situation would be an ideal application example for the 2T2D smart error sum (cf. Section 8.3).

In the case of $K_2Cr_2O_7$ we did not cut the single crystal into a cube because its natural faces were not that far away from being orthogonal to each other. Certainly, it is then necessary to adapt the rotation matrices correspondingly when the reflectance is calculated by the 4×4 matrix formalism. In addition, for low angles of incidences, a speed-increasing measure is to use both, s- and p-polarized light since the formalism always calculates both spectra. This also has the advantage that instead of rotating the sample by 90° around its face normal, the polarizer is rotated, something which can usually be done with higher accuracy (note that the angle of incidence was 8° and was considered properly, although for this angle of incidence, rotating the polarizer instead of the sample does not lead to very different spectra). The experimental spectra together with the best fits are depicted in Fig. 14.23. From the spectra it is obvious that $K_2Cr_2O_7$ is not pseudo-orthorhombic since the spectra taken from two faces with polarization direction along the same edge differ much stronger than those of blue vitriol.

Up to now (February 2024), to my best knowledge, only 4 correctly implemented dispersion analyses of triclinic crystals in the infrared have been carried out [26–29]. In addition to those, there exists one dispersion analysis of $K_2Cr_2O_7$ of ellipsometry data in the UV-vis [30]. The smaller number of successful analyses is an obvious consequence of the much higher complexity compared to those of monoclinic crystals. While the spectra of a monoclinic crystal can be analyzed within a couple of hours, the analysis of the spectra of a triclinic crystal can take months. However, even when much more interventions of the operator are necessary, most of the effort is computing time, which is in my opinion well spent.

14.6 Generalized dispersion analysis

The idea behind generalized dispersion analysis is to carry out dispersion analysis without any a priori knowledge of crystal symmetry and orientation. It goes back to my former Post-doc Sonja Höfer who also implemented every step of the method. Sonja's basic idea was to use the same procedure that is applied to a triclinic crystal for arbitrary crystals regardless of symmetry and orientation. Accordingly, the crystal of interest would be cut into a cube (under preservation of a prominent surface) and exactly the same measurements as for triclinic crystals would be performed. If a crystal has triclinic symmetry, the orientations of the transition moments would be uncorrelated, whereas any correlation among the orientations would indicate a crystal of higher symmetry. On the other hand, if all 12 spectra would be identical, the symmetry would be cubic. Therefore, the two cases, triclinic and cubic, can easily be identified. In the next step, uniaxial, orthorhombic and monoclinic symmetry needs to be distinguished and the particular orientation and the dispersion parameters (= oscillator parameters + dielectric background tensor) need to be determined.

Originally, we separated the problem in two parts and showed first that when we a priori only know the crystal symmetry, but not the orientation, we can determine the dispersion parameters unambiguously, which we proved separately for all three symmetries [31–33].

In a uniaxial crystal the original fit reveals two different kinds of orientations, cf. Fig. 14.24. The first kind belongs to oscillators along the optical axis which all have pairs of Φ, Θ which are similar while the other oscillator parameters are different. The second kind is oscillators that have in pairs the same oscillator parameters because the oscillators are twofold degenerated. This also means that the orientation of their transition moment is both perpendicular to each other and to the optical axis. This relationship is not directly revealed, but it can be exploited by either using corresponding orthogonality relations as Sonja preferred or using generalized oscillator models (cf. Eq. 13.14) which is what I favor.

In any way, both methods work reliably. The accuracy of the determined principal dielectric functions and the dispersion parameters are certainly not as good as if a suitable principal cut would be employed, but overall the agreement between generalized and common dispersion analysis is quite satisfactorily [31]. For orthorhombic and monoclinic crystals similar orthogonality relations and generalized dispersion formulas exist (cf. Eqs. 13.16, 13.17) [32,33]. Unfortunately, these conditions are only necessary, but not sufficient to allow generalized dispersion analysis due to the vast parameter space in these cases. After some struggling with the problem, we adopted an idea that I had earlier in relation with determining the orientation in oriented glass ceramics, where I used in addition to one polarizer experiments measurements with crossed polarizers to determine orientation (cf. Section 15.3) [34]. Put into concrete terms, it is necessary to support the twelve one polarizer experiments with another twelve measurements with the same sample orientations, but employing

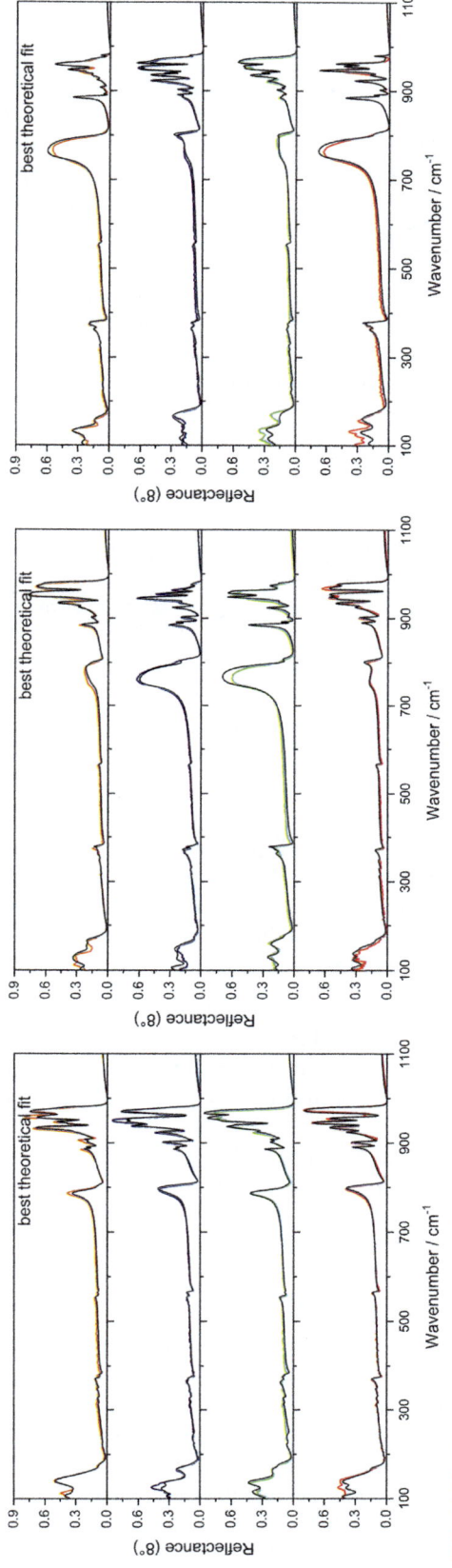

FIG. 14.23 The twelve experimental reflectance spectra and corresponding modeled spectra (*in black*) for $K_2Cr_2O_7$.

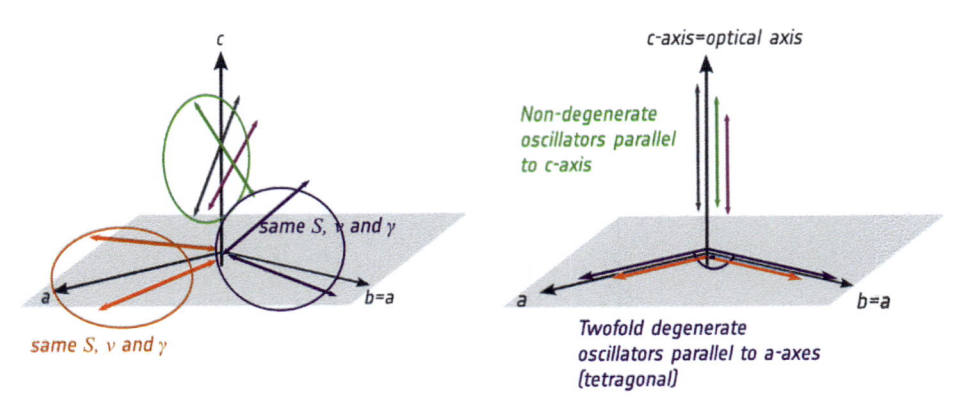

FIG. 14.24 Orientation of oscillators resulting from an unconstrained fit *(left)* and from a constrained fit *(right)*. Orthogonality condition or the generalized oscillator model allow to determine the orientation of the optical axis reliably [31].

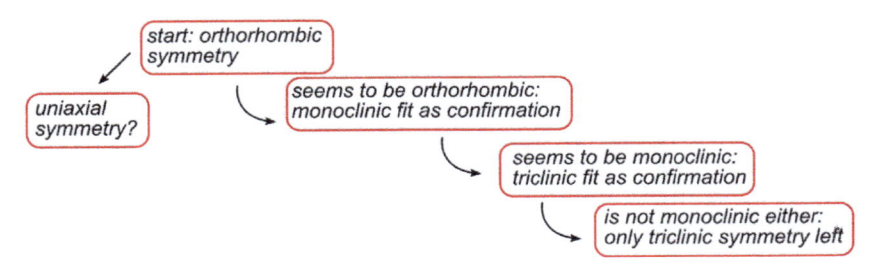

SCHEME 14.1 Illustration of the fitting procedure to obtain a priori unknown crystal symmetry and orientation [35].

crossed polarizers. This procedure enabled generalized dispersion analysis for orthorhombic and monoclinic crystals with about the same accuracy as for the uniaxial case [32,33]. Based on the additional cross-polarization spectra it was possible to perform dispersion analysis when the orientation was unknown. What still was missing was how to determine if a crystal is optically uniaxial, orthorhombic or monoclinic when this is not known a priori. The complete solution Sonja came up finally with is the following (cf. Scheme 14.1) [35]:

1. Perform dispersion analysis in the same way as for an orthorhombic crystal using the full set of 24 measurements (12 one polarizer measurements and 12 cross polarization measurements).
2. If some of the oscillators appear in pairs, i.e., with about the same oscillator strengths, damping constants and oscillator positions, the crystal is probably uniaxial. Confirm this by performing generalized dispersion analysis for uniaxial crystals.
3. If no oscillators appear in pairs, the crystal might be orthorhombic. To confirm this, perform generalized dispersion analysis under the assumption of monoclinic symmetry. If this less restrictive fit leads to the same orientations as the orthorhombic fit, this symmetry is confirmed.
4. If the orthorhombic symmetry is not confirmed, perform a final fit by releasing all constraints, i.e., perform an analysis as if the crystal is triclinic. If you obtain the same orientations as in the case of the monoclinic fit, the crystal is monoclinic; otherwise, it is triclinic.

Overall, we found that the orientation of the crystal can be determined with an accuracy of 2°–5° regarding the orientation angles, which depends on the optical anisotropy (the accuracy increases with the anisotropy). The same order of magnitude concerning the errors applies to the orientations of transition moments in general.

Depending on the quality of the spectra, dispersion parameters can be reproduced with errors that usually lie within about 5% and 20%. Note that, very much like for triclinic crystals, the surface quality of the faces can be expected to differ and correction factors have to be introduced (or, alternatively the employment of the 2T2D *smart error sum*, cf. Section 8.3, could be useful). Exemplary spectra for crystals of different symmetry and their best fits are shown in Fig. 14.25. Up to date (February 2024), nobody else has successfully employed generalized dispersion analysis.

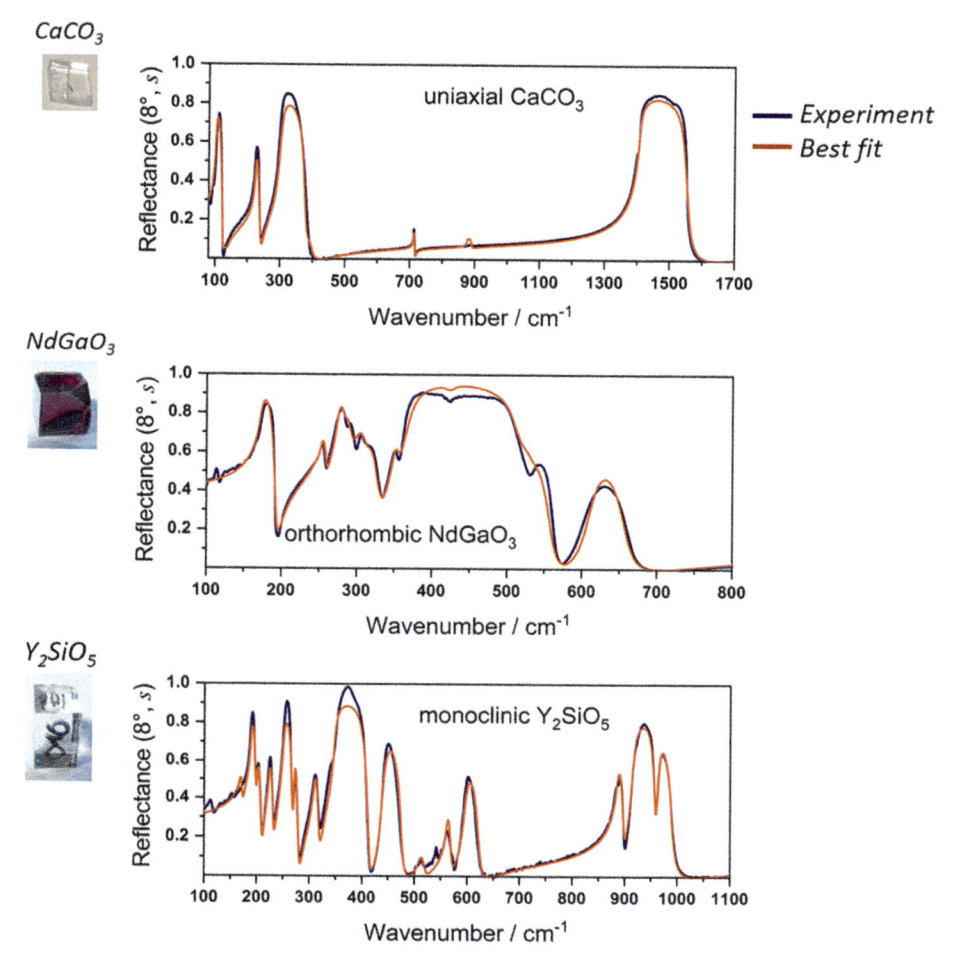

FIG. 14.25 Exemplary experimental spectra and their best fits obtained by generalized dispersion analysis.

14.7 Further reading

Unfortunately, there is no other literature that I could suggest as alternative.

References

[1] W. Spitzer, D. Kleinman, D. Walsh, Infrared properties of hexagonal silicon carbide, Phys. Rev. 113 (1959) 127–132.
[2] T.C. Damen, S.P.S. Porto, B. Tell, Raman effect in zinc oxide, Phys. Rev. 142 (1966) 570.
[3] T.G. Mayerhöfer, T. Höche, F. Schrempel, Infrared optical properties of Li- and Xe-irradiated KTiOPO$_4$, Appl. Phys. A Mater. Sci. Process. 78 (2004) 589–596.
[4] S. Höfer, R. Uecker, A. Kwasniewski, J. Popp, T.G. Mayerhöfer, Complete dispersion analysis of single crystal neodymium gallate, Vib. Spectrosc. 78 (2015) 17–22.
[5] V.F. Pavinich, M.V. Belousov, Dispersion analysis of the reflection spectra of monoclinic crystals, Opt. Spectrosc. 45 (1978) 881–883.
[6] V. Ivanovski, T.G. Mayerhöfer, J. Popp, Investigation of the peculiarities in the polarized reflectance spectra of some Tutton salt monoclinic single crystals using dispersion analysis, Vib. Spectrosc. 44 (2007) 369–374.
[7] V. Ivanovski, T.G. Mayerhöfer, J. Popp, Isosbestic-like point in the polarized reflectance spectra of monoclinic crystals – a quantitative approach, Spectrochim. Acta A Mol. Biomol. Spectrosc. 68 (2007) 632–638.
[8] V. Ivanovski, T.G. Mayerhöfer, J. Popp, V.M. Petrusevski, Polarized IR reflectance spectra of the monoclinic single crystal K$_2$Ni(SO$_4$)$_2$.6H$_2$O: dispersion analysis, dielectric and optical properties, Spectrochim. Acta A Mol. Biomol. Spectrosc. 69 (2008) 629–641.
[9] V. Ivanovski, T.G. Mayerhöfer, J. Popp, Dispersion analysis of polarized IR reflectance spectra of Tutton salts: the $\nu3(SO_4^{2-})$ frequency region, Vib. Spectrosc. 47 (2008) 91–98.
[10] T.G. Mayerhöfer, V. Ivanovski, J. Popp, Dispersion analysis of non-normal reflection spectra from monoclinic crystals, Vib. Spectrosc. 63 (2012) 396–403.

[11] V. Ivanovski, T.G. Mayerhöfer, Vibrational spectra and dispersion analysis of $K_2Ni(SeO_4)_2.6H_2O$ Tutton salt single crystal doped with $K_2Ni(SO_4)_2.6H_2O$, Spectrochim. Acta A Mol. Biomol. Spectrosc. 114 (2013) 553–562.

[12] S. Höfer, R. Uecker, A. Kwasniewski, J. Popp, T.G. Mayerhöfer, Complete dispersion analysis of single crystal yttrium orthosilicate, Vib. Spectrosc. 83 (2016) 151–158.

[13] S. Höfer, J. Popp, T.G. Mayerhöfer, Dispersion analysis of sodium dichromate dihydrate $Na_2Cr_2O_7.2H_2O$ single crystal, Spectrochim. Acta A Mol. Biomol. Spectrosc. 205 (2018) 243–250.

[14] S. Höfer, I. Uschmann, J. Popp, T.G. Mayerhöfer, Complete dispersion analysis of single crystal EDDt, Spectrochim. Acta A Mol. Biomol. Spectrosc. 206 (2018) 224–231.

[15] V. Ivanovski, T.G. Mayerhöfer, J. Stare, M.K. Gunde, J. Grdadolnik, Analysis of the polarized IR reflectance spectra of the monoclinic α-oxalic acid dihydrate, Spectrochim. Acta A Mol. Biomol. Spectrosc. 218 (2019) 1–8.

[16] T.G. Mayerhöfer, V. Ivanovski, J. Popp, Dispersion analysis with inverse dielectric function modelling, Spectrochim. Acta A Mol. Biomol. Spectrosc. 168 (2016) 212–217.

[17] J.L.P. Mosteller, F. Wooten, Optical properties and reflectance of uniaxial absorbing crystals, J. Opt. Soc. Am. 58 (1968) 511–518.

[18] F. Abelès, H.A. Washburn, H.H. Soonpaa, Calculating optical constants of anisotropic materials from reflectivity data, J. Opt. Soc. Am. 63 (1973) 104–105.

[19] T.G. Mayerhöfer, S. Weber, J. Popp, Dispersion analysis of perpendicular modes in anisotropic crystals and layers, J. Opt. Soc. Am. A 28 (2011) 2428–2435.

[20] J.P. Devlin, G. Pollard, R. Frech, ATR infrared spectra of uniaxial nitrate crystals, J. Chem. Phys. 53 (1970) 4147–4151.

[21] T.G. Mayerhöfer, I. Noda, J. Popp, Towards dispersion analysis of perpendicular modes using a hybrid 2T2D smart error sum, Appl. Spectrosc. (2024). Submitted for publication.

[22] C. Yang, W. Wang, A. Nefedov, Y. Wang, T.G. Mayerhöfer, C. Wöll, Polarization-dependent vibrational shifts on dielectric substrates, Phys. Chem. Chem. Phys. 22 (2020) 17129–17133.

[23] C. Yang, Y. Cao, P.N. Plessow, J. Wang, A. Nefedov, S. Heissler, F. Studt, Y. Wang, H. Idriss, T.G. Mayerhöfer, C. Wöll, N_2O adsorption and photochemistry on ceria surfaces, J. Phys. Chem. C 126 (2022) 2253–2263.

[24] J.R. Aronson, A.G. Emslie, P.F. Strong, Optical constants of triclinic anisotropic crystals: blue vitriol, Appl. Opt. 24 (1985) 1200–1203.

[25] A.G. Emslie, J.R. Aronson, Determination of the complex dielectric tensor of triclinic crystals: theory, J. Opt. Soc. Am. 73 (1983) 916–919.

[26] S. Höfer, J. Popp, T.G. Mayerhöfer, Determination of the dielectric tensor function of triclinic $CuSO_4 \cdot 5H_2O$, Vib. Spectrosc. 67 (2013) 44–54.

[27] S. Höfer, J. Popp, T.G. Mayerhöfer, Dispersion analysis of triclinic $K_2Cr_2O_7$, Vib. Spectrosc. 72 (2014) 111–118.

[28] R.M. Almeida, S. Höfer, T.G. Mayerhöfer, J. Popp, K. Krambrock, R.P.S.M. Lobo, A. Dias, R.L. Moreira, Optical phonon features of triclinic montebrasite: dispersion analysis and non-polar Raman modes, Vib. Spectrosc. 77 (2015) 10.

[29] C. Ye, M.J. Rucks, J.A. Arnold, T.D. Glotch, Mid-infrared optical constants of labradorite, a triclinic plagioclase mineral, earth and space, Science 6 (2019) 2410–2422.

[30] C. Sturm, S. Höfer, K. Hingerl, T.G. Mayerhöfer, M. Grundmann, Dielectric function decomposition by dipole interaction distribution: application to triclinic $K_2Cr_2O_7$, New J. Phys. 22 (2020) 073041.

[31] S. Höfer, R. Uecker, A. Kwasniewski, J. Popp, T.G. Mayerhöfer, Dispersion analysis of arbitrarily cut uniaxial crystals, Vib. Spectrosc. 78 (2015) 23–33.

[32] S. Höfer, V. Ivanovski, R. Uecker, A. Kwasniewski, J. Popp, T.G. Mayerhöfer, Dispersion analysis of arbitrarily cut orthorhombic crystals, Spectrochim. Acta A Mol. Biomol. Spectrosc. 180 (2017) 67–78.

[33] S. Höfer, V. Ivanovski, R. Uecker, A. Kwasniewski, J. Popp, T.G. Mayerhöfer, Generalized dispersion analysis of arbitrarily cut monoclinic crystals, Spectrochim. Acta A Mol. Biomol. Spectrosc. 185 (2017) 217–227.

[34] T.G. Mayerhöfer, J. Popp, Modelling IR spectra of polycrystalline materials in the large crystallites limit—quantitative determination of orientation, J. Opt. A Pure Appl. Opt. 8 (2006) 657–671.

[35] S. Höfer, J. Popp, T.G. Mayerhöfer, Generalized dispersion analysis of crystals with unknown symmetry and orientation, Spectrochim. Acta A Mol. Biomol. Spectrosc. 205 (2018) 348–363.

Chapter 15

Polycrystalline materials

15.1 How to calculate reflectance and transmittance for random orientation

In the preceding chapter, we have discussed how dispersion analysis can be performed on single crystals. Since the preparation of single crystals is often complex and costly, if not impossible, usually the corresponding polycrystalline sample is spectroscopically investigated instead. This has been done since ages. So, a polycrystalline sample surely must be a proper replacement for a single crystal. In general, this assumption implies that there might be conditions which, if applicable, may allow to some degree a quantitative analysis. The main problem is that to solve the inverse problem, which consists in determining the dispersion parameters from spectra of polycrystalline samples, the corresponding direct problem of forward calculating the spectra from single crystal data must be solved beforehand. If we consult the textbooks of optics, everything seems to be clear: For randomly oriented polycrystalline samples, the dielectric tensor reduces to a scalar and once we know this scalar (function), we can use the formalisms from the first part of this book to calculate the corresponding spectra. Unfortunately, there are a lot of problems concerning this assumption, the most important ones I present to you in the following.

The first example I chose is polycrystalline CuO. To understand the IR spectra of monoclinic CuO has once been seen as very important to comprehend the optical and, thereby, the structural properties of the at that time new high temperature superconductors, which are cuprates [1,2].

If we compare the spectra in Fig. 15.1, the lower wavenumber part seems to show some resemblance except that the peaks of the red spectrum are shifted to higher wavenumbers and that the relative intensities show some deviations. The high wavenumber region above $350\,cm^{-1}$, on the other hand, shows little similarity, also because the red spectrum seems to have only one peak and two shoulders whereas the black spectrum shows three peaks and one shoulder. The first idea that comes to one's mind is that the structures of CuO on the molecular level or, in this case, of the elementary cell must be different, but according to X-ray diffraction investigations both samples consist of the same monoclinic structure exclusively. The difference between the spectra was noticed by the authors of the later paper [2], as were the different conclusions with regard to physical properties like magnetic ordering etc., but no explanation for the differences was offered.

A further example that I came across during my PhD work and which I originally could not understand is exemplified in Fig. 15.2. The idea behind comparing spectra of polycrystalline Sr- with Ba-Fresnoite was to assign the vibrations where Barium is involved. From a theoretical point of view, replacing the heavier Barium by the lighter Strontium should result in some bands shifting to higher wavenumber (cf. Eq. 5.12), which would be those I was looking for (Ba-Fresnoite and Sr-Fresnoite are isostructural, which is certainly a precondition for this assumption). I think you can empathize with me and understand my surprise when I performed the comparison and found that nearly all bands of Sr-Fresnoite are shifted to higher wavenumbers. Does this mean that the Sr-atom is involved in nearly all vibrations? Certainly, in nonmolecular inorganic materials, all atoms are to some degree involved in all vibrations due to much smaller mass differences, so that the approximation of group vibrations does not apply, but this result was nevertheless unexpected. Not long after I made this comparison, we obtained a Sr-Fresnoite single crystal, and the corresponding spectra are compared with those of Ba-Fresnoite in Fig. 15.3.

I guess you can again easily imagine my surprise when I compared the single crystal spectra and found that indeed only some of the bands are blueshifted in Sr-Fresnoite, and, as originally expected, mostly those at lower wavenumbers. When I thought I had found an explanation, which connected the spectral changes to crystallite size in the polycrystalline samples, I was very excited. My models could even semiquantitatively predict why Sr-Fresnoite with its crystallites larger compared to the wavelength would have bands that were nearly all blueshifted compared to Ba-Fresnoite with its small (compared to the wavelength or the resolution limit) crystallites. Unfortunately, the reviewers would not accept my model, also because it did not lead to the holy grail of Linear Dichroism Theory, the magic angle.

Fortunately, I had a collaborator who was able to meet my special wishes for new samples. As a result, he was able to fabricate eventually also polycrystalline Ba-Fresnoite, with large crystallites, and the spectra clearly showed that there is indeed a crystallite size effect, cf. Fig. 15.4.

Wave Optics in Infrared Spectroscopy. https://doi.org/10.1016/B978-0-443-22031-9.00015-X
Copyright © 2024 Elsevier Inc. All rights reserved, including those for text and data mining, AI training, and similar technologies.

FIG. 15.1 Comparison of the spectra of polycrystalline CuO samples [3]. According to X-ray diffraction investigations, both samples consist of monoclinic randomly oriented CuO.

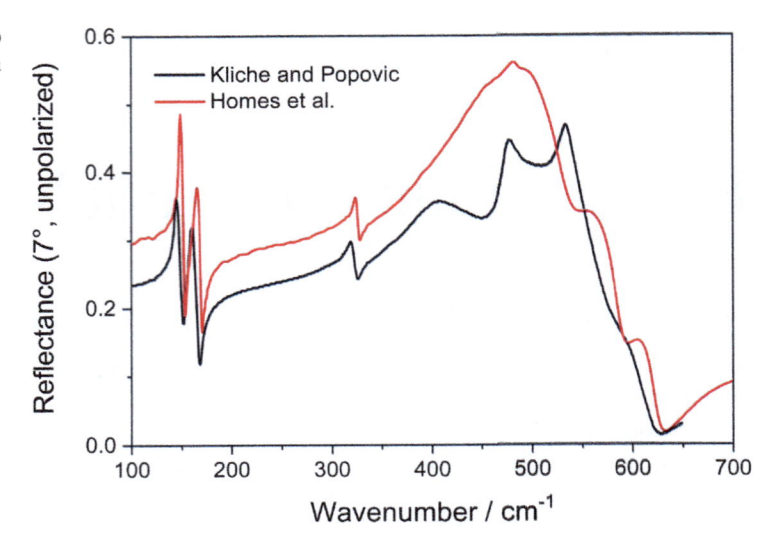

FIG. 15.2 Reflectance spectra of polycrystalline Ba- and Sr-Fresnoite. Both samples have the same structure.

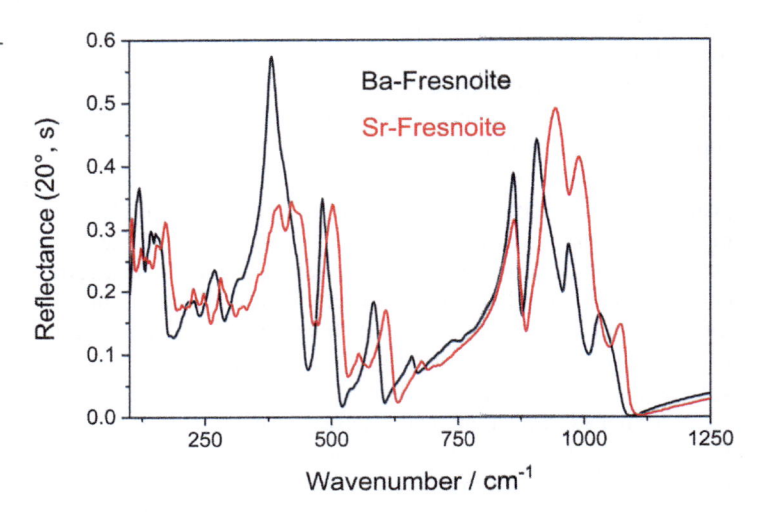

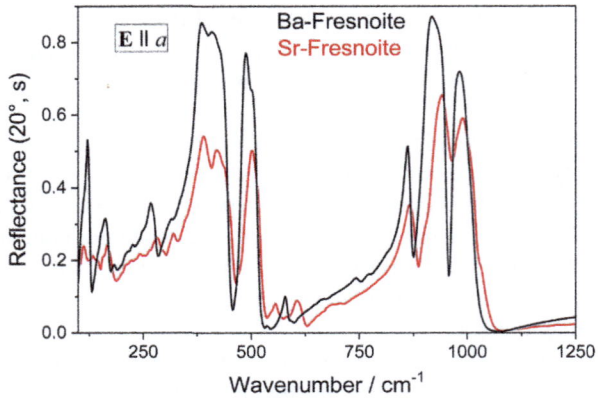

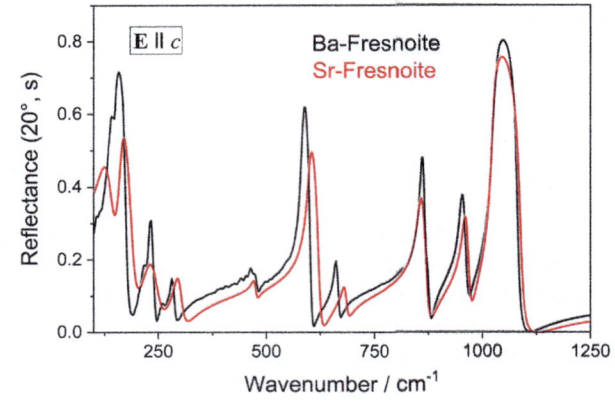

FIG. 15.3 Principal reflectance spectra of single crystal Ba- and Sr-Fresnoite.

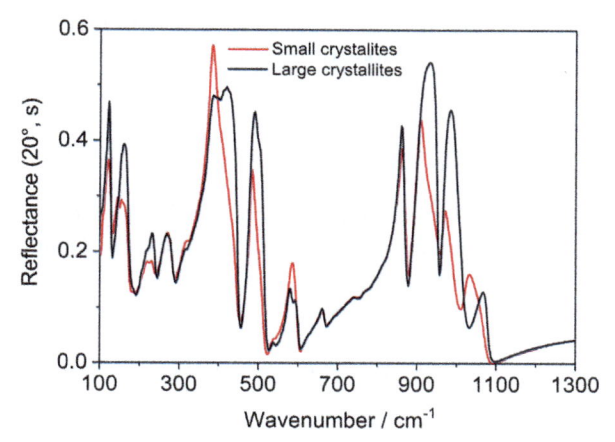

FIG. 15.4 Comparison of experimental spectra of polycrystalline Fresnoite with optically small crystallites *(red curve)* and optically large crystallites *(black curve)*.

Obviously, most of the peaks of the spectrum recorded from the sample with large crystallites are blueshifted. In addition, there are changes in the relative intensities and in the band shapes. Remarkably, there is also a region without noticeable changes between about 600 and 900 cm^{-1}. In this region, the single crystal is nearly isotropic, i.e., $\varepsilon_a \approx \varepsilon_c$. The corresponding conclusion is that the optical properties are differently averaged for small and large crystallites. In fact, for the sample with large crystallites, light registers the anisotropy of the individual crystallites. Therefore, it is obvious that such samples can be isotropic only on a macroscopic level. The proof of this assumption can simply be carried out by IR microscopy as can be seen in Fig. 15.5. For the sample with large crystallites, the spectra differ due to the different orientations of the crystallites, whereas for small crystallites, the number of crystallites in the measuring spot is so high that everywhere in the sample, the same averaged dielectric function results (the right side of Fig. 15.5 actually shows four different curves). The problem is now how can we develop a theory that seamlessly incorporates the extreme cases as well as samples that consist of both small and large crystallites? Since there were only a few articles about the spectroscopy of polycrystalline materials where colleagues were aware of this crystallite effect and even fewer who were not mainly interested in gaining knowledge of a special material's properties, only a patchwork of approaches existed. In the field of molecular spectroscopy as well as in the community dealing with astronomical applications usually crystallite size effects are either unknown or wrongly explained. In particular in the latter community, this is surprising, since, e.g., the famous appearance or disappearance of a hematite band, which is linked to the question had there been water on the Mars [5], is due to a crystallite size effect, and there are further examples [6].

Among those theories used to explain the reduction of the dielectric tensor to a scalar is the effective medium approximation (EMA, also known as Bruggeman's approximation, cf. Section 10.4). The basic assumption of EMA is closely related to the Lorentz-Lorenz formula and is based on the averaging of local field effects, except that for anisotropic

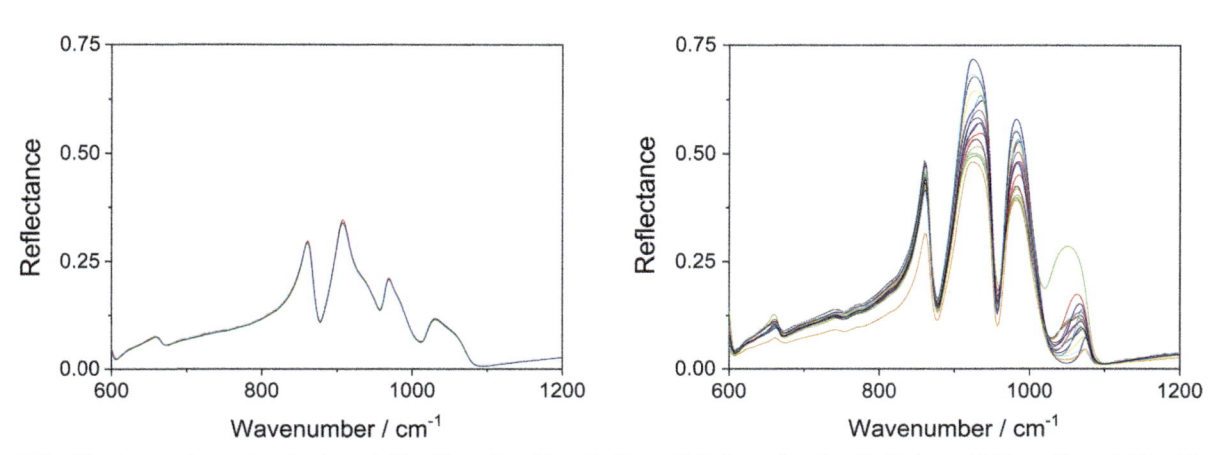

FIG. 15.5 IR-microscopic spectra of polycrystalline Fresnoite with optically small (left panel) and optically large (right panel) crystallites [4].

materials, it is assumed that the polycrystalline material consists of three different isotropic materials, each of which is characterized by one of the principal dielectric functions of the dielectric tensor function. Accordingly, the average $\langle \varepsilon \rangle$ is implicitly given by the condition that

$$\frac{1}{3}\frac{\varepsilon_a - \langle \varepsilon \rangle}{\varepsilon_a + 2\langle \varepsilon \rangle} + \frac{1}{3}\frac{\varepsilon_b - \langle \varepsilon \rangle}{\varepsilon_b + 2\langle \varepsilon \rangle} + \frac{1}{3}\frac{\varepsilon_c - \langle \varepsilon \rangle}{\varepsilon_c + 2\langle \varepsilon \rangle} = 0. \tag{15.1}$$

On first view this seems to make sense, but on further thinking there are a number of problems connected with Eq. (15.1). An immediate problem is what to do with monoclinic and triclinic materials. For monoclinic materials only ε_b exists, and for triclinic materials there is no principal dielectric function. There are certainly practical solutions. If we reformulate Eq. (15.1) and solve it, we obtain three different solutions [3]:

$$\begin{aligned}
\langle \varepsilon \rangle^3 &- \frac{1}{4}(\varepsilon_a\varepsilon_b + \varepsilon_b\varepsilon_c + \varepsilon_c\varepsilon_a)\langle \varepsilon \rangle + \frac{1}{4}\varepsilon_a\varepsilon_b\varepsilon_c = 0 \\
\langle \varepsilon \rangle_1 &= -\frac{3^{\frac{1}{3}}K_1 + K_3^{\frac{2}{3}}}{2 \times 3^{\frac{2}{3}}K_3^{\frac{1}{3}}} \\
\langle \varepsilon \rangle_2 &= -\frac{2(-3)^{\frac{1}{3}}K_1 + (1 - i\sqrt{3})K_3^{\frac{2}{3}}}{4 \times 3^{\frac{2}{3}}K_3^{\frac{1}{3}}} \\
\langle \varepsilon \rangle_3 &= -\frac{3^{\frac{1}{3}}(1 - i\sqrt{3})K_1 + (1 + i\sqrt{3})K_3^{\frac{2}{3}}}{4 \times 3^{\frac{2}{3}}K_3^{\frac{1}{3}}} \\
K_1 &= \varepsilon_a\varepsilon_b + \varepsilon_b\varepsilon_c + \varepsilon_c\varepsilon_a \\
K_2 &= \varepsilon_a\varepsilon_b\varepsilon_c \\
K_3 &= -9K_2 + \sqrt{3}\sqrt{27K_2^2 - K_1^3}
\end{aligned} \tag{15.2}$$

Instead of employing the principal dielectric functions, I suggested to use the three solutions of the following eigenvalue relation:

$$\text{Det}(\boldsymbol{\varepsilon} - \varepsilon_i \mathbf{I}) = 0 \quad i = 1, 2, 3, \tag{15.3}$$

where the eigenvalues are given by:

$$\begin{aligned}
\varepsilon_1 &= \frac{1}{3}(\varepsilon_{XX} + \varepsilon_{YY} + \varepsilon_{ZZ}) + \frac{2^{1/3}K_1}{3\left(K_2 + \sqrt{K_2^2 + 4K_1^3}\right)^{1/3}} - \frac{\left(K_2 + \sqrt{K_2^2 + 4K_1^3}\right)^{1/3}}{3 \times 2^{1/3}} \\
\varepsilon_2 &= \frac{1}{3}(\varepsilon_{XX} + \varepsilon_{YY} + \varepsilon_{ZZ}) - \frac{(1 + i\sqrt{3})K_1}{3 \times 2^{2/3}\left(K_2 + \sqrt{K_2^2 + 4K_1^3}\right)^{1/3}} + \frac{(1 - i\sqrt{3})\left(K_2 + \sqrt{K_2^2 + 4K_1^3}\right)^{1/3}}{6 \times 2^{1/3}} \\
\varepsilon_3 &= \frac{1}{3}(\varepsilon_{XX} + \varepsilon_{YY} + \varepsilon_{ZZ}) - \frac{(1 - i\sqrt{3})K_1}{3 \times 2^{2/3}\left(K_2 + \sqrt{K_2^2 + 4K_1^3}\right)^{1/3}} + \frac{(1 + i\sqrt{3})\left(K_2 + \sqrt{K_2^2 + 4K_1^3}\right)^{1/3}}{6 \times 2^{1/3}}
\end{aligned} \tag{15.4}$$

$$\begin{aligned}
K_1 &= -(\varepsilon_{XX} + \varepsilon_{YY} + \varepsilon_{ZZ})^2 - 3(\varepsilon_{XY}^2 + \varepsilon_{XZ}^2 + \varepsilon_{YZ}^2 - \varepsilon_{XX}\varepsilon_{YY} - \varepsilon_{XX}\varepsilon_{ZZ} - \varepsilon_{YY}\varepsilon_{ZZ}) \\
K_2 &= -2(\varepsilon_{XX}^3 + \varepsilon_{YY}^3 + \varepsilon_{ZZ}^3) + 3(\varepsilon_{XX}^2\varepsilon_{YY} + \varepsilon_{XX}^2\varepsilon_{ZZ} + \varepsilon_{YY}^2\varepsilon_{XX} + \varepsilon_{YY}^2\varepsilon_{ZZ} + \varepsilon_{ZZ}^2\varepsilon_{XX} + \varepsilon_{ZZ}^2\varepsilon_{YY}) \\
&\quad - 9(\varepsilon_{XX}\varepsilon_{XY}^2 + \varepsilon_{XX}\varepsilon_{XZ}^2 + \varepsilon_{YY}\varepsilon_{XY}^2 + \varepsilon_{YY}\varepsilon_{YZ}^2 + \varepsilon_{ZZ}\varepsilon_{XZ}^2 + \varepsilon_{ZZ}\varepsilon_{YZ}^2) + 18(\varepsilon_{XX}\varepsilon_{YZ}^2 + \varepsilon_{YY}\varepsilon_{XZ}^2 + \varepsilon_{ZZ}\varepsilon_{XY}^2) \\
&\quad - 12\,\varepsilon_{XX}\varepsilon_{YY}\varepsilon_{ZZ} - 54\,\varepsilon_{XY}\varepsilon_{XZ}\varepsilon_{YZ}
\end{aligned}$$

For crystals of orthorhombic or higher symmetry, the solutions are simply the principal dielectric functions. Eq. (15.4) solves the problem for monoclinic and triclinic materials. Even with this solution, however, there is still another problem that remains when EMA is used. EMA is based on local field effects. Are these effects out of a sudden no longer present when the crystallites become somewhat larger than the resolution limit? For large crystallites it was correspondingly a mosaiced surface approach that was employed which does not consider local field effects. Even more serious is however

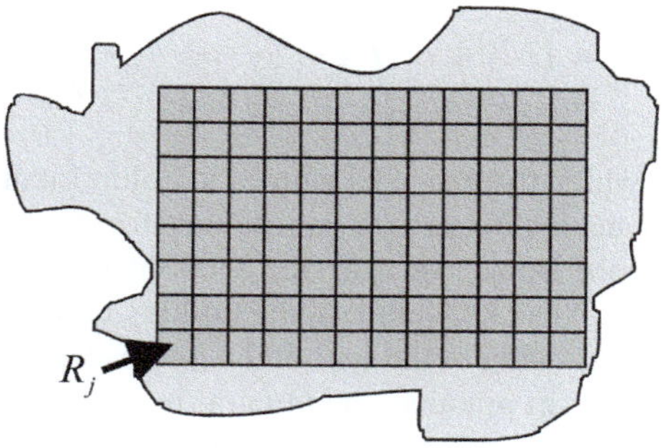

SCHEME 15.1 Grid on the surface of a sample [7].

that, as we know, geometric resonances are responsible for band shifts and shape changes, cf. Chapter 10. How would those be connected with the anisotropy of materials?

Because of these problems, I decided to develop an approach that still does not solve every related problem (e.g., that of samples of partly oriented small crystallites), but allows to treat randomly oriented polycrystalline samples with small and large crystallites consistently. The overall underlying concept is that reflectance and transmittance are quantities that are additive. So, if I lay a grid with grid constant $\lambda/10$ (or alternatively, the resolution limit of light) onto a sample surface, like in Scheme 15.1, different grid cells have in general different reflectances.

The overall reflectance/transmittance that one would measure from a microheterogeneous sample would be an average of the individual reflectances/transmittances from the microhomogenous squares, which could be formulated as [7],

$$\langle R \rangle = N^{-1} \sum_{j=1}^{N} R_j$$
$$\langle T \rangle = N^{-1} \sum_{j=1}^{N} T_j$$
(15.5)

if N would be the number of different microhomogenous squares. Note that this is consistent with the idea of microheterogeneous and microhomogeneous materials and corresponding mixing rules as discussed in Section 6.6 (in fact, those are special cases of the approach presented in this section). Correspondingly, the same idea also holds for transmittance, provided that the sample is homogeneous in the direction normal to the surface. For reflectance, at least for strongly absorbing materials, it does not matter in practice if there are other layers of differently oriented crystallites beyond the first layer, because their influence on the spectrum should be negligible. For transmittance to be nonzero, light has to pass through all crystallites. Overall, one would expect that averaging also leads to a spectrum that is not a function of the number of crystallite layers the light has to pass through. Since the shape of crystallites is usually irregular, interference effects should not play a role (most probably there is actually no well-defined second or third layer), but scattering may have an influence. On the other hand, scattering should be minimized since, in the absence of absorption, the material properties outside the individual crystallite may not be much different from its own, and the presence of absorption would lead to the fact that most light that is removed by scattering is nevertheless absorbed. In any way, as we will see in the following, the direct problem of calculating the spectra of polycrystalline samples when the single crystal data are known is already complicated, so that the inverse problem of determining the material properties from the spectra of randomly oriented polycrystalline samples is next to impossible to solve.

This is a statement with grave consequences, given the fact that usually only polycrystalline samples can be investigated. A fortunate coincidence makes it nevertheless possible that at least oscillator positions can be extracted if the sample consists of exclusively small crystallites, since for such materials the reflectance peaks are located at or close to the TO positions. This is an important difference with regard to cubic materials, even if, in the literature, both are pigeonholed simply as isotropic.

Coming back to Scheme 15.1, let us first assume that the crystallites are small and that there are a large number of crystallites within one of the squares. Under the assumption that local field effects do not play a larger role, we can obtain an averaged index of refraction by,

$$\langle n \rangle_j = \left(N^{(3)}\right)^{-1} \int_{\Omega^{(3)}} \left(\frac{n_1(\Omega)}{2} + \frac{n_2(\Omega)}{2}\right) d\Omega, \tag{15.6}$$

wherein Ω represents the orientation for which the use of a symmetric Euler orientation representation is especially useful except for uniaxial crystallites [8]. $n_1(\Omega)$ and $n_2(\Omega)$ are the two solutions for the forward traveling waves one obtains in the corresponding single crystal from solving the Booker quartic (see Sections 12.2/12.3). Accordingly, this is why this theory is called Average Refractive Index Theory (ARIT). Overall, in the limiting case of very small crystallites, $\langle n \rangle_j = \langle n \rangle$, and it can be expected that independent of the positioning of the spot of the microscope, everywhere the same spectrum results as this is the case in the left panel of Fig. 15.5. In practice, the integral in Eq. (15.6) would be replaced by a summation.

How well does this averaging scheme perform in practice? In the generally very demanding case of Fresnoite, it performs very well. I have averaged the indices of refraction for 8281 (91^2) different orientations and obtained an excellent resemblance between the experimental and forward calculated spectrum which is depicted in Fig. 15.6. The residual deviations can be explained with the assumptions of moderately higher damping constants in polycrystalline materials with a common factor for all bands, something that is not possible if the forward calculation is based on Eq. (15.1), albeit it resembles the original spectrum much better (cf. Fig. 15.5, although not as good as ARIT) than the famous but highly approximative and related $1/3 - 1/3 - 1/3$ approximation, which is only valid in the limit of vanishing absorption or anisotropy:

$$\langle \varepsilon \rangle = \left(N^{(3)}\right)^{-1} \int_{\Omega^{(3)}} \mathbf{A}(\Omega) \cdot \begin{pmatrix} \varepsilon_a & 0 & 0 \\ 0 & \varepsilon_b & 0 \\ 0 & 0 & \varepsilon_c \end{pmatrix} \cdot \mathbf{A}(\Omega)^{-1} d\Omega = 1/3\, \varepsilon_a + 1/3\, \varepsilon_b + 1/3\, \varepsilon_c. \tag{15.7}$$

Since this approximation results from an orientational average of the dielectric tensor itself, it again means that it is not directly applicable to monoclinic and triclinic materials. What proves that this approximation is in general not valid? A simple comparison with an experiment. This comparison, which is provided in Fig. 15.7, clearly shows that the $1/3 - 1/3 - 1/3$ approximation (which is a $2/3 - 1/3$ approximation for optically uniaxial materials) clearly overestimates the experimental reflectance, except in the region between 600 and 900 cm^{-1} where Fresnoite is nearly optically isotropic, i.e. where $\varepsilon_a \approx \varepsilon_c$. Apart from this, not only the absolute intensities but also the band shapes are different. Whereas the $1/3 - 1/3 - 1/3$ approximation obviously preserves the original Lorentz-oscillator band shapes, the experimental band shapes are much more asymmetric. In addition, the peaks are shifted toward the TO position. This is an important aspect, because it tells us that conventional dispersion analysis is not applicable for such samples. It may not be obvious, but by employing conventional dispersion analysis to the spectrum of a randomly oriented polycrystalline sample, one automatically assumes that the $1/3 - 1/3 - 1/3$ approximation holds:

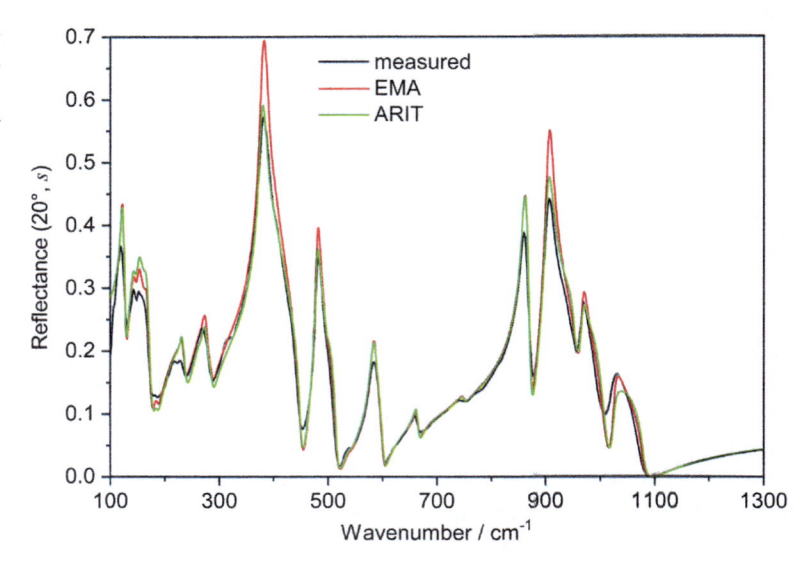

FIG. 15.6 Comparison between experimental and by ARIT and EMA forward calculated spectra of randomly oriented polycrystalline Fresnoite with exclusively small crystallites (for the forward calculations the single crystal data were used).

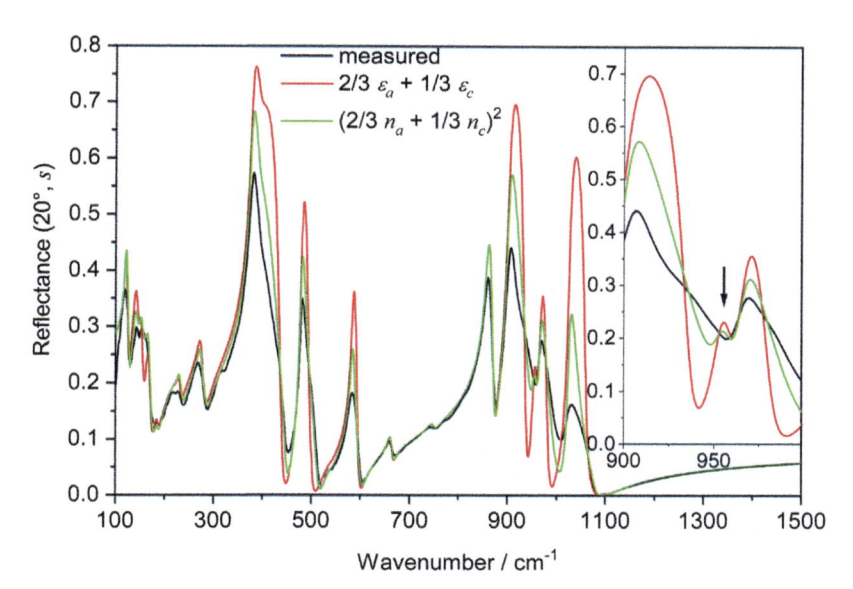

FIG. 15.7 Comparison between experimental and by two different $1/3 - 1/3 - 1/3$ approximations forward calculated spectra of randomly oriented polycrystalline Fresnoite with exclusively small crystallites [4].

$$\varepsilon(\tilde{\nu}) = \varepsilon_\infty + \sum_{j=1}^{N_a} \frac{S_{j,a}^2}{\tilde{\nu}_{j,a}^2 - \tilde{\nu}^2 - i\tilde{\nu}\,\gamma_{j,a}} + \sum_{j=1}^{N_b} \frac{S_{j,b}^2}{\tilde{\nu}_{j,b}^2 - \tilde{\nu}^2 - i\tilde{\nu}\,\gamma_{j,b}} + \sum_{j=1}^{N_c} \frac{S_{j,c}^2}{\tilde{\nu}_{j,c}^2 - \tilde{\nu}^2 - i\tilde{\nu}\,\gamma_{j,c}} = \frac{\varepsilon_a + \varepsilon_b + \varepsilon_c}{3}. \quad (15.8)$$

In Eq. (15.8), the oscillator strengths are certainly not the same as the ones one would obtain when the principal spectra of an orthorhombic crystal would be analyzed (in particular, it is not possible to determine from the spectrum of a polycrystalline sample to which crystallographic axis a transition moment is parallel to). The important point of Eq. (15.8) is that applying dispersion analysis to randomly oriented polycrystalline materials would assume a linear combination of the different principal dielectric functions.

The assumption of linear mixing is obviously incorrect, and this conclusion certainly holds for the spectra of both small and large crystallite samples. While for the latter, the whole procedure is obviously meaningless, for the former at least the obtained dielectric function is meaningful, whereas the so-obtained oscillator parameters lack any meaning. Nevertheless, since the band shapes deviate from the typical CDHO shapes, the more the stronger the oscillator, it makes much more sense to apply Poor Man's dispersion analysis (cf. Section 5.8) to such kind of samples to obtain the average dielectric function. This can be easily seen in Fig. 15.8. The peak at about $860\,\mathrm{cm}^{-1}$, for which $\varepsilon_a \approx \varepsilon_c$, shows the typical shape of a CDHO oscillator in the spectrum of a cubic material or of a principal spectrum of an anisotropic crystal, whereas the shape of the

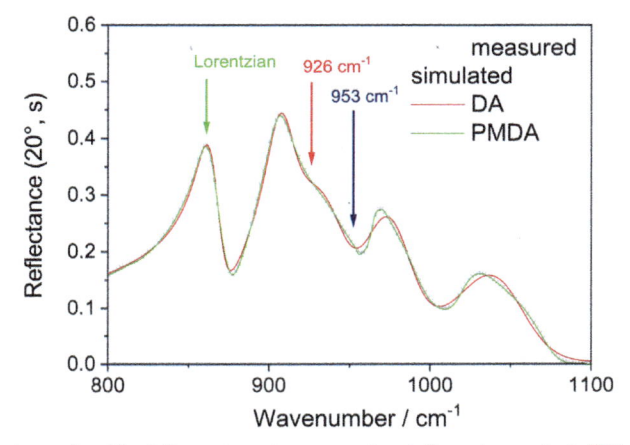

FIG. 15.8 Comparison between experimental and best fit spectra using conventional dispersion analysis (DA, *red curve*) and Poor Man's Dispersion Analysis (PDMA) to model the spectrum of randomly oriented polycrystalline Fresnoite with small crystals.

other bands more or less deviate from this typical shape. It is therefore not very surprising that these bands cannot be satisfactorily fitted with conventional dispersion analysis. Instead, as already stated, this is a typical application of Poor Man's Dispersion Analysis, since the best fit agrees with the experimental data within line thickness (which is intentionally very small in Fig. 15.8!) and the usual disadvantage that the dispersion parameters cannot be determined is not existent, since these parameters can anyway not be determined from the spectra of randomly oriented polycrystalline samples with small crystallites. The latter statement is certainly also true if the crystallites are not all small compared to the resolution limit of light, since then the spectrum can in any way not be described by an averaged dielectric function. Before we discuss this case, we first study another advantage of ARIT compared to EMA in more detail. This advantage that I have already mentioned above is that the experimental and the calculated spectra can be made congruent if the damping constants are multiplied by a common factor as can be seen in Fig. 15.9. This is a very appealing finding, since the damping constants are seen, as detailed in Section 5.4, as the inverse phonon lifetime. If the crystallites become smaller, so should the inverse lifetime and that is certainly by a common factor.

Before I continue with optically large crystallites, I want to shortly digress and discuss the optical and spectroscopic properties of glasses. From a historical point of view, there are two different theories to describe their structure. One theory is the so-called Crystallite theory and the other the Random Network theory [9]. According to the former, a glass consisted of very small crystallites. While nowadays it is mostly believed that there are no real crystallites in glasses, but rather a kind of medium range order, it is tempting to assume that in the spectra of glasses orientational averaging must somehow also play an important role. In fact, often glasses that have compositions like crystalline materials resemble the spectra of polycrystalline materials with small crystallites. So, if you have to do orientational averaging to simulate the spectra of polycrystalline materials, you might have to do the same for corresponding glasses. Nevertheless, one has to take into account that structural units in glasses are not necessarily the same as in polycrystalline materials. An example is Fresnoite glass, which also contains Ti units in which Ti is fourfold or sixfold coordinated by oxygen, in addition to the fivefold coordination known from crystalline Fresnoite [10]. For vitreous silica, one also has to factor in that crystalline SiO_2 exists in various modifications, of which it seems that the spectrum of polycrystalline tridymite resembles best that of vitreous silica with regard to the position and number of bands as well as their relative intensities. Unfortunately, it looks like the centimeter-sized single crystal tridymite is not available. Thus, its spectra have never been investigated by dispersion analysis, so that its dielectric tensor function is unknown. Therefore, I could not use proper oscillator data to prove by simulations that indeed orientational averaging is quite important to understand glass spectra. But one can, for simplicity, assume two oscillators in a uniaxial structural unit, so that one oscillator has its transition moment along the optical axis and for the other oscillator, the transition moment is perpendicular to this axis. Can those two oscillators describe the SiO_2-spectrum in the range of the asymmetric Si—O—Si stretching vibrations? Have a look at Fig. 15.10.

FIG. 15.9 Upper panel: Comparison between the dielectric function of randomly oriented polycrystalline Fresnoite as obtained by Poor Man's dispersion analysis and the averaged dielectric functions according to EMA and ARIT. Lower panel: Comparison between dielectric function of randomly oriented polycrystalline Fresnoite as obtained by Poor Man's dispersion analysis and ARIT if all damping constants of the single crystal are multiplied with the factor $f = 1.3$.

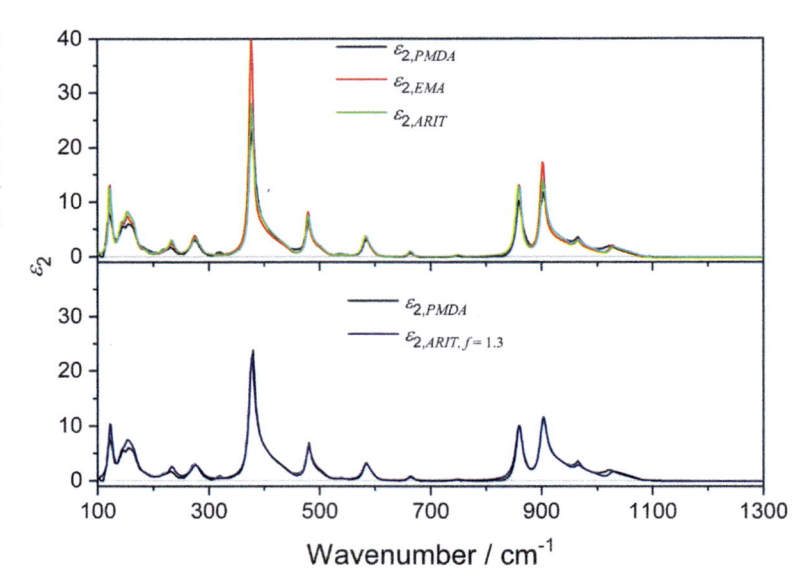

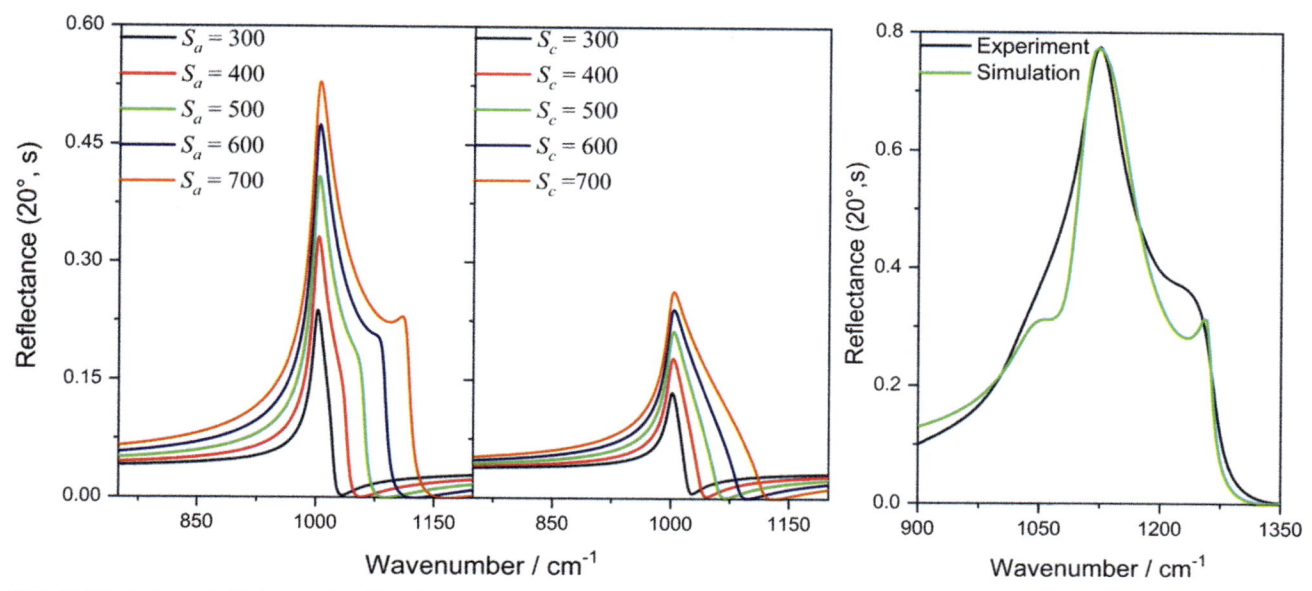

FIG. 15.10 Left panel: Variance of the IR reflectance spectrum of a classical damped harmonic oscillator with varying oscillator strength that is orientationally averaged (the medium range order is assumed to be optically uniaxial, the transition moment of the oscillator is oriented *perpendicular* to the optical axis, $\tilde{\nu}_{TO}/cm^{-1} = 1000$, $\gamma/cm^{-1} = 10$, $\varepsilon_{\infty,a}=2$, $\varepsilon_{\infty,c}=2$) [10]. Center panel: Variance of the IR reflectance spectrum of a classical damped harmonic oscillator with varying oscillator strength that is orientationally averaged (the medium range order is assumed to be optically uniaxial, the transition moment of the oscillator is oriented *parallel* to the optical axis, $\tilde{\nu}_{TO}/cm^{-1} = 1000$, $\gamma/cm^{-1} = 10$, $\varepsilon_{\infty,a}=2$, $\varepsilon_{\infty,c}=2$) [10]. Right panel: Experimental spectrum of vitreous silica compared with a simulated spectrum based on two classical damped harmonic oscillators, one with the transition moment perpendicular to the optical axis and the oscillator parameters $S/cm^{-1}=920$, $\tilde{\nu}_{TO}/cm^{-1}=1100$, $\gamma/cm^{-1}=10$, and the other with the transition moment parallel to the optical axis and the oscillator parameters $S/cm^{-1}=700$, $\tilde{\nu}_{TO}/cm^{-1}=1040$, $\gamma/cm^{-1}=40$ ($\varepsilon_{\infty,a}=2.25$, $\varepsilon_{\infty,c}=2$).

According to the results depicted in Fig. 15.10, the difference between an oscillator having its transition moment perpendicular to the optical axis and the one oriented along this axis is that the former obtains a pronounced shoulder when its oscillator strength increases due to its mixed TO-LO character, while the latter just broadens between the TO and LO positions. Both types of oscillators have a distinct maximum at the TO position, independent of the oscillator strength. In particular, the band shapes of the oscillator with a transition moment perpendicular to the optical axis remind strongly of the band due to the asymmetric Si—O—Si stretching vibrations of SiO_2, which can be simulated with two oscillators (cf. right panel in Fig. 15.10). Although the agreement is far from perfect, I would consider this a strong hint that orientational averaging is very important to understand the optical properties of a glass and that those properties agree less with those of crystals of cubic symmetry. Note that for SiO_2, the oscillator having its transition moment perpendicular to the optical axis has a comparably small damping constant. Usually, one would expect that the phonon lifetime in glasses is much smaller than in crystals due to their small dimensions. In fact, e.g., for Fresnoite glasses, we found that we would have to multiply the damping constants of the single crystals with factors of about 4 so that typical glass bands result. These factors are nevertheless much smaller than those that would have to be used if one assumed that glasses have cubic symmetry (cf. Section 5.5.5), a further hint that orientational averaging is important to understand the spectra of glasses. Finally, one can certainly use the convolution model together with orientational averaging. Since part of the intensity decreases relative to those for crystals due to orientational averaging, the standard distribution of the oscillator position would take on much more realistic values. Overall, the use of this type of oscillator, together with orientational averaging according to Eq. (15.6) (optically small crystallites), would in my opinion resemble the modern view on the structure of glasses, which is a kind of fusion of the crystallite and random network theory.

Coming back to polycrystalline materials, if the crystallites are large compared to the wavelength, the same crystallite may occupy more than one square (cf. Scheme 15.1). Correspondingly, and corresponding with the results of IR microspectroscopy depicted in Fig. 15.5, there will be a large number of different orientations only if we investigate a larger area of the sample. Then we could see such a sample as having a continuous distribution of orientations with regard to the area investigated and calculate the reflectance (and, correspondingly, the transmittance), by [7,11,12],

FIG. 15.11 Comparison between forward calculated and experimental spectra of randomly oriented polycrystalline Fresnoite with optically large crystallites [12].

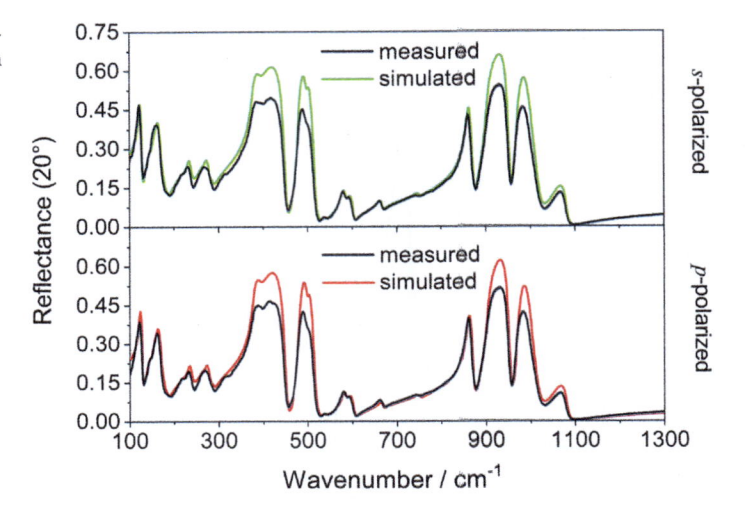

$$\langle R \rangle = \left(N^{(3)}\right)^{-1} \int_{\Omega^{(3)}} \left(\frac{R_s(\Omega)}{2} + \frac{R_p(\Omega)}{2}\right) d\Omega \rightarrow$$

$$\langle R_j \rangle = \left(N^{(3)}\right)^{-1} \int_{\Omega^{(3)}} R_j(\Omega) d\Omega \quad j = s, p$$

(15.9)

if naturally polarized incident light is used. R_s and R_p would have to be calculated by the 4×4 matrix formalism introduced in Chapter 12. I did this again using the Fresnoite single crystal data, and the comparison between the forward calculated and the experimental spectra is shown in Fig. 15.11.

Obviously, band positions, shapes, and relative intensities agree to a large degree. Only the absolute intensities are somewhat overestimated. This overestimation is an expected result since reflectance measurements depend strongly on the surface quality, which is due to the polycrystalline nature even after polishing, less good than that of the single crystal. Since the crystallites are large compared to the wavelength, the polishing will lead to some holes in the surface, which are also large compared to the wavelength, which will decrease the reflectance. Nonetheless, I think the result is so convincing that we can agree that the comparison proves the validity of Eq. (15.9).

There is, however, an additional proof for its correctness. As we have already discussed, Eq. (15.9) can be used for one-polarizer experiments, i.e., for experiments with a polarizer before the light hits the sample. Thanks to the 4×4 matrix formalism, we can also calculate the spectra of two polarizer experiments (cf. Fig. 12.1). Thereby, we can separately determine, e.g., R_{ss} and R_{sp}. In particular, the experiments with crossed polarizers are very interesting, since every reduction of a dielectric tensor to a scalar leads to the situation where the cross-polarized reflection or transmission becomes zero, since in a scalar material no polarization conversion can take place (this was the reason we can use 2×2 matrices in the first part of the book, cf. Section 4.5).

What happens if we average the cross-polarization terms R_{sp} and R_{ps}? Since the cross-polarization is zero only for principal orientations, the average will always be nonzero [11]:

$$\langle R_{sp} \rangle = \left(N^{(3)}\right)^{-1} \int_{\Omega^{(3)}} R_{sp}(\Omega) d\Omega > 0$$

$$\langle R_{ps} \rangle = \left(N^{(3)}\right)^{-1} \int_{\Omega^{(3)}} R_{ps}(\Omega) d\Omega > 0$$

(15.10)

The result is that an isotropic material can show a nonzero cross-polarization! For me, this was originally hard to accept. On the other hand, every mineralogist knows this from experiments with a microscope using crossed polarizers: All crystallites

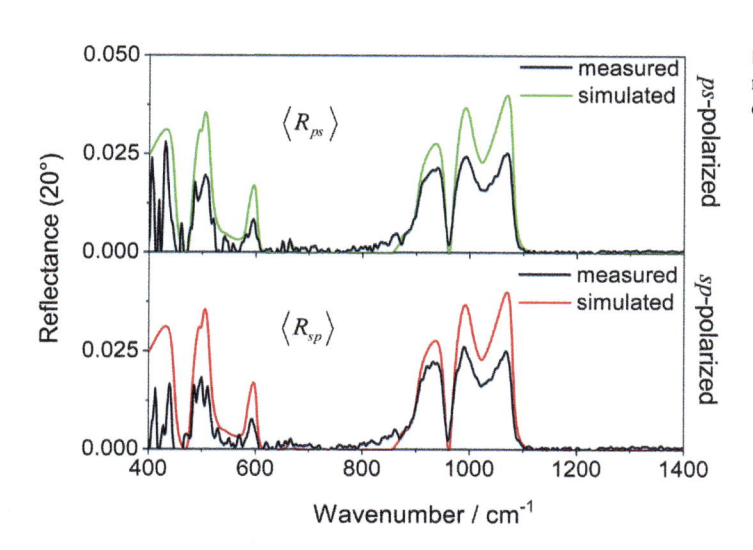

FIG. 15.12 Comparison between forward calculated and experimental spectra of randomly oriented polycrystalline Fresnoite with optically large crystallites for cross-polarization [12].

in nonprincipal orientations will show up. If the magnification of the microscope is so large that it originally shows only one crystallite, more will be visible once the magnification is reduced, and the image will not get dark, even when, in the end, the whole macroscopically isotropic sample with its randomly oriented crystallites would be visible. Nevertheless, even when Eq. (15.10) is obviously valid, let us first investigate the outcome for the above sample of Fresnoite, which is provided in Fig. 15.12. Even when the experimental spectra are somewhat noisy, which is not really surprising given the fact that the beam is weakened by two polarizers, I think we agree that the result is again convincing.

For randomly oriented samples with large crystallites it is expected for symmetry reasons that $\langle R_{ps}\rangle = \langle R_{sp}\rangle$ [12], which is also confirmed by the experiment. At this point I want to emphasize that this is again a result that cannot be explained by Linear Dichroism Theory, because, as already pointed out, it allows us to easily distinguish between random orientation, the magic angle (which is not magic at all in optical spectroscopy), and the perfect orientation of one third of each of the crystallographic axes along the polarization direction (the latter situation would lead to a zero cross-polarization, at least for orthorhombic symmetry, since only principal orientations are present with one of the crystallographic axes along the polarization direction).

Finally, I want to focus on the similarities between ARIT and the latter theory, which I call the Average Reflectance and Transmittance Theory (ARTT). While the averaging over all orientations and the negligence of local field corrections are obvious, also ARTT allows to correlate all damping constants of the oscillators in the polycrystalline material with one single factor. The expectation is certainly that this factor should be smaller than that of the small crystallite material, since the larger crystallite sizes should lead to increased phonon life times. Indeed, I found a factor of 1.1 in the large crystallite sample, which seems to fit considering the determined factor of 1.3 in the small crystallite sample. The corresponding comparison is provided in Fig. 15.13 [12]. Both theories, ARIT as well as ARTT can now be put together assuming that both small and large crystallites can be present, with the expectation that both theories can be linearly merged according to [7],

$$R = \phi_L \cdot R(d > \lambda/10) + (1 - \phi_L) \cdot R(d < \lambda/10)$$
$$T = \phi_L \cdot T(d > \lambda/10) + (1 - \phi_L) \cdot T(d < \lambda/10)$$
(15.11)

wherein ϕ_L is the volume fraction (or equivalently, the surface fraction) occupied by the large crystallites. The combined theory I called Unified Average Optical Properties Theory (UAOPT). I proved this theory again with a Fresnoite sample which consisted of upto 75% of optically small and 25% of optically large crystallites. The corresponding experimental and forward calculated spectra are depicted in Fig. 15.14.

Although this proves UAOPT, some questions remain. One of these is: what exactly is the resolution limit for a given wavenumber, or, in other words, how small is small and how large is large? Furthermore, is there an abrupt change in the spectrum at a particular crystallite/wavelength ratio, or is the change gradually from small to large (or vice-versa)? I can answer these questions provisionally since Sonja has done corresponding research in 2020 based on FDTD calculations (for some reasons, the results have not been published so far). In correspondence with the much simpler case of two different phases arranged like a chessboard, which we discussed in Section 10.6, also in this case the change is gradual as could be

FIG. 15.13 Comparison between forward calculated (for different factors f, γ_j(polycrystalline) $= f \cdot \gamma_j$(single crystal)) and experimental spectra of randomly oriented polycrystalline Fresnoite with optically large crystallites [12].

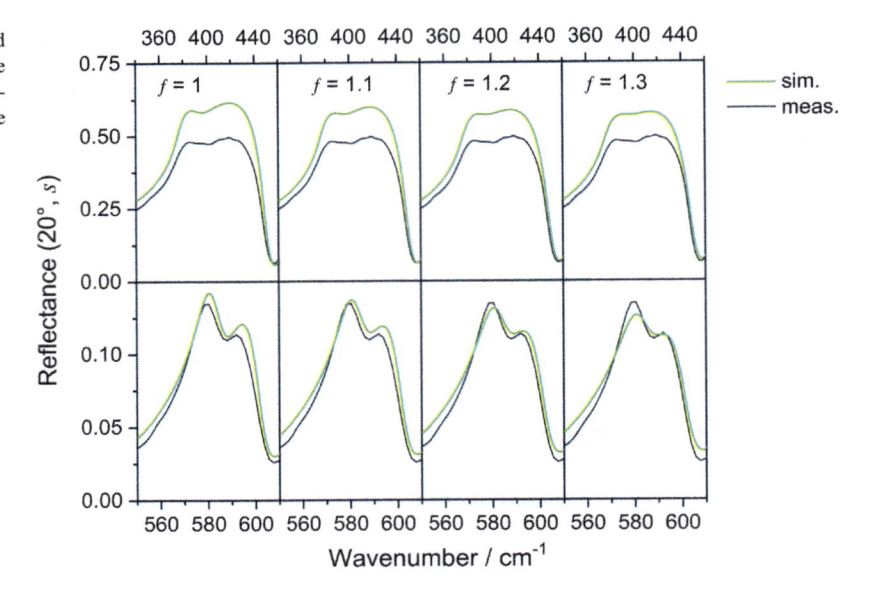

FIG. 15.14 Comparison between forward calculated and experimental spectra of randomly oriented polycrystalline Fresnoite with a 3:1-mixture of optically large and small crystallites. Upper panel: Simulated spectra were calculated from single crystal data. Center panel: Simulated spectra were calculated from the experimental spectra of the end members. Bottom panel: Comparison of the experimental spectra with the calculated cross-polarization spectra based on single crystal data [7].

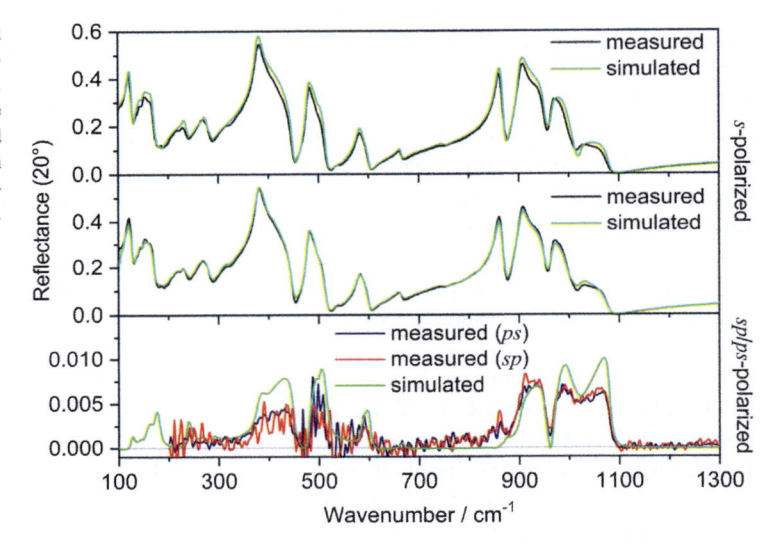

expected. For large crystallites, the FDTD results converge against those calculated by ARTT, while in the limit of small crystallites ARIT predicts the FDTD results more accurate than EMA, but the agreement is in general not perfect since local field effects definitely also play a role.

15.2 Optical properties of randomly oriented polycrystalline materials with large crystallites compared to those consisting of small crystallites

Generally, the term optical isotropy refers to the macroscopic optical properties of a solid that do not depend upon the orientation of the material. As is often the case, more or less subtle extensions to this definition have crept in over time, and nowadays optical isotropy refers to: (1) an isotropic medium that can be characterized by a scalar dielectric function and (2) the idea that the dispersion of the scalar dielectric function can be modeled in the infrared by the classic damped harmonic oscillator model (or oscillator models that are derived from that). As shown in the previous section, this is generally not possible if the crystallites themselves are anisotropic. In addition, if the dimensions of the anisotropic domains are not small compared to the wavelength of light (or are below the resolution limit), the optical properties of these materials cannot be characterized by a scalar dielectric function. This simply follows from the existence of nonzero cross-polarization terms. As a consequence, the term optical isotropy should be used only in the context of its original meaning.

The optical properties of a medium that can be properly characterized by a scalar dielectric function have been discussed extensively in the first part of this book. Optically isotropic media consisting of large domains show some interesting deviations from these properties, apart from the nonzero cross-polarization terms, of which the spectroscopist should be aware. The observation of these properties motivated me to investigate the optical properties of large- vs small-domain materials further. In particular, I think that in this respect, the angular dependence of the reflectance is of interest.

Therefore, the goal in the following is to determine this dependence and compare it with the well-known dependence of a scalar medium. I will focus on the case where the incidence medium has a high index of refraction (knowing that the material property is actually the dielectric function), so that, for nonabsorbing spectral regions, total internal reflection can be observed at higher angles of incidence. The reason is that the effects are in principle the same as for incidence from vacuum/air, but they are much larger for high-index incidence media due to the potential occurrence of (attenuated) total reflection. If you are interested in the full story, have a look at Ref. [4], or the original publication [13]. The latter case is, however, also of interest for incidence from vacuum or air since, as I will demonstrate, it can also occur in a conventional reflection experiment if one investigates the reflectance from a randomly oriented large domain medium at higher angles of incidence.

For the following simulations, we assume a semiinfinite medium consisting of randomly oriented and anisotropically ordered regions (the sample) with a smooth interface between it and the semiinfinite incidence medium. The ordered regions/domains shall be uniaxial with two principal indices of refraction, $n_a = 2.2$ and $n_c = 1.6$. For optically small ordered domains, we apply ARIT (see previous section). The result of this calculation according to Eq. (15.6) is that for our model system, the average index of refraction $\langle n \rangle = 1.96$, which is close to the result of the linear combination of the principal refractive indices according to the $2/3 - 1/3$ approximation (the small difference (1.96 vs 2) is certainly a consequence of the comparably small optical anisotropy of our sample). Next, we invoke the Fresnel equations, which allow us to relate $\langle n \rangle$ to R_s and R_p (for vacuum as an incidence medium, this would result in the well-known curves as displayed in Fig. 4.1).

In the opposite limit, where we assume large ordered domains, reflectance has to be averaged. Thus, we need to apply ARTT. For vacuum as an incidence medium, the results in dependence on the angle of incidence are not very different, although R_p does not become zero at Brewster's angle. But with $R_p = 0.005$ at the minimum, the difference would be hard to determine experimentally. Even harder to measure would be the cross-polarization terms with $R_{ps} = R_{sp} \approx 0.0015$. This changes fundamentally if we assume an incidence medium with a high refractive index. In the following, we will assume that $n_{\text{inc}} = 3$.

The corresponding simulations are shown in Fig. 15.15, where the results for two virtually semiinfinite exit media consisting of either large or small domains are compared. For the small domain medium, the typical and well-known curves result (cf. Fig. 4.4). Brewster's angle can be found at $\alpha_i = 33.3°$, and the onset of total reflection at $\alpha_i = 40.8°$. For the large domain sample, obvious differences can be noted. There is a considerable shift of the onset of total reflection, which is

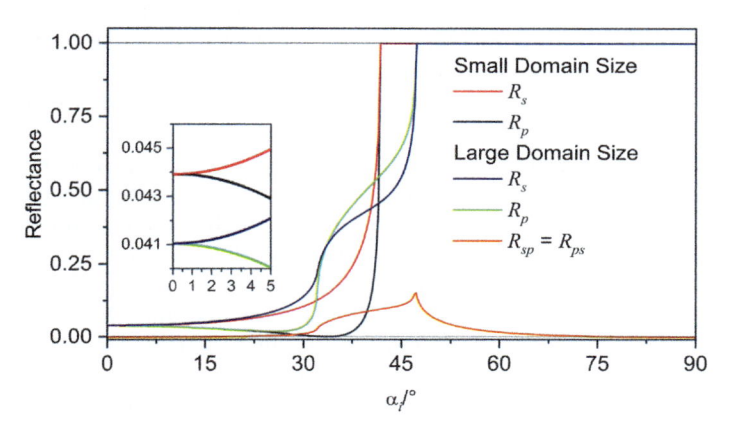

FIG. 15.15 Calculated angular dependence of the perpendicular and parallel polarized reflectance, R_s and R_p, from an isotropic medium consisting of optically small ($d \ll \lambda/10$) or large ($d \gg \lambda/10$) ordered uniaxial domains ($n_a = 2.2$ and $n_c = 1.6$) into a highly refractive incidence medium ($n_{\text{inc}} = 3$) [14].

located at $\alpha_i = 47.2°$. R_p still shows a minimum, but this minimum is not well-pronounced. A very striking difference is that there exists a range where R_p exceeds R_s, which is not possible for scalar media. This range extends from $\alpha_i = 33.8°$ up to the onset of total reflection at $\alpha_i = 47.2°$. With further simulations it can be shown, that its width depends on the difference between the principal indices of refraction, n_a and n_c and that it disappears at $n_a \approx 2.09$ and $n_c \approx 1.82$ for $n_{inc} = 3$. Further investigations showed that in the case of randomly oriented media with large domains, the onset of total reflection is always shifted to a higher angle of incidence compared with random media consisting of small domains. In particular, the onset of total reflection is determined by the largest of the principal indices of refraction of the anisotropic domains. It is equal to that of a homogenous and isotropic medium with this index of refraction [14]. Measuring the above differences between small and large domain media can still be easily performed even when $n_a = 2.02$ and $n_c = 1.96$ [14], which is a fairly small anisotropy.

If we allow the sample to be absorbing, again not much changes for vacuum or air as the incidence medium. Actually, the already small effects are further diminished [15]. Again, it is much more interesting to focus on high-index incidence media.

As pointed out above, the critical angle α_c is always shifted to higher values if the second medium consists of randomly oriented large domains, since the critical angle is then solely determined by the largest principal index of refraction of the domains. As a consequence, the shift of the onset increases with increasing anisotropy. In practice, this effect is important because it has to be taken into account in order, e.g., to choose a proper incidence medium and suitable angles of incidence for ATR experiments. Conditions that might be suitable for small domain media may easily be inappropriate for large domain media, in spite of domains having the same principal indices of refraction. This is in particular problematic, since to obtain larger penetration depths and signals, often angles of incidence are used that are only somewhat smaller than the critical angle, e.g., if ZnSe or diamond are used as ATR crystals together with $\alpha_i = 45°$.

Simulations of the angular dependence of the reflectance when absorption is nonzero are shown in Fig. 15.16. Besides the shift of α_c, we also notice comparably large cross-polarization terms, which show a marked increase near the angle $\alpha_i = \sin^{-1}(n_{lowest}/n_{inc})$, where n_{lowest} is the lowest of the principal indices of refraction of the domains, and an emphasized maximum in the vicinity of α_c. In addition, we again find ranges in which $R_p > R_s$ holds and R_p displays only a minimum but does not become zero.

These differences become increasingly reduced if we allow the polydomain medium to absorb and increase the indices of absorption k_i. If the absorption index values are comparable to those of the real parts n_i, the differences between the large and the small domain cases become very small, even if the ratio n_a/n_c, and therefore the anisotropy, is kept constant. This also applies to the cross-polarization terms, which are in such a case of the order of 1×10^{-3} and therefore almost negligible as in the case of incidence from vacuum or air. For organic materials, however, typically exhibiting low oscillator strengths and, thus, comparably low k_i values in the infrared, considerable misinterpretations may result, if the influence of the domain size is neglected.

In this context, I want to point out that the well-known relation $R_s^2(\alpha_i = 45°)/R_p(\alpha_i = 45°) = 1$ (cf., e.g., Ref. [16]) only holds for scalar media. Deviations from this relation may easily mislead to the assumption of, e.g., preferential orientation of the domains. In contrast, the occurrence of large domains might be responsible for $R_s^2(\alpha = 45°)/R_p(\alpha = 45°) \neq 1$. In this respect, please keep in mind that $\alpha_i = 45°$ is the angle of incidence most commonly employed in ATR accessories nowadays.

For the case of a medium consisting of small domains, it is usually not seen as a special case if the index of refraction of the incidence medium lies between the values of the principal indices of refraction of the polydomain medium, since its averaged index of refraction is either smaller or larger than that of the incidence medium and there is no third case. For media consisting of large domains, however, these conditions lead to a situation that deserves special attention as I will demonstrate in the following.

Again, we use the same set of optical constants as in the preceding sections, i.e., $n_a = 2.2$ and $n_c = 1.6$. But this time the refractive index of the incidence medium $n_{inc} = 2$. Since the averaged index of refraction (small domains) does not depend on the incidence medium, its value is still given by $\langle n \rangle = 1.96$ according to ARIT. Therefore, the small domain medium shows total reflection starting at $\alpha_c = 78.52°$ and, while the difference between R_p and R_s is very small, R_s is always larger than R_p (with the trivial exceptions at $\alpha_i = 0°$ and for $\alpha_i \geq \alpha_c$) as shown in the top panel of Fig. 15.17.

If we investigate the angular dependence of the reflectance for the large domain medium, we first note that total reflection cannot occur, since $n_a > n_{inc}$ and, hence, $\alpha_c = \sin^{-1}(n_{largest}/n_{inc})$ is a complex number. Indeed, it can be shown that total reflection is possible only for p-polarized light and domains having their optical axes oriented perpendicular to the surface. Additionally, this particular orientation results in $R_p > R_s$ if $\alpha_i \neq 0$ and $\alpha_i < \alpha_c$. Since there is a large variety of other orientations that also lead to $R_p > R_s$, at least over a limited range of α_i, we can expect to find an angular range where $R_p > R_s$

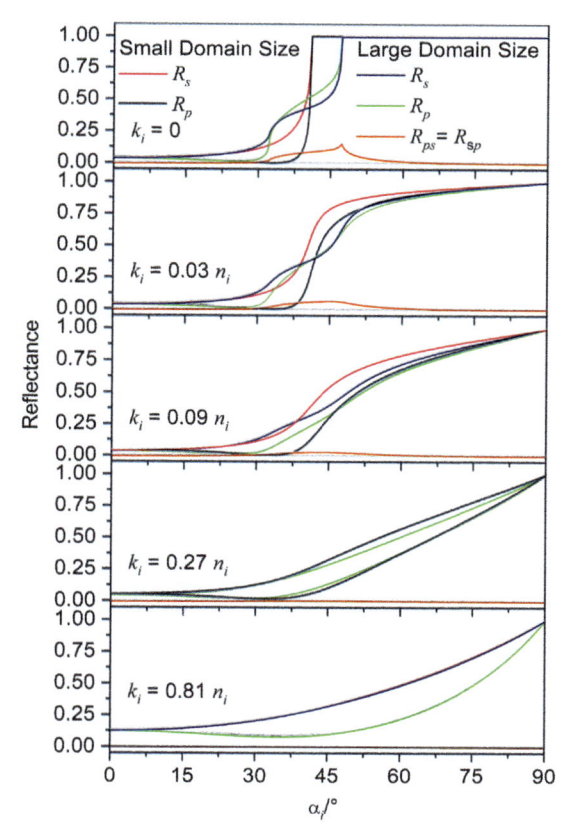

FIG. 15.16 Angular dependence of the reflectances R_s, R_p and the cross-polarization reflectance terms R_{sp} and R_{ps} ($R_{sp} = R_{ps}$) of an isotropic semiinfinite medium consisting of optically large uniaxial domains ($n_a = 2.2$ and $n_c = 1.6$) for several principal absorption indices ($k_i = 0$, ½ n_i, n_i, 2 n_i, $i = a, c$) and comparison with R_s and R_p of the corresponding small domain medium. The incidence medium is characterized by an index of refraction higher than any of the principal indices n_a and n_c of the polydomain medium ($n_{inc} = 3$) [15].

for the randomly oriented large domain medium. This is confirmed by the results of the simulations, which are shown in the top panel of Fig. 15.17. In this figure simulated cross-polarization terms are also presented. These possess again remarkable values in a range starting in the vicinity of $\alpha_i = \sin^{-1}(n_{lowest}/n_{inc}) = 53.1°$.

When we allow the domains to be absorptive while keeping the anisotropy constant, it can be once again noticed that the domain size-related effects are reduced by increasing absorption. However, at $k_i = 0.09\, n_i$, R_p is still slightly larger than R_s. Therefore, it can be assumed that it must be relatively easy to come across isotropic samples where R_p exceeds R_s. In line with that, Fresnoite can be employed as a test material to validate the importance of the above simulations. Its principal indices of refraction functions are shown in Fig. 15.18. These functions can be subdivided into several ranges according to the values of the real parts of the principal refractive indices in comparison with the index of refraction n_{inc} of the incidence medium ($n_{inc} = 1$) [17].

Accordingly, six ranges are obtained. From the visible spectral range down to 1110 cm^{-1} (I), the real parts of the principal refractive indices are both larger than n_{inc}. In the second range (II), between 1110 and 1063 cm^{-1}, the inequality $n_c < n_{inc} < n_a$ holds, while in the third range (III), which extends from 1063 to 1034 cm^{-1}, both n_a as well as n_c are smaller than n_{inc}. In the subsequent range (IV), which is situated between 1034 and 969 cm^{-1}, only n_c exceeds n_{inc}. In range V, starting from 969 cm^{-1} down to 957 cm^{-1}, there is a point, that is located between 961 and 962 cm^{-1}, where both the indices of refraction and the indices of absorption are equal, and therefore the optical properties of Fresnoite should resemble those of a material with cubic crystal symmetry. While in range V both real parts are larger than unity, n_a is again smaller than n_{inc} below 957 cm^{-1} in range VI.

The function of the average index of refraction $\langle n \rangle$ as determined from a Kramers-Kronig analysis (KKA) of a reflectance spectrum of randomly oriented Fresnoite consisting of small crystallites (cf. Fig. 15.19) shows that in the two ranges between 1030 and 1090 cm^{-1} and also between 976 and 996 cm^{-1}, an ATR experiment is performed if $\alpha_i \geq \alpha_c$. For a sample

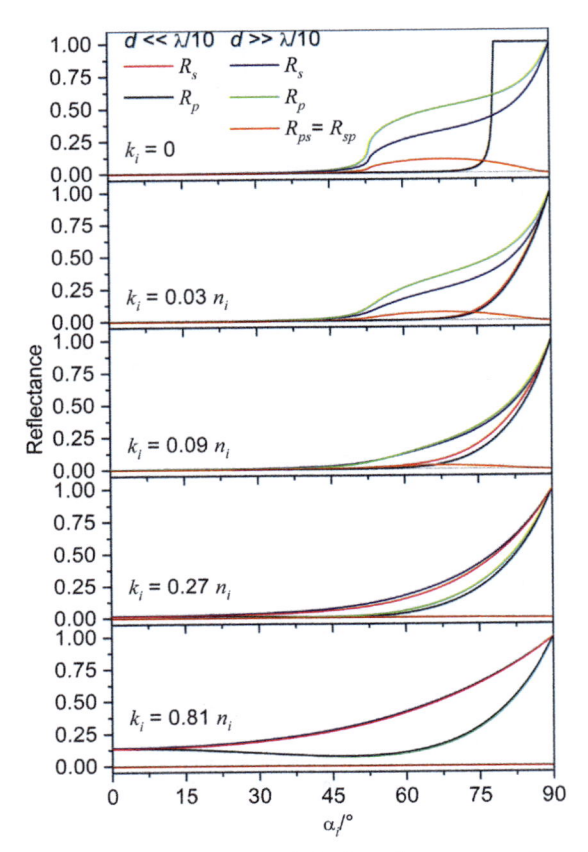

FIG. 15.17 Angular dependence of the reflectances R_s, R_p and the cross-polarization reflectance terms R_{sp} and R_{ps} ($R_{sp} = R_{ps}$) of an isotropic semiinfinite medium consisting of optically large uniaxial domains ($n_a = 2.2$ and $n_c = 1.6$) for several principal absorption indices ($k_i = 0, 0.03\, n_i, 0.09\, n_i, 0.27\, n_i, 0.81\, n_i, i = a, c$) and comparison with R_s and R_p of the corresponding small domain medium. The incidence medium is characterized by an index of refraction in between the principal indices n_a and n_c of the polydomain medium ($n_{\text{inc}} = 2$) [15].

FIG. 15.18 Real and imaginary parts of the principal indices of refraction of the Fresnoite single crystal obtained by Kramers-Kronig analyses [4,17].

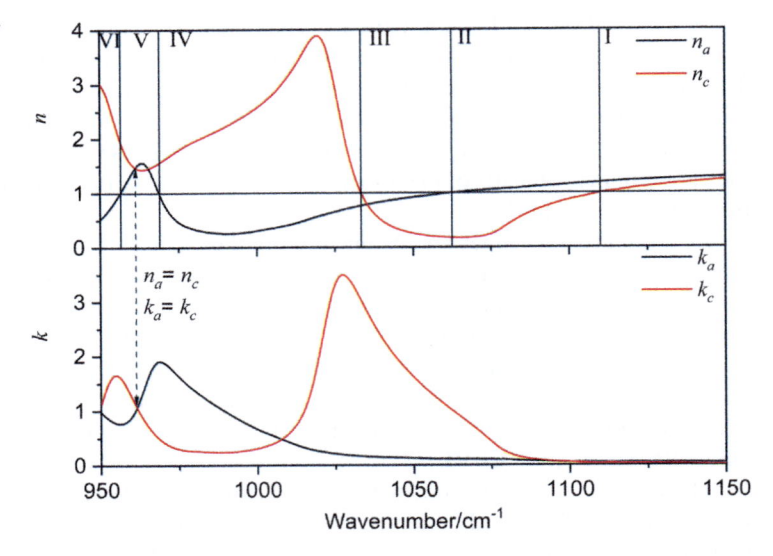

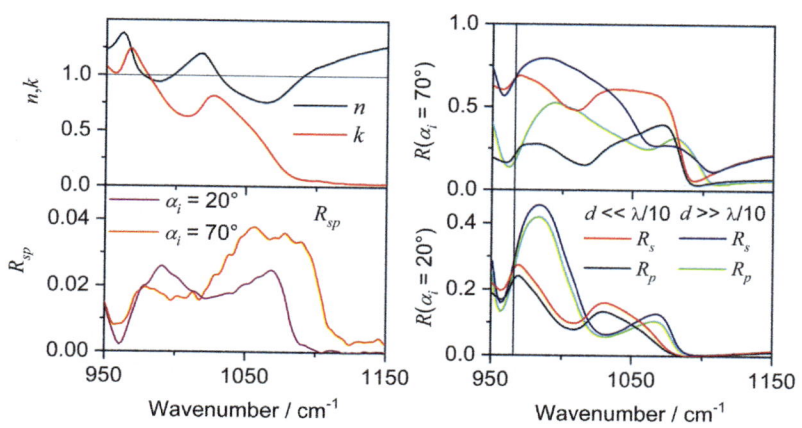

FIG. 15.19 Experimental reflectance spectra ($\alpha_i = 20°/70°$) of randomly oriented polycrystalline Fresnoite with either optically large or small crystallites. Left column, upper panel: wavenumber dependence of the averaged complex index of refraction $\langle n \rangle$ (small domain sample, derived from a KKA of $R_s(\alpha_i = 20°)$); left column, lower panel: cross-polarization terms of the reflectance R_{sp} for two different angles of incidence $\alpha_i = 20°$ and $\alpha_i = 70°$ (large domain sample). Right column: Comparison of the reflectance for s- and p-polarized incident radiation and two different angles of incidence $\alpha_i = 20°$ (lower panel) and $\alpha_i = 70°$ (upper panel).

with large crystallites, the situation is different. For such samples, the former range is identical to range III and therefore narrower (between 1034 and 1063 cm^{-1}), while the second range does not exist, since for a large domain medium, total reflection is suppressed in this range (vide supra). With respect to the angular dependence of the cross-polarization terms, at $\alpha_i = 70°$ the cross-polarized reflectance is higher down to 1020 cm^{-1} compared with its values at $\alpha_i = 20°$.

At $\alpha_i = 20°$, R_s and R_p show similar band shapes and fulfill the condition that $R_s > R_p$ over the whole spectral range for both types of polycrystalline samples. At $\alpha_i = 70°$ this inequality is always fulfilled only for the sample consisting of small crystallites. In contrast, for the large domain sample, $R_p > R_s$ holds between 1069 and 1096 cm^{-1}. Note that the latter range lies completely in the region where $n_c < n_{\text{inc}} < n_a$ (range II) and in which absorption is already comparatively low, in contrast to range III. In the latter range, absorption prevents R_p from being larger than R_s. Therefore, in particular at higher angles of incidence, take special care not to misinterpret the occurrence of spectral ranges where $R_p > R_s$, e.g., by assuming preferred orientation.

Fig. 15.20 presents a comparison between the experimental spectra of the large crystallite sample with those simulated by ARTT based on optical constant functions determined by KKA. Both sets of spectra show excellent agreement, in

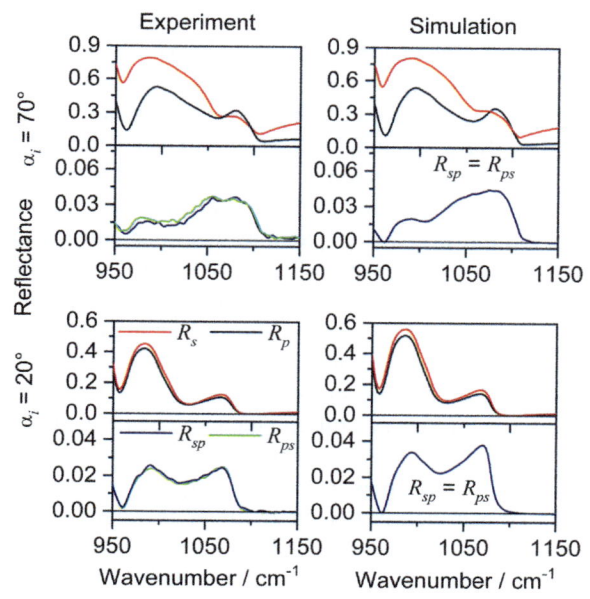

FIG. 15.20 Comparison of the experimental reflectance spectra of randomly oriented polycrystalline Fresnoite (large domains, first column from the left) with the simulated spectra based on single crystal data obtained by KKA (right column) at $\alpha_i = 70°$ (upper row) and $\alpha_i = 20°$ (lower row).

particular with regard to the cross-polarization terms. These terms are of special sensitivity to errors concerning the optical constants determined from the single crystal [17]. This example shows that a comparison between the experimental spectra of respective polycrystalline materials with large crystallites and the spectra simulated by ARTT can be used to assess the accuracy of optical constants of single crystals.

From the resemblance between measured and simulated spectra, we can also draw the conclusion that ARTT is valid at higher angles of incidence as well as at near normal incidence. Besides the quantitative prediction of cross-polarized reflectance terms R_{sp} and R_{ps}, ARTT is also able to predict band shapes, positions and relative intensities of bands quite satisfactorily. With regard to absolute intensities, the errors are larger due to the lower surface quality of the polycrystalline samples compared with that of a single crystal, the optical constants of which were used in the simulation.

Overall, I have to state that it is very unfortunate that most ATR accessories sold nowadays do not allow to change the angle of incidence, and to use a polarizer and analyzer. Certainly, not every purpose of application would justify the additional costs, but in many cases, it would be very beneficial to have the corresponding options.

15.3 Large crystallites and nonrandom orientation

As I admitted already, it is not yet clear to me how small crystallites and nonrandom orientations can be treated. I am at the moment not even convinced that this is something that is practically relevant, because if some neighboring crystallites share the same orientation, light would register them as one large crystallite. Anyway, at least for the latter case, it is comparably easy to extend ARTT so that it can be used to describe partly oriented samples. While for randomly oriented samples, every orientation has a priori the same probability, this is no longer the case for oriented samples. Accordingly, we have to introduce weights, which we do by employing a so-called orientation distribution function (ODF) which is normalized. Accordingly, we obtain:

$$\langle R \rangle = \int_{\Omega^{(3)}} \text{ODF}(\Omega) R(\Omega) d\Omega$$

$$\langle T \rangle = \int_{\Omega^{(3)}} \text{ODF}(\Omega) T(\Omega) d\Omega \quad (15.12)$$

$$\text{with} \int_{\Omega^{(3)}} \text{ODF}(\Omega) d\Omega = 1$$

Key elements to determine the orientation are the cross-polarization terms and a nonzero angle of incidence. For normal incidence, certain symmetry relations lead to the fact that, e.g., for a uniaxially oriented sample the orientations (φ, θ) and $(\varphi, -\theta)$ cannot be distinguished, since the reflectance and transmittance terms for the one and two polarizer experiments are all equal [18]. Nonnormal incidence breaks the symmetry, however only because of the cross-polarization terms, which means that $R_s(\varphi, \theta) \approx R_s(\varphi, -\theta)$, and the differences may be too small to be detected. Therefore, experiments with crossed polarizers are very important. If the cross-polarization terms are very small, it is useful to increase the angle of incidence. For Fresnoite, the differences are large enough to be realized even at moderate angles of incidence like $\alpha_i = 20°$. We have demonstrated this for a special kind of sample that was prepared by electrochemically induced nucleation: In a Pt crucible, which served as the anode, a mixture with the composition $Ba_2TiSi_2O_8 + 0.75$ SiO_2 was melted. After melting, a Pt wire, which served later as a cathode, was put in the center of the crucible. Next, the melt was supercooled, and a current was initiated. Subsequently, a glass ceramic with the above composition began to grow around the Pt wire. Before the ceramic reached the walls of the crucible, it was withdrawn. Afterwards, the ceramics (cf. Fig. 15.21) was cut into disks, which underwent investigations by X-ray diffraction and IR spectroscopy. Both methods confirmed that the growth direction agreed to a high degree with the orientation of the crystallographic c-axis. Interestingly, what only IR spectroscopy could reveal was that the remaining SiO_2, which was still glassy, was co-oriented to the Fresnoite crystallites [19]. This would certainly be nothing out of the ordinary for organic glasses, but for inorganic glasses usually orientation is not possible for the bulk; only at the surfaces can orientation effects be detected [20]. With regard to the disks, a measurement in the center should result in a two-dimensional randomly oriented sample. The c-axis is nearly parallel to the surface, which means that θ is close to 90°, while φ takes on random values. According to Linear Dichroism Theory, it is impossible to distinguish an

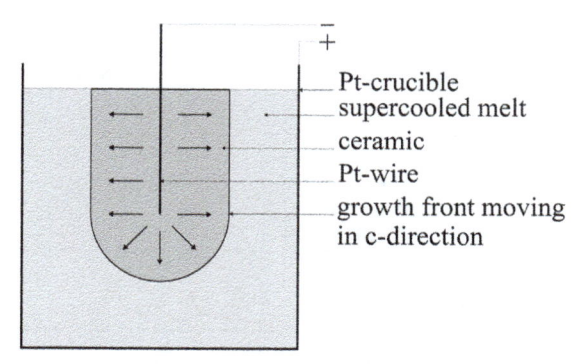

FIG. 15.21 Preparation of the glass ceramic. The *arrows* indicate not only the growth direction, but also the orientation of the *c*-axis [19].

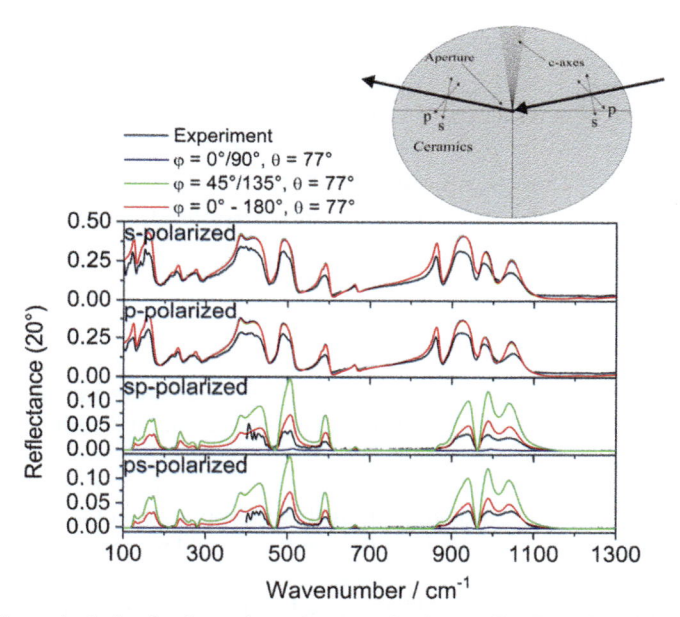

FIG. 15.22 Comparison between forward calculated and experimental spectra of polycrystalline Fresnoite oriented by the method of electrochemically induced nucleation. Spectra are taken at the center of the disk [18].

orientation distribution from a particular orientation having the average orientation. In Fig. 15.22 we compare the experimental spectra with those resulting from three different orientation distributions, which all have the same average orientation. The first consists of two orientations for which φ takes on the values 0° and 90°. The second also features only two orientations, but this time $\varphi = 45°$ and 135°. Finally, the third consists of a continuous distribution of φ-values between 0° and 180°. The first two panels of Fig. 15.22 from above show the one-polarizer experiments and the corresponding forward calculations, all of which agree within line thickness. Based on these simulations, it is not possible to decide which orientation distribution would be the one that better reflects the one in the sample. In stark contrast, the cross-polarization spectra differ strongly with regard to their intensities. The one that is most compatible with the experimental spectra is the continuous distribution of φ-values between 0° and 180°. Still, there might be orientation distributions that may have similar values of the cross-polarization terms, but it is again obvious from the results that LDT is nothing but a first approximation of the actual situation.

The more we move the aperture to the edge of the disk, the smaller becomes the orientation distribution with regard to the angle φ. At the edge, the distribution is so narrow that the spectra resemble strongly those of oriented single crystal spectra. This is obvious from the results presented in Fig. 15.23. Nevertheless, the comparison of the experimental

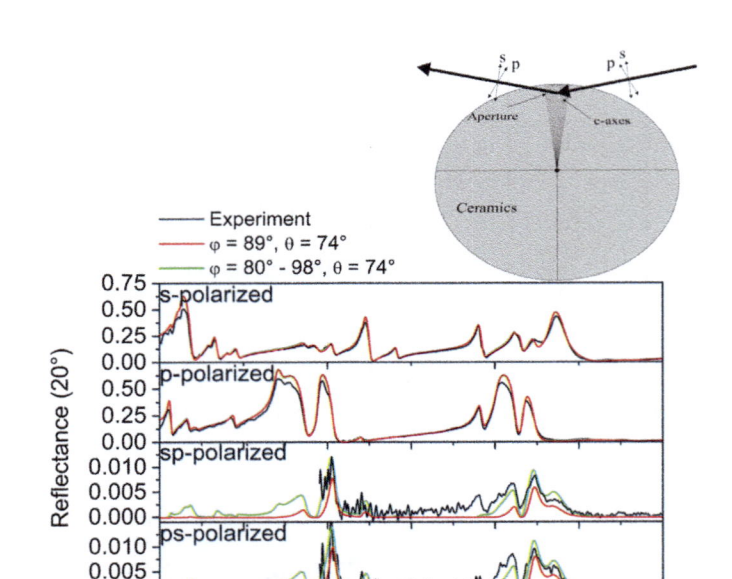

FIG. 15.23 Comparison between forward calculated and experimental spectra of polycrystalline Fresnoite oriented by the method of electrochemically induced nucleation. Spectra are taken at the edge of the disk [18].

cross-polarization spectra and those forward calculated for a particular orientation and an orientation distribution reveals that the latter two can still be distinguished and that the spectrum of the orientation distribution is clearly more similar to the experimental one than that for the particular orientation.

These results again show how important and fruitful the use of wave optics is to fully understand and exploit the possibilities IR spectroscopy offers to those who apply it. Still, wave optics is comparably sophisticated and effortful. While the application of two polarizers seems to be a requirement, we could recently show that the use of 2D-correlation spectroscopy might be able to remove this requirement and help in situations where not even the cross-polarization spectra are helpful. An example are disks that have been cut below the end of the Pt-wire (cf. Fig. 15.21). The spectra measured in the center of such disks are influenced more or less exclusively by the a-axis component of the dielectric tensor. If the aperture is moved toward the edge of the disk and the polarization direction is perpendicular to the radius, the spectra virtually do not change. Thanks to 2T2D-correlation spectroscopy (cf. Section 8.3) such spectra nevertheless contain exploitable information about the orientation and even the orientation distribution [21]. This is shown by the comparison between experimental and simulated asynchronous 2T2D-correlation spectra of the Fresnoite glass ceramic in Fig. 15.24.

While the differences are subtle, they nevertheless allow to clearly differentiate between the different orientations and orientation distributions. E.g., if you look at the 2D spectra in Fig. 15.24 at the brighter stripe between about 1050 and 1080 cm^{-1} above the diagonal (y-coordinates), then you see no brighter spot in the left spectra, while there is one from about 900 to 920 cm^{-1} (x-coordinates) in the center spectra and an additional one around 1000 cm^{-1} (x-coordinates) in the right spectra. Note that the simulated spectra are actually based on single crystal data. Therefore, the agreement is comparably excellent.

Overall, I hope I could convince you that based on wave optics in combination with dispersion theory, even subtle effects in spectra can be convincingly explained. Although this combination is much more demanding than Linear Dichroism theory, I think the added value clearly justifies the increase in complexity.

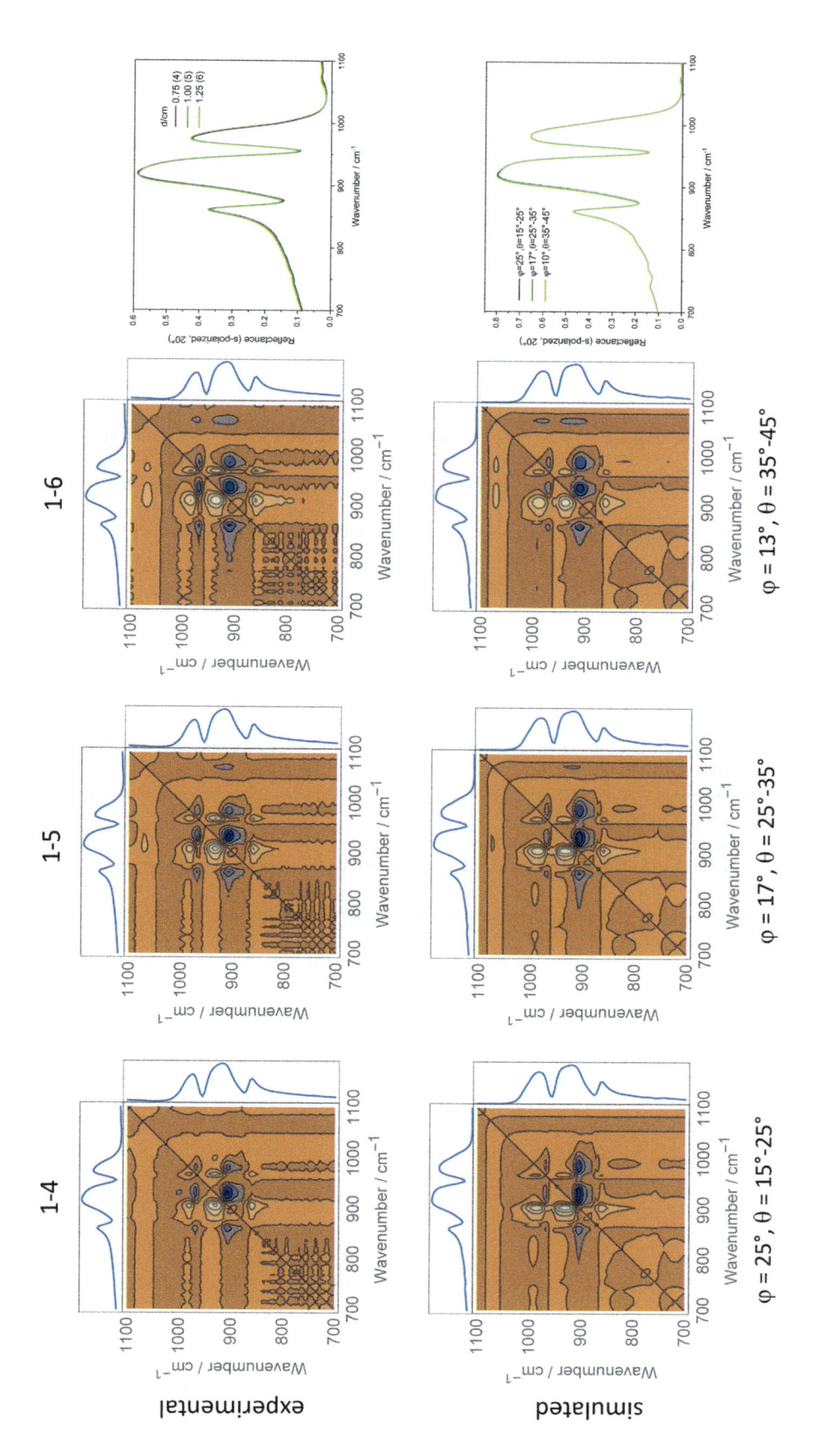

FIG. 15.24 Experimental and simulated asynchronous 2T2D-correlation spectra of disk 8 of the $Ba_2TiSi_2O_8 + 0.75$ SiO_2 sample with s-polarized light and 20° angle of incidence (plane of incidence along the radius). The experimental spectra consisted of the spectra for $d = 0.75$, $d = 1.0$, and $d = 1.25$ cm and are compared on the upper panel. For the simulations, we assumed wave optics to be valid and employed as a reference the spectrum of the single crystal for $\varphi = 0°$, $\theta = 0°$. For the second spectra we used averages of the single crystal spectra according to $\varphi = 25°$, $\theta = 15°$–$25°$, $\varphi = 17°$, $\theta = 25°$–$35°$, and $\varphi = 13°$, $\theta = 35°$–$45°$, which represent the perturbation for the simulated spectra. These spectra are depicted on the lower panel [21].

15.4 Further reading

Unfortunately, there is no other literature that I could suggest as an alternative.

References

[1] G. Kliche, Z.V. Popovic, Far-infrared spectroscopic investigations on CuO, Phys. Rev. B 42 (1990) 10060–10066.
[2] C. Homes, M. Ziaei, B. Clayman, J. Irwin, J. Franck, Softening of a reststrahlen band in CuO near the Néel transition, Phys. Rev. B 51 (1995) 3140–3150.
[3] T.G. Mayerhöfer, J. Popp, Employing spectra of polycrystalline materials for the verification of optical constants obtained from corresponding low-symmetry single crystals, Appl. Opt. 46 (2007) 327–334.
[4] T.G. Mayerhöfer, Optics and IR-spectroscopy of polydomain materials, in: Habilitation, Friedrich Schiller University Jena, 2006, https://doi.org/10.13140/RG.2.1.3880.7443.
[5] T.G. Mayerhöfer, J. Popp, The 390cm^{-1} feature of polycrystalline hematite—an optical crystallite size effect, Icarus 203 (2009) 303–309.
[6] V.E. Hamilton, C.W. Haberle, T.G. Mayerhöfer, Effects of small crystallite size on the thermal infrared (vibrational) spectra of minerals, Am. Mineral. 105 (2020) 1756–1760.
[7] T.G. Mayerhöfer, Modelling IR-spectra of single-phase polycrystalline materials with random orientation—a unified approach, Vib. Spectrosc. 35 (2004) 67–76.
[8] T.G. Mayerhöfer, Symmetric Euler orientation representations for orientational averaging, Spectrochim. Acta A Mol. Biomol. Spectrosc. 61 (2005) 2611–2621.
[9] A.C. Wright, The great crystallite versus random network controversy: a personal perspective, Int. J. Appl. Glas. Sci. 5 (2014) 31–56.
[10] T.G. Mayerhöfer, H.H. Dunken, R. Keding, C. Rüssel, Interpretation and modeling of IR-reflectance spectra of glasses considering medium range order, J. Non-Cryst. Solids 333 (2004) 172–181.
[11] T.G. Mayerhöfer, Modelling IR spectra of single-phase polycrystalline materials with random orientation in the large crystallites limit—extension to arbitrary crystal symmetry, J. Opt. A Pure Appl. Opt. 4 (2002) 540.
[12] T.G. Mayerhöfer, Z. Shen, R. Keding, T. Höche, Modelling IR-spectra of single-phase polycrystalline materials with random orientation—supplementations and refinements for optically uniaxial crystallites, Optik 114 (2003) 351–359.
[13] T.G. Mayerhöfer, J.L. Musfeldt, Angular dependence of the specular reflectance from an isotropic polydomain medium with large domains: surprising results regarding Brewster's angle, J. Opt. Soc. Am. A 22 (2005) 185–189.
[14] T.G. Mayerhöfer, J. Popp, Angular dependence of the reflectance from an isotropic polydomain medium: effect of large domain size on total reflection, J. Opt. Soc. Am. A 22 (2005) 569–573.
[15] T.G. Mayerhöfer, J. Popp, Angular dependence of the reflectance from an isotropic polydomain medium: the influence of absorption, J. Opt. Soc. Am. A Opt. Image Sci. Vis. 22 (2005) 2557–2563.
[16] F.M. Mirabella, Internal reflection spectroscopy, Appl. Spectrosc. Rev. 21 (1985) 45–178.
[17] T.G. Mayerhöfer, J. Popp, Angular dependence of the reflectance from an isotropic polydomain medium: experimental verification, J. Opt. A Pure Appl. Opt. 9 (2007) 581–585.
[18] T.G. Mayerhöfer, J. Popp, Modelling IR spectra of polycrystalline materials in the large crystallites limit—quantitative determination of orientation, J. Opt. A Pure Appl. Opt. 8 (2006) 657–671.
[19] T.G. Mayerhöfer, H.H. Dunken, R. Keding, C. Rüssel, Determination of the crystallographic orientation of oriented fresnoite glass ceramics by infrared reflectance spectroscopy, Phys. Chem. Glasses 42 (2001) 353–357.
[20] V. Ivanovski, T.G. Mayerhöfer, A. Kriltz, J. Popp, IR-ATR investigation of surface anisotropy in silicate glasses, Spectrochim. Acta A Mol. Biomol. Spectrosc. 173 (2017) 608–617.
[21] T.G. Mayerhöfer, I. Noda, J. Popp, The footprint of linear dichroism in infrared 2D-correlation spectra, Spectrochim. Acta A Mol. Biomol. Spectrosc. 304 (2024) 123311.

Chapter 16

Vibrational circular dichroism

16.1 Introduction

When you scan through the available textbooks about infrared spectroscopy or vibrational circular dichroism (VCD), do you get any idea how the actual microscopic mechanism behind VCD works? What I personally knew from Berreman's paper about the 4×4 matrix formalism [1] was the macroscopic reason for chirality, which is that the electric field and the magnetic field somehow couple in optically active materials—something which is not possible in achiral materials where electric and magnetic fields are always perpendicular to each other. So, the wave optics-related part was more or less clear to me, more or less, because it seems there is still some uncertainty concerning the actual form of the constitutive relations in the case of chiral media (cf. Section 12.8.5). However, what do the peaks in a VCD spectrum stand for, and why are some positive and some negative? Why do the peaks have the same positions as those in the corresponding conventional IR spectrum and why do the bands have an approximately Lorentzian shape? Is there any rule about the order of positive and negative peaks or about how many of each sign are present? Is there anything I can learn from these spectra except that I can determine the absolute configuration by the comparison of experimental spectra with the results of quantum mechanical calculations (which is certainly very important)? [2].

In search of answers, I read through the corresponding chapters and books, like the one of Laurence Nafie, Vibrational Optical Activity: Principles and Applications [2], and others, but to no avail. It is my impression that nowadays only the measurement principle, i.e., measuring the difference of the negative decadic logarithms of the transmittance of left and right circularly polarized light, ΔA, is explained and then the texts jump to how to calculate the VCD spectra based on quantum mechanics (is it just me or is there a big gap in-between?). Basically, this means that I would not be able to report anything in this chapter because neither theory on the level of absorbance nor quantum mechanics are topics of this book.

Usually, it helps to go back into history and study how a field developed to understand its present state. Talking about VCD measurements, it took a long time before the first successful one was reported in 1974, in contrast to CD measurements in the UV/Vis, which could be carried out much earlier. The reason is that measurements of conventional spectra in the IR are already more demanding than those in the UV/Vis because of less brilliant light sources, less sensitive detectors, and weaker oscillator strengths (for organic or biologic substances; note that this situation certainly is about to change, thanks to quantum cascade lasers [3] and, hopefully, supercontinuum light sources). But VCD effects are even some orders of magnitude less intense in relation to conventional IR absorptions than their UV/Vis counterparts relative to UV/Vis absorptions (there is a dependence on frequency; with increasing frequency, the CD signals become more intense). This means that a potential classical explanation for the occurrence of (V)CD had long been forgotten before the first spectra had been recorded.

The first explanation of the origin of CD bands was given by Paul Drude and provided in his textbook about optics in 1900 [4]. Unfortunately, while the formulas derived were useful, his assumptions were not correct as shown later by Werner Kuhn [5], which is why I will not go into detail concerning Drude's explanations. The first reasonable explanation was provided by Max Born in 1915 and partly corrected and extended in 1918 [6]. This explanation is based on an extended oscillator model with coupled oscillators, something which we have already been discussing in Section 7.3. Since there are some subtleties that I did not touch upon in that section, I will explain them in detail in this one. In addition, Werner Kuhn should be mentioned again because he also has great merits in developing the classical idea further, see, e.g., [7]. At the same time, Kuhn worked on the classical model, the quantum mechanical model was already developed [8]. In the following years, the focus was on the latter model, and the classical model was forgotten until the advent of chiral plasmonics, where it is known as the Born-Kuhn model [9]. Unfortunately, this rediscovery focused on a single pair of oscillators with a shared eigenfrequency, and it was generally disregarded that both Born and Kuhn developed their models for an unlimited number of different oscillators that are coupled with each other. Two oscillators with a shared eigenfrequency are a severe restriction, even for chiral plasmonics, which is why extended models have been introduced where this restriction has been removed [10,11]. These latter works are of great interest to us since they show how the CD spectrum can be influenced by changing the configuration of atoms (actually that of four metallic disks and one metallic bar in [10]), and how relatively accurately this spectrum can be modeled by the classical model. This brought me to the idea that it should also be possible to perform a kind of chiral dispersion analysis of VCD spectra. Investigating this idea is one main topic of this chapter, while the exploration of coupling effects has already been performed in Section 7.3.2. So, we will only briefly repeat the discussion of such effects together with chiral dispersion analysis.

16.2 Calculating the spectra of chiral materials

Conventional analysis of IR and corresponding VCD spectra is based on interpreting the absorbance and the difference of absorbance between left and right circularly polarized light and on comparing the result with quantum mechanical calculations. In contrast, chiral dispersion analysis uses a suitable oscillator model to generate the dielectric function and the chirality admittance function to calculate the conventional IR spectrum and the VCD spectrum based on wave optics. The modeled spectra are compared with the measured spectra and iteratively optimized.

In principle, a VCD spectrometer records the transmittance with left and right circularly polarized light in fast alternation and calculates a conventional absorbance and a VCD spectrum. The conventional absorbance spectrum could be reconverted to transmittance and used to determine the dielectric function. An alternative would be to use the ATR spectra after calibration of an ATR accessory (i.e., determining the polarization angle, cf. Section 6.7.1) and determine the dielectric function from this spectrum (in [12], we used an iterative formalism to evaluate the ATR spectra according to Section 6.7.1. Nowadays, I would suggest using a much faster algorithm based on a direct Kramers-Kronig analysis according to Section 5.8.2). Based on this dielectric function, I can calculate how a conventional transmittance absorbance spectrum of, e.g., (+)-α-pinene should look like and compare it to the one determined by the VCD spectrometer. Such a comparison is shown in Fig. 16.1. While peak shapes and positions agree, the relative absorbance values are different, as are the absolute values. Around $3000\,\text{cm}^{-1}$, absorbance is higher than 10. From this finding, it can be concluded that the ΔA spectrum will also not be reliable in this spectral region (note that the absorbance becomes unreliable already for absorbances larger than 1.5 in the experimental spectrum since not enough photons reach the detector anymore). In any case, it seems that the absorbance spectrum provided together with the ΔA spectrum cannot be used to determine the dielectric function of a chiral substance.

After the dielectric function is determined, we can employ the 4×4 matrix formalism using the simplification provided in Section 12.8.5 to determine the chirality admittance function g employing the same principal wave optics-based iterative correction scheme as originally applied to the ATR measurement and also to ΔA (cf. Section 6.7.1) [12]. The scheme first requires an estimation for g. How do we obtain such an estimation? We make use of the fact that the (complex) eigenvalues for normal incidence are given by (cf. Eq. 12.89):

FIG. 16.1 Comparison between the simulated absorbance spectrum of (+)-α-pinene based on the dielectric function obtained from ATR measurements and the one determined by the VCD spectrometer [12].

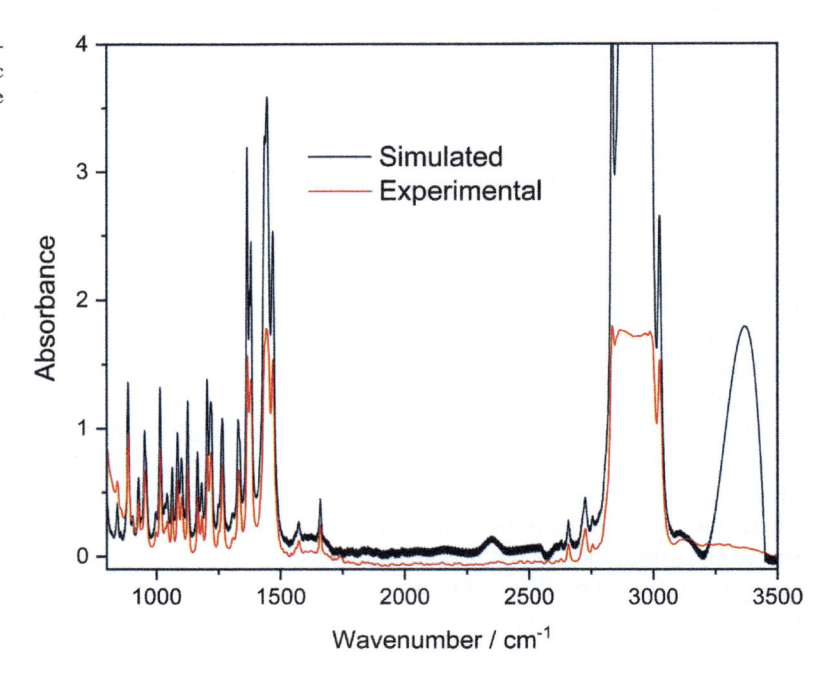

$$n_+ = n + g$$
$$n_- = n - g,$$
(16.1)

Taking the imaginary part of the difference $n_+ - n_-$ allows to calculate ΔA in dependence of g (d is the optical path length or the thickness of the cuvette):

$$\Delta A = 4\pi(\log_{10}e)d\tilde{\nu}\text{Im}(n_+ - n_-) = 8\pi(\log_{10}e)d\tilde{\nu}g'',$$
(16.2)

Therefore, the first estimation of the imaginary part of g is

$$g'' = \frac{\Delta A}{8\pi(\log_{10}e)d\tilde{\nu}}$$
(16.3)

When you apply the iterative formalism and use the above equation as the first estimation, there is no more improvement, and this first iteration is the final result. When this happened, my first thought was that this was an error. However, this thought was not correct. More investigations showed that ΔA is indeed a linear function of the absorbance difference and that it is independent of the dielectric function of the chiral compound and even of the cuvette material (since the formulas are very complex, we checked this by applying 2D-correlation analysis with the thickness as perturbation). In other words, it seems that Eq. (16.2) is exact on the level of wave optics.

This result was quite surprising to me, having spent a lot of time (and space in this book) showing that the Bouguer-Beer-Lambert (BBL) law is nothing but an approximation of the results obtained by using wave optics. On a second view, this finding might be less surprising, taking into account that the differences between n_+, n_- and n are very small. On the other hand, even if these differences are not small, as for metamaterials, Eq. (16.2) is still valid, at least for free-standing films [13]. In [13], Eq. (16.2) was not derived directly, but Eq. (15b) in this reference can easily be converted into Eq. (16.2). This is another proof that it is indeed not necessary to use the 4×4 matrix formalism and that VCD spectra can directly be quantitatively evaluated by Eq. (16.2) (in contrast to the related IR spectra!). Note that this finding refers only to the Bouguer-Lambert part of the BBL approximation. Beer's part, that is, the linear concentration dependence, is yet to be checked. My guess is that this part would not hold, although I might easily be wrong a second time.

I did not discuss this in Section 5.8, but the real part and the imaginary part of g, g' and g'' are also related by the Kramers-Kronig relation (cf. Section 5.8) [14]:

$$g'(\tilde{\nu}) = \frac{2}{\pi}\wp \int_0^\infty \frac{g''(\tilde{\nu}')\tilde{\nu}'}{\tilde{\nu}'^2 - \tilde{\nu}^2}d\tilde{\nu}'$$

$$g''(\tilde{\nu}) = -\frac{2\tilde{\nu}}{\pi}\wp \int_0^\infty \frac{g'(\tilde{\nu}')}{\tilde{\nu}'^2 - \tilde{\nu}^2}d\tilde{\nu}'$$
(16.4)

A distinctive feature is that g' starts at zero for very high frequencies/wavenumbers and returns to this value for low frequencies/wavenumbers. As I will show later, the area of the product $\tilde{\nu}g''$ above zero and below zero are equal (and the same holds true for g'.). In any way, we can represent g'' by the triangular function base and analytically calculate g' from it (cf. Section 5.8 and Section 6.7.1). Fig. 16.2 shows real and imaginary parts of the chirality admittance for Propylene oxide.

16.3 Chiral dispersion analysis

In this section, we come back to the oscillator model which included coupling by springs (and not by honeypots), as discussed in Section 7.3.2. There is, however, one important difference that we have to introduce to understand why a material would be optically active. In Section 7.3.2, we did not make an assumption about how the mutual coupling would come into effect. It could be the same interaction that is described by the force constant, but if this would be the case, why or how would we notice it? If we would in a thought experiment introduce such an interaction in an existing molecule, it would change its systems of normal vibrations, but if we would not be able to investigate it before the change takes place, we would not be able to detect it. Only if it would be different and if it would be something we could switch on and off, would we be able to spot the difference. We certainly cannot do this in the case of chiral molecules, but we can invert the direction of the coupling by choosing one enantiomer or the other. In addition, we can change the handedness of the circularly polarized light. The circular polarization is also key to the understanding of what may be actually going on. So far, we

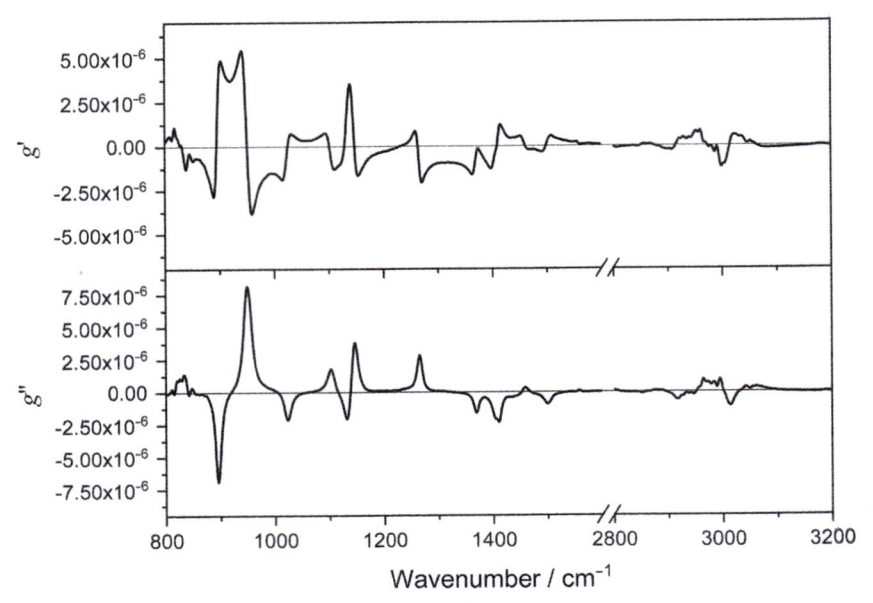

FIG. 16.2 Real and imaginary part of the chirality admittance of S-(−)-Propylene oxide. [12].

always assumed that all oscillators in a material are hit by electromagnetic waves at the same time and with the same phase at every location. This is an assumption that we have to modify to understand circular dichroism. Say a plane wave is traveling along the z-direction. If there are two oscillators that have their transition moments perpendicular to z at different z-values with the difference d, there will be a phase difference. In particular, we will assume that one oscillator is oriented along the x-axis while the other is oriented along the y-axis (cf. Fig. 16.3).

If these two oscillators are coupled, the equations of motion will be the following:

$$\frac{d^2x}{dt^2} + \gamma_1 \frac{dx}{dt} + \omega_{0,1}^2 x + v_{12}y = q_1 E_x \exp(i(k \cdot (z+d/2) - \omega \cdot t))$$
$$\frac{d^2y}{dt^2} + \gamma_2 \frac{dy}{dt} + \omega_{0,2}^2 y + v_{12}x = q_2 E_x \exp(i(k \cdot (z-d/2) - \omega \cdot t))$$
(16.5)

As already mentioned, the difference in relation to Eq. (7.11) is that we still have to consider the spatial dependence. Additionally, here we assume two oscillators that have perpendicular transition moments. Note that we also implicitly assume that the amplitude of the electric field does not change when traveling from $z+d/2$ to $z-d/2$, so that the magnitude of the wavevector k is a real number and is equal to $k = 2\pi/\lambda = 2\pi\tilde{v}$.

Remember that v_{12} is the coupling constant and that the term involving the coupling constant should read $v_{12}(y-x)$. In the above equation, I have already considered that $\omega_{TO,1}^2 - v_{12} = \omega_{0,1}^2$ and $\omega_{TO,2}^2 - v_{12} = \omega_{0,2}^2$, where $\omega_{TO,j}$ is the eigenfrequency of oscillator j.

After having carried out the derivations, and having introduced the wavenumber, we obtain:

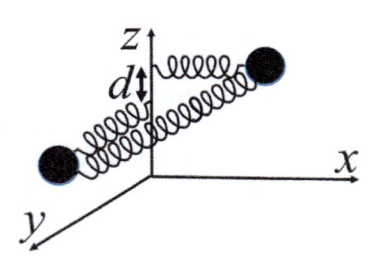

FIG. 16.3 Two coupled oscillators at a distance d.

$$\begin{pmatrix} A_1 & v_{12} \\ v_{12} & A_2 \end{pmatrix} \begin{pmatrix} x \\ y \end{pmatrix} = (q_1 E_x \exp(ik \cdot (z+d/2)) q_2 E_y \exp(ik \cdot (z-d/2)))$$

$$A_1 = \tilde{\nu}_{0,1}^2 - \tilde{\nu}^2 - i\tilde{\nu}\gamma_1$$

$$A_2 = \tilde{\nu}_{0,2}^2 - \tilde{\nu}^2 - i\tilde{\nu}\gamma_2$$

(16.6)

d, the z-distance between coupling oscillators is very small. We can therefore develop the spatial dependence in a series according to $\exp(ik\cdot(z+d/2)) \approx 1 + ik\cdot(z+d/2)$ and $\exp(ik\cdot(z-d/2)) \approx 1 - ik\cdot(z+d/2)$. Note that if only the first term is kept, we are back to the approximation of assuming that all parts of the molecule vibrate in phase and that we neglect all retardation effects; sometimes this is called the local approximation. If we keep the second term, we keep first-order spatial dispersion phenomena.

When we compare the result of the series development with the first of the constitutive relations as provided by Jaggard and Sun [15], in a somewhat different form:

$$\mathbf{D} = \varepsilon_0 \varepsilon \mathbf{E} + ig/c\mathbf{H}$$
$$\mathbf{B} = \mu_0 \mu \mathbf{H} - ig/c\mathbf{E},$$

(16.7)

then, based on the first term of the first relation, our result for the components of the dielectric tensor function is [16]

$$\varepsilon_{xx}(\tilde{\nu}) = 1 + \frac{S_1^2(\tilde{\nu}_{0,2}^2 - \tilde{\nu}^2 - i\tilde{\nu}\gamma_2)}{(\tilde{\nu}_{0,1}^2 - \tilde{\nu}^2 - i\tilde{\nu}\gamma_1)(\tilde{\nu}_{0,2}^2 - \tilde{\nu}^2 - i\tilde{\nu}\gamma_2) - v_{12}^2}$$

$$\varepsilon_{yy}(\tilde{\nu}) = 1 + \frac{S_2^2(\tilde{\nu}_{0,1}^2 - \tilde{\nu}^2 - i\tilde{\nu}\gamma_1)}{(\tilde{\nu}_{0,1}^2 - \tilde{\nu}^2 - i\tilde{\nu}\gamma_1)(\tilde{\nu}_{0,2}^2 - \tilde{\nu}^2 - i\tilde{\nu}\gamma_2) - v_{12}^2}$$

$$\varepsilon_{xy}(\tilde{\nu}) = \varepsilon_{yx}(\tilde{\nu}) = \frac{-2S_1 S_2 v_{12}}{(\tilde{\nu}_{0,1}^2 - \tilde{\nu}^2 - i\tilde{\nu}\gamma_1)(\tilde{\nu}_{0,2}^2 - \tilde{\nu}^2 - i\tilde{\nu}\gamma_2) - v_{12}^2}$$

$$\varepsilon_{zz}(\tilde{\nu}) = 1$$

$$\varepsilon_{xz}(\tilde{\nu}) = \varepsilon_{xz}(\tilde{\nu}) = \varepsilon_{yz}(\tilde{\nu}) = \varepsilon_{yz}(\tilde{\nu}) = 0$$

(16.8)

Certainly, in a real molecule, we have more than one pair of oscillators, oriented in the most general case in arbitrary directions. In a liquid or a polycrystalline solid, for small oscillator strengths, the components of the dielectric tensor are given by $\varepsilon_{ij} = \varepsilon\delta_{ij}$ with $\varepsilon = (\varepsilon_{xx} + \varepsilon_{yy} + \varepsilon_{zz})/3$, so that the mixed terms vanish. While the general applicability of this mixing formula is questionable (see preceding chapter), the mixed terms must vanish; otherwise, the enantiomers would not have the same dielectric functions. The mixed term, however, leads to an expression for $g(\tilde{\nu})$, based on the comparison with the constitutive relations above (Eq. 16.7) [16]:

$$g(\tilde{\nu}) = 2\pi\tilde{\nu}d \frac{2S_1 S_2 v_{12}}{(\tilde{\nu}_{0,1}^2 - \tilde{\nu}^2 - i\tilde{\nu}\gamma_1)(\tilde{\nu}_{0,2}^2 - \tilde{\nu}^2 - i\tilde{\nu}\gamma_2) - v_{12}^2}.$$

(16.9)

Note that we have somewhat simplified the problem for the moment insofar as we assume that we only have a pair of coupled vibrations [17]. In general, one would have to assume that every vibration couples with each other, an assumption that has been made also by Max Born and Werner Kuhn as discussed above. Accordingly, in chiral plasmonics, this model is named the Born-Kuhn model as already mentioned above [9,10,18]. We also do not assume that both oscillators have the same eigenfrequencies, damping constants, or oscillator strengths, which would be a drastic limitation for the use of the model for dispersion analysis of VCD spectra. Where we follow the modern model, we also do not try to find the new system of normal modes that considers the coupling and which leads to chirality but keep explicitly the coupling constants instead. In addition, we assume that Eq. (7.13) and Eq. (16.9) are also valid for our bi-isotropic and scalar systems. This is not necessarily the case as I showed in the preceding chapter, but for organic and biologic molecules and their comparably small VCD intensities, this assumption may nevertheless be a very good approximation. Furthermore, since the main goal is to give an illustrative approach to understand VCD, it may be advisable to keep things simple.

FIG. 16.4 Real and imaginary part of the sum of the two principal dielectric functions of two coupled oscillators *(red lines)* in comparison with that of the mixed term *(green lines)*. S_1 and S_2 have the same sign.

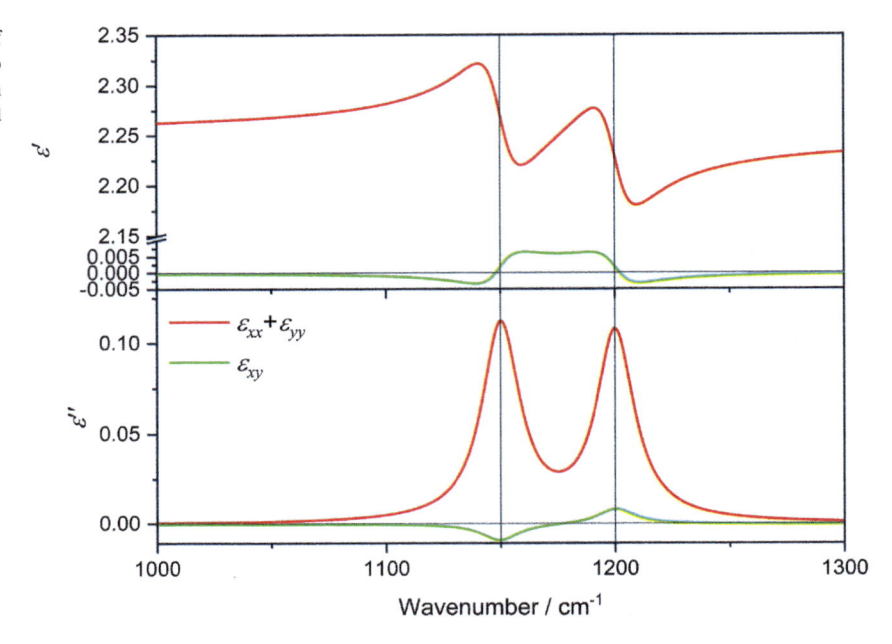

How do the predicted dielectric functions look, and how do they compare to the same systems without coupling? A comparison is given in Fig. 16.4.

For Fig. 16.4, I have chosen the oscillator parameters to describe comparably weak bands with oscillator strengths and $S_1 = S_2 = 50 \, \text{cm}^{-1}$. For such weak oscillators and a comparably small coupling constant of $v_{12} = 10,000 \, \text{cm}^{-1}$, the mixed fraction $-g(\tilde{v})/(2\pi\tilde{v}d)$ is practically the difference between the dielectric functions for coupled and uncoupled oscillators since the oscillators are nearly not shifted. The main effect of coupling in the conventional IR spectra, which would be a redshift of the lower and a blueshift of the higher wavenumber mode, is therefore not detectable. Also, the band shapes of the bands in the imaginary part of the dielectric function resemble those of the absorption index function, and thereby, those in absorbance spectra; the bands in the mixed fraction resemble those of the absorption difference spectra of left and right circularly polarized light.

It is quite instructive to investigate and compare how the imaginary parts of the dielectric function and the chirality admittance vary with the position of the second oscillator and with the coupling strength. The result of this investigation is shown in Fig. 16.5. The upper left panel is not equivalent to what I showed you in Section 7.3.2, because there the sum of the principal and the mixed components was depicted. For a coupling strength of $10,000 \, \text{cm}^{-1}$, it seems that the first oscillator always stays at $1150 \, \text{cm}^{-1}$, since the individual bands can no longer be resolved once the second oscillator comes close to the position of the first. That this is actually not the case, and that the oscillators even show anticrossing behavior for weak coupling can be seen when the change of the corresponding imaginary part of the chirality admittance is scrutinized. The second case is at least as much interesting. In this case, we have two degenerated oscillators with exactly the same position, strength, and damping constant. If we now introduce coupling and increase the coupling constant from zero to finally $100,000 \, \text{cm}^{-1}$, an increasing gap between the oscillators becomes visible. In contrast, two bands are seen in the $-g(\tilde{v})/(2\pi\tilde{v}d)$ spectrum. At least the case of small coupling constants, where the maximum and the minimum are close to each other, is known in the literature. This feature in CD spectra is often called a couplet.

It is also quite useful to look at the damping constant, which is approximately the full width at half height (FWHH). The peaks in the VCD spectra have the same FWHH as the parent dielectric function, which makes it sometimes easy to determine if one has two degenerated oscillators or not. Note also that dark modes, i.e., modes with zero oscillator strength cannot contribute to $g(\tilde{v})$ according to Eq. (16.9). On the other hand, if one oscillator changes the sign of S, e.g., because the direction of the transition dipole moment is inverted, then the VCD spectrum is multiplied by the factor -1. In other words, for the other enantiomer, the VCD spectrum is reversed, as can be expected. This makes sense if you assume a prochiral planar molecule where you add one atom from either side of the plane to make it chiral.

To further understand the properties of $g(\tilde{v})$, it may be interesting to investigate the limiting behavior of low and high wavenumbers as well as the sum rules that result from compliance to the Kramers-Kronig relation (Eq. 16.4). Away from absorptions, $g(\tilde{v})$ is in the form of:

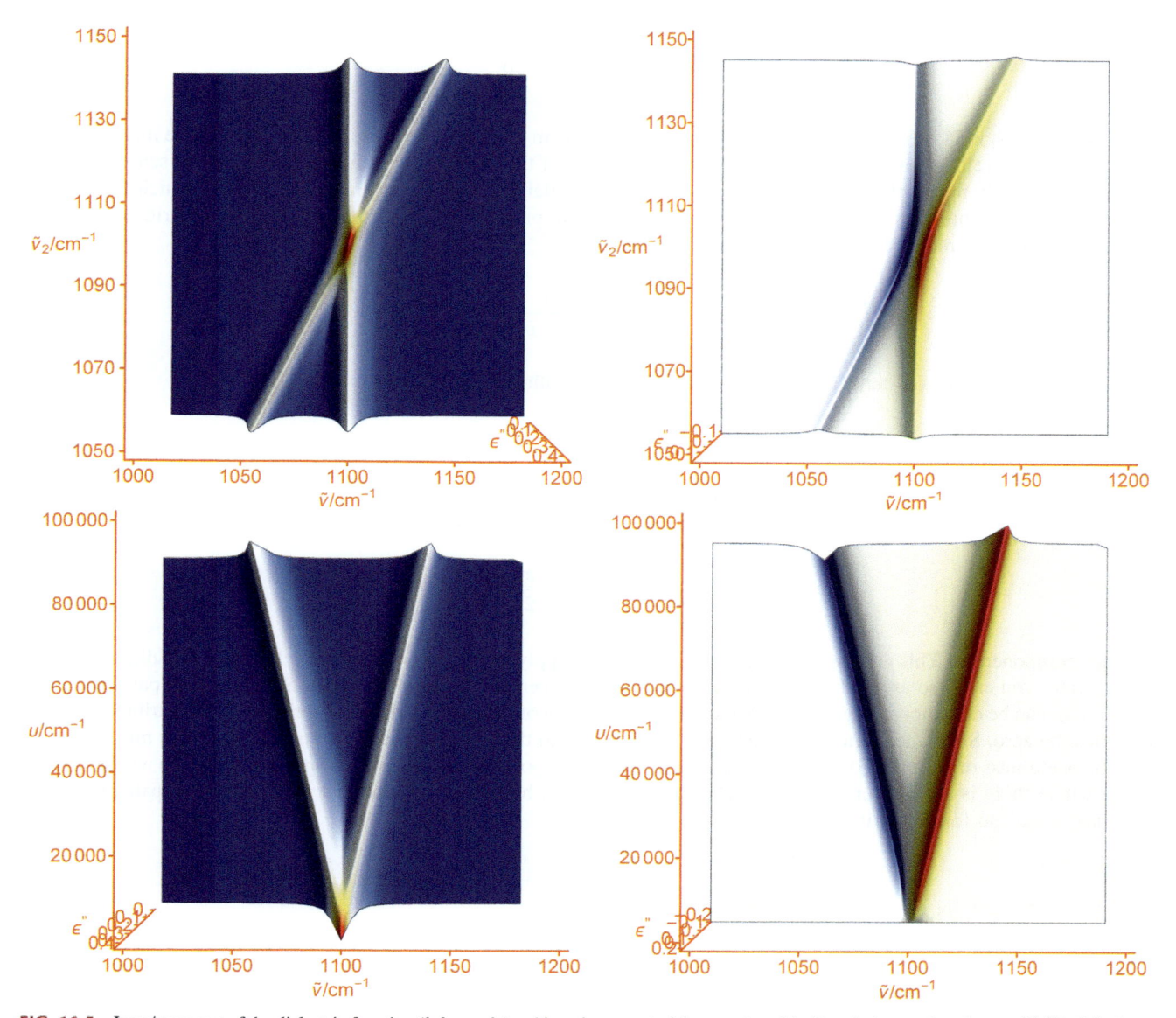

FIG. 16.5 Imaginary part of the dielectric function *(left panels)* and imaginary part of the negative chirality admittance function $-g(\tilde{\nu})/(2\pi\tilde{\nu}d)$ of two coupled oscillators *(right panels)* for varied oscillator position of the second oscillator *(top panels)* and varied coupling strength *(bottom panels)*.

$$g(\tilde{\nu}) = 2\pi\tilde{\nu} \sum_i d_{iab} \frac{2S_{ia}S_{ib}v_{iab}}{\left(\tilde{\nu}_{0,ia}^2 - \tilde{\nu}^2\right)\left(\tilde{\nu}_{0,ib}^2 - \tilde{\nu}^2\right) - v_{iab}^2}. \tag{16.10}$$

For very high wavenumbers, we find

$$g(\infty) = 2\pi \lim_{\tilde{\nu}\to\infty} \sum_i d_{iab} \frac{2S_{ia}S_{ib}v_{iab}}{\tilde{\nu}^3} = 0, \tag{16.11}$$

and the same results for very low wavenumbers,

$$g(0) = \lim_{\tilde{\nu}\to\infty} \left(\sum_i 2\pi\tilde{\nu} d_{iab} \frac{2S_{ia}S_{ib}v_{iab}}{\tilde{\nu}_{0,ia}^2\tilde{\nu}_{0,ib}^2 - v_{iab}^2} \right) = 0. \tag{16.12}$$

Concerning the sum rules, the same approach that we took in Section 5.8.3, which is to use the limiting behavior of the dispersion relation for very high wavenumbers and relate it to the Kramers-Kronig relation for the real part seems to be impossible due to the complexity of the dispersion relation. On the other hand, we can make use of the fact that $-g(\tilde{\nu})/(2\pi\tilde{\nu}d)$ is one of the terms of the dielectric function. For the imaginary part of the dielectric function,

$$\int_0^\infty \tilde{\nu}\varepsilon''(\tilde{\nu})d\tilde{\nu} = \frac{\pi}{2}S^2 \qquad (16.13)$$

holds with $S^2 = S_1^2 + S_2^2$, which is independent of the analytical form of $\varepsilon''(\tilde{\nu})$. Therefore, it must hold for the imaginary part of Eq. (7.13) as well. In particular, Eq. (16.13) is independent of the coupling constant v_{12}. This makes sense, since for $v_{12} = 0$. We simply regain the original form of the dispersion relation for noncoupling oscillators for which Eq. (16.13) was derived. Therefore, it is safe to conclude that the mixed term proportional to $-2S_1S_2v_{12}$ does not contribute anything, independent of what model is used for orientational averaging:

$$\int_0^\infty \tilde{\nu}\,\text{Im}\left(\frac{2S_1S_2v_{12}}{(\tilde{\nu}_{0,1}^2 - \tilde{\nu}^2 - i\tilde{\nu}\gamma_1)(\tilde{\nu}_{0,2}^2 - \tilde{\nu}^2 - i\tilde{\nu}\gamma_2) - v_{12}^2}\right)d\tilde{\nu} = 0. \qquad (16.14)$$

Therefore, since this mixed term is equal to $-g(\tilde{\nu})/(2\pi\tilde{\nu}d)$, it follows that

$$\int_0^\infty g''(\tilde{\nu})d\tilde{\nu} = 0. \qquad (16.15)$$

Note that for a dielectric tensor instead of a scalar, Eq. (16.13) is of the form

$$\int_0^\infty \tilde{\nu}\varepsilon_{ij}''(\tilde{\nu})d\tilde{\nu} = \frac{\pi}{2}S^2\delta_{ij}, \qquad (16.16)$$

for each component ε_{ij}. This was shown by the authors of [19] by applying super convergence techniques. In addition, to prove Eq. (16.16), you can also argue that the trace of a matrix is independent of the orientation. Since in principal orientation, Eq. (16.16) can be decomposed again to three equations of the form of Eq. (16.13), the cross terms cannot contribute anything and must be zero. Since the chirality admittance is proportional to the cross-polarization terms, Eq. (16.15) must hold.

The usefulness of Eq. (16.15) seems limited as long as it is not possible to record a CD spectrum over the whole spectral range where there is absorption. On the other hand, since it must hold if we assume that we have only a pair of coupled oscillators, we can formulate it as

$$\int_{\tilde{\nu}_e}^{\tilde{\nu}_f} g''(\tilde{\nu})d\tilde{\nu} = 0, \qquad (16.17)$$

where $\tilde{\nu}_e$ is a wavenumber well below $\tilde{\nu}_{0,1}$ and $\tilde{\nu}_f$ is a wavenumber well above $\tilde{\nu}_{0,2}$ so that $g''(\tilde{\nu}_e) \approx g''(\tilde{\nu}_f) \approx 0$. This should help to assign pairs of coupled oscillators, which is actually not easy in VCD spectra due to the usually large number of bands. Eq. (16.17) should also allow us to see if the simple model of two coupling oscillators holds or if the coupling of more oscillators needs to be assumed. Lastly, it should also help to assess the accuracy of a measurement if $\tilde{\nu}_e$ and $\tilde{\nu}_f$ are the limits of the measurement range and if this measurement range is limited below and above by transparency ranges. Note that accuracy also means, in this case, the position of the baseline, since the integration carried out in Eq. (16.17) is naturally very sensitive to this position.

Concerning the real part of the chirality admittance function, we can argue that for very high wavenumbers, $\tilde{\nu}_f$, $g''(\tilde{\nu}_f) = 0$. Thus, based on the other Kramers-Kronig relation, Eq. (16.4), we find that

$$g''(\tilde{\nu}_f) = -\frac{2\tilde{\nu}_f}{\pi}\wp\int_0^\infty \frac{g'(\tilde{\nu}')}{\tilde{\nu}'^2 - \tilde{\nu}_f^2}d\tilde{\nu}' = 0 \rightarrow$$

$$-\frac{2\tilde{\nu}_f}{\pi}\int_0^\infty \frac{g'(\tilde{\nu}')}{-\tilde{\nu}_f^2}d\tilde{\nu}' = 0 \rightarrow \qquad (16.18)$$

$$\int_0^\infty g'(\tilde{\nu}')d\tilde{\nu}' = 0$$

because the right side of Eq. (16.18) can only be zero if the integral is zero. Thus, also for $g'(\tilde{\nu}')$, the area above must be the same as the area below zero.

Finally, I will come back to dispersion analysis. As practiced in this book, it is usually carried out directly on the recorded spectra. Since we had to use two different methods, i.e., ATR and difference absorbance (the difference of the negative decadic logarithms of the transmittances), to determine the dielectric function and the chirality admittance function, I thought it would be easier to carry out dispersion analysis in a different way. In particular, as shown in Section 16.2, it is relatively straightforward to determine both functions, in particular the chirality admittance. So, instead of using the spectra, I directly fitted the imaginary parts of $\varepsilon(\tilde{\nu})$ and $g(\tilde{\nu})$ using coupled oscillator pairs in the form of:

$$\varepsilon(\tilde{\nu}) = \varepsilon_\infty + \sum_{j=1}^{M/2} \frac{S_{j-1}^2 A_j + S_j^2 A_{j-1}}{A_{j-1} A_j - v_{(j-1)j}^2}$$

$$g(\tilde{\nu}) = 2\pi\tilde{\nu} \sum_{j=1}^{M/2} d_{(j-1)j} \frac{2 S_{j-1} S_j v_{(j-1)j}}{A_{j-1} A_j - v_{(j-1)j}^2}$$

(16.19)

The results for (+)-α-pinene can be found in Fig. 16.6. Even though only the fingerprint region was analyzed, the effort was comparably large. Not only because I had to use 28 oscillators but also because the signs of the oscillator strengths had to be considered (actually only the sign of the product $S_{j-1} S_j$).

Overall, considering that $\varepsilon(\tilde{\nu})$ and $g(\tilde{\nu})$ originate from two different measurements, the result is satisfying. The fit of $\varepsilon''(\tilde{\nu})$ looks somewhat better than the one for $g''(\tilde{\nu})$. This seems to be natural, given that the errors should be relatively larger in the VCD than in the ATR spectrum. To a certain degree, the fit quality can be influenced by the weighting between the individual residual sum of the squares. Overall, the model of pairs of oscillators seems to be adequate, but certain deviations can be noted. If we integrate the band areas of the function $g''(\tilde{\nu})$, we find that the result is not zero, but negative. This is the reason why the fit quality is not good in the case of the strong negative bands at 997 and 1214 cm^{-1}. This is something that could not be improved even if we include oscillator models that allow more than two oscillators to couple. In a few regions, the negative contributions of one oscillator pair compensate partly for the positive contribution of the other, e.g., around 890 cm^{-1} (or the other way around). In other regions, the contribution of two pairs is needed to explain the overall intensity, cf. Fig. 16.6. This can lead to very complex patterns that are not possible in conventional dispersion analysis, where the contributions of two oscillators are always additive concerning the dielectric function (in the absence of local field effects). Due to this complex interplay, it is impossible to guarantee that the result presented in Fig. 16.6 is indeed the best possible, although I have tried a lot of combinations to arrive at this solution. Apart from the problem of equal areas above and below zero for $g''(\tilde{\nu})$, three explanations are possible for the shortcomings [17]:

1. The assumption of pairs of coupled oscillators while more or all oscillators may be coupled.
2. Assumption of orthogonality of the electric transition moments—usually this assumption is not valid for chiral molecules. Accordingly, mixed terms may contribute to the dielectric function, albeit these terms must be independent of the enantiomeric excess.

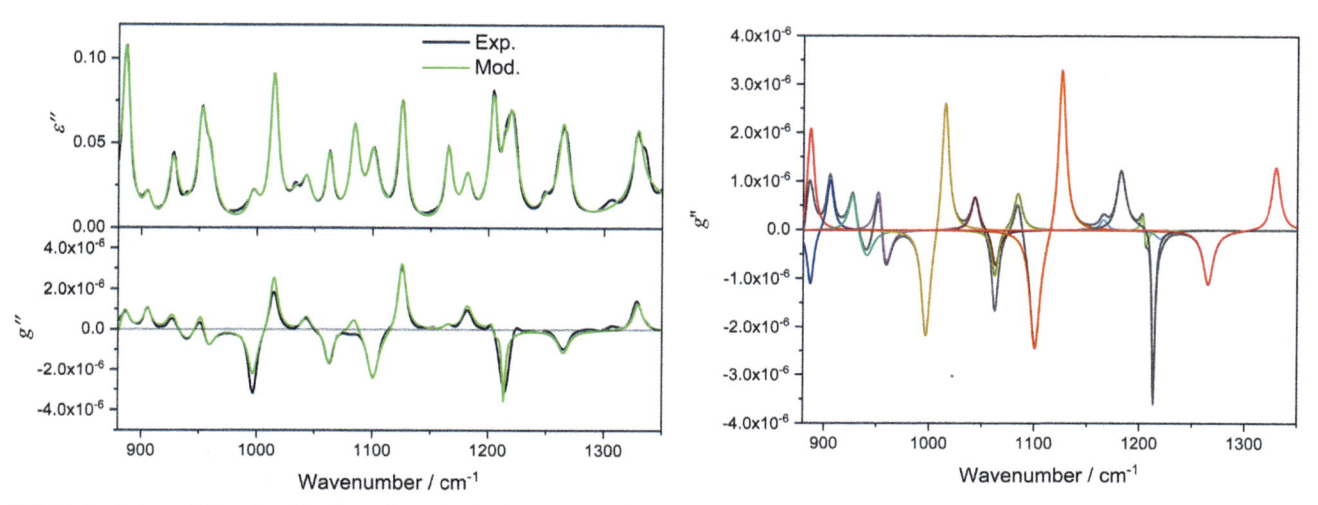

FIG. 16.6 *Left panel*: Results of the dispersion analysis of (+)-α-pinene in the fingerprint region. *Right panel*: Sum and individual curves of the simulated g'' [17].

3. The model assumes harmonic oscillations, while anharmonicity may contribute.

Points 2 and 3 will not be pursued further, but it may be worth to investigate if a model assuming three coupled oscillators will be of benefit. In general, we could describe a system of M coupling oscillators simply by extending Eq. (16.6): [17].

$$\begin{pmatrix} A_1 & v_{12} & \cdot & v_{1M-1} & v_{1M} \\ v_{12} & A_2 & & & \cdot \\ \cdot & & \cdot & & \cdot \\ v_{1M-1} & & & A_{M-1} & v_{(M-1)M} \\ v_{1M} & \cdot & \cdot & v_{(M-1)M} & A_M \end{pmatrix} \begin{pmatrix} x_1 \\ x_2 \\ \cdot \\ x_{M-1} \\ x_M \end{pmatrix} = \begin{pmatrix} q_1 E_{x_1} \exp(ik \cdot (z+d_1)) \\ q_2 E_{x_2} \exp(ik \cdot (z+d_2)) \\ \cdot \\ q_{M-1} E_{x_{M-1}} \exp(ik \cdot (z+d_{M-1})) \\ q_M E_M \exp(ik \cdot (z+d_M)) \end{pmatrix}. \quad (16.20)$$

It is obvious that by increasing M, the complexity of the solutions would quickly become immense. As in the case of the honeypot-coupled oscillators (Section 5.4), the outcome would most probably not be worth the effort. Nevertheless, for three oscillators, there still might be some benefit. To simplify the situation, I will assume that the first oscillator has an eigenfrequency in between those of oscillators 2 and 3 and that there is no coupling between these two oscillators. In addition, I assume that $d_1 = d/2$ and $d_2 = d_3 = -d/2$ (which explains the missing coupling between oscillators 2 and 3, since they are in the same z-plane). With these restrictions, the solutions are given by:

$$\varepsilon(\tilde{\nu}) = \varepsilon_\infty + \frac{S_1^2 A_2 A_3 + S_2^2 \left(A_1 A_3 - v_{12}^2 A_3/A_2 - v_{13}^2\right) + S_3^2 \left(A_1 A_2 - v_{12}^2 - v_{13}^2 A_2/A_3\right)}{A_1 A_2 A_3 - v_{12}^2 A_3 - v_{13}^2 A_2}$$

$$g(\tilde{\nu}) = 2\pi\tilde{\nu}d \frac{S_1 (S_2 v_{12} A_3 + S_3 v_{13} A_2)}{A_1 A_2 A_3 - v_{12}^2 A_3 - v_{13}^2 A_2}$$

(16.21)

After a lot of tests, I found that the fit of the fingerprint region of (+)-α-pinene profits most from the use of one term with three oscillators. For all other bands, I used only terms with two oscillators. Accordingly, the fit only improves in the region of the three oscillators, which is depicted in Fig. 16.7.

While the fit of $\varepsilon''(\tilde{\nu})$ between 1050 and 1150 cm^{-1} is overall of comparable quality, when the corresponding two pairs of two coupled oscillators are replaced by one set of three coupled oscillators following Eq. (16.21), the changes are stronger for $g''(\tilde{\nu})$. Whereas the wrongly positive feature at 1084 cm^{-1} can be inverted, the rule that the integration of $\tilde{\nu}g''(\tilde{\nu})$ must give a zero result for each set of coupled oscillators is one of the factors that leads to a less good fit of the features at 1100 and 1125 cm^{-1} (the second constraint is certainly that $\varepsilon''(\tilde{\nu})$ must also fit, and there is also some loss of flexibility by the fact that overall three instead of four oscillators are used).

All in all, it is obvious that the results are just hints that in general sets of pairs of coupled oscillators may be sufficient to describe $\varepsilon(\tilde{\nu})$ and $g(\tilde{\nu})$ sufficiently well (even if we did not check the real parts, we can be sure because of the

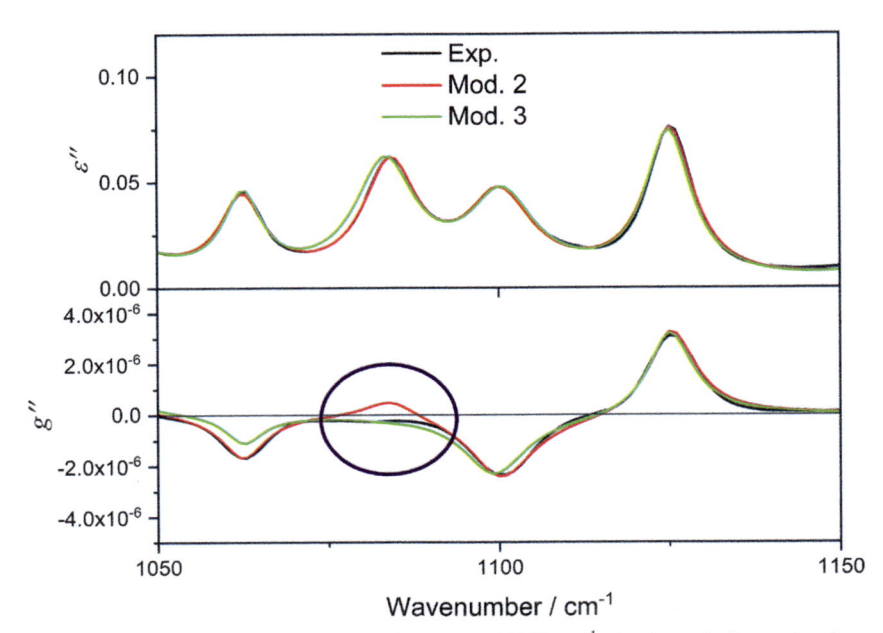

FIG. 16.7 Comparison of the results of dispersion analysis of (+)-α-pinene around 1100 cm^{-1} when instead of two pairs of two coupled oscillators one set of three coupled oscillators according to Eq. (16.21) is used [17].

Kramers-Kronig relations that the same as for the imaginary parts is true for the real parts of $\varepsilon(\tilde{\nu})$ and $g(\tilde{\nu})$). Note that while for electronic excitations, one-electron contributions to the CD may be important, [14] corresponding effects caused by a single vibration can hardly be imagined and can, in my opinion be excluded for VCD.

Apart from dispersion analysis, the addition of a third oscillator that couples with one of a pair of coupled oscillators adds some interesting features that I want to discuss as a last point in this chapter. In Fig. 16.8, you see two coupled oscillators with the same oscillator parameters. The oscillator at the higher wavenumber position additionally couples with a

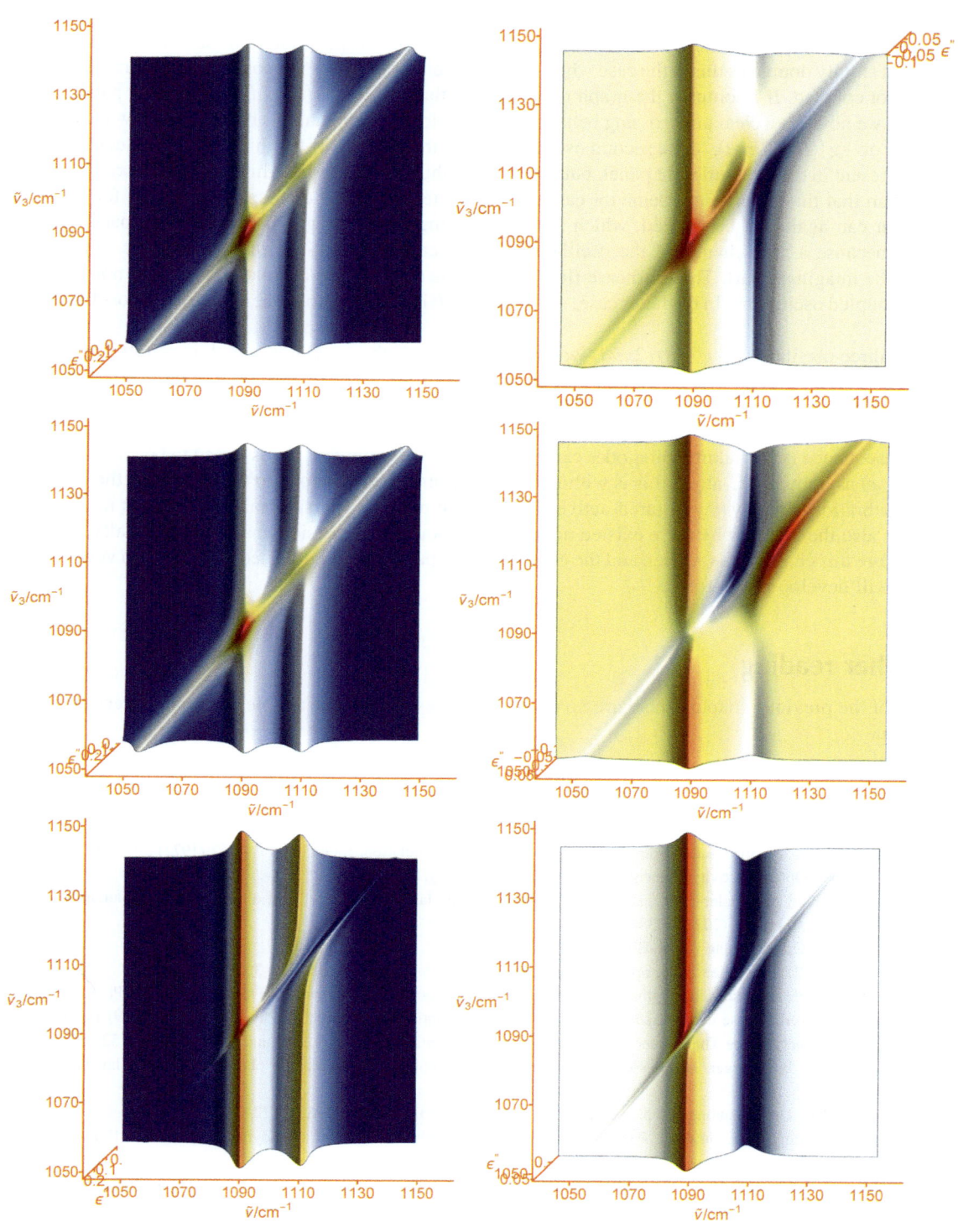

FIG. 16.8 Comparison of the imaginary parts of the dielectric functions $\varepsilon''(\tilde{\nu})$ *(left panels)* and $-g''(\tilde{\nu})/(2\pi\tilde{\nu}d)$ *(right panels)* which result for two coupled oscillators, if the oscillator at the higher wavenumber position is coupled to a third oscillator with altering wavenumber position.

third oscillator. The wavenumber of this third oscillator is varied. In principle, I find the comparison interesting for three different values of the oscillator strength of the third:

1. It has the same oscillator strength as the other two.
2. The oscillator strength of the third oscillator has the same value but with a negative sign.
3. The oscillator's strength is zero.

For the last case, I have additionally decreased the damping constant of the third oscillator so that the effect has increased visibility. Since the values of both coupling constants are $10,000\,\text{cm}^{-1}$ comparably small, an anticrossing behavior is not visible in the upper panel of the left side when the third oscillator crosses the position of the second oscillator, but the value of $\varepsilon''(\tilde{\nu})$ does not simply double as this is the case when the third oscillator crosses the position of the first oscillator, because those two are not coupled. If I compare the graph of $\varepsilon''(\tilde{\nu})$ with that of $-g''(\tilde{\nu})/(2\pi\tilde{\nu}d)$ (upper right panel), I find it very remarkable that we now see a clear anticrossing behavior. The obvious reason that this behavior is now visible, is that there is a sign change of $-g''(\tilde{\nu})/(2\pi\tilde{\nu}d)$, of the second oscillator, when the position of the third oscillator decreases. Such a sign change is also present in the medium right panel, but now it is the third oscillator for which $-g''(\tilde{\nu})/(2\pi\tilde{\nu}d)$ changes its sign. This means again that the anticrossing behavior can easily be detected. In the corresponding graph for $\varepsilon''(\tilde{\nu})$, this anticrossing behavior can again not be noticed, which is not surprising since this graph (medium left panel) is the same as the one above because a sign change of the oscillator strength cannot be noticed in the dielectric function, neither in the real nor in the imaginary part. The third case (lower panels) is dedicated to a situation that does not occur in the case of a set of two coupled oscillators. In the latter case, $-g''(\tilde{\nu})/(2\pi\tilde{\nu}d)$ is simply zero if one of the oscillators has zero oscillator strength.

For a set of three oscillators, $-g''(\tilde{\nu})/(2\pi\tilde{\nu}d)$ would only be zero if the oscillator, which is coupled to two other oscillators would have zero oscillator strength. This means that one of the other two oscillators can be a dark mode, which nevertheless shows up in a spectrum since it is fed by a bright mode. In this case, the anticrossing behavior is visible in both the $\varepsilon''(\tilde{\nu})$ and in the $-g''(\tilde{\nu})/(2\pi\tilde{\nu}d)$ spectrum which is also a consequence of the assumed small damping constant. The overall consequence is that additional modes can appear in the spectrum that are assumed to be noninfrared active. For molecules with an asymmetric C-atom, this is without relevance since those belong to the class with the lowest symmetry (C_1) where all vibrations are anyway infrared-active. This is what makes dispersion analysis for such molecules complex. Possibly, this is also the reason why there existed no literature about the dispersion analysis of optically active molecules and crystals before this chapter was written and the corresponding papers were published [12,17]. I am very curious to learn how this field will develop in the future.

16.4 Further reading

As in the case of the previous chapter, there are no other references or books available, so I cannot recommend any other literature to you.

References

[1] D.W. Berreman, Optics in stratified and anisotropic media: 4 × 4-matrix formulation, J. Opt. Soc. Am. 62 (1972) 502–510.
[2] L.A. Nafie, Vibrational Optical Activity: Principles and Applications, Wiley, 2011.
[3] D.R. Hermann, G. Ramer, M. Kitzler-Zeiler, B. Lendl, Quantum cascade laser-based vibrational circular dichroism augmented by a balanced detection scheme, Anal. Chem. 94 (2022) 10384–10390.
[4] P. Drude, Lehrbuch der Optik, Verlag von S, Hirzel, 1906.
[5] W. Kuhn, Über die DRUDEsche Theorie der optischen Aktivität, Z. Phys. Chem. 20B (1933) 325–332.
[6] M. Born, Elektronentheorie des natürlichen optischen Drehungsvermögens isotroper und anisotroper Flüssigkeiten, Ann. Phys. 360 (1918) 177–240.
[7] W. Kuhn, Quantitative Verhältnisse und Beziehungen bei der natürlichen optischen Aktivität, Z. Phys. Chem. 4B (1929) 14–36.
[8] L. Rosenfeld, Quantenmechanische Theorie der natürlichen optischen Aktivität von Flüssigkeiten und Gasen, Z. Phys. 52 (1929) 161–174.
[9] X. Yin, M. Schäferling, B. Metzger, H. Giessen, Interpreting chiral nanophotonic spectra: the plasmonic Born-Kuhn model, Nano Lett. 13 (2013) 6238–6243.
[10] X. Duan, S. Yue, N. Liu, Understanding complex chiral plasmonics, Nanoscale 7 (2015) 17237–17243.
[11] T.G. Mayerhöfer, A.K. Singh, J.-S. Huang, J. Popp, interpreting chiral nanophotonic spectra: employing the plasmonic Born-Kuhn model for dispersion analysis, Submitted for publication, 2024.
[12] T.G. Mayerhöfer, A.K. Singh, J.-S. Huang, C. Krafft, J. Popp, Unveiling chiral optical constants of α-pinene and propylene oxide through ATR and VCD spectroscopy in the mid-infrared range, Spectrochim. Acta A Mol. Biomol. Spectrosc. 302 (2023) 123136.

[13] R. Zhao, T. Koschny, C.M. Soukoulis, Chiral metamaterials: retrieval of the effective parameters with and without substrate, Opt. Express 18 (2010) 14553–14567.
[14] A. Moscowitz, Theoretical aspects of optical activity part one: small molecules, Adv. Chem. Phys. 4 (1962) 67–112.
[15] D.L. Jaggard, X. Sun, Theory of chiral multilayers, J. Opt. Soc. Am. A 9 (1992) 804–813.
[16] Y.P. Svirko, N.I. Zheludev, Polarization of Light in Nonlinear Optics, Wiley, 1998.
[17] T.G. Mayerhöfer, A.K. Singh, J.-S. Huang, C. Krafft, J. Popp, Quantitative evaluation of IR and corresponding VCD spectra, Spectrochim. Acta A Mol. Biomol. Spectrosc. 305 (2024) 123549.
[18] H. Kurosawa, S.-I. Inoue, Born-Kuhn model for magnetochiral effects, Phys. Rev. A 98 (2018) 053805.
[19] M. Altarelli, D.L. Dexter, H.M. Nussenzveig, D.Y. Smith, Superconvergence and sum rules for the optical constants, Phys. Rev. B 6 (1972) 4502–4509.

Index

Note: Page numbers followed by *f* indicate figures, *t* indicate tables, and *s* indicate schemes.

A

Absorbance, 34, 159, 166
   absorptance *vs.*, 287–288, 287–288*f*
   quantity, 4
Absorption, 57, 88–89, 104–105
   band fitting/deconvolution, 90
   and integrated absorbance, concentration dependence of, 108, 108*f*
   Lorentz profiles, 105
   UV-vis absorption, 91–92
Absorption index function, 27–29, 27–29*f*, 85, 99–100*f*, 101, 102*f*, 104, 142–145, 148, 154, 154*f*, 159, 162, 162–163*f*, 188–189
Absorption spectroscopy, 197
Acoustic vibrations, of chain, 92–93, 92*f*
Ampere's circuital law, 16
Anisotropic oscillator models
   generalized oscillator models, 322–323
   monoclinic crystals, 321, 321*f*
   orthorhombic crystals systems, 320, 320*f*
   tetragonal, hexagonal, and trigonal crystal systems, 318–320, 319*f*
   triclinic crystals, 321–322
Anomalous dispersion, 89
Anticrossing behavior, VCD, 372, 378
Apparent absorbance, 184–189, 185–186*f*, 187*s*, 188*f*
Asynchronous sums of squares (ARSS), 232, 238
Attenuated total reflection (ATR) spectroscopy, 57, 159, 178–182, 180–181*f*
   dispersion analysis, 331–333, 332–334*f*, 336
   vibrational circular dichroism (VCD), 368–369, 368*f*, 375–376
Attenuated total reflection technique, 108–109
Average electric field intensity, 166
Average index of refraction function, 162
Average reflectance and transmittance theory (ARTT), 355, 361–362
Average refractive index theory (ARIT), 349–352, 350*f*, 352*f*
Avogadro constant, 102

B

Band fitting/deconvolution, 90, 98, 101
Beer's approximation, 13–14, 26–28, 198, 199*f*, 200, 243–245, 247–249, 251–253, 252*f*
   dispersion relations and, 101–109
   absorbance, 104–105

absorption function, index of, 104
benzene-carbon tetrachloride, spectra of mixtures in, 108–109, 109*f*
benzene-cyclohexane, spectra of mixtures in, 108–109, 109*f*
benzene-toluene, spectra of mixtures in, 108–109, 109–110*f*
dielectric constant, 103–104
dielectric function, real and imaginary part, 103–104
dispersion theory, 104–105
integrated absorbance, concentration dependence of, 108, 108*f*
Kramers-Kronig sum rules, 105
Lorentz-Lorenz equation, 106–109
Loschmidt's constant, 106–107
macroscopic polarization, 102–103
medium strong oscillator, 104, 105*f*
molar concentration, 102–103
molar oscillator strength, 103
molecular dynamics simulations, 109
oscillator strength, 104–105
refraction function, index of, 104
relative dielectric function, 103
Beer's law, 5–6
Benzene-carbon tetrachloride system, 108–109, 109*f*
Benzene-cyclohexane system, 108–109, 109*f*
Benzene-toluene system, 108–109, 109–110*f*, 229, 229*f*
Bergman representation, 263–274
   many-particle systems, polarization in, 263–267, 266–267*f*
   microheterogeneity, 275–278, 275*f*
   microhomogeneity, 275, 275*f*
   percolation, 268–269
   polycrystalline materials, 276
   spectral densities (*see* Spectral density)
Berreman effect, 9
Berreman's formalism
   Maxwell equations, and constitutive relations, 294–295
Berreman-Unterwal model, 116–117, 116*f*, 118*f*, 129, 336
Biisotropic medium, 313–314
Blue vitriol, 337–338, 337*f*, 339*f*, 340
Boltzmann distribution, 97–98
Booker quartic, 295, 298, 310
   coefficients, 306, 308

Born-Kuhn model, 367–368, 371
Bouguer-Beer-Lambert (BBL) approximation, 4–5, 8–9, 11, 44, 369
   absorption index functions, 27–29, 27–29*f*
   attenuated total reflection (ATR), 159, 178–182, 180–181*f*
   electromagnetic waves, absorption on, 30–32, 31*f*
   errors, determination of, 159, 160*s*
   free-standing film embedded in vacuum/air, transmittance of, 164–168, 165*s*, 165*f*, 167*f*
   harmonic molecular vibrations, 21–28, 24*f*, 33*s*
   infrared spectroscopy, measurement methods and geometries for, 159, 160*s*
   Kramers-Kronig relations, 28–30
   longitudinal *vs.* transversal wave, 20–21, 20*f*
   matrix formalism, incoherent layers, 78
   mixing rules, 182–184, 183*f*
   phase shifts, 19
   reflection at interface, 32–34, 32*f*
   reflection measurements, 159–160
   semi-infinite medium, 32, 36, 41*s*
   slab embedded in vacuum/air, transmittance of, 161–164, 162*s*, 162–163*f*
   transflection technique, 168–171, 170–172*f*
   transmission
      at interface, 32–34, 32*f*
      thick film and free-standing film, 159, 160*s*
      through layer on substrate, 38–40, 39*s*
      through thick slab, 34–36, 35*s*
      through thin slab, 36–38, 37*s*
   transparent substrate, transmission of layer on absorptance, 173–174, 174*f*
      field maps for nonabsorbing films on Si, ZnSe, and CaF_2 substrates, 173–174, 174*f*
      fringes, removal of, 174–175, 175–176*f*
      PMMA layers on CaF_2, apparent absorption coefficient spectra of, 175–177, 176*f*
      PMMA layers on Si, simulated transmittance spectra of, 174–175, 175*f*
   two-dimensional correlation spectroscopy (2DCOS), 227–230
Boundary conditions, 48–49
Brendel oscillator, 117–119, 123–124
Brewster's angle, 66

Bruggeman approximation, 347–348
Buried metal layer infrared reflection absorption spectroscopy (BML-IRRAS), 202

C

Calcite, 284–286, 286f
Cauchy principal value, 29
Cauchy's formulas, 152–153
Chemometrics, 230, 243–245
   benzene-toluene system, 244–245, 245f
   classical least squares (CLS) regression, 245–249, 246f, 248f
   inverse least squares (ILS) regression, 249, 250f
   multivariate curve resolution-alternating least squares (MCR-ALS), 252–255, 254f
   principal component analysis (PCA), 249–252
   principal component regression (PCR), 249–252, 253f
Chirality admittance function, VCD, 368–369, 370f, 372, 373f, 374–375
Chiral materials
   dispersion analysis, 369–378, 370f, 372–373f
   spectra of, 368–369
Circularly polarized light, 54, 54f
Classical damped harmonic oscillator (CDHO) model, 98–99, 99f, 101, 114–115f, 117–119, 118–119f, 122, 226, 227f, 317–322
Classical least squares (CLS) regression, 245–249, 246f, 248f
Clausius-Mossotti equation, 5
Combined 2D correlation residual sum of squares (C2DCRSS), 232
Complex oscillator strength, 120–122, 122f
Complex refractive index function, 99, 101, 138–139, 147, 186–187, 189
Constitutive relations, Berreman's formalism, 294–295
Convolution model, 123–124, 123f
Coupled oscillator model
   chiral dispersion analysis, 367–368, 370f, 372–373f, 374, 376–378, 376–377f
   dispersion relations, 110–115, 110s, 112–115f, 114s
Coupling effects, 207–221
Cramer's rule, 250, 298–299, 309
Critical angle, 68
Critical point transitions, 130–131
Cross-polarization, 354–355, 355–356f, 357–364, 359f, 361f
   nonzero, 357
   vibrational circular dichroism (VCD), 374
Crystallite theory, 123, 352
Crystal symmetry, 317, 318f
   4×4 matrix formalism, 304–305, 304–305f
   generalized dispersion analysis, 340, 342s
Cubic crystal system, 317–318

D

Damped harmonic oscillator model, 98, 208
Damping, 88–89, 95, 117, 130
Damping constant, 89–90, 101, 116–119, 121, 123–124, 127, 129, 150
Decadic absorption coefficient, 165
Dielectric constants, 58s
Dielectric function, 20, 85–86, 89, 96–97, 126, 184, 189–190
   inverse of, 125
   pole and zero of, 116–117, 116f
   real and imaginary part changes, 85, 86f, 103–104
   of semiconductors, in visible spectral range, 117–119
   tensor, 89–90
   two coupled oscillators, 111, 112–113f
Dielectric loss function, 117
Dielectric material, nonmagnetic behavior
   anisotropic, 305–307
   monoclinic, 307–309
   orthorhombic, 312
   uniaxial, 309–312
Dielectric tensor, 273
   crystal symmetries and, 317, 318f
   function, 325, 326f, 333, 371
Differential refractive index, 198–199, 199f
Dipole moment, 85–86, 88, 94, 103
Dirac delta function, 272–273
Direct coupling, 214–217
Dispersion analysis, 90–91, 96, 117, 122, 152–153, 160–161, 189–194, 190–192f, 191t, 193t, 194f, 325
   ATR spectroscopy, 331–333, 332–334f, 336
   blue vitriol, 337–338, 337f, 339f, 340
   chiral, 369–378, 370f, 372–373f
   conventional, 325
   generalized, 340–342, 342–343f, 342s
   monoclinic crystals, 328, 329–330f
   optically uniaxial crystals, 325–327
   orthorhombic crystals, 327–328, 327–328f
   perpendicular modes, 329–336
   reflectance spectra (see Reflectance spectrum)
   triclinic crystals, 337–340, 337–339f, 341f
Dispersion relations
   Beer's approximation, 101–109, 105f, 107–110f
   coupled oscillator model, 110–115, 110s, 112–115f, 114s
   dielectric function, 85–86, 86f
   dielectric/refractive index background, 152–154, 153–155f
   dipole moment, 85–86
   Drude model, 130–131, 131–132f
   electromagnetic fields in matter, solution for, 86
   inverse dielectric function model, 124–130, 124f, 128–129f
   Kramers-Kronig relations (KKR) (see Kramers-Kronig relations (KKR))
   Kramers-Kronig sum rules, 105, 111, 125, 127, 148–152
   Lorentz profile vs. oscillator, 98–101, 99–100f
   plate capacitor, electric field, 86, 87f

refraction function, index of, 85
resonance wavenumber, 86
semi-empirical four parameter models
   Berreman-Unterwal model, 116–117, 116f, 118f
   complex oscillator strength, classical model with, 120–122, 122f
   convolution model, 123–124, 123f
   frequency-dependent damping constant, classical model with, 119–120, 120f
   Kim oscillator, 117–119, 118–119f
Dispersion theory, 4–5, 9–10, 9f, 97–98, 104–105, 147
Drude-Lorentz model, 95
Drude model, 130–131, 131–132f
Drude's theory, 9, 284–285
Dynamical matrix, 293, 297, 299, 303, 306

E

Effective dielectric function, 269–274, 269–270f
Effective medium approximation (EMA), 347–349, 350f, 351–352, 352f
Effective thickness, 178
Eigenfrequency, 88–89, 110
Electric displacement, 11–14, 14f
Electric field, 11–14, 12f, 57–58, 62, 71–72, 74–75, 81, 86, 87f, 88, 94
   Bouguer-Beer-Lambert (BBL) approximation, 11–14, 12f
   to incoming electric field strengths, 7, 7f
   intensity, 69, 69–70f, 166
   plate capacitor and point charge, 41, 41f
   strengths
      coherent layers, 77–78
      mixed coherent and incoherent multilayers, 81–83, 83f
Electric field intensity distribution, 202–203, 203–207f
Electric field standing wave (EFSW) effect, 4, 164
   in freestanding polyethylene film, 166–167, 167–168f
   in nonabsorbing layer on gold, 169, 170f
   in transmission, 165, 165f
   in weakly absorbing layer on gold, 169–171, 170f
Electric wave complex, 18
Electromagnetic field, 86
   boundary conditions, 48–49, 48–49s
   energy density and flux, 49–50
   Maxwell's relations, 47–48
   polarized waves, 52–54, 53s, 54f
   wave equation, 50–52
Electromagnetic theory, 26–27
Electromagnetic waves, 30–32, 31f
Enantiomers, 369–372
Energy density and flux, 49–50
Equations of motion, 110
Euclidian distance, 231–232
Euler's formula, 18
Extended multiplicative scatter correction (EMSC), 278

F

Fano-like profiles, 121–122
Faraday's law of induction, 16
Fermi's golden rule, 3
Finite-difference time-domain (FDTD) simulations, 277, 277–278f
4×4 matrix formalism, 293
  Berreman's formalism, 294–297
  crystal symmetry, 304–305, 304–305f
  monoclinic crystal, 293–294, 305, 307, 309
  nonmagnetic behavior, 305–312
    biisotropic medium, 313–314
    dielectric anisotropic material, 305–307
    dielectric monoclinic material, 307–309
    dielectric orthorhombic material, 312
    dielectric uniaxial material, 309–312
    optical model and coordinate system, 309–310, 311f
  propagation matrix, 314
  reflectance and transmittance coefficients, 303
  simplifications for special cases, 304–314
  singularities, treatment of, 300–303
    degenerate eigenvalues, 300–302
    dynamical matrix, 303
  transfer matrix, 299–300
  triclinic crystals, 307
  Yeh's formalism, 297–298
Four-parameter model, 119
Frequency-dependent damping constant, 119–120, 120f
Fresnel's equation, 145, 189, 211–212, 284–285
Fresnel's law, 61, 63, 68
Fresnoite, 124–125, 124f, 137, 137f
  dielectric tensor function, 325, 326f
  principal reflectance spectra, 325, 326f
  Raman spectra of, 325–327, 327f
Fröhlich mode, 266

G

Gauss divergence theorem, 48, 48s
Gaussian distribution, 187
Gaussian profile, 117–119
Gauss-Lorentz switch, 117–119, 210, 210f
Gauss's law, 15–16
Generalized dispersion analysis, 322, 340–342, 342–343f, 342s
Generalized oscillator models, 322–323

H

Harmonic molecular vibrations, 21–28, 24f, 33s
Harmonic waves, BBL approximation, 17–21
Hexagonal crystal systems, 318–320, 319f
Hilbert-Noda transformation matrix, 226
Honeypot coupling, 214, 216
Hooke's law, 21–22, 95
Humlíček's equation, 125

I

Indirect coupling, 207–214
Infrared refraction spectroscopy, 197–200
Interaction damping constant, 110
Interference effects, 71, 164–165, 167–168, 171–172, 174–177, 181–183, 192
Interference-enhanced ATR spectroscopy, 203, 205–206
Interference-enhanced infrared spectroscopy (IEIRS), 202
Interference-enhanced Raman spectroscopy, 201
Inverse dielectric function, 124–130, 124f, 128–129f, 151, 205, 211–213
Inverse least squares (ILS) regression, 249, 250f
Isosbestic points, 183–184

K

Kim-Lorentz profiles, 210–211, 210–211f
Kim oscillator, 117–119, 118–119f
Kirchhoff's law, 105, 150–151
Kramers-Kronig analysis (KKA), 143–144, 144f, 152–153, 351–352, 360f, 361–362, 368–369, 372–377
Kramers-Kronig conformity, 117–119, 121–122
Kramers-Kronig constrained variational analysis (KKCVA), 137
Kramers-Kronig relations (KKR), 91–92, 101, 119, 185, 189, 198, 200, 210–211
  a-axis of Fresnoite, reflectance spectrum of, 137, 137–138f, 139
  Cauchy principal value, 132
  classical damped harmonic oscillator model, 135–137, 136f
  consequences, 132
  dispersion theory, 132–133
  Hilbert transform pairs, 132
  inverse dielectric function, real and imaginary parts of, 132–133
  Kuzmenko's idea, advantages and drawback, 137
  lineshape/dielectric function model, 137
  Maclaurin's formula, 134–137, 134f, 136f, 139
  optical constants from transmittance/reflectance, determination of, 140–148, 143–146f, 148–149f
  piecewise dielectric function model, 134–135
  Poor Man's Kramers-Kronig analysis (PMKKA), 135–136, 136f
  practical problems, 133–134
  transparency regions, 133–134
  triangular functions, 134–135, 135–136f, 137
Kramers-Kronig sum rules, 105, 111, 125, 127, 148–152
Kretschmann configuration, 205–206, 205f

L

Lattice $vs.$ photon, dispersion of, 93, 93f
Least squares (LS) regression
  classical, 245–249, 246f, 248f
  inverse, 249, 250f
Light-matter interaction, 207
Linear dichroism theory (LDT), 9, 320, 362–363
  absorptance $vs.$ absorbance, 287–288, 287–288f
  optical model, 285, 285f
  polycrystalline fresnoite $vs.$ large/small crystallites, 290, 290f
  stretched PET (polyethylene terephthalate) foil, 284, 284f
  transmittance, 291, 291f
Linear polarized light, 52–53, 54f
Local field effect, 7, 94, 96–97, 105, 108–109, 257
Longitudinal dielectric function, 126
Longitudinal optical (LO)
  damping constant, 125–126
  mode, 117, 127–129
  oscillator strength, 125, 127–128, 148, 151, 212–213, 212–213f
  vibrations, 125
  wavenumber position, 128–129
Longitudinal plasmon-polaritons (LPP), 213–214
Lorentz-Lorenz equation, 4–6, 94, 96–97, 106–109, 183, 209, 209f, 211, 211f, 226–227, 229–230, 248–249, 248f, 252–255, 252f
Lorentz-Lorenz theory, 13, 258–259, 259f, 260s
Lorentz oscillator, 27–28
Lorentz profiles, 208–210, 209f
Lorentz profile $vs.$ oscillator, 98–101, 99–100f
Loschmidt's constant, 106–107
Lyddane-Sachs-Teller (LST) relation, 126, 217

M

Mach-Zehnder interferometer, 197
Maclaurin's formula, 134–137, 134f, 136f, 139, 142
Magnetic field, 62, 64, 71–72
  tangential components of, 33
Malus' law, 187–188
Many-particle systems, polarization in, 263–267, 266–267f
Matrix formalism, 70–83
  arbitrary number of layers, 76–77, 76s
  coherent layers, 77–78
  combined matrix formulation, for waves at single interface, 73–74
  incoherent layers, 78–80, 79s
  mixed coherent and incoherent layers, 80–81, 80s
  p-polarized waves, at single interface, 72–73, 73s
  s-polarized waves, at single interface, 71–72, 72s
  two infinite media, layer sandwiched by, 74–76, 74s
Maxwell-Garnett (MG) approximation, 259–261, 261f, 263f
Maxwell's equations, 11, 40, 50, 58, 61, 85–86, 208, 218, 284–285
  Ampere's circuital law, 16
  Berreman's formalism, 294–295
  Faraday's law of induction, 16
  Gauss's law, 15–16
  magnetic fields and, 15
  in simplified form, 15–16
  vector calculations, 52

Maxwell's wave equation, 5
Microheterogeneity, Bergman representation, 275–278, 275f
Microhomogeneity, Bergman representation, 275, 275f
Mie theory, 7–8
Mixing rules, 257
   benzene-toluene systems, 259f
   Bergman representation, 263–274
   Bouguer-Beer-Lambert (BBL) approximation, 182–184, 183f
   Bruggeman approximation, 261–262
   extended multiplicative scatter correction (EMSC), 278
   Lorentz-Lorenz theory, 258–259, 259f, 260s
   Maxwell-Garnett (MG) approximation, 259–261, 261f, 263f
   Wigner bounds, 262, 262s
Molar concentration, 102–103
Molar decadic absorption coefficient, 166
Molar oscillator strength, 103, 150
Molecular dynamics simulations, 109
Monoclinic crystals, 321, 321f
   dispersion analysis, 328, 329–330f
   4×4 matrix formalism, 293–294, 305, 307, 309
Multivariate curve resolution-alternating least squares (MCR-ALS), 252–255, 254f

N

Napierian absorption coefficient, 3, 31, 161, 166
Nearest neighbor model, 115, 119, 121
Newton's second law, 87
Non-IR active oscillators, 111
Nonnormal incidence, reflection and transmission
   two scalar infinite media, 60–66, 60s
      angle of incidence, 60–61
      angle of reflection and refraction, 60
      dependence of reflectance from angle of incidence, 66, 66f
      Fresnel's law, 61
      Maxwell's equations, 61
      plane of incidence, 60
      p-polarized light, 60, 63–65, 63s
      reflectance and transmittance, calculation of, 65–66
      s-polarized light, 60–63, 61s
      wave vectors, 61
Normal incidence, two scalar media, 57–60, 58s

O

Octadecanethiol (ODT), 201–202
One-dimensional waves, BBL approximation, 17–21
Optical branch, 93–94, 116
Optically uniaxial crystals, dispersion analysis, 325–327
Optically uniaxial, perpendicular modes, 329–336, 330f
Optical model, 285, 285f
Optical vibrations, of chain, 92f, 93
Orientational averaging, 350–353, 373–374
Orientation distribution function (ODF), 362
Orthorhombic crystals, 320, 320f, 327–328, 327–328f
Oscillator position, 96–97, 99–101, 105, 107–108, 113, 117, 137, 149, 151
Oscillators, direct coupling of, 214–217
Oscillator strength, 88–90, 95–98, 104–105, 111, 130–131, 147, 150, 152–153, 159–160, 166, 198, 199f
Otto configuration, 205–206, 206f

P

PCA. See Principal component analysis (PCA)
PCR. See Principal component regression (PCR)
Pellet method, 108–109
Percolation, Bergman representation, 268–269
Perpendicular modes, dispersion analysis, 329–336, 335f
   experimental and modeled reflectance, 336, 336f
   optically uniaxial crystal, 329–330, 330f
   reflectance spectrum, 331, 331–332f, 333, 334f
Phase angle, 236–238
Phase shifts, 19
Phonon-polariton dispersion, 218, 218f
Phonons, 90–91, 123
Plane of incidence, 60
Plane waves, 16, 41, 44, 50–53
   harmonic, 42, 42f
   reflection and transmission of (see Reflection and transmission, of plane waves)
   through two semi-infinite medium, 41–42, 41s
Plasma frequency, 130
Plasmonics, 201, 206
Plate capacitor, 86, 87f
PMKKA. See Poor Man's Kramers-Kronig analysis (PMKKA)
Polaritons, 217–221
Polarization
   direction, calculation of
      Berreman's formalism, 295–297
      Yeh's formalism, 298–299
   in many-particle systems, 263–267, 266–267f
   of matter by light, 4
Polycrystalline materials, 345–356, 346f
   angular dependence, 357–361, 357f, 359–360f
   average reflectance and transmittance theory (ARTT), 355, 361–362
   Bergman representation, 276
   cross-polarization, 354–355, 355–356f, 357–364, 359f, 361f
   experimental spectra, 345, 347f
   high-index incidence media, 358
   IR-microscopic spectra, 347, 347f
   IR reflectance spectrum, 352, 353f
   large crystallites and nonrandom orientation, 362–365
   optical properties, 356–362
   Poor Man's dispersion analysis (PDMA), 351–352, 351–352f
   randomly oriented, 356–362
   reflectance spectra, 345, 346f
   unified average optical properties theory (UAOPT), 355–356
Poly(methyl methacrylate) (PMMA), infrared spectra of, 201–202, 218–221, 219–221f
Poor Man's dispersion analysis (PDMA), 101, 160–161, 351–352, 351–352f
Poor Man's Kramers-Kronig analysis (PMKKA), 135–136, 136f, 185–186
Poynting's vector, 49–50, 52, 57–58
Principal component analysis (PCA), 6, 249–252
Principal component regression (PCR), 249–252, 253f
Principal orientations, 189–190

Q

Quantum cascade laser-based Mach-Zehnder interferometer, 106–107
Quantum field theory (QFT), 207, 217
Quantum mechanical model, 211, 367–368

R

Raman spectroscopy, 116–117, 125, 284
Random network theory, 123, 352
Raoult's law, 108–109, 244
Reduced mass, 87–88
Reflectance spectrum
   angle of incidence, 331, 332f
   ATR spectrum, 331–333, 332f
   blue vitriol, 337–338, 339f, 340
   Fresnoite, 325, 326f, 333, 334–335f, 335
   Tuton-salts, 328, 330f
Reflection and transmission, of plane waves
   nonnormal incidence, two scalar infinite media, 60–66, 60s
      angle of incidence, 60–61
      angle of reflection and refraction, 60
      dependence of reflectance from angle of incidence, 66, 66f
      Fresnel's law, 61
      Maxwell's equations, 61
      plane of incidence, 60
      p-polarized light, 60, 63–65, 63s
      reflectance and transmittance, calculation of, 65–66
      s-polarized light, 60–63, 61s
      wave vectors, 61
   nonnormal incidence, two scalar media
      absorbing media, 67, 68f
      matrix formalism (see Matrix formalism)
      total/internal reflection, 68–70, 68–70f
   normal incidence, two scalar media, 57–60, 58s
Reflection coefficient, 34, 62–65, 72, 75, 79–80, 124–125, 140–141, 166, 173
Refraction spectroscopy, 145, 197
Refractive index
   Berreman's formalism, 295–297
   Bouguer-Beer-Lambert (BBL) approximation, 27–31, 28–29f
   dispersion relations, 152–154, 153–155f
   Yeh's formalism, 295–297

Refractive index function, 85, 89–90, 99, 99f, 101, 103f, 104, 136f, 139, 142–145, 143f, 148, 151–152, 154, 154–155f, 159–160, 188–189, 197–198, 200f, 208–209, 220–221
Refractometry, 199–200
Relative dielectric constant, 86
Relative dielectric function, 103
 dispersion of, 88–89, 89–90f
 tensor, 89–90
Residual sum of squares (RSS), 239–240
Resonance disaster, 88–89
Resonance frequency, 125–126
Resonance wavenumber, 86, 89–90, 94, 121–124, 150–152
Rigorous coupled-wave analysis (RCWA), 206–207

S

Scalar fields, BBL approximation, 40–44
Schrödinger's equation, 40
Schwarz's theorem, 44
SEIRA. *See* Surface-enhanced infrared absorption (SEIRA)
Semi-empirical four parameter models
 Berreman-Unterwal model, 116–117, 116f, 118f
 classical model with
  complex oscillator strength, 120–122, 122f
  frequency-dependent damping constant, 119–120, 120f
 convolution model, 123–124, 123f
 Kim oscillator, 117–119, 118–119f
Semi-infinite medium, BBL approximation, 32, 36, 41s
Silent mode/oscillator, 111
Sinusoidal wave, BBL approximation, 17, 18f
Smart error sum
 asynchronous maps, antisymmetric, 231–232
 combined 2D correlation residual sum of squares (C2DCRSS), 232
 dispersion analysis, 232, 235
 minimum of, 235
 multiplicative errors, 235
 PMMA, corrected and uncorrected reflectance spectra of, 232, 233f
 PMMA layers on CaF$_2$, transmittance spectra of, 235
 PMMA layers on gold
  average corrected experimental and best-fit reflectance spectra of, 235–236, 237f
  average experimental and best-fit reflectance spectra of, 235, 236f
  reference spectrum, change of, 232, 234f
 synchronous and asynchronous maps, 232–234, 233f
 two-trace two-dimensional (2T2D) smart error sum
  fastest convergence, 238
  null spectrum, 236
  phase angle, 236–238
  PMMA on gold, average experimental and best-fit reflectance spectra of, 238–239, 239–240f
  PMMA on gold, conventional dispersion analysis of corrected reflectance spectra, 239, 240f
  residual sum of squares (RSS), 239–240
  series-based smart error sum, 236–238
  synchronous and asynchronous hybrid maps, 236–238
  synchronous and asynchronous spectra, 236
Spectral density
 of Bruggeman approximation, 271, 271f
 concrete, effective dielectric function on, 269–274, 269–270f
 properties of, 267–268, 272–274f
 size dependence of, 275–278, 276f
Spectral mixing rules, 257
 Bergman representation, 263–274
 Bruggeman approximation, 261–262
 extended multiplicative scatter correction (EMSC), 278
 ideal systems benzene-toluene, 259f
 Lorentz-Lorenz theory, 258–259, 259f, 260s
 Maxwell-Garnett (MG) approximation, 259–261, 261f, 263f
 Wigner bounds, 262, 262s
Spring coupling, 214, 216
Standing wave artifacts, 168–169
Stoke's theorem, 49, 49s
Strong coupling, 207–208, 210, 214–221, 215–216f, 218f
Surface-enhanced infrared absorption (SEIRA), 201–207
Surface-enhanced IR spectroscopy, 121–122
Surface plasmons, 205
Synchronous sums of squares (SRSS), 232, 238

T

Tetragonal crystal systems, 318–320, 319f
Transfer matrix, 201, 299–300
Transflection technique, 130–131, 168–171, 170–172f
Transition moment
 generalized oscillator models, 322–323
 monoclinic crystals, 321, 321f
 orthorhombic crystals systems, 320
 tetragonal, hexagonal, and trigonal crystal systems, 318–320
 triclinic crystals, 321–322
Transmission
 Bouguer-Beer-Lambert (BBL) approximation
  at interface, 32–34, 32f
  through layer on substrate, 38–40, 39s
  through thick slab, 34–36, 35s
  through thin slab, 36–38, 37s
 coefficient, 34, 62–65, 72, 75, 79–80, 140–141, 173
 of plane waves (*see* Reflection and transmission, of plane waves)
Transversal optical dielectric function, 126
Transversal optical (TO)
 mode, 117, 127–129
 oscillator strength, 127–128, 148, 151, 212–213
 vibrations, 125

Triclinic crystals, dispersion analysis, 337–340, 337–339f, 341f
Trigonal crystal systems, 318–320, 319f
Two-dimensional correlation analysis, 364
 Euclidian distance, 231–232
 hetero 2D correlation analysis, 230
 hybrid 2D correlation analysis, 230–232
 Minkowski distance, 231–232
 smart error sum (*see* Smart error sum)
 synchronous and asynchronous hybrid maps, 230–232, 231f
 synchronous and asynchronous sums of squares (SRSS and ARSS), 232
Two-dimensional correlation spectroscopy (2DCOS), 6
 absorbance, 225–226, 229
 asynchronous null spectrum, 227–229
 asynchronous spectrum, 226
 auto(correlation)peaks, 226
 Beer's approximation, 227–230
 benzene-toluene system
  experimental *vs.* forward calculated spectra, 229, 229f
  synchronous and asynchronous spectra of, 229, 230f
 binary mixtures, 225
 caprolactam in chloroform
  absorbance spectrum of, 227–229, 228f
  synchronous and asynchronous spectra of, 227–229, 228f
 chemometrics, 230
 cross-peaks, 226–227
 dynamic spectra, series of, 225
 Hilbert-Noda transformation matrix, 226
 history, 225
 local fields, concept of, 225
 Lorentz-Lorenz equation, 226–227, 229–230
 matrices, based on, 225–226
 nonlinear effects, sensitive to, 225
 perturbation, 225
 reflectance/transmittance, 225–226
 simulated synchronous and asynchronous spectra, of two oscillators
  classical damped harmonic (Lorentz) oscillators, 226, 227f
  Lorentz oscillators and local field of Lorentz, 226–227, 227f
  Lorentz profiles, 226, 226f
 synchronous spectrum, 226
 variance-covariance matrix, 226
Two-trace two-dimensional correlation analysis (2T2D-COS), smart error sum, 187, 236
 experimental and simulated spectrum, 236–238
 fastest convergence, 238
 null spectrum, 236
 phase angle, 236–238
 PMMA layers on gold
  average experimental and best-fit reflectance spectra of, 238–239, 239–240f
  corrected reflectance spectra, conventional dispersion analysis of, 239, 240f
  residual sum of squares (RSS), 239–240

Two-trace two-dimensional correlation analysis (2T2D-COS), smart error sum *(Continued)*
   synchronous and asynchronous hybrid maps, 236–238
   synchronous and asynchronous spectra, 236
Two-trace two-dimensional (2T2D) correlation spectroscopy, 364, 365*f*

U

Uniaxial crystals systems, 318–320, 319*f*
Unified average optical properties theory (UAOPT), 355–356
Univariate regression, 245

V

Vacuum refractive index, 151–152
Variance-covariance matrix, 226
Vector fields, BBL approximation, 40–44
Vibrational circular dichroism (VCD), 314, 367–368
   anticrossing behavior, 372, 378
   chiral dispersion analysis, 369–378, 370*f*, 372–373*f*
   chirality admittance function, 368–369, 370*f*, 372, 373*f*, 374–375
   Kramers-Kronig analysis, 368–369, 372–377
   measurement principle, 367
   spectra of chiral materials, 368–369

W

Wave equation, 15, 17, 50–52
Wave vector, 18, 41–42
Weak coupling, 207–208, 214–215, 215–216*f*, 217, 218*f*
Wigner bound, 262, 262*s*

Y

Yeh's formalism
   constitutive relations, 297–298
   Maxwell equations, 297–298
   polarization direction, 298–299
   refractive indices, 295–297

CPI Antony Rowe
Eastbourne, UK
June 06, 2024